Global Transversality, Resonance and Chaotic Dynamics

Global Transversality, Resonance and Chaotic Dynamics

Albert C J Luo
Southern Illinois University Edwardsville, USA

World Scientific

NEW JERSEY · LONDON · SINGAPORE · BEIJING · SHANGHAI · HONG KONG · TAIPEI · CHENNAI

Published by

World Scientific Publishing Co. Pte. Ltd.
5 Toh Tuck Link, Singapore 596224
USA office: 27 Warren Street, Suite 401-402, Hackensack, NJ 07601
UK office: 57 Shelton Street, Covent Garden, London WC2H 9HE

British Library Cataloguing-in-Publication Data
A catalogue record for this book is available from the British Library.

GLOBAL TRANSVERSALITY, RESONANCE AND CHAOTIC DYNAMICS

ISBN-13 978-981-277-111-7
ISBN-10 981-277-111-5

Printed by Fulsland Offset Printing (S) Pte Ltd, Singapore

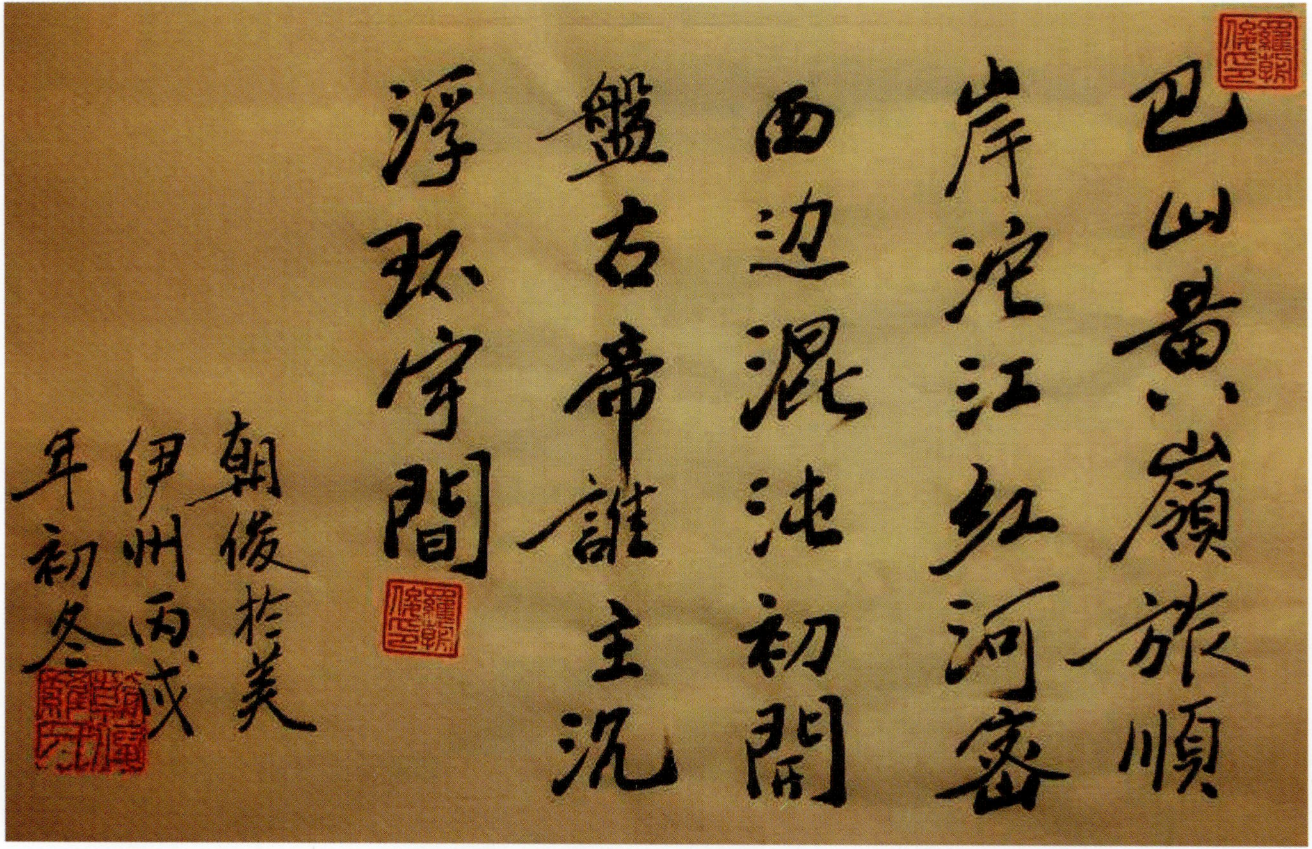
巴山黃嶺旅順岸
沱江紅河谷西边
混沌初開盤古帝
誰主沉浮环宇間
朝俊於美
伊州丙戌
年初冬

Preface

The modern theory of dynamical systems originates from the Poincarè's qualitative analysis, which focuses on the complexity and stability of motions in such dynamical systems. In 1892, Poincarè discovered that the motion of nonlinear coupled oscillators is sensitive to the initial condition and qualitatively presented that the inherent characteristics of the motion in vicinity of unstable fixed points of nonlinear oscillation systems may be *stochastic* under regular applied forces. This is because the separatrix exists in nonlinear dynamical systems. Such a separatrix connected with hyperbolic points is generic, which is called the generic separatrix. However, another kind of separatrix is generated by the nonlinear resonance between the system and periodic forcing or between the two oscillation interactions. Such a separatrix is called the resonant separatrix. The natural frequency of the separatrix connected with hyperbolic points is zero and the natural frequencies of motion on both sides of the separatrix are different, which makes such a stochastic motion exists in the neighborhood of separatrix. This book will discuss the global transversality of a flow to the separatrix from a domain to another domain in order to understand the mechanism for the onset, growth and destruction of chaos in vicinity of separatrix. The author would like to present a different point of view in order to look into a fundamental theory on global transversality, resonance and chaotic dynamics in nonlinear dynamic systems. The ideas presented in this book are less formal and rigorous in an informal and lively manner. The author hopes the initial ideas may give some inspirations in the field of nonlinear dynamics.

To measure the complex behaviors in a nonlinear dynamical system, the corresponding well-behaved dynamical system is employed. For doing so, the differential geometrical relations of two flows in the two nonlinear dynamical systems are presented. Based on such differential geometric relations, this book presents a theory of global transversality, resonance and chaos in n-dimensional nonlinear dynamics. The history and recent development of nonlinear dynamics is briefly discussed first, and then the global transversality of a flow to the separatrix is investigated to determine motion complexity in n-dimensional dynamical systems. The resonant mechanism of chaos in n-dimensional dynamical systems is discussed in general. Further, the resonant theory of the stochastic layer in 2-dimensional dynamical systems is presented. In addition, the stochasticity of the resonant separatrix layers for $2n$-dimensional, nonlinear Hamiltonian systems is also presented, and nonlinear dynamics on a $(2n\text{-}1)$-dimensional equi-energy surface is briefly discussed. For dissipative, nonlinear dynamical systems, the stability and grazing bifurcation are addressed. The global dynamics of 2-dimensional dissipative dynamical systems is presented. Finally, the switchability of a flow from a domain to its adjacent domain in discontinuous dynamical systems is discussed. The objective of this book is to throw out some original ideas on global transversality, resonant dynamics and chaos in nonlinear dynamics. The author believes that some ideas may not be very mature and some typos may exist in the book. The author sincerely hopes that readers can forgive such unavoidable errors here and where in the book. The author really appreciates readers for providing suggestions and comments to improve the theory presented in this book.

This book is dedicated to people who are challenging difficult and unsolved problems in natural science. Finally, I also dedicate this book to my wife (Sherry X. Huang) for support and to my lovely children (Yanyi Luo, Robin Ruo-Bing Luo, and Robert Zong-Yuan Luo) for their happiness to stimulate my inspiration.

Albert C.J. Luo
Edwardsville, Illinois

Contents

Chapter 1

Introduction

In order to understand complex phenomena in nature and social sciences, people desire to achieve the most accurate description and modeling of such complex phenomena and to understand the corresponding physics. Nonlinear dynamics, as an engine of modern science, drives investigations on nonlinear science in interdisciplinary fields, and dynamics is the most fascinating field in the modern science. The modern science began in the 17th century, and since then, one has paid a great deal of attentions on dynamics. The work of Galileo, Newton, Laplace, Clausius, Raylaigh lay on developing the fundamental physical principles and applications of dynamics. The work of Lagrange, Hamilton and Jacobi provided the further and formal developments of dynamics theory. In the 19th century, Hill and Poincare provided the qualitative analysis of dynamical systems to realize the remarkable significance of dynamics for scientific thought. The three-body problem in celestial mechanics is one of the origins for one to keep interests and to develop the fundamental theories and methodologies in dynamics. The ordinary differential equation theory in the real domain is a core to play an important role in dynamics. In the 20th century, one followed the Poincaré's idea to develop and apply the qualitative theory to understand the complexity in dynamical systems. Birkhoff was a key person to push the further development and applications of the Poincare's qualitative theory. The Taylor series expansion and perturbation analyses play a central role in qualitative and quantitative analyses. However, the Taylor series expansion analysis is valid in the finite domain under certain convergent conditions, and the perturbation analysis based on the small parameters, as an approximate estimate, is only acceptable for a very small domain during a short time period. Therefore, the qualitative theory based on the Taylor series and perturbation analysis cannot provide enough useful tools to understand the global complexity in nonlinear systems in the infinite time interval. The author would like to present a different view to look into the fundamental theory in dynamics. The ideas presented in this book are less formal and rigorous in an informal and lively manner. The author hopes the new ideas can give some inspirations in the field of nonlinear dynamics. In the first chapter of this book, the brief development history of the modern theory of dynamics will be presented. Further, chaotic dynamics in nonlinear Hamiltonian systems will be discussed and the corresponding existing methodologies for approximate predictions of the onset, growth and destruction of chaotic layers will be reviewed. The current status of researches on dissipative, nonlinear dynamical systems will be addressed and the crucial, unsolved problems in nonlinear dynamical systems will be briefly discussed. Finally, the summary and layout of contents in this book will be given.

1.1. A brief history of dynamics

The modern theory of dynamics originates from the Poincare's qualitative analysis. Poincare (1892) discovered that the motion of nonlinear coupled oscillator is sensitive to the initial condition, and qualitatively presented that the inherent characteristics of the motion in the vicinity of unstable fixed points of nonlinear oscillation systems may be *stochastic* under regular applied forces. In addition, Poincare developed the perturbation theory for periodic motions in planar dynamical systems. Birkhoff (1913) continued the Poincaré's work, and provided a proof of Poincare's geometric theorem. Birkhooff (1927) showed that both stable and unstable fixed points of nonlinear oscillation systems with two degrees of freedom must exist whenever their frequency ratio (or called *resonance*) is rational. The sub-resonances in periodic motions of such systems change the topological structures of phase trajectories, and the island chains are obtained when the dynamical systems renormalized with fine scales are used. The work of Poincaré and Birkhoff implies that the complexity of topological structures in phase space exists for nonlinear dynamic systems. The question is whether the complicated trajectory can fill the entire phase space or not. The formal and normal forms in the vicinity of equilibrium are developed through the Taylor series to investigate the complexity of trajectory in the neighborhood of the equilibrium. Since the trajectory complexity exists in the vicinity of hyperbolic points, one focused on investigating the dynamics in such vicinity of hyperbolic points.

From a topological point of view, the Smale's horseshoe was presented in Smale (1967). Further, a differentiable dynamical system theory was developed. Such a theory has been extensively used to interpret the homoclinic tangle phenomenon in nonlinear dynamics. Smale found the infinite, many periodic motions, and a perfect minimal Cantor set near a homoclinic motion can be formed. However, Smale's results cannot apply to Hamiltonian systems with more than two-degrees of freedom. Because the differentiable dynamical system theory is based on the linearization of dynamical systems at hyperbolic points, it may not be adequate to explain the complexity of chaotic motions in nonlinear dynamical systems. To continue the Birkhoff' formal stability, Glimm (1963) investigated the formal stability of an equilibrium (or a periodic solution) of Hamiltonian systems through the rational functions instead of the power series expansion. Such an investigation just gave another kind of approximation. Though those theories are extensively applied in nonlinear dynamical systems, such analyses based on the formal and normal forms are still the local analyses in the vicinity of equilibrium. Those theories cannot be applied for the global behaviors of nonlinear dynamical systems.

To understand the complexity of motion in nonlinear Hamiltonian systems, based on the non-rigorous theory of perturbation, Kolmogorov (1954) postulated the KAM theorem. In the KAM theorem, Kolmogorov suggested a procedure which ultimately led to the stability proof of the periodic solutions of the Hamiltonian systems with two-degrees of freedom. This problem is intimately connected with the difficulty of small divisors. The aforementioned theorem was proved under different restrictions in Arnold (1963) and Moser (1962). Further, Arnold (1964) investigated the instability of dynamical systems with several degrees of freedom, and the diffusion of motion along the generic separatrix was discussed. The results of Arnold (1964) extended the Kolmogorov's results to the Hamiltonian system with several degrees of freedom system. The stability in the sense of Lyapunov cannot be inferred. The KAM theory is based on the separable oscillators with weak interactions. In fact, once the perturbation exists, the dynamics of the perturbed Hamiltonian systems may not be well-behaved to the separable dynamical systems. In physical systems, the interaction between two oscillators in a nonlinear dynamical system cannot be very small. The KAM theorem may provide an acceptable prediction only when the interaction perturbation is very weak. The KAM theory is based on separable, integrable Hamiltonian systems. In fact, the complexity of motions in non-integrable, nonlinear Hamiltonian systems is much beyond what the KAM theory stated.

The instability zone (or *stochastic layer*) of Hamiltonian systems, as investigated in Arnold (1964), is a domain of chaotic motion in the vicinity of the generic separatrix. Even if the width of the separatrix splitting was estimated, the dynamics of the separatrix splitting was not developed. Henon and Heiles (1964) gave a numerical investigation on the nonlinear Hamiltonian system with two-degrees of freedom in order to determine whether or not a well-behaved constant of the motion exists for such a Hamiltonian. Izrailev and Chirikov (1964) first pointed out that the periodically forced, nonlinear Hamiltonian system with one degree of freedom exhibits a KAM instability leading to the stochastic behavior (or stochastic and resonance layers). Walker and Ford (1969) investigated the amplitude instability and ergodic behavior for nonlinear Hamiltonian systems with two degrees of freedom to develop the verifiable scheme for prediction of the onset of the amplitude instability. Isolated resonance and double resonance were investigated and the resonance was determined through the transformed coordinates. Such ergodic behavior in nonlinear Hamiltonian system originates from Birkhoff (1927). In other words, to investigate the enormous complexity of non-special motions in dynamical systems from geodesic flows, Birkhoff (1927) presented that the set of non-special motions (or chaotic motions) is measurable in the sense of Lebesgue, and the set of the special motions (or regular motion) is of zero measure. Furthermore, the ergodic theory had been developed in the 20th century and it is as a fundamental base for fractal theory. The thorough study of the geodesic flows in the ergodic theory can be found in Hopf's book (1937). Those ideas were generalized by Anosov (1962) to study a class of differential equations, which can be also referenced to Sinai (1976). Even though the ergodic theory is a foundation for fractality of chaotic motions in nonlinear dynamical systems, such a theory still cannot provide enough hopes to understand the complexity of chaotic motions in nonlinear dynamics.

For a nonlinear Hamiltonian system with n-degrees of freedom, it is very difficult to understand the mechanism of chaotic motions. To date, such a problem is unsolved. Around 1960, one considered extremely simple, nonlinear Hamiltonian systems to investigate such a mechanism. Melnikov (1962) used the concept of Poincaré (1892) to investigate the behavior of trajectories of perturbed systems near autonomous Hamiltonian systems. Melnikov (1963) further investigated the behavior of trajectories of perturbed Hamiltonian systems and the width of the separatrix splitting were approximately estimated. The width gives the domain of the chaotic motion in the vicinity of the generic separatrix. Even if the width of the separatix splitting was approximately estimated, the dynamics of the separatrix splitting was not developed. From a physical point of view, Chirikov (1960) investigated the resonance processes in magnetic traps, and the resonance overlap was presented initially. Zaslavsky and Chirikov (1964) discussed the mechanism of one-dimensional Fermi acceleration and determined the stochastic property of such a system. Rosenblut et al (1966) investigated the appearance of a stochastic instability (or chaotic motion) of trapped particles in the magnetic field of a traveling wave under a perturbation. Filonenko et al (1967) further discussed the destruction of magnetic surface generated by the resonance harmonics of perturbation. The destruction of such a magnetic surface demonstrates the formation and destruction of the resonant surface. Zaslavsky and Filonenko (1968) gave a systematic investigation of the stochastic instability of trapped particles through the separatrix map (or whisker motion in Arnold (1964)), and the fractional equation for diffusion was developed. Zaslavsky and Chirikov (1972) further presented the stochastic instability of nonlinear oscillations. Chirikov (1979) refined the resonance overlap criterion to predict the onset of chaos in stochastic layers. In addition, the most important achievements for prediction of the appearance of chaotic motions were summarized. Escande and Doveil (1981) used the resonance overlap concept and gave a criterion through a renormalization group method (also see, Escande, 1985). The details for the resonance overlap theory and renormalization group scheme can be referred to references (e.g., Lichtenberg and Lieberman, 1992; Reichl, 1992). Though the resonant overlap criterion can provide a rough prediction of the onset of chaotic motion in the stochastic layers, the mechanism of the chaotic motion in the stochastic layers still cannot be fully understood until now.

Luo (1995) proposed the resonance theory for chaotic motions in the vicinity of generic separatrix in nonlinear Hamiltonian system (also see, Luo and Han, 2001), and it was asserted that chaotic motions in nonlinear Hamiltonian systems are caused by the resonant interaction. Furthermore, the mechanism for the formation, growth and destruction of stochastic layers in nonlinear Hamiltonian systems was discussed in Luo and Han (2001). In Luo et al (1999), the resonant webs formed in the stochastic layer were presented, and it was observed that the webs are similar to the stochastic layer of the parametrically forced pendulum system. The recent investigations (e.g., Han and Luo, 1998; Luo, 2001b, 2001c, 2002) discovered that the resonance interaction generates the resonant separatrix, and the chaotic motion forms in vicinity of such a resonant separatrix. The corresponding criteria were presented for analytical predictions of chaotic motions in 1-DOF nonlinear Hamiltonian systems with periodic perturbations. The maximum and minimum energy spectrum methods were developed for numerical predictions of chaotic motions in nonlinear Hamiltonian systems (Luo et al, 1999; and Luo, 2002). The energy spectrum approach is applicable not only for small perturbations but for the large perturbation. The recent achievements for stochastic layers in periodically forced Hamiltonians with one-degree of freedom were summarized in Luo (2004a). Luo (2006a) investigated quasi-periodic and chaotic motions in n-dimensional nonlinear Hamiltonian systems. The energy spectrum method was systematically presented for arbitrary interactions of the integrable nonlinear Hamiltonian systems. The internal resonance was discussed analytically for weak interactions, and the chaotic and quasi-periodic motions can be predicted. From a theory for discontinuous dynamical system in Luo(2006b), Luo (2007a) presented a general theory for n-dimensional nonlinear dynamical systems. The global tangency and trans-versality to the separatrix were discussed from the first integral quantity. The first integral quantity increment was introduced to investigate the periodic and chaotic flows. In the following section, the basic theory and methodology for nonlinear Hamiltonian system will be addressed.

1.2. Nonlinear Hamiltonian systems

The *stochastic layer* in nonlinear Hamiltonian systems is a domain of chaotic motion in the vicinity of separatrices (e.g., Chirikov, 1979; Lichtenberg and Lieberman, 1992). The chaotic motion is generated through resonance interaction in nonlinear Hamiltonian systems. The separatrices of a Hamiltonian system divide the system phase space into different domains, and from numerical simulations, resonance interactions in all the domains are distinguishing themselves (e.g., Han and Luo, 1998; Luo and Han, 1999). Therefore, the resonance interaction in stochastic layers is a key for a better understanding of the mechanism of such a chaotic motion in the vicinity of separatrix. Poincare (1890) described qualitatively such a chaotic motion formed through the separatrix splitting of Hamiltonian systems for the first time (also see, Poincare, 1892), and Melnikov (1963) continued the Poincaré's investigation and computed the width of the separatrix splitting. Since then, the separatrix splitting has been further investigated quantitatively (e.g., Melnikov, 1963; Holmes et al, 1988; Lazutkin et al, 1989; Gelfreich et al, 1991; Gelfreich et al, 1994; Treschev, 1995, 1998), and the separatrix and standard map approaches have been developed for chaotic motions in the stochastic layer (e.g., Filonenko et al, 1967; Zaslavsky and Filonenko, 1968; Luo and Han, 1999; Luo, 1995). After the KAM torus in vicinity of separatrix is destroyed, the resonance overlap, generated by the interaction of resonance between the unperturbed and perturbed orbits, occurs in the stochastic layer in Chirikov (1979). However, such a resonant overlap mechanism of the stochastic layer needs to be further investigated. Therefore, the resonant characterization of stochastic layers in nonlinear systems was extensively investigated in recent years.

Consider a nonlinear Hamiltonian system in Chirikov (1979) as

$$H(I,\varphi)=H_0(I)+\mu H_1(I,\varphi,\Omega t,) \tag{1.1}$$

where $(I,\varphi)=(I_1,\cdots,I_n,\varphi_1,\cdots,\varphi_n)\,(n>1)$ are canonical variables, and t is time. The unperturbed integrable Hamiltonian $H_0(I)$ and the $2\pi/\Omega$-periodic perturbation $H_1(I,\varphi,\Omega t)$ are real analytic function (i.e., C^k ($k\geq 2$) continuous function) and the small parameter $\mu>0$ is the magnitude of perturbation, and the parameter Ω is perturbation frequency. As in Chirikov (1979), the Fourier series expansion of Eq.(1.1) for given I is

$$H(I,\varphi)=H_0(I)+\mu\sum_{m,n}^{\infty}H_{mn}(I)\exp(\mathrm{i}[n\varphi+m\Omega(t+t_0)]) \tag{1.2}$$

where m,n are integer and $\mathrm{i}=\sqrt{-1}$. $H_{mn}(I)$ is the Fourier series coefficient of $H_1(I,\varphi,\Omega t)$. From Eq.(1.2) the resonant condition is

$$n\omega(I)+m\Omega=0. \tag{1.3}$$

Due to different Ω, the following forcing will be discussed: (a) fast forcing ($\Omega=1/\varepsilon$ and $0<\varepsilon<<1$), (b) slow forcing ($\Omega=\varepsilon$ and $0<\varepsilon<<1$), (c) regular forcing (μ and Ω not small). Before discussing the nonlinear dynamics of stochastic layers in nonlinear Hamiltonian systems with at least a separatrix, the separatrix splitting should be discussed first.

1.2.1. Separatrix splitting

Poincare (1892) gave the measure of the splitting being exponentially small in μ (i.e., $\exp(-\frac{\pi}{\sqrt{2\mu}})$). The series solutions for limiting trajectories of Eq.(1.1) include the terms of $\exp(-\frac{\alpha}{\sqrt{\mu}})$ and $\alpha>0$ (e.g.,Poincare,1892; Melnikov,1962,1963). Arnold (1964) investigated the diffusion in the stochastic layer (or instability zone) through the separatrix splitting, and the average velocity for the shift of action variables is of the order of $\exp(-\frac{\alpha}{\sqrt{\mu}})$. Some explicit upper bounds for the speed of the Arnold diffusion in steep unperturbed Hamiltonian were presented (see also, Nekhoroshev, 1977; Neishtadt, 1984). For small μ, the stochastic layer is a very narrow domain in vicinity of the separatrix, and both the width and angle are of the order $\exp(-\frac{\alpha}{\sqrt{\mu}})$ in Neishtadt (1984). Holmes et al (1988) gave both upper and lower estimates of the exponentially small splitting of separatrices for rapidly forced systems (e.g., rapid forcing $\mu\sin(t/\varepsilon)$ and $0<\varepsilon<<1$) through the splitting distance computed by the Melnikov function. As in Treschev (1995), the Melnikov method gives the correct asymptotics of the splitting only when the parameter μ is sufficiently small. The other exponentially small effects were investigated in Slutskin (1964). Such effects are generated by a very slow perturbation (or slow forcing $\mu\sin(\varepsilon t)$ and $0<\varepsilon<<1$) rather than a very rapid perturbation of an integrable system.

Consider a 2-dimensional dynamic system possessing at least a homoclinic or heteroclinic orbit and such a system is perturbed by a periodic excitation, which is given by

$$\left.\begin{aligned}\dot{x}&=f_1(x,y)+\mu g_1(x,y,t),\\ \dot{y}&=f_2(x,y)+\mu g_2(x,y,t);\end{aligned}\right\}\tag{1.4}$$

where $f_1(x,y)=\partial H_0/\partial y$, $f_2(x,y)=-\partial H_0/\partial x$. The unperturbed Hamiltonian $H_0=H_0(x,y)$, $g_1(x,y,t)$ and $g_2(x,y,t)$ are analytic. In Melnikov (1962), the limiting trajectories of Eq.(1.4) were expressed by a series in the power of $\sqrt{\mu}$ and the trajectories converge for all the $\sqrt{\mu}$-terms. Consider an oscillatory frequency $\omega=\frac{1}{\sqrt{\mu}}$, and then the terms of $\exp(-\frac{\alpha}{\sqrt{\mu}})$ with ($\alpha>0$) should exist in the series solution of the foregoing equation. To investigate the separatrix splitting of the Hamiltonian system in Eq.(1.4), from Melnikov (1963), the asymptotic solution to Eq.(1.4) is determined by

$$\left.\begin{aligned}x_\varepsilon(t,\omega,t_0)&=x_0(t-t_0)+\sum_{n=1}^{\infty}x_n(t,\omega,t_0)\mu^n,\\ y_\varepsilon(t,\omega,t_0)&=y_0(t-t_0)+\sum_{n=1}^{\infty}y_n(t,\omega,t_0)\mu^n;\end{aligned}\right\}\tag{1.5}$$

where $\dot{x}_0=f_1(x_0,y_0)$ and $\dot{y}_0=f_2(x_0,y_0)$ gives a solution of the unperturbed system (i.e., $x_0(t-t_0)$ and $y_0(t-t_0)$). Through the perturbation analysis, equations for variables (x_n,y_n) can be obtained from the order of ε^n, and the corresponding perturbed solution (i.e., $x_n(t,\omega,t_0)$ and $y_n(t,\omega,t_0)$) is determined by

$$\begin{Bmatrix} \dot{x}_n \\ \dot{y}_n \end{Bmatrix} = \begin{bmatrix} \frac{\partial f_1(x,y)}{\partial x} & \frac{\partial f_1(x,y)}{\partial y} \\ \frac{\partial f_2(x,y)}{\partial x} & \frac{\partial f_2(x,y)}{\partial y} \end{bmatrix}_{(x_0,y_0)} \begin{Bmatrix} x_n \\ y_n \end{Bmatrix} + \begin{Bmatrix} P_n(x_0,\cdots,x_{n-1},y_0,\cdots,y_{n-1}) \\ Q_n(x_0,\cdots,x_{n-1},y_0,\cdots,y_{n-1}) \end{Bmatrix}. \tag{1.6}$$

From Eq.(1.6), Melnikov (1963) developed an analytical method to predict the gap (or width, splitting distance) between two branches of separatrix splitting, and the gap was computed by

$$\Delta_\mu(t_0,\omega) = \sum_{n=1}^{\infty} \Delta_n(t_0,\omega)\mu^n, \tag{1.7}$$

where

$$\begin{aligned} \Delta_n(t_0,\omega) = \int_{-\infty}^{\infty} &\left[Q_n(x_0,\cdots,y_{n-1}) f_1(x_0,y_0) - P_n(x_0,\cdots,y_{n-1}) f_2(x_0,y_0)\right] \\ &\times \exp\left(-\int_{t_0}^{t-t_0} \left[\frac{\partial f_1(x,y)}{\partial x} + \frac{\partial g_1(x,y)}{\partial y}\right]_{(x_0(\xi),y_0(\xi))} d\xi\right) dt. \end{aligned} \tag{1.8}$$

For $P_1(x_0,y_0) = f_2(x_0,y_0)$ and $Q_1(x_0,y_0) = g_2(x_0,y_0)$ in Eq.(1.8), the quantity $\Delta_n(t_0,\omega)$ becomes $\Delta_1(t_0,\omega)$ which is called the Melnikov function. Equation (1.7) gives the asymptotic behavior of the function $\Delta_\mu(t_0,\omega)$ as $\omega \to \pm\infty$.

In recent years, analytic estimates of the exponentially small effect of separatrix splitting were very active through the standard map (e.g., Lazutkin et al, 1989; Gelfreich et al, 1994; Treschev,1998) and for continuous Hamiltonian systems (e.g., Gelfreich et al, 1991; Treschev, 1995). When the homoclinic orbit is splitting in perturbed Hamiltonian systems with a small parameter μ, its stable and unstable manifolds intersect each other, and the tangent vectors of the two manifolds are nonzero intersection angles. The splitting of homoclinic orbit is characterized by the maximum angle of intersection and the width of the stochastic layer. The angle and width are of the order $\exp(-\frac{\alpha}{\sqrt{\mu}})$ $(\alpha > 0)$ in Poincaré (1890) (also see, Melnikov, 1962). For the exponentially small splitting of a homoclinic orbit γ, Gelfreich et al (1991) introduced the homoclinic invariant instead of the maximum angle of intersection since this angle is not symplectically invariant. When the stable and unstable manifolds $\varphi^{s,u}(t)$ related to the hyperbolic fixed point z_0 are used, the homoclinic invariant is

$$\omega(\gamma) = \left\| \frac{d\varphi^u(t)}{dt}\Big|_{z_h} \right\| \times \left\| \frac{d\varphi^s(t)}{dt}\Big|_{z_h} \right\| \sin\phi, \tag{1.9}$$

where ϕ is the intersection angle of two manifolds $\varphi^{s,u}(t)$ at z_h and $\|\cdot\|$ is a Riemanian metric. A generalized standard map in Gelfreich et al (1991) is

$$x_{n+1} = x_n + y_{n+1}, \text{ and } y_{n+1} = y_n + \mu\sum_{k=1}^{\infty} a_k \sin(kx_n + b_k). \tag{1.10}$$

The corresponding homoclinic invariant is

$$\omega(\gamma)=\frac{4\pi|\Theta|}{n^2\mu}\exp(-\frac{2\pi\rho}{\sqrt{\mu}})\sum_{k=1}^{m}c_k\cos\frac{2\pi\beta_k}{\sqrt{\mu}}+o\left(\mu^{-1+\delta}\exp(\frac{2\pi\rho}{\sqrt{\mu}})\right). \quad (1.11)$$

The constant in the o-estimate depends on the choice of parameters. $\delta\in(0,\frac{1}{2(n+1)})$ and $z_k=\beta_k$ ($k=0,1,\cdots,m$). $c_k=1$ if $\beta_k=0$ and $c_k=2$ if $\beta_k\neq 0$. $|\Theta|=1118.827\ 705\ 95\cdots$. If $k=1$, we have $\omega(\gamma)=4\phi$ and Eq.(1.11) reduces to the maximum intersection angle in Lazutkin et al (1989)

$$\phi=\frac{\pi|\Theta|}{\mu}\exp(-\frac{\pi^2}{\sqrt{\mu}})\left[1+o\left(\mu^{1/8-\delta}\right)\right]. \quad (1.12)$$

The other exponentially small effects, generated by a very slow perturbation (or slow forcing $\mu\sin(\varepsilon t)$ and $0<\varepsilon<<1$) rather than the very rapid forcing of an integrable system, were investigated in Slutskin (1964). Analytical estimates for the exponentially small splitting of separatrices in the standard Chirikov map were presented (e.g., Lazutkin et al, 1989; Gelfreich et al, 1991, 1994; Treschev, 1995). Treschev(1995) investigated the separatrix splitting in nonlinear Hamiltonian systems, which is exponentially close to integrable Hamiltonian systems through the averaging method in Slutskin (1964), and such a separatrix splitting includes all the aforementioned separatrix splitting. Treschev (1998) estimated the width of stochastic layers in near-integrable two-dimensional symplectic maps, and the upper and lower bounds for both the width and area of the stochastic layers were also estimated through the separatrix splitting.

The separatrix splitting for slow forcing ($\mu>1$) and usual forcing is still not estimated because the aforementioned method is based on the perturbation analysis. In addition, only the separatrix is used for estimates of the separatrix splitting without the periodic solutions of an unperturbed Hamiltonian system in the vicinity of the separatrix. The exponentially small separatrix splitting is crucial to understand the behavior and characteristics of trajectories in the stochastic layer, and the widths and/or intersection angle of the separatrix splitting in the stochastic layer have been discussed. However, no any criteria associated with the widths for transition to global stochasticity is provided as in the standard map (e.g., Chirikov, 1979). Therefore, the amplitudes (or strengths) of external excitation cannot be determined when a primary resonance is absorbed in the stochastic layer.

1.2.2. Standard and whisker maps

Zaslavsky and Filonenko (1968) first introduced the separatrix map to investigate the one-dimensional motion of a charged particle in the field of a traveling wave nder a perturbation. Such a whisker map is based on the following three assumptions:

(i) The approximation of the first complete elliptic integral gives an expression $K(k) \approx \log(4/\sqrt{1-k^2})$ when the periodic orbits of an unperturbed system are close to the separatrix (e.g., Chirikov, 1979).
(ii) Only the period of libration is used for the estimate of phase change without the rotation.
(iii) The energy change (ΔH_0) for one period of a periodic orbit of an unperturbed Hamiltonian system in vicinity of separatrix in Eq.(1.1) is approximated by the one (ΔH_0^{sx}) along the separatrix (e.g., Chirikov, 1979; Gelfreich et al, 1994; Arnold, 1964; Reichl, 1992), i.e.,

$$\begin{aligned}\Delta H_0 \approx \Delta H_0^{sx} &= \mu \int_{-\infty}^{\infty} \left.\frac{dH_0}{dt}\right|_{(I^{sx},\varphi^{sx})} dt \\ &= \mu \int_{-\infty}^{\infty} \left[H_0\left(I^{sx}\right), H_1\left(I^{sx}, \varphi^{sx}, \Omega t\right)\right] dt\end{aligned} \tag{1.13}$$

where I^{sx} and φ^{sx} are the separatrix solution, and $[\cdot,\cdot]$ is the Poisson bracket. The whisker map for the forced pendulum (e.g., Zaslavsky and Filonenko, 1968; Chirikov, 1979; Lichtenberg and Lieberman, 1992) is

$$w_{i+1} = w_i + W \sin \varphi_i, \text{ and } \varphi_{i+1} = \varphi_i + \tfrac{\Omega}{\omega} \log(\tfrac{32}{|w_{i+1}|}) \bmod 2\pi; \tag{1.14}$$

where $w = E/\omega^2 - 1$. W is related to the magnitude of energy change ΔH_0. Both E and ω are the Hamiltonian energy and natural frequency of the unperturbed system, respectively. Ω is excitation frequency.

For the stochastic instability of trapped particles, Zaslavsky and Filonenko (1968) also linearized the whisker map to obtain a standard map in the neighborhood of an assumed resonance (e.g.,Chirikov,1979; Lichtenberg and Lieberman, 1992; Reichl,1992). Vecheslavov (1996) used a similar procedure to investigate chaotic motions in vicinity of the separatrix under high-frequency excitations. In addition to the above-mentioned three assumptions, two more assumptions are used in derivation of the standard map:

(iv) The whisker map is approximated at its period-1 fixed point based on a primary resonance.
(v) The primary resonant condition is assumed but not derived from Eq.(1.1) (i.e., equation (1.3) is not used).

From Assumptions (iv)-(v), the standard map (e.g., Chirikov, 1979; Lichtenberg and Lieberman, 1992) is

$$I_{i+1} = I_i + K \sin \varphi_i \text{, and } \varphi_{i+1} \approx \varphi_i + I_{i+1}; \tag{1.15}$$

where $I = -(w - w_r)\Omega/\omega w_r$ and $K = -\Omega W/w_r\omega$. w_r is the resonant value based on the arbitrary resonant condition. Greene (1968, 1979) developed a method to numerically determine the strength of the stochasticity (i.e., $K^* \approx 0.9716\cdots$) when the transition to global stochasticity for Eq.(1.15) occurs. Other estimates for the strength of stochasticity parameter can be referred to the references (e.g., Chirikov, 1979 and Lichtenberg and Lieberman, 1992). Treschev (1998) presented a qualitative relationship between the width (w) of the stochastic layer and the width (d) of the lobe domain, such as $w/d \sim 1/\log \mu$ and μ is a multiplier at the corresponding hyperbolic point. From the energy analysis and numerical simulations, the resonance in regions separated by the separatrix in phase space are distinguishing themselves (e.g., Luo, 1995; Han and Luo, 1998; Luo and Han, 1999). Therefore, the Assumption (v) is inadequate for the derivation of the standard map. Further, the established standard map approach cannot provide a satisfactory prediction of the stochastic layer for the original system in Eq.(1.1). Although the approximate separatrix map (i.e., Eq.(1.14)) is not accurate to model Hamiltonian systems, it provides a better prediction of the stochastic layer than the standard map. Therefore, in recent years, the approximate whisker map was investigated extensively (e.g., Rom-Keder, 1990, 1994, 1995; Zaslavsky and Abdullaev, 1995; Abdullaev and Zaslavsky, 1995, 1996; Ahn et al, 1996; Iomin and Fishman, 1996). Rom-Kedar (1990) derived the approximate whisker map through the Melnikov function, and used such an approximate map to predict transport rates statistically and numerically. The escape rates in the vicinity of homoclinic tangles were investigated. However, the approximate whisker map in Rom-Kedar (1994) was further investigated. Rom-Kedar (1995) determined the secondary homoclinic bifurcation in the stochastic layer by setting the Hamiltonian energy at the second iteration of the approximate whisker map to zero. For an improved computation of energy change, Zaslavsky and Abdullaev (1995) introduced a shifted separatrix map. The procedure of the conventional derivation of whisker and standard maps was given in Luo and Han (1999) (also see, Zaslavsky and Filonenko, 1968; Luo, 2001). Even if so, the aforementioned approach cannot provide an appropriate prediction of the onset of resonance in the stochastic layer. Therefore, Luo (2001) developed an improved standard map approach based on the accurate whisker map for such a prediction.

To provide a more adequate prediction of the stochastic and resonance layers, Luo (1995) presented the incremental energy method. This method is based on the accurate whisker mapping rather than the standard mapping technique (also see, Han and Luo, 1998; Luo and Han, 2001). The criterion presented in Luo and Han (2001) can be expressed through the action variables and natural frequency of the unperturbed Hamiltonian systems, as in Arnold (1989). The onset of a specified resonance in the stochastic layer is predicted through the incremental energy approach. This approach is also applicable for strong excitations when the energy increments still maintain in good accuracy. For a better prediction of resonance interactions in the stochastic layer, a new computational method for the incremental energy should be developed. For the numerical prediction of stochastic layer, Luo et al (1999) developed an energy spectrum technique, and the resonant characteristics in stochastic layers were investigated through the energy spectra. This technique computes the maximum and minimum energies of the Poincaré mapping points. A comparison of analytical and numerical predictions was presented in Luo and Han (2001). From numerical results, the standard mapping approach developed in Luo and Han (2001) cannot provide a very good prediction for strong excitations.

1.2.3. Chirikov resonance overlap criterion

Chirikov (1979) presented a resonant overlap criterion, and pointed out that the transverse, homoclinic intersections, caused by the resonant overlap, transfer the energy from one resonant orbit to another one. For the transition of local stochasticity to chaos (or global stochasticity), Chirikov postulated that the last KAM surface between two lowest-order primary resonance is destroyed when the sum of the half-widths of the two island separatrices formed by the resonance is equal to the distance between the two resonance. Using this criterion, Reichl and Zheng (1984a,b) estimated the width of the stochastic layer (or excitation strength) for the undamped Duffing oscillator, and such estimates quantitatively agree with the predictions of excitation strength through the established standard map. The Chirikov overlap criterion usually gives the correct order of magnitude for excitation strength though the self-similarity of resonance is ignored. Consider a Hamiltonian pertaining to the standard map (e.g., Zaslavsky and Filonenko,1968; Zaslvasky, 1998), i.e.,

$$H = \tfrac{1}{2}y^2 + K\cos\theta + 2K\cos\theta\cos(\Omega t), \tag{1.16}$$

where $y = \dot{\theta}$. The third term of Eq.(1.16) is considered as a perturbation. The unperturbed Hamiltonian of Eq.(1.16) is

$$H_0 = \tfrac{1}{2}y^2 + K\cos\theta. \tag{1.17}$$

The half-width of the island separatrix at $\cos\theta = -1$ is

$$\Delta y_{\max} = 2\sqrt{K}. \tag{1.18}$$

Therefore, three overlap criteria are as follows:

(i) For the simple overlap, the critical condition is obtained by

$$2\Delta y_{\max} = 2\pi \Rightarrow K = \tfrac{\pi^2}{4} \approx 2.46740. \tag{1.19}$$

(ii) For the first and second-order resonance overlap, the critical value is

$$2\sqrt{K} + \tfrac{1}{2}K = \pi \quad \Rightarrow K \approx 1.46. \tag{1.20}$$

(iii) For the improved overlap, the critical value is improved as

$$2\left[\sqrt{K} + \tfrac{(2\pi)^2}{K^2}\exp(-\tfrac{\pi^2}{\sqrt{K}})\right] + \tfrac{K}{2} = \pi \Rightarrow K \approx 1.2. \tag{1.21}$$

Consider a perturbed Hamiltonian system, the rescaled energy near the resonant orbit is:

$$H = \tfrac{1}{2}y^2 - M\cos\theta - P\cos\nu\left(\theta - \Omega t\right). \tag{1.22}$$

For $P = 0$, the half-width of the resonance at $y = 0$ is $\Delta y_{\max} = 2\sqrt{M}$. Similarly, when $M = 0$, the half-width of the resonance at $y = 1$ is $\Delta y_{\max} = 2\sqrt{P}$. The Chirikov overlap criterion in Reichl (1992) is

$$2\sqrt{M} + 2\sqrt{P} = 1. \tag{1.23}$$

Using the foregoing equation, Reichl and Zheng (1984a,b) estimated the widths of the stochastic layers (actually, excitation strengths) for the undamped Duffing oscillator (also see, Luo and Han, 1999, 2000). Luo and Han (1999) used the Chirikov overlap criterion to predict the onset of the resonant layer. Compared to the numerical prediction, this criterion gives a rough estimate of the resonant layer.

1.2.4. Renormalization group technique

Owing to the self-similar structure of resonance, the renormalization group theory (e.g., Escande and Doveil, 1981; Escande, 1985) was developed for prediction of the excitation strength when the last KAM torus surface between two lowest-order primary resonances is destroyed. Following such a theory in Escande (1985) and assuming $M > P$, the energy H in Eq. (1.22) is separated into two parts as

$$H_0 = \tfrac{1}{2}y^2 - M\cos\theta = E_0, H_1 = -P\cos\nu\left(\theta - \Omega t\right); \tag{1.24}$$

where E_0 is constant. The first one of the forgoing equation gives

$$\theta = 2\mathrm{am}(\tfrac{K(k)\varphi}{\pi}, k), y = \pm 2k\sqrt{M}\,\mathrm{cn}(\tfrac{2K(k)\varphi}{\pi}, k); \tag{1.25}$$

where am and cn are the elliptic amplitude and the Jacobi elliptic functions, and

$$k^2 = \frac{2M}{E_0 + M},\ J = \frac{4\sqrt{M}}{\pi k} K(k). \tag{1.26}$$

Substitution of Eq. (1.25) into Eq. (1.22) yields

$$H = H_0(J) - P\sum_{-\infty}^{\infty} V_n(J)\cos\left[(\nu + n)\varphi - \nu\Omega t\right], \tag{1.27}$$

where $V_n(J)$ is a coefficient of the nth-order term in the Fourier series. From Eq.(1.27), we have the nth-order resonant condition

$$\omega = \frac{\pi\sqrt{M}}{kK(k)} = \frac{\nu\Omega}{\nu + n + \delta}, \tag{1.28}$$

where $\delta = \{0,1\}$. The energy renormalization in *Eq.*(1.27) for the nth *-order* resonance yields,

$$\bar{H} = \frac{\bar{y}^2}{2} - \bar{M}\cos\bar{\theta} - \bar{P}\cos\bar{\nu}(\bar{\theta} - \bar{\Omega}t), \tag{1.29}$$

where

$$\left.\begin{aligned}
\bar{M} &= \frac{PV_{n+\delta}\,\omega(\nu+n)^2(\nu+n+1)^2}{\nu^2},\\
\bar{P} &= \frac{PV_{n+1-\delta}\,\omega(\nu+n)^2(\nu+n+1)^2}{\nu^2},\\
\bar{\nu} &= \frac{(\nu+n+1-\delta)}{\nu+n+\delta},\ \bar{\Omega} = \frac{(2\delta-1)\nu\Omega}{\nu+n+1-\delta},\\
\bar{\theta} &= (\nu+n+\delta)\varphi - \nu\Omega t.
\end{aligned}\right\} \tag{1.30}$$

Solving Eq.(1.30) as the self-similar structures gives

$$2\sqrt{M} + 2\sqrt{P} \approx 0.7. \tag{1.31}$$

Note that Eqs.(1.23) and (1.31) have a similar form, but only two of the infinite primary resonances are modeled (e.g., Luo, 1995; Luo and Han, 1999; Luo et al, 1995). Han and Luo(1998) developed an improved standard map approach for predicting the onset and destruction of resonant separatrix layers. The Chirikov overlap criterion in Luo and Han (1999) and the renormalization group technique (e.g.,Reichl and Zheng, 1984; Lin and Reichl,1986; Luo et al, 1995) were modified for the prediction of the appearance of resonant layers. The renormalization group technique can give a good prediction of the resonance interaction when the excitation perturbation is very weak. For strong perturbations, the renormalization technique may not be adequate. Therefore, Luo (1995) presented an incremental energy method to give an approximate prediction (also see, Han and Luo, 1998), and the corresponding energy spectrum technique was developed in Luo (2002). Using the energy spectrum method, the appearance and disappearance of the resonant layer can be obtained.

1.3. Dissipative nonlinear systems

For dissipative nonlinear dynamical systems, the perturbation analysis is an important tool to determine the periodic motion behaviors. The earliest approximation method is the method of averaging, and the idea of averaging originates from Lagrange (1788) to investigate the three-body problem. In the end of the 19th century, Poincaré (1892) used the similar ideas to develop the perturbation theory in 2-dimensional dynamic systems, one has applied the Poincaré perturbation methods for periodic motions in nonlinear dynamical systems. The classic perturbation methods for nonlinear oscillators were presented (e.g. Stoker 1950, Minorsky, 1962; Hayashi, 1964). Because the time for periodic motions in nonlinear oscillation is finite, the perturbation analysis can provide an approximate estimate of periodic motions in nonlinear oscillators. However, the time for chaotic motions in nonlinear dynamical systems is infinite, so the perturbation method may not provide an adequate analysis of chaotic motions. In recent decades, chaos in dissipative nonlinear dynamical system was extensively investigated. For instance, Ueda (1980) gave the numerical simulation of regular and chaotic motions in the damped Duffing oscillator. So far, chaotic motions in dissipated nonlinear systems with one degree of freedom are analytically predicted by the Melnikov method. The detailed discussion of the Melnikov method can be referred to Guckenheimer and Holmes (1983). In fact, the Melnikov function (1963) can only give an acceptable, approximate estimate of the width for the separatrix splitting for a weak perturbation. Such a function may not be adequate for the global transversality of flows to the homoclinic or heteroclinic orbit in nonlinear dynamical systems. In this book, the author would like to discuss this question.

Based on the work of Melnikov (1963), Greenspan (1981) extended the Melnikov's ideas to the dissipative dynamical systems (also see, Greenspan and Holmes, 1982; Guckenheimer and Holmes, 1983). Further, the Melnikov method was developed for the global transversality in dissipative nonlinear systems. Once the global transversality to the separatrix exists, one thought that the Smale horseshoe presented in Smale (1967) may exist, and further chaos in such a nonlinear dynamical system may occur. In recent years, many researches were carried out from such a thought. However, from such a prediction based on the

Melnikov method, one cannot observe the global transversality in nonlinear dynamical systems. The Smale horseshoe theory may not be adequate for nonlinear dynamical systems rather than the topological structure, the author does not proceed to discuss whether the Smale horseshoe theory is right or wrong in nonlinear dynamical systems. From the perturbation analysis, the Melnikov function was obtained for Hamiltonian systems with a small perturbation. One used such a function to analytically predict global behaviors (e.g., chaos) in Hamiltonian systems with a small perturbation. Due to the perturbation analysis, the Melnikov method can give a reasonable analysis of the global behavior only when the perturbation is very small and close to zero. However, the perturbation is very small to zero, chaos in the nonlinear dynamical systems may not occur. So the Melnikov method may not help us understand the global behaviors of nonlinear dynamical systems. Luo (1995) used the Chirikov criterion to determine Hamiltonian chaos and applied the Melnikov function to investigate the global transversality (also see, Luo and Han, 1999). It was found that the Melnikov method cannot provide an adequate prediction of chaotic motions in the dissipative system. For a better understanding of the Melnikov method, the paper of Melnikov (1963) should be revisited. In 1963, Melnikov presented a perturbation analysis to estimate the width of the separatrix splitting. Indeed, the width of the separatrix can be approximately estimated, but it cannot be used for prediction of the existence of chaos. In fact, the Melnikov function is an approximate energy increment during a certain time period, which can be found in references (e.g., Arnold, 1964; Chirikov,1979; Luo and Han, 2001). If the Melnikov function is zero, from physical point of view, it implies that the system energy is conserved during a certain time period. Such a zero value of the Melnikov function does not imply that the flow has any global transversality to the separatrix. To resolve this puzzle, Luo (2007a) used the concepts in discontinuous dynamical systems in Luo (2005) (also see, Luo, 2006a) to investigate the global transversality and tangency to the separatrix in nonlinear dynamical systems with a continuous vector field. The concepts of global tangency and transversality to the separatrix are introduced. It is very easy for us to observe that complex behaviors in nonlinear systems are involved with the separatrix. This is because the singularity originates from separatrix with zero frequency. The phase space of a dynamical system is divided by the separatrix into many sub-domains. In each individual sub-domain, the dynamical system has the similar dynamic behaviors which, however, are different in the other subdomains. The differences of dynamic behaviors in different sub-domains cause the complexity of motion near the separatrix in nonlinear dynamical systems. Luo (2007a) developed a general theory for the global tangency and transversality of flows in n-dimensional nonlinear dynamical systems. Such a general theory is applied to a dissipative dynamical system (e.g., a periodically forced, damped Duffing oscillator) in Luo (2007b), and the global tangency and transversality of the periodic and chaotic motions to the separatrix were discussed.

The detailed discussion was given in Luo (2007c). It was observed that the transversality and tangency to the separatrix in nonlinear dynamical systems are independent of the Melnikov function (or the energy increments). Furthermore, the conditions for the global transversality and tangency of periodic and chaotic motions were given, which may help us understand the complexity of motions in nonlinear dynamic systems with separatrix.

In dynamic systems, the stability of equilibrium and periodic solutions is a very important issue. This is another key to understand the complexity of flows in nonlinear dynamical systems correctly. The stability by linearization is extensively used, which is completed through the eigenvalue analysis of the linearized system of equilibriums or periodic orbits in the corresponding neighborhoods. Such a stability analysis is local. To determine the non-local stability of nonlinear dynamical systems, the Lyapunov method should be employed. Lyapunov (1907) generalized the work of the Torricelli, Huygens and Lagrange about the ideas of the stability in mechanics, and developed the Lyapunov direct method based on differential equations rather than the potential energy or an energy-alike quantity in general. From the stability definition, the stability conditions in the Lyapunov direct method are too strong. So far, except for numerical simulations of dissipative systems, almost no analytical methods can be used to predict chaotic motions. The most popular method is the method of Lyapunov exponents, which is based on the theory of linear differential equations. To use the linearization of instantaneous point of a given trajectory of the nonlinear dynamical system, the characteristic exponents are computed. One thought that such a trajectory is chaotic if the maximum exponent is greater than zero. In other words, to use the ideas of the local linearization, the Lyapunov exponents of a given trajectory characterize the mean exponential rate of divergence of the trajectories in the vicinity of such a given trajectory. Honen and Heiles (1964) introduced the divergence of near trajectories to investigate the stochasticity of a trajectory in phase space. Zaslvasky and Chirikov (1972) gave the further studies of such stochasticity of a given trajectory in nonlinear oscillation systems. The connection between Lyapunov exponents and exponential divergence was presented by Benettin et al (1976), and the comprehensive presentation about the Lyaponuv characteristic exponents can be found in Benettin et al (1980a,b). Wolf et al (1985) presented a simplest computational scheme to determine the Lyapunov exponents from a time series. The continuous orthogalization methods were properly implemented for computing the Lyapunov exponents of continuous dynamical systems (e.g., Goldhirsch et al, 1987; Dieci et al, 1997), which is much better than the method of Wolf et al (1985). Based on the QR method, the further improvement of methods for computing Lyapunov exponents can be found in references (e.g., Dieci et al, 1994, 1995; Udwadia and von Bremen, 2001). Bartler (1999) used the idea of tracking Lyapunov vectors to determine Lyapunov exponents. From Bartler's inspiration, Yang et al (2005) modified the Lyapunov vector method. Lu et al (2005) gave the further mathematic development of the Lyapunov vector method, and a numerical integration scheme was derived that can automatically preserve the orthogonality between any two consecutive vectors. In such a method, the vectors are orthogonal but not necessary orthonormal. Except for the numerical method based on the linearized equation of the dynamical system, Luo (2007a) pointed out the G-function to the first integral manifold can be used to determine stability of the system. This stability conditions similar to the Lyapunov stability is very strong. Therefore, it was suggested that the first integral quantity increment is used to determine the stability of periodic motions in nonlinear dynamical systems in Luo(2007a). The chaotic motions in nonlinear dynamical systems can be investigated through the corresponding iterations given by the first integral quantity increment. For this method, the key is to determine the first integral manifold, which can be done from Lie group analysis. Sometimes, it is very difficult to obtain the first integral surface of a specific, nonlinear dynamical system. In recent years, Ao (2004) constructed the potential function for stochastic differential equations (see, Ao, 2005, Yin and Ao, 2006). Once the potential function can be found, the first integral quantity for stochastic dynamical systems can be developed. Such an issue needs further discussion.

1.4. Book layout

This book will systematically present a theory for n-dimensional nonlinear dynamics from a different point of view. The history and recent development of dynamics will be presented first. The differential geometrical relations between two flows in two different dynamical systems will be presented through G-functions. Based on the G-functions, the global transversality in nonlinear dynamical systems will be investigated. A theory for chaotic layer dynamics for nonlinear Hamiltonian systems will be discussed, which includes the resonant theory of stochastic layers and the stochasticity of resonant layers. The nonlinear dynamics on the ($2n$-1)-dimensional equi-energy surface will be discussed. For dissipative nonlinear dynamical systems, the stability and grazing bifurcations will be discussed in general. Finally, the global dynamics of 2-dimensional dynamical systems will be systematically discussed as an example. Switchability of a flow from a domain to an adjacent domain in discontinuous dynamical systems will be also discussed through the G-functions, and the first incremental increment for discontinuous dynamical system will be given. All the materials are scattered in ten chapters, as summarized as follows.

In Chapter 1, the brief development history of the modern theory of dynamics will be presented. The recent development of chaotic dynamics in nonlinear Hamiltonian systems was discussed first and the corresponding existing methodologies for approximate predictions of the onset, growth and destruction of chaotic layers were reviewed. The current researches on dissipative, nonlinear dynamical systems were addressed and the crucial, unsolved problems existing in nonlinear dynamical systems were briefly discussed. Finally, the summary and layout of contents in this book are given.

In Chapter 2, in two dynamical systems, the concepts for both *compared* and *reference* flows will be introduced to determine flow complexity. It is assumed that the flows of the reference dynamic system always exist on the certain reference surfaces. The time-change rate of the normal distance between two flows in the normal direction of the reference surface will be measured by a new function (i.e., *G*-function). Based on the reference flow, the kth-order *G*-function for the non-contact and l^{th}-order contact flows in two different dynamical systems will be introduced in the normal direction of the reference surface. Through the new functions, the geometric relations between two flows in two dynamical systems will be presented for two flows with and without contact. Finally, the brief discussion of applications will be given. Such G-function quantity will be extensively used to measure the complexity properties of a flow in nonlinear dynamical systems.

In Chapter 3, the first integral quantity for nonlinear dynamical systems will be introduced for the global transversality of a flow to the separatrix. From the first integral quantity, the separatrix surface and initial sets of flows will be introduced, and the local and global flows to the separatrix surface will be discussed. The global transversality of flows to the separatrix surface in nonlinear dynamical systems will be discussed, and the corresponding necessary and sufficient conditions will be presented. The global transversality of a flow to the $2n$-dimensional Hamiltonian systems will be discussed comprehensively. The global transversality of a flow to the generalized separatrix will be also discussed. The global transversality of a flow in nonlinear dynamical system is a basic element to understand the trajectory complexity in phase space, which can very clearly explain the complexity of gradient dynamical systems.

In Chapter 4, a theory for chaotic layer dynamics will be presented. The small domain of flows to a specific, first integral manifold will be introduced in phase space. The first integral quantity increment will be introduced for periodic and chaotic flows in nonlinear dynamic systems. Based on different reference surfaces, all possible expressions for the first integral quantity increment will be given for different applications. The relations of the periodic and chaotic flows with the first integral quantity increments will be presented. The criteria for resonances in the stochastic and resonant layers will be presented through the first integral quantity increment. The first integral quantity increment for periodic flows in nonlinear dynamical systems is zero.

In Chapter 5, a resonant mechanism for stochastic layers in nonlinear Hamiltonian systems will be presented. The mathematical description of the stochastic layers will be given to help one understand the complexity of trajectory in the stochastic layer, and the theory and methodology for approximate predictions of the onset and vanishing of the resonances in stochastic layers will be discussed in detail. Several examples will be presented for demonstrating how to approximately predict the onset, growth and destruction of resonance in the stochastic layers. The maximum and minimum energy spectrum techniques will be presented, and the width computation of the stochastic layers will be estimated. From such a presentation, it will be observed that chaos in the stochastic layer of nonlinear Hamiltonian systems is caused by the resonant separatrix structure instead of the Smale's horseshoe structure.

In Chapter 6, the stochasticity of the resonant separatrix layers (or resonant layers) in nonlinear Hamiltonian systems will be discussed. The renormalization procedure in the vicinity of the resonant separatrix will be presented and the approximate conditions for the onset, growth and disappearance of the resonant layers will be given as well. The maximum and minimum energy increment spectrum method will be presented for numerically detecting resonant layers. Chaos in the resonant layer of nonlinear Hamiltonian systems is caused by the sub-resonant separatrix structure and resonance overlap.

In Chapter 7, the nonlinear dynamics on the (2*n*-1)-dimensional equi-energy space will be presented for 2*n*-dimensional nonlinear Hamiltonian systems. A general methodology for quasi-periodic motions and chaos in such a Hamiltonian system will be presented. Such an idea can be extended to the dynamics on (*n*-1)-dimensional first integral invariant surface for *n*-dimensional dynamical systems. A nonlinear Hamiltonian system with two degrees of freedom will be investigated as a sampled problem.

In Chapter 8, the stability of equilibriums and periodic flows in dissipative dynamical systems will be discussed through the first integral quantity increment. Compared to the Lyapunov stability conditions, the weak stability conditions for equilibriums and periodic flows will be developed. Using the first integral quantity increment, the limit cycle in 2-dimensional nonlinear systems will be briefly discussed. Grazing bifurcation and the mapping structures based on the first integral manifold surface will be discussed for periodic and chaotic flows. The invariant set fragmentation of chaotic motions, caused by the grazing bifurcation, will be presented.

In Chapter 9, the analytical conditions for the global transversality of 2-dimensional, time-dependent, nonlinear dynamical systems will be presented. The first integral quantity increment (i.e., the energy increment) for a certain time interval will be achieved for periodic flows and chaos in the 2-dimensional nonlinear dynamical systems. Under perturbation assumptions and convergent conditions, the Melnikov function can be recovered from the first integral quantity increment. An example will be presented to show how to apply the newly developed theory.

In Chapter 10, the switchability of a flow from one domain to its adjacent domain in discontinuous dynamical systems will be presented. The *G*-function for discontinuous dynamical systems will be introduced. The imaginary flow in discontinuous dynamical systems will be employed. Based on such a *G*-function, the passability conditions for a flow from a domain to the adjacent domains will be investigated. Because of the discontinuity, the nonpassable flow to the separation boundary will be investigated. The tangential flow to the separation boundary will be addressed. Further, the switching bifurcation between the passable and non-passable flows to the separation boundary will be discussed. Finally, the first integral quantity increment for discontinuous dynamical systems will be presented instead of the approximate function (e.g., the Melnikov function) to develop the iterative mapping relations.

Chapter 2

Differential Geometry of Flows

In this Chapter, to investigate the complexity of a flow in a nonlinear dynamical system, a well-behaved dynamical system will be employed. A flow for such a well-behaved dynamical system is called *the reference flow.* It is assumed that the reference flow always lies on a specific surface defined by the well-behaved dynamical system, and such a surface is called *the reference surface.* A flow in an investigated dynamical system is called *the compared flow.* For a small time interval, the time-change rate of a normal distance between the reference and compared flows to the reference surface is measured by a new function (i.e., *G*-function). Based on the reference flow on the reference surface, the kth -order G-function for the non-contact and the lth -order contact of the compared and reference flows will be introduced. A compared flow without contacting a reference flow, *passing through*, *returning back from* and *paralleling to* the corresponding reference surface, will be discussed first. The *tangency* and *passability* of a compared flow to a reference surface with the lth -order contact will be presented. Finally, the brief discussion of applications will be given. The kth-order contact of the compared flow to the reference surface indicates that the compared flow to the reference surface is of the kth -order singularity. However, the compared flow with the kth -order singularity to the reference surface does not mean that the compared flow to the reference surface is of the kth-order contact.

2.1. Normal distance and G-functions

To investigate the differential geometry relations of two flows in two dynamical systems, a concept for the comparison between two flows in two dynamical systems is introduced. Consider two *n*-dimensional dynamical systems

$$\dot{\mathbf{x}} = \mathbf{F}(\mathbf{x}, t, \mathbf{p}) \quad \mathbf{x} \in \mathbb{R}^n, \mathbf{p} \in \mathbb{R}^m \tag{2.1}$$

and

$$\dot{\bar{x}} = \mathbf{f}(\bar{\mathbf{x}}, t, \boldsymbol{\mu}) \quad \bar{\mathbf{x}} \in \mathbb{R}^n, \boldsymbol{\mu} \in \mathbb{R}^{m_1} \tag{2.2}$$

The vector field function $\mathbf{F}(\mathbf{x}, t, \mathbf{p})$ is a C^r -smooth vector function ($r \geq 1$) defined on two open subsets $U_{\mathbf{x}} \subseteq \mathbb{R}^n$, $U_{\mathbf{p}} \subseteq \mathbb{R}^m$ and $U_t \subseteq \mathbb{R}$. With an initial condition $(t_0, \mathbf{x}_0)$, equation (2.1) gives a unique solution $\mathbf{x}(t) = \mathbf{\Phi}(t_0, \mathbf{x}_0, \mathbf{p}, t)$ in $U_{\mathbf{x}}$ and $\mathbf{x}_0 \in U_{\mathbf{x}}$ for $t \in I_{12}$ and $t_0 \in I_{12} \equiv (t_1, t_2) \subseteq \mathbb{R}$. The flow has the semi-group property (i.e., $\mathbf{\Phi}(t_0, \mathbf{x}_0, \mathbf{p}, t+s) = \mathbf{\Phi}(t, \mathbf{\Phi}(t_0, \mathbf{x}_0, \mathbf{p}, t), \mathbf{p}, s)$ with $\mathbf{x}(t_0) = \mathbf{\Phi}(t_0, \mathbf{x}_0, \mathbf{p}, t_0)$).

Similarly, the vector field functions $\mathbf{f}(\bar{\mathbf{x}}, t, \boldsymbol{\mu})$ is a C^r-smooth vector function ($r \geq 1$) defined on two open subsets $U_{\bar{\mathbf{x}}} \subseteq \mathbb{R}^n$, $U_{\boldsymbol{\mu}} \subseteq \mathbb{R}^{m_1}$ and $U_t \subseteq \mathbb{R}$. With an initial condition $(t_0, \bar{\mathbf{x}}_0)$, equation (2.2) has unique solution $\bar{\mathbf{x}}(t) = \mathbf{\Phi}(t_0, \bar{\mathbf{x}}_0, \boldsymbol{\mu}, t)$ in $U_{\bar{\mathbf{x}}}$ and $\bar{\mathbf{x}}_0 \in U_{\bar{\mathbf{x}}}$ for $t \in I_{12}$ and $t_0 \in I_{12} \equiv (t_1, t_2) \subseteq \mathbb{R}$. The semi-group property of such a flow gives $\bar{\mathbf{\Phi}}(t_0, \bar{\mathbf{x}}_0, \boldsymbol{\mu}, t+s) = \bar{\mathbf{\Phi}}(t, \bar{\mathbf{\Phi}}(t_0, \bar{\mathbf{x}}_0, \boldsymbol{\mu}, t), \boldsymbol{\mu}, s)$ with $\bar{\mathbf{x}}(t_0) = \bar{\mathbf{\Phi}}(t_0, \bar{\mathbf{x}}_0, \boldsymbol{\mu}, t_0)$. Suppose there is an ($n$-1)-dimensional surface S_t on which the flow $\bar{\mathbf{x}}_t$ lies for time t, and the surface S_t is determined by

$$\left.\begin{aligned} & F(\bar{\mathbf{x}}, \bar{\mathbf{x}}_0, t) = 0 \text{ and } \nabla F \bullet \dot{\bar{\mathbf{x}}} + \tfrac{\partial F}{\partial t} = 0, \\ & \tfrac{d\bar{x}_i}{f_i} = \tfrac{d\bar{x}_j}{f_j} = \tfrac{dF}{-\partial F/\partial t} \text{ for } i, j = 1, 2, \cdots, n \text{ and } i \neq j; \\ & \mathbf{n}_{S_t} = (\tfrac{\partial F}{\partial \bar{x}_1}, \tfrac{\partial F}{\partial \bar{x}_2}, \cdots, \tfrac{\partial F}{\partial \bar{x}_n})^T. \end{aligned}\right\} \tag{2.3}$$

To determine the dynamical property of the flow in Eq.(2.1) to the surface of flow in Eq.(2.2), the dynamical system in Eq.(2.2) is considered as a referenced dynamical system, and the flow $\bar{\mathbf{x}}_t \equiv \bar{\mathbf{x}}(t) = \mathbf{\Phi}(t_0, \bar{\mathbf{x}}_0, \boldsymbol{\mu}, t)$ for time $t \in [t_0, \infty)$ is termed *the reference flow* on the reference surface. Consider a dynamical system in Eq.(2.1) as a compared dynamical system to be investigated, and the corresponding flow $\mathbf{x}_t \equiv \mathbf{x}(t) = \mathbf{\Phi}(t_0, \mathbf{x}_0, \mathbf{p}, t)$ for time $t \in [t_0, \infty)$ is termed *the compared flow*.

Consider two infinitesimal time intervals $[t-\varepsilon, t)$ and $(t, t+\varepsilon]$ for an arbitrarily small $\varepsilon > 0$. The two flows for two systems in Eqs.(2.1) and (2.2) are sketched in Fig.2.1. The vector difference between the two flows for three time instants are given by $\mathbf{x}_{t-\varepsilon} - \bar{\mathbf{x}}_{t-\varepsilon}$, $\mathbf{x}_t - \bar{\mathbf{x}}_t$ and $\mathbf{x}_{t+\varepsilon} - \bar{\mathbf{x}}_{t+\varepsilon}$. The normal vector of the reference surface and the tangential vectors of the reference flow, at three locations, are expressed by ($\mathbf{n}_{S_{t-\varepsilon}}$, $\mathbf{n}_{S_t}$ and $\mathbf{n}_{S_{t+\varepsilon}}$) and ($\mathbf{t}_{\bar{\mathbf{x}}_{t-\varepsilon}}$, $\mathbf{t}_{\bar{\mathbf{x}}_t}$ and $\mathbf{t}_{\bar{\mathbf{x}}_{t+\varepsilon}}$), respectively. From the normal vectors of the reference surface S_t, the normal component of the position vector difference between the reference and compared flows is defined by

$$d_{t-\varepsilon} = \mathbf{n}^T_{S_{t-\varepsilon}} \cdot (\mathbf{x}_{t-\varepsilon} - \bar{\mathbf{x}}_{t-\varepsilon}),\ d_t = \mathbf{n}^T_{S_t} \cdot (\mathbf{x}_t - \bar{\mathbf{x}}_t),\ d_{t+\varepsilon} = \mathbf{n}^T_{S_{t+\varepsilon}} \cdot (\mathbf{x}_{t+\varepsilon} - \bar{\mathbf{x}}_{t+\varepsilon}); \tag{2.4}$$

where the normal vector of the reference surface S_t at a point $\mathbf{x}_t$ is given by $\mathbf{n}_{S_t} = (\frac{\partial F}{\partial \bar{x}_1}, \cdots, \frac{\partial F}{\partial \bar{x}_n})^{\mathrm{T}}$ for time t and the normal component is the distance of the two points of two flows in the normal direction of the reference surface. Herein, we focus on the change rate of the distance in the normal direction.

DEFINITION 2.1. Consider two dynamical systems in Eqs.(2.1) and (2.2) which have a compared flow $\mathbf{x}_t = \mathbf{\Phi}(t_0, \mathbf{x}_0, \mathbf{p}, t)$ with an initial condition $(t_0, \mathbf{x}_0)$ and a reference flow $\overline{\mathbf{x}}_t = \mathbf{\Phi}(t_0, \overline{\mathbf{x}}_0, \boldsymbol{\mu}, t)$ with an initial condition $(t_0, \overline{\mathbf{x}}_0)$, respectively. Suppose the reference flow $\overline{\mathbf{x}}_t$ lies on a reference surface S_t with $F(\overline{\mathbf{x}}, \overline{\mathbf{x}}_0, t) = 0$ and $\nabla F \bullet \dot{\overline{\mathbf{x}}} + \frac{\partial F}{\partial t} = 0$ for all time t. For an arbitrarily small $\varepsilon > 0$, there are two time intervals $[t-\varepsilon, t)$ and $(t, t+\varepsilon]$. Two *left* and *right* G-functions (G_- and G_+) of the compared flow $\mathbf{x}_t$ to the reference flow $\overline{\mathbf{x}}_t$ in the normal direction of the reference surface S_t are defined as

$$\left.\begin{aligned} G_-(\overline{\mathbf{x}}_t, t, \mathbf{x}_t) &= \lim_{\varepsilon \to 0} \frac{1}{\varepsilon}\left[\mathbf{n}_{S_t}^T \cdot (\mathbf{x}_t - \overline{\mathbf{x}}_t) - \mathbf{n}_{S_{t-\varepsilon}}^T \cdot (\mathbf{x}_{t-\varepsilon} - \overline{\mathbf{x}}_{t-\varepsilon})\right], \\ G_+(\overline{\mathbf{x}}_t, t, \mathbf{x}_t) &= \lim_{\varepsilon \to 0} \frac{1}{\varepsilon}\left[\mathbf{n}_{S_{t+\varepsilon}}^T \cdot (\mathbf{x}_{t+\varepsilon} - \overline{\mathbf{x}}_{t+\varepsilon}) - \mathbf{n}_{S_t}^T \cdot (\mathbf{x}_t - \overline{\mathbf{x}}_t)\right]. \end{aligned}\right\} \tag{2.5}$$

From Eq.(2.5), since $\mathbf{x}_t$ and $\overline{\mathbf{x}}_t$ are the solutions of Eqs.(2.1) and (2.2), their derivatives exist. Further, by use of the Taylor series expansion, equation (2.5) gives

$$\left.\begin{aligned} G_-(\overline{\mathbf{x}}_t, t, \mathbf{x}_t) &= D_{\overline{\mathbf{x}}_t} \mathbf{n}_{S_t}^T \cdot (\mathbf{x}_{t-} - \overline{\mathbf{x}}_t) + \mathbf{n}_{S_t}^T \cdot (\dot{\mathbf{x}}_{t-} - \dot{\overline{\mathbf{x}}}_t), \\ G_+(\overline{\mathbf{x}}_t, t, \mathbf{x}_t) &= D_{\overline{\mathbf{x}}_t} \mathbf{n}_{S_t}^T \cdot (\mathbf{x}_{t+} - \overline{\mathbf{x}}_t) + \mathbf{n}_{S_t}^T \cdot (\dot{\mathbf{x}}_{t+} - \dot{\overline{\mathbf{x}}}_t) \end{aligned}\right\} \tag{2.6}$$

where the total derivative $D_{\overline{\mathbf{x}}_t}(\cdot) = \frac{\partial(\cdot)}{\partial \overline{\mathbf{x}}_t}\dot{\overline{\mathbf{x}}}_t + \frac{\partial(\cdot)}{\partial t}$. Due to the continuity, $G_- = G_+ = G$. Thus, the *G-function* of the compared flow $\mathbf{x}_t$ to the reference flow $\overline{\mathbf{x}}_t$ on the normal direction of the reference surface S_t is defined as

$$\begin{aligned} G(\overline{\mathbf{x}}_t, t, \mathbf{x}_t) &= D_{\overline{\mathbf{x}}_t} \mathbf{n}_{S_t}^T \cdot (\mathbf{x}_t - \overline{\mathbf{x}}_t) + \mathbf{n}_{S_t}^T \cdot (\dot{\mathbf{x}}_t - \dot{\overline{\mathbf{x}}}_t) \\ &= \left[\frac{\partial \mathbf{n}_{S_t}^T}{\partial \overline{\mathbf{x}}_t}\dot{\overline{\mathbf{x}}}_t + \frac{\partial \mathbf{n}_{S_t}^T}{\partial t}\right] \cdot (\mathbf{x}_t - \overline{\mathbf{x}}_t) + \mathbf{n}_{S_t}^T \cdot (\dot{\mathbf{x}}_t - \dot{\overline{\mathbf{x}}}_t). \end{aligned} \tag{2.7}$$

Notice that if the reference surface S_t is a hyper-surface for discontinuous dynamical systems, the *left* and *right* G-functions (i.e., $G_-(\overline{\mathbf{x}}_t, t, \mathbf{x}_t)$ and $G_+(\overline{\mathbf{x}}_t, t, \mathbf{x}_t)$) are different. Using Eqs.(2.1) and (2.2), the *G-function* in Eq.(2.7) becomes

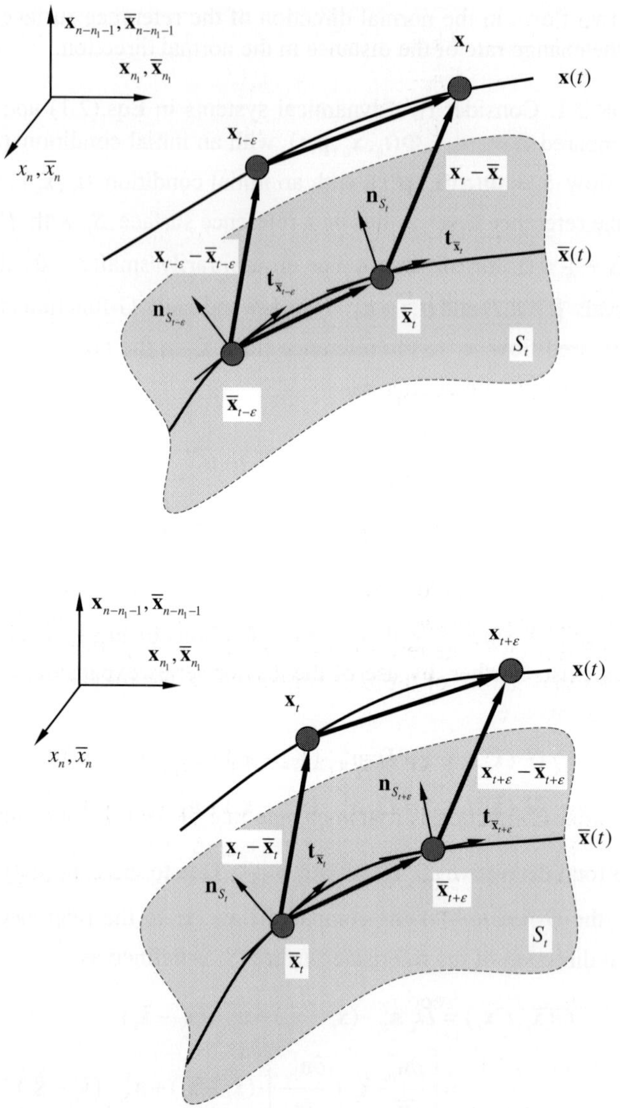

(a)

(b)

Figure 2.1. Differential flows of the two dynamical system changes for: (a) infinitesimal time interval $[t-\varepsilon, t]$ and (b) infinitesimal time interval $[t, t+\varepsilon]$. The referenced flow is $\bar{\mathbf{x}}(t)$ and the compared flow is $\mathbf{x}(t)$. The normal vectors $\mathbf{n}_{\bar{\mathbf{x}}_{t-\varepsilon}}$, $\mathbf{n}_{\bar{\mathbf{x}}_t}$ and $\mathbf{n}_{\bar{\mathbf{x}}_{t+\varepsilon}}$ on the referenced flow are expressed.

$$\begin{aligned}G(\bar{\mathbf{x}}_t,t,\mathbf{x}_t) &= D_{\bar{\mathbf{x}}_t}\mathbf{n}_{S_t}^T\cdot(\mathbf{x}_t-\bar{\mathbf{x}}_t)+\mathbf{n}_{S_t}^T\cdot[\mathbf{F}(\mathbf{x}_t,t,\mathbf{p})-\mathbf{f}(\bar{\mathbf{x}}_t,t,\boldsymbol{\mu})]\\ &=\left[\frac{\partial\mathbf{n}_{S_t}^T}{\partial\bar{\mathbf{x}}_t}\mathbf{f}(\bar{\mathbf{x}}_t,t,\boldsymbol{\mu})+\frac{\partial\mathbf{n}_{S_t}^T}{\partial t}\right]\cdot(\mathbf{x}_t-\bar{\mathbf{x}}_t)\\ &\quad+\mathbf{n}_{S_t}^T\cdot[\mathbf{F}(\mathbf{x}_t,t,\mathbf{p})-\mathbf{f}(\bar{\mathbf{x}}_t,t,\boldsymbol{\mu})].\end{aligned}\tag{2.8}$$

The formulas in Eqs.(2.7) and (2.8) are given for the non-contact between the compared and reference flows. As in Kreyszig (1959), consider the zero-order contact of the compared and reference flows at the instant time t, i.e.,

$$\mathbf{x}_t=\bar{\mathbf{x}}_t.\tag{2.9}$$

The corresponding G-functions for the zero-order contact between the reference and compared flows become

$$\begin{aligned}G_0(\bar{\mathbf{x}}_t,t,\mathbf{x}_t)=G(\bar{\mathbf{x}}_t,t,\mathbf{x}_t)\big|_{\mathbf{x}_t=\bar{\mathbf{x}}_t} &=\mathbf{n}_{S_t}^T\cdot(\dot{\mathbf{x}}_t-\dot{\bar{\mathbf{x}}}_t)\\ &=\mathbf{n}_{S_t}^T\cdot[\mathbf{F}(\mathbf{x}_t,t,\mathbf{p})-\mathbf{f}(\bar{\mathbf{x}}_t,t,\boldsymbol{\mu})].\end{aligned}\tag{2.10}$$

From the foregoing equation, the G-function for the zero-order contact of the flows is the normal component of the corresponding vector field difference on the normal direction of the reference surface. Because the derivatives to the vector fields in Eq.(2.8) are of the zero-order, such a *G-function* is called the *zero-order G-function*, i.e.,

$$G^{(0)}(\bar{\mathbf{x}}_t,t,\mathbf{x}_t)=G(\bar{\mathbf{x}}_t,t,\mathbf{x}_t).\tag{2.11}$$

The zero-order G-function is a time-change rate of the relative distance between the compared and reference flows on the normal direction of the reference surface S_t. Therefore, it can be called *the relative speed* in the normal direction of the reference surface S_t. To investigate the time-change rate of such a speed, the *first order, left and right G-functions* of the compared flow $\mathbf{x}_t$ to the reference flow $\bar{\mathbf{x}}_t$ on the normal direction of the reference surface S_t are defined as

$$\left.\begin{aligned}&G_-^{(1)}(\bar{\mathbf{x}}_t,t,\mathbf{x}_t)\\ &=-\lim_{\varepsilon\to0}\frac{1}{\varepsilon^2}\left[\mathbf{n}_{S_t}^T\cdot(\mathbf{x}_t-\bar{\mathbf{x}}_t)-\mathbf{n}_{S_{t-\varepsilon}}^T\cdot(\mathbf{x}_{t-\varepsilon}-\bar{\mathbf{x}}_{t-\varepsilon})-G_-^{(0)}(\bar{\mathbf{x}}_t,t,\mathbf{x}_t)\varepsilon\right],\\ &G_+^{(1)}(\bar{\mathbf{x}}_t,t,\mathbf{x}_t)\\ &=\lim_{\varepsilon\to0}\frac{1}{\varepsilon^2}\left[\mathbf{n}_{S_{t+\varepsilon}}^T\cdot(\mathbf{x}_{t+\varepsilon}-\bar{\mathbf{x}}_{t+\varepsilon})-\mathbf{n}_{S_t}^T\cdot(\mathbf{x}_t-\bar{\mathbf{x}}_t)-G_+^{(0)}(\bar{\mathbf{x}}_t,t,\mathbf{x}_t)\varepsilon\right].\end{aligned}\right\}\tag{2.12}$$

Using the Taylor series, the foregoing definition gives

$$\left.\begin{aligned} G_{-}^{(1)}(\bar{\mathbf{x}}_t,t,\mathbf{x}_t) &= D_{\bar{\mathbf{x}}_t}^{(2)}\mathbf{n}_{S_t}^T\cdot(\mathbf{x}_{t-}-\bar{\mathbf{x}}_t)+2D_{\bar{\mathbf{x}}_t}\mathbf{n}_{S_t}^T\cdot(\dot{\mathbf{x}}_{t-}-\dot{\bar{\mathbf{x}}}_t)\\ &\quad+\mathbf{n}_{S_t}^T\cdot(\ddot{\mathbf{x}}_{t-}-\ddot{\bar{\mathbf{x}}}_t),\\ G_{+}^{(1)}(\bar{\mathbf{x}}_t,t,\mathbf{x}_t) &= D_{\bar{\mathbf{x}}_t}^{(2)}\mathbf{n}_{S_t}^T\cdot(\mathbf{x}_{t+}-\bar{\mathbf{x}}_t)+2D_{\bar{\mathbf{x}}_t}\mathbf{n}_{S_t}^T\cdot(\dot{\mathbf{x}}_{t+}-\dot{\bar{\mathbf{x}}}_t)\\ &\quad+\mathbf{n}_{S_t}^T\cdot(\ddot{\mathbf{x}}_{t+}-\ddot{\bar{\mathbf{x}}}_t). \end{aligned}\right\} \tag{2.13}$$

Using Eqs.(2.1) and (2.2), we have the first-order G-functions as

$$\left.\begin{aligned} G_{-}^{(1)}(\bar{\mathbf{x}}_t,t,\mathbf{x}_t) &= D_{\bar{\mathbf{x}}_t}^{(2)}\mathbf{n}_{S_t}^T\cdot(\mathbf{x}_{t-}-\bar{\mathbf{x}}_t)\\ &\quad+2D_{\bar{\mathbf{x}}_t}\mathbf{n}_{S_t}^T\cdot(\mathbf{F}(\mathbf{x}_{t-},t,\mathbf{p})-\mathbf{f}(\bar{\mathbf{x}}_t,t,\boldsymbol{\mu}))\\ &\quad+\mathbf{n}_{S_t}^T\cdot(D_{\mathbf{x}_t}\mathbf{F}(\mathbf{x}_{t-},t,\mathbf{p})-D_{\bar{\mathbf{x}}_t}\mathbf{f}(\bar{\mathbf{x}}_t,t,\boldsymbol{\mu})),\\ G_{+}^{(1)}(\bar{\mathbf{x}}_t,t,\mathbf{x}_t) &= D_{\bar{\mathbf{x}}_t}^{(2)}\mathbf{n}_{S_t}^T\cdot(\mathbf{x}_{t+}-\bar{\mathbf{x}}_t)\\ &\quad+2D_{\bar{\mathbf{x}}_t}\mathbf{n}_{S_t}^T\cdot(\mathbf{F}(\mathbf{x}_{t+},t,\mathbf{p})-\mathbf{f}(\bar{\mathbf{x}}_t,t,\boldsymbol{\mu}))\\ &\quad+\mathbf{n}_{S_t}^T\cdot(D_{\mathbf{x}_t}\mathbf{F}(\mathbf{x}_{t+},t,\mathbf{p})-D_{\bar{\mathbf{x}}_t}\mathbf{f}(\bar{\mathbf{x}}_t,t,\boldsymbol{\mu})). \end{aligned}\right\} \tag{2.14}$$

where

$$\left.\begin{aligned} D_{\mathbf{x}_t}(\cdot) &= \tfrac{\partial(\cdot)}{\partial \mathbf{x}_t}\dot{\mathbf{x}}_t+\tfrac{\partial(\cdot)}{\partial t} = \tfrac{\partial(\cdot)}{\partial \mathbf{x}_t}\mathbf{F}(\mathbf{x}_t,t,\mathbf{p})+\tfrac{\partial(\cdot)}{\partial t} \text{ and}\\ D_{\bar{\mathbf{x}}_t}(\cdot) &= \tfrac{\partial(\cdot)}{\partial \bar{\mathbf{x}}_t}\dot{\bar{\mathbf{x}}}_t+\tfrac{\partial(\cdot)}{\partial t} = \tfrac{\partial(\cdot)}{\partial \bar{\mathbf{x}}_t}\mathbf{f}(\bar{\mathbf{x}}_t,t,\boldsymbol{\mu})+\tfrac{\partial(\cdot)}{\partial t}. \end{aligned}\right\} \tag{2.15}$$

Due to the continuity, $G_{-}^{(1)} = G_{+}^{(1)} = G^{(1)}$. The first-order G-functions is the time change rate of the relative, normal speed in the normal direction of the reference surface, and such a change rate is the relative acceleration in the normal direction of the reference surface S_t. The extension of the above idea gives the definition of the kth-order, *left* and *right* *G*-functions of the compared flow $\mathbf{x}_t$ to the reference flow $\bar{\mathbf{x}}_t$ on the normal direction of the reference surface S_t as follows:

DEFINITION 2.2. Consider two dynamical systems in Eqs.(2.1) and (2.2) which have a compared flow $\mathbf{x}_t = \boldsymbol{\Phi}(t_0,\mathbf{x}_0,\mathbf{p},t)$ with an initial condition $(t_0,\mathbf{x}_0)$ and a reference flow $\bar{\mathbf{x}}_t = \boldsymbol{\Phi}(t_0,\bar{\mathbf{x}}_0,\boldsymbol{\mu},t)$ with an initial condition $(t_0,\bar{\mathbf{x}}_0)$, respectively. Suppose the reference flow $\bar{\mathbf{x}}_t$ lies on a reference surface S_t with $F(\bar{\mathbf{x}},\bar{\mathbf{x}}_0,t)=0$ and $\nabla F\bullet\dot{\bar{\mathbf{x}}}+\frac{\partial F}{\partial t}=0$ for all time *t*. For an arbitrarily small $\varepsilon>0$, there are two time intervals $[t-\varepsilon,t)$ and $(t,t+\varepsilon]$. The vector fields $\mathbf{F}(\mathbf{x},t,\mathbf{p})$ and $\mathbf{f}(\bar{\mathbf{x}},t,\boldsymbol{\mu})$ are $C^r_{[t-\varepsilon,t+\varepsilon]}$-continuous ($r\ge k$) for time *t*. The flow $\mathbf{x}_t$ and $\bar{\mathbf{x}}_t$ are $C^r_{[t-\varepsilon,t+\varepsilon]}$-continuous ($r\ge k+1$) for time *t*, $\| d^{r+1}\mathbf{x}_t/dt^{r+1}\|<\infty$ and $\| d^{r+1}\bar{\mathbf{x}}_t/dt^{r+1}\|<\infty$.

The kth-*order, left and right G-functions* of the compared flow $\mathbf{x}_t$ to the reference flow $\overline{\mathbf{x}}_t$ in the normal direction of the reference surface S_t are defined as

$$\left.\begin{aligned}
G_-^{(k)}(\overline{\mathbf{x}}_t,t,\mathbf{x}_{t-}) &= \lim_{\varepsilon\to 0}\frac{(-1)^{k+2}}{\varepsilon^{k+1}}\Big[\mathbf{n}_{S_t}^T\cdot(\mathbf{x}_{t-}-\overline{\mathbf{x}}_t)-\mathbf{n}_{S_{t-\varepsilon}}^T\cdot(\mathbf{x}_{t-\varepsilon}-\overline{\mathbf{x}}_{t-\varepsilon})\\
&\quad+\sum_{r=0}^{k-1}G_-^{(r)}(\overline{\mathbf{x}}_t,t,\mathbf{x}_{t-})(-\varepsilon)^{r+1}\Big],\\
G_+^{(k)}(\overline{\mathbf{x}}_t,t,\mathbf{x}_{t+}) &= \lim_{\varepsilon\to 0}\frac{1}{\varepsilon^{k+1}}\Big[\mathbf{n}_{S_{t+\varepsilon}}^T\cdot(\mathbf{x}_{t+\varepsilon}-\overline{\mathbf{x}}_{t+\varepsilon})-\mathbf{n}_{S_t}^T\cdot(\mathbf{x}_{t+}-\overline{\mathbf{x}}_t)\\
&\quad-\sum_{r=0}^{k-1}G_+^{(r)}(\overline{\mathbf{x}}_t,t,\mathbf{x}_{t+})\varepsilon^{r+1}\Big].
\end{aligned}\right\}\tag{2.16}$$

Again, application of the Taylor series expansion of $\mathbf{n}_{S_{t-\varepsilon}}^T\cdot(\mathbf{x}_{t-\varepsilon}-\overline{\mathbf{x}}_{t-\varepsilon})$ and $\mathbf{n}_{S_{t+\varepsilon}}^T\cdot(\mathbf{x}_{t+\varepsilon}-\overline{\mathbf{x}}_{t+\varepsilon})$ to the foregoing equation yields

$$\left.\begin{aligned}
G_-^{(k)}(\overline{\mathbf{x}}_t,t,\mathbf{x}_{t-}) &= \sum_{r=0}^{k+1}C_{k+1}^r D_{\overline{\mathbf{x}}_t}^{k+1-r}\mathbf{n}_{S_t}^T\cdot\Big(\frac{d^r\mathbf{x}_t}{dt^r}-\frac{d^r\overline{\mathbf{x}}_t}{dt^r}\Big)\Big|_{(\overline{\mathbf{x}}_t,t,\mathbf{x}_{t-})},\\
G_+^{(k)}(\overline{\mathbf{x}}_t,t,\mathbf{x}_{t+}) &= \sum_{r=0}^{k+1}C_{k+1}^r D_{\overline{\mathbf{x}}_t}^{k+1-r}\mathbf{n}_{S_t}^T\cdot\Big(\frac{d^r\mathbf{x}_t}{dt^r}-\frac{d^r\overline{\mathbf{x}}_t}{dt^r}\Big)\Big|_{(\overline{\mathbf{x}}_t,t,\mathbf{x}_{t+})}.
\end{aligned}\right\}\tag{2.17}$$

Due to the continuity, $G_-^{(k)}=G_+^{(k)}=G^{(k)}$. The kth-*order G-function* at time t is defined as

$$G^{(k)}(\overline{\mathbf{x}}_t,t,\mathbf{x}_t)=\sum_{r=0}^{k+1}C_{k+1}^r D_{\overline{\mathbf{x}}_t}^{k+1-r}\mathbf{n}_{S_t}^T\cdot\Big(\frac{d^r\mathbf{x}_t}{dt^r}-\frac{d^r\overline{\mathbf{x}}_t}{dt^r}\Big).\tag{2.18}$$

Using Eqs.(2.1) and (2.2), the kth-order G-function of the compared flow $\mathbf{x}_t$ to the reference flow $\overline{\mathbf{x}}_t$ on the normal direction of the reference surface S_t becomes

$$\begin{aligned}
G^{(k)}(\overline{\mathbf{x}}_t,t,\mathbf{x}_t) &= \sum_{r=1}^{k+1}C_{k+1}^r D_{\overline{\mathbf{x}}_t}^{k+1-r}\mathbf{n}_{S_t}^T\cdot(D_{\mathbf{x}_t}^{r-1}\mathbf{F}(\mathbf{x}_t,t,\mathbf{p})-D_{\overline{\mathbf{x}}_t}^{r-1}\mathbf{f}(\overline{\mathbf{x}}_t,t,\boldsymbol{\mu}))\\
&\quad+D_{\overline{\mathbf{x}}_t}^{k+1}\mathbf{n}_{S_t}^T\cdot(\mathbf{x}_t-\overline{\mathbf{x}}_t),
\end{aligned}\tag{2.19}$$

with $C_{k+1}^r=\frac{(k+1)k(k-1)\cdots(k+1-r)}{r!}$ and $C_{k+1}^0=1$ and $r!=1\times 2\times\cdots\times r$.

From Differential Geometry in Kreyszig(1959), the lth-*order* contact between the reference and compared flows requires

$$\frac{d^s\mathbf{x}_t}{dt^s}=\frac{d^s\overline{\mathbf{x}}_t}{dt^s}\ \text{for}\ s=0,1,2,\cdots,l;\tag{2.20}$$

or

$$\left.\begin{aligned}&\mathbf{x}_t = \overline{\mathbf{x}}_t \text{ and } D_{\mathbf{x}_t}^{s-1}\mathbf{F}(\mathbf{x}_t,t,\mathbf{p}) = D_{\overline{\mathbf{x}}_t}^{s-1}\mathbf{f}(\overline{\mathbf{x}}_t,t,\boldsymbol{\mu})\\ &\text{for } s = 1,2,\cdots,l.\end{aligned}\right. \tag{2.21}$$

Further, the kth-*order* G-function in Eqs.(2.18) and (2.19) for the kth-*order* contact becomes

$$\begin{aligned}G^{(k)}(\overline{\mathbf{x}}_t,t,\mathbf{x}_t) &= \mathbf{n}_{S_t}^T\cdot\left(\frac{d^{k+1}\mathbf{x}_t}{dt^{k+1}} - \frac{d^{k+1}\overline{\mathbf{x}}_t}{dt^{k+1}}\right)\\ &= \mathbf{n}_{S_t}^T\cdot\left(D_{\mathbf{x}_t}^{k}\mathbf{F}(\mathbf{x}_t,t,\mathbf{p}) - D_{\overline{\mathbf{x}}_t}^{k}\mathbf{f}(\overline{\mathbf{x}}_t,t,\boldsymbol{\mu})\right).\end{aligned} \tag{2.22}$$

Therefore, the theory for discontinuous dynamical systems in Luo (2005a,2006b) is applicable only for the lth-order ($l \geq 0$) contact between the separation boundary and flows, which is because the lth-order singularity is different from the lth-order contact. Similarly, in Luo (2007a), the theory for n-dimensional, nonlinear dynamics on continuous vector fields is also applicable only for the lth-order contact between the unperturbed and perturbed flows. The lth-order contact between two flows must be the lth-order singularity. However, the lth-order singularity between two flows may not be the lth-order contact. To make such a concept clear, the kth-order G-function for the compared and reference flows with the lth-order contact is introduced as follows.

DEFINITION 2.3. Consider two dynamical systems in Eqs.(2.1) and (2.2) which have a compared flow $\mathbf{x}_t = \boldsymbol{\Phi}(t_0,\mathbf{x}_0,\mathbf{p},t)$ with an initial condition $(t_0,\mathbf{x}_0)$ and a reference flow $\overline{\mathbf{x}}_t = \boldsymbol{\Phi}(t_0,\overline{\mathbf{x}}_0,\boldsymbol{\mu},t)$ with an initial condition $(t_0,\overline{\mathbf{x}}_0)$, respectively. Suppose the reference flow $\overline{\mathbf{x}}_t$ lies on a reference surface S_t with $F(\overline{\mathbf{x}},\overline{\mathbf{x}}_0,t) = 0$ and $\nabla F \bullet \dot{\overline{\mathbf{x}}} + \frac{\partial F}{\partial t} = 0$ for all time t. For an arbitrarily small $\varepsilon > 0$, there are two time intervals $[t-\varepsilon,t)$ and $(t,t+\varepsilon]$. The vector fields $\mathbf{F}(\mathbf{x},t,\mathbf{p})$ and $\mathbf{f}(\overline{\mathbf{x}},t,\boldsymbol{\mu})$ are $C_{[t-\varepsilon,t+\varepsilon]}^{r}$-continuous ($r \geq k$) for time t. The flow $\mathbf{x}_t$ and $\overline{\mathbf{x}}_t$ are $C_{[t-\varepsilon,t+\varepsilon]}^{r}$-continuous ($r \geq k+1$) for t, $\| d^{r+1}\mathbf{x}_t/dt^{r+1} \| < \infty$ and $\| d^{r+1}\overline{\mathbf{x}}_t/dt^{r+1} \| < \infty$. For the lth-order ($l \geq 0$) contact of the compared and reference flows, the kth-order G-function of a compared flow $\mathbf{x}_t$ to a reference flow $\overline{\mathbf{x}}_t$ on the normal direction of the reference surface S_t is defined as

$$\left.\begin{aligned}&\frac{d^s\mathbf{x}_t}{dt^s} = \frac{d^s\overline{\mathbf{x}}_t}{dt^s} \text{ for } s = 0,1,2,\cdots,l\\ &G_l^{(k)}(\overline{\mathbf{x}}_t,t,\mathbf{x}_t) = \sum\nolimits_{s=l}^{k+1} C_{k+1}^{s} D_{\overline{\mathbf{x}}_t}^{k+1-s}\mathbf{n}_{S_t}^T\cdot\left(\frac{d^s\mathbf{x}_t}{dt^s} - \frac{d^s\overline{\mathbf{x}}_t}{dt^s}\right)\end{aligned}\right\} \tag{2.23}$$

or

$$\left.\begin{array}{l}\mathbf{x}_t = \bar{\mathbf{x}}_t, \text{ and } D_{\mathbf{x}_t}^{s-1}\mathbf{F}(\mathbf{x}_t,t,\mathbf{p}) = D_{\bar{\mathbf{x}}_t}^{s-1}\mathbf{f}(\bar{\mathbf{x}}_t,t,\boldsymbol{\mu}) \\ \quad \text{for } s = 0,1,2,\cdots,l \\ G_l^{(k)}(\bar{\mathbf{x}}_t,t,\mathbf{x}_t) = \sum_{s=l}^{k+1} C_{k+1}^s D_{\bar{\mathbf{x}}_t}^{k+1-s}\mathbf{n}_{S_t}^T \cdot (D_{\mathbf{x}_t}^{s-1}\mathbf{F}(\mathbf{x}_t,t,\mathbf{p}) - D_{\bar{\mathbf{x}}_t}^{s-1}\mathbf{f}(\bar{\mathbf{x}}_t,t,\boldsymbol{\mu})).\end{array}\right\} \quad (2.24)$$

If $l = k$, $G_l^{(k)}(\bar{\mathbf{x}}_t,t,\mathbf{x}_t)$ gives

$$\left.\begin{array}{l}G_k^{(s)}(\bar{\mathbf{x}}_t,t,\mathbf{x}_t) = 0 \text{ for s} = 0,1,\cdots,k\text{-}1, \\ G_k^{(k)}(\bar{\mathbf{x}}_t,t,\mathbf{x}_t) = \mathbf{n}_{S_t}^T \cdot \left(\dfrac{d^{k+1}\mathbf{x}_t}{dt^{k+1}} - \dfrac{d^{k+1}\bar{\mathbf{x}}_t}{dt^{k+1}}\right) \\ \qquad = \mathbf{n}_{S_t}^T \cdot (D_{\mathbf{x}_t}^k\mathbf{F}(\mathbf{x}_t,t,\mathbf{p}) - D_{\bar{\mathbf{x}}_t}^k\mathbf{f}(\bar{\mathbf{x}}_t,t,\boldsymbol{\mu})).\end{array}\right\} \quad (2.25)$$

2.2. Non-contact flows

In this section, a geometric relation of a compared flow to the reference surface without any contact between the reference and compared flows in two different dynamical systems will be discussed through G-function. The basic concepts for a flow passing through, returning back from and paralleling to the reference surface S_t will be introduced first.

DEFINITION 2.4. Consider two dynamical systems in Eqs.(2.1) and (2.2) which have a compared flow $\mathbf{x}_t = \boldsymbol{\Phi}(t_0,\mathbf{x}_0,\mathbf{p},t)$ with an initial condition $(t_0,\mathbf{x}_0)$ and a reference flow $\bar{\mathbf{x}}_t = \boldsymbol{\Phi}(t_0,\bar{\mathbf{x}}_0,\boldsymbol{\mu},t)$ with an initial condition $(t_0,\bar{\mathbf{x}}_0)$, respectively. The flow $\mathbf{x}_t$ does not contact with the reference flow $\bar{\mathbf{x}}_t$. Suppose the reference flow $\bar{\mathbf{x}}_t$ lies on a reference surface S_t with $F(\bar{\mathbf{x}},\bar{\mathbf{x}}_0,t) = 0$ and $\nabla F \bullet \dot{\bar{\mathbf{x}}} + \frac{\partial F}{\partial t} = 0$ for time *t*. For an arbitrarily small $\varepsilon > 0$, there are two time intervals $[t_m - \varepsilon, t_m)$ and $(t_m, t_m + \varepsilon]$.

(i) The flow $\mathbf{x}_t$ to the reference flow $\bar{\mathbf{x}}_t$ is *leaving* from the reference surface S_t at time t_m with $\mathbf{x}_{t_m} \notin S_t$ if for $t \in (t_m, t_m + \varepsilon]$,

$$\left.\begin{array}{l}\mathbf{n}_{S_{t_{m+\varepsilon}}}^T \cdot (\mathbf{x}_{t_m+\varepsilon} - \bar{\mathbf{x}}_{t_m+\varepsilon}) - \mathbf{n}_{S_{t_m}}^T \cdot (\mathbf{x}_{t_m} - \bar{\mathbf{x}}_{t_m}) > 0 \text{ for } \mathbf{n}_{S_t}^T \cdot (\mathbf{x}_t - \bar{\mathbf{x}}_t) > 0 \text{ or} \\ \mathbf{n}_{S_{t_{m+\varepsilon}}}^T \cdot (\mathbf{x}_{t_m+\varepsilon} - \bar{\mathbf{x}}_{t_m+\varepsilon}) - \mathbf{n}_{S_{t_m}}^T \cdot (\mathbf{x}_{t_m} - \bar{\mathbf{x}}_{t_m}) < 0 \text{ for } \mathbf{n}_{S_t}^T \cdot (\mathbf{x}_t - \bar{\mathbf{x}}_t) < 0.\end{array}\right\} \quad (2.26)$$

(ii) The flow $\mathbf{x}_t$ to the reference flow $\bar{\mathbf{x}}_t$ is *approaching* the reference surface S_t at time t_m with $\mathbf{x}_{t_m} \notin S_t$ if for $t \in [t_m - \varepsilon, t_m)$,

$$\left.\begin{array}{l}\mathbf{n}^T_{S_{t_m}}\cdot(\mathbf{x}_{t_m}-\bar{\mathbf{x}}_{t_m})-\mathbf{n}^T_{S_{t_{m-\varepsilon}}}\cdot(\mathbf{x}_{t_{m-\varepsilon}}-\bar{\mathbf{x}}_{t_{m-\varepsilon}})>0 \quad \text{for } \mathbf{n}^T_{S_t}\cdot(\bar{\mathbf{x}}_t-\mathbf{x}_t)>0 \text{ or}\\ \mathbf{n}^T_{S_{t_m}}\cdot(\mathbf{x}_{t_m}-\bar{\mathbf{x}}_{t_m})-\mathbf{n}^T_{S_{t_{m-\varepsilon}}}\cdot(\mathbf{x}_{t_{m-\varepsilon}}-\bar{\mathbf{x}}_{t_{m-\varepsilon}})<0 \quad \text{for } \mathbf{n}^T_{\bar{\mathbf{x}}_t}\cdot(\bar{\mathbf{x}}_t-\mathbf{x}_t)<0.\end{array}\right\} \tag{2.27}$$

(iii) The flow $\mathbf{x}_t$ to the reference flow $\bar{\mathbf{x}}_t$ is *paralleling to* the reference surface S_t at time t_m with $\mathbf{x}_{t_m} \notin S_t$ if

$$\left.\begin{array}{l}\mathbf{n}^T_{S_{t_m}}\cdot(\mathbf{x}_{t_m}-\bar{\mathbf{x}}_{t_m})=\mathbf{n}^T_{S_{t_{m-\varepsilon}}}\cdot(\mathbf{x}_{t_{m-\varepsilon}}-\bar{\mathbf{x}}_{t_{m-\varepsilon}})\neq 0 \text{ for } t\in[t_m-\varepsilon,t_m),\\ \mathbf{n}^T_{S_{t_{m+\varepsilon}}}\cdot(\mathbf{x}_{t_{m+\varepsilon}}-\bar{\mathbf{x}}_{t_{m+\varepsilon}})=\mathbf{n}^T_{S_{t_m}}\cdot(\mathbf{x}_{t_m}-\bar{\mathbf{x}}_{t_m})\neq 0 \text{ for } t\in(t_m,t_m+\varepsilon].\end{array}\right\} \tag{2.28}$$

DEFINITION 2.5. Consider two dynamical systems in Eqs.(2.1) and (2.2) which have a compared flow $\mathbf{x}_t=\mathbf{\Phi}(t_0,\mathbf{x}_0,\mathbf{p},t)$ with an initial condition $(t_0,\mathbf{x}_0)$ and a reference flow $\bar{\mathbf{x}}_t=\mathbf{\Phi}(t_0,\bar{\mathbf{x}}_0,\boldsymbol{\mu},t)$ with an initial condition $(t_0,\bar{\mathbf{x}}_0)$, respectively. The flow $\mathbf{x}_t$ does not contact with the reference flow $\bar{\mathbf{x}}_t$. Suppose the reference flow $\bar{\mathbf{x}}_t$ lies on a reference surface S_t with $F(\bar{\mathbf{x}},\bar{\mathbf{x}}_0,t)=0$ and $\nabla F\bullet\dot{\bar{\mathbf{x}}}+\frac{\partial F}{\partial t}=0$ for time t. For an arbitrarily small $\varepsilon>0$, there are two time intervals $[t_{m-\varepsilon},t_m)$ and $(t_m,t_{m+\varepsilon}]$.

(i) The flow $\mathbf{x}_t$ at point $(t_m,\mathbf{x}_{t_m})$ with $\mathbf{x}_{t_m}=\bar{\mathbf{x}}'_{t_m}\in S_t$ is *passing through* the reference surface S_t if

$$\left.\begin{array}{l}\mathbf{n}^T_{S_{t_m}}\cdot(\mathbf{x}_{t_m}-\bar{\mathbf{x}}_{t_m})-\mathbf{n}^T_{S_{t_{m-\varepsilon}}}\cdot(\mathbf{x}_{t_{m-\varepsilon}}-\bar{\mathbf{x}}_{t_{m-\varepsilon}})>0,\\ \mathbf{n}^T_{S_{t_{m+\varepsilon}}}\cdot(\mathbf{x}_{t_{m+\varepsilon}}-\bar{\mathbf{x}}_{t_{m+\varepsilon}})-\mathbf{n}^T_{S_{t_m}}\cdot(\mathbf{x}_{t_m}-\bar{\mathbf{x}}_{t_m})>0\end{array}\right\} \tag{2.29}$$

with $\mathbf{n}^T_{S_t}\cdot(\bar{\mathbf{x}}_t-\mathbf{x}_t)>0$ for $t\in[t_m-\varepsilon,t_m)$ or

$$\left.\begin{array}{l}\mathbf{n}^T_{S_{t_m}}\cdot(\bar{\mathbf{x}}_{t_m}-\mathbf{x}_{t_m})-\mathbf{n}^T_{S_{t_{m-\varepsilon}}}\cdot(\mathbf{x}_{t_{m-\varepsilon}}-\bar{\mathbf{x}}_{t_{m-\varepsilon}})<0,\\ \mathbf{n}^T_{S_{t_{m+\varepsilon}}}\cdot(\mathbf{x}_{t_{m+\varepsilon}}-\bar{\mathbf{x}}_{t_{m+\varepsilon}})-\mathbf{n}^T_{S_{t_m}}\cdot(\mathbf{x}_{t_m}-\bar{\mathbf{x}}_{t_m})<0.\end{array}\right\} \tag{2.30}$$

with $\mathbf{n}^T_{S_t}\cdot(\bar{\mathbf{x}}_t-\mathbf{x}_t)<0$ for $t\in[t_m-\varepsilon,t_m)$.

(ii) The flow $\mathbf{x}_t$ at point $(t_m,\mathbf{x}_{t_m})$ with $\mathbf{x}_{t_m}\notin S_t$ is *returning back* and away from the reference surface S_t if

$$\left.\begin{array}{l}\mathbf{n}^T_{S_{t_m}}\cdot(\mathbf{x}_{t_m}-\bar{\mathbf{x}}_{t_m})-\mathbf{n}^T_{S_{t_{m-\varepsilon}}}\cdot(\mathbf{x}_{t_{m-\varepsilon}}-\bar{\mathbf{x}}_{t_{m-\varepsilon}})>0,\\ \mathbf{n}^T_{S_{t_{m+\varepsilon}}}\cdot(\mathbf{x}_{t_{m+\varepsilon}}-\bar{\mathbf{x}}_{t_{m+\varepsilon}})-\mathbf{n}^T_{S_{t_m}}\cdot(\mathbf{x}_{t_m}-\bar{\mathbf{x}}_{t_m})<0;\end{array}\right\} \tag{2.31}$$

with $\mathbf{n}_{S_t}^T \cdot (\bar{\mathbf{x}}_t - \mathbf{x}_t) > 0$ for $t \in [t_m - \varepsilon, t_m)$ or

$$\left.\begin{aligned} &\mathbf{n}_{S_{t_m}}^T \cdot (\mathbf{x}_{t_m} - \bar{\mathbf{x}}_{t_m}) - \mathbf{n}_{S_{t_{m-\varepsilon}}}^T \cdot (\mathbf{x}_{t_{m-\varepsilon}} - \bar{\mathbf{x}}_{t_{m-\varepsilon}}) < 0, \\ &\mathbf{n}_{S_{t_{m+\varepsilon}}}^T \cdot (\mathbf{x}_{t_{m+\varepsilon}} - \bar{\mathbf{x}}_{t_{m+\varepsilon}}) - \mathbf{n}_{S_{t_m}}^T \cdot (\mathbf{x}_{t_m} - \bar{\mathbf{x}}_{t_m}) > 0 \end{aligned}\right\} \tag{3.32}$$

with $\mathbf{n}_{S_t}^T \cdot (\bar{\mathbf{x}}_t - \mathbf{x}_t) < 0$ for $t \in [t_m - \varepsilon, t_m)$.

The foregoing definitions for the flow $\mathbf{x}_t$ *passing through* and *returning back from* the reference surface S_t for the flow $\bar{\mathbf{x}}_t$ are illustrated in Fig.2.2. The difference for two kinds of flows is the compared flow turning the direction after the compared flow $\mathbf{x}_t$ arrives to some point $\mathbf{x}_{t_m}$ at time t_m. For a flow passing through the surface S_t, $\mathbf{x}_{t_m} = \bar{\mathbf{x}}'_{t_m} \in S_t$ but $\bar{\mathbf{x}}_{t_m} \neq \bar{\mathbf{x}}'_{t_m}$. From the previous two definitions, the corresponding theorems are given as follows.

THEOREM 2.1. *Consider two dynamical systems in Eqs.*(2.1) *and* (2.2) *which have a compared flow* $\mathbf{x}_t = \mathbf{\Phi}(t_0, \mathbf{x}_0, \mathbf{p}, t)$ *with an initial condition* $(t_0, \mathbf{x}_0)$ *and a reference flow* $\bar{\mathbf{x}}_t = \mathbf{\Phi}(t_0, \bar{\mathbf{x}}_0, \boldsymbol{\mu}, t)$ *with an initial condition* $(t_0, \bar{\mathbf{x}}_0)$, *respectively. The flow* $\mathbf{x}_t$ *does not contact with the reference flow* $\bar{\mathbf{x}}_t$. *Suppose the reference flow* $\bar{\mathbf{x}}_t$ *lies on a reference surface* S_t *with* $F(\bar{\mathbf{x}}, \bar{\mathbf{x}}_0, t) = 0$ *and* $\nabla F \bullet \dot{\bar{\mathbf{x}}} + \frac{\partial F}{\partial t} = 0$ *for time t. For an arbitrarily small* $\varepsilon > 0$, *there are two time intervals* $[t_{m-\varepsilon}, t_m)$ *and* $(t_m, t_{m+\varepsilon}]$. *The vector fields* $\mathbf{F}(\mathbf{x}, t, \mathbf{p})$ *and* $\mathbf{f}(\bar{\mathbf{x}}, t, \boldsymbol{\mu})$ *are* $C^r_{[t_{m-\varepsilon}, t_{m+\varepsilon}]}$*-continuous* ($r \geq 1$) *for time t. The flow* $\mathbf{x}_t$ *and* $\bar{\mathbf{x}}_t$ *are* $C^r_{[t_{m-\varepsilon}, t_{m+\varepsilon}]}$*-continuous* ($r \geq 2$) *for time t,* $\| d^{r+1}\mathbf{x}_t / dt^{r+1} \| < \infty$ *and* $\| d^{r+1}\bar{\mathbf{x}}_t / dt^{r+1} \| < \infty$.

(i) *The flow* $\mathbf{x}_t$ *at point* $(t_m, \mathbf{x}_{t_m})$ *with* $\mathbf{x}_{t_m} \notin S_t$ *is leaving from the reference surface* S_t *iff for* $t \in (t_m, t_m + \varepsilon]$

$$\left.\begin{aligned} &G(\bar{\mathbf{x}}_{t_m}, t_m, \mathbf{x}_{t_m}) > 0 \text{ for } \mathbf{n}_{S_t}^T \cdot (\mathbf{x}_t - \bar{\mathbf{x}}_t) > 0 \text{ or} \\ &G(\bar{\mathbf{x}}_{t_m}, t_m, \mathbf{x}_{t_m}) < 0 \text{ for } \mathbf{n}_{S_t}^T \cdot (\mathbf{x}_t - \bar{\mathbf{x}}_t) < 0. \end{aligned}\right\} \tag{2.33}$$

(ii) *The flow* $\mathbf{x}_t$ *at point* $(t_m, \mathbf{x}_{t_m})$ *with* $\mathbf{x}_{t_m} \notin S_t$ *is approaching the reference surface* S_t *for the flow* $\bar{\mathbf{x}}_t$ *iff for* $t \in [t_m - \varepsilon, t_m)$

$$\left.\begin{aligned} &G(\bar{\mathbf{x}}_{t_m}, t_m, \mathbf{x}_{t_m}) > 0 \text{ for } \mathbf{n}_{S_t}^T \cdot (\bar{\mathbf{x}}_t - \mathbf{x}_t) > 0, \text{or} \\ &G(\bar{\mathbf{x}}_{t_m}, t_m, \mathbf{x}_{t_m}) < 0 \text{ for } \mathbf{n}_{S_t}^T \cdot (\bar{\mathbf{x}}_t - \mathbf{x}_t) < 0. \end{aligned}\right\} \tag{2.34}$$

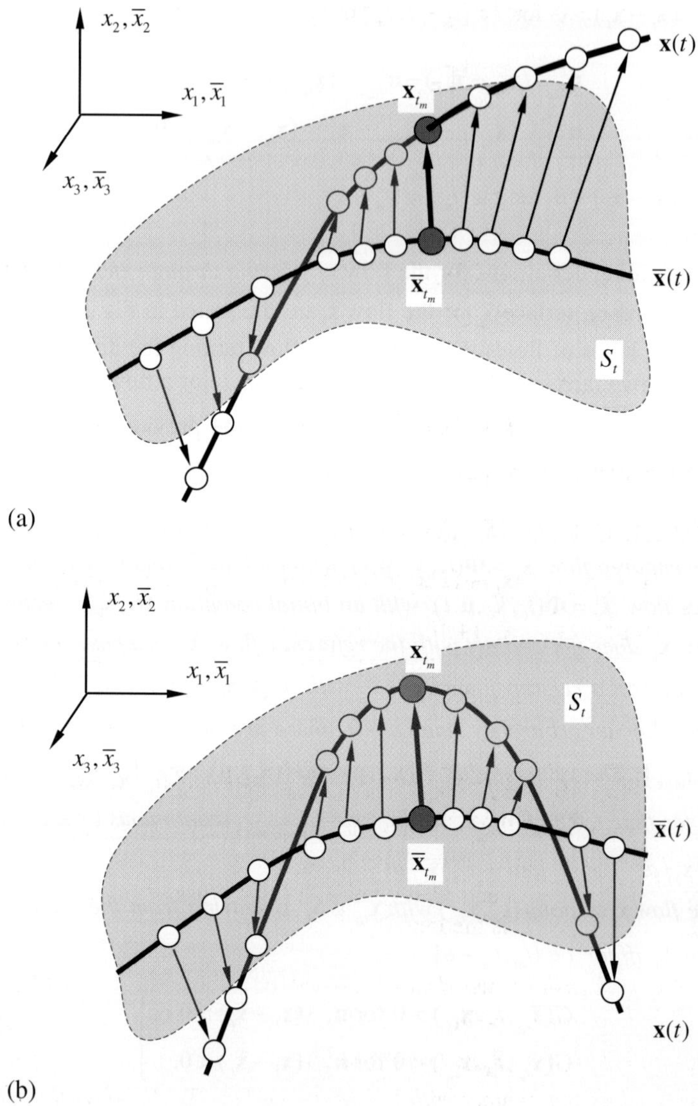

Figure 2.2. A 3-dimesional view of (a) the flow $\mathbf{x}_t$ passing through the reference surface for the reference flow $\bar{\mathbf{x}}_t$ and (b) the flow $\mathbf{x}_t$ returning back from the reference surface for the reference flow $\bar{\mathbf{x}}_t$.

PROOF. From Eq.(2.5), the left and right G-functions at point $(t_m, \bar{\mathbf{x}}_m)$ are

$$\left.\begin{aligned} G_-(\bar{\mathbf{x}}_{t_m}, t_m, \mathbf{x}_{t_m}) &= \lim_{\varepsilon\to 0}\frac{1}{\varepsilon}\left[\mathbf{n}^T_{S_{t_m}}\cdot(\mathbf{x}_{t_m}-\bar{\mathbf{x}}_{t_m}) - \mathbf{n}^T_{S_{t_m-\varepsilon}}\cdot(\mathbf{x}_{t_m-\varepsilon}-\bar{\mathbf{x}}_{t_m-\varepsilon})\right],\\ G_+(\bar{\mathbf{x}}_{t_m}, t_m, \mathbf{x}_{t_m}) &= \lim_{\varepsilon\to 0}\frac{1}{\varepsilon}\left[\mathbf{n}^T_{S_{t_m+\varepsilon}}\cdot(\mathbf{x}_{t_m+\varepsilon}-\bar{\mathbf{x}}_{t_m+\varepsilon}) - \mathbf{n}^T_{S_{t_m}}\cdot(\mathbf{x}_{t_m}-\bar{\mathbf{x}}_{t_m})\right].\end{aligned}\right\}$$

Owing to the continuity, $G_-(\bar{\mathbf{x}}_{t_m}, t_m, \mathbf{x}_{t_m}) = G_+(\bar{\mathbf{x}}_{t_m}, t_m, \mathbf{x}_{t_m})$. Equations (2.26) and (2.27) give Eqs.(2.33) and (2.34). With Eq.(2.6), the Taylor series expansions of $\mathbf{n}^T_{S_{t_m-\varepsilon}}\cdot(\mathbf{x}_{t_m-\varepsilon}-\bar{\mathbf{x}}_{t_m-\varepsilon})$ and $\mathbf{n}^T_{S_{t_m+\varepsilon}}\cdot(\mathbf{x}_{t_m+\varepsilon}-\bar{\mathbf{x}}_{t_m+\varepsilon})$ up to the ε -term give

$$\begin{aligned} &\mathbf{n}^T_{S_{t_m-\varepsilon}}\cdot(\mathbf{x}_{t_m-\varepsilon}-\bar{\mathbf{x}}_{t_m-\varepsilon})\\ &= \mathbf{n}^T_{S_{t_m}}\cdot(\mathbf{x}_{t_m}-\bar{\mathbf{x}}_{t_m}) - (D\mathbf{n}^T_{S_t}\cdot(\mathbf{x}_t-\bar{\mathbf{x}}_t) + \mathbf{n}^T_{S_t}\cdot(\dot{\mathbf{x}}_t-\dot{\bar{\mathbf{x}}}_t))\varepsilon + o(\varepsilon)\\ &= \mathbf{n}^T_{S_{t_m}}\cdot(\mathbf{x}_{t_m}-\bar{\mathbf{x}}_{t_m}) - G(\bar{\mathbf{x}}_{t_m}, t_m, \mathbf{x}_{t_m})\varepsilon + o(\varepsilon),\\ &\mathbf{n}^T_{S_{t_m+\varepsilon}}\cdot(\mathbf{x}_{t_m+\varepsilon}-\bar{\mathbf{x}}_{t_m+\varepsilon})\\ &= \mathbf{n}^T_{S_{t_m}}\cdot(\mathbf{x}_{t_m}-\bar{\mathbf{x}}_{t_m}) + (D\mathbf{n}^T_{S_t}\cdot(\mathbf{x}_t-\bar{\mathbf{x}}_t) + \mathbf{n}^T_{S_t}\cdot(\dot{\mathbf{x}}_t-\dot{\bar{\mathbf{x}}}_t))\varepsilon + o(\varepsilon)\\ &= \mathbf{n}^T_{S_{t_m}}\cdot(\mathbf{x}_{t_m}-\bar{\mathbf{x}}_{t_m}) + G(\bar{\mathbf{x}}_{t_m}, t_m, \mathbf{x}_{t_m})\varepsilon + o(\varepsilon).\end{aligned}$$

Once the ε^2- and higher-order terms are ignores, the deformation of the foregoing two equations gives

$$\begin{aligned} \mathbf{n}^T_{S_{t_m}}\cdot(\mathbf{x}_{t_m}-\bar{\mathbf{x}}_{t_m}) - \mathbf{n}^T_{S_{t_m-\varepsilon}}\cdot(\mathbf{x}_{t_m-\varepsilon}-\bar{\mathbf{x}}_{t_m-\varepsilon}) &= G(\bar{\mathbf{x}}_{t_m}, t_m, \mathbf{x}_{t_m})\varepsilon,\\ \mathbf{n}^T_{S_{t_m+\varepsilon}}\cdot(\mathbf{x}_{t_m+\varepsilon}-\bar{\mathbf{x}}_{t_m+\varepsilon}) - \mathbf{n}^T_{S_{t_m}}\cdot(\mathbf{x}_{t_m}-\bar{\mathbf{x}}_{t_m}) &= G(\bar{\mathbf{x}}_{t_m}, t_m, \mathbf{x}_{t_m})\varepsilon.\end{aligned}$$

Application of Eqs.(2.33) and (2.34) to the foregoing equations gives Eqs.(2.26) and (2.27), vice versa. This theorem is proved. ■

THEOREM 2.2. *Consider two dynamical systems in Eqs.*(2.1) *and* (2.2) *which have a compared flow* $\mathbf{x}_t = \mathbf{\Phi}(t_0, \mathbf{x}_0, \mathbf{p}, t)$ *with an initial condition* $(t_0, \mathbf{x}_0)$ *and a reference flow* $\bar{\mathbf{x}}_t = \mathbf{\Phi}(t_0, \bar{\mathbf{x}}_0, \boldsymbol{\mu}, t)$ *with an initial condition* $(t_0, \bar{\mathbf{x}}_0)$, *respectively. The flow* $\mathbf{x}_t$ *does not contact with the reference flow* $\bar{\mathbf{x}}_t$. *Suppose the reference flow* $\bar{\mathbf{x}}_t$ *lies on a reference surface* S_t *with* $F(\bar{\mathbf{x}}, \bar{\mathbf{x}}_0, t) = 0$ *and* $\nabla F \bullet \dot{\bar{\mathbf{x}}} + \frac{\partial F}{\partial t} = 0$ *for time t. For an arbitrarily small* $\varepsilon > 0$, *there are two time intervals* $[t_{m-\varepsilon}, t_m)$ *and* $(t_m, t_{m+\varepsilon}]$. *The vector fields* $\mathbf{F}(\mathbf{x}, t, \mathbf{p})$ *and* $\mathbf{f}(\bar{\mathbf{x}}, t, \boldsymbol{\mu})$ *are* $C^r_{[t_{m-\varepsilon}, t_{m+\varepsilon}]}$*-continuous* ($r \geq 1$) *for time t. The flows* $\mathbf{x}_t$ *and* $\bar{\mathbf{x}}_t$ *are* $C^r_{[t_{m-\varepsilon}, t_{m+\varepsilon}]}$*-continuous* ($r \geq 2$) *for time* *t with* $\| d^{r+1}\mathbf{x}_t / dt^{r+1} \| < \infty$ *and* $\| d^{r+1}\bar{\mathbf{x}}_t / dt^{r+1} \| < \infty$. *The flow* $\mathbf{x}_t$ *at point* $(t_m, \mathbf{x}_{t_m})$ *with* $\mathbf{x}_{t_m} = \bar{\mathbf{x}}'_{t_m} \in S_t$ *is passing through the reference surface* S_t *iff for* $t \in [t_m - \varepsilon, t_m)$

$$\left.\begin{array}{l} G(\bar{\mathbf{x}}_{t_m}, t_m, \mathbf{x}_{t_m}) > 0 \text{ for } \mathbf{n}_{S_t}^T \cdot (\bar{\mathbf{x}}_t - \mathbf{x}_t) > 0 \text{ or} \\ G(\bar{\mathbf{x}}_{t_m}, t_m, \mathbf{x}_{t_m}) < 0 \text{ for } \mathbf{n}_{S_t}^T \cdot (\bar{\mathbf{x}}_t - \mathbf{x}_t) < 0. \end{array}\right\} \tag{2.35}$$

PROOF. The proof is the same as in Proof of THEOREM 2.1. ■

THEOREM 2.3. *Consider two dynamical systems in Eqs.(2.1) and (2.2) which have a compared flow* $\mathbf{x}_t = \mathbf{\Phi}(t_0, \mathbf{x}_0, \mathbf{p}, t)$ *with an initial condition* $(t_0, \mathbf{x}_0)$ *and a reference flow* $\bar{\mathbf{x}}_t = \mathbf{\Phi}(t_0, \bar{\mathbf{x}}_0, \boldsymbol{\mu}, t)$ *with an initial condition* $(t_0, \bar{\mathbf{x}}_0)$, *respectively. The flow* $\mathbf{x}_t$ *does not contact with the reference flow* $\bar{\mathbf{x}}_t$. *Suppose the reference flow* $\bar{\mathbf{x}}_t$ *lies on a reference surface* S_t *with* $F(\bar{\mathbf{x}}, \bar{\mathbf{x}}_0, t) = 0$ *and* $\nabla F \bullet \dot{\bar{\mathbf{x}}} + \frac{\partial F}{\partial t} = 0$ *for time t. For an arbitrarily small* $\varepsilon > 0$, *there are two time intervals* $[t_{m-\varepsilon}, t_m)$ *and* $(t_m, t_{m+\varepsilon}]$. *The vector fields* $\mathbf{F}(\mathbf{x}, t, \mathbf{p})$ *and* $\mathbf{f}(\bar{\mathbf{x}}, t, \boldsymbol{\mu})$ *are* $C^r_{[t_{m-\varepsilon}, t_{m+\varepsilon}]}$*-continuous* ($r \geq 1$) *for time t. The flows* $\mathbf{x}_t$ *and* $\bar{\mathbf{x}}_t$ *are* $C^r_{[t_{m-\varepsilon}, t_{m+\varepsilon}]}$*-continuous* ($r \geq 2$) *for time t with* $\| d^{r+1}\mathbf{x}_t / dt^{r+1} \| < \infty$ *and* $\| d^{r+1}\bar{\mathbf{x}}_t / dt^{r+1} \| < \infty$. *The flow* $\mathbf{x}_t$ *at point* $(t_m, \mathbf{x}_{t_m})$ *with* $\mathbf{x}_{t_m} \notin S_t$ *is returning back and away from the reference surface* S_t *iff for* $t \in [t_{m-\varepsilon}, t_m)$

$$G(\bar{\mathbf{x}}_{t_m}, t_m, \mathbf{x}_{t_m}) \equiv G^{(0)}(\bar{\mathbf{x}}_{t_m}, t_m, \mathbf{x}_{t_m}) = 0, \tag{2.36}$$

$$\left.\begin{array}{l} G^{(1)}(\bar{\mathbf{x}}_{t_m}, t_m, \mathbf{x}_{t_m}) < 0 \text{ for } \mathbf{n}_{S_t}^T \cdot (\bar{\mathbf{x}}_t - \mathbf{x}_t) > 0 \text{ or} \\ G^{(1)}(\bar{\mathbf{x}}_{t_m}, t_m, \mathbf{x}_{t_m}) > 0 \text{ for } \mathbf{n}_{S_t}^T \cdot (\bar{\mathbf{x}}_t - \mathbf{x}_t) < 0. \end{array}\right\} \tag{2.37}$$

PROOF. From Eq.(2.5), we have at point $(t_m, \bar{\mathbf{x}}_m)$

$$\left.\begin{array}{l} G_-(\bar{\mathbf{x}}_{t_m}, t_m, \mathbf{x}_{t_m}) = \lim\limits_{\varepsilon \to 0} \dfrac{1}{\varepsilon} \Big[\mathbf{n}_{S_{t_m}}^T \cdot (\mathbf{x}_{t_{m-}} - \bar{\mathbf{x}}_{t_m}) - \mathbf{n}_{S_{t_{m-\varepsilon}}}^T \cdot (\mathbf{x}_{t_m - \varepsilon} - \bar{\mathbf{x}}_{t_m - \varepsilon}) \Big], \\ G_+(\bar{\mathbf{x}}_{t_m}, t_m, \mathbf{x}_{t_m}) = \lim\limits_{\varepsilon \to 0} \dfrac{1}{\varepsilon} \Big[\mathbf{n}_{S_{t_{m+\varepsilon}}}^T \cdot (\mathbf{x}_{t_m + \varepsilon} - \bar{\mathbf{x}}_{t_m + \varepsilon}) - \mathbf{n}_{S_{t_m}}^T \cdot (\mathbf{x}_{t_{m+}} - \bar{\mathbf{x}}_{t_m}) \Big]. \end{array}\right\}$$

For $\mathbf{n}_{S_t}^T \cdot (\bar{\mathbf{x}}_t - \mathbf{x}_t) > 0$ for $t \in [t_{m-\varepsilon}, t_m)$, with Eq.(2.31), we have

$$G_{-}(\overline{\mathbf{x}}_{t_m}, t_m, \mathbf{x}_{t_m}) \geq 0 \text{ and } G_{+}(\overline{\mathbf{x}}_{t_m}, t_m, \mathbf{x}_{t_m}) \leq 0.$$

The continuity of flows in Eqs.(2.1) and (2.2) at time t_m gives $G_{-} = G_{+}$. Thus,

$$G(\overline{\mathbf{x}}_{t_m}, t_m, \mathbf{x}_{t_m}) = G_{-}(\overline{\mathbf{x}}_{t_m}, t_m, \mathbf{x}_{t_m}) = G_{+}(\overline{\mathbf{x}}_{t_m}, t_m, \mathbf{x}_{t_m}) = 0.$$

Using the foregoing equation, equation (2.12) gives

$$\left.\begin{aligned} G_{-}^{(1)}(\overline{\mathbf{x}}_t, t, \mathbf{x}_t) &= -\lim_{\varepsilon \to 0} \frac{1}{\varepsilon^2}\left[\mathbf{n}_{S_t}^T \cdot (\mathbf{x}_t - \overline{\mathbf{x}}_t) - \mathbf{n}_{S_{t-\varepsilon}}^T \cdot (\mathbf{x}_{t-\varepsilon} - \overline{\mathbf{x}}_{t-\varepsilon})\right], \\ G_{+}^{(1)}(\overline{\mathbf{x}}_t, t, \mathbf{x}_t) &= \lim_{\varepsilon \to 0} \frac{1}{\varepsilon^2}\left[\mathbf{n}_{S_{t+\varepsilon}}^T \cdot (\mathbf{x}_{t+\varepsilon} - \overline{\mathbf{x}}_{t+\varepsilon}) - \mathbf{n}_{S_t}^T \cdot (\mathbf{x}_t - \overline{\mathbf{x}}_t)\right]. \end{aligned}\right\}$$

Application of Eq.(2.31) to the foregoing equations yields

$$G_{-}^{(1)}(\overline{\mathbf{x}}_{t_m}, t_m, \mathbf{x}_{t_m}) < 0 \text{ and } G_{+}^{(1)}(\overline{\mathbf{x}}_{t_m}, t_m, \mathbf{x}_{t_m}) < 0.$$

Therefore, based on the continuity at time t_m, the following condition exists,

$$G^{(1)}(\overline{\mathbf{x}}_{t_m}, t_m, \mathbf{x}_{t_m}) = G_{-}^{(1)}(\overline{\mathbf{x}}_{t_m}, t_m, \mathbf{x}_{t_m}) = G_{+}^{(1)}(\overline{\mathbf{x}}_{t_m}, t_m, \mathbf{x}_{t_m}) < 0.$$

Similarly, for $\mathbf{n}_{\overline{\mathbf{x}}_t}^T \cdot (\overline{\mathbf{x}}_t - \mathbf{x}_t) < 0$ for time $t \in [t_{m\text{-}\varepsilon}, t_m]$, equation (2.32) gives

$$G_{-}(\overline{\mathbf{x}}_{t_m}, t_m, \mathbf{x}_{t_m}) \leq 0 \text{ and } G_{+}(\overline{\mathbf{x}}_{t_m}, t_m, \mathbf{x}_{t_m}) \geq 0.$$

The continuity condition gives

$$G(\overline{\mathbf{x}}_{t_m}, t_m, \mathbf{x}_{t_m}) = G_{-}(\overline{\mathbf{x}}_{t_m}, t_m, \mathbf{x}_{t_m}) = G_{+}(\overline{\mathbf{x}}_{t_m}, t_m, \mathbf{x}_{t_m}) = 0.$$

Using the foregoing equation, equation (2.12) gives

$$\left.\begin{aligned} G_{-}^{(1)}(\overline{\mathbf{x}}_{t_m}, t_m, \mathbf{x}_{t_m}) &= -\lim_{\varepsilon \to 0} \frac{1}{\varepsilon^2}\left[\mathbf{n}_{S_{t_m}}^T \cdot (\mathbf{x}_{t_m} - \overline{\mathbf{x}}_{t_m}) - \mathbf{n}_{S_{t_m-\varepsilon}}^T \cdot (\mathbf{x}_{t_m-\varepsilon} - \overline{\mathbf{x}}_{t_m-\varepsilon})\right], \\ G_{+}^{(1)}(\overline{\mathbf{x}}_{t_m}, t_m, \mathbf{x}_{t_m}) &= \lim_{\varepsilon \to 0} \frac{1}{\varepsilon^2}\left[\mathbf{n}_{S_{t_m+\varepsilon}}^T \cdot (\mathbf{x}_{t_m+\varepsilon} - \overline{\mathbf{x}}_{t_m+\varepsilon}) - \mathbf{n}_{S_{t_m}}^T \cdot (\mathbf{x}_{t_m} - \overline{\mathbf{x}}_{t_m})\right]. \end{aligned}\right\}$$

Application of Eq.(2.32) to the foregoing equations yields

$$G_{-}^{(1)}(\overline{\mathbf{x}}_{t_m}, t_m, \mathbf{x}_{t_m}) > 0 \text{ and } G_{+}^{(1)}(\overline{\mathbf{x}}_{t_m}, t_m, \mathbf{x}_{t_m}) > 0.$$

Therefore, the continuity gives

$$G^{(1)}(\overline{\mathbf{x}}_{t_m}, t_m, \mathbf{x}_{t_m}) = G_{-}^{(1)}(\overline{\mathbf{x}}_{t_m}, t_m, \mathbf{x}_{t_m}) = G_{+}^{(1)}(\overline{\mathbf{x}}_{t_m}, t_m, \mathbf{x}_{t_m}) > 0.$$

With Eq.(2.6) and (2.13), the Taylor series expansions of $\mathbf{n}_{S_{t_m-\varepsilon}}^T \cdot (\mathbf{x}_{t_m-\varepsilon} - \overline{\mathbf{x}}_{t_m-\varepsilon})$

and $\mathbf{n}^T_{S_{t_{m+\varepsilon}}} \cdot (\mathbf{x}_{t_m+\varepsilon} - \bar{\mathbf{x}}_{t_m+\varepsilon})$ up to ε^2-terms give

$$\begin{aligned}
\mathbf{n}^T_{S_{t_{m-\varepsilon}}} \cdot (\mathbf{x}_{t_m-\varepsilon} - \bar{\mathbf{x}}_{t_m-\varepsilon}) &= \mathbf{n}^T_{S_{t_m}} \cdot (\mathbf{x}_{t_m} - \bar{\mathbf{x}}_{t_m}) - G(\bar{\mathbf{x}}_{t_m}, t_m, \mathbf{x}_{t_m})\varepsilon \\
&\quad + G^{(1)}(\bar{\mathbf{x}}_{t_m}, t_m, \mathbf{x}_{t_m})\varepsilon^2 + o(\varepsilon^2), \\
\mathbf{n}^T_{S_{t_{m+\varepsilon}}} \cdot (\mathbf{x}_{t_m+\varepsilon} - \bar{\mathbf{x}}_{t_m+\varepsilon}) &= \mathbf{n}^T_{S_{t_m}} \cdot (\mathbf{x}_{t_m} - \bar{\mathbf{x}}_{t_m}) + G(\bar{\mathbf{x}}_{t_m}, t_m, \mathbf{x}_{t_m})\varepsilon \\
&\quad + G^{(1)}(\bar{\mathbf{x}}_{t_m}, t_m, \mathbf{x}_{t_m})\varepsilon^2 + o(\varepsilon^2).
\end{aligned}$$

With Eq.(2.36), the foregoing equations give

$$\begin{aligned}
\mathbf{n}^T_{S_{t_m}} \cdot (\mathbf{x}_{t_m} - \bar{\mathbf{x}}_{t_m}) - \mathbf{n}^T_{S_{t_{m-\varepsilon}}} \cdot (\mathbf{x}_{t_m-\varepsilon} - \bar{\mathbf{x}}_{t_m-\varepsilon}) &= -G^{(1)}(\bar{\mathbf{x}}_{t_m}, t_m, \mathbf{x}_{t_m})\varepsilon, \\
\mathbf{n}^T_{S_{t_{m+\varepsilon}}} \cdot (\mathbf{x}_{t_m+\varepsilon} - \bar{\mathbf{x}}_{t_m+\varepsilon}) - \mathbf{n}^T_{S_{t_m}} \cdot (\mathbf{x}_{t_m} - \bar{\mathbf{x}}_{t_m}) &= G^{(1)}(\bar{\mathbf{x}}_{t_m}, t_m, \mathbf{x}_{t_m})\varepsilon.
\end{aligned}$$

Application of Eqs.(2.37) to the foregoing equations gives Eqs.(2.31) and (2.32), vice versa. This theorem is proved. ■

THEOREM 2.4. *Consider two dynamical systems in Eqs.*(2.1) *and* (2.2) *which have a compared flow* $\mathbf{x}_t = \mathbf{\Phi}(t_0, \mathbf{x}_0, \mathbf{p}, t)$ *with an initial condition* $(t_0, \mathbf{x}_0)$ *and a reference flow* $\bar{\mathbf{x}}_t = \mathbf{\Phi}(t_0, \bar{\mathbf{x}}_0, \boldsymbol{\mu}, t)$ *with an initial condition* $(t_0, \bar{\mathbf{x}}_0)$, *respectively. The flow* $\mathbf{x}_t$ *does not contact with the reference flow* $\bar{\mathbf{x}}_t$. *Suppose the reference flow* $\bar{\mathbf{x}}_t$ *lies on a reference surface* S_t *with* $F(\bar{\mathbf{x}}, \bar{\mathbf{x}}_0, t) = 0$ *and* $\nabla F \bullet \dot{\bar{\mathbf{x}}} + \frac{\partial F}{\partial t} = 0$ *for time t. For an arbitrarily small* $\varepsilon > 0$, *there are two time intervals* $[t_{m-\varepsilon}, t_m)$ *and* $(t_m, t_{m+\varepsilon}]$. *The vector fields* $\mathbf{F}(\mathbf{x}, t, \mathbf{p})$ *and* $\mathbf{f}(\bar{\mathbf{x}}, t, \boldsymbol{\mu})$ *are* $C^r_{[t_{m-\varepsilon}, t_{m+\varepsilon}]}$*-continuous* ($r \geq k$) *only.* $D^{(l)}_{\mathbf{x}}\mathbf{F}(\mathbf{x}, t, \mathbf{p}) = 0$ *and* $D^{(l)}_{\bar{\mathbf{x}}}\mathbf{f}(\bar{\mathbf{x}}, t, \boldsymbol{\mu}) = 0$ *for* ($l > k$). *The flow* $\mathbf{x}_t$ *and* $\bar{\mathbf{x}}_t$ *are* $C^r_{[t_{m-\varepsilon}, t_{m+\varepsilon}]}$ *-continuous* ($r \geq k+2$) *for time t. The flow* $\mathbf{x}_t$ *at point* $(t_m, \mathbf{x}_{t_m})$ *with* $\mathbf{x}_{t_m} \notin S_t$ *is paralleling to the reference surface* S_t *iff*

$$\left.\begin{aligned}
&\mathbf{n}^T_{S_{t_m}} \cdot (\mathbf{x}_{t_m} - \bar{\mathbf{x}}_{t_m}) \neq 0, \\
&G^{(r)}(\bar{\mathbf{x}}_{t_m}, t_m, \mathbf{x}_{t_m}) = 0 \text{ for } r = 0, 1, 2, \cdots, k.
\end{aligned}\right\} \tag{2.38}$$

PROOF. From Eq.(2.5), we have at point $(t_m, \bar{\mathbf{x}}_m)$

$$G_-(\bar{\mathbf{x}}_{t_m}, t_m, \mathbf{x}_{t_m}) = \lim_{\varepsilon \to 0} \frac{1}{\varepsilon}\Big[\mathbf{n}^T_{S_{t_m}} \cdot (\mathbf{x}_{t_m} - \bar{\mathbf{x}}_{t_m}) - \mathbf{n}^T_{S_{t_{m-\varepsilon}}} \cdot (\mathbf{x}_{t_m-\varepsilon} - \bar{\mathbf{x}}_{t_m-\varepsilon})\Big],$$

$$G_+(\bar{\mathbf{x}}_{t_m}, t_m, \mathbf{x}_{t_m}) = \lim_{\varepsilon \to 0} \frac{1}{\varepsilon}\Big[\mathbf{n}^T_{S_{t_{m+\varepsilon}}} \cdot (\mathbf{x}_{t_m+\varepsilon} - \bar{\mathbf{x}}_{t_m+\varepsilon}) - \mathbf{n}^T_{S_{t_m}} \cdot (\mathbf{x}_{t_m} - \bar{\mathbf{x}}_{t_m})\Big].$$

With Eq.(2.27), the foregoing equation gives

$$G_-(\bar{\mathbf{x}}_{t_m}, t_m, \mathbf{x}_{t_m}) = G_+(\bar{\mathbf{x}}_{t_m}, t_m, \mathbf{x}_{t_m}) = 0.$$

Because of the continuity, we have

$$G^{(0)}(\bar{\mathbf{x}}_{t_m}, t_m, \mathbf{x}_{t_m}) = G(\bar{\mathbf{x}}_{t_m}, t_m, \mathbf{x}_{t_m}) = 0.$$

From Eq.(2.15), for $k = 1$, using the foregoing equation,

$$\left.\begin{aligned} G_-^{(1)}(\bar{\mathbf{x}}_{t_m}, t_m, \mathbf{x}_{t_{m-}}) &= \lim_{\varepsilon \to 0} \frac{(-1)}{\varepsilon^2}\left[\mathbf{n}_{S_{t_m}}^T \cdot (\mathbf{x}_{t_{m-}} - \bar{\mathbf{x}}_{t_m}) - \mathbf{n}_{S_{t_{m-\varepsilon}}}^T \cdot (\mathbf{x}_{t_{m-\varepsilon}} - \bar{\mathbf{x}}_{t_{m-\varepsilon}})\right], \\ G_+^{(1)}(\bar{\mathbf{x}}_{t_m}, t_m, \mathbf{x}_{t_{m+}}) &= \lim_{\varepsilon \to 0} \frac{1}{\varepsilon^2}\left[\mathbf{n}_{S_{t_{m+\varepsilon}}}^T \cdot (\mathbf{x}_{t_{m+\varepsilon}} - \bar{\mathbf{x}}_{t_{m+\varepsilon}}) - \mathbf{n}_{S_t}^T \cdot (\mathbf{x}_{t_{m+}} - \bar{\mathbf{x}}_{t_m})\right]. \end{aligned}\right\}$$

Using Eq.(2.28) again, we have

$$G_-^{(1)}(\bar{\mathbf{x}}_{t_m}, t_m, \mathbf{x}_{t_{m-}}) = G_+^{(1)}(\bar{\mathbf{x}}_{t_m}, t_m, \mathbf{x}_{t_{m+}}) = 0$$

which implies $G^{(1)}(\bar{\mathbf{x}}_{t_m}, t_m, \mathbf{x}_{t_m}) = 0$. Similarly, for $s = 0, 1, 2, \cdots, k-1$

$$G_-^{(s)}(\bar{\mathbf{x}}_{t_m}, t_m, \mathbf{x}_{t_{m-}}) = G_+^{(s)}(\bar{\mathbf{x}}_{t_m}, t_m, \mathbf{x}_{t_{m+}}) = 0,$$

and furthermore, $G^{(r)}(\bar{\mathbf{x}}_{t_m}, t_m, \mathbf{x}_{t_m}) = 0$ holds. For the case of $r = k$, with the foregoing equation, equation (2.15) gives

$$\left.\begin{aligned} G_-^{(k)}(\bar{\mathbf{x}}_{t_m}, t_m, \mathbf{x}_{t_{m-}}) &= \lim_{\varepsilon \to 0} \frac{(-1)^{k+2}}{\varepsilon^{k+1}}\left[\mathbf{n}_{S_{t_m}}^T \cdot (\mathbf{x}_{t_{m-}} - \bar{\mathbf{x}}_{t_m}) - \mathbf{n}_{S_{t_{m-\varepsilon}}}^T \cdot (\mathbf{x}_{t_{m-\varepsilon}} - \bar{\mathbf{x}}_{t_{m-\varepsilon}})\right] \\ G_+^{(k)}(\bar{\mathbf{x}}_{t_m}, t_m, \mathbf{x}_{t_{m+}}) &= \lim_{\varepsilon \to 0} \frac{1}{\varepsilon^{k+1}}\left[\mathbf{n}_{S_{t_{m+\varepsilon}}}^T \cdot (\mathbf{x}_{t_{m+\varepsilon}} - \bar{\mathbf{x}}_{t_{m+\varepsilon}}) - \mathbf{n}_{S_t}^T \cdot (\mathbf{x}_{t_{m+}} - \bar{\mathbf{x}}_{t_m})\right]. \end{aligned}\right\}$$

Again, equation (2.28) applying to the above equation yields

$$G_-^{(k)}(\bar{\mathbf{x}}_{t_m}, t_m, \mathbf{x}_{t_{m-}}) = G_+^{(k)}(\bar{\mathbf{x}}_{t_m}, t_m, \mathbf{x}_{t_{m+}}) = 0.$$

Therefore,

$$G^{(k)}(\bar{\mathbf{x}}_{t_m}, t_m, \mathbf{x}_{t_m}) = 0.$$

With Eq.(2.18), the Taylor series expansions of the two normal distances $\mathbf{n}_{S_{t_{m-\varepsilon}}}^T \cdot (\mathbf{x}_{t_{m-\varepsilon}} - \bar{\mathbf{x}}_{t_{m-\varepsilon}})$ and $\mathbf{n}_{S_{t_{m+\varepsilon}}}^T \cdot (\mathbf{x}_{t_{m+\varepsilon}} - \bar{\mathbf{x}}_{t_{m+\varepsilon}})$ up to ε^{k+1}-terms give

$$\mathbf{n}^T_{S_{t_{m-\varepsilon}}}\cdot(\mathbf{x}_{t_m-\varepsilon}-\overline{\mathbf{x}}_{t_m-\varepsilon})=\mathbf{n}^T_{S_{t_m}}\cdot(\mathbf{x}_{t_m-}-\overline{\mathbf{x}}_{t_m})+\sum\nolimits_{r=0}^{k}G_{-}^{(r)}(\overline{\mathbf{x}}_{t_m},t_m,\mathbf{x}_{t_m-})(-\varepsilon)^{r+1},$$
$$\mathbf{n}^T_{S_{t_{m-\varepsilon}}}\cdot(\mathbf{x}_{t_m+\varepsilon}-\overline{\mathbf{x}}_{t_m+\varepsilon})=\mathbf{n}^T_{S_{t_m}}\cdot(\mathbf{x}_{t_m+}-\overline{\mathbf{x}}_{t_m})+\sum\nolimits_{r=0}^{k}G_{+}^{(r)}(\overline{\mathbf{x}}_{t_m},t_m,\mathbf{x}_{t_m+})\varepsilon^{r+1}.$$

Because $\mathbf{F}(\mathbf{x},t,\mathbf{p})$ and $\mathbf{f}(\overline{\mathbf{x}},t,\boldsymbol{\mu})$ with the $C^r_{[t_{m-\varepsilon},t_{m+\varepsilon}]}$-continuity ($r\geq k$) for time t exist only and $D^{(l)}_{\mathbf{x}}\mathbf{F}(\mathbf{x},t,\mathbf{p})=0$ and $D^{(l)}_{\overline{\mathbf{x}}}\mathbf{f}(\overline{\mathbf{x}},t,\boldsymbol{\mu})=0$ for ($l>k$), it implies that $G^{(r)}(\overline{\mathbf{x}}_{t_m},t_m,\mathbf{x}_{t_m})=0$ for $r>k$. Using Eq.(2.38), we have

$$\mathbf{n}^T_{S_{t_{m-\varepsilon}}}\cdot(\mathbf{x}_{t_m-\varepsilon}-\overline{\mathbf{x}}_{t_m-\varepsilon})=\mathbf{n}^T_{S_{t_m}}\cdot(\mathbf{x}_{t_m-}-\overline{\mathbf{x}}_{t_m}),$$
$$\mathbf{n}^T_{S_{t_{m-\varepsilon}}}\cdot(\mathbf{x}_{t_m+\varepsilon}-\overline{\mathbf{x}}_{t_m+\varepsilon})=\mathbf{n}^T_{S_{t_m}}\cdot(\mathbf{x}_{t_m+}-\overline{\mathbf{x}}_{t_m}).$$

This theorem is proved. ■

Notice that from this theorem, the compared flow paralleling to the reference surface requires all the order G-function be zero.

DEFINITION 2.6. Consider two dynamical systems in Eqs.(2.1) and (2.2) which have a compared flow $\mathbf{x}_t=\boldsymbol{\Phi}(t_0,\mathbf{x}_0,\mathbf{p},t)$ with an initial condition $(t_0,\mathbf{x}_0)$ and a reference flow $\overline{\mathbf{x}}_t=\boldsymbol{\Phi}(t_0,\overline{\mathbf{x}}_0,\boldsymbol{\mu},t)$ with an initial condition $(t_0,\overline{\mathbf{x}}_0)$, respectively. The flow $\mathbf{x}_t$ does not contact with the reference flow $\overline{\mathbf{x}}_t$. Suppose the reference flow $\overline{\mathbf{x}}_t$ lies on a reference surface S_t with $F(\overline{\mathbf{x}},\overline{\mathbf{x}}_0,t)=0$ and $\nabla F\bullet\dot{\overline{\mathbf{x}}}+\frac{\partial F}{\partial t}=0$ for time t. For an arbitrarily small $\varepsilon>0$, there are two time intervals $[t_{m-\varepsilon},t_m)$ and $(t_m,t_{m+\varepsilon}]$. The vector fields $\mathbf{F}(\mathbf{x},t,\mathbf{p})$ and $\mathbf{f}(\overline{\mathbf{x}},t,\boldsymbol{\mu})$ are $C^r_{[t_{m-\varepsilon},t_{m+\varepsilon}]}$-continuous ($r\geq k+1$) for time t. The flows $\mathbf{x}_t$ and $\overline{\mathbf{x}}_t$ are $C^r_{[t_{m-\varepsilon},t_{m+\varepsilon}]}$-continuous ($r\geq k+2$) for time t, $\|d^{r+1}\mathbf{x}_t/dt^{r+1}\|<\infty$ and $\|d^{r+1}\overline{\mathbf{x}}_t/dt^{r+1}\|<\infty$.

$$G^{(s)}(\overline{\mathbf{x}}_{t_m},t_m,\mathbf{x}_{t_m})=0 \text{ for } s=0,1,2,\cdots,k. \tag{2.39}$$

(i) The flow $\mathbf{x}_t$ at point $(t_m,\mathbf{x}_{t_m})$ with $\mathbf{x}_{t_m}\notin S_t$ is *leaving from* the reference surface S_t with the kth-order singularity if for $t\in(t_m,t_m+\varepsilon]$

$$\left.\begin{aligned}&\mathbf{n}^T_{S_{t_{m+\varepsilon}}}\cdot(\mathbf{x}_{t_m+\varepsilon}-\overline{\mathbf{x}}_{t_m+\varepsilon})-\mathbf{n}^T_{S_{t_m}}\cdot(\mathbf{x}_{t_m}-\overline{\mathbf{x}}_{t_m})>0 \text{ for } \mathbf{n}^T_{S_t}\cdot(\mathbf{x}_t-\overline{\mathbf{x}}_t)>0,\\&\mathbf{n}^T_{S_{t_{m+\varepsilon}}}\cdot(\mathbf{x}_{t_m+\varepsilon}-\overline{\mathbf{x}}_{t_m+\varepsilon})-\mathbf{n}^T_{S_{t_m}}\cdot(\mathbf{x}_{t_m}-\overline{\mathbf{x}}_{t_m})<0 \text{ for } \mathbf{n}^T_{S_t}\cdot(\mathbf{x}_t-\overline{\mathbf{x}}_t)<0.\end{aligned}\right\}\tag{2.40}$$

(ii) The flow $\mathbf{x}_t$ at point $(t_m,\mathbf{x}_{t_m})$ with $\mathbf{x}_{t_m}\notin S_t$ is *approaching* the reference surface S_t with the kth-order singularity if for $t\in[t-\varepsilon,t)$

$$\left.\begin{aligned} &\mathbf{n}^T_{S_{t_m}}\cdot(\mathbf{x}_{t_m}-\overline{\mathbf{x}}_{t_m})-\mathbf{n}^T_{S_{t_m-\varepsilon}}\cdot(\mathbf{x}_{t_m-\varepsilon}-\overline{\mathbf{x}}_{t_m-\varepsilon})>0 \quad \text{for } \mathbf{n}^T_{S_t}\cdot(\overline{\mathbf{x}}_t-\mathbf{x}_t)>0,\\ &\mathbf{n}^T_{S_{t_m}}\cdot(\mathbf{x}_{t_m}-\overline{\mathbf{x}}_{t_m})-\mathbf{n}^T_{S_{t_m-\varepsilon}}\cdot(\mathbf{x}_{t_m-\varepsilon}-\overline{\mathbf{x}}_{t_m-\varepsilon})<0 \quad \text{for } \mathbf{n}^T_{S_t}\cdot(\overline{\mathbf{x}}_t-\mathbf{x}_t)<0. \end{aligned}\right\} \tag{2.41}$$

(iii) The flow $\mathbf{x}_t$ at point $(t_m, \mathbf{x}_{t_m})$ with $\mathbf{x}_{t_m} \notin S_t$ is *paralleling to* the reference surface S_t with the kth-order singularity if

$$\left.\begin{aligned} &\mathbf{n}^T_{S_{t_m}}\cdot(\mathbf{x}_{t_m}-\overline{\mathbf{x}}_{t_m})-\mathbf{n}^T_{S_{t_m-\varepsilon}}\cdot(\mathbf{x}_{t_m-\varepsilon}-\overline{\mathbf{x}}_{t_m-\varepsilon})=0 \quad \text{for } t\in[t_m-\varepsilon,t_m),\\ &\mathbf{n}^T_{S_{t_m+\varepsilon}}\cdot(\mathbf{x}_{t_m+\varepsilon}-\overline{\mathbf{x}}_{t_m+\varepsilon})-\mathbf{n}^T_{S_{t_m}}\cdot(\mathbf{x}_{t_m}-\overline{\mathbf{x}}_{t_m})=0 \quad \text{for } t\in(t_m,t_m+\varepsilon]. \end{aligned}\right\} \tag{2.42}$$

DEFINITION 2.7. Consider two dynamical systems in Eqs.(2.1) and (2.2) which have a compared flow $\mathbf{x}_t = \mathbf{\Phi}(t_0, \mathbf{x}_0, \mathbf{p}, t)$ with an initial condition $(t_0, \mathbf{x}_0)$ and a reference flow $\overline{\mathbf{x}}_t = \mathbf{\Phi}(t_0, \overline{\mathbf{x}}_0, \boldsymbol{\mu}, t)$ with an initial condition $(t_0, \overline{\mathbf{x}}_0)$, respectively. The flow $\mathbf{x}_t$ does not contact with the reference flow $\overline{\mathbf{x}}_t$. Suppose the reference flow $\overline{\mathbf{x}}_t$ lies on a reference surface S_t with $F(\overline{\mathbf{x}}, \overline{\mathbf{x}}_0, t) = 0$ and $\nabla F \bullet \dot{\overline{\mathbf{x}}} + \frac{\partial F}{\partial t} = 0$ for time t. For an arbitrarily small $\varepsilon > 0$, there are two time intervals $[t_{m-\varepsilon}, t_m)$ and $(t_m, t_{m+\varepsilon}]$. The vector fields $\mathbf{F}(\mathbf{x}, t, \mathbf{p})$ and $\mathbf{f}(\overline{\mathbf{x}}, t, \boldsymbol{\mu})$ are $C^r_{[t_{m-\varepsilon}, t_{m+\varepsilon}]}$-continuous ($r \geq 2k$) for time t. The flows $\mathbf{x}_t$ and $\overline{\mathbf{x}}_t$ are $C^r_{[t_{m-\varepsilon}, t_{m+\varepsilon}]}$-continuous ($r \geq 2k+1$) for time t with $\| d^{r+1}\mathbf{x}_t / dt^{r+1} \| < \infty$ and $\| d^{r+1}\overline{\mathbf{x}}_t / dt^{r+1} \| < \infty$. The flow $\mathbf{x}_t$ at point $(t_m, \mathbf{x}_{t_m})$ with $\mathbf{x}_{t_m} = \overline{\mathbf{x}}'_{t_m} \in S_t$ is *passing through* the reference surface S_t with the $(2k)^{\text{th}}$-order singularity if

$$G^{(s)}(\overline{\mathbf{x}}_{t_m}, t_m, \mathbf{x}_{t_m}) = 0 \text{ for } s = 0, 1, 2, \cdots, 2k-1, \tag{2.43}$$

$$\left.\begin{aligned} &\mathbf{n}^T_{S_{t_m}}\cdot(\mathbf{x}_{t_m}-\overline{\mathbf{x}}_{t_m})-\mathbf{n}^T_{S_{t_m-\varepsilon}}\cdot(\mathbf{x}_{t_m-\varepsilon}-\overline{\mathbf{x}}_{t_m-\varepsilon})>0,\\ &\mathbf{n}^T_{S_{t_m+\varepsilon}}\cdot(\mathbf{x}_{t_m+\varepsilon}-\overline{\mathbf{x}}_{t_{m+\varepsilon}})-\mathbf{n}^T_{S_{t_m}}\cdot(\mathbf{x}_{t_m}-\overline{\mathbf{x}}_{t_m})>0; \end{aligned}\right\} \tag{2.44}$$

with $\mathbf{n}^T_{S_t}\cdot(\overline{\mathbf{x}}_t-\mathbf{x}_t)>0$ for $t\in[t_m-\varepsilon,t_m)$ or

$$\left.\begin{aligned} &\mathbf{n}^T_{S_{t_m}}\cdot(\overline{\mathbf{x}}_{t_m}-\mathbf{x}_{t_m})-\mathbf{n}^T_{S_{t_m-\varepsilon}}\cdot(\mathbf{x}_{t_m-\varepsilon}-\overline{\mathbf{x}}_{t_m-\varepsilon})<0,\\ &\mathbf{n}^T_{S_{t_m+\varepsilon}}\cdot(\mathbf{x}_{t_m+\varepsilon}-\overline{\mathbf{x}}_{t_{m+\varepsilon}})-\mathbf{n}^T_{S_{t_m}}\cdot(\mathbf{x}_{t_m}-\overline{\mathbf{x}}_{t_m})<0 \end{aligned}\right\} \tag{2.45}$$

with $\mathbf{n}^T_{S_t}\cdot(\overline{\mathbf{x}}_t-\mathbf{x}_t)<0$ for $t\in[t_m-\varepsilon,t_m)$.

DEFINITION 2.8. Consider two dynamical systems in Eqs.(2.1) and (2.2) which have a compared flow $\mathbf{x}_t = \mathbf{\Phi}(t_0, \mathbf{x}_0, \mathbf{p}, t)$ with an initial condition $(t_0, \mathbf{x}_0)$ and a reference flow $\overline{\mathbf{x}}_t = \mathbf{\Phi}(t_0, \overline{\mathbf{x}}_0, \boldsymbol{\mu}, t)$ with an initial condition $(t_0, \overline{\mathbf{x}}_0)$, respectively. The flow $\mathbf{x}_t$ does not contact with the reference flow $\overline{\mathbf{x}}_t$. Suppose the reference flow $\overline{\mathbf{x}}_t$ lies on a reference surface S_t with $F(\overline{\mathbf{x}}, \overline{\mathbf{x}}_0, t) = 0$ and $\nabla F \bullet \dot{\overline{\mathbf{x}}} + \frac{\partial F}{\partial t} = 0$ for time t. For an arbitrarily small $\varepsilon > 0$, there are two time intervals $[t_{m-\varepsilon}, t_m)$ and $(t_m, t_{m+\varepsilon}]$. The vector fields $\mathbf{F}(\mathbf{x}, t, \mathbf{p})$ and $\mathbf{f}(\overline{\mathbf{x}}, t, \boldsymbol{\mu})$ are $C^r_{[t_{m-\varepsilon}, t_{m+\varepsilon}]}$-continuous ($r \geq 2k-1$) for time t. The flows $\mathbf{x}_t$ and $\overline{\mathbf{x}}_t$ are $C^r_{[t_{m-\varepsilon}, t_{m+\varepsilon}]}$-continuous ($r \geq 2k$) for time t with $\| d^{r+1}\mathbf{x}_t / dt^{r+1} \| < \infty$ and $\| d^{r+1}\overline{\mathbf{x}}_t / dt^{r+1} \| < \infty$. The flow $\mathbf{x}_t$ at point $(t_m, \mathbf{x}_{t_m})$ with $\mathbf{x}_{t_m} \notin S_t$ is *returning back* and away from the reference surface S_t with the $(2k\text{-}1)^{\text{th}}$-order singularity if

$$G^{(s)}(\overline{\mathbf{x}}_{t_m}, t_m, \mathbf{x}_{t_m}) = 0 \text{ for } s = 0, 1, 2, \cdots, 2k-2; \tag{2.46}$$

$$\left.\begin{aligned} &\mathbf{n}^T_{S_{t_m}} \cdot (\mathbf{x}_{t_m} - \overline{\mathbf{x}}_{t_m}) - \mathbf{n}^T_{S_{t_{m-\varepsilon}}} \cdot (\mathbf{x}_{t_m-\varepsilon} - \overline{\mathbf{x}}_{t_m-\varepsilon}) > 0 \\ &\mathbf{n}^T_{S_{t_{m+\varepsilon}}} \cdot (\mathbf{x}_{t_m+\varepsilon} - \overline{\mathbf{x}}_{t_{m+\varepsilon}}) - \mathbf{n}^T_{S_{t_m}} \cdot (\mathbf{x}_{t_m} - \overline{\mathbf{x}}_{t_m}) < 0 \end{aligned}\right\} \tag{2.47}$$

with $\mathbf{n}^T_{S_t} \cdot (\overline{\mathbf{x}}_t - \mathbf{x}_t) > 0$ for $t \in [t_m - \varepsilon, t_m)$ or

$$\left.\begin{aligned} &\mathbf{n}^T_{S_{t_m}} \cdot (\mathbf{x}_{t_m} - \overline{\mathbf{x}}_{t_m}) - \mathbf{n}^T_{S_{t_{m-\varepsilon}}} \cdot (\mathbf{x}_{t_m-\varepsilon} - \overline{\mathbf{x}}_{t_m-\varepsilon}) < 0, \\ &\mathbf{n}^T_{S_{t_{m+\varepsilon}}} \cdot (\mathbf{x}_{t_m+\varepsilon} - \overline{\mathbf{x}}_{t_{m+\varepsilon}}) - \mathbf{n}^T_{S_{t_m}} \cdot (\mathbf{x}_{t_m} - \overline{\mathbf{x}}_{t_m}) > 0 \end{aligned}\right\} \tag{2.48}$$

with $\mathbf{n}^T_{S_t} \cdot (\overline{\mathbf{x}}_t - \mathbf{x}_t) < 0$ for $t \in [t_m - \varepsilon, t_m)$.

THEOREM 2.5. *Consider two dynamical systems in Eqs.*(2.1) *and* (2.2) *which have a compared flow* $\mathbf{x}_t = \mathbf{\Phi}(t_0, \mathbf{x}_0, \mathbf{p}, t)$ *with an initial condition* $(t_0, \mathbf{x}_0)$ *and a reference flow* $\overline{\mathbf{x}}_t = \mathbf{\Phi}(t_0, \overline{\mathbf{x}}_0, \boldsymbol{\mu}, t)$ *with an initial condition* $(t_0, \overline{\mathbf{x}}_0)$, *respectively. The flow* $\mathbf{x}_t$ *does not contact with the reference flow* $\overline{\mathbf{x}}_t$. *Suppose the reference flow* $\overline{\mathbf{x}}_t$ *lies on a reference surface* S_t *with* $F(\overline{\mathbf{x}}, \overline{\mathbf{x}}_0, t) = 0$ *and* $\nabla F \bullet \dot{\overline{\mathbf{x}}} + \frac{\partial F}{\partial t} = 0$ *for time t. For an arbitrarily small* $\varepsilon > 0$, *there are two time intervals* $[t_{m-\varepsilon}, t_m)$ *and* $(t_m, t_{m+\varepsilon}]$. *The vector fields* $\mathbf{F}(\mathbf{x}, t, \mathbf{p})$ *and* $\mathbf{f}(\overline{\mathbf{x}}, t, \boldsymbol{\mu})$ *are* $C^r_{[t_{m-\varepsilon}, t_{m+\varepsilon}]}$*-continuous* ($r \geq k+1$) *for time t. The flows* $\mathbf{x}_t$ *and* $\overline{\mathbf{x}}_t$ *are* $C^r_{[t_{m-\varepsilon}, t_{m+\varepsilon}]}$*-continuous* ($r \geq k+2$) *for time t with* $\| d^{r+1}\mathbf{x}_t / dt^{r+1} \| < \infty$ *and* $\| d^{r+1}\overline{\mathbf{x}}_t / dt^{r+1} \| < \infty$.

$$G^{(s)}(\overline{\mathbf{x}}_{t_m}, t_m, \mathbf{x}_{t_m}) = 0 \text{ for } s = 0, 1, 2, \cdots, k-1. \tag{2.49}$$

(i) *The flow* $\mathbf{x}_t$ *at point* $(t_m, \mathbf{x}_m)$ *with* $\mathbf{x}_{t_m} \notin S_t$ *is leaving from the reference surface* S_t *with the* k^{th}*-order singularity iff for* $t \in (t_m, t_m + \varepsilon]$

$$\left.\begin{array}{l} G^{(k)}(\bar{\mathbf{x}}_{t_m}, t_m, \mathbf{x}_{t_m}) > 0 \text{ for } \mathbf{n}_{S_t}^T \cdot (\mathbf{x}_t - \bar{\mathbf{x}}_t) > 0 \text{ or} \\ G^{(k)}(\bar{\mathbf{x}}_{t_m}, t_m, \mathbf{x}_{t_m}) < 0 \text{ for } \mathbf{n}_{S_t}^T \cdot (\mathbf{x}_t - \bar{\mathbf{x}}_t) < 0. \end{array}\right\} \tag{2.50}$$

(ii) *The flow* $\mathbf{x}_t$ *at point* $(t_m, \bar{\mathbf{x}}_m)$ *with* $\mathbf{x}_{t_m} \notin S_t$ *is approaching the reference surface* S_t *with the* k^{th}*-order singularity iff for* $t \in [t_m - \varepsilon, t_m)$

$$\left.\begin{array}{l} \left.\begin{array}{l} G^{(k)}(\bar{\mathbf{x}}_{t_m}, t_m, \mathbf{x}_{t_m}) > 0 \quad k = 2m, \\ G^{(k)}(\bar{\mathbf{x}}_{t_m}, t_m, \mathbf{x}_{t_m}) < 0 \quad k = 2m-1 \end{array}\right\} \text{for } \mathbf{n}_{S_t}^T \cdot (\bar{\mathbf{x}}_t - \mathbf{x}_t) > 0, \text{ or} \\ \left.\begin{array}{l} G^{(k)}(\bar{\mathbf{x}}_{t_m}, t_m, \mathbf{x}_{t_m}) < 0 \quad k = 2m, \\ G^{(k)}(\bar{\mathbf{x}}_{t_m}, t_m, \mathbf{x}_{t_m}) > 0 \quad k = 2m-1 \end{array}\right\} \text{for } \mathbf{n}_{S_t}^T \cdot (\bar{\mathbf{x}}_t - \mathbf{x}_t) < 0 \end{array}\right\} \tag{2.51}$$

PROOF. From Eq.(2.39), equation (2.49) exists, i.e.,

$$G^{(s)}(\bar{\mathbf{x}}_{t_m}, t_m, \mathbf{x}_{t_m}) = 0 \text{ for } s = 0, 1, 2, \cdots, k;$$

and vice versa.

(i) For the case of $r = k$, equation (2.16) gives

$$G_+^{(k)}(\bar{\mathbf{x}}_{t_m}, t_m, \mathbf{x}_{t_{m+}}) = \lim_{\varepsilon \to 0} \frac{1}{\varepsilon^{k+1}} \left[\mathbf{n}_{S_{t_m+\varepsilon}}^T \cdot (\mathbf{x}_{t_m+\varepsilon} - \bar{\mathbf{x}}_{t_m+\varepsilon}) - \mathbf{n}_{S_t}^T \cdot (\mathbf{x}_{t_{m+}} - \bar{\mathbf{x}}_{t_m}) \right]$$

From Eq.(2.40), the following equation exists

$$\left.\begin{array}{l} G^{(k+1)}(\bar{\mathbf{x}}_{t_m}, t_m, \mathbf{x}_{t_m}) = G_+^{(k+1)}(\bar{\mathbf{x}}_{t_m}, t_m, \mathbf{x}_{t_m}) > 0 \text{ for } \mathbf{n}_{S_t}^T \cdot (\bar{\mathbf{x}}_t - \mathbf{x}_t) \geq 0, \\ G^{(k+1)}(\bar{\mathbf{x}}_{t_m}, t_m, \mathbf{x}_{t_m}) = G_+^{(k+1)}(\bar{\mathbf{x}}_{t_m}, t_m, \mathbf{x}_{t_m}) < 0 \text{ for } \mathbf{n}_{S_t}^T \cdot (\bar{\mathbf{x}}_t - \mathbf{x}_t) \leq 0. \end{array}\right\}$$

With Eq.(2.18), the Taylor series of $\mathbf{n}_{S_{t_m+\varepsilon}}^T \cdot (\mathbf{x}_{t_m+\varepsilon} - \bar{\mathbf{x}}_{t_m+\varepsilon})$ up to ε^{k+2}-terms gives

$$\begin{aligned} \mathbf{n}_{S_{t_m+\varepsilon}}^T \cdot (\mathbf{x}_{t_m+\varepsilon} - \bar{\mathbf{x}}_{t_m+\varepsilon}) &= \mathbf{n}_{S_{t_m}}^T \cdot (\mathbf{x}_{t_{m+}} - \bar{\mathbf{x}}_{t_m}) + \sum_{s=0}^{k-1} G_+^{(s)}(\bar{\mathbf{x}}_{t_m}, t_m, \mathbf{x}_{t_{m+}}) \varepsilon^{s+1} \\ &\quad + G_+^{(k)}(\bar{\mathbf{x}}_{t_m}, t_m, \mathbf{x}_{t_{m+}}) \varepsilon^{k+1} \end{aligned}$$

Application of Eq.(2.49) to the foregoing equation gives

$$\mathbf{n}_{S_{t_m+\varepsilon}}^T \cdot (\mathbf{x}_{t_m+\varepsilon} - \bar{\mathbf{x}}_{t_m+\varepsilon}) - \mathbf{n}_{S_{t_m}}^T \cdot (\mathbf{x}_{t_{m+}} - \bar{\mathbf{x}}_{t_m}) = G_+^{(k)}(\bar{\mathbf{x}}_{t_m}, t_m, \mathbf{x}_{t_{m+}}) \varepsilon^{k+1}.$$

Equation (2.40) is employed in the above equation, equation (2.50) holds.

(ii) Using the similar procedure, for the case of $s = k$, equation (2.16) gives

$$G_{-}^{(k)}(\bar{\mathbf{x}}_{t_m}, t_m, \mathbf{x}_{t_m -}) = \lim_{\varepsilon\to 0}\frac{(-1)^{k+2}}{\varepsilon^{k+1}}\left[\mathbf{n}_{S_t}^T\cdot(\mathbf{x}_{t_m+} - \bar{\mathbf{x}}_{t_m}) - \mathbf{n}_{S_{t_m-\varepsilon}}^T\cdot(\mathbf{x}_{t_m-\varepsilon} - \bar{\mathbf{x}}_{t_m-\varepsilon})\right].$$

From Eq.(2.41), the following equation is obtained

$$\left.\begin{array}{ll} G^{(k)}(\bar{\mathbf{x}}_{t_m}, t_m, \mathbf{x}_{t_m}) > 0 & k = 2m, \\ G^{(k)}(\bar{\mathbf{x}}_{t_m}, t_m, \mathbf{x}_{t_m}) < 0 & k = 2m-1 \end{array}\right\}\text{for } \mathbf{n}_{S_t}^T\cdot(\bar{\mathbf{x}}_t - \mathbf{x}_t) > 0, \text{ or}$$

$$\left.\begin{array}{ll} G^{(k)}(\bar{\mathbf{x}}_{t_m}, t_m, \mathbf{x}_{t_m}) < 0 & k = 2m, \\ G^{(k)}(\bar{\mathbf{x}}_{t_m}, t_m, \mathbf{x}_{t_m}) > 0 & k = 2m-1 \end{array}\right\}\text{for } \mathbf{n}_{S_t}^T\cdot(\bar{\mathbf{x}}_t - \mathbf{x}_t) < 0$$

With Eq.(2.18),the Taylor series expansion of $\mathbf{n}_{S_{t_m-\varepsilon}}^T\cdot(\mathbf{x}_{t_m-\varepsilon} - \bar{\mathbf{x}}_{t_m-\varepsilon})$ up to ε^{k+2}-terms gives

$$\mathbf{n}_{S_{t_m-\varepsilon}}^T\cdot(\mathbf{x}_{t_m-\varepsilon} - \bar{\mathbf{x}}_{t_m-\varepsilon}) = \mathbf{n}_{S_{t_m}}^T\cdot(\mathbf{x}_{t_m -} - \bar{\mathbf{x}}_{t_m}) + \sum_{r=0}^{k-1} G_{-}^{(r)}(\bar{\mathbf{x}}_{t_m}, t_m, \mathbf{x}_{t_m -})(-\varepsilon)^{r+1}$$
$$+ G_{-}^{(k)}(\bar{\mathbf{x}}_{t_m}, t_m, \mathbf{x}_{t_m -})(-\varepsilon)^{k+1}.$$

Application of eq.(2.49) to the foregoing equation gives

$$\mathbf{n}_{S_{t_m}}^T\cdot(\mathbf{x}_{t_m -} - \bar{\mathbf{x}}_{t_m}) - \mathbf{n}_{S_{t_m-\varepsilon}}^T\cdot(\mathbf{x}_{t_m-\varepsilon} - \bar{\mathbf{x}}_{t_m-\varepsilon}) = (-1)^{k+2} G_{-}^{(k)}(\bar{\mathbf{x}}_{t_m}, t_m, \mathbf{x}_{t_m -})\varepsilon^{k+1}.$$

When Eq.(2.41) is employed in the above equation, equation (2.51) holds, vice versa. ■

THEOREM 2.6. *Consider two dynamical systems in Eqs.(2.1) and (2.2) which have a compared flow $\mathbf{x}_t = \mathbf{\Phi}(t_0, \mathbf{x}_0, \mathbf{p}, t)$ with an initial condition $(t_0, \mathbf{x}_0)$ and a reference flow $\bar{\mathbf{x}}_t = \mathbf{\Phi}(t_0, \bar{\mathbf{x}}_0, \boldsymbol{\mu}, t)$ with an initial condition $(t_0, \bar{\mathbf{x}}_0)$, respectively. The flow $\mathbf{x}_t$ does not contact with the reference flow $\bar{\mathbf{x}}_t$. Suppose the reference flow $\bar{\mathbf{x}}_t$ lies on a reference surface S_t with $F(\bar{\mathbf{x}}, \bar{\mathbf{x}}_0, t) = 0$ and $\nabla F\bullet\dot{\bar{\mathbf{x}}} + \frac{\partial F}{\partial t} = 0$ for time t. For an arbitrarily small $\varepsilon > 0$, there are two time intervals $[t_{m-\varepsilon}, t_m)$ and $(t_m, t_{m+\varepsilon}]$. The vector fields $\mathbf{F}(\mathbf{x}, t, \mathbf{p})$ and $\mathbf{f}(\bar{\mathbf{x}}, t, \boldsymbol{\mu})$ are $C^r_{[t_{m-\varepsilon}, t_{m+\varepsilon}]}$-continuous ($r \geq 2k$) for time t. The flows $\mathbf{x}_t$ and $\bar{\mathbf{x}}_t$ are $C^r_{[t_{m-\varepsilon}, t_{m+\varepsilon}]}$-continuous ($r \geq 2k+1$) for time t with $\| d^{r+1}\mathbf{x}_t/dt^{r+1} \| < \infty$ and $\| d^{r+1}\bar{\mathbf{x}}_t/dt^{r+1} \| < \infty$. The flow $\mathbf{x}_t$ at point $(t_m, \mathbf{x}_m)$ with $\mathbf{x}_{t_m} = \bar{\mathbf{x}}'_{t_m} \in S_t$ is passing through the reference surface S_t with the $(2k)^{th}$-order singularity iff for $t \in [t_m - \varepsilon, t_m)$*

$$G^{(s)}(\overline{\mathbf{x}}_{t_m}, t_m, \mathbf{x}_{t_m}) = 0 \text{ for } s = 0,1,2,\cdots,2k-1 \tag{2.52}$$

$$\left.\begin{aligned} G^{(2k)}(\overline{\mathbf{x}}_{t_m}, t_m, \mathbf{x}_{t_m}) > 0 \ \text{ for } \ \mathbf{n}_{S_t}^T \cdot (\overline{\mathbf{x}}_t - \mathbf{x}_t) > 0 \ \text{ or} \\ G^{(2k)}(\overline{\mathbf{x}}_{t_m}, t_m, \mathbf{x}_{t_m}) < 0 \ \text{ for } \mathbf{n}_{S_t}^T \cdot (\overline{\mathbf{x}}_t - \mathbf{x}_t) < 0 \end{aligned}\right\} \tag{2.53}$$

PROOF. The proof is the same as in Proof of Theorem 2.5. Equation (2.43) is identical to Eq.(2.52), i.e.,

$$G^{(s)}(\overline{\mathbf{x}}_{t_m}, t_m, \mathbf{x}_{t_m}) = 0 \text{ for } s = 0,1,2,\cdots,2k-1;$$

Consider the case of $r = 2k$. Equation (2.16) gives

$$G_-^{(2k)}(\overline{\mathbf{x}}_{t_m}, t_m, \mathbf{x}_{t_{m-}}) = \lim_{\varepsilon\to 0} \frac{(-1)^{2k+2}}{\varepsilon^{2k+1}} \left[\mathbf{n}_{S_t}^T \cdot (\mathbf{x}_{t_{m-}} - \overline{\mathbf{x}}_{t_m}) - \mathbf{n}_{S_{t_m-\varepsilon}}^T \cdot (\mathbf{x}_{t_m-\varepsilon} - \overline{\mathbf{x}}_{t_m-\varepsilon}) \right]$$

$$G_+^{(2k)}(\overline{\mathbf{x}}_{t_m}, t_m, \mathbf{x}_{t_{m+}}) = \lim_{\varepsilon\to 0} \frac{1}{\varepsilon^{2k+1}} \left[\mathbf{n}_{S_{t_m+\varepsilon}}^T \cdot (\mathbf{x}_{t_m+\varepsilon} - \overline{\mathbf{x}}_{t_m+\varepsilon}) - \mathbf{n}_{S_t}^T \cdot (\mathbf{x}_{t_{m+}} - \overline{\mathbf{x}}_{t_m}) \right]$$

From Eqs.(2.44) and (2.45), the following equation given in Eq.(2.53) exists

$$\left.\begin{aligned} G^{(2k)}(\overline{\mathbf{x}}_{t_m}, t_m, \mathbf{x}_{t_m}) > 0 \ \text{ for } \ \mathbf{n}_{S_t}^T \cdot (\overline{\mathbf{x}}_t - \mathbf{x}_t) > 0 \ \text{ or} \\ G^{(2k)}(\overline{\mathbf{x}}_{t_m}, t_m, \mathbf{x}_{t_m}) < 0 \ \text{ for } \mathbf{n}_{S_t}^T \cdot (\overline{\mathbf{x}}_t - \mathbf{x}_t) < 0 \end{aligned}\right\}$$

With Eq.(2.18), the Taylor series expansions of the two normal distances $\mathbf{n}_{S_{t_m-\varepsilon}}^T \cdot (\mathbf{x}_{t_m-\varepsilon} - \overline{\mathbf{x}}_{t_m-\varepsilon})$ and $\mathbf{n}_{S_{t_m+\varepsilon}}^T \cdot (\mathbf{x}_{t_m+\varepsilon} - \overline{\mathbf{x}}_{t_m+\varepsilon})$ up to ε^{k+2} terms give

$$\begin{aligned} \mathbf{n}_{S_{t_m-\varepsilon}}^T \cdot (\mathbf{x}_{t_m-\varepsilon} - \overline{\mathbf{x}}_{t_m-\varepsilon}) = \mathbf{n}_{S_{t_m}}^T \cdot (\mathbf{x}_{t_{m-}} - \overline{\mathbf{x}}_{t_m}) + \sum\nolimits_{s=0}^{2k-1} G_-^{(s)}(\overline{\mathbf{x}}_{t_m}, t_m, \mathbf{x}_{t_{m-}})(-\varepsilon)^{s+1} \\ + G_-^{(2k)}(\overline{\mathbf{x}}_{t_m}, t_m, \mathbf{x}_{t_{m-}})(-\varepsilon)^{2k+1} \end{aligned}$$

$$\begin{aligned} \mathbf{n}_{S_{t_m+\varepsilon}}^T \cdot (\mathbf{x}_{t_m+\varepsilon} - \overline{\mathbf{x}}_{t_m+\varepsilon}) = \mathbf{n}_{S_{t_m}}^T \cdot (\mathbf{x}_{t_{m+}} - \overline{\mathbf{x}}_{t_m}) + \sum\nolimits_{s=0}^{2k-1} G_+^{(s)}(\overline{\mathbf{x}}_{t_m}, t_m, \mathbf{x}_{t_{m+}})\varepsilon^{s+1} \\ + G_+^{(2k)}(\overline{\mathbf{x}}_{t_m}, t_m, \mathbf{x}_{t_{m+}})\varepsilon^{2k+1} \end{aligned}$$

Application of Eq.(2.52) to the foregoing equations gives

$$\mathbf{n}_{S_{t_m}}^T \cdot (\mathbf{x}_{t_{m-}} - \overline{\mathbf{x}}_{t_m}) - \mathbf{n}_{S_{t_m-\varepsilon}}^T \cdot (\mathbf{x}_{t_m-\varepsilon} - \overline{\mathbf{x}}_{t_m-\varepsilon}) = -G_-^{(2k)}(\overline{\mathbf{x}}_{t_m}, t_m, \mathbf{x}_{t_{m-}})(-\varepsilon)^{2k+1}$$

$$\mathbf{n}_{S_{t_m+\varepsilon}}^T \cdot (\mathbf{x}_{t_m+\varepsilon} - \overline{\mathbf{x}}_{t_m+\varepsilon}) - \mathbf{n}_{S_{t_m}}^T \cdot (\mathbf{x}_{t_{m+}} - \overline{\mathbf{x}}_{t_m}) = G_+^{(2k)}(\overline{\mathbf{x}}_{t_m}, t_m, \mathbf{x}_{t_{m+}})\varepsilon^{2k+1}$$

When Eq.(2.53) is employed in the above equation, equations (2.44) and (2.45) hold, vice versa. ∎

THEOREM 2.7. *Consider two dynamical systems in Eqs.*(2.1) *and* (2.2) *which have a compared flow* $\mathbf{x}_t = \mathbf{\Phi}(t_0, \mathbf{x}_0, \mathbf{p}, t)$ *with an initial condition* $(t_0, \mathbf{x}_0)$ *and a reference flow* $\bar{\mathbf{x}}_t = \mathbf{\Phi}(t_0, \bar{\mathbf{x}}_0, \boldsymbol{\mu}, t)$ *with an initial condition* $(t_0, \bar{\mathbf{x}}_0)$, *respectively. The flow* $\mathbf{x}_t$ *does not contact with the reference flow* $\bar{\mathbf{x}}_t$. *Suppose the reference flow* $\bar{\mathbf{x}}_t$ *lies on a reference surface* S_t *with* $F(\bar{\mathbf{x}}, \bar{\mathbf{x}}_0, t) = 0$ *and* $\nabla F \bullet \dot{\bar{\mathbf{x}}} + \frac{\partial F}{\partial t} = 0$ *for time t. For an arbitrarily small* $\varepsilon > 0$, *there are two time intervals* $[t_{m-\varepsilon}, t_m)$ *and* $(t_m, t_{m+\varepsilon}]$. *The vector fields* $\mathbf{F}(\mathbf{x}, t, \mathbf{p})$ *and* $\mathbf{f}(\bar{\mathbf{x}}, t, \boldsymbol{\mu})$ *are* $C^r_{[t_{m-\varepsilon}, t_{m+\varepsilon}]}$*-continuous* ($r \geq 2k-1$) *for time t. The flows* $\mathbf{x}_t$ *and* $\bar{\mathbf{x}}_t$ *are* $C^r_{[t_{m-\varepsilon}, t_{m+\varepsilon}]}$*-continuous* ($r \geq 2k$) *for time t with* $\| d^{r+1}\mathbf{x}_t / dt^{r+1} \| < \infty$ *and* $\| d^{r+1}\bar{\mathbf{x}}_t / dt^{r+1} \| < \infty$. *The flow* $\mathbf{x}_t$ *at point* $(t_m, \mathbf{x}_m)$ *with* $\mathbf{x}_{t_m} \notin S_t$ *is returning back and away from the reference surface* S_t *with the* (2*k*-1)*th-order singularity iff for* $t \in [t_{m\text{-}\varepsilon}, t_m)$

$$G^{(s)}(\bar{\mathbf{x}}_{t_m}, t_m, \mathbf{x}_{t_m}) = 0 \text{ for } s = 0, 1, 2, \cdots, 2k-2; \tag{2.54}$$

$$\left.\begin{aligned} &G^{(2k-1)}(\bar{\mathbf{x}}_{t_m}, t_m, \mathbf{x}_{t_m}) < 0 \text{ with } \mathbf{n}^T_{S_t} \cdot (\bar{\mathbf{x}}_t - \mathbf{x}_t) > 0 \text{ or} \\ &G^{(2k-1)}(\bar{\mathbf{x}}_{t_m}, t_m, \mathbf{x}_{t_m}) > 0 \text{ with } \mathbf{n}^T_{S_t} \cdot (\bar{\mathbf{x}}_t - \mathbf{x}_t) < 0. \end{aligned}\right\} \tag{2.55}$$

PROOF. The proof is the same as in Proof of THEOREM 2.5. Equation (2.46) is identical to Eq.(2.54), i.e.,

$$G^{(s)}(\bar{\mathbf{x}}_{t_m}, t_m, \mathbf{x}_{t_m}) = 0 \text{ for } s = 0, 1, 2, \cdots, 2k;$$

Consider the case of $s = 2k$, equation (2.16) gives

$$G^{(2k+1)}_{-}(\bar{\mathbf{x}}_{t_m}, t_m, \mathbf{x}_{t_m-}) = \lim_{\varepsilon \to 0} \frac{(-1)^{2k+3}}{\varepsilon^{2k+2}} \left[\mathbf{n}^T_{S_t} \cdot (\mathbf{x}_{t_m-} - \bar{\mathbf{x}}_{t_m}) - \mathbf{n}^T_{S_{t_m-\varepsilon}} \cdot (\mathbf{x}_{t_m-\varepsilon} - \bar{\mathbf{x}}_{t_m-\varepsilon}) \right],$$

$$G^{(2k+1)}_{+}(\bar{\mathbf{x}}_{t_m}, t_m, \mathbf{x}_{t_m+}) = \lim_{\varepsilon \to 0} \frac{1}{\varepsilon^{2k+2}} \left[\mathbf{n}^T_{S_{t_m+\varepsilon}} \cdot (\mathbf{x}_{t_m+\varepsilon} - \bar{\mathbf{x}}_{t_m+\varepsilon}) - \mathbf{n}^T_{S_t} \cdot (\mathbf{x}_{t_m+} - \bar{\mathbf{x}}_{t_m}) \right].$$

From Eq.(2.47), the equations in Eq.(2.55) exist, i.e.,

$$\left.\begin{aligned} &G^{(2k+1)}(\bar{\mathbf{x}}_{t_m}, t_m, \mathbf{x}_{t_m}) < 0 \text{ for } \mathbf{n}^T_{S_t} \cdot (\bar{\mathbf{x}}_t - \mathbf{x}_t) > 0 \text{ or} \\ &G^{(2k+1)}(\bar{\mathbf{x}}_{t_m}, t_m, \mathbf{x}_{t_m}) > 0 \text{ for } \mathbf{n}^T_{S_t} \cdot (\bar{\mathbf{x}}_t - \mathbf{x}_t) < 0 \end{aligned}\right\}$$

With Eq.(2.18), the Taylor series expansion of the two normal components $\mathbf{n}^T_{S_{t_m-\varepsilon}} \cdot \mathbf{n}^T_{S_{t_m-\varepsilon}} \cdot (\mathbf{x}_{t_m-\varepsilon} - \bar{\mathbf{x}}_{t_m-\varepsilon})$ and $\mathbf{n}^T_{S_{t_m+\varepsilon}} \cdot (\mathbf{x}_{t_m+\varepsilon} - \bar{\mathbf{x}}_{t_m+\varepsilon})$ in the normal direction of the reference surface up to ε^{k+2} terms give

$$\mathbf{n}^T_{S_{t_m-\varepsilon}}\cdot(\mathbf{x}_{t_m-\varepsilon}-\overline{\mathbf{x}}_{t_m-\varepsilon})=\mathbf{n}^T_{S_{t_m}}\cdot(\mathbf{x}_{t_m-}-\overline{\mathbf{x}}_{t_m})+\sum\nolimits_{s=0}^{2k-2}G_-^{(s)}(\overline{\mathbf{x}}_{t_m},t_m,\mathbf{x}_{t_m-})(-\varepsilon)^{s+1}$$
$$+G_-^{(2k-1)}(\overline{\mathbf{x}}_{t_m},t_m,\mathbf{x}_{t_m-})(-\varepsilon)^{2k},$$

$$\mathbf{n}^T_{S_{t_m+\varepsilon}}\cdot(\mathbf{x}_{t_m+\varepsilon}-\overline{\mathbf{x}}_{t_m+\varepsilon})=\mathbf{n}^T_{S_{t_m}}\cdot(\mathbf{x}_{t_m+}-\overline{\mathbf{x}}_{t_m})+\sum\nolimits_{s=0}^{2k-2}G_+^{(s)}(\overline{\mathbf{x}}_{t_m},t_m,\mathbf{x}_{t_m+})\varepsilon^{s+1}$$
$$+G_+^{(2k-1)}(\overline{\mathbf{x}}_{t_m},t_m,\mathbf{x}_{t_m+})\varepsilon^{2k}.$$

Application of Eq.(2.46) to the foregoing equations gives

$$\mathbf{n}^T_{S_{t_m}}\cdot(\mathbf{x}_{t_m-}-\overline{\mathbf{x}}_{t_m})-\mathbf{n}^T_{S_{t_m-\varepsilon}}\cdot(\mathbf{x}_{t_m-\varepsilon}-\overline{\mathbf{x}}_{t_m-\varepsilon})=-G_-^{(2k-1)}(\overline{\mathbf{x}}_{t_m},t_m,\mathbf{x}_{t_m-})(-\varepsilon)^{2k},$$
$$\mathbf{n}^T_{S_{t_m+\varepsilon}}\cdot(\mathbf{x}_{t_m+\varepsilon}-\overline{\mathbf{x}}_{t_m+\varepsilon})-\mathbf{n}^T_{S_{t_m}}\cdot(\mathbf{x}_{t_m+}-\overline{\mathbf{x}}_{t_m})=G_+^{(2k-1)}(\overline{\mathbf{x}}_{t_m},t_m,\mathbf{x}_{t_m+})\varepsilon^{2k}.$$

If Eq.(2.55) is employed in the above equations, equations (2.47) and (2.48) hold, vice versa. ■

2.3. Contact flows

As in the previous section, the singularity for the two flows with contact at point $(t_m,\overline{\mathbf{x}}_{t_m})$ will be presented in this section. Consider the zero-order contact of the reference and compared flows first, which requires $\mathbf{x}_{t_m}=\overline{\mathbf{x}}_{t_m}$ at time t_m.

DEFINITION 2.9. Consider two dynamical systems in Eqs.(2.1) and (2.2) which have a compared flow $\mathbf{x}_t=\mathbf{\Phi}(t_0,\mathbf{x}_0,\mathbf{p},t)$ with an initial condition $(t_0,\mathbf{x}_0)$ and a reference flow $\overline{\mathbf{x}}_t=\mathbf{\Phi}(t_0,\overline{\mathbf{x}}_0,\boldsymbol{\mu},t)$ with an initial condition $(t_0,\overline{\mathbf{x}}_0)$, respectively. At time t_m, $\mathbf{x}_m=\overline{\mathbf{x}}_m$. Suppose the reference flow $\overline{\mathbf{x}}_t$ lies on a reference surface S_t with $F(\overline{\mathbf{x}},\overline{\mathbf{x}}_0,t)=0$ and $\nabla F\boldsymbol{\cdot}\dot{\overline{\mathbf{x}}}+\frac{\partial F}{\partial t}=0$ for time *t*. For an arbitrarily small $\varepsilon>0$, there are two time intervals $[t_{m-\varepsilon},t_m)$ and $(t_m,t_{m+\varepsilon}]$.

(i) The flow $\mathbf{x}_t$ is *leaving from* the reference surface S_t at point $(t_m,\overline{\mathbf{x}}_m)$ if for $t\in(t_m,t_m+\varepsilon]$

$$\left.\begin{aligned}&\mathbf{n}^T_{S_{t_m+\varepsilon}}\cdot(\mathbf{x}_{t_m+\varepsilon}-\overline{\mathbf{x}}_{t_m+\varepsilon})>0\text{ for }\mathbf{n}^T_{S_t}\cdot(\mathbf{x}_t-\overline{\mathbf{x}}_t)>0\text{ or}\\&\mathbf{n}^T_{S_{t_m+\varepsilon}}\cdot(\mathbf{x}_{t_m+\varepsilon}-\overline{\mathbf{x}}_{t_m+\varepsilon})<0\text{ for }\mathbf{n}^T_{S_t}\cdot(\mathbf{x}_t-\overline{\mathbf{x}}_t)<0.\end{aligned}\right\}\tag{2.56}$$

(ii) The flow $\mathbf{x}_t$ is *approaching* the reference surface S_t at point $(t_m,\overline{\mathbf{x}}_m)$ if for $t\in[t_m-\varepsilon,t_m)$

$$\left.\begin{array}{l} -\mathbf{n}^T_{S_{t_{m-\varepsilon}}} \cdot (\mathbf{x}_{t_m-\varepsilon} - \overline{\mathbf{x}}_{t_m-\varepsilon}) > 0 \;\; \text{for } \mathbf{n}^T_{S_t} \cdot (\overline{\mathbf{x}}_t - \mathbf{x}_t) > 0 \text{ or} \\ -\mathbf{n}^T_{S_{t_{m-\varepsilon}}} \cdot (\mathbf{x}_{t_m-\varepsilon} - \overline{\mathbf{x}}_{t_m-\varepsilon}) < 0 \;\; \text{for } \mathbf{n}^T_{S_t} \cdot (\overline{\mathbf{x}}_t - \mathbf{x}_t) < 0. \end{array}\right\} \tag{2.57}$$

(iii) The compared flow $\mathbf{x}_t$ contacting with the reference flow $\overline{\mathbf{x}}_t$ at point $(t_m, \overline{\mathbf{x}}_m)$ are on the reference surface if for $t \in [t_m - \varepsilon, t_m + \varepsilon]$

$$\mathbf{n}^T_{S_{t_{m-\varepsilon}}} \cdot (\mathbf{x}_{t_m-\varepsilon} - \overline{\mathbf{x}}_{t_m-\varepsilon}) = \mathbf{n}^T_{S_{t_{m+\varepsilon}}} \cdot (\mathbf{x}_{t_m+\varepsilon} - \overline{\mathbf{x}}_{t_m+\varepsilon}) = 0. \tag{2.58}$$

DEFINITION 2.10. Consider two dynamical systems in Eqs.(2.1) and (2.2) which have a compared flow $\mathbf{x}_t = \mathbf{\Phi}(t_0, \mathbf{x}_0, \mathbf{p}, t)$ with an initial condition $(t_0, \mathbf{x}_0)$ and a reference flow $\overline{\mathbf{x}}_t = \mathbf{\Phi}(t_0, \overline{\mathbf{x}}_0, \boldsymbol{\mu}, t)$ with an initial condition $(t_0, \overline{\mathbf{x}}_0)$, respectively. At time t_m, $\mathbf{x}_m = \overline{\mathbf{x}}_m$. Suppose the reference flow $\overline{\mathbf{x}}_t$ lies on a reference surface S_t with $F(\overline{\mathbf{x}}, \overline{\mathbf{x}}_0, t) = 0$ and $\nabla F \bullet \dot{\overline{\mathbf{x}}} + \frac{\partial F}{\partial t} = 0$ for time t. For an arbitrarily small $\varepsilon > 0$, there are two time intervals $[t_{m-\varepsilon}, t_m)$ and $(t_m, t_{m+\varepsilon}]$.

(i) The flow $\mathbf{x}_t$ is *passing* through the reference surface S_t at point $(t_m, \overline{\mathbf{x}}_m)$ if

$$-\mathbf{n}^T_{S_{t_{m-\varepsilon}}} \cdot (\mathbf{x}_{t_m-\varepsilon} - \overline{\mathbf{x}}_{t_m-\varepsilon}) > 0 \text{ and } \mathbf{n}^T_{S_{t_{m+\varepsilon}}} \cdot (\mathbf{x}_{t_m+\varepsilon} - \overline{\mathbf{x}}_{t_{m+\varepsilon}}) > 0 \tag{2.59}$$

with $\mathbf{n}^T_{S_t} \cdot (\overline{\mathbf{x}}_t - \mathbf{x}_t) > 0$ for $t \in [t_m - \varepsilon, t_m)$ or

$$-\mathbf{n}^T_{S_{t_{m-\varepsilon}}} \cdot (\mathbf{x}_{t_m-\varepsilon} - \overline{\mathbf{x}}_{t_m-\varepsilon}) < 0 \text{ and } \mathbf{n}^T_{S_{t_{m+\varepsilon}}} \cdot (\mathbf{x}_{t_m+\varepsilon} - \overline{\mathbf{x}}_{t_{m+\varepsilon}}) < 0 \tag{2.60}$$

with $\mathbf{n}^T_{S_t} \cdot (\overline{\mathbf{x}}_t - \mathbf{x}_t) < 0$ for $t \in [t_m - \varepsilon, t_m)$.

(ii) The flow $\mathbf{x}_t$ is *returning back* from the reference surface S_t at point $(t_m, \overline{\mathbf{x}}_m)$ if

$$-\mathbf{n}^T_{S_{t_{m-\varepsilon}}} \cdot (\mathbf{x}_{t_m-\varepsilon} - \overline{\mathbf{x}}_{t_m-\varepsilon}) > 0 \text{ and } \mathbf{n}^T_{S_{t_{m+\varepsilon}}} \cdot (\mathbf{x}_{t_m+\varepsilon} - \overline{\mathbf{x}}_{t_{m+\varepsilon}}) < 0 \tag{2.61}$$

with $\mathbf{n}^T_{S_t} \cdot (\overline{\mathbf{x}}_t - \mathbf{x}_t) > 0$ for $t \in [t_m - \varepsilon, t_m)$ or

$$-\mathbf{n}^T_{S_{t_{m-\varepsilon}}} \cdot (\mathbf{x}_{t_m-\varepsilon} - \overline{\mathbf{x}}_{t_m-\varepsilon}) < 0 \text{ and } \mathbf{n}^T_{S_{t_{m+\varepsilon}}} \cdot (\mathbf{x}_{t_m+\varepsilon} - \overline{\mathbf{x}}_{t_{m+\varepsilon}}) > 0 \tag{2.62}$$

with $\mathbf{n}^T_{S_t} \cdot (\overline{\mathbf{x}}_t - \mathbf{x}_t) < 0$ for $t \in [t_m - \varepsilon, t_m)$.

Since the flow $\overline{\mathbf{x}}_t$ is on the surface S_t, the flow $\mathbf{x}_t$ passing through the reference surface is called the flow $\mathbf{x}_t$ *transversal to* such a surface S_t at point $(t_m, \overline{\mathbf{x}}_m)$. Because the flow $\mathbf{x}_t$ is continuous at the contact point of $\overline{\mathbf{x}}_t$, the flow $\mathbf{x}_t$ returning back from the reference surface S_t at point $(t_m, \bar{\mathbf{x}}_m)$ is called the flow $\mathbf{x}_t$ *tangential to* the reference surface S_t at point $(t_m, \bar{\mathbf{x}}_m)$. This case is extensively used in many dynamical systems. In Eq.(2.10), due to $\mathbf{n}_{S_t}^T \cdot \dot{\bar{\mathbf{x}}}_t = 0$ for a time-independent reference surface, the zero-order, G-function for the zero-order contact between the compared and reference flows can be re-written as

$$G_0^{(0)}(\bar{\mathbf{x}}_t, t, \mathbf{x}_t) = G(\bar{\mathbf{x}}_t, t, \mathbf{x}_t)\big|_{\mathbf{x}_t = \bar{\mathbf{x}}_t} = \mathbf{n}_{S_t}^T \cdot \dot{\mathbf{x}}_t\big|_{\mathbf{x}_t = \bar{\mathbf{x}}_t} = \mathbf{n}_{S_t}^T \cdot \mathbf{F}(\mathbf{x}_t, t, \mathbf{p})\big|_{\mathbf{x}_t = \bar{\mathbf{x}}_t}. \quad (2.63)$$

Accordingly, from Eqs.(2.18) and (2.19), the kth-order G-function for the zero-order contact can be re-written as

$$\begin{aligned} G_0^{(k)}(\bar{\mathbf{x}}_t, t, \mathbf{x}_t) &= G^{(k)}(\bar{\mathbf{x}}_t, t, \mathbf{x}_t)\big|_{\mathbf{x}_t = \bar{\mathbf{x}}_t} \\ &= \sum_{r=1}^{k+1} C_{k+1}^r D_{\bar{\mathbf{x}}_t}^{k+1-r} \mathbf{n}_{S_t}^T \cdot \left(\tfrac{d^r \mathbf{x}_t}{dt^r} - \tfrac{d^r \bar{\mathbf{x}}_t}{dt^r}\right)\Big|_{\mathbf{x}_t = \bar{\mathbf{x}}_t} \\ &= \sum_{r=1}^{k+1} C_{k+1}^r D_{\bar{\mathbf{x}}_t}^{k+1-r} \mathbf{n}_{S_t}^T \cdot \left(D_{\mathbf{x}_t}^{r-1} \mathbf{F}(\mathbf{x}_t, t, \mathbf{p}) - D_{\bar{\mathbf{x}}_t}^{r-1} \mathbf{f}(\bar{\mathbf{x}}_t, t, \boldsymbol{\mu})\right)\Big|_{\mathbf{x}_t = \bar{\mathbf{x}}_t}. \end{aligned} \quad (2.64)$$

For $k = 0$, we have

$$G_0^{(0)}(\bar{\mathbf{x}}_t, t, \mathbf{x}_t) = G_0(\bar{\mathbf{x}}_t, t, \mathbf{x}_t) = G^{(0)}(\bar{\mathbf{x}}_t, t, \mathbf{x}_t)\big|_{\mathbf{x}_t = \bar{\mathbf{x}}_t}. \quad (2.65)$$

Notice that for the contact case, the following relation holds.

$$G_0^{(k)}(\bar{\mathbf{x}}_t, t, \mathbf{x}_t) = \left[\tfrac{d}{dt} G_0^{(k-1)}(\bar{\mathbf{x}}_t, t, \mathbf{x}_t)\right]\Big|_{\mathbf{x}_t = \bar{\mathbf{x}}_t} \neq \tfrac{d}{dt}\left[G_0^{(k-1)}(\bar{\mathbf{x}}_t, t, \mathbf{x}_t)\big|_{\mathbf{x}_t = \bar{\mathbf{x}}_t}\right]. \quad (2.66)$$

With the zero-order contact between the reference and compared flows, the higher order *G-function* can be obtained through the corresponding non-contact G-function plus the flow contact conditions. The compared flow $\mathbf{x}_t$ passing through and tangential to the reference surface for the flow $\bar{\mathbf{x}}_t$ at point $(t_m, \mathbf{x}_{t_m})$ are sketched in Fig. 2.3.

THEOREM 2.8. *Consider two dynamical systems in Eqs.*(2.1) *and* (2.2) *which have a compared flow* $\mathbf{x}_t = \mathbf{\Phi}(t_0, \mathbf{x}_0, \mathbf{p}, t)$ *with an initial condition* $(t_0, \mathbf{x}_0)$ *and a reference flow* $\bar{\mathbf{x}}_t = \mathbf{\Phi}(t_0, \bar{\mathbf{x}}_0, \boldsymbol{\mu}, t)$ *with an initial condition* $(t_0, \bar{\mathbf{x}}_0)$, *respectively. At time* t_m, $\mathbf{x}_m = \bar{\mathbf{x}}_m$. *Suppose the reference flow* $\bar{\mathbf{x}}_t$ *lies on a reference surface* S_t *with* $F(\bar{\mathbf{x}}, \bar{\mathbf{x}}_0, t) = 0$ *and* $\nabla F \bullet \dot{\bar{\mathbf{x}}} + \frac{\partial F}{\partial t} = 0$ *for time t. For an arbitrarily small* $\varepsilon > 0$, *there are two time intervals* $[t_{m-\varepsilon}, t_m)$ *and* $(t_m, t_{m+\varepsilon}]$. *The vector fields* $\mathbf{F}(\mathbf{x}, t, \mathbf{p})$ *and* $\mathbf{f}(\bar{\mathbf{x}}, t, \boldsymbol{\mu})$ *are* $C^r_{[t_{m-\varepsilon}, t_{m+\varepsilon}]}$*-continuous* ($r \geq 1$) *for time t. The flows* $\mathbf{x}_t$ *and* $\bar{\mathbf{x}}_t$ *are*

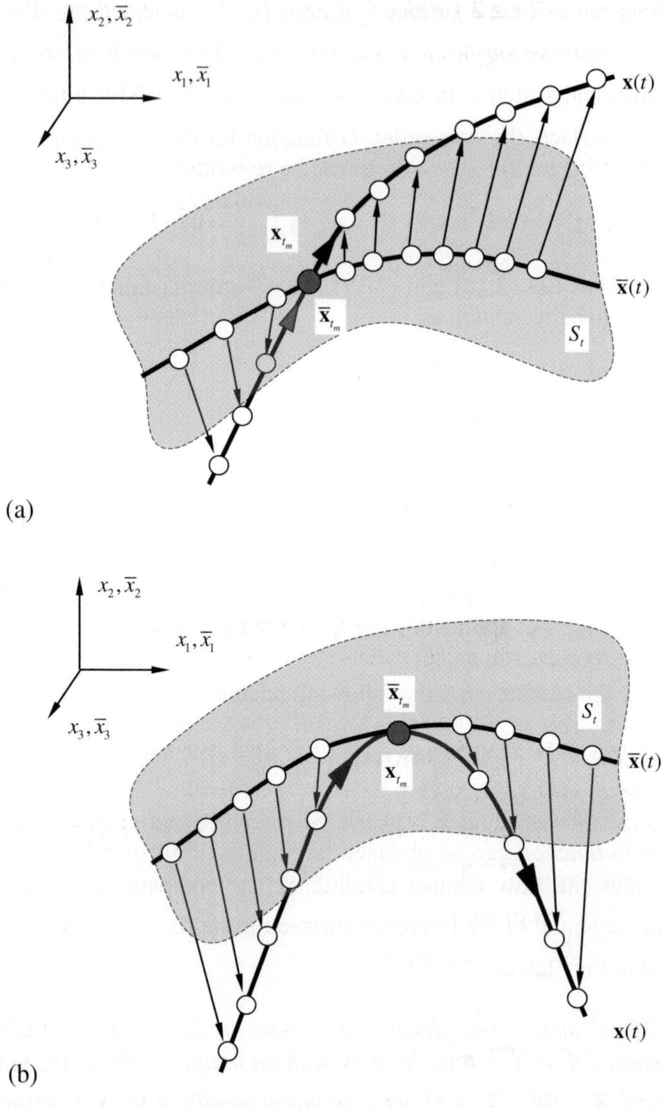

Figure 2.3. A 3-dimesional view of (a) the flow $\mathbf{x}_t$ passing over the reference surface S_t at $(t_m, \bar{\mathbf{x}}_m)$ and (b) the flow $\mathbf{x}_t$ tangential to the reference surface S_t at $(t_m, \bar{\mathbf{x}}_m)$. The zero-order contact point is $\mathbf{x}_{t_m} = \bar{\mathbf{x}}_{t_m}$ at time t_m.

$C^r_{[t_{m-\varepsilon},t_{m+\varepsilon}]}$*-continuous* ($r \geq 2$) *with* $\| d^{r+1}\mathbf{x}_t / dt^{r+1} \| < \infty$ *and* $\| d^{r+1}\bar{\mathbf{x}}_t / dt^{r+1} \| < \infty$.

(i) *The flow* $\mathbf{x}_t$ *is leaving from the reference surface* S_t *at point* $(t_m, \bar{\mathbf{x}}_m)$ *iff for* $t \in (t_m, t_m + \varepsilon]$

$$\left.\begin{aligned} G_0^{(0)}(\bar{\mathbf{x}}_{t_m}, t_m, \mathbf{x}_{t_m}) = G_0(\bar{\mathbf{x}}_{t_m}, t_m, \mathbf{x}_{t_m}) > 0 \text{ for } \mathbf{n}_{S_t}^T \cdot (\mathbf{x}_t - \bar{\mathbf{x}}_t) > 0 \\ G_0^{(0)}(\bar{\mathbf{x}}_{t_m}, t_m, \mathbf{x}_{t_m}) = G_0(\bar{\mathbf{x}}_{t_m}, t_m, \mathbf{x}_{t_m}) < 0 \text{ for } \mathbf{n}_{S_t}^T \cdot (\mathbf{x}_t - \bar{\mathbf{x}}_t) < 0 \end{aligned}\right\} \tag{2.67}$$

(ii) *The flow* $\mathbf{x}_t$ *is approaching the reference surface* S_t *at point* $(t_m, \bar{\mathbf{x}}_m)$ *iff for* $t \in [t_m - \varepsilon, t_m]$

$$\left.\begin{aligned} G_0^{(0)}(\bar{\mathbf{x}}_{t_m}, t_m, \mathbf{x}_{t_m}) = G_0(\bar{\mathbf{x}}_{t_m}, t_m, \mathbf{x}_{t_m}) > 0 \text{ for } \mathbf{n}_{S_t}^T \cdot (\bar{\mathbf{x}}_t - \mathbf{x}_t) \geq 0, \text{or} \\ G_0^{(0)}(\bar{\mathbf{x}}_{t_m}, t_m, \mathbf{x}_{t_m}) = G_0(\bar{\mathbf{x}}_{t_m}, t_m, \mathbf{x}_{t_m}) < 0 \text{ for } \mathbf{n}_{S_t}^T \cdot (\bar{\mathbf{x}}_t - \mathbf{x}_t) \leq 0. \end{aligned}\right\} \tag{2.68}$$

PROOF. The proof is the same as in Theorem 2.1. ■

The foregoing theorem is application of Theorem 2.1 for the reference and compared flows with the zero-contact. Only the G-function is different. However, this case has a lot of application, such a theorem is restated. For a flow passing through a special surface with a zero-order contact, the theorem is given:

THEOREM 2.9. *Consider two dynamical systems in Eqs.*(2.1) *and* (2.2) *which have a compared flow* $\mathbf{x}_t = \mathbf{\Phi}(t_0, \mathbf{x}_0, \mathbf{p}, t)$ *with an initial condition* $(t_0, \mathbf{x}_0)$ *and a reference flow* $\bar{\mathbf{x}}_t = \mathbf{\Phi}(t_0, \bar{\mathbf{x}}_0, \boldsymbol{\mu}, t)$ *with an initial condition* $(t_0, \bar{\mathbf{x}}_0)$, *respectively. At time* t_m, $\mathbf{x}_m = \bar{\mathbf{x}}_m$. *Suppose the reference flow* $\bar{\mathbf{x}}_t$ *lies on a reference surface* S_t *with* $F(\bar{\mathbf{x}}, \bar{\mathbf{x}}_0, t) = 0$ *and* $\nabla F \bullet \dot{\bar{\mathbf{x}}} + \frac{\partial F}{\partial t} = 0$ *for time t. For an arbitrarily small* $\varepsilon > 0$, *there are two time intervals* $[t_{m-\varepsilon}, t_m)$ *and* $(t_m, t_{m+\varepsilon}]$. *The vector fields* $\mathbf{F}(\mathbf{x}, t, \mathbf{p})$ *and* $\mathbf{f}(\bar{\mathbf{x}}, t, \boldsymbol{\mu})$ *are* $C^r_{[t_{m-\varepsilon},t_{m+\varepsilon}]}$*-continuous* ($r \geq 1$) *for time t. The flows* $\mathbf{x}_t$ *and* $\bar{\mathbf{x}}_t$ *are* $C^r_{[t_{m-\varepsilon},t_{m+\varepsilon}]}$*-continuous* ($r \geq 2$) *with* $\| d^{r+1}\mathbf{x}_t / dt^{r+1} \| < \infty$ *and* $\| d^{r+1}\bar{\mathbf{x}}_t / dt^{r+1} \| < \infty$. *The flow* $\mathbf{x}_t$ *is passing through the reference surface* S_t *at point* $(t_m, \bar{\mathbf{x}}_m)$ *iff* $t \in [t_m - \varepsilon, t_m)$

$$\left.\begin{aligned} G_0^{(0)}(\bar{\mathbf{x}}_{t_m}, t_m, \mathbf{x}_{t_m}) = G_0(\bar{\mathbf{x}}_{t_m}, t_m, \mathbf{x}_{t_m}) > 0 \text{ for } \mathbf{n}_{S_t}^T \cdot (\bar{\mathbf{x}}_t - \mathbf{x}_t) > 0 \text{ or} \\ G_0^{(0)}(\bar{\mathbf{x}}_{t_m}, t_m, \mathbf{x}_{t_m}) = G_0(\bar{\mathbf{x}}_{t_m}, t_m, \mathbf{x}_{t_m}) < 0 \text{ for } \mathbf{n}_{S_t}^T \cdot (\bar{\mathbf{x}}_t - \mathbf{x}_t) < 0. \end{aligned}\right\} \tag{2.69}$$

PROOF. The proof is the same as in Proof of THEOREM 2.2. ■

For a flow tangential to a specific surface with a zero-order contact, the corresponding theorem is given as follows:

THEOREM 2.10. *Consider two dynamical systems in Eqs.*(2.1) *and* (2.2) *which have a compared flow* $\mathbf{x}_t = \mathbf{\Phi}(t_0, \mathbf{x}_0, \mathbf{p}, t)$ *with an initial condition* $(t_0, \mathbf{x}_0)$ *and a reference flow* $\bar{\mathbf{x}}_t = \mathbf{\Phi}(t_0, \bar{\mathbf{x}}_0, \boldsymbol{\mu}, t)$ *with an initial condition* $(t_0, \bar{\mathbf{x}}_0)$, *respectively. At time* t_m, $\mathbf{x}_m = \bar{\mathbf{x}}_m$. *Suppose the reference flow* $\bar{\mathbf{x}}_t$ *lies on a reference surface* S_t *with* $F(\bar{\mathbf{x}}, \bar{\mathbf{x}}_0, t) = 0$ *and* $\nabla F \bullet \dot{\bar{\mathbf{x}}} + \frac{\partial F}{\partial t} = 0$ *for time t. For an arbitrarily small* $\varepsilon > 0$, *there are two time intervals* $[t_{m-\varepsilon}, t_m)$ *and* $(t_m, t_{m+\varepsilon}]$. *The vector fields* $\mathbf{F}(\mathbf{x}, t, \mathbf{p})$ *and* $\mathbf{f}(\bar{\mathbf{x}}, t, \boldsymbol{\mu})$ *are* $C^r_{[t_{m-\varepsilon}, t_{m+\varepsilon}]}$*-continuous* ($r \geq 2$) *for time t. The flows* $\mathbf{x}_t$ *and* $\bar{\mathbf{x}}_t$ *are* $C^r_{[t_{m-\varepsilon}, t_{m+\varepsilon}]}$*-continuous* ($r \geq 3$) *with* $\| d^{r+1}\mathbf{x}_t / dt^{r+1} \| < \infty$ *and* $\| d^{r+1}\bar{\mathbf{x}}_t / dt^{r+1} \| < \infty$. *The flow* $\mathbf{x}_t$ *is tangential to the reference surface* S_t *at point* $(t_m, \bar{\mathbf{x}}_m)$ *iff for* $t \in [t_{m-\varepsilon}, t_m)$

$$G_0^{(0)}(\bar{\mathbf{x}}_{t_m}, t_m, \mathbf{x}_{t_m}) \equiv G_0(\bar{\mathbf{x}}_{t_m}, t_m, \mathbf{x}_{t_m}) = 0, \tag{2.70}$$

$$\left.\begin{aligned} G_0^{(1)}(\bar{\mathbf{x}}_{t_m}, t_m, \mathbf{x}_{t_m}) < 0 \ \text{ with } \mathbf{n}_{S_t}^T \cdot (\bar{\mathbf{x}}_t - \mathbf{x}_t) > 0 \text{ or} \\ G_0^{(1)}(\bar{\mathbf{x}}_{t_m}, t_m, \mathbf{x}_{t_m}) > 0 \ \text{ with } \mathbf{n}_{S_t}^T \cdot (\bar{\mathbf{x}}_t - \mathbf{x}_t) < 0. \end{aligned}\right\} \tag{2.71}$$

PROOF. The proof is the same as in Theorem 2.3. ∎

THEOREM 2.11. *Consider two dynamical systems in Eqs.*(2.1) *and* (2.2) *which have a compared flow* $\mathbf{x}_t = \mathbf{\Phi}(t_0, \mathbf{x}_0, \mathbf{p}, t)$ *with an initial condition* $(t_0, \mathbf{x}_0)$ *and a reference flow* $\bar{\mathbf{x}}_t = \mathbf{\Phi}(t_0, \bar{\mathbf{x}}_0, \boldsymbol{\mu}, t)$ *with an initial condition* $(t_0, \bar{\mathbf{x}}_0)$, *respectively. At time* t_m, $\mathbf{x}_{t_m} = \bar{\mathbf{x}}_{t_m}$. *Suppose the reference flow* $\bar{\mathbf{x}}_t$ *lies on a reference surface* S_t *with* $F(\bar{\mathbf{x}}, \bar{\mathbf{x}}_0, t) = 0$ *and* $\nabla F \bullet \dot{\bar{\mathbf{x}}} + \frac{\partial F}{\partial t} = 0$ *for time t. For an arbitrarily small* $\varepsilon > 0$, *there are two time intervals* $[t_{m-\varepsilon}, t_m)$ *and* $(t_m, t_{m+\varepsilon}]$. *The vector fields* $\mathbf{F}(\mathbf{x}, t, \mathbf{p})$ *and* $\mathbf{f}(\bar{\mathbf{x}}, t, \boldsymbol{\mu})$ *are* $C^r_{[t_{m-\varepsilon}, t_{m+\varepsilon}]}$*-continuous* ($r \geq k$) *for time t only.* $D_{\mathbf{x}}^{(l)}\mathbf{F}(\mathbf{x}, t, \mathbf{p}) = 0$ *and* $D_{\bar{\mathbf{x}}}^{(l)}\mathbf{f}(\bar{\mathbf{x}}, t, \boldsymbol{\mu}) = 0$ *for* ($l > k$). *The flow* $\mathbf{x}_t$ *at point* $(t_m, \mathbf{x}_{t_m})$ *to the reference* S_t *is of the* k^{th}*-order singularity iff*

$$G_0^{(s)}(\bar{\mathbf{x}}_{t_m}, t_m, \mathbf{x}_{t_m}) = 0 \text{ for } s = 0, 1, 2, \cdots, k-1. \tag{2.72}$$

PROOF. The proof of this Theorem is the same as in Theorem 2.4 ∎

DEFINITION 2.11. Consider two dynamical systems in Eqs.(2.1) and (2.2) which have a compared flow $\mathbf{x}_t = \mathbf{\Phi}(t_0, \mathbf{x}_0, \mathbf{p}, t)$ with an initial condition $(t_0, \mathbf{x}_0)$ and a reference flow $\overline{\mathbf{x}}_t = \mathbf{\Phi}(t_0, \overline{\mathbf{x}}_0, \boldsymbol{\mu}, t)$ with an initial condition $(t_0, \overline{\mathbf{x}}_0)$, respectively. At time t_m, $\mathbf{x}_m = \overline{\mathbf{x}}_m$. Suppose the reference flow $\overline{\mathbf{x}}_t$ lies on a reference surface S_t with $F(\overline{\mathbf{x}}, \overline{\mathbf{x}}_0, t) = 0$ and $\nabla F \bullet \dot{\overline{\mathbf{x}}} + \frac{\partial F}{\partial t} = 0$ for time t. For an arbitrarily small $\varepsilon > 0$, there are two time intervals $[t_{m-\varepsilon}, t_m)$ and $(t_m, t_{m+\varepsilon}]$. The vector fields $\mathbf{F}(\mathbf{x}, t, \mathbf{p})$ and $\mathbf{f}(\overline{\mathbf{x}}, t, \boldsymbol{\mu})$ are $C^r_{[t_{m-\varepsilon}, t_{m+\varepsilon}]}$-continuous ($r \geq k+1$) for time t. The flows $\mathbf{x}_t$ and $\overline{\mathbf{x}}_t$ are $C^r_{[t_{m-\varepsilon}, t_{m+\varepsilon}]}$-continuous ($r \geq k+2$) for time t with $\| d^{r+1}\mathbf{x}_t / dt^{r+1} \| < \infty$ and $\| d^{r+1}\overline{\mathbf{x}}_t / dt^{r+1} \| < \infty$. The kth-order singularity requires

$$G_0^{(s)}(\overline{\mathbf{x}}_{t_m}, t_m, \mathbf{x}_{t_m}) = 0 \text{ for } s = 0, 1, 2, \cdots, k-1. \tag{2.73}$$

(i) The flow $\mathbf{x}_t$ contacting with the flow $\overline{\mathbf{x}}_t$ at point $(t_m, \overline{\mathbf{x}}_m)$ is leaving from the reference surface S_t with the kth-order singularity if for $t \in (t_m, t_m + \varepsilon]$

$$\left.\begin{array}{l} \mathbf{n}^T_{S_{t_m+\varepsilon}} \cdot (\mathbf{x}_{t_m+\varepsilon} - \overline{\mathbf{x}}_{t_m+\varepsilon}) > 0 \text{ for } \mathbf{n}^T_{S_t} \cdot (\mathbf{x}_t - \overline{\mathbf{x}}_t) > 0 \text{ or} \\ \mathbf{n}^T_{S_{t+\varepsilon}} \cdot (\mathbf{x}_{t_m+\varepsilon} - \overline{\mathbf{x}}_{t_m+\varepsilon}) < 0 \text{ for } \mathbf{n}^T_{S_t} \cdot (\mathbf{x}_t - \overline{\mathbf{x}}_t) < 0. \end{array}\right\} \tag{2.74}$$

(ii) The flow $\mathbf{x}_t$ contacting with the flow $\overline{\mathbf{x}}_t$ at point $(t_m, \overline{\mathbf{x}}_m)$ is approaching the reference surface S_t with the kth-order singularity if $t \in [t_m - \varepsilon, t_m)$

$$\left.\begin{array}{l} \mathbf{n}^T_{S_{t_m-\varepsilon}} \cdot (\mathbf{x}_{t_m-\varepsilon} - \overline{\mathbf{x}}_{t_m-\varepsilon}) < 0 \quad \text{for } \mathbf{n}^T_{S_t} \cdot (\overline{\mathbf{x}}_t - \mathbf{x}_t) > 0 \text{ or} \\ \mathbf{n}^T_{S_{t_m-\varepsilon}} \cdot (\mathbf{x}_{t_m-\varepsilon} - \overline{\mathbf{x}}_{t_m-\varepsilon}) > 0 \quad \text{for } \mathbf{n}^T_{S_t} \cdot (\overline{\mathbf{x}}_t - \mathbf{x}_t) < 0. \end{array}\right\} \tag{2.75}$$

(iii) The flow $\mathbf{x}_t$ contacting with the reference flow $\overline{\mathbf{x}}_t$ at point $(t_m, \overline{\mathbf{x}}_m)$ lies on the reference surface S_t if

$$\mathbf{n}^T_{S_{t_m-\varepsilon}} \cdot (\mathbf{x}_{t_m-\varepsilon} - \overline{\mathbf{x}}_{t_m-\varepsilon}) = \mathbf{n}^T_{S_{t_m+\varepsilon}} \cdot (\mathbf{x}_{t_m+\varepsilon} - \overline{\mathbf{x}}_{t_m+\varepsilon}) = 0. \tag{2.76}$$

DEFINITION 2.12. Consider two dynamical systems in Eqs.(2.1) and (2.2) which have a compared flow $\mathbf{x}_t = \mathbf{\Phi}(t_0, \mathbf{x}_0, \mathbf{p}, t)$ with an initial condition $(t_0, \mathbf{x}_0)$ and a reference flow $\overline{\mathbf{x}}_t = \mathbf{\Phi}(t_0, \overline{\mathbf{x}}_0, \boldsymbol{\mu}, t)$ with an initial condition $(t_0, \overline{\mathbf{x}}_0)$, respectively. At time t_m, $\mathbf{x}_m = \overline{\mathbf{x}}_m$. Suppose the reference flow $\overline{\mathbf{x}}_t$ lies on a reference surface S_t with $F(\overline{\mathbf{x}}, \overline{\mathbf{x}}_0, t) = 0$ and $\nabla F \bullet \dot{\overline{\mathbf{x}}} + \frac{\partial F}{\partial t} = 0$ for time t. For an arbitrarily small $\varepsilon > 0$, there are two time intervals $[t_{m-\varepsilon}, t_m)$ and $(t_m, t_{m+\varepsilon}]$. The vector fields $\mathbf{F}(\mathbf{x}, t, \mathbf{p})$

and $\mathbf{f}(\overline{\mathbf{x}},t,\boldsymbol{\mu})$ are $C^r_{[t_{m-\varepsilon},t_{m+\varepsilon}]}$-continuous ($r \ge 2k$). The flow $\mathbf{x}_t$ and $\overline{\mathbf{x}}_t$ are $C^r_{[t_{m-\varepsilon},t_{m+\varepsilon}]}$-continuous ($r \ge 2k+1$) for time t with $\| d^{r+1}\mathbf{x}_t/dt^{r+1} \| < \infty$ and $\| d^{r+1}\overline{\mathbf{x}}_t/dt^{r+1} \| < \infty$. The flow $\mathbf{x}_t$ contacting with the flow $\overline{\mathbf{x}}_t$ at point $(t_m, \overline{\mathbf{x}}_m)$ is passing through the reference surface S_t with the $(2k)^{\text{th}}$-order singularity if

$$G_0^{(s)}(\overline{\mathbf{x}}_{t_m}, t_m, \mathbf{x}_{t_m}) = 0 \text{ for } s = 0,1,2,\cdots,2k-1. \tag{2.77}$$

$$-\mathbf{n}^T_{S_{t_{m-\varepsilon}}} \cdot (\mathbf{x}_{t_m-\varepsilon} - \overline{\mathbf{x}}_{t_m-\varepsilon}) > 0 \text{ and } \mathbf{n}^T_{S_{t_{m+\varepsilon}}} \cdot (\mathbf{x}_{t_m+\varepsilon} - \overline{\mathbf{x}}_{t_{m+\varepsilon}}) > 0 \tag{2.78}$$

with $\mathbf{n}^T_{S_t} \cdot (\overline{\mathbf{x}}_t - \mathbf{x}_t) > 0$ for $t \in [t_m - \varepsilon, t_m)$ or

$$-\mathbf{n}^T_{S_{t_{m-\varepsilon}}} \cdot (\mathbf{x}_{t_m-\varepsilon} - \overline{\mathbf{x}}_{t_m-\varepsilon}) < 0 \text{ and } \mathbf{n}^T_{S_{t_{m+\varepsilon}}} \cdot (\mathbf{x}_{t_m+\varepsilon} - \overline{\mathbf{x}}_{t_{m+\varepsilon}}) < 0. \tag{2.79}$$

with $\mathbf{n}^T_{S_t} \cdot (\overline{\mathbf{x}}_t - \mathbf{x}_t) < 0$ for $t \in [t_m - \varepsilon, t_m)$.

DEFINITION 2.13. Consider two dynamical systems in Eqs.(2.1) and (2.2) which have a compared flow $\mathbf{x}_t = \boldsymbol{\Phi}(t_0, \mathbf{x}_0, \mathbf{p}, t)$ with an initial condition $(t_0, \mathbf{x}_0)$ and a reference flow $\overline{\mathbf{x}}_t = \boldsymbol{\Phi}(t_0, \overline{\mathbf{x}}_0, \boldsymbol{\mu}, t)$ with an initial condition $(t_0, \overline{\mathbf{x}}_0)$, respectively. At time t_m, $\mathbf{x}_m = \overline{\mathbf{x}}_m$. Suppose the reference flow $\overline{\mathbf{x}}_t$ lies on a reference surface S_t with $F(\overline{\mathbf{x}}, \overline{\mathbf{x}}_0, t) = 0$ and $\nabla F \bullet \dot{\overline{\mathbf{x}}} + \frac{\partial F}{\partial t} = 0$ for time t. For an arbitrarily small $\varepsilon > 0$, there are two time intervals $[t_{m-\varepsilon}, t_m)$ and $(t_m, t_{m+\varepsilon}]$. The vector fields $\mathbf{F}(\mathbf{x},t,\mathbf{p})$ and $\mathbf{f}(\overline{\mathbf{x}},t,\boldsymbol{\mu})$ are $C^r_{[t_{m-\varepsilon},t_{m+\varepsilon}]}$-continuous ($r \ge 2k+1$) for time t. The flow $\mathbf{x}_t$ and $\overline{\mathbf{x}}_t$ are $C^r_{[t_{m-\varepsilon},t_{m+\varepsilon}]}$-continuous ($r \ge 2k+2$) for time t and $\| d^{r+1}\mathbf{x}_t/dt^{r+1} \| < \infty$ and $\| d^{r+1}\overline{\mathbf{x}}_t/dt^{r+1} \| < \infty$. The flow $\mathbf{x}_t$ contacting with the flow $\overline{\mathbf{x}}_t$ at point $(t_m, \overline{\mathbf{x}}_m)$ is tangential to the reference surface S_t with the $(2k+1)^{\text{th}}$-order singularity if

$$G_0^{(s)}(\overline{\mathbf{x}}_{t_m}, t_m, \mathbf{x}_{t_m}) = 0 \text{ for } s = 0,1,2,\cdots,2k \tag{2.80}$$

$$-\mathbf{n}^T_{S_{t_{m-\varepsilon}}} \cdot (\mathbf{x}_{t_m-\varepsilon} - \overline{\mathbf{x}}_{t_m-\varepsilon}) < 0 \text{ and } \mathbf{n}^T_{S_{t_{m+\varepsilon}}} \cdot (\mathbf{x}_{t_m+\varepsilon} - \overline{\mathbf{x}}_{t_{m+\varepsilon}}) < 0 \tag{2.81}$$

with $\mathbf{n}^T_{S_t} \cdot (\mathbf{x}_t - \overline{\mathbf{x}}_t) > 0$ for $t \in [t_m - \varepsilon, t_m)$ or

$$-\mathbf{n}^T_{S_{t_{m-\varepsilon}}} \cdot (\mathbf{x}_{t_m-\varepsilon} - \overline{\mathbf{x}}_{t_m-\varepsilon}) > 0 \text{ and } \mathbf{n}^T_{S_{t_{m+\varepsilon}}} \cdot (\mathbf{x}_{t_m+\varepsilon} - \overline{\mathbf{x}}_{t_{m+\varepsilon}}) > 0 \tag{2.82}$$

with $\mathbf{n}^T_{S_t} \cdot (\mathbf{x}_t - \overline{\mathbf{x}}_t) < 0$ for $t \in [t_m - \varepsilon, t_m)$.

THEOREM 2.12. *Consider two dynamical systems in Eqs.*(2.1) *and* (2.2) *which have a compared flow* $\mathbf{x}_t = \mathbf{\Phi}(t_0, \mathbf{x}_0, \mathbf{p}, t)$ *with an initial condition* $(t_0, \mathbf{x}_0)$ *and a reference flow* $\bar{\mathbf{x}}_t = \mathbf{\Phi}(t_0, \bar{\mathbf{x}}_0, \boldsymbol{\mu}, t)$ *with an initial condition* $(t_0, \bar{\mathbf{x}}_0)$, *respectively. At time* t_m, $\mathbf{x}_m = \bar{\mathbf{x}}_m$. *Suppose the reference flow* $\bar{\mathbf{x}}_t$ *lies on a reference surface* S_t *with* $F(\bar{\mathbf{x}}, \bar{\mathbf{x}}_0, t) = 0$ *and* $\nabla F \bullet \dot{\bar{\mathbf{x}}} + \frac{\partial F}{\partial t} = 0$ *for time t. For an arbitrarily small* $\varepsilon > 0$, *there are two time intervals* $[t_{m-\varepsilon}, t_m)$ *and* $(t_m, t_{m+\varepsilon}]$. *The vector fields* $\mathbf{F}(\mathbf{x}, t, \mathbf{p})$ *and* $\mathbf{f}(\bar{\mathbf{x}}, t, \boldsymbol{\mu})$ *are* $C^r_{[t_{m-\varepsilon}, t_{m+\varepsilon}]}$*-continuous* $(r \geq k+1)$ *for time t. The flows* $\mathbf{x}_t$ *and* $\bar{\mathbf{x}}_t$ *are* $C^r_{[t_{m-\varepsilon}, t_{m+\varepsilon}]}$*-continuous* $(r \geq k+2)$ *for time t with* $\| d^{r+1}\mathbf{x}_t / dt^{r+1} \| < \infty$ *and* $\| d^{r+1}\bar{\mathbf{x}}_t / dt^{r+1} \| < \infty$. *The* k^{th}*-order singularity is*

$$G_0^{(s)}(\bar{\mathbf{x}}_{t_m}, t_m, \mathbf{x}_{t_m}) = 0 \text{ for } s = 0, 1, 2, \cdots, k-1. \tag{2.83}$$

(i) *The flow* $\mathbf{x}_t$ *contacting with the flow* $\bar{\mathbf{x}}_t$ *at point* $(t_m, \bar{\mathbf{x}}_m)$ *is leaving from the reference surface* S_t *with the* k^{th}*-order singularity iff for* $t \in (t_m, t_m + \varepsilon]$

$$\left.\begin{aligned} G_0^{(k)}(\bar{\mathbf{x}}_{t_m}, t_m, \mathbf{x}_{t_m}) > 0 \text{ for } \mathbf{n}_{S_t}^T \cdot (\mathbf{x}_t - \bar{\mathbf{x}}_t) > 0 \\ G_0^{(k)}(\bar{\mathbf{x}}_{t_m}, t_m, \mathbf{x}_{t_m}) < 0 \text{ for } \mathbf{n}_{S_t}^T \cdot (\mathbf{x}_t - \bar{\mathbf{x}}_t) < 0. \end{aligned}\right\} \tag{2.84}$$

(ii) *The flow* $\mathbf{x}_t$ *contacting with the flow* $\bar{\mathbf{x}}_t$ *at point* $(t_m, \bar{\mathbf{x}}_m)$ *is approaching the reference surface* S_t *with the* k^{th}*-order singularity iff* $t \in [t_m - \varepsilon, t_m)$

$$\left.\begin{aligned} &\left.\begin{aligned} G_0^{(k)}(\bar{\mathbf{x}}_{t_m}, t_m, \mathbf{x}_{t_m}) > 0 \quad k = 2m, \\ G_0^{(k)}(\bar{\mathbf{x}}_{t_m}, t_m, \mathbf{x}_{t_m}) < 0 \quad k = 2m-1 \end{aligned}\right\} \text{for } \mathbf{n}_{S_t}^T \cdot (\bar{\mathbf{x}}_t - \mathbf{x}_t) > 0, \\ &\text{or} \\ &\left.\begin{aligned} G_0^{(k)}(\bar{\mathbf{x}}_{t_m}, t_m, \mathbf{x}_{t_m}) < 0 \quad k = 2m, \\ G_0^{(k)}(\bar{\mathbf{x}}_{t_m}, t_m, \mathbf{x}_{t_m}) > 0 \quad k = 2m-1 \end{aligned}\right\} \text{for } \mathbf{n}_{S_t}^T \cdot (\bar{\mathbf{x}}_t - \mathbf{x}_t) < 0. \end{aligned}\right\} \tag{2.85}$$

PROOF. The proof is the same as in Theorem 2.5. ■

THEOREM 2.13. *Consider two dynamical systems in Eqs.*(2.1) *and* (2.2) *which have a compared flow* $\mathbf{x}_t = \mathbf{\Phi}(t_0, \mathbf{x}_0, \mathbf{p}, t)$ *with an initial condition* $(t_0, \mathbf{x}_0)$ *and a reference flow* $\bar{\mathbf{x}}_t = \mathbf{\Phi}(t_0, \bar{\mathbf{x}}_0, \boldsymbol{\mu}, t)$ *with an initial condition* $(t_0, \bar{\mathbf{x}}_0)$, *respectively. At time* t_m, $\mathbf{x}_m = \bar{\mathbf{x}}_m$. *Suppose the reference flow* $\bar{\mathbf{x}}_t$ *lies on a reference surface* S_t *with* $F(\bar{\mathbf{x}}, \bar{\mathbf{x}}_0, t) = 0$ *and* $\nabla F \bullet \dot{\bar{\mathbf{x}}} + \frac{\partial F}{\partial t} = 0$ *for time t. For an arbitrarily small* $\varepsilon > 0$, *there are two time intervals* $[t_{m-\varepsilon}, t_m)$ *and* $(t_m, t_{m+\varepsilon}]$. *The vector fields* $\mathbf{F}(\mathbf{x}, t, \mathbf{p})$

and $\mathbf{f}(\overline{\mathbf{x}},t,\boldsymbol{\mu})$ *are* $C^r_{[t_{m-\varepsilon},t_{m+\varepsilon}]}$*-continuous* ($r \geq 2k$) *for time t. The flow* $\mathbf{x}_t$ *and* $\overline{\mathbf{x}}_t$ *are* $C^r_{[t_{m-\varepsilon},t_{m+\varepsilon}]}$*-continuous* ($r \geq 2k+1$) *with* $\| d^{r+1}\mathbf{x}_t / dt^{r+1} \| < \infty$ *and* $\| d^{r+1}\overline{\mathbf{x}}_t / dt^{r+1} \| < \infty$ *for time t. The flow* $\mathbf{x}_t$ *is passing through the reference surface* S_t *with the* $(2k)^{th}$*-order singularity at point* $(t_m, \overline{\mathbf{x}}_m)$ *iff* $t \in [t_m - \varepsilon, t_m)$

$$G_0^{(s)}(\overline{\mathbf{x}}_{t_m}, t_m, \mathbf{x}_{t_m}) = 0 \text{ for } s = 0,1,2,\cdots,2k-1; \tag{2.86}$$

$$\left.\begin{array}{l} G_0^{(2k)}(\overline{\mathbf{x}}_{t_m}, t_m, \mathbf{x}_{t_m}) > 0 \text{ for } \mathbf{n}^T_{\overline{\mathbf{x}}_t} \cdot (\overline{\mathbf{x}}_t - \mathbf{x}_t) > 0 \text{ or} \\ G_0^{(2k)}(\overline{\mathbf{x}}_{t_m}, t_m, \mathbf{x}_{t_m}) < 0 \text{ for } \mathbf{n}^T_{\overline{\mathbf{x}}_t} \cdot (\overline{\mathbf{x}}_t - \mathbf{x}_t) < 0. \end{array}\right\} \tag{2.87}$$

PROOF. The proof is the same as in Theorem 2.6 ■

THEOREM 2.14. *Consider two dynamical systems in Eqs.*(2.1) *and* (2.2) *which have a compared flow* $\mathbf{x}_t = \boldsymbol{\Phi}(t_0, \mathbf{x}_0, \mathbf{p}, t)$ *with an initial condition* $(t_0, \mathbf{x}_0)$ *and a reference flow* $\overline{\mathbf{x}}_t = \boldsymbol{\Phi}(t_0, \overline{\mathbf{x}}_0, \boldsymbol{\mu}, t)$ *with an initial condition* $(t_0, \overline{\mathbf{x}}_0)$, *respectively. At time* t_m, $\mathbf{x}_m = \overline{\mathbf{x}}_m$. *Suppose the reference flow* $\overline{\mathbf{x}}_t$ *lies on a reference surface* S_t *with* $F(\overline{\mathbf{x}}, \overline{\mathbf{x}}_0, t) = 0$ *and* $\nabla F \bullet \dot{\overline{\mathbf{x}}} + \frac{\partial F}{\partial t} = 0$ *for time t. For an arbitrarily small* $\varepsilon > 0$, *there are two time intervals* $[t_{m-\varepsilon}, t_m)$ *and* $(t_m, t_{m+\varepsilon}]$. *The vector fields* $\mathbf{F}(\mathbf{x},t,\mathbf{p})$ *and* $\mathbf{f}(\overline{\mathbf{x}},t,\boldsymbol{\mu})$ *are* $C^r_{[t_{m-\varepsilon},t_{m+\varepsilon}]}$*-continuous* ($r \geq 2k+1$) *for time t. The flow* $\mathbf{x}_t$ *and* $\overline{\mathbf{x}}_t$ *are* $C^r_{[t_{m-\varepsilon},t_{m+\varepsilon}]}$*-continuous* ($r \geq 2k+2$)*for time t with* $\| d^{r+1}\mathbf{x}_t / dt^{r+1} \| < \infty$ *and* $\| d^{r+1}\overline{\mathbf{x}}_t / dt^{r+1} \| < \infty$. *The flow* $\mathbf{x}_t$ *contacting with the flow* $\overline{\mathbf{x}}_t$ *at point* $(t_m, \overline{\mathbf{x}}_m)$ *is tangential to the reference surface* S_t *with the* $(2k+1)^{th}$*-order singularity iff for* $t \in [t_{m-\varepsilon}, t_m]$

$$G_0^{(s)}(\overline{\mathbf{x}}_{t_m}, t_m, \mathbf{x}_{t_m}) = 0 \text{ for } s = 0,1,2,\cdots,2k; \tag{2.88}$$

$$\left.\begin{array}{l} G_0^{(2k+1)}(\overline{\mathbf{x}}_{t_m}, t_m, \mathbf{x}_{t_m}) < 0 \text{ with } \mathbf{n}^T_{\overline{\mathbf{x}}_t} \cdot (\overline{\mathbf{x}}_t - \mathbf{x}_t) > 0, \text{ or} \\ G_0^{(2k+1)}(\overline{\mathbf{x}}_{t_m}, t_m, \mathbf{x}_{t_m}) > 0 \text{ with } \mathbf{n}^T_{\overline{\mathbf{x}}_t} \cdot (\overline{\mathbf{x}}_t - \mathbf{x}_t) < 0. \end{array}\right\} \tag{2.89}$$

PROOF. The proof is the same as in Theorem 2.7 ■

The contact and singularity between two flows are different. The kth-order contact of the two flows implies that the two flows are of the kth-order singularity. If the two flows are of the kth-order singularity, it does not mean that the two flows are of the kth-order contact.

DEFINITION 2.14. Consider two dynamical systems in Eqs.(2.1) and (2.2) which have a compared flow $\mathbf{x}_t = \mathbf{\Phi}(t_0, \mathbf{x}_0, \mathbf{p}, t)$ with an initial condition $(t_0, \mathbf{x}_0)$ and a reference flow $\bar{\mathbf{x}}_t = \mathbf{\Phi}(t_0, \bar{\mathbf{x}}_0, \boldsymbol{\mu}, t)$ with an initial condition $(t_0, \bar{\mathbf{x}}_0)$, respectively. Suppose the reference flow $\bar{\mathbf{x}}_t$ lies on a reference surface S_t with $F(\bar{\mathbf{x}}, \bar{\mathbf{x}}_0, t) = 0$ and $\nabla F \bullet \dot{\bar{\mathbf{x}}} + \frac{\partial F}{\partial t} = 0$ for time t. For an arbitrarily small $\varepsilon > 0$, there are two time intervals $[t_{m-\varepsilon}, t_m)$ and $(t_m, t_{m+\varepsilon}]$. The vector fields $\mathbf{F}(\mathbf{x}, t, \mathbf{p})$ and $\mathbf{f}(\bar{\mathbf{x}}, t, \boldsymbol{\mu})$ are $C^r_{[t_{m-\varepsilon}, t_{m+\varepsilon}]}$-continuous ($r \geq k+1$) for time t. The flow $\mathbf{x}_t$ and $\bar{\mathbf{x}}_t$ are $C^r_{[t_{m-\varepsilon}, t_{m+\varepsilon}]}$-continuous ($r \geq k+2$) for time t with $\| d^{r+1}\mathbf{x}_t / dt^{r+1} \| < \infty$ and $\| d^{r+1}\bar{\mathbf{x}}_t / dt^{r+1} \| < \infty$. At time t_m, the two flows are of the lth-order contact ($l < k$), i.e.,

$$\left. \frac{d^s \mathbf{x}_t}{dt^s} \right|_{(t_m, \mathbf{x}_m)} = \left. \frac{d^s \bar{\mathbf{x}}_t}{dt^s} \right|_{(t_m, \bar{\mathbf{x}}_m)} \quad \text{for } s = 0, 1, 2, \cdots, l. \tag{2.90}$$

The kth-order singularity is

$$G_l^{(s)}(\bar{\mathbf{x}}_{t_m}, t_m, \mathbf{x}_{t_m}) = 0 \text{ for } s = l, l+1, \cdots, k. \tag{2.91}$$

(i) The flow $\mathbf{x}_t$ contacting the flow $\bar{\mathbf{x}}_t$ with the lth-order at point $(t_m, \bar{\mathbf{x}}_m)$ is leaving from the reference surface S_t with the kth-order singularity ($k \geq l$) if for $t \in (t_m, t_m + \varepsilon]$

$$\left. \begin{aligned} \mathbf{n}^T_{S_{t+\varepsilon}} \cdot (\mathbf{x}_{t+\varepsilon} - \bar{\mathbf{x}}_{t+\varepsilon}) > 0 \text{ for } \mathbf{n}^T_{S_t} \cdot (\mathbf{x}_t - \bar{\mathbf{x}}_t) > 0 \text{ or} \\ \mathbf{n}^T_{S_{t+\varepsilon}} \cdot (\mathbf{x}_{t+\varepsilon} - \bar{\mathbf{x}}_{t+\varepsilon}) < 0 \text{ for } \mathbf{n}^T_{S_t} \cdot (\mathbf{x}_t - \bar{\mathbf{x}}_t) < 0. \end{aligned} \right\} \tag{2.92}$$

(ii) The flow $\mathbf{x}_t$ contacting the flow $\bar{\mathbf{x}}_t$ with the lth-order at point $(t_m, \bar{\mathbf{x}}_m)$ is approaching the reference surface S_t with the kth-order singularity ($k \geq l$) if $t \in [t_m - \varepsilon, t_m)$

$$\left. \begin{aligned} -\mathbf{n}^T_{S_{t_m-\varepsilon}} \cdot (\mathbf{x}_{t_m-\varepsilon} - \bar{\mathbf{x}}_{t_m-\varepsilon}) > 0 \quad \text{for } \mathbf{n}^T_{S_t} \cdot (\bar{\mathbf{x}}_t - \mathbf{x}_t) > 0 \text{ or} \\ -\mathbf{n}^T_{S_{t_m-\varepsilon}} \cdot (\mathbf{x}_{t_m-\varepsilon} - \bar{\mathbf{x}}_{t_m-\varepsilon}) < 0 \quad \text{for } \mathbf{n}^T_{S_t} \cdot (\bar{\mathbf{x}}_t - \mathbf{x}_t) < 0. \end{aligned} \right\} \tag{2.93}$$

(iii) The flow $\mathbf{x}_t$ contacting the flow $\bar{\mathbf{x}}_t$ with the lth-order at point $(t_m, \bar{\mathbf{x}}_m)$ lies in the reference surface S_t if

$$\mathbf{n}^T_{S_{t-\varepsilon}} \cdot (\mathbf{x}_{t_m-\varepsilon} - \bar{\mathbf{x}}_{t_m-\varepsilon}) = \mathbf{n}^T_{S_{t_m+\varepsilon}} \cdot (\mathbf{x}_{t_m+\varepsilon} - \bar{\mathbf{x}}_{t_m+\varepsilon}) = 0. \tag{2.94}$$

DEFINITION 2.15. Consider two dynamical systems in Eqs.(2.1) and (2.2) which have a compared flow $\mathbf{x}_t = \mathbf{\Phi}(t_0, \mathbf{x}_0, \mathbf{p}, t)$ with an initial condition $(t_0, \mathbf{x}_0)$ and a reference flow $\bar{\mathbf{x}}_t = \mathbf{\Phi}(t_0, \bar{\mathbf{x}}_0, \boldsymbol{\mu}, t)$ with an initial condition $(t_0, \bar{\mathbf{x}}_0)$, respectively. Suppose the reference flow $\bar{\mathbf{x}}_t$ lies on a reference surface S_t with $F(\bar{\mathbf{x}}, \bar{\mathbf{x}}_0, t) = 0$ and $\nabla F \bullet \dot{\bar{\mathbf{x}}} + \frac{\partial F}{\partial t} = 0$ for time t. For an arbitrarily small $\varepsilon > 0$, there are two time intervals $[t_{m-\varepsilon}, t_m)$ and $(t_m, t_{m+\varepsilon}]$. The vector fields $\mathbf{F}(\mathbf{x}, t, \mathbf{p})$ and $\mathbf{f}(\bar{\mathbf{x}}, t, \boldsymbol{\mu})$ are $C^r_{[t_{m-\varepsilon}, t_{m+\varepsilon}]}$-continuous ($r \geq 2k$) for time t. The flow $\mathbf{x}_t$ and $\bar{\mathbf{x}}_t$ are $C^r_{[t_{m-\varepsilon}, t_{m+\varepsilon}]}$-continuous ($r \geq 2k+1$) for time t with $\| d^{r+1}\mathbf{x}_t / dt^{r+1} \| < \infty$ and $\| d^{r+1}\bar{\mathbf{x}}_t / dt^{r+1} \| < \infty$. At time t_m, the two flows are of the lth-order contact ($l < 2k$), i.e.,

$$\left. \frac{d^s \mathbf{x}_t}{dt^s} \right|_{(t_m, \mathbf{x}_m)} = \left. \frac{d^s \bar{\mathbf{x}}_t}{dt^s} \right|_{(t_m, \bar{\mathbf{x}}_m)} \quad \text{for } s = 0, 1, 2, \cdots, l. \tag{2.95}$$

The flow $\mathbf{x}_t$ contacting the flow $\bar{\mathbf{x}}_t$ with the lth-order at point $(t_m, \bar{\mathbf{x}}_m)$ is passing through the reference surface S_t with the $(2k)^{\text{th}}$-order singularity if

$$G_l^{(s)}(\bar{\mathbf{x}}_{t_m}, t_m, \mathbf{x}_{t_m}) = 0 \text{ for } s = l, l+1, \cdots, 2k-1; \tag{2.96}$$

$$-\mathbf{n}^T_{S_{t_{m-\varepsilon}}} \cdot (\mathbf{x}_{t_m-\varepsilon} - \bar{\mathbf{x}}_{t_m-\varepsilon}) > 0 \text{ and } \mathbf{n}^T_{S_{t_{m+\varepsilon}}} \cdot (\mathbf{x}_{t_m+\varepsilon} - \bar{\mathbf{x}}_{t_{m+\varepsilon}}) > 0 \tag{2.97}$$

with $\mathbf{n}^T_{S_t} \cdot (\bar{\mathbf{x}}_t - \mathbf{x}_t) > 0$ for $t \in [t_m - \varepsilon, t_m)$ or

$$-\mathbf{n}^T_{S_{t_{m-\varepsilon}}} \cdot (\mathbf{x}_{t_m-\varepsilon} - \bar{\mathbf{x}}_{t_m-\varepsilon}) < 0 \text{ and } \mathbf{n}^T_{S_{t_{m+\varepsilon}}} \cdot (\mathbf{x}_{t_m+\varepsilon} - \bar{\mathbf{x}}_{t_{m+\varepsilon}}) < 0 \tag{2.98}$$

with $\mathbf{n}^T_{S_t} \cdot (\mathbf{x}_t - \bar{\mathbf{x}}_t) < 0$ for $t \in [t_m - \varepsilon, t_m)$.

DEFINITION 2.16. Consider two dynamical systems in Eqs.(2.1) and (2.2) which have a compared flow $\mathbf{x}_t = \mathbf{\Phi}(t_0, \mathbf{x}_0, \mathbf{p}, t)$ with an initial condition $(t_0, \mathbf{x}_0)$ and a reference flow $\bar{\mathbf{x}}_t = \mathbf{\Phi}(t_0, \bar{\mathbf{x}}_0, \boldsymbol{\mu}, t)$ with an initial condition $(t_0, \bar{\mathbf{x}}_0)$, respectively. Suppose the reference flow $\bar{\mathbf{x}}_t$ lies on a reference surface S_t with $F(\bar{\mathbf{x}}, \bar{\mathbf{x}}_0, t) = 0$ and $\nabla F \bullet \dot{\bar{\mathbf{x}}} + \frac{\partial F}{\partial t} = 0$ for time t. For an arbitrarily small $\varepsilon > 0$, there are two time intervals $[t_{m-\varepsilon}, t_m)$ and $(t_m, t_{m+\varepsilon}]$. The vector fields $\mathbf{F}(\mathbf{x}, t, \mathbf{p})$ and $\mathbf{f}(\bar{\mathbf{x}}, t, \boldsymbol{\mu})$ are $C^r_{[t_{m-\varepsilon}, t_{m+\varepsilon}]}$-continuous ($r \geq 2k-1$) for time t. The flow $\mathbf{x}_t$ and $\bar{\mathbf{x}}_t$ are $C^r_{[t_{m-\varepsilon}, t_{m+\varepsilon}]}$-continuous ($r \geq 2k$) for time t with $\| d^{r+1}\mathbf{x}_t / dt^{r+1} \| < \infty$ and $\| d^{r+1}\bar{\mathbf{x}}_t / dt^{r+1} \| < \infty$. At time t_m, the two flows are of the lth-order contact ($l < 2k$), i.e.,

$$\left.\frac{d^s\mathbf{x}_t}{dt^s}\right|_{(t_m,\mathbf{x}_m)}=\left.\frac{d^s\overline{\mathbf{x}}_t}{dt^s}\right|_{(t_m,\overline{\mathbf{x}}_m)} \quad \text{for } s=0,1,2,\cdots,l. \tag{2.99}$$

The flow $\mathbf{x}_t$ contacting the flow $\overline{\mathbf{x}}_t$ with the lth-order at point $(t_m,\overline{\mathbf{x}}_m)$ is tangential to the reference surface with the $(2k-1)^{th}$-order singularity at point $(t_m,\overline{\mathbf{x}}_m)$ if

$$G_l^{(s)}(\overline{\mathbf{x}}_{t_m},t_m,\mathbf{x}_{t_m})=0 \text{ for } s=l,l+1,\cdots,2k-2; \tag{2.100}$$

$$-\mathbf{n}^T_{S_{t_m-\varepsilon}}\cdot(\mathbf{x}_{t_m-\varepsilon}-\overline{\mathbf{x}}_{t_m-\varepsilon})>0 \text{ and } \mathbf{n}^T_{S_{t_m+\varepsilon}}\cdot(\mathbf{x}_{t_m+\varepsilon}-\overline{\mathbf{x}}_{t_{m+\varepsilon}})<0 \tag{2.101}$$

with $\mathbf{n}^T_{S_t}\cdot(\mathbf{x}_t-\overline{\mathbf{x}}_t)>0$ for $t\in[t_m-\varepsilon,t_m)$ or

$$-\mathbf{n}^T_{S_{t_m-\varepsilon}}\cdot(\mathbf{x}_{t_m-\varepsilon}-\overline{\mathbf{x}}_{t_m-\varepsilon})<0 \text{ and } \mathbf{n}^T_{S_{t_m+\varepsilon}}\cdot(\mathbf{x}_{t_m+\varepsilon}-\overline{\mathbf{x}}_{t_{m+\varepsilon}})>0 \tag{2.102}$$

with $\mathbf{n}^T_{S_t}\cdot(\mathbf{x}_t-\overline{\mathbf{x}}_t)<0$ for $t\in[t_m-\varepsilon,t_m)$.

THEOREM 2.15. *Consider two dynamical systems in Eqs.*(2.1) *and* (2.2) *which have a compared flow* $\mathbf{x}_t=\mathbf{\Phi}(t_0,\mathbf{x}_0,\mathbf{p},t)$ *with an initial condition* $(t_0,\mathbf{x}_0)$ *and a reference flow* $\overline{\mathbf{x}}_t=\mathbf{\Phi}(t_0,\overline{\mathbf{x}}_0,\boldsymbol{\mu},t)$ *with an initial condition* $(t_0,\overline{\mathbf{x}}_0)$, *respectively. Suppose the reference flow* $\overline{\mathbf{x}}_t$ *lies on a reference surface* S_t *with* $F(\overline{\mathbf{x}},\overline{\mathbf{x}}_0,t)=0$ *and* $\nabla F\bullet\dot{\overline{\mathbf{x}}}+\frac{\partial F}{\partial t}=0$ *for time t. For an arbitrarily small* $\varepsilon>0$, *there are two time intervals* $[t_{m-\varepsilon},t_m)$ *and* $(t_m,t_{m+\varepsilon}]$. *The vector fields* $\mathbf{F}(\mathbf{x},t,\mathbf{p})$ *and* $\mathbf{f}(\overline{\mathbf{x}},t,\boldsymbol{\mu})$ *are* $C^r_{[t_{m-\varepsilon},t_{m+\varepsilon}]}$*-continuous* ($r\geq k+1$) *for time t. The flow* $\mathbf{x}_t$ *and* $\overline{\mathbf{x}}_t$ *are* $C^r_{[t_{m-\varepsilon},t_{m+\varepsilon}]}$*-continuous* ($r\geq k+2$) *for time t with* $\| d^{r+1}\mathbf{x}_t/dt^{r+1}\|<\infty$ *and* $\| d^{r+1}\overline{\mathbf{x}}_t/dt^{r+1}\|<\infty$. *At time* t_m, *the two flows are of the* l^{th}*-order contact* ($l<k$), *i.e.,*

$$\left.\frac{d^s\mathbf{x}_t}{dt^s}\right|_{(t_m,\mathbf{x}_m)}=\left.\frac{d^s\overline{\mathbf{x}}_t}{dt^s}\right|_{(t_m,\overline{\mathbf{x}}_m)} \quad \text{for } s=0,1,2,\cdots,l. \tag{2.103}$$

The kth*-order singularity is*

$$G_l^{(s)}(\overline{\mathbf{x}}_{t_m},t_m,\mathbf{x}_{t_m})=0 \text{ for } s=l,l+1,\cdots,k-1. \tag{2.104}$$

(i) *The flow* $\mathbf{x}_t$ *contacting the flow* $\overline{\mathbf{x}}_t$ *with the* l^{th}*-order at point* $(t_m,\overline{\mathbf{x}}_m)$ *is leaving from the reference surface* S_t *with the* k^{th}*-order singularity* ($k\geq l$) *iff for* $t\in(t_m,t_m+\varepsilon]$

$$\left.\begin{array}{l} G_l^{(k+1)}(\bar{\mathbf{x}}_{t_m},t_m,\mathbf{x}_{t_m})>0 \text{ for } \mathbf{n}_{S_t}^T\cdot(\mathbf{x}_t-\bar{\mathbf{x}}_t)>0 \\ G_l^{(k+1)}(\bar{\mathbf{x}}_{t_m},t_m,\mathbf{x}_{t_m})<0 \text{ for } \mathbf{n}_{S_t}^T\cdot(\mathbf{x}_t-\bar{\mathbf{x}}_t)<0. \end{array}\right\} \tag{2.105}$$

(ii) *The flow* $\mathbf{x}_t$ *contacting the flow* $\bar{\mathbf{x}}_t$ *with the* l^{th}*-order at point* $(t_m,\bar{\mathbf{x}}_m)$ *is approaching the reference surface* S_t *with the* k^{th}*-order singularity* ($k\geq l$) *iff* $t\in[t_m-\varepsilon,t_m)$

$$\left.\begin{array}{l} \left.\begin{array}{l} G_l^{(k)}(\bar{\mathbf{x}}_{t_m},t_m,\mathbf{x}_{t_m})>0 \;\; k=2m, \\ G_l^{(k)}(\bar{\mathbf{x}}_{t_m},t_m,\mathbf{x}_{t_m})<0 \;\; k=2m-1 \end{array}\right\} \text{for } \mathbf{n}_{S_t}^T\cdot(\bar{\mathbf{x}}_t-\mathbf{x}_t)>0, \\ \left.\begin{array}{l} G_l^{(k)}(\bar{\mathbf{x}}_{t_m},t_m,\mathbf{x}_{t_m})<0 \;\; k=2m, \\ G_l^{(k)}(\bar{\mathbf{x}}_{t_m},t_m,\mathbf{x}_{t_m})>0 \;\; k=2m-1 \end{array}\right\} \text{for } \mathbf{n}_{S_t}^T\cdot(\bar{\mathbf{x}}_t-\mathbf{x}_t)<0. \end{array}\right\} \tag{2.106}$$

PROOF. The proof is the same as in Theorem 2.5. ■

THEOREM 2.16. *Consider two dynamical systems in Eqs.(2.1) and (2.2) which have a compared flow* $\mathbf{x}_t=\mathbf{\Phi}(t_0,\mathbf{x}_0,\mathbf{p},t)$ *with an initial condition* $(t_0,\mathbf{x}_0)$ *and a reference flow* $\bar{\mathbf{x}}_t=\mathbf{\Phi}(t_0,\bar{\mathbf{x}}_0,\boldsymbol{\mu},t)$ *with an initial condition* $(t_0,\bar{\mathbf{x}}_0)$, *respectively. Suppose the reference flow* $\bar{\mathbf{x}}_t$ *lies on a reference surface* S_t *with* $F(\bar{\mathbf{x}},\bar{\mathbf{x}}_0,t)=0$ *and* $\nabla F\bullet\dot{\bar{\mathbf{x}}}+\frac{\partial F}{\partial t}=0$ *for time t. For an arbitrarily small* $\varepsilon>0$, *there are two time intervals* $[t_{m-\varepsilon},t_m)$ *and* $(t_m,t_{m+\varepsilon}]$. *The vector fields* $\mathbf{F}(\mathbf{x},t,\mathbf{p})$ *and* $\mathbf{f}(\bar{\mathbf{x}},t,\boldsymbol{\mu})$ *are* $C^r_{[t_{m-\varepsilon},t_{m+\varepsilon}]}$*-continuous* ($r\geq 2k$) *for time t. The flow* $\mathbf{x}_t$ *and* $\bar{\mathbf{x}}_t$ *are* $C^r_{[t_{m-\varepsilon},t_{m+\varepsilon}]}$*-continuous* ($r\geq 2k+1$) *for time t with* $\|d^{r+1}\mathbf{x}_t/dt^{r+1}\|<\infty$ *and* $\|d^{r+1}\bar{\mathbf{x}}_t/dt^{r+1}\|<\infty$. *At time* t_m, *the two flows are of the* l^{th}*-order contact* ($l<2k$), *i.e.,*

$$\left.\frac{d^s\mathbf{x}_t}{dt^s}\right|_{(t_m,\mathbf{x}_m)}=\left.\frac{d^s\bar{\mathbf{x}}_t}{dt^s}\right|_{(t_m,\bar{\mathbf{x}}_m)} \quad \text{for } s=0,1,2,\cdots,l. \tag{2.107}$$

The flow $\mathbf{x}_t$ *contacting with the flow* $\bar{\mathbf{x}}_t$ *with the* l^{th}*-order at point* $(t_m,\bar{\mathbf{x}}_m)$ *is passing through the reference surface* S_t *with the (2k)*th*-order singularity if*

$$G_l^{(s)}(\bar{\mathbf{x}}_{t_m},t_m,\mathbf{x}_{t_m})=0 \text{ for } s=l,l+1,\cdots,2k-1; \tag{2.108}$$

$$\left.\begin{array}{l} G_l^{(2k)}(\bar{\mathbf{x}}_{t_m},t_m,\mathbf{x}_{t_m})>0 \text{ for } \mathbf{n}_{S_t}^T\cdot(\bar{\mathbf{x}}_t-\mathbf{x}_t)>0 \text{ or} \\ G_l^{(2k)}(\bar{\mathbf{x}}_{t_m},t_m,\mathbf{x}_{t_m})<0 \text{ for } \mathbf{n}_{S_t}^T\cdot(\bar{\mathbf{x}}_t-\mathbf{x}_t)<0. \end{array}\right\} \tag{2.109}$$

PROOF. The proof is the same as in Theorem 2.6. ■

THEOREM 2.17. *Consider two dynamical systems in Eqs.*(2.1) *and* (2.2) *which have a compared flow* $\mathbf{x}_t = \mathbf{\Phi}(t_0, \mathbf{x}_0, \mathbf{p}, t)$ *with an initial condition* $(t_0, \mathbf{x}_0)$ *and a reference flow* $\overline{\mathbf{x}}_t = \mathbf{\Phi}(t_0, \overline{\mathbf{x}}_0, \boldsymbol{\mu}, t)$ *with an initial condition* $(t_0, \overline{\mathbf{x}}_0)$, *respectively. Suppose the reference flow* $\overline{\mathbf{x}}_t$ *lies on a reference surface* S_t *with* $F(\overline{\mathbf{x}}, \overline{\mathbf{x}}_0, t) = 0$ *and* $\nabla F \bullet \dot{\overline{\mathbf{x}}} + \frac{\partial F}{\partial t} = 0$ *for time t. For an arbitrarily small* $\varepsilon > 0$, *there are two time intervals* $[t_{m-\varepsilon}, t_m)$ *and* $(t_m, t_{m+\varepsilon}]$. *The vector fields* $\mathbf{F}(\mathbf{x}, t, \mathbf{p})$ *and* $\mathbf{f}(\overline{\mathbf{x}}, t, \boldsymbol{\mu})$ *are* $C^r_{[t_{m-\varepsilon}, t_{m+\varepsilon}]}$*-continuous* ($r \geq 2k-1$) *for time t. The flow* $\mathbf{x}_t$ *and* $\overline{\mathbf{x}}_t$ *are* $C^r_{[t_{m-\varepsilon}, t_{m+\varepsilon}]}$*-continuous* ($r \geq 2k$) *for time t with* $\| d^{r+1}\mathbf{x}_t / dt^{r+1} \| < \infty$ *and* $\| d^{r+1}\overline{\mathbf{x}}_t / dt^{r+1} \| < \infty$. *At time* t_m, *the two flows are of the* l^{th}*-order contact* ($l < 2k$), *i.e.,*

$$\left. \frac{d^s \mathbf{x}_t}{dt^s} \right|_{(t_m, \mathbf{x}_m)} = \left. \frac{d^s \overline{\mathbf{x}}_t}{dt^s} \right|_{(t_m, \overline{\mathbf{x}}_m)} \quad \text{for } s = 0, 1, 2, \cdots, l. \tag{2.110}$$

The flow $\mathbf{x}_t$ *contacting the flow* $\overline{\mathbf{x}}_t$ *with the* l^{th}*-order at point* $(t_m, \overline{\mathbf{x}}_m)$ *is tangential to the reference surface with the* $(2k-1)^{th}$*-order singularity if*

$$G_l^{(s)}(\overline{\mathbf{x}}_{t_m}, t_m, \mathbf{x}_{t_m}) = 0 \text{ for } s = l, l+1, \cdots, 2k-2; \tag{2.111}$$

$$\left. \begin{aligned} &G_l^{(2k-1)}(\overline{\mathbf{x}}_{t_m}, t_m, \mathbf{x}_{t_m}) < 0 \ \text{ with } \mathbf{n}_{S_t}^T \cdot (\overline{\mathbf{x}}_t - \mathbf{x}_t) > 0 \ \text{ or} \\ &G_l^{(2k-1)}(\overline{\mathbf{x}}_{t_m}, t_m, \mathbf{x}_{t_m}) > 0 \ \text{ with } \mathbf{n}_{S_t}^T \cdot (\overline{\mathbf{x}}_t - \mathbf{x}_t) < 0. \end{aligned} \right\} \tag{2.112}$$

PROOF. The proof is the same as in Theorem 2.7. ■

2.4. Concluding remarks

To investigate the complexity of flows in a dynamical system, the well-behavior dynamical system in the same dimensional phase space can be selected as a reference dynamical system. Using this idea, the stability of flow in an investigated dynamical system can be easily developed. For example, consider the reference dynamical system to be conservative, the complexity and stability for the corresponding non-conservative dynamical system can be investigated. When a reference flow is on a discontinuous boundary, the passability of the compared flow to the separation boundary can be discussed through such a *G*-function. The left and right *G*-functions are different owing to the discontinuity. In addition, the contact and singularity between the compared and reference flows are different.

If the two flows are of the kth -order contact,

$$\left.\frac{d^s \mathbf{x}_t}{dt^s}\right|_{(t_m,\mathbf{x}_m)} = \left.\frac{d^s \overline{\mathbf{x}}_t}{dt^s}\right|_{(t_m,\overline{\mathbf{x}}_m)} \quad \text{for } s = 0,1,2,\cdots,k; \tag{2.113}$$

then, the kth -order singularity between the two flows requires

$$G^{(l-1)}(\overline{\mathbf{x}}_t,t,\mathbf{x}_t) = \sum_{r=0}^{l} C_l^r D_{\overline{\mathbf{x}}_t}^{l-r} \mathbf{n}_{S_t}^T \cdot \left(\frac{d^r \mathbf{x}_t}{dt^r} - \frac{d^r \overline{\mathbf{x}}_t}{dt^r}\right) = 0 \ (l = 1,2,\cdots,k). \tag{2.114}$$

From the foregoing equations, the kth-order contact of the two flows implies that the compared flow to the reference surface S_t is of the kth -order singularity. However, if the condition in Eq.(2.114) for all the $(l-1)^{th}$ -order G-function ($l = 1,2,\cdots,k-1$) holds, it is not necessary that the conditions in Eq.(2.113) for the kth -order contact hold. So if the compared flow to the reference surface is of the kth -order singularity, it does not mean that the compared and reference flows are of the kth-order contact.

Chapter 3

Global Transversality in Continuous Dynamical Systems

In this Chapter, the global tangency and transversality of a perturbed flow in dynamical systems with at least one separatrix will be discussed. The local and global flows in perturbed dynamical systems will be introduced through the geometrical intuition. The global and tangential flows will be defined mathematically, and the sufficient and necessary conditions of such global and tangential flows to the separatrix will be given as well. The $G^{(k)}$-function ($k = 0,1,2,\cdots$) for a perturbed flow to a specified, first integral surface will be introduced for simplicity. Based on the global tangency and transversality of a perturbed flow to the separatrix, the complexity of flows in nonlinear dynamic systems can be investigated. The perturbed Hamiltonian system will be discussed for how to obtain such necessary and sufficient conditions of the global tangency and transversality of a perturbed flow to the separatrix. To explain the physical meaning of analytical conditions, the global transversality and tangentcy of a perturbed flow to the separatrix in a two-dimensional nonlinear dynamical system will be discussed. A periodically forced, damped Duffing oscillator with a separatrix will be presented, and the corresponding analytical conditions for the global transversality and tangency to the separatrix will be obtained, and the global and tangential flows are illustrated as well. Finally, the generalized separatrix in nonlinear dynamical systems will be introduced, and the corresponding conditions for the global transversality and tangency will be presented. The van der Pol oscillator will be discussed as an example.

3.1. Nonlinear dynamical systems

Consider an n-dimensional nonlinear dynamical system

$$\dot{\mathbf{x}} = \mathbf{F}(\mathbf{x},t,\mathbf{p}) \equiv \mathbf{f}(\mathbf{x},\boldsymbol{\mu}) + \mathbf{g}(\mathbf{x},t,\boldsymbol{\pi}) \tag{3.1}$$

with $\mathbf{x} \equiv \mathbf{x}(t) \in \mathbb{R}^n$, $\mathbf{p} \in \mathbb{R}^{m_1+m_2}$, $\boldsymbol{\mu} \in \mathbb{R}^{m_1}$ and $\boldsymbol{\pi} \in \mathbb{R}^{m_2}$ where the integrable, time-independent vector field function $\mathbf{f}(\mathbf{x},\boldsymbol{\mu})$ is a C^r-smooth vector function ($r \geq 1$) defined on two open subsets $U_{\mathbf{x}} \subseteq \mathbb{R}^n$ and $U_{\boldsymbol{\mu}} \subseteq \mathbb{R}^{m_1}$. The non-integrable, vector field function $\mathbf{g}(\mathbf{x},t,\boldsymbol{\pi})$ is a C^r-smooth vector function ($r \geq 1$) defined on three open subsets $U_{\mathbf{x}} \subseteq \mathbb{R}^n$, $U_{\boldsymbol{\mu}} \subseteq \mathbb{R}^{m_2}$ and $U_t \subseteq \mathbb{R}$. For a parameter vector $\mathbf{p} \in U_{\mathbf{p}}$, the vector field $\mathbf{F}(\mathbf{x}, t, \mathbf{p})$ in Eq.(3.1) with an initial condition $(t_0, \mathbf{x}_0)$ generates a flow $\mathbf{x}(t) = \mathbf{\Phi}(t_0, \mathbf{x}_0, \mathbf{p}, t)$ in $U_{\mathbf{x}}$ and $\mathbf{x}_0 \in U_{\mathbf{x}}$ for $t \in I_{12}$ and $t_0 \in I_{12} \equiv (t_1, t_2) \subseteq \mathbb{R}$. The flow possesses the semi-group property, i.e., $\mathbf{\Phi}(t, \mathbf{\Phi}(t_0, \mathbf{x}_0, \mathbf{p}, t), \mathbf{p}, s) = \mathbf{\Phi}(t_0, \mathbf{x}_0, \mathbf{p}, t+s)$ with $\mathbf{x}(t_0) = \mathbf{\Phi}(t_0, \mathbf{x}_0, \mathbf{p}, t_0)$.

To restrict our discussion, the following assumptions should be introduced.

A1: The time-independent dynamical system of Eq.(3.1),

$$\dot{\bar{\mathbf{x}}} = \mathbf{f}(\bar{\mathbf{x}}, \boldsymbol{\mu}), \quad \bar{\mathbf{x}} \equiv \bar{\mathbf{x}}(t) \in \mathbb{R}^n \text{ and } \boldsymbol{\mu} \in \mathbb{R}^{m_1}, \tag{3.2}$$

is integrable on the two open subsets $U_{\mathbf{x}} \subseteq \mathbb{R}^n$ and $U_{\boldsymbol{\mu}} \subseteq \mathbb{R}^{m_1}$. For given $\boldsymbol{\mu} \in U_{\boldsymbol{\mu}}$, there is a scalar function $F(\bar{\mathbf{x}}, \boldsymbol{\mu})$ satisfying the following condition

$$DF(\bar{\mathbf{x}}, \boldsymbol{\mu}) = \dot{\bar{\mathbf{x}}} \bullet \nabla F(\bar{\mathbf{x}}, \boldsymbol{\mu}) = 0 \tag{3.3}$$

where the total differentiation operator $D = \sum_i^n \dot{\bar{x}}_i \frac{\partial}{\partial \bar{x}_i} + \frac{\partial}{\partial t}$ and $\nabla = \frac{\partial}{\partial \bar{x}_i} \mathbf{i}_i$.

A2: The time-independent dynamical system of Eq.(3.2) possesses at least one saddle point p_0. There is an (n-1)-dimensional separatrix surface to connect such a saddle point p_0, and this separatrix surface divides the phase space with m sub-domain in phase spaces $\Omega_i \subset \mathbb{R}^n$ ($i = 1, 2, \cdots, m$).

A3: For a dynamical system in Eq.(3.1), in sub-domain Ω_i, the corresponding dynamical system is re-expressed as

$$\dot{\mathbf{x}}^{(i)} = \mathbf{F}(\mathbf{x}^{(i)}, t, \mathbf{p}) \equiv \mathbf{f}(\mathbf{x}^{(i)}, \boldsymbol{\mu}) + \mathbf{g}(\mathbf{x}^{(i)}, t, \boldsymbol{\pi}), \tag{3.4}$$

with $\mathbf{x}^{(i)} \in \Omega_i \subset \mathbb{R}^n$, $\mathbf{p} \in \mathbb{R}^{m_1+m_2}$, $\boldsymbol{\mu} \in \mathbb{R}^{m_1}$ and $\boldsymbol{\pi} \in \mathbb{R}^{m_2}$. There is an open initial set $D_i \subset \Omega_i$ and the flow for some initial set can approach the separatrix surface.

From Assumption (**A1**), equation (3.3) gives the first integral of Eq.(3.2), i.e.,

$$F(\bar{\mathbf{x}}, \boldsymbol{\mu}) = F(\bar{x}_1, \bar{x}_2, \cdots, \bar{x}_n, \boldsymbol{\mu}) = E = \text{const.} \tag{3.5}$$

With different constant E, the first integral is different. Under a specified constant, all the trajectories of motion will be on such an (n-1)-dimensional surface. Therefore, this (n-1)-dimensional surface is called *the first integral manifold*, as shown in Fig.3.1. With different constant E_σ ($\sigma \in [0,1]$), the first integral manifold cannot be intersected, which is determined by the first integral invariant. An

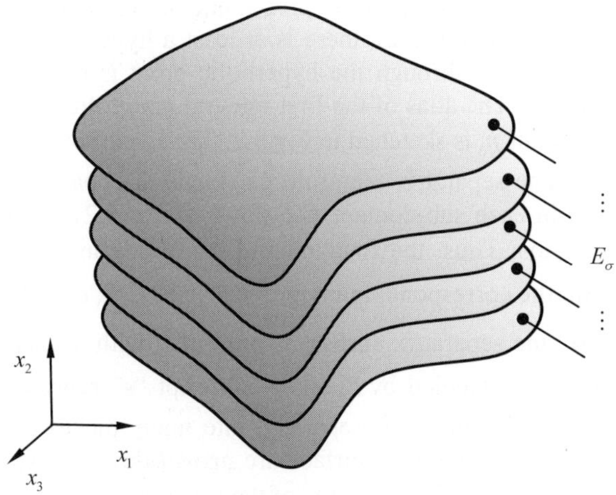

Figure 3.1. An atlas of 3-dimensional first integral manifolds with different constants.

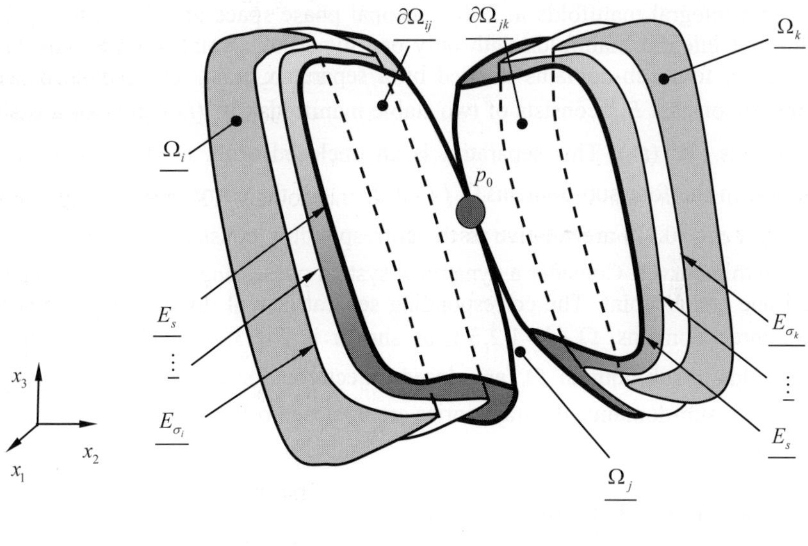

Figure 3.2. A atlas of the first integral manifolds in the vicinity of the hyperbolic point p_0. The separatrix manifold is presented with a thick boundary, and labeled by the integral constant E_s. The first integral manifolds in Ω_i and Ω_k will be sketched with labels E_{σ_α} ($\sigma_\alpha \in (0,1]$ for $\alpha = i, j, k$).

atlas of the first integral manifolds is presented through the different integral constants. From Assumption (**A2**), there is at least a hyperbolic point. The first integral manifold passing through the hyperbolic point is called *the separatrix surface* (or manifold). The atlas of the first integral manifold surfaces in vicinity of the hyperbolic point p_0 is sketched in Fig.3.2. The hyperbolic point is depicted by a filled circle. The separatrix manifold partitions the phase space into many sub-domains, and in each sub-domain, the flows of Eq.(3.2) are different in the adjacent sub-domains. Thus, the first integral manifolds in domains Ω_i and Ω_k are presented with the corresponding integral constant E_{σ_α} ($\sigma_\alpha \in (0,1]$)and $\alpha = i, j,k$). For $\sigma_\alpha = 0$, the separatrix surface is presented with a thick curve, which the integral constant is labeled by E_s. The subscript "s" represents "separatrix surface". The separatrix surface is separated into many pieces by sub-domains. The small pieces of the separatrix surface are named through two adjacent sub-domains. So, the two pieces $\partial\Omega_{ij}$ and $\partial\Omega_{kj}$ of the separatrix surface are depicted in Fig.3.2. In the sub-domain Ω_j, the first integral manifolds for Eq.(3.2) is not presented in order to make the view clearer. To help one understand this concept, the first integral manifolds in 2-dimensional phase space are shown in Fig.3.3. The first integral manifolds with only one hyperbolic point are presented. For this case, four sub-domains divided by a separatrix exist. The separatrix with integral constant $E_s^{(0)}$ consists of two stable manifolds ($W_s^s(t)$) and two unstable manifolds ($W_s^u(t)$). This separatrix is an unclosed orbit, as depicted by dark curves. In the four sub-domains Ω_i ($i = 1,2,\cdots,4$), the corresponding trajectories $\mathbf{x}_\sigma^{(i)}(t)$ ($\sigma \in (0,1]$) are relative to the corresponding constants $E_\sigma^{(i)}$, as sketched by the thin curves. Consider a dynamical system possessing one hyperbolic point and one center point. The corresponding separatrix will divide the phase space into three domains Ω_i ($i = 1,2,3$), as shown in Fig.3.3(b). The first integral manifolds in sub-domain Ω_1 are closed trajectories for all integral constants. In other two sub-domains, the trajectories are unclosed orbits with different integral constants.

3.2. Local and global flows

Because the non-integrable vector function exists in the dynamical system of Eq. (3.1), its flow cannot keep on the first integral manifold of the integrable dynamical system in Eq.(3.2) anymore. The flows of dynamical systems in Eq.(3.1) will move from one first integral manifold surface to another one with time varying. Such a flow placing in the atlas of the first integral manifold surfaces is sketched in Fig. 3.4. A flow intersecting with a specific integral manifold surface

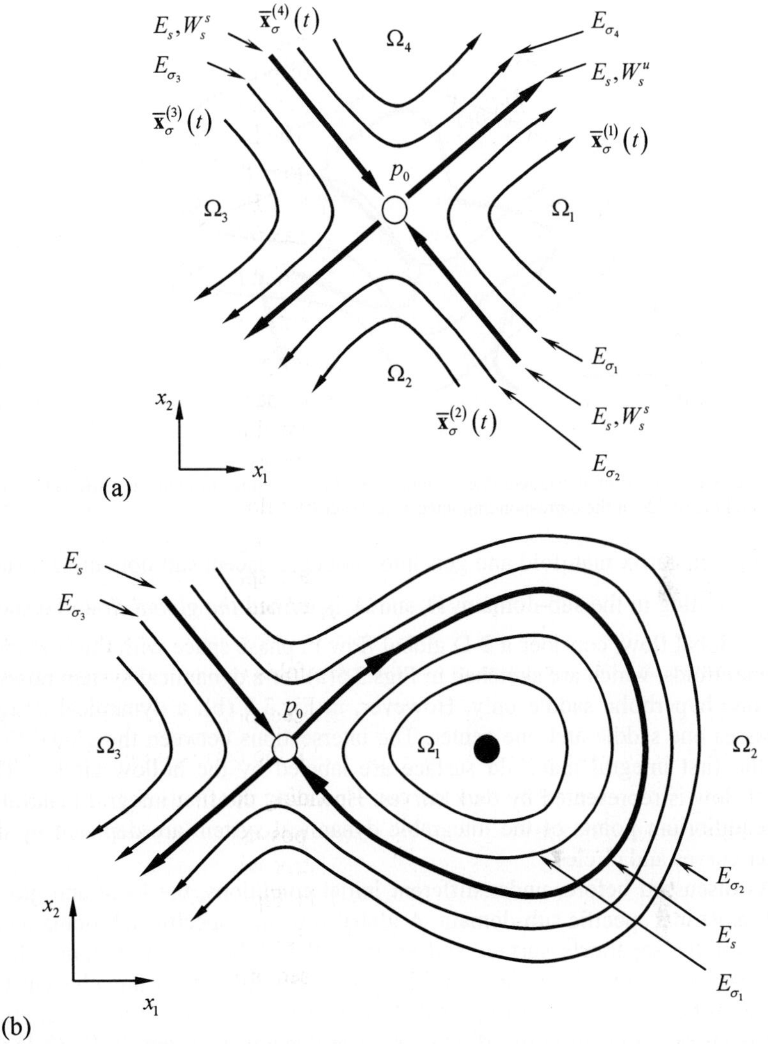

Figure 3.3. Phase portraits of the first integral manifolds on 2-D phase space: (a) only one hyperbolic point and (b) one saddle and one center.

is labeled circles. Under a special condition, the flow of the dynamical system in Eq.(3.1) may stay only in a specific sub-domain Ω_i ($i = 1, 2, \cdots, m$). Such a flow existing in the sub-domain Ω_i only is termed *the local flow* in such a domain. In this chapter, we are not be interested in the local flow in such a sub-domain. If a flow of the dynamical system in Eq.(3.1) starting in the sub-domain Ω_i passes

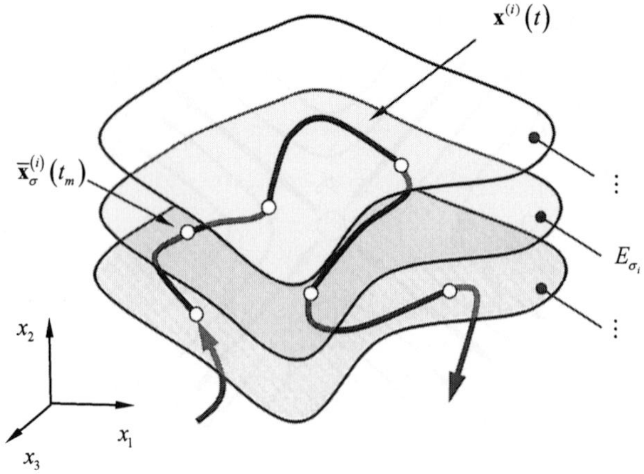

Figure 3.4. A flow of non-integrable system in Eq.(3.1) passing through the different first integral manifolds of the corresponding integrable system.

over the separatrix manifold and gets into another adjacent sub-domain Ω_j, such a flow existing in the sub-domains Ω_i and Ω_j is termed *the global flow*. To show such a global flow, consider a 2-D global flow in phase space with the first integral manifolds, which are sketched in Figs.3.5(a) for a dynamical system possessing one hyperbolic saddle only. However, in Fig.3.5 (b), a dynamical system possesses one saddle and one center. The intersections between the global flow and the first integral manifold surface are labeled by the hollow circles. The global flow is represented by dark curves. However, the first integral manifolds and equilibrium points of the integrable dynamical system are depicted by the lighter curves and circles.

As discussed before, under different initial conditions, the local and global flows exist in a specific sub-domain. A global flow in a specific sub-domain can pass over the separatrix surface and enter another adjacent sub-domain. However, the local flow will be only in this special sub-domain. In this chapter, the flow transition from a local to global flow is of great interest. So, Assumption (A3) is inserted to restrict our discussion. The dynamical system in Eq.(3.1) requires an initial set in the sub-domain Ω_i to make the corresponding flow approach its separatrix. The separation boundary between two sub-domains is formed by the separatrix surface of Eq.(3.2), and the corresponding boundary is defined through the first integral manifold of integrable system.

DEFINITION 3.1. For a dynamical system in Eq.(3.1), its integrable dynamical system in Eq.(3.2) possesses a first integral manifold relative to saddle points (i.e., $F(\bar{\mathbf{x}}^{(s)},\boldsymbol{\mu}) = E_s$), which partitions the phase space into many sub-domains

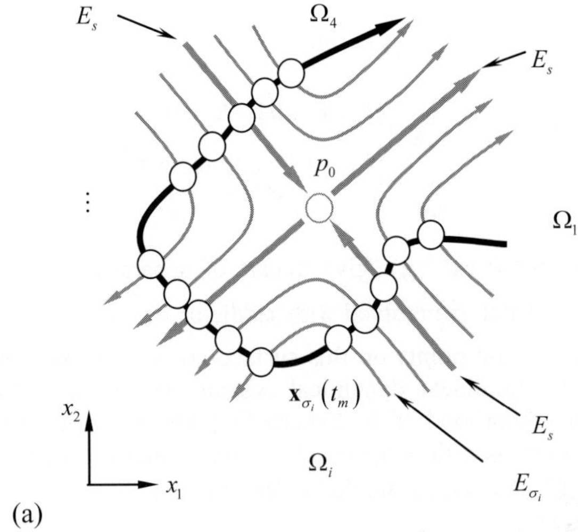

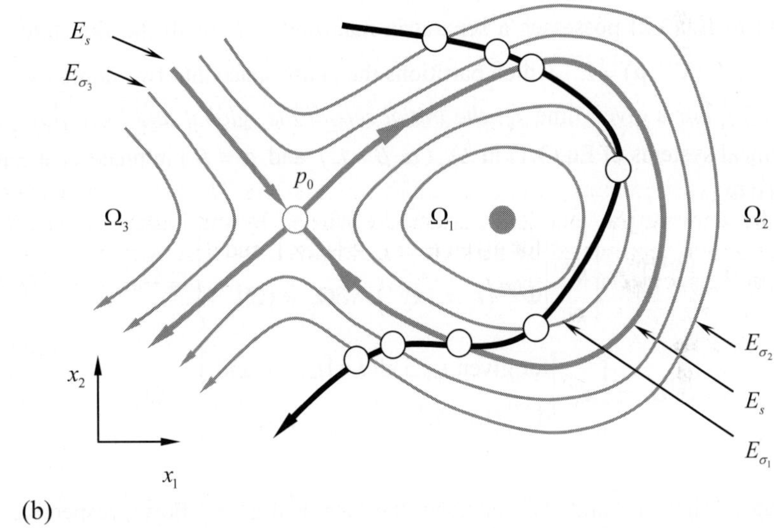

Figure 3.5. The 2-D global flows in the first integral with: (a) only one hyperbolic saddle and (b) one saddle and one center.

Ω_α ($\alpha \in \{1, 2, \cdots, m\}$). The first integral manifold surface pertaining to the saddle point is called *the separatrix* of the dynamical system in Eq.(3.1), defined as

$$S_s = \left\{ \bar{\mathbf{x}}^{(s)} \,\middle|\, F\left(\bar{\mathbf{x}}^{(s)}, \boldsymbol{\mu}\right) - E_s = 0,\ F \text{ is } C^r\text{-continuous } (r \geq 1) \right\} \subset \Re^{n-1} \quad (3.6a)$$

and the sub-manifold surface (or sub-separatrix) pertaining to two domains Ω_α ($\alpha=\{i,j\}$ and $i,j=\{1,2,\cdots,m\}$) is defined by

$$\partial\Omega_{ij}\equiv S_s\cap\bar{\Omega}_i\cap\bar{\Omega}_j=\left\{\bar{\mathbf{x}}^{(s)}\left|\begin{array}{l}F_{ij}\left(\bar{\mathbf{x}}^{(s)},\boldsymbol{\mu}\right)-E_s=0,\\ F_{ij}\text{ is }C^r\text{-continuous }(r\geq 1)\end{array}\right.\right\}\subset S_s. \tag{3.6b}$$

The scalar function $F_{ij}(\bar{\mathbf{x}}^{(s)},\boldsymbol{\mu})$ give a part of separatrix manifold surface between Ω_i and Ω_j, which is involved with saddle or sink and source. The flows connected with hyperbolic points on this surface are varied exponentially with time t. Responses in nonlinear dynamical systems are sensitive to the initial conditions in the neighborhood of the hyperbolic point. From Assumption (A3), to make sure that there is a flow approaching the separatrix, the initial sets for time t_0 in domain Ω_α ($\alpha=i,j$) should be defined, i.e.,

DEFINITION 3.2. For a dynamical system in Eq.(3.1), its integrable dynamical system in Eq.(3.2) possesses a separatrix manifold $\partial\Omega_{ij}$ with the first integral manifold $F_{ij}(\bar{\mathbf{x}}^{(s)},\boldsymbol{\mu})=E_s$, which partitions the phase space into two domains Ω_α ($\alpha=i,j$). For a given time t_0, *the initial sets of the global and local flows* of dynamical systems of Eq.(3.1) in Ω_α ($\alpha,\beta=i,j$ and $\alpha\neq\beta$) in phase space are defined as

$$ {}^{(G_\beta)}_{t_0}D_\alpha=\left\{\mathbf{x}_0^{(\alpha)}\left|\begin{array}{l}\text{for given }t_0,\ \exists\ t_m\in(t_0,\infty),\text{ and}\\ \Phi^{(\alpha)}\left(t_m,t_0,\mathbf{x}_0^{(\alpha)}\right)\cap\partial\Omega_{ij}\neq\{\varnothing\}\end{array}\right.\right\}\subseteq\Omega_\alpha, \tag{3.7a}$$

$$ {}^{(L)}_{t_0}D_\alpha=\left\{\mathbf{x}_0^{(\alpha)}\left|\begin{array}{l}\text{for given }t_0,\ \exists\text{ all }t\in[t_0,\infty),\text{ and}\\ \Phi^{(\alpha)}\left(t,t_0,\mathbf{x}_0^{(\alpha)}\right)\cap\partial\Omega_{ij}=\{\varnothing\}\end{array}\right.\right\}\subseteq\Omega_\alpha. \tag{3.7b}$$

Superscripts "L" and "G" represent the local and global flows, respectively. Based on an initial set ${}^{(G_\beta)}_{t_0}D_\alpha$ at time t_0 in domain Ω_α, a flow $\Phi^{(\alpha)}(t,t_0,\mathbf{x}_0)$ can approach the separatrix $\partial\Omega_{ij}$ and enter the domain Ω_β ($\beta=i,j$ and $\alpha\neq\beta$). However, from an initial set ${}^{(L)}_{t_0}D_\alpha$ at an initial time t_0, all the flows $\Phi^{(\alpha)}(t,t_0,\mathbf{x}_0)$ cannot approach the separatrix $\partial\Omega_{ij}$, and they form a family of local flows in the domain Ω_α. From the foregoing definition, the domain Ω_α can be expressed through the initial sets

$$\Omega_\alpha = \cup_\beta {}^{(G_\beta)}_{t_0}D_\alpha \cup {}^{(L)}_{t_0}D_\alpha \cup {}_{t_0}\Gamma_{\alpha\beta} \tag{3.8}$$

where ${}_{t_0}\Gamma_{\alpha\beta} = \cup_\beta {}^{(G_\beta)}_{t_0}\bar{D}_\alpha \cap {}^{(L)}_{t_0}\bar{D}_\alpha$ and the closures of the global and local initial sets are ${}^{(G_\beta)}_{t_0}\bar{D}_\alpha$ and ${}^{(L)}_{t_0}\bar{D}_\alpha$, respectively. For the separatrix surface $\partial\Omega_{\alpha\beta}$ ($\beta \in \{i, j, \cdots, k\}$ and $\alpha \neq \beta$), the boundary of sub-domain Ω_α is composed of many different pieces of separatrix ($\partial\Omega_{\alpha\beta}$). To demonstrate the global and local flows from the initial sets, consider two initial sets ${}^{(G_j)}_{t_0}D_i$ and ${}^{(L)}_{t_0}D_i$ for an initial time t_0 in sub-domain Ω_i. From such initial sets, the global and local flows in Ω_i are sketched in Figs.3.6(a) and (b), respectively. If the boundary ${}_{t_0}\Gamma_{\alpha\beta}$ between ${}^{(G_j)}_{t_0}D_i$ and ${}^{(L)}_{t_0}D_i$ is as an initial set, the flows in Ω_i will be tangential to the separatrix $\partial\Omega_{ij}$. The gray-filled, initial set is ${}^{(G_j)}_{t_0}D_i$ on which the corresponding global flow will go into the domains Ω_j. The shaded domains represent the initial sets to map the global flow to the other domains. The starting and ending points of the global and local flows are represented by circles. The flows are depicted by the curves with arrows. The local flow cannot arrive to any separatrix boundary of Ω_j. The global flow to the separatrix manifold $\partial\Omega_{ij}$ in Ω_i can pass over this boundary. If this passing flow just gets into a local initial set in Ω_j at a suitable time, the flow will stay in such a domain forever. If this passing flow gets into a global initial set in Ω_j, it will continue to enter another domain. If the arrived set is an global initial set in Ω_j from which the global flow returns back into the global initial set in Ω_i, the passable, global regular and irregular flows may exist to pass over the separatrix surface $\partial\Omega_{ij}$. To determine such flows, the initial sets should be determined first. In addition, the tangential flow is a key to distinguish the global and local flows for a given time t_0. Therefore, the global transversal and tangential flows to the separatrix manifold $\partial\Omega_{\alpha\beta}$ will be investigated in the following sections.

3.3. Global transversality

From the previous discussion, if a flow $\mathbf{x}^{(\alpha)}(t) = \Phi^{(\alpha)}(t, t_0, \mathbf{x}_0)$ in Ω_α is global to the separatrix manifold surface $\partial\Omega_{\alpha\beta}$, the initial condition for time t_0 should be chosen from the initial set ${}^{(G_\beta)}_{t_0}D_\alpha$. To investigate the global transversality of a flow, the mathematical definition of the global flow is given as follows:

DEFINITION 3.3. For a dynamical system in Eq.(3.1), its integrable dynamical system in Eq.(3.2) possesses a separatrix surface $\partial\Omega_{ij}$ ($i,j\in\{1,2,\cdots,m\}$) with the first integral manifold $F_{ij}(\overline{\mathbf{x}}^{(s)},\boldsymbol{\mu})=E_s$, which partitions its phase space into two domains Ω_α ($\alpha=i,j$). A flow $\mathbf{x}^{(\alpha)}(t)$ of the dynamical system in Eq.(3.1) in the two domains possesses different solution characteristics. Choose $\overline{\mathbf{x}}^{(s)}(t_m)\equiv\mathbf{x}_m\in\partial\Omega_{ij}$ at time t_m. Suppose $\mathbf{x}^{(i)}(t_{m-})=\mathbf{x}_m=\mathbf{x}^{(j)}(t_{m+})$. For an arbitrarily small $\varepsilon>0$, there are two time intervals $[t_{m-\varepsilon},t_m)$ and $(t_m,t_{m+\varepsilon}]$. The flow $\mathbf{x}^{(\alpha)}(t)$ ($\alpha\in\{i,j\}$) with the initial set ${}^{(G_\beta)}_{t_0}D_\alpha$ ($\beta\in\{i,j\}$ and $\alpha\neq\beta$) to the separatrix surface $\partial\Omega_{ij}$ is *global* if the following properties exist

$$\left.\begin{array}{ll}\text{either} & \left.\begin{array}{l}\mathbf{n}^T_{\partial\Omega_{ij}}(\overline{\mathbf{x}}^{(s)}_{m-\varepsilon})\bullet\left[\overline{\mathbf{x}}^{(s)}_{m-\varepsilon}-\mathbf{x}^{(\alpha)}_{m-\varepsilon}\right]>0 \text{ and}\\ \mathbf{n}^T_{\partial\Omega_{ij}}(\overline{\mathbf{x}}^{(s)}_{m+\varepsilon})\bullet\left[\mathbf{x}^{(\beta)}_{m+\varepsilon}-\overline{\mathbf{x}}^{(s)}_{m+\varepsilon}\right]>0\end{array}\right\}\text{for } \mathbf{n}_{\partial\Omega_{ij}}\to\Omega_\beta,\\ \text{or} & \left.\begin{array}{l}\mathbf{n}^T_{\partial\Omega_{ij}}(\overline{\mathbf{x}}^{(s)}_{m-\varepsilon})\bullet\left[\overline{\mathbf{x}}^{(s)}_{m-\varepsilon}-\mathbf{x}^{(\alpha)}_{m-\varepsilon}\right]<0 \text{ and}\\ \mathbf{n}^T_{\partial\Omega_{ij}}(\overline{\mathbf{x}}^{(s)}_{m+\varepsilon})\bullet\left[\mathbf{x}^{(\beta)}_{m+\varepsilon}-\overline{\mathbf{x}}^{(s)}_{m+\varepsilon}\right]<0\end{array}\right\}\text{for } \mathbf{n}_{\partial\Omega_{ij}}\to\Omega_\alpha\end{array}\right\} \tag{3.9}$$

where the normal vector of the boundary $\partial\Omega_{ij}$ is

$$\mathbf{n}_{\partial\Omega_{ij}}(\overline{\mathbf{x}}^{(s)}_m)=\nabla F_{ij}\left(\overline{\mathbf{x}}^{(s)}_m\right)=\left(\frac{\partial F_{ij}}{\partial x_1},\frac{\partial F_{ij}}{\partial x_2},\cdots,\frac{\partial F_{ij}}{\partial x_n}\right)^{\mathrm{T}}_{(t_m,\overline{\mathbf{x}}^{(s)}_m)}. \tag{3.10}$$

Notice that the expression ($\mathbf{n}_{\partial\Omega_{ij}}\to\Omega_\alpha$) means that the normal vector points to domain Ω_α ($\alpha\in\{i,j\}$).The notations $t_{m\pm\varepsilon}=t_m\pm\varepsilon$ and $t_{m\pm}=t_m\pm0$ are used. Consider a flow in Eq.(3.1) from a domain Ω_i into domain Ω_j through the separatrix surface $\partial\Omega_{ij}$. At time t_m, the flow $\mathbf{x}^{(\alpha)}(t)=\Phi^{(\alpha)}(t_0,\mathbf{x}_0,t)$ arrives to the separatrix $\partial\Omega_{ij}$, and there is a small neighborhood $(t_{m-\varepsilon},t_{m+\varepsilon})$ of the time t_m, which is arbitrarily selected. As $\varepsilon\to0$, the time increment $\Delta t\equiv\varepsilon\to0$. Before the flow reaches to the separatrix, the point $\mathbf{x}^{(i)}(t_{m-\varepsilon})$ lies in the domain Ω_i. The point $\mathbf{x}_m$ is for the flow on the separatrix. After the flow passes through the boundary, the point $\mathbf{x}^{(j)}(t_{m+\varepsilon})$ is in the neighborhood of separatrix on the side of domain Ω_j. The input and output flow vectors are $\mathbf{x}^{(i)}(t_m)-\mathbf{x}^{(i)}(t_{m-\varepsilon})$ and $\mathbf{x}^{(j)}(t_{m+\varepsilon})-\mathbf{x}^{(j)}(t_m)$, respectively. The global flow $\mathbf{x}^{(i)}(t)$ with the initial sets ${}^{(G_j)}_{t_0}D_i$ passes through the separatrix $\partial\Omega_{ij}$ from Ω_i to Ω_j, as shown in Fig.3.7. The two cases ($\mathbf{n}_{\partial\Omega_{ij}}\to\Omega_j$

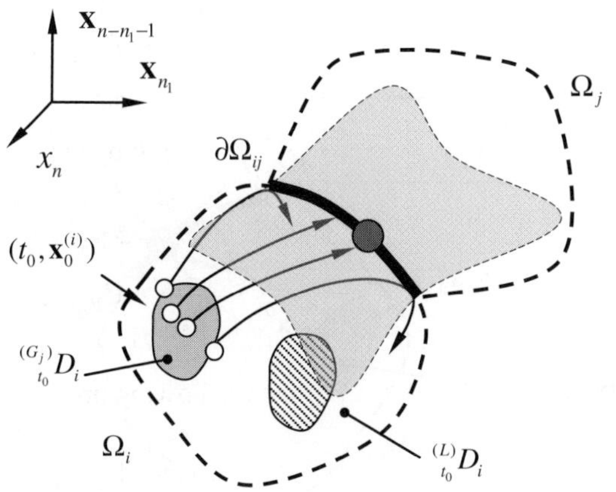

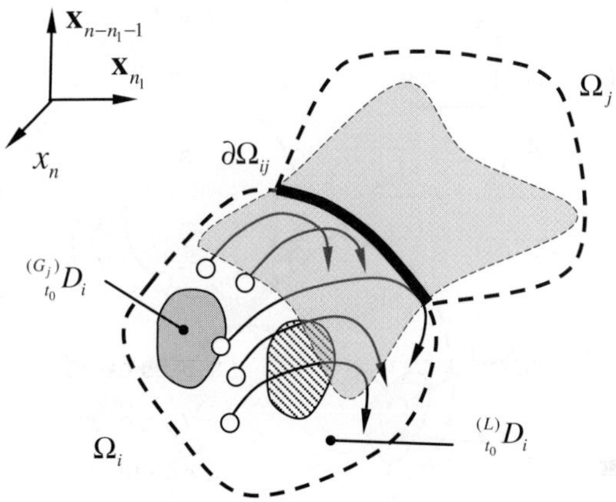

Figure 3.6. (a) Global and (b) local flows in sub-domain Ω_i starting from the initial sets. The tangential flows on the initial sets boundary of the local and global flows are sketched.

and $\mathbf{n}_{\partial\Omega_{ij}} \to \Omega_i$) are presented. Three points $\mathbf{x}^{(i)}(t_{m-\varepsilon})$, $\mathbf{x}^{(j)}(t_{m+\varepsilon})$ and $\mathbf{x}_m$ are in domains Ω_i and Ω_j on the separatrix $\partial\Omega_{ij}$, respectively. The vector ($\overline{\mathbf{x}}^{(s)}(t_{m-\varepsilon})-\mathbf{x}^{(i)}(t_{m-\varepsilon})$) (or $\mathbf{x}^{(j)}(t_{m+\varepsilon})-\overline{\mathbf{x}}^{(s)}(t_{m+\varepsilon})$) is the difference between flows $\mathbf{x}^{(i)}(t)$ and $\overline{\mathbf{x}}^{(s)}(t)$ at time $t_{m-\varepsilon}$ (or flows $\mathbf{x}^{(j)}(t)$ and $\overline{\mathbf{x}}^{(s)}(t)$ at time $t_{m+\varepsilon}$). The input and output vector differences are depicted through the dotted arrows. Two vectors $\mathbf{n}_{\partial\Omega_{ij}}$

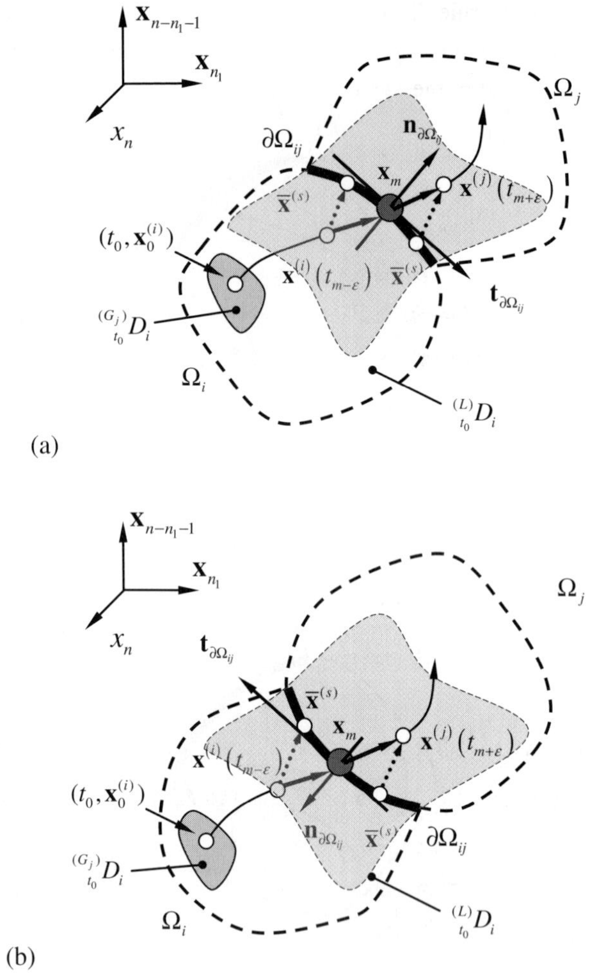

Figure 3.7. The global flow $\mathbf{x}^{(i)}(t)$ with the initial sets ${}^{(G_j)}_{t_0}D_i$, passing through separatrix $\partial\Omega_{ij}$ from domain Ω_i to Ω_j : (a) $\mathbf{n}_{\partial\Omega_{ij}} \to \Omega_j$ and (b) $\mathbf{n}_{\partial\Omega_{ij}} \to \Omega_i$. Three points $\mathbf{x}^{(i)}(t_{m-\varepsilon})$, $\mathbf{x}^{(j)}(t_{m+\varepsilon})$ and $\mathbf{x}_m$ are in the domains Ω_i and Ω_j , and on the separatrix $\partial\Omega_{ij}$, respectively. Two vectors $\mathbf{n}_{\partial\Omega_{ij}}$ and $\mathbf{t}_{\partial\Omega_{ij}}$ are the normal and tangential vectors of $\partial\Omega_{ij}$.

and $\mathbf{t}_{\partial\Omega_{ij}}$ are the normal and tangential vectors of $\partial\Omega_{ij}$, which are determined by $F_{ij}(\overline{\mathbf{x}}^{(s)},\boldsymbol{\mu}) = E_s$. The direction of $\mathbf{t}_{\partial\Omega_{ij}} \times \mathbf{n}_{\partial\Omega_{ij}}$ is the positive direction of the coordinate by the right-hand rule. To determine existence of the global flows in Ω_α ($\alpha = i, j$), the theorem can be easily obtained from the foregoing definition because of the continuity of the dynamical systems in Eq.(3.1).

THEOREM 3.1. *For a dynamical system in Eq.*(3.1), *its integrable dynamical system in Eq.*(3.2) *possesses a separatrix surface* $\partial\Omega_{ij}$ *with the first integral manifold* $F_{ij}(\bar{\mathbf{x}}^{(s)}, \boldsymbol{\mu}) = E_s$*, which partitions its phase space into domains* Ω_α ($\alpha = i, j$). *Two flows* $\mathbf{x}^{(\alpha)}(t)$ ($\alpha = i, j$) *of the dynamical system in Eq.*(3.1) *in the corresponding domains have different solution characteristics. Choose* $\bar{\mathbf{x}}^{(s)}(t_m) \equiv \mathbf{x}_m \in \partial\Omega_{ij}$ *at time* t_m. *Suppose* $\mathbf{x}^{(i)}(t_{m-}) = \mathbf{x}_m = \mathbf{x}^{(j)}(t_{m+})$. *For an arbitrarily small* $\varepsilon > 0$, *there are two time intervals* $[t_{m-\varepsilon}, t_m)$ *and* $(t_m, t_{m+\varepsilon}]$. *Both* $\mathbf{x}^{(i)}(t)$ *and* $\mathbf{x}^{(j)}(t)$ *are* $C^r_{[t_{m-\varepsilon}, t_m)}$ *and* $C^r_{(t_m, t_{m+\varepsilon}]}$*-continuous* ($r \ge 2$) *for time t, respectively and* $\| d^r\mathbf{x}^{(\alpha)}/dt^r \| < \infty$ ($\alpha \in \{i, j\}$). *The flow* $\mathbf{x}^{(i)}(t)$ *with the initial set* ${}^{(G_j)}_{t_0}D_i$ *to the separatrix boundary* $\partial\Omega_{ij}$ *is global from the domain* Ω_i *to* Ω_j *iff*

$$\begin{aligned} &\text{either} \quad \mathbf{n}^T_{\partial\Omega_{ij}}(\bar{\mathbf{x}}_m)\bullet\dot{\mathbf{x}}^{(i)}_{m-} > 0 \text{ and } \mathbf{n}^T_{\partial\Omega_{ij}}(\bar{\mathbf{x}}_m)\bullet\dot{\mathbf{x}}^{(j)}_{m+} > 0 \quad \text{for } \mathbf{n}_{\partial\Omega_{ij}} \to \Omega_j, \\ &\text{or} \quad \mathbf{n}^T_{\partial\Omega_{ij}}(\bar{\mathbf{x}}_m)\bullet\dot{\mathbf{x}}^{(i)}_{m-} < 0 \text{ and } \mathbf{n}^T_{\partial\Omega_{ij}}(\bar{\mathbf{x}}_{t_m})\bullet\dot{\mathbf{x}}^{(j)}_{m+} < 0 \quad \text{for } \mathbf{n}_{\partial\Omega_{ij}} \to \Omega_i. \end{aligned} \tag{3.11}$$

PROOF. For a point $\mathbf{x}_m \in \partial\Omega_{ij}$, suppose $\mathbf{x}^{(i)}(t_{m-}) = \mathbf{x}_m$ and $\mathbf{x}_m = \mathbf{x}^{(j)}(t_{m+})$. Both $\mathbf{x}^{(i)}(t)$ and $\mathbf{x}^{(j)}(t)$ are $C^r_{[t_{m-\varepsilon}, t_m)}$ and $C^r_{(t_m, t_{m+\varepsilon}]}$-continuous ($r \ge 2$) for time t, respectively. $\| \ddot{\mathbf{x}}^{(\alpha)}(t) \| < \infty$ ($\alpha \in \{i, j\}$) for $0 < \varepsilon \ll 1$. Consider $a \in [t_{m-\varepsilon}, t_{m-})$ and $b \in (t_{m-}, t_{m+\varepsilon}]$. The Taylor series expansion of $\mathbf{x}^{(\alpha)}(t_{m\pm\varepsilon})$ with $t_{m\pm\varepsilon} = t_m \pm \varepsilon$ ($\alpha \in \{i, j\}$) to $\mathbf{x}^{(\alpha)}(a)$ and $\mathbf{x}^{(\alpha)}(b)$ gives

$$\begin{aligned} \mathbf{x}^{(i)}_{m-\varepsilon} &\equiv \mathbf{x}^{(i)}(t_{m-\varepsilon}) \equiv \mathbf{x}^{(i)}(t_{m-} - \varepsilon) \\ &= \mathbf{x}^{(i)}(a) + \dot{\mathbf{x}}^{(i)}(a)(t_{m-} - \varepsilon - a) + o((t_{m-} - \varepsilon - a)), \\ \mathbf{x}^{(j)}_{m+\varepsilon} &= \mathbf{x}^{(j)}(t_{m+\varepsilon}) \equiv \mathbf{x}^{(j)}(t_{m+} + \varepsilon) \\ &= \mathbf{x}^{(j)}(b) + \dot{\mathbf{x}}^{(j)}(t_{m+} + \varepsilon - b)\varepsilon + o((t_{m+} + \varepsilon - b)). \end{aligned}$$

Let $a \to t_{m-}$ and $b \to t_{m+}$, the limits of the foregoing equations lead to

$$\begin{aligned} \mathbf{x}^{(i)}_{m-\varepsilon} &\equiv \mathbf{x}^{(i)}(t_{m-} - \varepsilon) = \mathbf{x}^{(i)}_{m-} - \dot{\mathbf{x}}^{(i)}_{m-}\varepsilon + o(\varepsilon); \\ \mathbf{x}^{(j)}_{m+\varepsilon} &\equiv \mathbf{x}^{(j)}(t_{m+} + \varepsilon) = \mathbf{x}^{(j)}_{m+} + \dot{\mathbf{x}}^{(j)}_{m+}\varepsilon + o(\varepsilon). \end{aligned}$$

Due to $0<\varepsilon<<1$, the ε^2 and higher order terms of the foregoing equations is ignored. In a similar fashion, we have

$$\begin{aligned}
&\bar{\mathbf{x}}_{m\pm\varepsilon}^{(s)} \equiv \mathbf{x}^{(s)}(t_m \pm \varepsilon) = \bar{\mathbf{x}}_m^{(s)} \pm \dot{\bar{\mathbf{x}}}_m^{(s)}\varepsilon + o(\varepsilon);\\
&\mathbf{n}_{\partial\Omega_{ij}}(\bar{\mathbf{x}}_{m\pm\varepsilon}^{(s)}) \equiv \mathbf{n}_{\partial\Omega_{ij}}(\bar{\mathbf{x}}_m^{(s)}) \pm D_{\bar{\mathbf{x}}}\mathbf{n}_{\partial\Omega_{ij}}(\bar{\mathbf{x}}_m^{(s)})\varepsilon + o(\varepsilon).
\end{aligned}$$

Further, we have

$$\begin{aligned}
\mathbf{n}_{\partial\Omega_{ij}}^T(\bar{\mathbf{x}}_{m-\varepsilon}^{(s)})\bullet\left[\bar{\mathbf{x}}_{m-\varepsilon}^{(s)} - \mathbf{x}_{m-\varepsilon}^{(\alpha)}\right] &= -D_{\bar{\mathbf{x}}}\mathbf{n}_{\partial\Omega_{ij}}^T(\bar{\mathbf{x}}_m^{(s)})\cdot\left[\bar{\mathbf{x}}_m^{(s)} - \mathbf{x}_{m-}^{(\alpha)}\right]\\
&\quad -\mathbf{n}_{\partial\Omega_{ij}}^T(\bar{\mathbf{x}}_m^{(s)})\cdot\left[\dot{\bar{\mathbf{x}}}_m^{(s)} - \dot{\mathbf{x}}_{m-}^{(\alpha)}\right]\varepsilon,\\
\mathbf{n}_{\partial\Omega_{ij}}^T(\bar{\mathbf{x}}_{m+\varepsilon}^{(s)})\bullet\left[\mathbf{x}_{m+\varepsilon}^{(\beta)} - \bar{\mathbf{x}}_{m+\varepsilon}^{(s)}\right] &= D_{\bar{\mathbf{x}}}\mathbf{n}_{\partial\Omega_{ij}}^T(\bar{\mathbf{x}}_m^{(s)})\cdot\left[\mathbf{x}_{m+}^{(\beta)} - \bar{\mathbf{x}}_m^{(s)}\right]\\
&\quad +\mathbf{n}_{\partial\Omega_{ij}}^T(\bar{\mathbf{x}}_m^{(s)})\cdot\left[\mathbf{x}_{m+}^{(\beta)} - \dot{\bar{\mathbf{x}}}_m^{(s)}\right]\varepsilon.
\end{aligned}$$

Owing to $\mathbf{x}^{(\alpha)}(t_{m-}) = \bar{\mathbf{x}}^{(s)}(t_m) = \mathbf{x}_m = \mathbf{x}^{(\beta)}(t_{m+})$, the foregoing equations give

$$\begin{aligned}
\mathbf{n}_{\partial\Omega_{ij}}^T(\bar{\mathbf{x}}_{m-\varepsilon}^{(s)})\bullet\left[\bar{\mathbf{x}}_{m-\varepsilon}^{(s)} - \mathbf{x}_{m-\varepsilon}^{(\alpha)}\right] &= -\mathbf{n}_{\partial\Omega_{ij}}^T(\bar{\mathbf{x}}_m^{(s)})\cdot\left[\dot{\bar{\mathbf{x}}}_m^{(s)} - \dot{\mathbf{x}}_{m-}^{(\alpha)}\right]\varepsilon,\\
\mathbf{n}_{\partial\Omega_{ij}}^T(\bar{\mathbf{x}}_{m+\varepsilon}^{(s)})\bullet\left[\mathbf{x}_{m+\varepsilon}^{(\beta)} - \bar{\mathbf{x}}_{m+\varepsilon}^{(s)}\right] &= \mathbf{n}_{\partial\Omega_{ij}}^T(\bar{\mathbf{x}}_m^{(s)})\cdot\left[\mathbf{x}_{m+}^{(\beta)} - \dot{\bar{\mathbf{x}}}_m^{(s)}\right]\varepsilon.
\end{aligned}$$

Owing to $\mathbf{n}_{\partial\Omega_{ij}}^T(\bar{\mathbf{x}}_m^{(s)})\cdot\dot{\bar{\mathbf{x}}}_m^{(s)} = 0$, the following equations can be obtained,

$$\begin{aligned}
\mathbf{n}_{\partial\Omega_{ij}}^T(\bar{\mathbf{x}}_{m-\varepsilon}^{(s)})\bullet\left[\bar{\mathbf{x}}_{m-\varepsilon}^{(s)} - \mathbf{x}_{m-\varepsilon}^{(\alpha)}\right] &= \mathbf{n}_{\partial\Omega_{ij}}^T(\bar{\mathbf{x}}_m^{(s)})\cdot\dot{\mathbf{x}}_{m-}^{(\alpha)}\varepsilon\\
\mathbf{n}_{\partial\Omega_{ij}}^T(\bar{\mathbf{x}}_{m+\varepsilon}^{(s)})\bullet\left[\mathbf{x}_{m+\varepsilon}^{(\beta)} - \bar{\mathbf{x}}_{m+\varepsilon}^{(s)}\right] &= \mathbf{n}_{\partial\Omega_{ij}}^T(\bar{\mathbf{x}}_m^{(s)})\cdot\mathbf{x}_{m+}^{(\beta)}\varepsilon
\end{aligned}$$

Applying Eq.(3.9) to the above equations, equation (3.11) is obtained, vice versa. This theorem is proved. ■

THEOREM 3.2. *For a dynamical system in Eq.*(3.1), *its integrable dynamical system in Eq.*(3.2) *possesses a separatrix surface* $\partial\Omega_{ij}$ *with the first integral manifold* $F_{ij}(\bar{\mathbf{x}}^{(s)},\boldsymbol{\mu}) = E_s$, *which partitions its phase space into domains* Ω_α ($\alpha = i, j$). *Two flows* $\mathbf{x}^{(\alpha)}(t)$ ($\alpha = i, j$) *of the dynamical system in Eq.(3.1) in the corresponding domains have different solution characteristics. Choose* $\bar{\mathbf{x}}^{(s)}(t_m) \equiv \mathbf{x}_m \in \partial\Omega_{ij}$ *at time* t_m. *Suppose* $\mathbf{x}^{(i)}(t_{m-}) = \mathbf{x}_m = \mathbf{x}^{(j)}(t_{m+})$. *For an arbitrarily small* $\varepsilon > 0$, *there are two time intervals* $[t_{m-\varepsilon}, t_m)$ *and* $(t_m, t_{m+\varepsilon}]$. $\mathbf{F}(\mathbf{x}, t, \mathbf{p})$ *is* $C^r_{[t_{m-\varepsilon}, t_{m+\varepsilon}]}$*-continuous* ($r \geq 1$) *for time t, and* $\| d^{r+1}\mathbf{x}^{(\alpha)}/dt^{r+1} \| < \infty$ ($\alpha \in \{i, j\}$). *The flow* $\mathbf{x}^{(i)}(t)$ ($\alpha \in \{i, j\}$) *with the initial set* ${}^{(G_i)}_{t_0}D_i$ *to the separatrix* $\partial\Omega_{ij}$ *is* *global from the domain* Ω_i *to* Ω_j *iff*

$$\begin{aligned}
\mathbf{n}_{\partial\Omega_{ij}}^T(\mathbf{x}_m)\bullet\mathbf{g}(\mathbf{x}_m,t_m,\boldsymbol{\pi}) = \nabla F_{ij}(\mathbf{x}_m)\bullet\mathbf{g}(\mathbf{x}_m,t_m,\boldsymbol{\pi}) > 0 \quad \text{for } \mathbf{n}_{\partial\Omega_{ij}} \to \Omega_j,\\
\mathbf{n}_{\partial\Omega_{ij}}^T(\mathbf{x}_m)\bullet\mathbf{g}(\mathbf{x}_m,t_m,\boldsymbol{\pi}) = \nabla F_{ij}(\mathbf{x}_m)\bullet\mathbf{g}(\mathbf{x}_m,t_m,\boldsymbol{\pi}) < 0 \quad \text{for } \mathbf{n}_{\partial\Omega_{ij}} \to \Omega_i.
\end{aligned} \tag{3.12}$$

PROOF. Although the flows in each domain are different, they must satisfy the dynamical systems in Eq.(3.1), as stated in Eq.(3.4). For $t = t_m$, equation (3.1) is written as

$$\dot{\mathbf{x}}^{(\alpha)}(t_{m\pm}) = \mathbf{F}(\mathbf{x}^{(\alpha)}(t_{m\pm}),t_m,\mathbf{p}) \equiv \mathbf{f}(\mathbf{x}^{(\alpha)}(t_{m\pm}),\boldsymbol{\mu}) + \mathbf{g}(\mathbf{x}^{(\alpha)}(t_{m\pm}),t_m,\boldsymbol{\pi}).$$

For a point $\mathbf{x}_m \in \partial\Omega_{ij}$, the relationship $\mathbf{x}^{(i)}(t_{m-}) = \mathbf{x}_m = \mathbf{x}^{(j)}(t_{m+})$ exists. The foregoing equation becomes

$$\dot{\mathbf{x}}^{(\alpha)}(t_{m\pm}) = \mathbf{F}(\mathbf{x}_m,t_m,\mathbf{p}) \equiv \mathbf{f}(\mathbf{x}_m,\boldsymbol{\mu}) + \mathbf{g}(\mathbf{x}_m,t_m,\boldsymbol{\pi}).$$

Using $\mathbf{n}_{\partial\Omega_{ij}}^T = \nabla F_{ij}$, multiplication of both sides of the foregoing equation gives

$$\begin{aligned}
\mathbf{n}_{\partial\Omega_{ij}}^T\bullet\dot{\mathbf{x}}^{(\alpha)}(t_{m\pm}) &= \mathbf{n}_{\partial\Omega_{ij}}^T\bullet\mathbf{f}(\mathbf{x}_m,\boldsymbol{\mu}) + \mathbf{n}_{\partial\Omega_{ij}}^T\bullet\mathbf{g}(\mathbf{x}_m,t_m,\boldsymbol{\pi})\\
&= \nabla F_{ij}(\mathbf{x}_m,\boldsymbol{\mu})\bullet\mathbf{f}(\mathbf{x}_m,\boldsymbol{\mu}) + \nabla F_{ij}(\mathbf{x}_m,\boldsymbol{\mu})\bullet\mathbf{g}(\mathbf{x}_m,t_m,\boldsymbol{\pi}).
\end{aligned}$$

Because $\mathbf{x}_m \in \partial\Omega_{ij}$ is on the separatrix, the separatrix is the first integral manifold relative to the hyperbolic-saddle. So Eq.(3.3) should be satisfied, i.e.,

$$\nabla F_{ij}(\mathbf{x}_m,\boldsymbol{\mu})\bullet\mathbf{f}(\mathbf{x}_m,\boldsymbol{\mu}) = 0.$$

Using the foregoing equation, we have

$$\mathbf{n}_{\partial\Omega_{ij}}^T\bullet\dot{\mathbf{x}}^{(\alpha)}(t_{m\pm}) = \nabla F_{ij}(\mathbf{x}_m,\boldsymbol{\mu})\bullet\mathbf{g}(\mathbf{x}_m,t_m,\boldsymbol{\pi}).$$

Using Eq.(3.12), equation (3.11) is obtained, and vice versa. From Theorem 3.1, this theorem is proved. ■

As in Chapter 2, the G-function is introduced to measure the relation between two flows in two different dynamical systems. To investigate the global transversality of a perturbed flow to the separatrix, the perturbed flow given by Eq.(3.1) is considered as a compared flow. The integrable flow given by Eq.(3.2) on the separatrix is used as a reference flow. For the global transversality of a perturbed flow to the separatrix at point $(t_m, \bar{\mathbf{x}}_m)$, the flows $\mathbf{x}^{(\alpha)}(t)$ and $\bar{\mathbf{x}}^{(s)}(t)$ at least are the zero-order contact. So $\mathbf{x}^{(\alpha)}(t_m) = \bar{\mathbf{x}}^{(s)}(t_m) = \bar{\mathbf{x}}_m$. From Definition 2.3, the G-

function for the zero-order contact between a perturbed flow in Eqs.(3.1) and an integrable flow on the separatrix in Eq.(3.2) can be determined. Herein, the *G*-function for a perturbed flow to the separatrix is defined as follows.

DEFINITION 3.4. For a dynamical system in Eq.(3.1), its integrable dynamical system in Eq.(3.2) possesses a separatrix surface $\partial\Omega_{ij}$ with the first integral manifold $F_{ij}(\bar{\mathbf{x}}^{(s)},\boldsymbol{\mu}) = E_s$, which partitions its phase space into two domains Ω_α ($\alpha = i, j$). The *G*-function of the kth-order $G^{(k)}_{\partial\Omega_{ij}}(\bullet)$ for a perturbed flow to the first integral surface $\partial\Omega_{ij} \subset S$ at point $\bar{\mathbf{x}}^{(s)}_m = \mathbf{x}^{(\alpha)}(t_m)$ is defined for ($k = 0,1,2,\cdots$)

$$
\begin{aligned}
G^{(k)}_{\partial\Omega_{ij}}(\bar{\mathbf{x}}^{(s)}_m,\mathbf{x}^{(\alpha)}_m,t_m,\mathbf{p}) &= \sum_{r=1}^{k+1} C^r_{k+1} D^{k+1-r}_{\bar{\mathbf{x}}}\mathbf{n}^T_{\partial\Omega_{ij}}(\bar{\mathbf{x}})\bullet\left[\frac{d^r\mathbf{x}}{dt^r}-\frac{d^r\bar{\mathbf{x}}}{dt^r}\right]\Bigg|_{(\bar{\mathbf{x}}^{(s)}_m,\mathbf{x}^{(\alpha)}_m,t_m)}\\
&= \sum_{r=1}^{k+1} C^r_{k+1} D^{k+1-r}_{\bar{\mathbf{x}}}\mathbf{n}^T_{\partial\Omega_{ij}}(\bar{\mathbf{x}})\bullet\Big[D^{r-1}_{\mathbf{x}}\mathbf{F}(\mathbf{x},t,\mathbf{p})\\
&\quad -D^{r-1}_{\bar{\mathbf{x}}}\mathbf{f}(\bar{\mathbf{x}},\boldsymbol{\mu})\Big]\Big|_{(\bar{\mathbf{x}}^{(s)}_m,\mathbf{x}^{(\alpha)}_m,t_m)}.
\end{aligned}
\tag{3.13}
$$

For the special case ($k = 0$), the normal component function is

$$
\begin{aligned}
G^{(0)}_{\partial\Omega_{ij}}(\bar{\mathbf{x}}^{(s)}_m,\mathbf{x}^{(\alpha)}_m,t_m,\mathbf{p}) &= G_{\partial\Omega_{ij}}(\bar{\mathbf{x}}^{(s)}_m,\mathbf{x}^{(\alpha)}_m,t_m,\mathbf{p}) = \mathbf{n}^T_{\partial\Omega_{ij}}(\bar{\mathbf{x}})\bullet(\dot{\mathbf{x}}-\dot{\bar{\mathbf{x}}})\Big|_{(\bar{\mathbf{x}}^{(s)}_m,\mathbf{x}^{(\alpha)}_m,t_m)}\\
&= \mathbf{n}^T_{\partial\Omega_{ij}}(\bar{\mathbf{x}})\bullet[\mathbf{F}(\mathbf{x},t,\mathbf{p})-\mathbf{f}(\bar{\mathbf{x}},\boldsymbol{\mu})]\Big|_{(\bar{\mathbf{x}}^{(s)}_m,\mathbf{x}^{(\alpha)}_m,t_m)}.
\end{aligned}
\tag{3.14}
$$

$$
G_{\partial\Omega_{ij}}(\bar{\mathbf{x}}^{(s)}_m,\mathbf{x}^{(\alpha)}_m,t_m,\mathbf{p}) = \mathbf{n}^T_{\partial\Omega_{ij}}(\bar{\mathbf{x}}_m)\bullet\dot{\mathbf{x}}_m = \mathbf{n}^T_{\partial\Omega_{ij}}(\bar{\mathbf{x}}^{(s)}_m)\bullet\mathbf{g}(\mathbf{x}^{(\alpha)}_m,t_m,\boldsymbol{\pi}). \tag{3.15}
$$

The superscript “0” in Eq.(3.14) has been dropped in Eq.(3.15). From Eq.(3.15), Theorems 3.1 and 3.2 can be expressed through G-function. Using Eq.(3.13), the $(2k)^{th}$-order *global* flow from the domain Ω_i to Ω_j is defined.

DEFINITION 3.5. For a dynamical system in Eq.(3.1), there is an integrable dynamical system in Eq.(3.2) possessing a separatrix manifold $\partial\Omega_{ij}$ with the first integral manifold $F_{ij}(\bar{\mathbf{x}}^{(s)},\boldsymbol{\mu}) = E_s$, and the separatrix partitions the phase space into two domains Ω_α ($\alpha = i, j$). The dynamical system of Eq.(3.1) in the two domains has different flows $\mathbf{x}^{(\alpha)}(t)$ ($\alpha = i, j$). Choose $\bar{\mathbf{x}}^{(s)}(t_m) \equiv \mathbf{x}_m \in \partial\Omega_{ij}$ at time t_m. Suppose $\mathbf{x}^{(i)}(t_{m-}) = \mathbf{x}_m = \mathbf{x}^{(j)}(t_{m+})$. For an arbitrarily small $\varepsilon > 0$, there are two time intervals $[t_{m-\varepsilon}, t_m)$ and $(t_m, t_{m+\varepsilon}]$. A flow $\mathbf{x}^{(\alpha)}(t)$ is $C^r_{[t_{m-\varepsilon},t_m)}$ and $C^r_{(t_m,t_{m+\varepsilon}]}$-continuous ($r \geq 2l+1$) for time *t*. The perturbed flow $\mathbf{x}^{(i)}(t)$ with the initial set ${}^{(G_j)}_{t_0}D_i$ in Ω_i is the $(2k)^{th}$-order *global* flow from the domain Ω_i to Ω_j if the three conditions hold:

$$G^{(r)}_{\partial\Omega_{ij}}(\overline{\mathbf{x}}^{(s)}_{t_m},\mathbf{x}^{(\alpha)}_{t_m},t_m,\mathbf{p})=0 \text{ for } (r=0,1,\cdots,2k-1), \tag{3.16}$$

$$G^{(2k)}_{\partial\Omega_{ij}}(\overline{\mathbf{x}}^{(s)}_{t_m},\mathbf{x}^{(\alpha)}_{t_m},t_m,\mathbf{p})\neq 0, \tag{3.17}$$

$$\left.\begin{array}{ll}\text{either} & \left.\begin{array}{l}\mathbf{n}^T_{\partial\Omega_{ij}}(\overline{\mathbf{x}}^{(i)}_{m-\varepsilon})\bullet\left[\overline{\mathbf{x}}^{(s)}_{m-\varepsilon}-\mathbf{x}^{(i)}_{m-\varepsilon}\right]>0\\ \mathbf{n}^T_{\partial\Omega_{ij}}(\overline{\mathbf{x}}^{(s)}_{m+\varepsilon})\bullet\left[\mathbf{x}^{(j)}_{m+\varepsilon}-\overline{\mathbf{x}}^{(s)}_{m+\varepsilon}\right]>0\end{array}\right\}\text{ for }\mathbf{n}_{\partial\Omega_{ij}}\to\Omega_j\\ \text{or} & \left.\begin{array}{l}\mathbf{n}^T_{\partial\Omega_{ij}}(\overline{\mathbf{x}}^{(i)}_{m-\varepsilon})\bullet\left[\overline{\mathbf{x}}^{(s)}_{m-\varepsilon}-\mathbf{x}^{(i)}_{m-\varepsilon}\right]<0\\ \mathbf{n}^T_{\partial\Omega_{ij}}(\overline{\mathbf{x}}^{(s)}_{m+\varepsilon})\bullet\left[\mathbf{x}^{(j)}_{m+\varepsilon}-\overline{\mathbf{x}}^{(s)}_{m+\varepsilon}\right]<0\end{array}\right\}\text{ for }\mathbf{n}_{\partial\Omega_{ij}}\to\Omega_i.\end{array}\right\} \tag{3.18}$$

THEOREM 3.3. *For a dynamical system in Eq.*(3.1)*, its integrable dynamical system in Eq.*(3.2) *possesses a separatrix surface $\partial\Omega_{ij}$ with the first integral manifold $F_{ij}(\overline{\mathbf{x}}^{(s)},\boldsymbol{\mu})=E_s$, and the separatrix surface partitions phase space into two domains Ω_α ($\alpha=i,j$). The dynamical system of Eq.*(3.1) *in the two domains has different flows $\mathbf{x}^{(\alpha)}(t)$. Choose $\overline{\mathbf{x}}^{(s)}(t_m)\equiv\mathbf{x}_m\in\partial\Omega_{ij}$ at time t_m. Suppose $\mathbf{x}^{(i)}(t_{m-})$ $=\mathbf{x}_m=\mathbf{x}^{(j)}(t_{m+})$. For an arbitrarily small $\varepsilon>0$, there are two time intervals $[t_{m-\varepsilon},t_m)$ and $(t_m,t_{m+\varepsilon}]$. The flow $\mathbf{x}^{(\alpha)}(t)$ is $C^r_{[t_{m-\varepsilon},t_m)}$ and $C^r_{(t_m,t_{m+\varepsilon}]}$-continuous ($r\geq$ $2k+1$) for time t with $\| d^{r+1}\mathbf{x}^{(\alpha)}/dt^{r+1}\|<\infty$. The flow $\mathbf{x}^{(i)}(t)$ with an initial set ${}^{(G_j)}_{t_0}D_i$ in Ω_i is the $(2k)^{th}$-order global flow from domain Ω_i to Ω_j iff*

$$G^{(r)}_{\partial\Omega_{ij}}(\overline{\mathbf{x}}^{(s)}_m,\mathbf{x}^{(\alpha)}_m,t_m,\mathbf{p})=0 \quad \text{for } (r=0,1,\cdots,2k-1), \tag{3.19}$$

$$\left.\begin{array}{lll}\text{either} & G^{(2k)}_{\partial\Omega_{ij}}(\overline{\mathbf{x}}^{(s)}_m,\mathbf{x}_m,t_m,\mathbf{p})>0 & \text{for } \mathbf{n}_{\partial\Omega_{ij}}\to\Omega_j\\ \text{or} & G^{(2k)}_{\partial\Omega_{ij}}(\overline{\mathbf{x}}^{(s)}_m,\mathbf{x}_m,t_m,\mathbf{p})<0 & \text{for } \mathbf{n}_{\partial\Omega_{ij}}\to\Omega_i.\end{array}\right\} \tag{3.20}$$

PROOF. Equation (3.19) is identical to Eq.(3.16), thus the condition in Eq.(3.16) is satisfied, and vice versa. Suppose $\mathbf{x}^{(\alpha)}(t_{m\pm})=\mathbf{x}_m$ ($\alpha\in\{i,j\}$) and $\mathbf{x}^{(\alpha)}(t)$ are $C^r_{[t_{m-\varepsilon},t_{m+\varepsilon}]}$-continuous (2$k$+1) for time t and $\| d^r\mathbf{x}^{(\alpha)}/dt^r\|<\infty$ ($\alpha\in\{i,j\}$). For a $\in[t_{m-\varepsilon},t_m)$ or $a\in(t_m,t_{m+\varepsilon}]$, the Taylor series expansion of $\mathbf{x}^{(\alpha)}(t_{m\pm\varepsilon})$ to $\mathbf{x}^{(\alpha)}(a)$ up to the ε^{2k+1}-order term gives

$$\mathbf{x}_{m\pm\varepsilon}^{(\alpha)} \equiv \mathbf{x}^{(\alpha)}\left(t_{m\pm} \pm \varepsilon\right) = \mathbf{x}^{(\alpha)}(a) + \sum_{r=1}^{2k} \tfrac{d^r}{dt^r}\mathbf{x}^{(\alpha)}(a)\left(t_{m\pm} \pm \varepsilon - a\right)^r$$
$$+ \tfrac{d^{2k+1}}{dt^{2k+1}}\mathbf{x}^{(\alpha)}(a)\left(t_{m\pm} \pm \varepsilon - a\right)^{2k+1} + o(\left(t_{m\pm} \pm \varepsilon - a\right)^{2k+1}),$$

As $a \to t_{m\pm}$, taking the limit of the foregoing equation leads to

$$\mathbf{x}_{m\pm\varepsilon}^{(\alpha)} \equiv \mathbf{x}^{(\alpha)}(t_m \pm \varepsilon) = \mathbf{x}_{m\pm}^{(\alpha)} + \sum_{r=1}^{2k} \tfrac{d^r}{dt^r}\mathbf{x}_{m\pm}^{(\alpha)}(\pm\varepsilon)^r + \tfrac{d^{2k+1}}{dt^{2k+1}}\mathbf{x}_{m\pm}^{(\alpha)}(\pm\varepsilon)^{2k+1} + o((\pm\varepsilon)^{2k+1}).$$

In a similar fashion, we have

$$\overline{\mathbf{x}}_{m\pm\varepsilon}^{(s)} \equiv \overline{\mathbf{x}}^{(s)}\left(t_m \pm \varepsilon\right) = \overline{\mathbf{x}}_{m\pm}^{(s)} + \sum_{r=1}^{2k} \tfrac{d^r}{dt^r}\overline{\mathbf{x}}_{m\pm}^{(s)}(\pm\varepsilon)^r + \tfrac{d^{2k+1}}{dt^{2k+1}}\overline{\mathbf{x}}_{m\pm}^{(s)}(\pm\varepsilon)^{2k+1} + o\left((\pm\varepsilon)^{2k+1}\right),$$

$$\mathbf{n}_{\partial\Omega_{ij}}\left(\overline{\mathbf{x}}_{m\pm\varepsilon}^{(s)}\right) = \mathbf{n}_{\partial\Omega_{ij}}\left(\overline{\mathbf{x}}_m^{(s)}\right) + \sum_{r=1}^{2k} D_{\overline{\mathbf{x}}}^r \mathbf{n}_{\partial\Omega_{ij}}\left(\overline{\mathbf{x}}_m^{(s)}\right)(\pm\varepsilon)^r$$
$$+ D_{\overline{\mathbf{x}}}^{2k+1}\mathbf{n}_{\partial\Omega_{ij}}\left(\overline{\mathbf{x}}_m^{(s)}\right)(\pm\varepsilon)^{2k+1} + o\left((\pm\varepsilon)^{2k+1}\right).$$

The ignorance of the ε^{2k+2} and high order terms, the deformation of the above equation and left multiplication of $\mathbf{n}_{\partial\Omega_{ij}}$ gives

$$\mathbf{n}_{\partial\Omega_{ij}}^T(\overline{\mathbf{x}}_{m-\varepsilon}^{(s)})\bullet\left[\mathbf{x}_{m-\varepsilon}^{(i)} - \overline{\mathbf{x}}_{m-\varepsilon}^{(s)}\right] = \mathbf{n}_{\partial\Omega_{ij}}^T(\overline{\mathbf{x}}_m^{(s)})\bullet\left[\mathbf{x}_{m-}^{(i)} - \overline{\mathbf{x}}_m^{(s)}\right]$$
$$+\sum_{r=1}^{2k}(-\varepsilon)^r G_{\partial\Omega_{ij}}^{(r-1)}(\overline{\mathbf{x}}_m^{(s)}, \mathbf{x}_m^{(i)}, t_m, \mathbf{p}) - \varepsilon^{2k+1}G_{\partial\Omega_{ij}}^{(2k)}(\overline{\mathbf{x}}_m^{(s)}, \mathbf{x}_m^{(i)}, t_m, \mathbf{p}),$$

$$\mathbf{n}_{\partial\Omega_{ij}}^T(\overline{\mathbf{x}}_{m+\varepsilon}^{(s)})\bullet\left[\mathbf{x}_{m+\varepsilon}^{(j)} - \overline{\mathbf{x}}_{m+\varepsilon}^{(s)}\right] = \mathbf{n}_{\partial\Omega_{ij}}^T(\overline{\mathbf{x}}_m^{(s)})\bullet\left[\mathbf{x}_{m+}^{(j)} - \overline{\mathbf{x}}_m^{(s)}\right]$$
$$+\sum_{r=1}^{2k}\varepsilon^r G_{\partial\Omega_{ij}}^{(r-1)}(\overline{\mathbf{x}}_m^{(s)}, \mathbf{x}_m^{(j)}, t_m, \mathbf{p}) + \varepsilon^{2k+1}G_{\partial\Omega_{ij}}^{(2k)}(\overline{\mathbf{x}}_m^{(s)}, \mathbf{x}_m^{(j)}, t_m, \mathbf{p}),$$

Because of $\mathbf{x}_{m\pm}^{(\alpha)} = \overline{\mathbf{x}}_m^{(s)}$ and $G_{\partial\Omega_{ij}}^{(r)}(\overline{\mathbf{x}}_m^{(s)}, \mathbf{x}_m^{(\alpha)}, t_m, \mathbf{p}) = 0$ for $r = 0, 1, \cdots, 2k-1$, the foregoing equation becomes

$$\mathbf{n}_{\partial\Omega_{ij}}^T(\overline{\mathbf{x}}_{m-\varepsilon}^{(s)})\bullet\left[\overline{\mathbf{x}}_{m-\varepsilon}^{(s)} - \mathbf{x}_{m-\varepsilon}^{(i)}\right] = \varepsilon^{2k+1}G_{\partial\Omega_{ij}}^{(2k)}(\overline{\mathbf{x}}_m^{(s)}, \mathbf{x}_m^{(i)}, t_m, \mathbf{p}),$$
$$\mathbf{n}_{\partial\Omega_{ij}}^T(\overline{\mathbf{x}}_{m+\varepsilon}^{(s)})\bullet\left[\mathbf{x}_{m+\varepsilon}^{(j)} - \overline{\mathbf{x}}_{m+\varepsilon}^{(s)}\right] = \varepsilon^{2k+1}G_{\partial\Omega_{ij}}^{(2k)}(\overline{\mathbf{x}}_m^{(s)}, \mathbf{x}_m^{(j)}, t_m, \mathbf{p}).$$

With Eq.(3.20), the forgoing equation gives Eq.(3.18). However, with Eq.(3.18), the foregoing equation gives Eq.(3.20). The proof is completed. ■

3.4. Global tangency

To investigate the global transversality from a local flow to global flow, the global tangency of a flow to the separatrix is very crucial. The mathematical description of the global tangency is given through the following definition, i.e.,

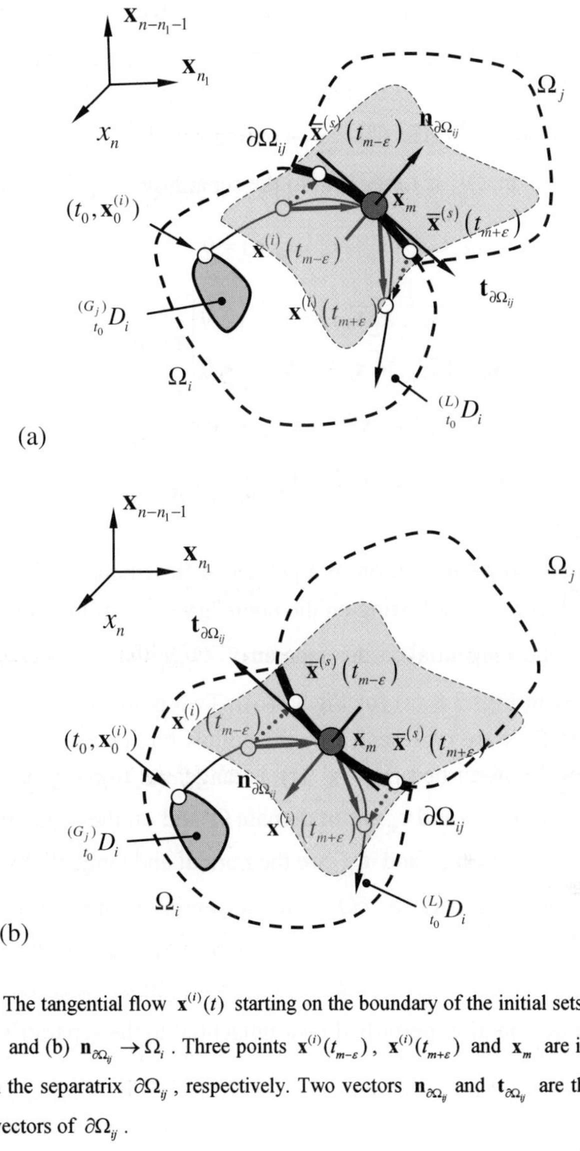

Figure 3.8. The tangential flow $\mathbf{x}^{(i)}(t)$ starting on the boundary of the initial sets ${}^{(G_j)}_{t_0}D_i$ for (a) $\mathbf{n}_{\partial\Omega_{ij}} \to \Omega_j$ and (b) $\mathbf{n}_{\partial\Omega_{ij}} \to \Omega_i$. Three points $\mathbf{x}^{(i)}(t_{m-\varepsilon})$, $\mathbf{x}^{(i)}(t_{m+\varepsilon})$ and $\mathbf{x}_m$ are in the domains Ω_i and on the separatrix $\partial\Omega_{ij}$, respectively. Two vectors $\mathbf{n}_{\partial\Omega_{ij}}$ and $\mathbf{t}_{\partial\Omega_{ij}}$ are the normal and tangential vectors of $\partial\Omega_{ij}$.

DEFINITION 3.6. For a dynamical system in Eq.(3.1), its integrable dynamical system in Eq.(3.2) possessing a separatrix surface $\partial\Omega_{ij}$ with the first integral manifold $F_{ij}(\overline{\mathbf{x}}^{(s)},\boldsymbol{\mu}) = E_s$, and the separatrix surface partitions the phase space into two domains Ω_α ($\alpha = i, j$). The dynamical system of Eq.(3.1) in the two domains has different flows $\mathbf{x}^{(\alpha)}(t)$ ($\alpha = i, j$). Choose $\overline{\mathbf{x}}^{(s)}(t_m) \equiv \mathbf{x}_m \in \partial\Omega_{ij}$ at time t_m. For an arbitrarily small $\varepsilon > 0$, there is a time interval $[t_{m-\varepsilon}, t_{m+\varepsilon}]$. Suppose $\mathbf{x}^{(\alpha)}(t_{m\pm}) = \mathbf{x}_m$, a flow $\mathbf{x}^{(\alpha)}(t)$ is $C^r_{[t_{m-\varepsilon}, t_{m+\varepsilon}]}$-continuous ($r \geq 2$) for time *t*. The perturbed flow $\mathbf{x}^{(\alpha)}(t)$ in Ω_α is *tangential* to the separatrix $\partial\Omega_{ij}$ if

$$G^{(0)}_{\partial\Omega_{ij}}(\overline{\mathbf{x}}^{(s)}_m, \mathbf{x}^{(\alpha)}_m, t_m, \mathbf{p}) = 0; \tag{3.21}$$

$$\left.\begin{array}{ll} \text{either} & \left.\begin{array}{l} \mathbf{n}^T_{\partial\Omega_{ij}}(\overline{\mathbf{x}}^{(s)}_{m-\varepsilon})\bullet\left[\overline{\mathbf{x}}^{(s)}_{m-\varepsilon} - \mathbf{x}^{(\alpha)}_{m-\varepsilon}\right] > 0 \\ \mathbf{n}^T_{\partial\Omega_{ij}}(\overline{\mathbf{x}}^{(s)}_{m+\varepsilon})\bullet\left[\mathbf{x}^{(\alpha)}_{m+\varepsilon} - \overline{\mathbf{x}}^{(s)}_{m+\varepsilon}\right] < 0 \end{array}\right\} \text{for } \mathbf{n}_{\partial\Omega_{ij}} \to \Omega_\beta \\ \text{or} & \left.\begin{array}{l} \mathbf{n}^T_{\partial\Omega_{ij}}(\overline{\mathbf{x}}^{(s)}_{m-\varepsilon})\bullet\left[\overline{\mathbf{x}}^{(s)}_{m-\varepsilon} - \mathbf{x}^{(\alpha)}_{m-\varepsilon}\right] < 0 \\ \mathbf{n}^T_{\partial\Omega_{ij}}(\overline{\mathbf{x}}^{(s)}_{m+\varepsilon})\bullet\left[\mathbf{x}^{(\alpha)}_{m+\varepsilon} - \overline{\mathbf{x}}^{(s)}_{m+\varepsilon}\right] > 0 \end{array}\right\} \text{for } \mathbf{n}_{\partial\Omega_{ij}} \to \Omega_\alpha. \end{array}\right\} \tag{3.22}$$

To explain the above definition of a perturbed flow tangential to the separatrix, a perturbed flow $\mathbf{x}^{(i)}(t)$ starting on the boundary ${}_{t_0}\Gamma_{ij}$ of the initial set ${}^{(G_j)}_{t_0}D_i$ in domain Ω_i, and tangential to the separatrix $\partial\Omega_{ij}$ with $\mathbf{n}_{\partial\Omega_{ij}} \to \Omega_j$ (or $\mathbf{n}_{\partial\Omega_{ij}} \to \Omega_i$), is sketched in Figs.3.8 (a) (or Fig.3.8(b)). Two points ($\mathbf{x}^{(i)}_{m\pm\varepsilon} = \mathbf{x}^{(i)}(t_m \pm \varepsilon)$) on the perturbed flow are represented by the hollow circles, and they are located before and after the perturbed flow $\mathbf{x}^{(i)}(t)$ is tangential to the separatrix. Three points $\mathbf{x}^{(i)}(t_{m-\varepsilon})$, $\mathbf{x}^{(i)}(t_{m+\varepsilon})$ and $\mathbf{x}_m$ are in domain Ω_i and on the separatrix $\partial\Omega_{ij}$, respectively. Two vectors $\mathbf{n}_{\partial\Omega_{ij}}$ and $\mathbf{t}_{\partial\Omega_{ij}}$ are the normal and tangential vectors of an integrable flow on the separatrix $\partial\Omega_{ij}$. The tangential point $\mathbf{x}_m$ on the separatrix $\partial\Omega_{ij}$ is depicted by a large circle. The dotted arrows denote the flow differences $\overline{\mathbf{x}}^{(s)}(t_{m-\varepsilon}) - \mathbf{x}^{(i)}(t_{m-\varepsilon})$ and $\mathbf{x}^{(i)}(t_{m+\varepsilon}) - \overline{\mathbf{x}}^{(s)}(t_{m+\varepsilon})$. The solid arrows represent such differences before and after perturbed flow tangential to the separatrix.

THEOREM 3.4. *For a dynamical system in Eq.*(3.1)*, its integrable dynamical system in Eq.*(3.2) *possesses a separatrix surface* $\partial\Omega_{ij}$ *with the first integral manifold* $F_{ij}(\overline{\mathbf{x}}^{(s)}, \boldsymbol{\mu}) = E_s$*, and the separatrix surface partitions the phase space into two domains* Ω_α ($\alpha = i, j$)*. The dynamical system of Eq.*(3.1) *in the two domains has different flows* $\mathbf{x}^{(\alpha)}(t)$ ($\alpha = i, j$)*. For* $\overline{\mathbf{x}}^{(s)}(t_m) \equiv \mathbf{x}_m \in \partial\Omega_{ij}$ *at time* t_m*, suppose* $\mathbf{x}^{(\alpha)}(t_{m\pm}) = \mathbf{x}_m$*. For an arbitrarily small* $\varepsilon > 0$*, there is a time interval* $[t_{m-\varepsilon}$,

$t_{m+\varepsilon}]$. *The flow* $\mathbf{x}^{(\alpha)}(t)$ *is* $C^r_{[t_{m-\varepsilon},t_{m+\varepsilon}]}$*-continuous* ($r \geq 2$) *for time t. A flow* $\mathbf{x}^{(\alpha)}(t)$ *in* Ω_α *is tangential to the separatrix* $\partial\Omega_{ij}$ *iff*

$$G^{(0)}\left(\overline{\mathbf{x}}_m, \mathbf{x}_m^{(\alpha)}, t_m, \mathbf{p}\right) = 0; \tag{3.23}$$

$$G^{(0)}\left(\overline{\mathbf{x}}_{m-\varepsilon}, \mathbf{x}_{m-\varepsilon}^{(\alpha)}, t_{m-\varepsilon}, \mathbf{p}\right) \times G^{(0)}\left(\overline{\mathbf{x}}_{m+\varepsilon}, \mathbf{x}_{m+\varepsilon}^{(\alpha)}, t_{m+\varepsilon}, \mathbf{p}\right) < 0, \tag{3.24}$$

where

$$\begin{aligned} &G^{(0)}(\overline{\mathbf{x}}_{m\pm\varepsilon}, \mathbf{x}_{m\pm\varepsilon}^{(\alpha)}, t_{m\pm\varepsilon}, \mathbf{p}) \\ &= D_{\overline{\mathbf{x}}_t}\mathbf{n}^T_{\overline{\mathbf{x}}_t}(\overline{\mathbf{x}}_{m\pm\varepsilon})\cdot(\mathbf{x}_{m\pm\varepsilon}^{(\alpha)} - \overline{\mathbf{x}}_{m\pm\varepsilon}) + \mathbf{n}^T_{\overline{\mathbf{x}}_t}(\overline{\mathbf{x}}_{m\pm\varepsilon})\cdot\left[\dot{\mathbf{x}}_{m\pm\varepsilon}^{(\alpha)} - \dot{\overline{\mathbf{x}}}_{m\pm\varepsilon}\right] \\ &= \left\{\frac{\partial \mathbf{n}^T_{\overline{\mathbf{x}}_t}}{\partial \overline{\mathbf{x}}_t}\mathbf{f}(\overline{\mathbf{x}}_t, \boldsymbol{\mu})\cdot(\mathbf{x}_t - \overline{\mathbf{x}}_t) + \mathbf{n}^T_{\overline{\mathbf{x}}_t}\cdot[\mathbf{F}(\mathbf{x}_t, t, \mathbf{p}) - \mathbf{f}(\overline{\mathbf{x}}_t, \boldsymbol{\mu})]\right\}\Big|_{(\overline{\mathbf{x}}_{m\pm\varepsilon}, \mathbf{x}_{m\pm\varepsilon}^{(\alpha)}, t_{m\pm\varepsilon}, \mathbf{p})}. \end{aligned} \tag{3.25}$$

PROOF. Since Eq.(3.23) is identical to Eq.(3.21), the condition in Eq.(3.21) is satisfied. The Taylor series expansions of $\mathbf{x}_{t_{m\pm}}^{(\alpha)}$, $\overline{\mathbf{x}}_{t_{m\pm}}^{(s)}$ and $\mathbf{n}_{\partial\Omega_{ij}}(\overline{\mathbf{x}}_m^{(s)})$ give

$$\begin{aligned} \mathbf{x}_{t_{m\pm}}^{(\alpha)} &= \mathbf{x}_{t_{m\pm\varepsilon}}^{(\alpha)} \mp \varepsilon\dot{\mathbf{x}}_{t_{m\pm\varepsilon}}^{(\alpha)} + o(\varepsilon), \\ \overline{\mathbf{x}}_{t_{m\pm}}^{(s)} &= \overline{\mathbf{x}}_{t_{m\pm\varepsilon}}^{(s)} \mp \varepsilon\dot{\overline{\mathbf{x}}}_{t_{m\pm\varepsilon}}^{(s)} + o(\varepsilon), \\ \mathbf{n}_{\partial\Omega_{ij}}(\overline{\mathbf{x}}_m^{(s)}) &\equiv \mathbf{n}_{\partial\Omega_{ij}}(\overline{\mathbf{x}}_{m\pm\varepsilon}^{(s)}) \mp \varepsilon D_{\overline{\mathbf{x}}}\mathbf{n}_{\partial\Omega_{ij}}(\overline{\mathbf{x}}_{m\pm\varepsilon}^{(s)}) + o(\varepsilon). \end{aligned}$$

For $0 < \varepsilon << 1$, the higher order terms in the above equation are ignored. Thus,

$$\begin{aligned} \mathbf{n}^T_{\partial\Omega_{ij}}(\overline{\mathbf{x}}_m)\bullet\left[\mathbf{x}_{m-}^{(\alpha)} - \overline{\mathbf{x}}_m^{(s)}\right] &= \mathbf{n}^T_{\partial\Omega_{ij}}(\overline{\mathbf{x}}_{m-\varepsilon})\bullet\left[\mathbf{x}_{m-\varepsilon}^{(\alpha)} - \overline{\mathbf{x}}_{m-\varepsilon}^{(s)}\right] \\ &\quad +\varepsilon G^{(0)}\left(\overline{\mathbf{x}}_{m-\varepsilon}, \mathbf{x}_{m-\varepsilon}^{(\alpha)}, t_{m-\varepsilon}, \mathbf{p}\right), \\ \mathbf{n}^T_{\partial\Omega_{ij}}(\overline{\mathbf{x}}_m)\bullet\left[\mathbf{x}_{m+}^{(\alpha)} - \overline{\mathbf{x}}_m^{(s)}\right] &= \mathbf{n}^T_{\partial\Omega_{ij}}(\overline{\mathbf{x}}_{m+\varepsilon})\bullet\left[\mathbf{x}_{m+\varepsilon}^{(\alpha)} - \overline{\mathbf{x}}_{m+\varepsilon}^{(s)}\right] \\ &\quad -\varepsilon G^{(0)}\left(\overline{\mathbf{x}}_{m+\varepsilon}, \mathbf{x}_{m+\varepsilon}^{(\alpha)}, t_{m+\varepsilon}, \mathbf{p}\right). \end{aligned}$$

Because of $\overline{\mathbf{x}}_m^{(s)} = \mathbf{x}_{m-}^{(\alpha)}$ and $\overline{\mathbf{x}}_m^{(s)} = \mathbf{x}_m^{(\alpha)}$, we have

$$\begin{aligned} \mathbf{n}^T_{\partial\Omega_{ij}}(\overline{\mathbf{x}}_{m-\varepsilon})\bullet\left[\overline{\mathbf{x}}_{m-\varepsilon}^{(s)} - \mathbf{x}_{m-\varepsilon}^{(\alpha)}\right] &= \varepsilon G^{(0)}\left(\overline{\mathbf{x}}_{m-\varepsilon}, \mathbf{x}_{m-\varepsilon}^{(\alpha)}, t_{m-\varepsilon}, \mathbf{p}\right), \\ \mathbf{n}^T_{\partial\Omega_{ij}}(\overline{\mathbf{x}}_{m+\varepsilon})\bullet\left[\mathbf{x}_{m+\varepsilon}^{(\alpha)} - \overline{\mathbf{x}}_{m+\varepsilon}^{(s)}\right] &= \varepsilon G^{(0)}\left(\overline{\mathbf{x}}_{m+\varepsilon}, \mathbf{x}_{m+\varepsilon}^{(\alpha)}, t_{m+\varepsilon}, \mathbf{p}\right). \end{aligned}$$

From Eq.(3.22), the above equations give Eq.(3.24). Using Eq.(3.24), the above equation gives Eq.(3.22). This theorem is proved. ■

THEOREM 3.5. *For a dynamical system in Eq.*(3.1)*, its integrable dynamical system in Eq.*(3.2) *possesses a separatrix surface* $\partial\Omega_{ij}$ *with the first integral manifold* $F_{ij}(\overline{\mathbf{x}}^{(s)},\boldsymbol{\mu}) = E_s$*, and the separatrix surface partitions the phase space into two domains* Ω_α ($\alpha = i, j$)*. The dynamical system of Eq.*(3.1) *in the two domains has different flows* $\mathbf{x}^{(\alpha)}(t)$ ($\alpha = i, j$)*. Choose* $\overline{\mathbf{x}}^{(s)}(t_m) \equiv \mathbf{x}_m \in \partial\Omega_{ij}$ *at time* t_m*. For an arbitrarily small* $\varepsilon > 0$*, there is a time interval* $[t_{m-\varepsilon}, t_{m+\varepsilon}]$*. Suppose* $\mathbf{x}^{(\alpha)}(t_{m\pm}) = \mathbf{x}_m$*. The flow* $\mathbf{x}^{(\alpha)}(t)$ *is* $C^r_{[t_{m-\varepsilon}, t_{m+\varepsilon}]}$*-continuous* ($r \geq 3$) *and* $\| d^{r+1}\mathbf{x}^{(\alpha)}/dt^{r+1} \| < \infty$ *for time t. The flow* $\mathbf{x}^{(\alpha)}(t)$ *in* Ω_α *is tangential to the separatrix* $\partial\Omega_{ij}$ *iff*

$$G^{(0)}_{\partial\Omega_{ij}}(\overline{\mathbf{x}}^{(s)}_m, \mathbf{x}^{(\alpha)}_m, t_m, \mathbf{p}) = 0; \tag{3.26}$$

$$\begin{aligned} &\text{either} \quad G^{(1)}_{\partial\Omega_{ij}}(\overline{\mathbf{x}}^{(s)}_{t_m}, \mathbf{x}^{(\alpha)}_{t_m}, t_m, \mathbf{p}) < 0 \quad \text{for } \mathbf{n}_{\partial\Omega_{ij}} \to \Omega_\beta \\ &\text{or} \quad G^{(1)}_{\partial\Omega_{ij}}(\overline{\mathbf{x}}^{(s)}_{t_m}, \mathbf{x}^{(\alpha)}_{t_m}, t_m, \mathbf{p}) > 0 \quad \text{for } \mathbf{n}_{\partial\Omega_{ij}} \to \Omega_\alpha. \end{aligned} \tag{3.27}$$

with $\beta \in \{i, j\}, \beta \neq \alpha$.

PROOF. Equation (3.26) is identical to Eq.(3.21), thus the condition in Eq.(3.21) is satisfied, and vice versa. Suppose $\mathbf{x}^{(\alpha)}(t_{m\pm}) = \mathbf{x}_m$ ($\alpha \in \{i, j\}$) and $\mathbf{x}^{(\alpha)}(t)$ are $C^r_{[t_{m-\varepsilon}, t_{m+\varepsilon}]}$ -continuous ($r \geq 3$) for time t and $\| d^r\mathbf{x}^{(\alpha)}/dt^r \| < \infty$ ($\alpha \in \{i, j\}$). For $a \in [t_{m-\varepsilon}, t_m)$ or $a \in (t_m, t_{m+\varepsilon}]$, the Taylor series expansion of $\mathbf{x}^{(\alpha)}(t_{m\pm\varepsilon})$ to $\mathbf{x}^{(\alpha)}(a)$ up to the third-order term gives

$$\begin{aligned} \mathbf{x}^{(\alpha)}_{m\pm\varepsilon} &\equiv \mathbf{x}^{(\alpha)}(t_{m\pm} \pm \varepsilon) = \mathbf{x}^{(\alpha)}(a) + \dot{\mathbf{x}}^{(\alpha)}(a)(t_{m\pm} \pm \varepsilon - a) \\ &+ \ddot{\mathbf{x}}^{(\alpha)}(a)(t_{m\pm} \pm \varepsilon - a)^2 + o((t_{m\pm} \pm \varepsilon - a)^2), \end{aligned}$$

As $a \to t_{m\pm}$, taking the limit of the foregoing equation leads to

$$\mathbf{x}^{(\alpha)}_{m\pm\varepsilon} \equiv \mathbf{x}^{(\alpha)}(t_m \pm \varepsilon) = \mathbf{x}^{(\alpha)}_{m\pm} \pm \dot{\mathbf{x}}^{(\alpha)}_{m\pm}\varepsilon + \ddot{\mathbf{x}}^{(\alpha)}_{m\pm}\varepsilon^2 + o(\varepsilon^2).$$

In a similar fashion, we have

$$\begin{aligned} &\overline{\mathbf{x}}^{(s)}_{m\pm\varepsilon} \equiv \overline{\mathbf{x}}^{(s)}(t_m \pm \varepsilon) = \overline{\mathbf{x}}^{(s)}_{m\pm} \pm \dot{\overline{\mathbf{x}}}^{(s)}_{m\pm}\varepsilon + \ddot{\overline{\mathbf{x}}}^{(s)}_{m\pm}\varepsilon^2 + o(\varepsilon^2), \\ &\mathbf{n}_{\partial\Omega_{ij}}(\overline{\mathbf{x}}^{(s)}_{m\pm\varepsilon}) \equiv \mathbf{n}_{\partial\Omega_{ij}}(\overline{\mathbf{x}}^{(s)}_m) \pm D_{\overline{\mathbf{x}}}\mathbf{n}_{\partial\Omega_{ij}}(\overline{\mathbf{x}}^{(s)}_m)\varepsilon + D^2_{\overline{\mathbf{x}}}\mathbf{n}_{\partial\Omega_{ij}}(\overline{\mathbf{x}}^{(s)}_m)\varepsilon^2 + o(\varepsilon^2). \end{aligned}$$

The ignorance of the ε^3 and high order terms, the deformation of the above equation and left multiplication of $\mathbf{n}_{\partial\Omega_{ij}}$ gives

$$\mathbf{n}^T_{\partial\Omega_{ij}}(\overline{\mathbf{x}}^{(s)}_{m-\varepsilon})\bullet\left[\mathbf{x}^{(\alpha)}_{m-\varepsilon}-\overline{\mathbf{x}}^{(s)}_{m-\varepsilon}\right]=\mathbf{n}^T_{\partial\Omega_{ij}}(\overline{\mathbf{x}}^{(s)}_{m})\bullet\left[\mathbf{x}^{(\alpha)}_{m-}-\overline{\mathbf{x}}^{(s)}_{m}\right]-\varepsilon G^{(0)}_{\partial\Omega_{ij}}(\overline{\mathbf{x}}^{(s)}_{m},\mathbf{x}^{(\alpha)}_{m},t_m,\mathbf{p})$$
$$+\varepsilon^2 G^{(1)}_{\partial\Omega_{ij}}(\overline{\mathbf{x}}^{(s)}_{m},\mathbf{x}^{(\alpha)}_{m},t_m,\mathbf{p}),$$
$$\mathbf{n}^T_{\partial\Omega_{ij}}(\overline{\mathbf{x}}^{(s)}_{m+\varepsilon})\bullet\left[\mathbf{x}^{(\alpha)}_{m+\varepsilon}-\overline{\mathbf{x}}^{(s)}_{m+\varepsilon}\right]=\mathbf{n}^T_{\partial\Omega_{ij}}(\overline{\mathbf{x}}^{(s)}_{m})\bullet\left[\mathbf{x}^{(\alpha)}_{m+}-\overline{\mathbf{x}}^{(s)}_{m}\right]+\varepsilon G^{(0)}_{\partial\Omega_{ij}}(\overline{\mathbf{x}}^{(s)}_{m},\mathbf{x}^{(\alpha)}_{m},t_m,\mathbf{p})$$
$$+\varepsilon^2 G^{(1)}_{\partial\Omega_{ij}}(\overline{\mathbf{x}}^{(s)}_{m},\mathbf{x}^{(\alpha)}_{m},t_m,\mathbf{p}).$$

Due to $\mathbf{x}^{(\alpha)}_{m\pm}=\overline{\mathbf{x}}^{(s)}_{m}$ and $G^{(0)}_{\partial\Omega_{ij}}(\overline{\mathbf{x}}^{(s)}_{m},\mathbf{x}^{(\alpha)}_{m},t_m,\mathbf{p})=0$, the foregoing equation becomes

$$\mathbf{n}^T_{\partial\Omega_{ij}}(\overline{\mathbf{x}}^{(s)}_{m-\varepsilon})\bullet\left[\overline{\mathbf{x}}^{(s)}_{m-\varepsilon}-\mathbf{x}^{(\alpha)}_{m-\varepsilon}\right]=-\varepsilon^2 G^{(1)}_{\partial\Omega_{ij}}(\overline{\mathbf{x}}^{(s)}_{m},\mathbf{x}^{(\alpha)}_{m},t_m,\mathbf{p}),$$
$$\mathbf{n}^T_{\partial\Omega_{ij}}(\overline{\mathbf{x}}^{(s)}_{m+\varepsilon})\bullet\left[\mathbf{x}^{(\alpha)}_{m+\varepsilon}-\overline{\mathbf{x}}^{(s)}_{m+\varepsilon}\right]=\varepsilon^2 G^{(1)}_{\partial\Omega_{ij}}(\overline{\mathbf{x}}^{(s)}_{m},\mathbf{x}^{(\alpha)}_{m},t_m,\mathbf{p}).$$

Using Eq.(3.27), equation (3.21) is obtained. On the other hand, using Eq. (3.21), equation (3.27) is achieved. Therefore, this theorem is proved. ■

DEFINITION 3.7. For a dynamical system in Eq.(3.1), its integrable dynamical system in Eq.(3.2) possesses a separatrix surface $\partial\Omega_{ij}$ with the first integral manifold $F_{ij}(\overline{\mathbf{x}}^{(s)},\boldsymbol{\mu})=E_s$, and the separatrix surface partitions its phase space into two domains Ω_α ($\alpha=i,j$). The dynamical system of Eq.(3.1) in the two domains has different flows $\mathbf{x}^{(\alpha)}(t)$ ($\alpha=i,j$). Choose $\overline{\mathbf{x}}^{(s)}(t_m)\equiv\mathbf{x}_m\in\partial\Omega_{ij}$ at time t_m. Suppose $\mathbf{x}^{(\alpha)}(t_{m\pm})=\mathbf{x}_m$. For an arbitrarily small $\varepsilon>0$, there is a time interval $[t_{m-\varepsilon},t_{m+\varepsilon}]$. The flow $\mathbf{x}^{(\alpha)}(t)$ is $C^r_{[t_{m-\varepsilon},t_{m+\varepsilon}]}$-continuous ($r\geq 2k$) for time t. A flow $\mathbf{x}^{(\alpha)}(t)$ in Ω_i is tangential to the separatrix $\partial\Omega_{ij}$ with the $(2k-1)^{th}$-order if

$$G^{(r)}_{\partial\Omega_{ij}}(\overline{\mathbf{x}}^{(s)}_{m},\mathbf{x}^{(\alpha)}_{m},t_m,\mathbf{p})=0 \text{ for } r=0,1,\cdots,2k-2; \tag{3.28}$$

$$G^{(2k-1)}_{\partial\Omega_{ij}}(\overline{\mathbf{x}}^{(s)}_{m},\mathbf{x}^{(\alpha)}_{m},t_m,\mathbf{p})\neq 0 \tag{3.29}$$

$$\left.\begin{array}{ll}\text{either} & \left.\begin{array}{l}\mathbf{n}^T_{\partial\Omega_{ij}}(\overline{\mathbf{x}}^{(s)}_{m-\varepsilon})\bullet\left[\overline{\mathbf{x}}^{(s)}_{m-\varepsilon}-\mathbf{x}^{(\alpha)}_{m-\varepsilon}\right]>0\\ \mathbf{n}^T_{\partial\Omega_{ij}}(\overline{\mathbf{x}}^{(s)}_{m+\varepsilon})\bullet\left[\mathbf{x}^{(\alpha)}_{m+\varepsilon}-\overline{\mathbf{x}}^{(s)}_{m+\varepsilon}\right]<0\end{array}\right\}\text{for } \mathbf{n}_{\partial\Omega_{ij}}\to\Omega_\beta\\ \text{or} & \left.\begin{array}{l}\mathbf{n}^T_{\partial\Omega_{ij}}(\overline{\mathbf{x}}^{(s)}_{m-\varepsilon})\bullet\left[\overline{\mathbf{x}}^{(s)}_{m-\varepsilon}-\mathbf{x}^{(\alpha)}_{m-\varepsilon}\right]<0\\ \mathbf{n}^T_{\partial\Omega_{ij}}(\overline{\mathbf{x}}^{(s)}_{m+\varepsilon})\bullet\left[\mathbf{x}^{(\alpha)}_{m+\varepsilon}-\overline{\mathbf{x}}^{(s)}_{m+\varepsilon}\right]>0\end{array}\right\}\text{for } \mathbf{n}_{\partial\Omega_{ij}}\to\Omega_\alpha.\end{array}\right\} \tag{3.30}$$

THEOREM 3.6. *For a dynamical system in Eq.*(3.1), *its integrable dynamical system in Eq.*(3.2) *possesses a separatrix surface* $\partial\Omega_{ij}$ *with the first integral mani-*

fold $F_{ij}(\bar{\mathbf{x}}^{(s)},\boldsymbol{\mu})=E_s$*, and the separatrix surface partitions its phase space into two domains* Ω_α ($\alpha=i,j$)*. The dynamical system of Eq.*(3.1) *in two domains has different flows* $\mathbf{x}^{(\alpha)}(t)$ ($\alpha=i,j$)*. For* $\bar{\mathbf{x}}^{(s)}(t_m)\equiv\mathbf{x}_m\in\partial\Omega_{ij}$*, suppose* $\mathbf{x}^{(\alpha)}(t_{m\pm})=\mathbf{x}_m$. *at time* t_m*. For an arbitrarily small* $\varepsilon>0$*, there is a time interval* $[t_{m-\varepsilon},t_{m+\varepsilon}]$*. The flow* $\mathbf{x}^{(\alpha)}(t)$ *is* $C^r_{[t_{m-\varepsilon},t_{m+\varepsilon}]}$*-continuous* ($r\ge 2k+1$) *and* $\|d^r\mathbf{x}^{(\alpha)}/dt^r\|<\infty$ *for time t. A flow* $\mathbf{x}^{(\alpha)}(t)$ *in* Ω_α *is tangential to the separatrix* $\partial\Omega_{ij}$ *with the* $(2k-1)^{th}$*-order iff*

$$G^{(r)}_{\partial\Omega_{ij}}(\bar{\mathbf{x}}^{(s)}_m,\mathbf{x}^{(\alpha)}_m,t_m,\mathbf{p})=0 \text{ for } r=0,1,\cdots,2k-2; \tag{3.31}$$

$$\left.\begin{array}{ll} \textit{either} & G^{(2k-1)}_{\partial\Omega_{ij}}(\bar{\mathbf{x}}^{(s)}_m,\mathbf{x}^{(\alpha)}_m,t_m,\mathbf{p})<0 \text{ for } \mathbf{n}_{\partial\Omega_{ij}}\to\Omega_\beta \\ \textit{or} & G^{(2k-1)}_{\partial\Omega_{ij}}(\bar{\mathbf{x}}^{(s)}_m,\mathbf{x}^{(\alpha)}_m,t_m,\mathbf{p})>0 \text{ for } \mathbf{n}_{\partial\Omega_{ij}}\to\Omega_\alpha. \end{array}\right\} \tag{3.32}$$

PROOF. Equation (3.31) is identical to Eq.(3.28), thus the condition in Eq.(3.28) is satisfied, and vice versa. Suppose $\mathbf{x}^{(\alpha)}(t_{m\pm})=\mathbf{x}_m$ ($\alpha\in\{i,j\}$) and $\mathbf{x}^{(\alpha)}(t)$ are $C^r_{[t_{m-\varepsilon},t_{m+\varepsilon}]}$ -continuous ($r\ge 2k+1$) for time t and $\|d^r\mathbf{x}^{(\alpha)}/dt^r\|<\infty$ ($\alpha\in\{i,j\}$). For $a\in[t_{m-\varepsilon},t_m)$ or $a\in(t_m,t_{m+\varepsilon}]$, the Taylor series expansion of $\mathbf{x}^{(\alpha)}(t_{m\pm\varepsilon})$ to $\mathbf{x}^{(\alpha)}(a)$ up to the $(2k+1)^{\text{th}}$-order term gives

$$\begin{aligned}\mathbf{x}^{(\alpha)}_{m\pm\varepsilon}&\equiv\mathbf{x}^{(\alpha)}(t_{m\pm}\pm\varepsilon)=\mathbf{x}^{(\alpha)}(a)+\sum_{r=1}^{2k-1}\tfrac{d^r}{dt^r}\mathbf{x}^{(\alpha)}(a)(t_{m\pm}\pm\varepsilon-a)^r\\&\quad+\tfrac{d^{2k}}{dt^{2k}}\mathbf{x}^{(\alpha)}(a)(t_{m\pm}\pm\varepsilon-a)^{2k}+o((t_{m\pm}\pm\varepsilon-a)^{2k}).\end{aligned}$$

As $a\to t_{m\pm}$, taking the limit of the foregoing equation leads to

$$\mathbf{x}^{(\alpha)}_{m\pm\varepsilon}\equiv\mathbf{x}^{(\alpha)}(t_m\pm\varepsilon)=\mathbf{x}^{(\alpha)}_{m\pm}+\sum_{r=1}^{2k-1}\tfrac{d^r}{dt^r}\mathbf{x}^{(\alpha)}_{m\pm}(\pm\varepsilon)^r+\tfrac{d^{2k}}{dt^{2k}}\mathbf{x}^{(\alpha)}_{m\pm}(\pm\varepsilon)^{2k}+o(\varepsilon^{2k}).$$

In a similar fashion, we have

$$\bar{\mathbf{x}}^{(s)}_{m\pm\varepsilon}\equiv\bar{\mathbf{x}}^{(s)}(t_m\pm\varepsilon)=\bar{\mathbf{x}}^{(s)}_{m\pm}+\sum_{r=1}^{2k-1}\tfrac{d^r}{dt^r}\bar{\mathbf{x}}^{(s)}_{m\pm}(\pm\varepsilon)^r+\tfrac{d^{2k}}{dt^{2k}}\bar{\mathbf{x}}^{(s)}_{m\pm}\varepsilon^{2k}+o(\varepsilon^{2k}),$$

$$\begin{aligned}\mathbf{n}_{\partial\Omega_{ij}}(\bar{\mathbf{x}}^{(s)}_{m\pm\varepsilon})&=\mathbf{n}_{\partial\Omega_{ij}}(\bar{\mathbf{x}}^{(s)}_m)+\sum_{r=1}^{2k-1}D^r_{\bar{\mathbf{x}}}\mathbf{n}_{\partial\Omega_{ij}}(\bar{\mathbf{x}}^{(s)}_m)(\pm\varepsilon)^r\\&\quad+D^{2k}_{\bar{\mathbf{x}}}\mathbf{n}_{\partial\Omega_{ij}}(\bar{\mathbf{x}}^{(s)}_m)\varepsilon^{2k}+o(\varepsilon^{2k}).\end{aligned}$$

The ignorance of the ε^{2k+1} and high order terms, the deformation of the above equation and left multiplication of $\mathbf{n}_{\partial\Omega_{ij}}$ gives

$$\mathbf{n}^T_{\partial\Omega_{ij}}(\overline{\mathbf{x}}^{(s)}_{m-\varepsilon})\bullet\left[\mathbf{x}^{(\alpha)}_{m-\varepsilon}-\overline{\mathbf{x}}^{(s)}_{m-\varepsilon}\right]=\mathbf{n}^T_{\partial\Omega_{ij}}(\overline{\mathbf{x}}^{(s)}_{m})\bullet\left[\mathbf{x}^{(\alpha)}_{m-}-\overline{\mathbf{x}}^{(s)}_{m}\right]$$
$$+\sum_{r=1}^{2k-1}(-\varepsilon)^r G^{(r-1)}_{\partial\Omega_{ij}}(\overline{\mathbf{x}}^{(s)}_m,\mathbf{x}^{(\alpha)}_m,t_m,\mathbf{p})$$
$$+\varepsilon^{2k}G^{(2k-1)}_{\partial\Omega_{ij}}(\overline{\mathbf{x}}^{(s)}_m,\mathbf{x}^{(\alpha)}_m,t_m,\mathbf{p}),$$

$$\mathbf{n}^T_{\partial\Omega_{ij}}(\overline{\mathbf{x}}^{(s)}_{m+\varepsilon})\bullet\left[\mathbf{x}^{(\alpha)}_{m+\varepsilon}-\overline{\mathbf{x}}^{(s)}_{m+\varepsilon}\right]=\mathbf{n}^T_{\partial\Omega_{ij}}(\overline{\mathbf{x}}^{(s)}_{m})\bullet\left[\mathbf{x}^{(\alpha)}_{m+}-\overline{\mathbf{x}}^{(s)}_{m}\right]$$
$$+\sum_{r=1}^{2k-1}\varepsilon^r G^{(r-1)}_{\partial\Omega_{ij}}(\overline{\mathbf{x}}^{(s)}_m,\mathbf{x}^{(\alpha)}_m,t_m,\mathbf{p})$$
$$+\varepsilon^{2k}G^{(2k-1)}_{\partial\Omega_{ij}}(\overline{\mathbf{x}}^{(s)}_m,\mathbf{x}^{(\alpha)}_m,t_m,\mathbf{p}).$$

Because of $\mathbf{x}^{(\alpha)}_{m\pm}=\overline{\mathbf{x}}^{(s)}_m$ and $G^{(r)}_{\partial\Omega_{ij}}(\overline{\mathbf{x}}^{(s)}_m,\mathbf{x}^{(\alpha)}_m,t_m,\mathbf{p})=0$ for $r=0,1,\cdots,2k-2$, the foregoing equation becomes

$$\mathbf{n}^T_{\partial\Omega_{ij}}(\overline{\mathbf{x}}^{(s)}_{m-\varepsilon})\bullet\left[\overline{\mathbf{x}}^{(s)}_{m-\varepsilon}-\mathbf{x}^{(\alpha)}_{m-\varepsilon}\right]=-\varepsilon^{2k}G^{(2k-1)}_{\partial\Omega_{ij}}(\overline{\mathbf{x}}^{(s)}_m,\mathbf{x}^{(\alpha)}_m,t_m,\mathbf{p}),$$
$$\mathbf{n}^T_{\partial\Omega_{ij}}(\overline{\mathbf{x}}^{(s)}_{m+\varepsilon})\bullet\left[\mathbf{x}^{(\alpha)}_{m+\varepsilon}-\overline{\mathbf{x}}^{(s)}_{m+\varepsilon}\right]=\varepsilon^{2k}G^{(2k-1)}_{\partial\Omega_{ij}}(\overline{\mathbf{x}}^{(s)}_m,\mathbf{x}^{(\alpha)}_m,t_m,\mathbf{p}).$$

Using Eq.(3.32), equation (3.30) is obtained. However, using Eq.(3.30), equation (3.32) is obtained. Therefore, this theorem is proved. ■

The G-functions can be defined for a perturbed flow not only to the separatrix but also to any surface of the first integral manifold. The G-functions are the normal components relative to the perturbed vector fields and time change rates in the normal direction of the first integral surface S_α. Once such a concept is introduced, the properties of the perturbed flow in the local domain Ω_α ($\alpha=1,2,\cdots,m$) can be investigated.

DEFINITION 3.8. For a dynamical system in Eq.(3.1), its integrable dynamical system in Eq.(3.2) possesses a family of the first integral manifolds $F(\overline{\mathbf{x}},\boldsymbol{\mu})=E$. The G-function of the kth-order, $G^{(k)}_\alpha(\bullet)$, for a perturbed flow to the specified the first integral surface S_α at point $\mathbf{x}^{(\alpha)}_m=\mathbf{x}^{(\alpha)}(t_m)$ is defined for ($k=0,1,2,\cdots$)

$$\begin{aligned}G^{(k)}_\alpha(\mathbf{x}^{(\alpha)}_m,t_m,\mathbf{p})&=\sum_{r=1}^{k+1}C^r_{k+1}D^{k+1-r}_{\overline{\mathbf{x}}}\mathbf{n}^T_{S_\alpha}(\overline{\mathbf{x}})\bullet\left[\frac{d^r\mathbf{x}}{dt^r}-\frac{d^r\overline{\mathbf{x}}}{dt^r}\right]\Big|_{(\overline{\mathbf{x}}^{(\alpha)}_m,\mathbf{x}^{(\alpha)}_m,t_m)}\\&=\sum_{r=1}^{k+1}C^r_{k+1}D^{k+1-r}_{\overline{\mathbf{x}}}\mathbf{n}^T_{S_\alpha}(\overline{\mathbf{x}})\bullet\left[D^{r-1}_{\mathbf{x}}\mathbf{F}(\mathbf{x},t,\mathbf{p})\right.\\&\left.-D^{r-1}_{\overline{\mathbf{x}}}\mathbf{f}(\overline{\mathbf{x}},\boldsymbol{\mu})\right]\Big|_{(\overline{\mathbf{x}}^{(\alpha)}_m,\mathbf{x}^{(\alpha)}_m,t_m)}.\end{aligned}\tag{3.33}$$

For the special case ($k=0$), the zero-order G- function is

$$\begin{aligned} G_\alpha(\mathbf{x}_m^{(\alpha)}, t_m, \mathbf{p}) &= \mathbf{n}_{S_\alpha}^T(\overline{\mathbf{x}}_m^{(\alpha)}) \bullet \mathbf{F}(\mathbf{x}_m^{(\alpha)}, t_m, \mathbf{p}) = \nabla F(\overline{\mathbf{x}}_m^{(\alpha)}, \boldsymbol{\mu}) \bullet \mathbf{g}(\mathbf{x}_m^{(\alpha)}, t_m, \mathbf{p}) \\ &= \mathbf{n}_{S_\alpha}^T(\overline{\mathbf{x}}_m^{(\alpha)}) \bullet \mathbf{g}(\mathbf{x}_m^{(\alpha)}, t_m, \boldsymbol{\pi}) = \nabla F(\overline{\mathbf{x}}_m^{(\alpha)}, \boldsymbol{\mu}) \bullet \mathbf{g}(\mathbf{x}_m^{(\alpha)}, t_m, \boldsymbol{\pi}). \end{aligned} \tag{3.34}$$

From the foregoing equation, the G-function is the component of the vector field in the normal direction of the first integral surface. From the definitions in Chapter 2, the $G^{(k)}$-function is the time-change rate of the $G^{(k-1)}$-function in the normal direction of the first integral surface. If the G-function is defined in the normal direction of the separatrix, the global transversality and tangency of the flow can be expressed if $\overline{\mathbf{x}}_m^{(\alpha)} = \overline{\mathbf{x}}_m^{(s)}$. To determine the tangential bifurcation to the separatrix $\partial\Omega_{ij}$, the following conditions should be satisfied,

$$\tfrac{\partial}{\partial \mathbf{p}} G_{\partial\Omega_{ij}}^{(2k-1)}(\mathbf{x}_m^{(\alpha)}, t_m, \mathbf{p}) \neq 0. \tag{3.35}$$

If a local flow $\mathbf{x}^{(\alpha)}(t)$ is periodic and its tangency to the separatrix $\partial\Omega_{ij}$ occurs, this tangential bifurcation is called *the grazing bifurcation* of the local, periodic flow. Similarly, if the global flow $\mathbf{x}^{(i)}(t) \cup \mathbf{x}^{(j)}(t)$ to the separatrix $\partial\Omega_{ij}$ is periodic, the corresponding tangential bifurcation is a global bifurcation of the global, periodic flow. After the grazing bifurcation, the new global flow to the separatrix $\partial\Omega_{ij}$ may be periodic or chaotic.

3.5. Perturbed Hamiltonian systems

In this section, the global and tangential flows for perturbed nonlinear Hamiltonian systems will be discussed as an application in nonlinear dynamical systems. Consider a $2n$-dimenionsal perturbed nonlinear Hamiltonian system as

$$\left.\begin{aligned} \dot{\mathbf{x}} &= \mathbf{F}_1(\mathbf{x}, \mathbf{y}, t, \mathbf{p}) \equiv \mathbf{f}_1(\mathbf{x}, \mathbf{y}, \boldsymbol{\mu}) + \mathbf{g}_1(\mathbf{x}, \mathbf{y}, t, \boldsymbol{\pi}) \\ \dot{\mathbf{y}} &= \mathbf{F}_2(\mathbf{x}, \mathbf{y}, t, \mathbf{p}) \equiv \mathbf{f}_2(\mathbf{x}, \mathbf{y}, \boldsymbol{\mu}) + \mathbf{g}_2(\mathbf{x}, \mathbf{y}, t, \boldsymbol{\pi}) \end{aligned}\right\} \tag{3.36}$$

where $\mathbf{x} = (x_1, x_2, \cdots, x_n)^T$ and $\mathbf{y} = (y_1, y_2, \cdots, y_n)^T$. Without any perturbation, the Hamiltonian system is

$$\dot{\overline{\mathbf{x}}} = \mathbf{f}_1(\overline{\mathbf{x}}, \overline{\mathbf{y}}, \boldsymbol{\mu}) \text{ and } \dot{\overline{\mathbf{y}}} = \mathbf{f}_2(\overline{\mathbf{x}}, \overline{\mathbf{y}}, \boldsymbol{\mu}). \tag{3.37}$$

From the foregoing equation, its Hamiltonian is $H_0(\overline{\mathbf{x}}, \overline{\mathbf{y}}, \boldsymbol{\mu})$ and

$$\dot{\overline{\mathbf{x}}} = \tfrac{\partial H_0(\overline{\mathbf{x}}, \overline{\mathbf{y}}, \boldsymbol{\mu})}{\partial \overline{\mathbf{y}}} = \mathbf{f}_1(\overline{\mathbf{x}}, \overline{\mathbf{y}}, \boldsymbol{\mu}) \text{ and } \dot{\overline{\mathbf{y}}} = -\tfrac{\partial H_0(\overline{\mathbf{x}}, \overline{\mathbf{y}}, \boldsymbol{\mu})}{\partial \overline{\mathbf{x}}} = \mathbf{f}_2(\overline{\mathbf{x}}, \overline{\mathbf{y}}, \boldsymbol{\mu}); \tag{3.38}$$

or for ($i = 1, 2, \cdots, n$)

$$\dot{\bar{x}}_i = \frac{\partial H_0(\bar{\mathbf{x}},\bar{\mathbf{y}},\boldsymbol{\mu})}{\partial \bar{y}_i} = f_{1i}(\bar{\mathbf{x}},\bar{\mathbf{y}},\boldsymbol{\mu}) \text{ and } \dot{\bar{y}}_i = -\frac{\partial H_0(\bar{\mathbf{x}},\bar{\mathbf{y}},\boldsymbol{\mu})}{\partial \bar{x}_i} = f_{2i}(\bar{\mathbf{x}},\bar{\mathbf{y}},\boldsymbol{\mu}). \tag{3.39}$$

Application of Eq.(3.37) to Eq.(3.3) gives

$$\nabla_{\bar{\mathbf{x}}} H_0(\bar{\mathbf{x}},\bar{\mathbf{y}},\boldsymbol{\mu})\cdot\dot{\bar{\mathbf{x}}} + \nabla_{\bar{\mathbf{y}}} H_0(\bar{\mathbf{x}},\bar{\mathbf{y}},\boldsymbol{\mu})\cdot\dot{\bar{\mathbf{y}}} = 0. \tag{3.40}$$

Therefore, the Hamiltonian $H_0(\bar{\mathbf{x}},\bar{\mathbf{y}},\boldsymbol{\mu})$ is the first integral manifold of the dynamical system in Eq.(3.36), i.e., $H_0(\bar{\mathbf{x}},\bar{\mathbf{y}},\boldsymbol{\mu}) = F(\bar{\mathbf{x}},\bar{\mathbf{y}},\boldsymbol{\mu})$.

If the unperturbed Hamiltonian system in Eq.(3.36) has at least a hyperbolic point, then there is a separatrix surface including the hyperbolic point, given by $H_0(\bar{\mathbf{x}},\bar{\mathbf{y}},\boldsymbol{\mu}) = E_s$. The global and tangential flows to the separatrix manifold of the perturbed Hamiltonian system in Eq.(3.36) can be discussed. Based on this separatrix surface, the separatrix $\partial\Omega_{ij}$ for domains Ω_i and Ω_j can be defined. The necessary and sufficient conditions for the *global* flow to the separatrix boundary $\partial\Omega_{ij}$ in the perturbed Hamiltonian system of Eq.(3.36) can be determined by Eq.(3.12) of Theorem 3.2. With Eqs.(3.36) and (3.38), the zero-order G-function in Eq.(3.14) is determined by

$$\begin{aligned} G^{(0)}_{\partial\Omega_{ij}}(\mathbf{x}_m,\mathbf{y}_m,t_m,\mathbf{p}) &= \mathbf{f}_1(\mathbf{x}_m,\mathbf{y}_m,\boldsymbol{\mu})\cdot\mathbf{F}_2(\mathbf{x}_m,\mathbf{y}_m,t_m,\mathbf{p}) \\ &\quad -\mathbf{f}_2(\mathbf{x}_m,\mathbf{y}_m,\boldsymbol{\mu})\cdot\mathbf{F}_1(\mathbf{x}_m,\mathbf{y}_m,t_m,\mathbf{p}) \\ &= \mathbf{f}_1(\mathbf{x}_m,\mathbf{y}_m,\boldsymbol{\mu})\cdot\mathbf{g}_2(\mathbf{x}_m,\mathbf{y}_m,t_m,\boldsymbol{\pi}) \\ &\quad -\mathbf{f}_2(\mathbf{x}_m,\mathbf{y}_m,\boldsymbol{\mu})\cdot\mathbf{g}_1(\mathbf{x}_m,\mathbf{y}_m,t_m,\boldsymbol{\pi}) \end{aligned} \tag{3.41}$$

where $(\mathbf{x}_m,\mathbf{y}_m)\in\partial\Omega_{ij}$ and the $G^{(0)}_{\partial\Omega_{ij}}$-function is the component of the vector field of the perturbed Hamiltonian system on the normal direction of the separatrix surface. The normal vector always points to the domain Ω_α with $E_\alpha > E_s$ ($\alpha \in \{i,j\}$). Therefore, the corresponding conditions for the global transversality on the separatrix $\partial\Omega_{ij}$ are given by

$$\begin{aligned} &H_0(\mathbf{x}_m,\mathbf{y}_m,\boldsymbol{\mu}) = E_s, \\ &G^{(0)}_{\partial\Omega_{ij}}(\mathbf{x}_m,\mathbf{y}_m,t_m,\mathbf{p}) > 0 \quad \text{for } E_\alpha < E_s, \\ &G^{(0)}_{\partial\Omega_{ij}}(\mathbf{x}_m,\mathbf{y}_m,t_m,\mathbf{p}) < 0 \quad \text{for } E_\alpha > E_s. \end{aligned} \tag{3.42}$$

Once a flow $(\mathbf{x}^{(\alpha)}(t),\mathbf{y}^{(\alpha)}(t))$ ($\alpha = i,j$) arrives to the separatrix $\partial\Omega_{ij}$, if conditions in Eq.(3.42) are satisfied, this flow is a global flow from the domain Ω_α into Ω_β passing through the separatrix boundary $\partial\Omega_{ij}$.

The necessary and sufficient conditions for the *tangential* flow of the perturbed Hamiltonian system in Eq.(3.36) to the separatrix can be obtained from Eqs.(3.26) and (3.27) in Theorem 3.5. To determine the global tangency to the separatrix, the first-order G-function in the normal direction of the separatrix is determined from Eq.(3.13) for $k=1$ first. With Eqs.(3.36) and (3.38), the first-order G-function in Eq.(3.13) for $(\mathbf{x}_m,\mathbf{y}_m)=(\overline{\mathbf{x}}_m,\overline{\mathbf{y}}_m)\in\partial\Omega_{ij}$ becomes

$$\begin{aligned}G^{(1)}_{\partial\Omega_{ij}}(\mathbf{x}_m,\mathbf{y}_m,t_m,\mathbf{p})=&\,2D\mathbf{f}_1(\overline{\mathbf{x}}_m,\overline{\mathbf{y}}_m,\boldsymbol{\mu})\bullet\mathbf{g}_2(\mathbf{x}_m,\mathbf{y}_m,t_m,\boldsymbol{\pi})\\&-2D\mathbf{f}_2(\overline{\mathbf{x}}_m,\overline{\mathbf{y}}_m,\boldsymbol{\mu})\bullet\mathbf{g}_1(\mathbf{x}_m,\mathbf{y}_m,t_m,\boldsymbol{\pi})\\&+\mathbf{f}_1(\overline{\mathbf{x}}_m,\overline{\mathbf{y}}_m,\boldsymbol{\mu})\bullet\left[D\mathbf{F}_2(\mathbf{x}_m,\mathbf{y}_m,t_m,\mathbf{p})-D\mathbf{f}_2(\overline{\mathbf{x}}_m,\overline{\mathbf{y}}_m,\boldsymbol{\mu})\right]\\&-\mathbf{f}_2(\overline{\mathbf{x}}_m,\overline{\mathbf{y}}_m,\boldsymbol{\mu})\bullet\left[D\mathbf{F}_1(\mathbf{x}_m,\mathbf{y}_m,t_m,\mathbf{p})-D\mathbf{f}_1(\overline{\mathbf{x}}_m,\overline{\mathbf{y}}_m,\boldsymbol{\mu})\right]\end{aligned}\tag{3.43}$$

where for ($\sigma=1,2$ and $\alpha=i,j$) and ($p=1,2,\cdots,n$ and $q=1,2,\cdots,n$)

$$\begin{aligned}D\mathbf{F}_\sigma(\mathbf{x},\mathbf{y},t,\mathbf{p})=&\left[\frac{\partial F_{\sigma p}(\mathbf{x},\mathbf{y},t,\mathbf{p})}{\partial x_q}\right]_{n\times n}\bullet\mathbf{F}_1(\mathbf{x},\mathbf{y},t,\mathbf{p})\\&+\left[\frac{\partial F_{\sigma p}(\mathbf{x},\mathbf{y},t,\mathbf{p})}{\partial y_q}\right]_{n\times n}\bullet\mathbf{F}_2(\mathbf{x},\mathbf{y},t,\mathbf{p})\\&+\frac{\partial}{\partial t}\mathbf{g}_\sigma(\mathbf{x},\mathbf{y},t,\boldsymbol{\pi}),\end{aligned}\tag{3.44}$$

$$D\mathbf{f}_\sigma(\overline{\mathbf{x}},\overline{\mathbf{y}},\boldsymbol{\mu})=\left[\frac{\partial f_{\sigma p}(\overline{\mathbf{x}},\overline{\mathbf{y}},\boldsymbol{\mu})}{\partial \overline{x}_q}\right]_{n\times n}\bullet\mathbf{f}_1(\overline{\mathbf{x}},\overline{\mathbf{y}},\boldsymbol{\mu})+\left[\frac{\partial f_{\sigma p}(\overline{\mathbf{x}},\overline{\mathbf{y}},\boldsymbol{\mu})}{\partial \overline{y}_q}\right]_{n\times n}\bullet\mathbf{f}_2(\overline{\mathbf{x}},\overline{\mathbf{y}},\boldsymbol{\mu}).\tag{3.45}$$

Therefore, the corresponding sufficient and necessary conditions for the global tangency to the separatrix for $(\mathbf{x}^{(\alpha)}_m,\mathbf{y}^{(\alpha)}_m)=(\overline{\mathbf{x}}^{(\alpha)}_m,\overline{\mathbf{y}}^{(\alpha)}_m)\in\partial\Omega_{ij}$ are

$$\left.\begin{aligned}&H_0(\mathbf{x}^{(\alpha)}_m,\mathbf{y}^{(\alpha)}_m,\boldsymbol{\mu})=E_s,\\&G^{(0)}_{\partial\Omega_{ij}}(\mathbf{x}^{(\alpha)}_m,\mathbf{y}^{(\alpha)}_m,t_m,\mathbf{p})=0,\\&G^{(1)}_{\partial\Omega_{ij}}(\mathbf{x}^{(\alpha)}_m,\mathbf{y}^{(\alpha)}_m,t_m,\mathbf{p})<0\quad\text{for } E_\alpha<E_s,\\&G^{(1)}_{\partial\Omega_{ij}}(\mathbf{x}^{(\alpha)}_m,\mathbf{y}^{(\alpha)}_m,t_m,\mathbf{p})>0\quad\text{for } E_\alpha>E_s.\end{aligned}\right\}\tag{3.46}$$

The point $(\mathbf{x}_m,\mathbf{y}_m,t_m)$ must be on the separatrix. Once a flow $(\mathbf{x}^{(\alpha)}(t),\mathbf{y}^{(\alpha)}(t))$ ($\alpha=i,j$) arrives to the separatrix $\partial\Omega_{ij}$, if conditions in Eq.(3.46) are satisfied, the flow $(\mathbf{x}^{(\alpha)}(t),\mathbf{y}^{(\alpha)}(t))$ will be a tangential flow to the separatrix $\partial\Omega_{ij}$. From Eq.(3.46), the normal component of the vector field on the normal direction of the separatrix at point $(\mathbf{x}_m,\mathbf{y}_m,t_m)$ satisfies the second equation of Eq.(3.46). For $E_\alpha<E_s$, the time-change rate of the normal component of the vector field should be less than zero. However, for $E_\alpha>E_s$, such a time-change rate should be greater than zero. For this case, the flow will not pass through the separatrix.

In a similar fashion, with Eqs.(3.36) and (3.38), the kth-order G-function in the normal direction of the separatrix surface) for $(\mathbf{x}_m, \mathbf{y}_m) = (\overline{\mathbf{x}}_m, \overline{\mathbf{y}}_m) \in \partial\Omega_{ij}$ is determined by

$$\begin{aligned} G^{(k)}_{\partial\Omega_{ij}}(\mathbf{x}_m, \mathbf{y}_m, t_m, \mathbf{p}) = \sum_{r=1}^{k+1} C^r_{k+1} \Big\{ & D^{k+1-r}\mathbf{f}_1(\overline{\mathbf{x}}_m, \overline{\mathbf{y}}_m, \boldsymbol{\mu}) \bullet \big[D^{r-1}\mathbf{F}_2(\mathbf{x}_m, \mathbf{y}_m, t_m, \mathbf{p}) \\ & - D^{r-1}\mathbf{f}_2(\overline{\mathbf{x}}_m, \overline{\mathbf{y}}_m, \boldsymbol{\mu}) \big] - D^{k+1-r}\mathbf{f}_2(\overline{\mathbf{x}}_m, \overline{\mathbf{y}}_m, \boldsymbol{\mu}) \\ & \bullet \big[D^{r-1}\mathbf{F}_1(\mathbf{x}_m, \mathbf{y}_m, t_m, \mathbf{p}) - D^{r-1}\mathbf{f}_1(\overline{\mathbf{x}}_m, \overline{\mathbf{y}}_m, \boldsymbol{\mu}) \big] \Big\} \end{aligned} \tag{3.47}$$

for ($l = 1, 2, \cdots, k$), where for ($\sigma = 1, 2$)

$$\begin{aligned} D^r\mathbf{F}_\sigma(\mathbf{x}, \mathbf{y}, t, \mathbf{p}) &= D^{r-1}\big[D\mathbf{F}_\sigma(\mathbf{x}, \mathbf{y}, t, \mathbf{p}) \big] \\ &= D^{r-1}\Big\{ \Big[\tfrac{\partial F_{\sigma p}(\mathbf{x},\mathbf{y},t,\mathbf{p})}{\partial x_q} \Big]_{n\times n} \bullet \mathbf{F}_1(\mathbf{x}, \mathbf{y}, t, \mathbf{p}) \\ &\quad + \Big[\tfrac{\partial F_{\sigma p}(\mathbf{x},\mathbf{y},t,\mathbf{p})}{\partial y_q} \Big]_{n\times n} \bullet \mathbf{F}_2(\mathbf{x}, \mathbf{y}, t, \mathbf{p}) \\ &\quad + \tfrac{\partial}{\partial t}\mathbf{g}_\sigma(\mathbf{x}, \mathbf{y}, t, \boldsymbol{\pi}) \Big\}. \end{aligned} \tag{3.48}$$

$$\begin{aligned} D^r\mathbf{f}_\sigma(\overline{\mathbf{x}}, \overline{\mathbf{y}}, \boldsymbol{\mu}) &= D^{r-1}\big[D\mathbf{f}_\sigma(\overline{\mathbf{x}}, \overline{\mathbf{y}}, \boldsymbol{\mu}) \big] \\ &= D^{r-1}\Big\{ \Big[\tfrac{\partial f_{\sigma p}(\overline{\mathbf{x}},\overline{\mathbf{y}},t,\boldsymbol{\mu})}{\partial \overline{x}_q} \Big]_{n\times n} \bullet \mathbf{f}_1(\overline{\mathbf{x}}, \overline{\mathbf{y}}, \boldsymbol{\mu}) \\ &\quad + \Big[\tfrac{\partial f_{\sigma p}(\overline{\mathbf{x}},\overline{\mathbf{y}},t,\boldsymbol{\mu})}{\partial \overline{y}_q} \Big]_{n\times n} \bullet \mathbf{f}_2(\overline{\mathbf{x}}, \overline{\mathbf{y}}, \boldsymbol{\mu}) \Big\}. \end{aligned} \tag{3.49}$$

The necessary and sufficient conditions for the $(2k)^{th}$-order, global flow of the perturbed Hamiltonian systems to the separatrix surface can be developed from Theorem 3.3. Further, if a flow $(\mathbf{x}^{(\alpha)}(t), \mathbf{y}^{(\alpha)}(t))$ of the perturbed Hamiltonian system in Eq.(3.36) can pass over the separatrix, with Eqs.(3.47)-(3.49), the corresponding necessary and sufficient conditions for such a global flow must be satisfied for $(\mathbf{x}^{(\alpha)}_m, \mathbf{y}^{(\alpha)}_m) = (\overline{\mathbf{x}}^{(\alpha)}_m, \overline{\mathbf{y}}^{(\alpha)}_m) \in \partial\Omega_{ij}$ as follows:

$$\left. \begin{aligned} & H_0(\mathbf{x}^{(\alpha)}_m, \mathbf{y}^{(\alpha)}_m, \boldsymbol{\mu}) = E_s, \\ & G^{(r)}_{\partial\Omega_{ij}}(\mathbf{x}^{(\alpha)}_m, \mathbf{y}^{(\alpha)}_m, t_m, \mathbf{p}) = 0 \text{ for } r = 1, 2, \cdots 2k-1; \\ & G^{(2k)}_{\partial\Omega_{ij}}(\mathbf{x}^{(\alpha)}_m, \mathbf{y}^{(\alpha)}_m, t_m, \mathbf{p}) > 0 \quad \text{for } E_\alpha < E_s, \\ & G^{(2k)}_{\partial\Omega_{ij}}(\mathbf{x}^{(\alpha)}_m, \mathbf{y}^{(\alpha)}_m, t_m, \mathbf{p}) < 0 \quad \text{for } E_\alpha > E_s. \end{aligned} \right\} \tag{3.50}$$

Finally, the necessary and sufficient conditions for the $(2k-1)^{th}$-order, global, tangential flow $\mathbf{x}^{(\alpha)}(t)$ of a perturbed Hamiltonian system in Eq.(3.36) are given in Theorem 3.6. For the perturbed Hamiltonian system in Eq.(3.36), with help of Eqs.(3.47)-(3.49), the conditions of the global tangential flow to the separatrix for $(\mathbf{x}_m^{(\alpha)},\mathbf{y}_m^{(\alpha)})=(\bar{\mathbf{x}}_m^{(\alpha)},\bar{\mathbf{y}}_m^{(\alpha)})\in\partial\Omega_{ij}$ is

$$\left.\begin{aligned}
&H_0(\mathbf{x}_m^{(\alpha)},\mathbf{y}_m^{(\alpha)},\boldsymbol{\mu})=E_s,\\
&G_{\partial\Omega_{ij}}^{(r)}(\mathbf{x}_m^{(\alpha)},\mathbf{y}_m^{(\alpha)},t_m,\mathbf{p})=0 \text{ for } r=0,1,2,\cdots 2k-2;\\
&G_{\partial\Omega_{ij}}^{(2k-1)}(\mathbf{x}_m^{(\alpha)},\mathbf{y}_m^{(\alpha)},t_m,\mathbf{p})<0 \quad \text{for } E_\alpha<E_s,\\
&G_{\partial\Omega_{ij}}^{(2k-1)}(\mathbf{x}_m^{(\alpha)},\mathbf{y}_m^{(\alpha)},t_m,\mathbf{p})>0 \quad \text{for } E_\alpha>E_s.
\end{aligned}\right\} \tag{3.51}$$

3.6. Two-dimensional Hamiltonian systems

From the $2n$-dimensional dynamical system, the expressions for sufficient and necessary conditions are still not intuitive. To help one further understand the concepts of the global transversality and tangency to the separatrix, the geometrical interpretation of such global transversality and tangency will be presented in this section. Consider a two-dimensional nonlinear Hamiltonian system with perturbation as

$$\left.\begin{aligned}
\dot{x}=F_1(x,y,t,\mathbf{p})=f_1(x,y,\boldsymbol{\mu})+g_1(x,y,t,\boldsymbol{\pi})\\
\dot{y}=F_2(x,y,t,\mathbf{p})=f_2(x,y,\boldsymbol{\mu})+g_2(x,y,t,\boldsymbol{\pi})
\end{aligned}\right\} \tag{3.52}$$

where two parameter vectors $\boldsymbol{\mu}\in\mathbb{R}^{m_1}$ and $\boldsymbol{\pi}\in\mathbb{R}^{m_2}$. Without any perturbation, the Hamiltonian system is governed by

$$\dot{\bar{x}}=f_1(\bar{x},\bar{y},\boldsymbol{\mu}) \text{ and } \dot{\bar{y}}=f_2(\bar{x},\bar{y},\boldsymbol{\mu}). \tag{3.53}$$

The Hamiltonian of Eq.(3.53) is $H_0(\bar{x},\bar{y},\boldsymbol{\mu})$ from which we have

$$\dot{\bar{x}}=\tfrac{\partial H_0(\bar{x},\bar{y},\mu)}{\partial\bar{y}}=f_1(\bar{x},\bar{y},\boldsymbol{\mu}) \text{ and } \dot{\bar{y}}=-\tfrac{\partial H_0(\bar{x},\bar{y},\mu)}{\partial\bar{x}}=f_2(\bar{x},\bar{y},\boldsymbol{\mu}). \tag{3.54}$$

For a given initial condition $(\bar{x}_0,\bar{y}_0,t_0)$, the corresponding Hamiltonian is

$$H_0(\bar{x},\bar{y},\boldsymbol{\mu})=H_0(\bar{x}_0,\bar{y}_0,\boldsymbol{\mu})\equiv E_0. \tag{3.55}$$

Because of energy conservation, the change rate of non-perturbed Hamiltonian system with respect to time is zero. That is,

$$\frac{\partial H_0(\bar{x},\bar{y},\boldsymbol{\mu})}{\partial \bar{x}}\dot{\bar{x}}+\frac{\partial H_0(\bar{x},\bar{y},\boldsymbol{\mu})}{\partial \bar{y}}\dot{\bar{y}}=0. \tag{3.56}$$

Therefore, the Hamiltonian $H_0(\bar{x},\bar{y},\boldsymbol{\mu})$ is the first integral manifold of the dynamical system in Eq.(3.52), i.e., $H_0(\bar{x},\bar{y},\boldsymbol{\mu})=F(\bar{x},\bar{y},\boldsymbol{\mu})$. If the unperturbed Hamiltonian system in Eq.(3.52) has at least a hyperbolic point, then there is a separatrix manifold including the hyperbolic point, given by

$$H_0(\bar{x},\bar{y},\boldsymbol{\mu})=E_s. \tag{3.57}$$

The normal vector of the Hamiltonian energy surface S_α (i.e., $H_0(\bar{x},\bar{y},\boldsymbol{\mu})=E_\alpha$) is defined by

$$\mathbf{n}_{S_\alpha}=(\tfrac{\partial H_0}{\partial \bar{x}},\tfrac{\partial H_0}{\partial \bar{y}})^T=\left(-f_2(\bar{x},\bar{y}),f_1(\bar{x},\bar{y})\right)^T. \tag{3.58}$$

The perturbed vector field is

$$\mathbf{F}(x,y,t,\mathbf{p})=\left(F_1(x,y,t,\mathbf{p}),F_2(x,y,t,\mathbf{p})\right)^T \tag{3.59}$$

The component of the perturbed vector field in the normal direction of the Hamiltonian energy surface (or the first integral manifold surface) is for $\bar{x}=x$ and $y=\bar{y}$

$$\mathbf{n}_{S_\alpha}^T\bullet\mathbf{F}(x,y,t,\mathbf{p})=f_1(\bar{x},\bar{y},\boldsymbol{\mu})\bullet F_2(x,y,t,\mathbf{p})-f_2(\bar{x},\bar{y},\boldsymbol{\mu})\bullet F_1(x,y,t,\mathbf{p}) \tag{3.60}$$

As in Eq.(3.41), the zero-order G-function from for for $\bar{x}=x$ and $y=\bar{y}$ gives

$$\begin{aligned}G_\alpha^{(0)}(x,y,t,\mathbf{p})&=f_1(\bar{x},\bar{y},\boldsymbol{\mu})\bullet F_2(x,y,t,\mathbf{p})-f_2(\bar{x},\bar{y},\boldsymbol{\mu})\bullet F_1(x,y,t,\mathbf{p})\\&=f_1(\bar{x},\bar{y},\boldsymbol{\mu})\bullet g_2(x,y,t,\boldsymbol{\pi})-f_2(\bar{x},\bar{y},\boldsymbol{\mu})\bullet g_1(x,y,t,\boldsymbol{\pi})\end{aligned} \tag{3.61}$$

where $G_\alpha(x,y,t,\mathbf{p})\equiv G_\alpha^{(0)}(x,y,t,\mathbf{p})$. The zero-order G-function of a perturbed flow to the specified, first integral surface is a project of the vector field in the normal direction of the Hamiltonian energy surface (or the first integral surface). Such a quantity is illustrated in Fig.3.9. The Hamiltonian energy orbit is expressed by the gray curves. The normal vector is expressed through $\mathbf{n}_{S_\alpha}$. Such a quantity can be determined for an instantaneous time to measure the change of vector fields in the normal direction of the first integral surface. If the separatrix is selected as a reference surface, the normal vector can be expressed by $\mathbf{n}_{S_s}$ or $\mathbf{n}_{\partial\Omega_{ij}}$. The dashed lines on the Hamiltonian energy surface are in the tangential direction of the integrable flow. The normal component of the perturbed vector field is clearly illustrated. The saddle point is p_0. The separatrix is depicted by a thick

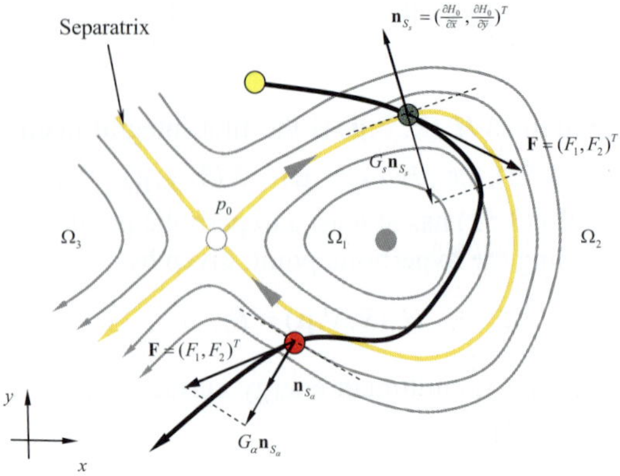

Figure 3.9. The normal component of the perturbed vector field on the normal direction of the Hamiltonian energy surface (or the first integral surface).

yellow curve. Because the global transversality means that the perturbed flow can pass over the separatrix from one domain to another adjacent domain. The necessary and sufficient condition for the *global* transversal flow to the separatrix $\partial\Omega_{ij}$ for the perturbed Hamiltonian system in Eq.(3.52) is determined for $(x_m^{(\alpha)}, y_m^{(\alpha)}) = (\overline{x}_m^{(\alpha)}, \overline{y}_m^{(\alpha)}) \in \partial\Omega_{ij}$ from

$$\begin{aligned} &H_0(x_m^{(\alpha)}, y_m^{(\alpha)}, \boldsymbol{\mu}) = E_s; \\ &G_{\partial\Omega_{ij}}^{(0)}\left(x_m^{(\alpha)}, y_m^{(\alpha)}, t_m, \mathbf{p}\right) > 0 \quad \text{for } E_\alpha < E_s, \\ &G_{\partial\Omega_{ij}}^{(0)}\left(x_m^{(\alpha)}, y_m^{(\alpha)}, t_m, \mathbf{p}\right) < 0 \quad \text{for } E_\alpha > E_s. \end{aligned} \tag{3.62}$$

where $G_\alpha(x, y, t, \mathbf{p}) \equiv G_\alpha^{(0)}(x, y, t, \mathbf{p})$. As in Eq.(3.43), the first-order G-function from for $\overline{x} = x$ and $y = \overline{y}$ gives

$$\begin{aligned} G_\alpha^{(1)}\left(x, y, t, p\right) = &\, 2D_{\overline{\mathbf{x}}} f_1\left(\overline{x}, \overline{y}, \boldsymbol{\mu}\right) \bullet g_2\left(x, y, t, \boldsymbol{\pi}\right) \\ &- 2D_{\overline{\mathbf{x}}} f_2\left(\overline{x}, \overline{y}, \boldsymbol{\mu}\right) \bullet g_1\left(x, y, t, \boldsymbol{\pi}\right) \\ &+ f_1\left(\overline{x}, \overline{y}, \boldsymbol{\mu}\right) \bullet \left[D_{\mathbf{x}} F_2\left(x, y, t, \mathbf{p}\right) - D_{\overline{\mathbf{x}}} f_2\left(\overline{x}, \overline{y}, \boldsymbol{\mu}\right)\right] \\ &- f_2\left(\overline{x}, \overline{y}, \boldsymbol{\mu}\right) \bullet \left[D_{\mathbf{x}} F_1\left(x, y, t, \mathbf{p}\right) - D_{\overline{\mathbf{x}}} f_1\left(\overline{x}, \overline{y}, \boldsymbol{\mu}\right)\right] \end{aligned} \tag{3.63}$$

where the total differentiations are for ($\sigma = 1, 2$)

$$D_{\overline{\mathbf{x}}} f_\sigma\left(\overline{x}, \overline{y}, \boldsymbol{\mu}\right) = \left[\frac{\partial f_\sigma\left(\overline{x}, \overline{y}, \boldsymbol{\mu}\right)}{\partial \overline{x}} f_1\left(\overline{x}, \overline{y}, \boldsymbol{\mu}\right) + \frac{\partial f_\sigma\left(\overline{x}, \overline{y}, \boldsymbol{\mu}\right)}{\partial \overline{y}} f_2\left(\overline{x}, \overline{y}, \boldsymbol{\mu}\right)\right], \tag{3.64a}$$

$$D_{\mathbf{x}}F_{\sigma}(x,y,t,\mathbf{p}) = \left[\frac{\partial F_{\sigma}(x,y,t,\mathbf{p})}{\partial x}F_1(x,y,t,\mathbf{p}) + \frac{\partial F_{\sigma}(x,y,t,\mathbf{p})}{\partial y}F_2(x,y,t,\mathbf{p}) + \frac{\partial g_{\sigma}(x,y,t,\boldsymbol{\pi})}{\partial t}\right]. \tag{3.64b}$$

From the definition in Chapter 2, the *first-order* G-function is the time rate of change for the zero-order G-function of the perturbed systems to the Hamiltonian system. In other words, the *first-order* G-function is the time rate of change of the component of the perturbed vector field in the normal direction of the Hamiltonian energy surface. The necessary and sufficient conditions for the global *tangential* flow of the perturbed Hamiltonian system to the separatrix in Eq.(3.52) are obtained for $(x_m^{(\alpha)}, y_m^{(\alpha)}) = (\overline{x}_m^{(\alpha)}, \overline{y}_m^{(\alpha)}) \in \partial\Omega_{ij}$, i.e.,

$$\left.\begin{aligned} &H_0(x_m^{(\alpha)}, y_m^{(\alpha)}, \boldsymbol{\mu}) = E_s;\\ &G_{\partial\Omega_{ij}}^{(0)}\left(x_m^{(\alpha)}, y_m^{(\alpha)}, t_m, \mathbf{p}\right) = 0;\\ &G_{\partial\Omega_{ij}}^{(1)}\left(x_m^{(\alpha)}, y_m^{(\alpha)}, t_m, \mathbf{p}\right) < 0 \quad \text{for } E_\alpha < E_s,\\ &G_{\partial\Omega_{ij}}^{(1)}\left(x_m^{(\alpha)}, y_m^{(\alpha)}, t_m, \mathbf{p}\right) > 0 \quad \text{for } E_\alpha > E_s. \end{aligned}\right\} \tag{3.65}$$

In the similar fashion, the $(2k)^{th}$-order, global transversality of a flow to the separatrix in Eq.(3.52) can be obtained, and the $(2k-1)^{th}$-order global tangency of a flow to the separatrix can be achieved as well. As in Fig.3.9, the global transversali-ty and tangency of the perturbed flow to the separatrix are sketched in Fig.3.10 (a) and (b).

3.7. A damped Duffing oscillator

To demonstrate the tangency and transversality of the perturbed flow to the separatrix, herein, consider a periodically forced, damped-Duffing oscillator with a twin-well potential

$$\ddot{x} + \delta\dot{x} - \alpha_1 x + \alpha_2 x^3 = Q_0\cos(\Omega t), \tag{3.66}$$

where system parameters $\alpha_1 > 0$ and $\alpha_2 > 0$ are relative to linear and nonlinear terms. Q_0 and Ω are excitation strength and frequency for the periodic forcing, respectively. Equation (3.66) can be expressed by state variables

$$\dot{x} = y,\ \ \dot{y} = \alpha_1 x - \alpha_2 x^3 + Q_0\cos(\Omega t) - \delta y \tag{3.67}$$

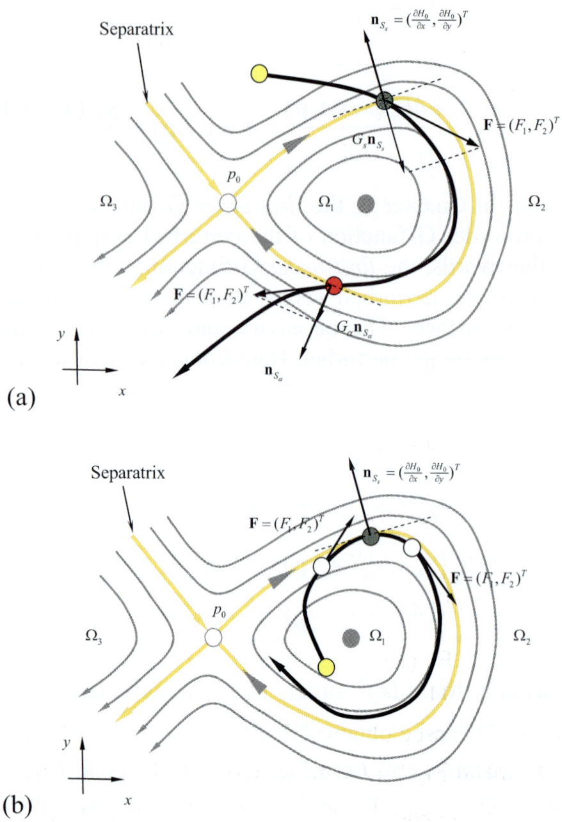

Figure 3.10. (a) The global transversality and (b) global tangency inside of the separatrix of a perturbed flow.

from which we have

$$\left.\begin{aligned} &f_1(x,y,\boldsymbol{\mu}) = y, f_2(x,y,\boldsymbol{\mu}) = \alpha_1 x - \alpha_2 x^3 \\ &g_1(x,y,\boldsymbol{\pi}) = 0, g_2(x,y,t,\boldsymbol{\mu}) = Q_0 \cos(\Omega t) - \delta y \\ &F_1(x,y,t,\mathbf{p}) = y, \\ &F_2(x,y,t,\mathbf{p}) = \alpha_1 x - \alpha_2 x^3 + Q_0 \cos(\Omega t) - \delta y \end{aligned}\right\} \tag{3.68}$$

and $\boldsymbol{\mu} = (\alpha_1, \alpha_2)^T$, $\boldsymbol{\pi} = (\delta, \Omega, Q_0)^T$ and $\mathbf{p} = (\alpha_1, \alpha_2, \delta, \Omega, Q_0)^T$.

The corresponding integrable system is given by

$$\dot{\overline{x}} = \overline{y} \text{ and } \dot{\overline{y}} = \alpha_1 \overline{x} - \alpha_2 \overline{x}^3 \tag{3.69}$$

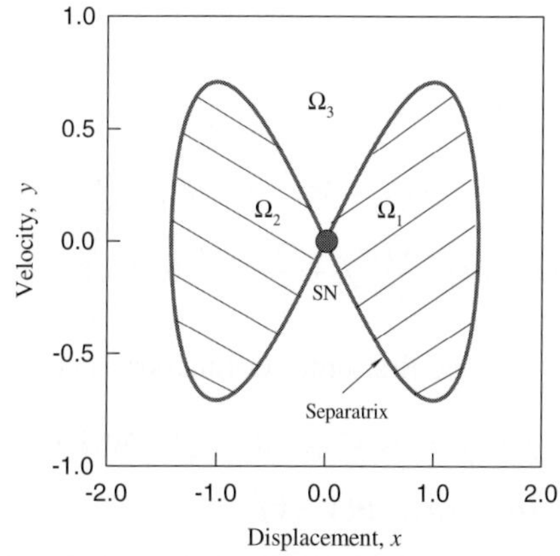

Figure 3.11. Separatrix and sub-domains of the integrable Duffing oscillator ($\alpha_1 = \alpha_2 = 1$).

with $f_1(\bar{x}, \bar{y}, \boldsymbol{\mu}) = \bar{y}$ and $f_2(\bar{x}, \bar{y}, \boldsymbol{\mu}) = \alpha_1 \bar{x} - \alpha_2 \bar{x}^3$, and the first integral manifold (or the Hamiltonian)

$$H_0 = \tfrac{1}{2}\bar{y}^2 - \tfrac{1}{2}\alpha_1 \bar{x}^2 + \tfrac{1}{4}\alpha_2 \bar{x}^4 = E_0. \tag{3.70}$$

The integrable system of Eq.(3.66) possesses a homoclinic separatrix relative to the saddle-node point $(0,0)$ for $H_0 = E_0 = 0$. This separatrix separates the domain of the integrable Duffing oscillator into three domains Ω_i ($i = 1,2,3$) in phase space, which is shown in Fig.3.11 with $\alpha_1 = \alpha_2 = 1$. For the same energy level, the motions of Eq.(3.69) in sub-domains Ω_1 and Ω_2 are of the skew symmetry. In sub-domain Ω_3, equation (3.70) will have the motion possessing a large orbit to enclose the separatrix. The separatrix is determined by

$$H_0 = \tfrac{1}{2}\bar{y}^2 - \tfrac{1}{2}\alpha_1 \bar{x}^2 + \tfrac{1}{4}\alpha_2 \bar{x}^4 = E_s = 0. \tag{3.71}$$

3.7.1. Transversal and tangential conditions

For any Hamiltonian energy surface $H_0(\bar{x}, \bar{y}, \boldsymbol{\mu}) = E$, the corresponding normal vector is given by $\mathbf{n}_{S_\alpha} = (-\alpha_1 \bar{x} + \alpha_2 \bar{x}^3, \bar{y})^T$. From Eqs.(3.61), (3.62) and (3.68), the zero-order G-function to the Hamiltonian energy surface is

$$\begin{aligned} G_\alpha^{(0)}(x,y,t,\mathbf{p}) &= f_1(\bar{x},\bar{y},\boldsymbol{\mu})\boldsymbol{\cdot} g_2(x,y,t,\boldsymbol{\pi}) - f_2(\bar{x},\bar{y},\boldsymbol{\mu})\boldsymbol{\cdot} g_1(x,y,t,\boldsymbol{\pi}) \\ &= \bar{y}\boldsymbol{\cdot}\left[Q_0 \cos(\Omega t) - \delta y\right] \underset{\bar{y}=y}{=} y Q_0 \cos(\Omega t) - \delta y^2. \end{aligned} \tag{3.72}$$

From Eqs.(3.64) and (3.68), the total derivatives of vector fields are

$$
\begin{aligned}
DF_2(x,y,t,\mathbf{p}) &= \frac{\partial F_2}{\partial x}F_1 + \frac{\partial F_2}{\partial y}F_2 + \frac{\partial F_2}{\partial t} \\
&= (\alpha_1 - 3\alpha_2 x^2)y - \delta\dot{y} - Q_0\Omega\sin(\Omega t), \\
Df_2(\bar{x},\bar{y},\boldsymbol{\mu}) &= \frac{\partial f_2}{\partial \bar{x}}f_1 + \frac{\partial f_2}{\partial \bar{y}}f_2 = (\alpha_1 - 3\alpha_2\bar{x}^2)\bar{y}.
\end{aligned}
\tag{3.73}
$$

From Eq.(3.63) and (3.68), the first-order G-function is computed by

$$
\begin{aligned}
G_\alpha^{(1)}(x,y,t,\mathbf{p}) &= \bar{y}\cdot\left[(\alpha_1 - 2\alpha_2 x^2)y - \delta\dot{y} - Q_0\Omega\sin\Omega t - (\alpha_1 - \alpha_2\bar{x}^2)\bar{y}\right] \\
&\quad + (\alpha_1\bar{x} - \alpha_2\bar{x}^3)\cdot(\dot{y} - \dot{\bar{y}}) \\
&\underbrace{=}_{\bar{y}=y,\bar{x}=x} -\delta Q_0 y\cos(\Omega t) + \delta^2 y^2 - yQ_0\Omega\sin(\Omega t) \\
&\quad + (\alpha_1 x - \alpha_2 x^3)[-2\delta y + Q_0\cos(\Omega t)].
\end{aligned}
\tag{3.74}
$$

With Eq.(3.70), the global transversality conditions of the perturbed flow in Eq. (3.67) starting with the initial energy ($H_0(\bar{x},\bar{y},\boldsymbol{\mu}) = E_0$) to the separatrix (i.e., $H_0(x,y,\boldsymbol{\mu}) = E_s = 0$) at point (x_m, y_m) with time t_m are

$$
\left.
\begin{aligned}
&\frac{1}{2}y_m^2 - \frac{1}{2}\alpha_1 x_m^2 + \frac{1}{4}\alpha_2 x_m^4 = E_s = 0 \\
&G_{\partial\Omega_{ij}}^{(0)}(x_m, y_m, t_m, \mathbf{p}) = y_m Q_0\cos(\Omega t_m) - \delta y_m^2 \begin{cases} > 0 \text{ for } E < E_s, \\ < 0 \text{ for } E > E_s. \end{cases}
\end{aligned}
\right\}
\tag{3.75}
$$

Introduce a critical parameter if $Q_0 > \delta\,|\,y_m\,|$

$$
\Phi_{cr} = \arccos\frac{\delta y_m}{Q_0}. \tag{3.76}
$$

The second equation in Eq.(3.75) gives for $Q_0 > \delta\,|\,y_m\,|$

$$
\operatorname{mod}(\Omega t_m, 2\pi) \in \begin{cases} [0,\Phi_{cr}) \cup (2\pi - \Phi_{cr}, 2\pi] & \text{if } y_m > 0 \\ (\Phi_{cr}, 2\pi - \Phi_{cr}) & \text{if } y_m < 0 \end{cases} \tag{3.77}
$$

for the case of $E > E_s = 0$,

$$
\operatorname{mod}(\Omega t_m, 2\pi) \in \begin{cases} [0,\Phi_{cr}) \cup (2\pi - \Phi_{cr}, 2\pi] & \text{if } y_m < 0 \\ (\Phi_{cr}, 2\pi - \Phi_{cr}) & \text{if } y_m > 0 \end{cases} \tag{3.78}
$$

for the case of $E < E_s = 0$. For all y_m satisfying $Q_0 < \delta \mid y_m \mid$, the second equation in Eq.(3.75) gives

$$\mathrm{mod}(\Omega t_m, 2\pi) \in \left[0, 2\pi\right] \text{ for } E > E_s, \tag{3.79}$$

which implies $G(x_m, y_m, t_m, \mathbf{p}) < 0$ always. However, if $G(x_m, y_m, t_m, \mathbf{p}) > 0$, the flow to the separatrix is transversal from domain Ω_i to Ω_3. With Eqs.(3.73) and (3.74), the global tangency conditions of a perturbed flow in Eq.(3.65) with the initial energy ($H_0(\overline{x}, \overline{y}, \boldsymbol{\mu}) = E_0$) to the separatrix surface (i.e., $H_0(\overline{x}, \overline{y}, \boldsymbol{\mu}) = E_s$) at point (x_m, y_m) with time t_m are

$$\left.\begin{aligned}
&\frac{1}{2} y_m^2 - \frac{1}{2}\alpha_1 x_m^2 + \frac{1}{4}\alpha_2 x_m^4 = E_s = 0, \\
&G_{\partial\Omega_{ij}}^{(0)}(x_m, y_m, t_m, \mathbf{p}) = y_m Q_0 \cos(\Omega t_m) - \delta y_m^2 = 0, \\
&G_{\partial\Omega_{ij}}^{(1)}(x_m, y_m, t_m, \mathbf{p}) = -y_m Q_0 \Omega \sin(\Omega t_m) + (\alpha_1 x_m - \alpha_2 x_m^3) \\
&\qquad\qquad \times[-2\delta y_m + Q_0 \cos(\Omega t_m)] \begin{cases} > 0 & \text{for } E > E_s, \\ < 0 & \text{for } E < E_s. \end{cases}
\end{aligned}\right\} \tag{3.80}$$

The second equation of Eq.(3.80) gives the necessary condition for the global tangency of a perturbed flow to the separatrix (i.e., $H_0(x, y, \boldsymbol{\mu}) = E_s$). The corresponding sufficient condition for such a flow tangential to the inside (or outside) of separatrix are $G^{(1)} < 0$ (or $G^{(1)} > 0$). The relation between the grazing velocity and phase can be obtained from Eq.(3.80). The conditions for the tangency are critical for the flow transversality. To illustrate the global tangency (or grazing) and transversality conditions, consider parameters $\alpha_1 = \alpha_2 = 1$ and the separatrix with $E_s = 0$. The notations $\overrightarrow{\partial\Omega_{ii}}$ and $\overrightarrow{\partial\Omega_{33}}$ represent the grazing occurrence at the separatrix in the domain Ω_i ($i = 1, 2$) and Ω_3, which are depicted by the circle and solid curves, respectively. The corresponding $G^{(1)}$-function for the global grazing relative to $\overrightarrow{\partial\Omega_{ii}}$ or $\overrightarrow{\partial\Omega_{33}}$ are less than or greater than zero (i.e., $G^{(1)} < 0$ or $G^{(1)} > 0$). The notation $\overleftarrow{\partial\Omega_{3i}}$ (or $\overrightarrow{\partial\Omega_{3i}}$) represents the flow globally transversal to the separatrix from domain Ω_i ($i = 1, 2$) to Ω_3 (or Ω_3 to Ω_i). The phase range of the global transversality will be given by Eqs.(3.77)-(3.79).

Consider the damping and excitation amplitude to be $\delta = 0.15$ and $Q_0 = 0.3$. From Eq.(3.80), we have $y_m \in [-2, 2]$, but from $H_0(x_m, y_m, \boldsymbol{\mu}) = E_s = 0$, the velocity on the separatrix satisfies $|y_m| < 2$. Hence, all values on the separatrix are possible for global tangency, and the grazing phase cannot take the entire range

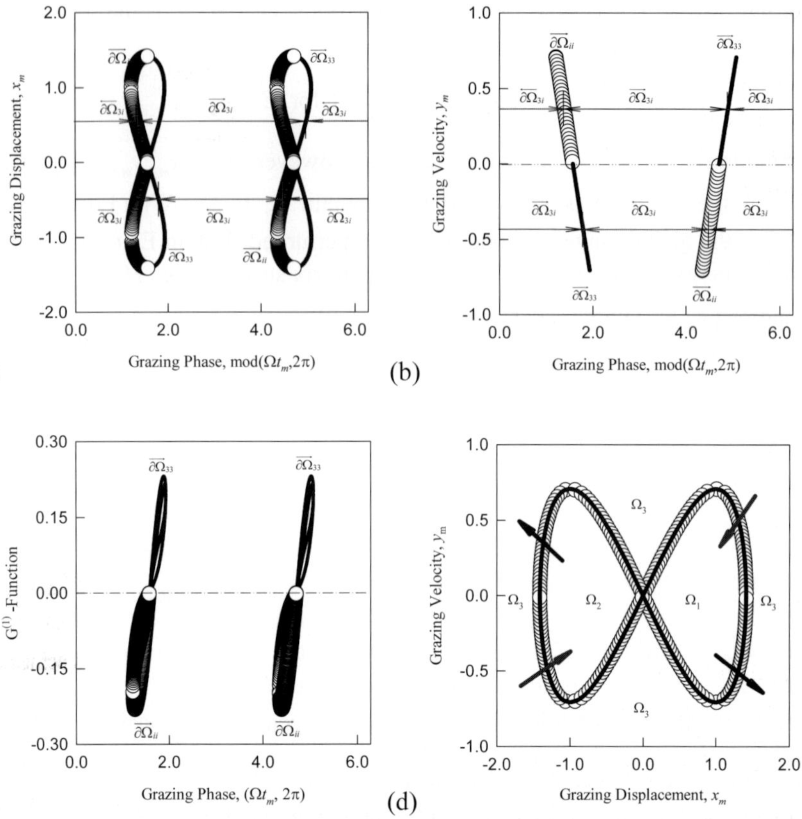

Figure 3.12. Global tangency and global transversality intervals on the separatrix: (a) grazing displacement versus grazing phase, (b) grazing velocity versus grazing phase, (c) $G^{(1)}$ varying with gazing phase and (d) inner and outer grazing on separatrix . ($\alpha_1 = \alpha_2 = 1$, $\delta = 0.15$, $Q_0 = 0.3$). $\overrightarrow{\partial\Omega_{ii}}$ ($i = 1,2$) and $\overrightarrow{\partial\Omega_{33}}$ represent the grazing occurrence at the separatrix in the domain Ω_i and Ω_3, which are depicted by the dotted and solid curves, respectively. $\overleftarrow{\partial\Omega_{3i}}$ ($i = 1,2$) (or $\overrightarrow{\partial\Omega_{3i}}$) represents the flow globally transversal to the separatrix from domain Ω_i to Ω_3 (or Ω_3 to Ω_i).

of $[0, 2\pi]$. The grazing displacement, grazing velocity and $G^{(1)}$-function against grazing phase are plotted in Figs.3.12(a)-(c), respectively. Note that $G^{(1)} = G^{(1)}_{\partial\Omega_{ij}}$. The intervals of the switching phase for the global transversality to the separatrix are labeled by $\overrightarrow{\partial\Omega_{3i}}$ and $\overleftarrow{\partial\Omega_{3i}}$ ($i = 1,2$). The skew symmetry of the global grazing and transversality is observed. The sum of the two phase intervals for the global transversality is $[0, 2\pi]$. In Fig.3.12(d), the grazing on separatrix in Ω_3 and Ω_i ($i = 1, 2$) are shown through the solid and dotted curves, respectively. The arrows in Fig.3.12(d) represent the global transversality on the separatrix from Ω_3 to Ω_i ($i = 1, 2$) and from Ω_i ($i = 1, 2$) to Ω_3 through the inward and outward arrows, respectively.

3.7.2. Global transversal and tangential flows

The analytical prediction of the global tangency and transversality of a perturbed flow to the separatrix is presented. The global tangency and transversality of perturbed flows to the separatrix are illustrated from the analytical predictions. The tangency to the perturbed flow in domain Ω_3 is presented first. Consider parameters ($\alpha_1 = \alpha_2 = 1$, $\delta = 0.15$, $Q_0 = 0.4$ and $\Omega = 1$). The corresponding transversality and tangency of perturbed flows to the separatrix are shown in Fig.3.13. The dash and solid curves represent the separatrix and global flows, respectively. A point $(x_m, y_m) \approx (0.5301, 0.4914)$ on the separatrix is arbitrarily chosen. The grazing phase on the separatrix is $\mathrm{mod}(\Omega t_m, 2\pi) \approx 4.899$ from the global grazing condition. Thus, the grazing phase $\Omega t_m \approx 4.899$ will be achieved for numerical simulation. The initial condition $(x_0, y_0) \approx (0.7640, -2.2492)$ for $t = 0$ is achieved from a time-inverse computation. Thus, a tangential flow to the separatrix in domain Ω_3 is presented for one period. The starting and ending points of the tangential flow to the separatrix are labeled by the two white circles. At the grazing point (x_m, y_m), the numerical computation gives $G = 0$ and $G^{(1)} = 0.1654 > 0$. Note that $G^{(1)} = G^{(1)}_{\partial\Omega_{ij}}$ and $G = G^{(0)}_{\partial\Omega_{ij}}$. However, at the same point on the separatrix, the different phase is chosen to look into the global transversality on the separatrix. For the switching phase $\Omega t_m = 4.0$, the inverse-time computation gives the initial condition $(x_0, y_0) \approx (0.0502, -1.9694)$ at $t = 0$. Using such an initial condition, it is observed that a flow possesses the global transversality from domain Ω_3 to Ω_1 at point $(x_m, y_m) \approx (0.5301, 0.4914)$ with $\Omega t_m = 4.0$. The starting and ending points of the global transversal flow are labeled by the two dark circles. The numerical computation gives $G \approx -0.1647 < 0$ and $G^{(1)} = 0.0176 > 0$. Similarly, consider the switching phase $\Omega t_m = 5.5$ at the same point of the separatrix. The initial condition $(x_0, y_0) \approx (-0.9262, -1.4245)$ at $t = 0$ is computed via the time-inverse computation. A perturbed flow passes through from domain Ω_1 to Ω_3 at point $(x_m, y_m) \approx (0.5301, 0.4914)$. The corresponding values of G and

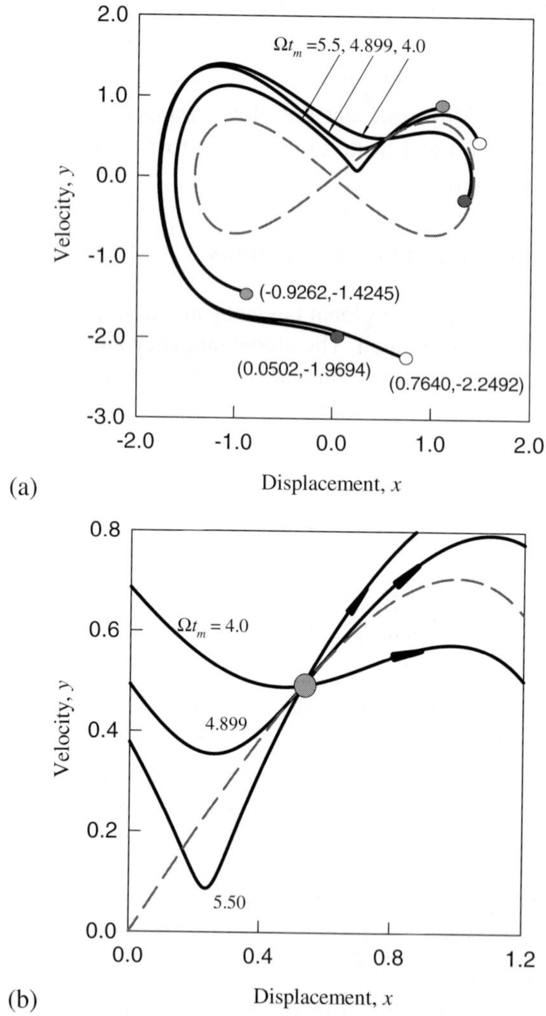

Figure 3.13. (a) Global grazing and transversal flows on the separatrix and (b) Zoomed view of the grazing and transversal point on the separatrix. ($\alpha_1 = \alpha_2 = 1$, $\delta = 0.15$, $Q_0 = 0.4$ and $\Omega = 1$). The dash and solid curves represent the separatrix and global flows. The gray, white and dark circles denote the starting and ending points, respectively, for the global transversal flow from Ω_1 to Ω_3, the global tangential flow in domain Ω_3, and the global transversal flow from Ω_3 to Ω_1 at point $(x_m, y_m) \approx (0.5344, 0.4947)$.

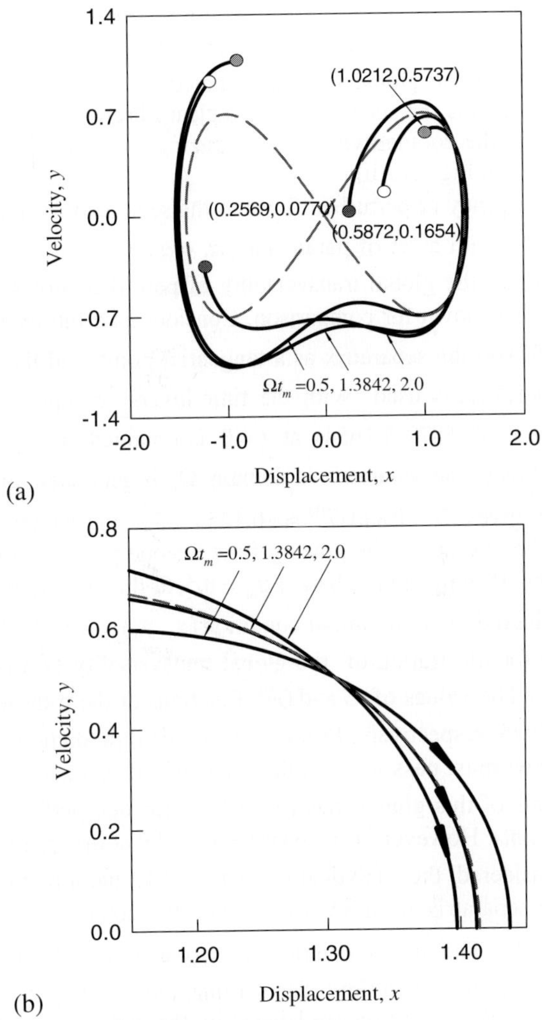

Figure 3.14. (a) Global grazing and transversal flows on the separatrix and (b) Zoomed view of the grazing and transversal point on the separatrix. ($\alpha_1 = \alpha_2 = 1$, $\delta = 0.15$, $Q_0 = 0.4$ and $\Omega = 1$). The dash and solid curves represent the separatrix and global flows. The gray, white and dark circles denote the starting and ending points, respectively, for the global transversal flow from Ω_1 to Ω_3, the global tangential flow in domain Ω_3, and the global transversal flow from Ω_3 to Ω_1 at point $(x_m, y_m) \approx (1.3094, 0.4947)$.

$G^{(1)}$ are 0.1301 and 0.1751. From the global transversality condition, the global flow must pass through the separatrix from Ω_1 to Ω_3 at that point. The starting and ending points of the global flow are labeled by the two gray circles. The dashed curve represents the separatrix. For a clear observation of global grazing and transversality, the zoomed view of the grazing and transversal point on the separatrix is given in Fig.3.13(b).

The global tangency of perturbed flows to the separatrix in domain Ω_1 is illustrated in Fig.3.14 with a set of parameter ($\alpha_1 = \alpha_2 = 1$, $\delta = 0.15$, $Q_0 = 0.4$ and $\Omega = 1$). In addition, the global transversality of perturbed flows near the tangential flow are also shown for comparison. Consider an arbitrary point (x_m, y_m) $\approx (1.3065, 0.4997)$ on the separatrix as a tangential point, and the grazing phase $\text{mod}(\Omega t_m, 2\pi) \approx 1.3821$ is used. With the time-inverse computation, the initial condition $(x_0, y_0) \approx (0.5872, 0.1654)$ at $t = 0$ is obtained. The global tangency of a perturbed flow to the separatrix in domain Ω_1 is generated, and the numerical computation gives $G = 0$ and $G^{(1)} \approx -0.1259 < 0$ at the tangential point. The starting and ending points of the grazing flow for one period are labeled by the two white circles. If the grazing phase $\Omega t_m = 0.5$ at the grazing point on the separatrix is considered, then the initial condition $(x_0, y_0) \approx (1.0212, 0.5737)$ at $t =$ 0 will be used for illustration of the global transversality of a perturbed flow from Ω_1 to Ω_3. The values of G and $G^{(1)}$-functions at the transversal point are 0.1380 and -0.3025, respectively. From the theoretical prediction, it implies that the perturbed flow must pass through the separatrix from Ω_1 to Ω_3. The starting and ending points of this global transversal flow for one period are labeled by the two green points. However, if the phase $\Omega t_m = 2.0$ at the grazing point on the separatrix is considered, the analytical prediction tells that a perturbed flow will pass though the separatrix from Ω_3 to Ω_1 with the condition $G = -0.1206 < 0$ and $G^{(1)} = -0.3176$. The initial condition $(x_0, y_0) \approx (0.2533, 0.0476)$ at $t = 0$ will be employed for this perturbed flow. The starting and ending points of the global transversal flow on the separatrix are labeled by the dark circles. In Fig.3.14(b), a zoom-ed view of the global grazing and transversal flows in the neighborhood of $(x_m, y_m) \approx (1.3065, 0.4997)$ is given.

3.8. Global transversality to a generalized separatrix

In beginning of this chapter, the separatrix is produced by the homoclinic or heteroclinic surface. In this section, a generalized separatrix surface will be discussed. Suppose there is an $(n-1)$-dimensional surface governed by an equation

$\varphi(\mathbf{x},\boldsymbol{\lambda})=0$ for $\mathbf{x}\in\mathbb{R}^n$ and $\boldsymbol{\lambda}\in\mathbb{R}^{m_3}$. This surface divided phase space into many sub-domains $\Omega_i\subset\mathbb{R}^n$ ($i=1,2,\cdots,m$) for Eq.(3.1). On each domain Ω_i, the dynamical behavior of Eq.(3.1) is different from the adjacent sub-domain Ω_j ($j\neq i$, $j\in\{1,2,\cdots,m\}$). So in each domain Ω_i, equation (3.4) can be used to determine the corresponding dynamic behaviors. This special surface is called *the generalized separatrix* for Eq.(3.1). The global tangency and transversality of the flow to the boundary $\partial\Omega_{ij}$ governed by $\varphi(\mathbf{x},\boldsymbol{\lambda})=0$ can be similarly determined as in Sections 3.3-3.5. The strict mathematical definition is given as follows

DEFINITION 3.9. Suppose there is an $(n-1)$-dimensional surface

$$S=\left\{\mathbf{x}\,\middle|\,\varphi(\mathbf{x},\boldsymbol{\lambda})=0 \text{ where } \varphi \text{ is } C^r\text{-continuous } (r\geq 1)\right\}\subset\Re^{n-1} \tag{3.81}$$

and the sub-manifold surface relative to two adjacent domains Ω_α ($\alpha=\{i,j\}$ and $i,j\in\{1,2,\cdots,m\}$) is defined by

$$\partial\Omega_{ij}=S\cap\bar{\Omega}_i\cap\bar{\Omega}_j=\left\{\mathbf{x}\,\middle|\,\begin{array}{l}\varphi_{ij}(\mathbf{x},\boldsymbol{\lambda})=0 \text{ where}\\ \varphi_{ij} \text{ is } C^r\text{-continuous } (r\geq 1)\end{array}\right\}\subset S. \tag{3.82}$$

In domain Ω_α ($\alpha=\{i,j\}$), the dynamic behavior is determined by

$$\dot{\mathbf{x}}^{(\alpha)}=\mathbf{F}(\mathbf{x}^{(\alpha)},t,\mathbf{p}) \tag{3.83}$$

and both $\mathbf{x}^{(i)}(t)=\boldsymbol{\Phi}(t_0,\mathbf{x}_0^{(i)},t,\mathbf{p})$ and $\mathbf{x}^{(j)}(t)=\boldsymbol{\Phi}(t_0,\mathbf{x}_0^{(j)},t,\mathbf{p})$ are different. On the boundary $\partial\Omega_{ij}$, $\mathbf{F}(\mathbf{x}^{(i)},t,\mathbf{p})$ and $\mathbf{F}(\mathbf{x}^{(j)},t,\mathbf{p})$ are C^r-continuous for all $r=0,1,2,\cdots$ and suppose the flow on the $(n-1)$-dimensional surface $\varphi(\mathbf{x}^{(0)},\boldsymbol{\lambda})=0$ is determined by

$$\dot{\mathbf{x}}^{(0)}=\mathbf{F}^{(0)}(\mathbf{x}^{(0)},\boldsymbol{\lambda}). \tag{3.84}$$

Therefore, the $(n-1)$-dimensional surface is termed *the generalized separatrix* for the dynamical system in Eq.(3.1). The surface $\varphi(\mathbf{x}^{(0)},\boldsymbol{\lambda})=0$ is the first integral manifold of Eq.(3.83). If Eq.(3.83) is replaced by Eq.(3.2) and the hyperbolic points of Eq.(3.83) exists on the generalized surface, then such a generalized separatrix becomes the generic separatrix with hyperbolic points.

The flow $\bar{\mathbf{x}}^{(s)}$ on the separatrix will be replaced by a flow $\mathbf{x}^{(0)}$ on the generalized separatrix and the corresponding normal vector of the separatrix boundary $\partial\Omega_{ij}$ is given at the point $(\mathbf{x}_m,t_m)$ by

$$\mathbf{n}_{\partial\Omega_{ij}} = \nabla\varphi_{ij} = (\frac{\partial\varphi_{ij}}{\partial x_1}, \frac{\partial\varphi_{ij}}{\partial x_2}, \cdots, \frac{\partial\varphi_{ij}}{\partial x_n})^T{}_{(\mathbf{x}_m, t_m)}. \tag{3.85}$$

Similar to Definition 3.8, the G-function of the kth -order to the generalized separatrix is defined for ($k = 0,1,2,\cdots$) as

$$\begin{aligned} G_{\partial\Omega_{ij}}^{(k)}(\mathbf{x}_m^{(\alpha)}, t_m, \mathbf{p}, \boldsymbol{\lambda}) &= \sum_{r=1}^{k+1} C_{k+1}^r D_{\bar{\mathbf{x}}}^{k+1-r} \mathbf{n}_{S_\alpha}^T(\mathbf{x}^{(0)}) \bullet \left[\frac{d^r\mathbf{x}}{dt^r} - \frac{d^r\mathbf{x}^{(0)}}{dt^r} \right] \Big|_{(\bar{\mathbf{x}}_m^{(0)}, \mathbf{x}_m^{(\alpha)}, t_m)} \\ &= \sum_{r=1}^{k+1} C_{k+1}^r D_{\bar{\mathbf{x}}}^{k+1-r} \mathbf{n}_{S_\alpha}^T(\mathbf{x}^{(0)}) \bullet \left[D_{\mathbf{x}}^{r-1}\mathbf{F}(\mathbf{x}, t, \mathbf{p}) \right. \\ &\quad \left. - D_{\bar{\mathbf{x}}}^{r-1}\mathbf{F}^{(0)}(\mathbf{x}^{(0)}, \boldsymbol{\lambda}) \right] \Big|_{(\mathbf{x}_m^{(0)}, \mathbf{x}_m^{(\alpha)}, t_m)}. \end{aligned} \tag{3.86}$$

Notice that $\mathbf{x}_m^{(\alpha)} \equiv \mathbf{x}_m^{(0)}$ on the generalized separatrix $\partial\Omega_{ij}$. For a special case ($k = 0$), owing to $\mathbf{n}_{\partial\Omega_{ij}}^T \bullet \mathbf{F}^{(0)}(\mathbf{x}_m^{(0)}, \boldsymbol{\lambda}) = \mathbf{n}_{\partial\Omega_{ij}}^T \bullet \dot{\mathbf{x}}_m^{(0)} = 0$, the G-function is

$$G_{\partial\Omega_{ij}}(\mathbf{x}_m^{(\alpha)}, t_m, \mathbf{p}, \boldsymbol{\lambda}) \equiv \mathbf{n}_{\partial\Omega_{ij}}^T \bullet \mathbf{F}(\mathbf{x}_m^{(\alpha)}, t_m, \mathbf{p}) = \nabla\varphi_{ij}(\mathbf{x}_m^{(0)}, \boldsymbol{\lambda}) \bullet \mathbf{F}(\mathbf{x}_m^{(\alpha)}, t_m, \mathbf{p}). \tag{3.87}$$

For a special case ($k = 1$), we have

$$\begin{aligned} G_{\partial\Omega_{ij}}^{(1)}(\mathbf{x}_m^{(\alpha)}, t_m, \mathbf{p}, \boldsymbol{\lambda}) &\equiv 2D\mathbf{n}_{\partial\Omega_{ij}}^T(\mathbf{x}_m^{(0)}, \boldsymbol{\lambda}) \bullet [\mathbf{F}(\mathbf{x}_m^{(\alpha)}, t_m, \mathbf{p}) - \mathbf{F}^{(0)}(\mathbf{x}_m^{(0)}, \boldsymbol{\lambda})] \\ &\quad + \mathbf{n}_{\partial\Omega_{ij}}^T(\mathbf{x}_m^{(0)}, \boldsymbol{\lambda}) \bullet [D\mathbf{F}(\mathbf{x}_m^{(\alpha)}, t_m, \mathbf{p}) - D\mathbf{F}^{(0)}(\mathbf{x}_m^{(0)}, \boldsymbol{\lambda})]. \end{aligned} \tag{3.88}$$

For $\mathbf{n}_{\partial\Omega_{ij}} \to \Omega_\alpha$ with $\alpha, \beta \in \{i, j\}$ and $\beta \neq \alpha$, the following cases are given:

(i) The necessary and sufficient conditions for a global tangential flow to the generalized separatrix $\partial\Omega_{ij}$ in Eq. (3.83) are

$$\begin{aligned} &G_{\partial\Omega_{ij}}(\mathbf{x}_m^{(\alpha)}, t_m, \mathbf{p}, \boldsymbol{\lambda}) = 0, \\ &G_{\partial\Omega_{ij}}^{(1)}(\mathbf{x}_m^{(\alpha)}, t_m, \mathbf{p}, \boldsymbol{\lambda}) < 0 \text{ and } G_{\partial\Omega_{ij}}^{(1)}(\mathbf{x}_m^{(\beta)}, t_m, \mathbf{p}, \boldsymbol{\lambda}) > 0. \end{aligned} \tag{3.89}$$

(ii) The necessary and sufficient conditions for a global transversal flow to the generalized separatrix $\partial\Omega_{ij}$ in Eq.(3.83) are

$$G_{\partial\Omega_{ij}}(\mathbf{x}_m^{(\alpha)}, t_m, \mathbf{p}, \boldsymbol{\lambda}) > 0 \text{ and } G_{\partial\Omega_{ij}}(\mathbf{x}_m^{(\beta)}, t_m, \mathbf{p}, \boldsymbol{\lambda}) < 0; \tag{3.90}$$

(iii) The necessary and sufficient conditions for the $(2k-1)^{th}$ -order global tangency of a flow to the generalized separatrix $\partial\Omega_{ij}$ in Eq.(3.83)are

$$G_{\partial\Omega_{ij}}^{(r)}(\mathbf{x}_m^{(\alpha)}, t_m, \mathbf{p}, \boldsymbol{\lambda}) = 0 \quad \text{for } r = 0,1,2,\cdots 2k-2, \tag{3.91}$$

$$G_{\partial\Omega_{ij}}^{(2k-1)}(\mathbf{x}_m^{(\alpha)}, t_m, \mathbf{p}, \boldsymbol{\lambda}) < 0 \text{ and } G_{\partial\Omega_{ij}}^{(2k-1)}(\mathbf{x}_m^{(\beta)}, t_m, \mathbf{p}, \boldsymbol{\lambda}) > 0. \tag{3.92}$$

(iv) The necessary and sufficient conditions for the $(2k)^{th}$-order global transversality of a flow to the generalized separatrix $\partial\Omega_{ij}$ in Eq.(3.83) are

$$G^{(r)}_{\partial\Omega_{ij}}(\mathbf{x}^{(\alpha)}_m, t_m, \mathbf{p}, \boldsymbol{\lambda}) = 0 \quad \text{for } r = 0, 1, 2, \cdots 2k-1 \tag{3.93}$$

$$G^{(2k)}_{\partial\Omega_{ij}}(\mathbf{x}^{(\alpha)}_m, t_m, \mathbf{p}, \boldsymbol{\lambda}) > 0 \text{ and } G^{(2k)}_{\partial\Omega_{ij}}(\mathbf{x}^{(\beta)}_m, t_m, \mathbf{p}, \boldsymbol{\lambda}) < 0. \tag{3.94}$$

If Eq.(3.83) becomes Eq.(3.2), the first integral manifold of Eq.(3.83) with hyperbolic points gives the generic separatrix. The G-functions in Eq.(3.33) and Eq.(3.34) can be recovered. If the generalized separatrix is controlled by an equation $\varphi(\mathbf{x}, t, \boldsymbol{\lambda}) = 0$, the corresponding dynamical system is given by

$$\dot{\mathbf{x}}^{(0)} = \mathbf{F}^{(0)}(\mathbf{x}^{(0)}, t, \boldsymbol{\lambda}). \tag{3.95}$$

The conditions $\mathbf{n}^T_{\partial\Omega_{ij}} \bullet \dot{\mathbf{x}}^{(\alpha)}(t_{m\pm}) = 0$ (or $\mathbf{n}^T_{\partial\Omega_{ij}} \bullet \mathbf{F}^{(\alpha)}(t_{m\pm}) = 0$) for the generalized separatrix $\varphi_{ij}(\mathbf{x}(t), \boldsymbol{\lambda}) = 0$ are replaced by $\mathbf{n}^T_{\partial\Omega_{ij}} \bullet \dot{\mathbf{x}}^{(\alpha)}(t_{m\pm}) + \frac{\partial\varphi_{ij}}{\partial t} = 0$ and $\mathbf{n}^T_{\partial\Omega_{ij}} \bullet \mathbf{F}^{(\alpha)}(t_{m\pm}) + \frac{\partial\varphi_{ij}}{\partial t} = 0$, respectively. The rest conditions can be kept as in the present form. In an alternative way, a new G-function is defined as

$$\begin{aligned} G^{(0)}_{\partial\Omega_{ij}}(\mathbf{x}^{(\alpha)}_m, t_m, \mathbf{p}, \boldsymbol{\lambda}) &\equiv \mathbf{n}^T_{\partial\Omega_{ij}}(\mathbf{x}^{(0)}, t, \boldsymbol{\lambda}) \bullet [\dot{\mathbf{x}}^{(\alpha)}(t) - \dot{\mathbf{x}}^{(0)}(t)]\big|_{(\mathbf{x}^{(0)}_m, \mathbf{x}^{(\alpha)}_m, t_{m\pm})} \\ &= \mathbf{n}^T_{\partial\Omega_{ij}}(\mathbf{x}^{(0)}, t, \boldsymbol{\lambda}) \bullet \dot{\mathbf{x}}^{(\alpha)}(t) + \frac{\partial\varphi_{ij}(\mathbf{x}^{(\alpha)}, t, \boldsymbol{\lambda})}{\partial t}\Big|_{(\mathbf{x}^{(0)}_m, \mathbf{x}^{(\alpha)}_m, t_{m\pm})} \\ &= \nabla\varphi_{ij}(\mathbf{x}^{(0)}, t, \boldsymbol{\lambda}) \bullet \dot{\mathbf{x}}^{(\alpha)}(t) + \frac{\partial\varphi_{ij}(\mathbf{x}^{(\alpha)}, t, \boldsymbol{\lambda})}{\partial t}\Big|_{(\mathbf{x}^{(0)}_m, \mathbf{x}^{(\alpha)}_m, t_{m\pm})} \end{aligned} \tag{3.96}$$

or

$$\begin{aligned} G^{(0)}_{\partial\Omega_{ij}}(\mathbf{x}^{(\alpha)}_m, t_m, \mathbf{p}, \boldsymbol{\lambda}) &\equiv \mathbf{n}^T_{\partial\Omega_{ij}}(\mathbf{x}^{(0)}, t) \bullet [\mathbf{F}(\mathbf{x}^{(\alpha)}, t, \mathbf{p}) - \mathbf{F}^{(0)}(\mathbf{x}^{(0)}, t, \boldsymbol{\lambda})]\big|_{(\mathbf{x}^{(0)}_m, \mathbf{x}^{(\alpha)}_m, t_{m\pm})} \\ &= \mathbf{n}^T_{\partial\Omega_{ij}}(\mathbf{x}^{(0)}, t) \bullet \mathbf{F}(\mathbf{x}^{(\alpha)}, t, \mathbf{p}) + \frac{\partial\varphi_{ij}(\mathbf{x}^{(0)}, t, \boldsymbol{\lambda})}{\partial t}\Big|_{(\mathbf{x}^{(0)}_m, \mathbf{x}^{(\alpha)}_m, t_{m\pm})} \\ &= \nabla\varphi_{ij}(\mathbf{x}^{(0)}, t, \boldsymbol{\lambda}) \bullet \mathbf{F}(\mathbf{x}^{(\alpha)}, t, \mathbf{p}) + \frac{\partial\varphi_{ij}(\mathbf{x}^{(0)}, t, \boldsymbol{\lambda})}{\partial t}\Big|_{(\mathbf{x}^{(0)}_m, \mathbf{x}^{(\alpha)}_m, t_{m\pm})}. \end{aligned} \tag{3.97}$$

The G-function of the kth-order is defined as

$$\begin{aligned} G^{(k)}_{\partial\Omega_{ij}}(\mathbf{x}^{(\alpha)}_m, t_m, \mathbf{p}, \boldsymbol{\lambda}) &= \sum_{r=1}^{k+1} C^r_{k+1} D^{k+1-r}_{\mathbf{x}^{(0)}} \mathbf{n}^T_{\partial\Omega_{ij}}(\mathbf{x}^{(0)}, t) \bullet \left[\frac{d^r\mathbf{x}}{dt^r} - \frac{d^r\mathbf{x}^{(0)}}{dt^r}\right]\Bigg|_{(\bar{\mathbf{x}}^{(0)}_m, \mathbf{x}^{(\alpha)}_m, t_m)} \\ &= \sum_{r=1}^{k+1} C^r_{k+1} D^{k+1-r}_{\mathbf{x}^{(0)}} \mathbf{n}^T_{\partial\Omega_{ij}}(\mathbf{x}^{(0)}, t) \bullet \Big[D^{r-1}_{\mathbf{x}} \mathbf{F}(\mathbf{x}, t, \mathbf{p}) \\ &\quad - D^{r-1}_{\mathbf{x}^{(0)}} \mathbf{F}^{(0)}(\mathbf{x}^{(0)}, t, \boldsymbol{\lambda})\Big]\Big|_{(\mathbf{x}^{(0)}_m, \mathbf{x}^{(\alpha)}_m, t_m)}. \end{aligned} \tag{3.98}$$

transversality can be determined by Eq.(3.88)–(3.94) once the new G-function replaced the old one.

Consider a periodically forced, van der Pol equation as an application

$$\ddot{x}+\delta(x^2-e^2)\dot{x}+\alpha x=Q_0\cos\Omega t \tag{3.99}$$

with $\alpha>0$, $e>0$ and $\delta>0$. Q_0 and Ω are forcing amplitude and frequency, respectively. The foregoing equation can be re-written as

$$\dot{\mathbf{x}}=\mathbf{F}(\mathbf{x},t,\mathbf{p}) \text{ with } \mathbf{x}=(x,y)^T \text{ and } \mathbf{F}=(F_1,F_2)^T \tag{3.100}$$

where

$$F_1=y \text{ and } F_2=-\delta(x^2-e^2)\dot{x}-\alpha x+Q_0\cos\Omega t. \tag{3.101}$$

For $|x|<e$, the system in (3.99) is unstable and for $|x|>e$, such a system is stable. From the definition of separatrix, there are two separatrices given by

$$\varphi(x,y)\equiv x\pm e=0. \tag{3.102}$$

Therefore, the phase space of Eq.(3.99) is divided into three regions, as shown in Fig.3.15. The three sub-domains are defined by

$$\left.\begin{aligned}\Omega_1&=\{(x,y)\,|\,x\in(e,\infty),y\in(-\infty,\infty)\},\\ \Omega_2&=\{(x,y)\,|\,x\in(-e,e),y\in(-\infty,\infty)\},\\ \Omega_3&=\{(x,y)\,|\,x\in(-\infty,-e),y\in(-\infty,\infty)\}.\end{aligned}\right\} \tag{3.103}$$

The entire phase space is given by

$$\Omega=\cup_{\alpha=1}^{3}\Omega_\alpha. \tag{3.104}$$

The corresponding separation boundaries are

$$\left.\begin{aligned}\partial\Omega_{12}&=\bar{\Omega}_1\cap\bar{\Omega}_2=\{(x,y)\,|\,\varphi_{12}(x,y)\equiv x-e=0\},\\ \partial\Omega_{23}&=\bar{\Omega}_2\cap\bar{\Omega}_3=\{(x,y)\,|\,\varphi_{23}(x,y)\equiv x+e=0\}.\end{aligned}\right\} \tag{3.105}$$

Such domains and the boundary are sketched in Fig.3.15. From the above definitions, equations (3.100) and (3.101) give

$$\dot{\mathbf{x}}^{(\alpha)}=\mathbf{F}\left(\mathbf{x}^{(\alpha)},t,\mathbf{p}\right) \text{ for } \alpha\in\{1,2,3\} \tag{3.106}$$

where for $\mathbf{x}^{(\alpha)}\in\Omega_\alpha$

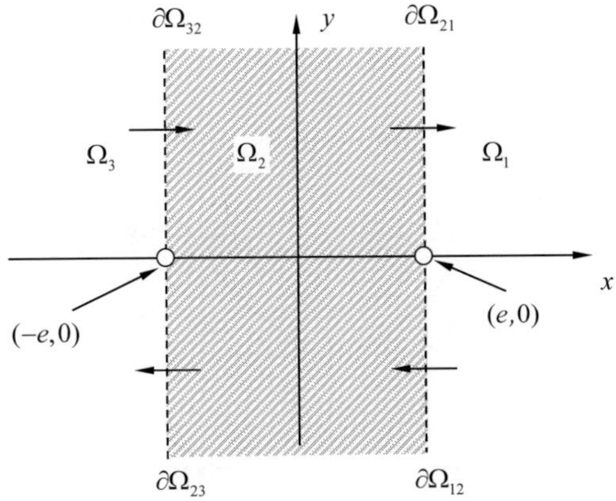

Figure 3.15. Sub-domains, boundaries and equilibriums $(\pm e, 0)$.

$$\mathbf{F}\left(\mathbf{x}^{(\alpha)}, t, \mathbf{p}\right) = \left(y^{(\alpha)}, -2\delta[(x^{(\alpha)})^2 - e^2]y^{(\alpha)} - \alpha x^{(\alpha)} + Q_0 \cos \Omega t\right)^T. \quad (3.107)$$

Note that $\mathbf{p} = (\Omega, Q_0, \alpha, \delta, e)^{\mathrm{T}}$. In Eq.(3.105), using $\mathbf{n}_{\partial\Omega_{ij}} = \nabla \varphi_{ij}$ gives

$$\mathbf{n}_{\partial\Omega_{12}} = \mathbf{n}_{\partial\Omega_{23}} = \left(1, 0\right)^{\mathrm{T}}. \quad (3.108)$$

Therefore, $D\mathbf{n}_{\partial\Omega_{ij}} = 0$, equations (3.107) and (3.108) give

$$G^{(0)}_{\partial\Omega_{\alpha\beta}}(\mathbf{x}^{(\alpha)}, t, \mathbf{p}) = \mathbf{n}^{\mathrm{T}}_{\partial\Omega_{\alpha\beta}} \bullet \mathbf{F}\left(\mathbf{x}^{(\alpha)}, t, \mathbf{p}\right) = y^{(\alpha)} \quad (3.109)$$

and

$$\begin{aligned} G^{(1)}_{\partial\Omega_{\alpha\beta}}(\mathbf{x}^{(\alpha)}, t, \mathbf{p}) &= \mathbf{n}^{\mathrm{T}}_{\partial\Omega_{\alpha\beta}} \bullet D\mathbf{F}(\mathbf{x}^{(\alpha)}, t, \mathbf{p}) \\ &= -2\delta[(x^{(\alpha)})^2 - e^2]y^{(\alpha)} - \alpha x^{(\alpha)} + Q_0 \cos \Omega t. \end{aligned} \quad (3.110)$$

The conditions for the global transversality to the generalized separatrix are

$$\begin{aligned} &y_m > 0 \quad \text{from } \Omega_3 \to \Omega_2 \text{ and } \Omega_2 \to \Omega_1, \text{ and} \\ &y_m < 0 \quad \text{from } \Omega_1 \to \Omega_2 \text{ and } \Omega_2 \to \Omega_3. \end{aligned} \quad (3.111)$$

From Eq.(3.109) and (3.110), the necessary and sufficient conditions for the global tangency to the generalized separatrix are

$$y_m = 0 \quad \text{on } \partial\Omega_{ij} \;\; i, j \in \{1, 2, 3\} \text{ and } i \neq j \quad (3.112)$$

and

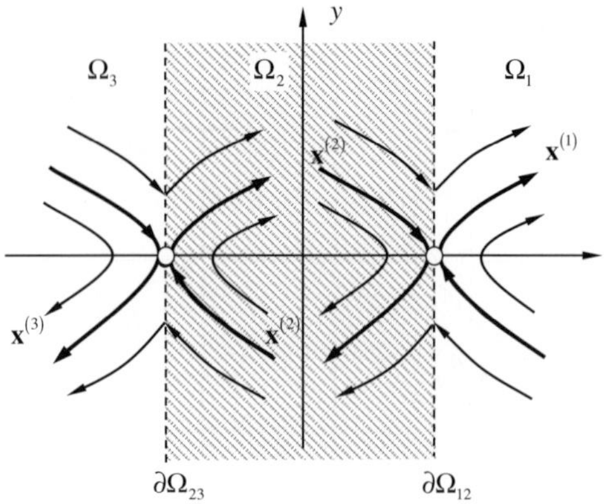

Figure 3.16. Phase portraits near equilibriums $(\pm e, 0)$ on the separatrix.

$$
\begin{aligned}
&-\alpha e+Q_0\cos\Omega t_m>0 \quad \text{on } \partial\Omega_{12} \text{ in } \Omega_1,\\
&-\alpha e+Q_0\cos\Omega t_m<0 \quad \text{on } \partial\Omega_{12} \text{ in } \Omega_2,\\
&\alpha e+Q_0\cos\Omega t_m>0 \quad \text{on } \partial\Omega_{23} \text{ in } \Omega_2,\\
&\alpha e+Q_0\cos\Omega t_m<0 \quad \text{on } \partial\Omega_{23} \text{ in } \Omega_3.
\end{aligned}
\tag{3.113}
$$

The global tangency and transversality of the van der Pol oscillator is sketched in Fig.3.16. The tangential and transversal flows on three separatrices from different domains are given. Let

$$
\Theta_{cr}=\arccos\frac{\alpha e}{Q_0}. \tag{3.114}
$$

From Eq.(3.113), the switching phase for the global grazing can be computed as

$$
\begin{aligned}
&\mathrm{mod}(\Omega t_m,2\pi)\in[0,\Theta_{cr})\cup(2\pi-\Theta_{cr},2\pi] && \text{on } \partial\Omega_{12} \text{ in } \Omega_1,\\
&\mathrm{mod}(\Omega t_m,2\pi)\in(\Theta_{cr},2\pi-\Theta_{cr}) && \text{on } \partial\Omega_{12} \text{ in } \Omega_2,\\
&\mathrm{mod}(\Omega t_m,2\pi)\in[0,\pi-\Theta_{cr})\cup(\pi+\Theta_{cr},2\pi] && \text{on } \partial\Omega_{23} \text{ in } \Omega_2,\\
&\mathrm{mod}(\Omega t_m,2\pi)\in(\pi-\Theta_{cr},\pi+\Theta_{cr}) && \text{on } \partial\Omega_{23} \text{ in } \Omega_3.
\end{aligned}
\tag{3.115}
$$

Owing to $|\cos\Omega t_m|<1$, if $Q_0<\alpha e$, the foregoing equation becomes

$$
\begin{aligned}
&\mathrm{mod}(\Omega t_m,2\pi)\in\{\varnothing\} && \text{on } \partial\Omega_{12} \text{ in } \Omega_1,\\
&\mathrm{mod}(\Omega t_m,2\pi)\in[0,2\pi] && \text{on } \partial\Omega_{12} \text{ in } \Omega_2,\\
&\mathrm{mod}(\Omega t_m,2\pi)\in[0,2\pi] && \text{on } \partial\Omega_{23} \text{ in } \Omega_2,\\
&\mathrm{mod}(\Omega t_m,2\pi)\in\{\varnothing\} && \text{on } \partial\Omega_{23} \text{ in } \Omega_3.
\end{aligned}
\tag{3.116}
$$

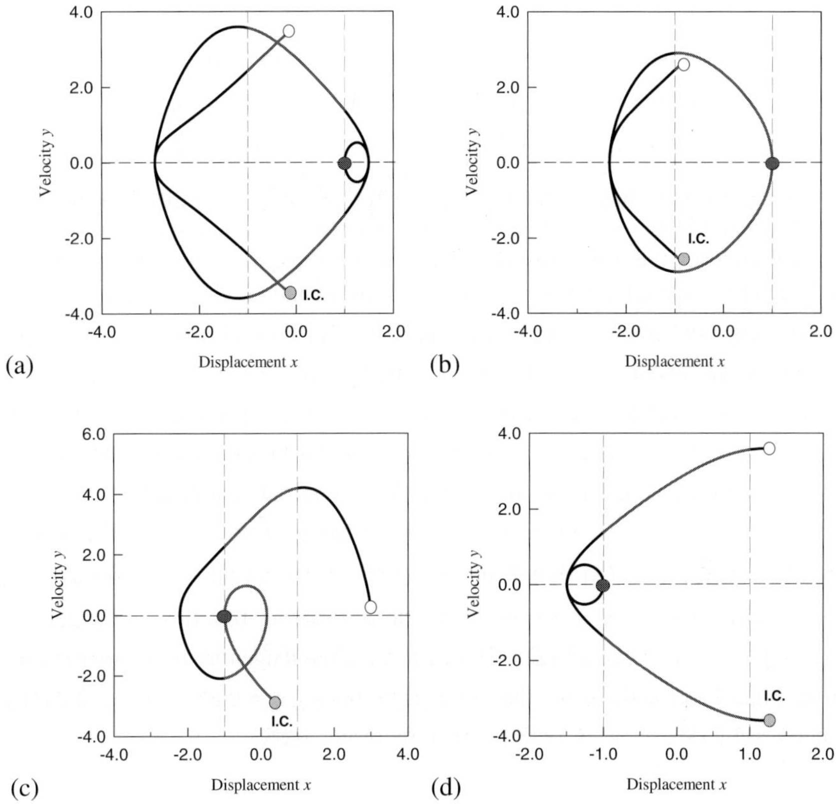

Figure 3.17. The global tangential flows at the boundary: (a) $\partial\Omega_{12}$ in Ω_1, (b) $\partial\Omega_{12}$ in Ω_2, (c) $\partial\Omega_{23}$ in Ω_2 and (d) $\partial\Omega_{23}$ in Ω_3. Acronym "I.C." represents the initial condition. Two vertical lines are the separatrices, and the horizontal line is reference line for global tangency.

If $\mathrm{mod}(\Omega t_m, 2\pi) \in \{\varnothing\}$, it means that no global tangency to the boundary $\partial\Omega_{12}$ in domain Ω_1 exists. For the following cases,

$$\begin{aligned} &\mathrm{mod}(\Omega t_m, 2\pi) = \{\Theta_{cr}, 2\pi - \Theta_{cr}\} \quad \text{on } \partial\Omega_{12} \text{ and} \\ &\mathrm{mod}(\Omega t_m, 2\pi) = \{\pi - \Theta_{cr}, \pi + \Theta_{cr}\} \text{ on } \partial\Omega_{23}, \end{aligned} \tag{3.117}$$

the high order G-function should be discussed. Consider a set of system parameter ($\alpha = 1.0$, $\delta = 0.5$, $e = 1.0$, $Q_0 = 2.0$, $\Omega = 1.0$). From Eq.(3.114), the critical switching phase is $\Theta_{cr} = \frac{\pi}{3}$. Hence, the switching phase for the global tangency can be chosen from

$$
\left.
\begin{array}{ll}
\mathrm{mod}(\Omega t_m, 2\pi) \in [0, \frac{\pi}{3}) \cup (\frac{5\pi}{3}, 2\pi] & \text{on } \partial\Omega_{12} \text{ in } \Omega_1, \\
\mathrm{mod}(\Omega t_m, 2\pi) \in (\frac{\pi}{3}, \frac{5\pi}{3}) & \text{on } \partial\Omega_{12} \text{ in } \Omega_2; \\
\mathrm{mod}(\Omega t_m, 2\pi) \in [0, \frac{2\pi}{3}) \cup (\frac{4\pi}{3}, 2\pi] & \text{on } \partial\Omega_{23} \text{ in } \Omega_2, \\
\mathrm{mod}(\Omega t_m, 2\pi) \in (\frac{2\pi}{3}, \frac{4\pi}{3}) & \text{on } \partial\Omega_{23} \text{ in } \Omega_3.
\end{array}
\right. \tag{3.118}
$$

The global tangency of flows in Eq.(3.99) to the two generalized separatrices are illustrated in Fig.3.17. Acronym "I.C." represents the initial condition. Two vertical lines are the generalized separatrices and the horizontal line is a reference line for the global tangency. From the above equation, choosing $\Omega t_m = 0$, the global tangential point $(x_m, y_m) = (1, 0)$ of a flow in Eq.(3.99) to the generalized separatrix $\partial\Omega_{12}$ occurs in Ω_1. However, for $\Omega t_m = \pi$, the global tangency of a flow to the generalized separatrix $\partial\Omega_{12}$ will occur in Ω_2. The initial condition ($t_0 = -2\pi$, $x_0 \approx -0.1792$ and $y_0 \approx -3.4122$) is used to the aforementioned grazing flow, as shown in Fig.3.17(a). In Fig.3.17(b), the initial condition ($t_0 = 0$, $x_0 \approx 0.9052$ and $y_0 \approx -2.5762$) is used. The global tangency of a flow to the boundary $\partial\Omega_{23}$ at $(x_m, y_m) = (-1, 0)$ will occur in Ω_2 with the switching phase $\Omega t_m = \frac{\pi}{3}$. If a switching phase $\Omega t_m = \pi$ is used, the global tangency of a flow to the generalized separatrix will occur in Ω_3. The corresponding tangential flows are presented in Fig.3.17(c) and (d) for the initial conditions ($x_0 \approx 0.3595$, $y_0 \approx -2.8218$) and ($x_0 \approx 1.2189$, $y_0 \approx -3.5948$) for $t_0 = 0$, respectively.

Chapter 4

Chaotic Layer Dynamics

In this Chapter, a domain in the vicinity of a specified, first integral manifold in phase space will be introduced through the first integral quantity. Such a domain occupied by chaotic motions of the non-integrable system in phase space is called *the chaotic layer* or *band* in vicinity of the specified, first integral manifold. The first integral quantity increment will be presented as an important quantity to investigate chaotic layer dynamics of the perturbed system. For different reference surfaces, all possible expressions of the first integral quantity increment will be given for perturbed systems. From the first integral quantity increment, the iterative mapping for chaotic motions in the non-integrable system will be presented. From an unperturbed flow on a specific first integral manifold, under certain convergent conditions, perturbed flows in nonlinear dynamical systems will be approximated via a perturbation expression. Further, the approximate, first integral quantity increment for the perturbed system will be derived, which can be used as an approximate estimate of the width of the chaotic layer in such a perturbed system. This approximate, first integral quantity increment will lead to the Melnikov function for the corresponding perturbed system. In addition, the resonance mechanism of the stochastic and resonant, chaotic layers in a perturbed system will be addressed.

4.1. Chaotic domains in phase space

In Chapter 3, the global tangency and transversality of a perturbed flow to the separatrix surface were discussed. The corresponding necessary and sufficient conditions for such global tangency and transversality were obtained. Once a perturbed flow crosses over the separatrix for infinite times, a chaotic motion will be formed in the vicinity of such a separatrix in phase space. The domain formed by such a chaotic motion in the vicinity of separatrix in phase space is termed *the stochastic layer or band.* However, the domain formed by such a chaotic motion in the vicinity of a specific, first integral manifold with resonance is termed *the resonant layer or band.* Both of the stochastic and resonant layers are called *the chaotic layers* in the perturbed system. In many perturbed dynamical systems without the separatrix, only the resonant layer exists. To discuss the chaotic layer dynamics in a perturbed system, besides Assumptions (A1)-(A3), the following assumption is inserted, i.e.,

A4. The integrable and non-integrable vector fields of Eq.(3.1) on phase space are bounded (or periodic), i.e., for arbitrary sub-domains $U_{\mathbf{x}} \subseteq \mathbb{R}^n$, $U_{\boldsymbol{\mu}} \subseteq \mathbb{R}^{m_1}$, $U_{\boldsymbol{\pi}} \subseteq \mathbb{R}^{m_2}$ and $U_t \subseteq \mathbb{R}$,

$$\|\mathbf{f}\| \le K_1 \text{ (const) and } \|\mathbf{g}\| \le K_2 \text{ (const)}. \tag{4.1}$$

The above assumption makes sure there is "no hole" in the domain $U_{\mathbf{x}} \subseteq \mathbb{R}^n$, $U_{\boldsymbol{\mu}} \subseteq \mathbb{R}^{m_1}$ and $U_{\boldsymbol{\pi}} \subseteq \mathbb{R}^{m_2}$. In addition, the non-integrable vector field in Eq.(3.1) is periodic with time t, i.e.,

$$\mathbf{g}(\mathbf{x}, t+T, \boldsymbol{\pi}) = \mathbf{g}(\mathbf{x}, t, \boldsymbol{\pi}). \tag{4.2}$$

4.1.1. Maximum and minimum first integral surfaces

From the above assumptions (A1)-(A4), once the initial sets are selected in the vicinity of a specified, first integral manifold surface, perturbed flows will stay in such vicinity. The first integral quantity for a specific non-perturbed flow is invariant, but the first integral quantity for a perturbed flow changes with time. With varying time, the instantaneous first integral quantity for a chaotic flow is chaotic. Such a chaotic motion can be measured through the instantaneous first integral quantity. To investigate the complicated behavior of perturbed flows in a vicinity of a specified, first integral manifold surface, point sets of unperturbed and perturbed flows in Eq.(3.1) are defined, and the corresponding first integral quantities are introduced as well.

DEFINITION 4.1. For a dynamical system in Eq.(3.1), its integrable dynamical system in Eq.(3.2) possesses a family of the first integral manifolds $F(\overline{\mathbf{x}}, \boldsymbol{\mu}) = E$. For a time $t_k \in (t_0, \infty)$ ($k = 1, 2, \cdots$), points $\overline{\mathbf{x}}(t_k)$ and $\mathbf{x}(t_k)$ are on the integrable and perturbed flows, respectively. The point sets and the corresponding first integral quantities of unperturbed and perturbed flows are

$$\Gamma_s = \left\{ \overline{\mathbf{x}}^{(s)}(t_k) \left| \begin{array}{l} \text{for } t_k \in (t_0, \infty), \\ E_s = F_{ij}\left(\overline{\mathbf{x}}^{(s)}(t_k), \boldsymbol{\mu}\right) \text{ and} \\ \overline{\mathbf{x}}^{(s)}(t_k) = \boldsymbol{\Phi}^{(s)}\left(\overline{\mathbf{x}}_0^{(s)}, t_k, t_0\right) \end{array} \right. \right\} \cup \{p_0\},$$
$$\text{and } E_s = F_{ij}\left(\overline{\mathbf{x}}^{(s)}(t), \boldsymbol{\mu}\right) \tag{4.3}$$

for a flow on the separatrix $\partial\Omega_{ij}$ of the integrable system of Eq.(3.2), and

$$\Gamma_{\sigma_\alpha} = \left\{ \overline{\mathbf{x}}^{(\alpha)}(t_k) \left| \begin{array}{l} E_{\sigma_\alpha} = F\left(\overline{\mathbf{x}}^{(\alpha)}(t_k), \boldsymbol{\mu}\right) \\ \text{for } \sigma_\alpha \in (0,1] \text{ and } t_k \in (t_0, \infty), \\ \text{and } \overline{\mathbf{x}}^{(\alpha)}(t_k) = \boldsymbol{\Phi}^{(\alpha)}\left(\overline{\mathbf{x}}_0^{(\alpha)}, t_k, t_0\right) \end{array} \right. \right\} \tag{4.4}$$

$$\text{and } E_{\sigma_\alpha} = F\left(\overline{\mathbf{x}}^{(\alpha)}(t), \boldsymbol{\mu}\right)$$

for a flow of the integrable system in Eq.(3.2) in domain Ω_α ($\alpha = 1,2,\cdots,m$)

$$\Gamma^{(\alpha)} = \left\{ \mathbf{x}^{(\alpha)}(t_k) \left| \begin{array}{l} \text{for } t_k \in (t_0, \infty), \\ \mathbf{x}^{(\alpha)}(t_k) = \boldsymbol{\Phi}^{(\alpha)}\left(\mathbf{x}_0^{(\alpha)}, t_k, t_0\right) \end{array} \right. \right\} \tag{4.5}$$

$$\text{and } E^{(\alpha)}(t) = F\left(\mathbf{x}^{(\alpha)}(t), \boldsymbol{\mu}\right)$$

for a flow of the non-integrable system in Eq.(3.1) in Ω_α ($\alpha = 1,2,\cdots,m$).

For the non-integrable system in Eq.(3.1), the first integral quantity $E^{(\alpha)}(t)$ is changed with time t. However, the first integral quantity E_{σ_α} of the integrable system is invariant with respect to time t. As $\sigma_\alpha \to 0$, the invariant quantity E_{σ_α} approaches E_s on the separatrix $\partial\Omega_{ij}$. To extend the separatrix manifold surface, the first integral manifold can be described as follows.

DEFINITION 4.2. For a dynamical system in Eq.(3.1), its integrable dynamical system in Eq.(3.2) possesses a family of the first integral manifolds $F(\overline{\mathbf{x}}, \boldsymbol{\mu}) = E$. The first integral manifold of a dynamical systems in Eq.(3.1) in Ω_α ($\alpha = 1,2,\cdots,$ m) is

$$S_{\sigma_\alpha} = \left\{ \overline{\mathbf{x}}^{(\alpha)} \left| \begin{array}{l} F\left(\overline{\mathbf{x}}^{(\alpha)}, \boldsymbol{\mu}\right) - E_{\sigma_\alpha} = 0 \\ \text{where } F \text{ is } C^r\text{-continuous } (r \geq 1) \end{array} \right. \right\} \subset \mathbb{R}^{n\text{-}1}. \tag{4.6}$$

From Definition 4.1, the point set on a flow of the integrable system of Eq. (3.1) for a specific initial set is the subset of the first integral manifold (i.e., $\Gamma_{\sigma_\alpha} \subseteq S_{\sigma_\alpha}$). Similarly, the G-function of the lth-order for a perturbed flow to the first integral manifold surface S_{σ_α} in Ω_α for the dynamical system in Eq.(3.1) is defined in Definition 3.8. The G-function will be employed to determine the maximum and minimum first integral manifolds of the domain of perturbed flows. So the Definition 3.8 is restated as follows:

DEFINITION 4.3. For a dynamical system in Eq.(3.1), its integrable dynamical system in Eq.(3.2) possesses a family of the first integral manifolds $F(\overline{\mathbf{x}},\boldsymbol{\mu})=E$. If there is an intersected point $\mathbf{x}^{(\alpha)}(t_k)$ between a perturbed flow $\mathbf{x}^{(\alpha)}(t)$ in Eq.(3.1) and the unperturbed flow $\overline{\mathbf{x}}^{(\alpha)}(t)$ on the first integral manifold surface S_{σ_σ} at time t_k (i.e., $\mathbf{x}_k^{(\alpha)}=\overline{\mathbf{x}}_k^{(\alpha)}$), the G-function $G_{\sigma_\alpha}^{(l)}(\bullet)$ of the lth-order for a perturbed flow $\mathbf{x}^{(\alpha)}(t)$ to the specified at the point $\overline{\mathbf{x}}^{(\alpha)}(t_k)=\mathbf{x}^{(\alpha)}(t_k)$ is defined for ($l=0,1,2,\cdots$) by

$$\begin{aligned}G_{\sigma_\alpha}^{(l)}(\mathbf{x}_k^{(\alpha)},t_k,\mathbf{p})=\sum\nolimits_{r=1}^{l+1}C_{l+1}^{r}D_{\overline{\mathbf{x}}}^{l+1-r}\mathbf{n}_{S_{\sigma_\alpha}}^{T}(\overline{\mathbf{x}})\bullet\big[D_{\mathbf{x}}^{r-1}\mathbf{F}(\mathbf{x},t,\mathbf{p})\\-D_{\overline{\mathbf{x}}}^{r-1}\mathbf{f}(\overline{\mathbf{x}},\boldsymbol{\mu})\big]\Big|_{(\overline{\mathbf{x}}_k^{(\alpha)},\mathbf{x}_k^{(\alpha)},t_k)}.\end{aligned}\tag{4.7}$$

For a special case ($l=0$), the G-function at point $\mathbf{x}^{(\alpha)}(t_k)$ is

$$G_{\sigma_\alpha}(\mathbf{x}_k^{(\alpha)},t_k,\mathbf{p})=\mathbf{n}_{S_{\sigma_\alpha}}^{T}\bullet\mathbf{g}\left(\mathbf{x}_k^{(\alpha)},t_k,\boldsymbol{\pi}\right)=\nabla F\left(\mathbf{x}_k^{(\alpha)},\boldsymbol{\mu}\right)\bullet\mathbf{g}\left(\mathbf{x}_k^{(\alpha)},t_k,\boldsymbol{\pi}\right).\tag{4.8}$$

The G-function is the normal component of a perturbed vector field at point $\mathbf{x}^{(\alpha)}(t_k)$ to the specified first integral manifold surface S_{σ_σ}. To bound perturbed flows in dynamical systems in Eq.(3.1), the maximum and minimum, first integral quantities should be introduced herein, and the corresponding first integral surfaces are described.

DEFINITION 4.4. The maximum and minimum of the first integral quantities of a perturbed flow $\mathbf{x}^{(\alpha)}(t)=\boldsymbol{\Phi}^{(\alpha)}(\mathbf{x}_0^{(\alpha)},t,t_0)$ in Ω_α of the dynamical system in Eq.(3.1) are defined as

$$\begin{aligned}E_{\max}^{(\alpha)}&=\max_{t_k\in[t_0,\infty)}\left\{E^{(\alpha)}(t_k)\left|\begin{array}{l}E^{(\alpha)}(t_k)=F\left(\mathbf{x}^{(\alpha)}(t_k),\boldsymbol{\mu}\right)\text{ and}\\ \mathbf{x}^{(\alpha)}(t)=\boldsymbol{\Phi}^{(\alpha)}\left(\mathbf{x}_0^{(\alpha)},t,t_0\right)\end{array}\right.\right\},\\E_{\min}^{(\alpha)}&=\min_{t_k\in[t_0,\infty)}\left\{E^{(\alpha)}(t_k)\left|\begin{array}{l}E^{(\alpha)}(t_k)=F\left(\mathbf{x}^{(\alpha)}(t_k),\boldsymbol{\mu}\right)\text{ and}\\ \mathbf{x}^{(\alpha)}(t)=\boldsymbol{\Phi}^{(\alpha)}\left(\mathbf{x}_0^{(\alpha)},t,t_0\right)\end{array}\right.\right\}.\end{aligned}\tag{4.9}$$

From the foregoing definition, all the first integrable quantity of a perturbed flow in Eq.(3.1) should be in such a range (i.e., $E^{(\alpha)}(t)\in[E_{\min}^{(\alpha)},E_{\max}^{(\alpha)}]$). Using the maximum and minimum integral quantities of the perturbed flow to the integrable system of Eq.(3.1), the two corresponding, first integral manifold surfaces

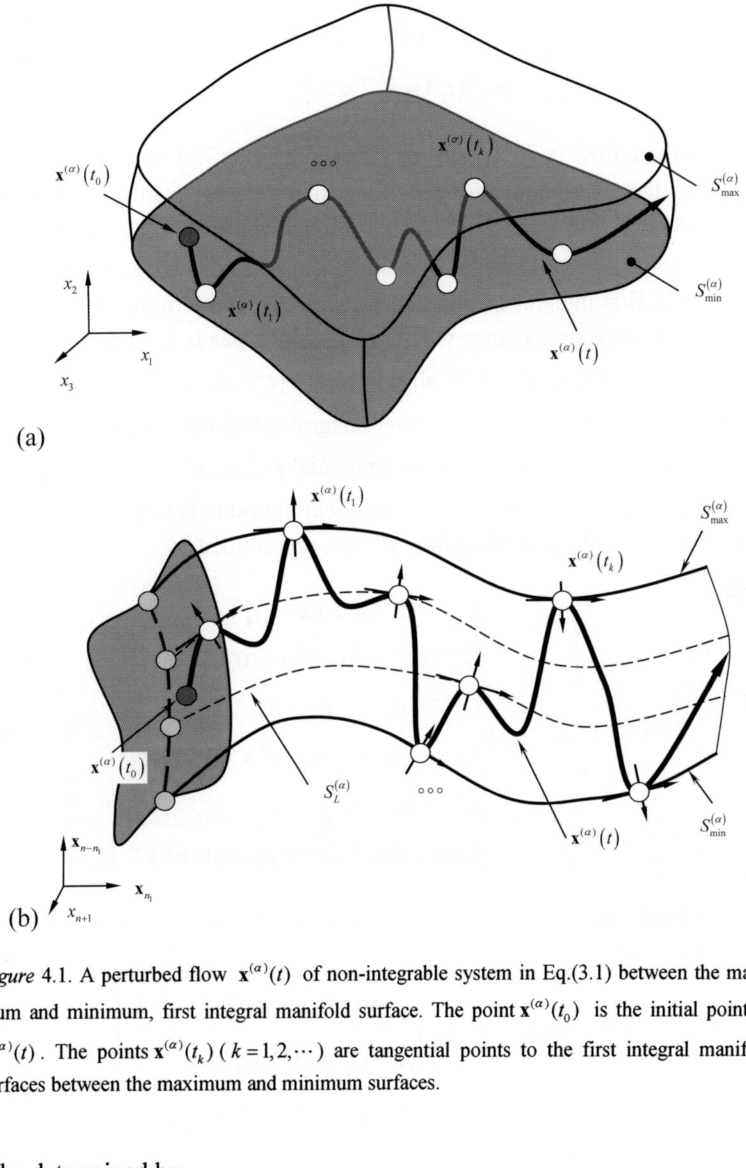

Figure 4.1. A perturbed flow $\mathbf{x}^{(\alpha)}(t)$ of non-integrable system in Eq.(3.1) between the maximum and minimum, first integral manifold surface. The point $\mathbf{x}^{(\alpha)}(t_0)$ is the initial point of $\mathbf{x}^{(\alpha)}(t)$. The points $\mathbf{x}^{(\alpha)}(t_k)$ ($k=1,2,\cdots$) are tangential points to the first integral manifold surfaces between the maximum and minimum surfaces.

will be determined by

$$S_{\max}^{(\alpha)} = \left\{ \overline{\mathbf{x}}^{(\alpha)} \left| \begin{array}{l} F\left(\overline{\mathbf{x}}^{(\alpha)}, \boldsymbol{\mu}\right) = E_{\max}^{(\alpha)} = F\left(\mathbf{x}^{(\alpha)}\left(t_k\right), \boldsymbol{\mu}\right), \\ \mathbf{x}^{(\alpha)}\left(t_k\right) = \overline{\mathbf{x}}_k^{(\alpha)} \in S_{\max}^{(\alpha)} \end{array} \right. \right\}, \tag{4.10a}$$

$$S_{\min}^{(\alpha)} = \left\{ \overline{\mathbf{x}}^{(\alpha)} \left| \begin{array}{l} F\left(\overline{\mathbf{x}}^{(\alpha)}, \boldsymbol{\mu}\right) = E_{\min}^{(\alpha)} = F\left(\mathbf{x}^{(\alpha)}\left(t_k\right), \boldsymbol{\mu}\right), \\ \mathbf{x}^{(\alpha)}\left(t_k\right) = \overline{\mathbf{x}}_k^{(\alpha)} \in S_{\min}^{(\alpha)} \end{array} \right. \right\}. \tag{4.10b}$$

A perturbed flow $\mathbf{x}^{(\alpha)}(t) = \boldsymbol{\Phi}^{(\alpha)}(\mathbf{x}_0^{(\alpha)}, t, t_0)$ of Eq.(3.1) for time $t \in [t_0, \infty)$ is bounded by the maximum and minimum, first integral manifold surfaces, as shown in Fig.4.1. The point $\mathbf{x}^{(\alpha)}(t_0)$ is an initial point of $\mathbf{x}^{(\alpha)}(t)$. Points $\mathbf{x}^{(\alpha)}(t_k)$ ($k = 1, 2, \cdots$) are tangential to the first integral manifolds between the maximum and minimum first integral quantity surfaces. How to determine such maximum and minimum surfaces becomes very crucial, which needs to find points $\mathbf{x}^{(\alpha)}(t_k)$ on both surfaces. Using Eq.(4.7), all points $\mathbf{x}^{(\alpha)}(t_k)$ should be found for a perturbed flow locally tangential to the first integral manifold surface $F(\overline{\mathbf{x}}^{(\alpha)}(t_k), \boldsymbol{\mu}) = E_{\sigma_\alpha}$ in domain Ω_α. For the locally tangential points, $\overline{\mathbf{x}}^{(\alpha)}(t_k) = \mathbf{x}^{(\alpha)}(t_k)$. Thus, a set that consists of such all the local tan-gential points is termed *the local tangential point set* of the perturbed flow $\mathbf{x}^{(\alpha)}(t)$, determined by

$$S_L^{(\alpha)}\left(t_k\right) = \left\{ \mathbf{x}^{(\alpha)}\left(t_k\right) \left| \begin{array}{l} E_{\sigma_\alpha} = E^{(\alpha)} = F\left(\mathbf{x}^{(\alpha)}\left(t_k\right), \boldsymbol{\mu}\right), \\ G_{\sigma_\alpha}^{(r)}(\mathbf{x}_k^{(\alpha)}, t_k, \mathbf{p}) = 0 \ (r = 0, 1, 2, \cdots, 2l-2) \\ G_{\sigma_\alpha}^{(2l-1)}(\mathbf{x}_k^{(\alpha)}, t_k, \mathbf{p}) > 0, \\ \text{for } \mathbf{n}_{S_{\sigma_\alpha}} \text{ on the side of } \mathbf{x}^{(\alpha)}\left(t_k\right) \text{ or} \\ G_{\sigma_\alpha}^{(2l-1)}(\mathbf{x}_k^{(\alpha)}, t_k, \mathbf{p}) < 0, \\ \text{for } \mathbf{n}_{S_{\sigma_\alpha}} \text{ on the opposite side of } \mathbf{x}^{(\alpha)}\left(t_k\right) \end{array} \right. \right\}. \tag{4.11}$$

From the local tangential point set, the maximum and minimum, first integral quantities are determined by

$$\left. \begin{array}{l} E_{\max}^{(\alpha)} = \max\limits_{t_k \in [t_0, \infty)} \left\{ E_{\sigma_\alpha}\left(t_k\right) \left| \begin{array}{l} E_{\sigma_\alpha}(t_k) = F\left(\mathbf{x}^{(\alpha)}\left(t_k\right), \boldsymbol{\mu}\right), \\ \mathbf{x}^{(\alpha)}\left(t_k\right) \in S_L^{(\alpha)}\left(t_k\right) \end{array} \right. \right\}, \\ E_{\min}^{(\alpha)} = \min\limits_{t_k \in [t_0, \infty)} \left\{ E_{\sigma_\alpha}\left(t_k\right) \left| \begin{array}{l} E_{\sigma_\alpha}(t_k) = F\left(\mathbf{x}^{(\alpha)}\left(t_k\right), \boldsymbol{\mu}\right), \\ \mathbf{x}^{(\alpha)}\left(t_k\right) \in S_L^{(\alpha)}\left(t_k\right) \end{array} \right. \right\}. \end{array} \right\} \tag{4.12}$$

Compared to Eq.(4.9), the maximum and minimum, first integral quantities can be easily computed through Eqs.(4.11) and (4.12). Further, from the maximum and minimum first integral quantities, the maximum and minimum first integral manifold surfaces are determined by

$$S_{\max}^{(\alpha)} = \left\{ \overline{\mathbf{x}}^{(\alpha)} \middle| E_{\max}^{(\alpha)} = F\left(\overline{\mathbf{x}}^{(\alpha)}, \boldsymbol{\mu}\right) \right\} \text{ and } S_{\min}^{(\alpha)} = \left\{ \overline{\mathbf{x}}^{(\alpha)} \middle| E_{\min}^{(\alpha)} = F\left(\overline{\mathbf{x}}^{(\alpha)}, \boldsymbol{\mu}\right) \right\}. \quad (4.13)$$

The two surfaces in Ω_α can form a sub-domain $D_\alpha \subseteq \Omega_\alpha$. From the above discussion, for $t \in [t_0, \infty)$, the perturbed flow $\mathbf{x}^{(\alpha)}(t)$ lies in such a sub-domain (i.e., $\mathbf{x}^{(\alpha)}(t) \in D_\alpha$), and the first integral quantity $E^{(\alpha)}(t) = F(\mathbf{x}^{(\alpha)}(t), \boldsymbol{\mu})$ is in the interval of $[E_{\min}^{(\alpha)}, E_{\max}^{(\alpha)}]$ (i.e., $E^{(\alpha)}(t) \in [E_{\min}^{(\alpha)}, E_{\max}^{(\alpha)}]$). To extend this idea, a δ-*set* in the vicinity of the first integral manifold surface $S_{\sigma_\alpha^0}$ ($\sigma_\alpha^0 \in (0,1]$) in Ω_α can be defined. If the first integral quantity interval $[E_{\min}^{(\alpha)}, E_{\max}^{(\alpha)}]$ of a perturbed flow $\mathbf{x}^{(\alpha)}(t)$ belongs to the first integral quantity interval relative to $E_{\sigma_\alpha^0}$, such a perturbed flow $\mathbf{x}^{(\alpha)}(t)$ exists in the neighborhood of the first integral manifold surface $S_{\sigma_\alpha^0}$.

4.1.2. First integral quantity intervals

Before discussing the flow characteristics of the $\mathbf{x}^{(\alpha)}(t)$ in Ω_α, from the maximum and minimum integral quantities, the corresponding δ-*set* of a specified first integral quantity is defined as follows:

DEFINITION 4.5. For a dynamical system in Eq.(3.1), its integrable dynamical system in Eq.(3.2) possesses a family of the first integral manifolds $F(\overline{\mathbf{x}}, \boldsymbol{\mu}) = E$. The δ-*set* of the first integral quantity in the neighborhood of a specified first integral manifold surface $S_{\sigma_\alpha^0}$ ($\sigma_\alpha^0 \in (0,1]$) in Ω_α is defined by

$$N_{\sigma_\alpha}^{\delta_\alpha}\left(E_{\sigma_\alpha^0}\right) = \left\{ E_{\sigma_\alpha} \middle| \left| E_{\sigma_\alpha} - E_{\sigma_\alpha^0} \right| \le \delta_\alpha \text{ for } \delta_\alpha > 0 \text{ and } \sigma_\alpha, \sigma_\alpha^0 \in (0,1] \right\}. \quad (4.14)$$

If the first integral manifold surface $S_{\sigma_\alpha^0}$ becomes the separatrix $\partial\Omega_{ij}$, the foregoing definition gives the δ-*set* of the first integral quantity in the neighborhood of the separatrix $\partial\Omega_{ij}$ in Ω_α ($\alpha = i, j$). For a global flow to the separatrix $\partial\Omega_{ij}$, the perturbed flows $\mathbf{x}^{(i)}(t)$ and $\mathbf{x}^{(j)}(t)$ should exist in the corresponding domains Ω_α ($\alpha = i, j$). Therefore, both the δ-*sets* of the first integral quantity relative to domain Ω_α ($\alpha = i, j$) in the vicinity of the separatrix $\partial\Omega_{ij}$ are defined. As in Definition 4.5, the corresponding δ-*sets* of the first integral quantity in the neighborhood of the separatrix $\partial\Omega_{ij}$ are described as follows.

DEFINITION 4.6. For a dynamical system in Eq.(3.1), its integrable dynamical system in Eq.(3.2) possesses a family of the first integral manifolds $F(\overline{\mathbf{x}},\boldsymbol{\mu})=E$. the δ-*set* of the first integral quantity in the neighborhood of the separatrix $\partial\Omega_{ij}$ in Ω_α is defined by

$$N_{\sigma_\alpha}^{\delta_\alpha}(E_s)=\left\{E_{\sigma_\alpha}\left|\left|E_{\sigma_\alpha}-E_s\right|\le\delta_\alpha \text{ for } \delta_\alpha>0 \text{ and } \sigma_\alpha\in(0,1]\right.\right\} \tag{4.15}$$

and the complete δ-*set* of the first integral quantity in the neighborhood of the separatrix $\partial\Omega_{ij}$ is

$$N^{\delta}(E_s)=\cup_\alpha N_{\sigma_\alpha}^{\delta_\alpha}(E_s)\cup\{E_s\}. \tag{4.16}$$

To explain the flow characteristics, two unperturbed flows on the first integral manifold surfaces are illustrated in Fig.4.2. For two different, first integral quantities (e.g., $E_{\sigma_\alpha^1}\neq E_{\sigma_\alpha^2}$), the two unperturbed flows ${}_{\sigma_\alpha^1}\overline{\mathbf{x}}^{(\alpha)}(t)$ and ${}_{\sigma_\alpha^2}\overline{\mathbf{x}}^{(\alpha)}(t)$ on the corresponding, first integral manifolds will never to be intersected. The two flows starting with the same cross-section surface (e.g., $\Xi_0^{(\alpha)}$) at time t_0 is sketched in Fig.4.2(a). For time $t\in(t_0,\infty)$, the two flows cannot arrive to the same final cross-section surface at time t. Even if the trajectories of two flows are very close in phase space, it does not imply that the time-responses of the two flows should be very close. For example, consider periodic flows relative to the first integral manifolds in a 2-dimensional, nonlinear dynamical system in Luo and Han (2001). The periodic flows on the different first integral manifolds possess the different frequencies. If they start with a same cross-section, for a final time t, the periodic orbits will have completely different time-responses of flows. To investigate the characteristics of two different unperturbed flows on the different first integral manifold surfaces, the two referenced, cross-section planes in phase space should be considered as in Fig.4.2(b). Two cross section surfaces are $\Xi_k^{(\alpha)}$ and $\Xi_{k+1}^{(\alpha)}$. The flows ${}_{\sigma_\alpha^1}\overline{\mathbf{x}}^{(\alpha)}(t)$ and ${}_{\sigma_\alpha^2}\overline{\mathbf{x}}^{(\alpha)}(t)$ possess the different time intervals $[t_0,t_1]$ and $[t_0,\overline{t}_1]$, respectively. For any nonlinear system, we have $t_1\neq\overline{t}_1$ but for linear systems, it may have $t_1=\overline{t}_1$. Therefore, the normal vectors of two unperturbed flows on the two first integral manifold surfaces $S_{\sigma_\alpha^1}$ and $S_{\sigma_\alpha^2}$ are defined from two points ${}_{\sigma_\alpha^1}\overline{\mathbf{x}}^{(\alpha)}\in\Gamma_{\sigma_\alpha^1}^{(\alpha)}$ and ${}_{\sigma_\alpha^2}\overline{\mathbf{x}}^{(\alpha)}\in\Gamma_{\sigma_\alpha^2}^{(\alpha)}$ on the same cross-section plane $\Xi_k^{(\alpha)}$. For two points ${}_{\sigma_\alpha^1}\overline{\mathbf{x}}^{(\alpha)}(t_k)\in\Xi_k^{(\alpha)}\cap S_{\sigma_\alpha^1}$ and ${}_{\sigma_\alpha^2}\overline{\mathbf{x}}^{(\alpha)}\in\Xi_k^{(\alpha)}\cap S_{\sigma_\alpha^2}$, the distance of the two points is computed by

$$\left\| {}_{\sigma_\alpha^2}\overline{\mathbf{x}}^{(\alpha)} - {}_{\sigma_\alpha^1}\overline{\mathbf{x}}^{(\alpha)} \right\|_{\Xi_k^{(\alpha)}} = \sqrt{\sum\nolimits_{i=1}^{n}\left({}_{\sigma_\alpha^2}\overline{x}_i^{(\alpha)} - {}_{\sigma_\alpha^1}\overline{x}_i^{(\alpha)} \right)^2}. \tag{4.17}$$

DEFINITION 4.7. For a dynamical system in Eq.(3.1), its integrable dynamical system in Eq.(3.2) possesses a family of the first integral manifolds $F(\overline{\mathbf{x}},\boldsymbol{\mu}) = E$. For a given point ${}_{\sigma_\alpha^1}\overline{\mathbf{x}}^{(\alpha)} \in S_{\sigma_\alpha^1}$ of the dynamical system in Eq.(3.2) on the cross-section plane $\Xi_k^{(\alpha)}$, there is a point ${}_{\sigma_\alpha^2}\overline{\mathbf{x}}^{(\alpha)} \in \Xi_k^{(\alpha)} \cap S_{\sigma_\alpha^2}$ to have the minimum distance from point ${}_{\sigma_\alpha^1}\overline{\mathbf{x}}^{(\alpha)}$ to the surface $S_{\sigma_\alpha^2}$. So the maximum distance of two integral manifold surfaces $S_{\sigma_\alpha^1}$ and $S_{\sigma_\alpha^2}$ is

$$\left\| {}_{\sigma_\alpha^2}\overline{\mathbf{x}}^{(\alpha)} - {}_{\sigma_\alpha^1}\overline{\mathbf{x}}^{(\alpha)} \right\| = \max_{k\in\mathbb{N}} \left\{ \min_{{}_{\sigma_\alpha^2}\overline{\mathbf{x}}^{(\alpha)} \in \Xi_k^{(\alpha)} \cap S_{\sigma_\alpha^2}} \left\| {}_{\sigma_\alpha^2}\overline{\mathbf{x}}^{(\alpha)} - {}_{\sigma_\alpha^1}\overline{\mathbf{x}}^{(\alpha)} \right\|_{\Xi_k^{(\alpha)}} \right\}. \tag{4.18}$$

The maximum distance of the two integral surfaces S_s and S_{σ_α} are defined by two points $\overline{\mathbf{x}}^{(s)} \in \Xi_k^{(s)} \cap S_s$ and ${}_{\sigma_\alpha}\overline{\mathbf{x}}^{(\alpha)} \in \Xi_k^{(s)} \cap S_{\sigma_\alpha}$ on the same cross-section surface $\Xi_k^{(s)}$ is

$$\left\| {}_{\sigma_\sigma}\overline{\mathbf{x}}^{(\alpha)} - \overline{\mathbf{x}}^{(s)} \right\| = \max_{k\in\mathbb{N}} \left\{ \min_{{}_{\sigma_\alpha}\overline{\mathbf{x}}^{(\alpha)} \in \Xi_k^{(\alpha)} \cap S_{\sigma_\alpha}} \left\| {}_{\sigma_\sigma}\overline{\mathbf{x}}^{(\alpha)} - \overline{\mathbf{x}}^{(s)} \right\|_{\Xi_k^{(s)}} \right\}. \tag{4.19}$$

LEMMA 4.1. *For the dynamical system in Eq.(3.1) with Assumptions (A1)-(A3) in Ω_α, consider the first integral surface $S_{\sigma_\alpha^0}$ as a reference one. For an arbitrary $\varepsilon_\alpha > 0$, there exists a $\delta_\alpha > 0$ and $E_{\sigma_\alpha} \in N_{\sigma_\alpha}^{\delta_\alpha}(E_{\sigma_\alpha^0})$. For ${}_{\sigma_\alpha}\overline{\mathbf{x}}^{(\alpha)} \in S_{\sigma_\alpha}$ and ${}_{\sigma_\alpha^0}\overline{\mathbf{x}}^{(\alpha)} \in S_{\sigma_\alpha^0}$, the condition $\| {}_{\sigma_\alpha}\overline{\mathbf{x}}^{(\alpha)} - {}_{\sigma_\alpha^0}\overline{\mathbf{x}}^{(\alpha)} \| \le \varepsilon_\alpha$ holds.*

PROOF. Choose ${}_{\sigma_\alpha^0}\overline{\mathbf{x}}^{(\alpha)} \in S_{\sigma_\alpha^0}$ ($k \in \mathbb{N}$)and consider a point ${}_{\sigma_\alpha^0}\overline{\mathbf{x}}^{(\alpha)} \in S_{\sigma_\alpha^0} \cap \Xi_k^{(\alpha)}$ on the referenced, first integral manifold. For an arbitrary $\varepsilon_\alpha > 0$, let $\delta_\alpha = \varepsilon \| F \| > 0$ satisfying

$$| E_{\sigma_\alpha} - E_{\sigma_\alpha^0} | = | F({}_{\sigma_\alpha}\overline{\mathbf{x}}^{(\alpha)},\boldsymbol{\mu}) - F({}_{\sigma_\alpha^0}\overline{\mathbf{x}}^{(\alpha)},\boldsymbol{\mu}) | = \| F \| \times \|_{\sigma_\alpha} \overline{\mathbf{x}}^{(\alpha)} - {}_{\sigma_\alpha^0}\overline{\mathbf{x}}^{(\alpha)} \| < \delta_\alpha$$

where

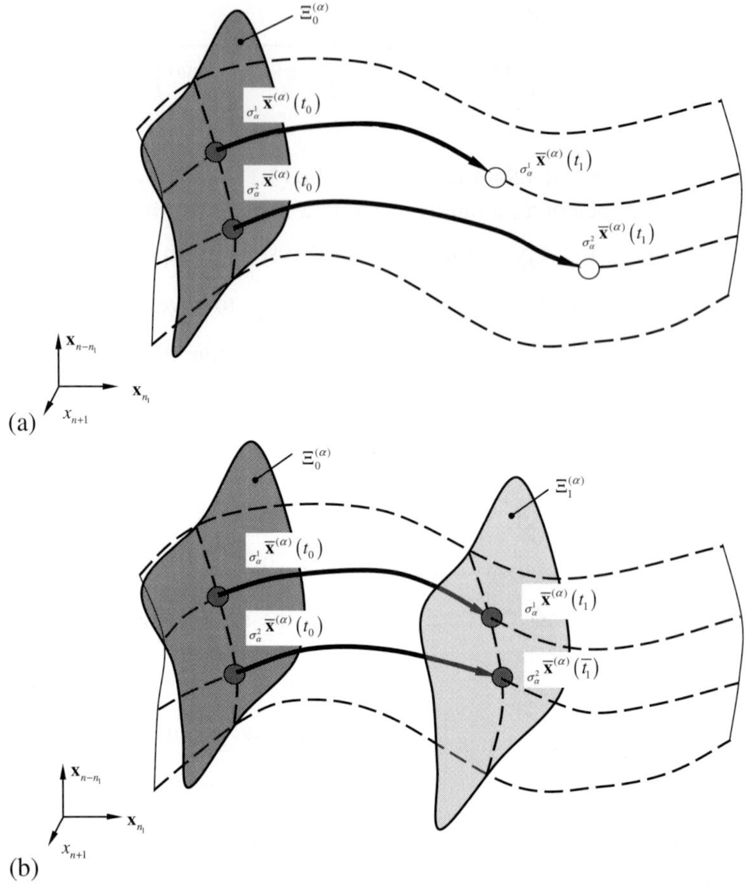

Figure 4.2. Flows on the first integral manifold surfaces: (a) flows with the same cross surface ($\Xi_0^{(\alpha)}$) from time t_0 to t_1; (b) flows between two cross section surfaces ($\Xi_0^{(\alpha)}$ and $\Xi_1^{(\alpha)}$).

$$\| F \| = \sup_{\sigma_\alpha \neq \sigma_\alpha^0} | F({}_{\sigma_\alpha}\overline{\mathbf{x}}^{(\alpha)}) - F_{\sigma_\alpha}({}_{\sigma_\alpha^0}\overline{\mathbf{x}}^{(\alpha)}) | \big/ \| {}_{\sigma_\alpha}\overline{\mathbf{x}}^{(\alpha)} - {}_{\sigma_\alpha^0}\overline{\mathbf{x}}^{(\alpha)} \|$$

because the first integral function $F(\overline{\mathbf{x}}, \boldsymbol{\mu})$ is C^r-smooth ($r \geq 3$) and $0 < \| F \| < \infty$ for ${}_{\sigma_\alpha}\overline{\mathbf{x}}^{(\alpha)}$ and ${}_{\sigma_\alpha^0}\overline{\mathbf{x}}^{(\alpha)}$. Therefore, we have

$$\| {}_{\sigma_\alpha}\overline{\mathbf{x}}^{(\alpha)} - {}_{\sigma_\alpha^0}\overline{\mathbf{x}}^{(\alpha)} \| < \delta_\alpha / \| F \| = \varepsilon_\alpha .$$

This lemma is proved. ■

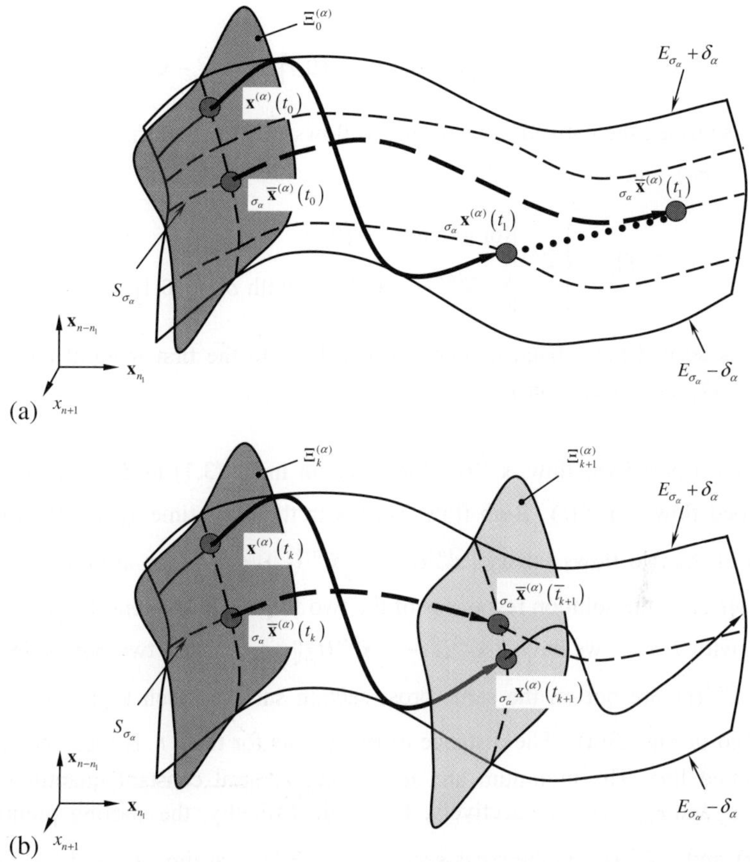

Figure 4.3. Non-integrable flow ($\mathbf{x}^{(\alpha)}(t)$) and integrable flow (${}_{\sigma_\alpha}\mathbf{x}^{(\alpha)}(t)$): (a) flows with the same cross surface ($\Xi_0^{(\alpha)}$) from time t_0 to t_1 ; (b) flows between two cross-section surfaces ($\Xi_k^{(\alpha)}$ and $\Xi_{k+1}^{(\alpha)}$).

If $S_{\sigma_\alpha^0} \equiv S_s$, the above lemma is applicable to the region of Ω_α ($\alpha = 1, 2, \cdots, m$) in vicinity of the separatrix. For any arbitrary $\varepsilon_\alpha > 0$, $\| {}_{\sigma_\alpha}\overline{\mathbf{x}}^{(\alpha)} - \overline{\mathbf{x}}^{(s)} \| \leq \varepsilon_\alpha$ holds for $E_{\sigma_\alpha} \in N_{\sigma_\alpha}^{\delta_\alpha}(E_s)$ and ${}_{\sigma_\alpha}\overline{\mathbf{x}}^{(\alpha)} \in S_{\sigma_\alpha}$.

DEFINITION 4.8. For a dynamical system in Eq.(3.1), its integrable dynamical system in Eq.(3.2) possesses a family of the first integral manifolds $F(\overline{\mathbf{x}}, \boldsymbol{\mu}) = E$. The ε -band for unperturbed flows of the integrable system in Eq.(3.1) is defined as

$$ {}_s\Omega_\alpha^\varepsilon = \left\{ {}_{\sigma_\alpha}\bar{\mathbf{x}}^{(\alpha)} \,\middle|\, \text{for all possible } \sigma_\alpha \in (0,1], \left\| {}_{\sigma_\alpha}\bar{\mathbf{x}}^{(\alpha)} - \bar{\mathbf{x}}^{(s)} \right\| < \varepsilon \right\} \tag{4.20} $$

which is termed the ε-band of unperturbed flows to the separatrix $\partial\Omega_{ij}$ in Ω_α ($\alpha = 1, 2, \cdots, m$), and

$$ {}_{\sigma_\alpha^0}\Omega_\alpha^\varepsilon = \left\{ \bar{\mathbf{x}}^{(\alpha)} \,\middle|\, \begin{array}{l} \text{for given } \sigma_\alpha^0 \in (0,1], \\ \left\| {}_{\sigma_\alpha}\bar{\mathbf{x}}^{(\alpha)} - {}_{\sigma_\alpha^0}\bar{\mathbf{x}}^{(\alpha)} \right\| < \varepsilon \text{ with } \sigma_\alpha \in (0,1] \end{array} \right\} \tag{4.21} $$

which is termed the ε-band of unperturbed flows to the first integral manifold $S_{\sigma_\alpha^0}$ in Ω_α ($\alpha = 1, 2, \cdots, m$).

Consider a perturbed flow $\mathbf{x}^{(\alpha)}(t)$ for a system in Eq.(3.1) in Ω_α near the unperturbed flow ${}_{\sigma_\alpha}\bar{\mathbf{x}}^{(\alpha)}(t)$. Both flows start with the same time t_0 and the initial points of the two flows satisfy $\| \mathbf{x}^{(\alpha)}(t_0) - {}_{\sigma_\alpha}\bar{\mathbf{x}}^{(\alpha)}(t_0) \| < \varepsilon_\alpha$ for an arbitrary $\varepsilon_\alpha > 0$. For $t \in [t_0, \infty)$, the solution responses of the two flows are different. For time $t \in [t_1, \infty)$ with $t_1 \neq t_0$, we have $\| \mathbf{x}^{(\alpha)}(t) - {}_{\sigma_\alpha}\bar{\mathbf{x}}^{(\alpha)}(t) \| < \varepsilon_\alpha$ and the two points $\mathbf{x}^{(\alpha)}(t)$ and ${}_{\sigma_\alpha}\bar{\mathbf{x}}^{(\alpha)}(t)$ are not on the same cross-section surface. Such a phenomena is sketched in Fig.4.3(a). The distance of two points for time t_1 is represented by the dotted line. The minimum and maximum, integral constant quantities are $E_{\sigma_\alpha} - \delta_\alpha$ and $E_{\sigma_\alpha} + \delta_\alpha$, respectively. For such difficulty, the starting points of $\mathbf{x}^{(\alpha)}(t)$ and ${}_{\sigma_\alpha}\bar{\mathbf{x}}^{(\alpha)}(t)$ on the cross-section plane $\Xi_k^{(\alpha)}$ is at time t_k, and $\| \mathbf{x}^{(\alpha)}(t_k) - {}_{\sigma_\alpha}\bar{\mathbf{x}}^{(\alpha)}(t_k) \| < \varepsilon_\alpha$. After next time $t_{k+1} > t_k$, the point $\mathbf{x}^{(\alpha)}(t_{k+1})$ of the flow $\mathbf{x}^{(\alpha)}(t)$ is as a base to form the cross-section plane $\Xi_{k+1}^{(\alpha)}$, as shown in Fig.4.3(b). The intersected point ${}_{\sigma_\alpha}\bar{\mathbf{x}}^{(\alpha)}(\bar{t}_{k+1}) \in {}_{\sigma_\alpha}\bar{\mathbf{x}}^{(\alpha)}(t) \cap \Xi_{k+1}^{(\alpha)}$ but $\bar{t}_{k+1} \neq t_{k+1}$. The two intersected points are on the same cross-section surface. If $| E_\alpha(t) - E_{\sigma_\alpha} | \leq \delta_\alpha$ is inserted to bound the flow of the non-integrable vector field, $\| \mathbf{x}^{(\alpha)}(t_{k+1}) - {}_{\sigma_\alpha}\bar{\mathbf{x}}^{(\alpha)}(\bar{t}_{k+1}) \| < \varepsilon_\alpha$ may not be satisfied. To explain this phenomenon between the non-integrable and integrable flows, a perturbed flow of system in Eq.(3.1) is projected on the referenced, first integral manifold surface S_{σ_α} with the first integral quantity E_{σ_α} is sketched in Fig.4.4. The perturbed flow of system in Eq.(3.1) is represented by the dark solid curve. The projected curve on the referenced surface is represented by the dotted curve.

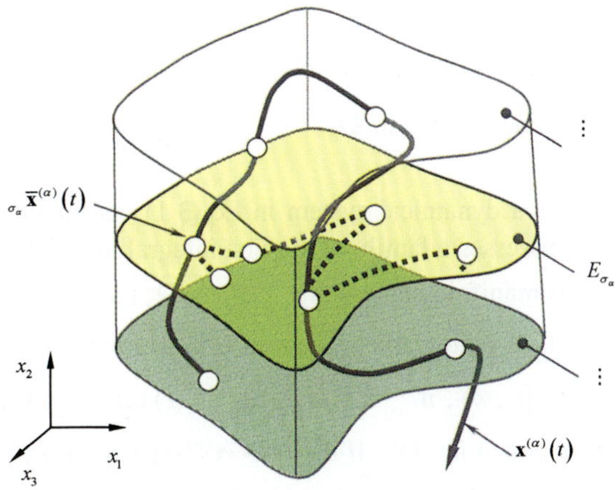

Figure 4.4. A flow of system in Eq.(3.1) projected on the referenced, first integral manifold surface S_{σ_α} with the first integral quantity E_{σ_α}. The flow of system in Eq.(3.1) is represented by the dark solid curve. The projected curve on the referenced surface is represented by the dotted curve.

The projected curve on the referenced, integral manifold surface will cross over a bundle of unperturbed flows with different initial conditions on such a surface. Thus, from the above two reasons, we cannot simply use $_{\sigma_\alpha}\bar{\mathbf{x}}^{(\alpha)}(t) \approx \mathbf{x}^{(\alpha)}(t)$ to compute any quantity. The first integral quantity will be used to define a flow band for perturbed flows of the non-integrable system in Eq.(3.1). The signific-ance of the flow band lies on determining the flow boundary, which is very easy for one to control perturbed flows of system in Eq.(3.1). If a flow of non-integrable system in Eq.(3.1) is bounded in the ε-band of unperturbed flows in the integrable system of Eq.(3.1).

4.1.3. Chaotic motion bands

From the ε-band of unperturbed flows, the band of perturbed flows in Eq.(3.1) can be further investigated. The mathematical description for the band of perturbed flows in the system of Eq.(3.1) is given as follows:

DEFINITION 4.9. For a dynamical system in Eq.(3.1), its integrable dynamical system in Eq.(3.2) possesses a family of the first integral manifolds $F(\bar{\mathbf{x}},\boldsymbol{\mu}) = E$. For the first integral manifold surface $S_{\sigma_\alpha^0}$ in Ω_α ($\alpha = 1,2,\cdots,m$), there is a perturbed flow $\mathbf{x}^{(\alpha)}(t)$ of non-integrable system in Eq.(3.1) with an initial set $_{t_0}^{(L)}D_\alpha$

$\subset {}_{\sigma_\alpha^0}\Omega_\alpha^\varepsilon$. For time $t \in [t_0, \infty)$, if ${}_{\sigma_\alpha}\bar{\mathbf{x}}^{(\alpha)} = S_{\sigma_\alpha} \cap \mathbf{x}^{(\alpha)}(t) \subset {}_{\sigma_\alpha^0}\Omega_\alpha^\varepsilon$, the flow $\mathbf{x}^{(\alpha)}(t)$ with $t \in [t_0, \infty)$ is bounded in ${}_{\sigma_\alpha^0}\Omega_\alpha^\varepsilon$. If the flow $\mathbf{x}^{(\alpha)}(t)$ is chaotic, then the band ${}_{\sigma_\alpha^0}\Omega_\alpha^\varepsilon$ is termed *the chaotic band* in Ω_α.

DEFINITION 4.10. For a dynamical system in Eq.(3.1), its integrable dynamical system in Eq.(3.2) possesses a family of the first integral manifolds $F(\bar{\mathbf{x}}, \boldsymbol{\mu}) = E$. For the first integral manifold surface $\partial\Omega_{ij} \subset S_s$ in Ω_α ($\alpha = i, j$), there is a perturbed flow $\mathbf{x}^{(\alpha)}(t)$ of non-integrable system in Eq.(3.1) with an initial set ${}^{(G)}_{t_0}D_\alpha \subset {}_s\Omega_\alpha^\varepsilon$. For time $t \in [t_0, \infty)$, if ${}_{\sigma_\alpha}\bar{\mathbf{x}}^{(\alpha)} = S_{\sigma_\alpha} \cap \mathbf{x}^{(\alpha)}(t) \subset {}_s\Omega_\alpha^\varepsilon$, the flow $\mathbf{x}^{(\alpha)}(t)$ with $t \in [t_0, \infty)$ is bounded in ${}_s\Omega_\alpha^\varepsilon$. If the flow $\mathbf{x}^{(\alpha)}(t)$ ($\alpha = i, j$) is chaotic, then the band ${}_s\Omega_\alpha^\varepsilon$ is termed *the chaotic band* in the vicinity of the separatrix $\partial\Omega_{ij}$ (or *the stochastic band*).

DEFINITION 4.11. If the non-integrable vector field of system in Eq.(3.1) is time-independent, the Poincare section is defined on a cross section Ξ_0 as

$$\Gamma^P = \left\{ \mathbf{x}(t_k) \middle| \begin{array}{l} \mathbf{x}(t_k) \in \mathbf{x}(t) \cap \Xi_0, \text{ the reference plane } \Xi_0 \subset \mathbb{R}^{n-1} \\ \text{and } \mathbf{x}(t) = \boldsymbol{\Phi}(\mathbf{x}_0, t, t_0) \text{ for } t \in (t_0, \infty) \end{array} \right\}. \tag{4.22}$$

If $\Gamma^P \subset \Xi_0$ is dense, the flow $\mathbf{x}(t)$ is chaotic. If Ξ_0 is a normal cross-section surface of $S_{\sigma_\alpha^0}$ in ${}_{\sigma_\alpha^0}\Omega_\alpha^\varepsilon$ and the dense set $\Gamma^P \subset \Xi_0 \cap {}_{\sigma_\alpha^0}\Omega_\alpha^\varepsilon$ exists, then the band ${}_{\sigma_\alpha^0}\Omega_\alpha^\varepsilon$ is termed the *chaotic band* in Ω_α. Otherwise, such a band is the periodic or quasi-periodic band in Ω_α. Similarly, if Ξ_0 is a normal cross-section plane of S_s in ${}_s\Omega_\alpha^\varepsilon$ ($\alpha = 1, 2, \cdots, m$) and the dense set $\Gamma^P \subset \Xi_0 \cap {}_s\Omega_\alpha^\varepsilon$ exists, then the band ${}_s\Omega_\alpha^\varepsilon$ is termed the *chaotic* band of the separatrix $\partial\Omega_{\alpha\beta}$ ($\beta \neq \alpha$, $\beta = 1, 2, \cdots, m$).

DEFINITION 4.12. If the non-integrable vector field in Eq.(3.1) has a periodically time-dependent term with period T, the Poincare section is defined as

$$\Gamma^P = \left\{ \mathbf{x}(t_N) \middle| \ \mathbf{x}(t_N) = \boldsymbol{\Phi}(\mathbf{x}_0, t_N, t_0),\ t_N = NT + t_0,\ N = 0, 1, 2, \ \cdots \right\}. \tag{4.23}$$

If $\Gamma^P \subset \Omega_\alpha$ is dense, the flow $\mathbf{x}(t)$ is chaotic. If the dense set $\Gamma^P \subset {}_{\sigma_\alpha^0}\Omega_\alpha^\varepsilon$ exists, then the band ${}_{\sigma_\alpha^0}\Omega_\alpha^\varepsilon$ is termed *the chaotic band* in Ω_α. Otherwise, the band is a periodic or quasi-periodic band in Ω_α. Similarly, if the dense set $\Gamma^P \subset {}_s\Omega_\alpha^\varepsilon$ (α

$=1,2,\cdots,m$) exists, then the band ${}_s\Omega_\alpha^\varepsilon$ is termed *the chaotic band* in vicinity of the separatrix $\partial\Omega_{\alpha\beta}$ ($\beta \neq \alpha$, $\beta = 1,2,\cdots,m$).

In the above chaotic bands, if the non-integrable vector field in Eq.(3.1) only includes periodic terms with period T, then the corresponding chaotic band in the vicinity of $\partial\Omega_{\alpha\beta}$ is also termed *the stochastic band* or *separatrix band.* If chaotic motion exists in the vicinity of the resonant separatrix, this chaotic band is called *the resonant band.* To determine the periodic, quasi-periodic or chaotic flows in Eq.(3.1), a new quantity should be introduced to continue our discussions.

4.2. First integral quantity increments

To determine the characteristics of a perturbed flow in a certain flow domain, consider the first integral quantity of a perturbed flow $\mathbf{x}^{(\alpha)}(t)$ in non-integrable system in Eq.(3.1) for time t in Ω_α , we have $E^{(\alpha)}(t) = F(\mathbf{x}^{(\alpha)}(t),\boldsymbol{\mu})$. The time-change rate of the first integral quantity is given by

$$\begin{aligned} \tfrac{dE^{(\alpha)}(t)}{dt} &= \nabla F\left(\mathbf{x}^{(\alpha)}(t),\boldsymbol{\mu}\right)\bullet\dot{\mathbf{x}}^{(\alpha)}(t) + \tfrac{\partial F}{\partial t} = \nabla F\left(\mathbf{x}^{(\alpha)}(t),\boldsymbol{\mu}\right)\bullet\dot{\mathbf{x}}^{(\alpha)}(t) \\ &= \nabla F\left(\mathbf{x}^{(\alpha)}(t),\boldsymbol{\mu}\right)\bullet\mathbf{F}\left(\mathbf{x}^{(\alpha)},t,\mathbf{p}\right) = \nabla F\left(\mathbf{x}^{(\alpha)}(t),\boldsymbol{\mu}\right)\bullet\mathbf{g}\left(\mathbf{x}^{(\alpha)},t,\boldsymbol{\pi}\right). \end{aligned} \tag{4.24}$$

From Chapter 2, such a time change rate of the first integral quantity is the zero-order G-function in the normal direction of the first integral manifold surface.

DEFINITION 4.13. For a dynamical system in Eq.(3.1), its integrable dynamical system in Eq.(3.2) possesses a family of the first integral manifolds $F(\overline{\mathbf{x}},\boldsymbol{\mu}) = E$, and the first integral quantity of a perturbed flow $\mathbf{x}^{(\alpha)}(t)$ is $E^{(\alpha)}(t) = F(\mathbf{x}^{(\alpha)}(t),\boldsymbol{\mu})$. The increment of the first integral quantity for a perturbed flow $\mathbf{x}^{(\alpha)}(t)$ for $t \in [t_k, t_{k+1}]$ is defined as

$$L^{(\alpha)}\left(t_k,t\right) \equiv \int_{t_k}^{t} G_\alpha^{(0)}\left(\mathbf{x}^{(\alpha)},t,\mathbf{p}\right)dt = \int_{t_k}^{t} \nabla F\left(\mathbf{x}^{(\alpha)},\boldsymbol{\mu}\right)\bullet\mathbf{g}\left(\mathbf{x}^{(\alpha)},t,\boldsymbol{\pi}\right)dt. \tag{4.25}$$

DEFINITION 4.14. A reference surface to a perturbed flow (i.e., $\gamma : \cup_\alpha \mathbf{x}^{(\alpha)}(t)$ ($\alpha \in \{\{m_1, m_2 \cdots m_l\} \subseteq \{1,2,\cdots,m\}$))of a time-independent system in Eq.(3.1) is defined through a continuous function f , i.e.,

$$\Xi_0 = \left\{\mathbf{x}_k^{(\alpha)} \middle| f\left(\mathbf{x}_k^{(\alpha)}\right) = 0 \text{ for } \mathbf{x}_k^{(\alpha)} \in \cup_\alpha \mathbf{x}^{(\alpha)}\right\}. \tag{4.26}$$

THEOREM 4.1. *For a dynamical system in Eq.*(3.1) *without any periodically time-dependent term, there is an* $(n-1)$ *-dimensional reference surface* Ξ_0 *perpendicular to an initial, first-integral manifold surface of Eq.*(3.1). *A perturbed flow* γ *is composed of* $\cup_\alpha \mathbf{x}^{(\alpha)}(t)$ ($\alpha \in \{m_1, m_2, \cdots m_l\} \subseteq \{1,2,\cdots,m\}$) *for* $t \in [t_0, t_1]$ *in all possible domains* Ω_α, *and the starting and ending points* ($\mathbf{x}_0^{(\alpha)}, \mathbf{x}_1^{(\alpha)} \in \Gamma^P$) *are on the reference surface* Ξ_0.

(i) *If a flow with one cycle to the reference surface* Ξ_0 *is periodic, the following conditions exist, i.e., the periodicity of the closed cycle is*

$$\mathbf{x}_1^{(\alpha)} = \mathbf{x}_0^{(\alpha)} \text{ for } t_1 \neq t_0 \tag{4.27}$$

and its first integral quantity increment is equal to zero, i.e.,

$$L^{(\alpha)}(t_0, t_1) = \int_{t_0}^{t_1} \nabla F\left(\mathbf{x}^{(\alpha)}, \boldsymbol{\mu}\right) \bullet \mathbf{g}\left(\mathbf{x}^{(\alpha)}, t, \boldsymbol{\pi}\right) dt = 0. \tag{4.28}$$

(ii) *If the flow with one cycle to the reference surface* Ξ_0 *is not periodic, then the corresponding condition becomes:*

$$\mathbf{x}_1^{(\alpha)} \neq \mathbf{x}_0^{(\alpha)} \text{ for } t_1 \neq t_0 \tag{4.29}$$

and the first integral quantity increment is the difference between the first integral quantities of the two points $(t_0, \mathbf{x}_0^{(\alpha)})$ *and* $(t_1, \mathbf{x}_1^{(\alpha)})$, *i.e.,*

$$L^{(\alpha)}(t_0, t_1) = \int_{t_0}^{t_1} \nabla F\left(\mathbf{x}^{(\alpha)}, \boldsymbol{\mu}\right) \bullet \mathbf{g}\left(\mathbf{x}^{(\alpha)}, t, \boldsymbol{\pi}\right) dt = F\left(\mathbf{x}_1^{(\alpha)}, \boldsymbol{\mu}\right) - F\left(\mathbf{x}_0^{(\alpha)}, \boldsymbol{\mu}\right). \tag{4.30}$$

PROOF. (i) If the perturbed flow $\cup_\alpha \mathbf{x}^{(\alpha)}(t)$ is periodic, it is a closed curve for $t \in [t_0, t_1]$ and $\mathbf{x}_1^{(\alpha)} = \mathbf{x}_0^{(\alpha)}$ for $t_1 \neq t_0$. For the first integral function $F(\mathbf{x}, \boldsymbol{\mu}) = E$, the corresponding first differentiation is exact, given by

$$dE = dF(\mathbf{x}, \boldsymbol{\mu}) = \nabla F(\mathbf{x}, \boldsymbol{\mu}) \bullet d\mathbf{x} = \sum_{i=1}^{n} \tfrac{\partial F(\mathbf{x}, \boldsymbol{\mu})}{\partial x_i} dx_i$$

Choosing

$$P_i = \tfrac{\partial F(\mathbf{x}, \boldsymbol{\mu})}{\partial x_i},$$

then the above expression is changed to

$$dE = dF(\mathbf{x}, \boldsymbol{\mu}) = \nabla F(\mathbf{x}, \boldsymbol{\mu}) \bullet d\mathbf{x} = \sum_{i=1}^{n} P_i dx_i.$$

Because the flow γ is periodic, the flow γ for the line integration is a closed orbit, so the first integral quantity increment is computed through the integration along a closed orbit. That is,

$$\Delta E(t_0,t_1) = \int_{\mathbf{x}_0}^{\mathbf{x}_1} \nabla F(\mathbf{x},\boldsymbol{\mu})\cdot d\mathbf{x} = \oint_{\gamma} \sum_{i=1}^{n} P_i dx_i$$

$$\underbrace{=}_{Stokes'Theorem} \iint_{\Omega_\gamma} \sum_{i=1}^{n} \sum_{j=i+1}^{n} \left(\frac{\partial P_j}{\partial x_i} - \frac{\partial P_i}{\partial x_j}\right) dx_i dx_j$$

$$= \iint_{\Omega_\gamma} \sum_{i=1}^{n} \sum_{j=i+1}^{n} \left(\frac{\partial^2 F}{\partial x_j \partial x_i} - \frac{\partial^2 F}{\partial x_i \partial x_j}\right) dx_i dx_j = 0.$$

Suppose the starting and ending points of a flow $\mathbf{x}^{(\alpha)}(t)$ are $(t_0, \mathbf{x}_0^{(\alpha)})$ and $(t_1, \mathbf{x}_1^{(\alpha)})$, respectively, changing the integration of the first integral quantity with space to time gives

$$\Delta E(t_0,t_1) = \int_{\mathbf{x}_0}^{\mathbf{x}_1} \nabla F(\mathbf{x}^{(\alpha)},\boldsymbol{\mu})\cdot d\mathbf{x} = \int_{t_0}^{t_1} \nabla F(\mathbf{x}^{(\alpha)},\boldsymbol{\mu})\cdot \dot{\mathbf{x}}^{(\alpha)} dt$$

$$= \int_{t_0}^{t_1} \nabla F(\mathbf{x}^{(\alpha)},\boldsymbol{\mu})\cdot \mathbf{g}(\mathbf{x}^{(\alpha)},t,\boldsymbol{\pi}) dt = L^{(\alpha)}(t_0,t_1).$$

If the flow $\mathbf{x}^{(\alpha)}(t)$ is periodic, equation (4.27) are satisfied. From the two foregoing equation, we have

$$L^{(\alpha)}(t_0,t_1) = \int_{t_0}^{t_1} \nabla F(\mathbf{x}^{(\alpha)},\boldsymbol{\mu})\cdot \mathbf{g}(\mathbf{x}^{(\alpha)},t,\boldsymbol{\pi}) dt = 0.$$

(ii) Because this flow is not periodic, it is a open curve for $t \in [t_0, t_1]$ and $\mathbf{x}_1^{(\alpha)} \neq \mathbf{x}_0^{(\alpha)}$ for $t_1 \neq t_0$. From Eq.(4.24), we have

$$L^{(\alpha)}(t_0,t_1) = \int_{t_0}^{t_1} \nabla F(\mathbf{x}^{(\alpha)},\boldsymbol{\mu})\cdot \mathbf{g}(\mathbf{x}^{(\alpha)},t,\boldsymbol{\pi}) dt$$

$$= \int_{\mathbf{x}_0^{(\alpha)}}^{\mathbf{x}_1^{(\alpha)}} \nabla F(\mathbf{x}^{(\alpha)},\boldsymbol{\mu})\cdot d\mathbf{x} = \int_{F(\mathbf{x}_0^{(\alpha)},\boldsymbol{\mu})}^{F(\mathbf{x}_1^{(\alpha)},\boldsymbol{\mu})} dF(\mathbf{x}^{(\alpha)},\boldsymbol{\mu})$$

$$= F(\mathbf{x}_1^{(\alpha)},\boldsymbol{\mu}) - F(\mathbf{x}_0^{(\alpha)},\boldsymbol{\mu}).$$

If Eq.(4.27) is applied, the first integral increment for the periodic flow is zero (i.e., $L^{(\alpha)}(t_0,t_1) = 0$), which is without application of the Stokes' theorem. So the domain Ω_γ enclosed by the periodic orbit should not be simply connected. The above theorem is proved. ■

REMARK. Compared Theorem 4.1 with the Bendixson's criterion, Bendixson's criterion requires the domain is simply connected, and it can be used only for 2-dimensional vector fields. However, the foregoing theorem can be applied for *n*-dimensional vector fields, and it does not have any requirement for domain to be simple and connected. Theorem 4.1 provides a necessary condition for periodic flows in n-dimensional systems of Eq.(3.1). On the other words, it means that $L^{(\alpha)}(t_0,t_1)=0$ does not imply the flow is periodic.

THEOREM 4.2. *For a dynamical system in Eq.*(3.1) *without any periodically time-dependent term, there is an* $(n-1)$*-dimensional reference surface* Ξ_0 *perpendicular to the initial first-integral manifold of Eq.*(3.1). *A perturbed flow* γ *is composed of* $\cup_\alpha \mathbf{x}^{(\alpha)}(t)$ ($\alpha \in \{m_1, m_2, \cdots m_l\} \subseteq \{1,2,\cdots,m\}$) *for* $t\in[t_0,t_k]$ *in all possible domains* Ω_α, *and the starting and ending points on the reference surface* Ξ_0 *are* $\mathbf{x}_l^{(\alpha)} \in \Gamma^P$ ($l=0,1,2,\cdots,k$) *at time* t_l.

(i) *If a flow* γ *with k-cycles to the reference* Ξ_0 *is periodic, then the following conditions exist for* $t_k > t_{k-1} > \cdots > t_1 > t_0$, *i.e.,*

$$\mathbf{x}_l^{(\alpha)} \neq \mathbf{x}_0^{(\alpha)} \text{ for } l=1,2,\cdots,k-1 \tag{4.31}$$

$$\mathbf{x}_k^{(\alpha)} = \mathbf{x}_0^{(\alpha)} \text{ for } t_k - t_0 \neq 0 \tag{4.32}$$

and the first integral quantity increment from time t_0 *to* t_l ($l=1,2,\cdots,k-1$) *is computed by*

$$L^{(\alpha)}(t_0,t_l) = \int_{t_0}^{t_l} \nabla F\left(\mathbf{x}^{(\alpha)},\boldsymbol{\mu}\right)\cdot\mathbf{g}\left(\mathbf{x}^{(\alpha)},t,\boldsymbol{\pi}\right)dt = F\left(\mathbf{x}_l^{(\alpha)},\boldsymbol{\mu}\right) - F\left(\mathbf{x}_0^{(\alpha)},\boldsymbol{\mu}\right), \tag{4.33}$$

and the first integral quantity increment from time t_0 *to* t_k *is zero, i.e.,*

$$L^{(\alpha)}(t_0,t_k) = \int_{t_0}^{t_k} \nabla F\left(\mathbf{x}^{(\alpha)},\boldsymbol{\mu}\right)\cdot\mathbf{g}\left(\mathbf{x}^{(\alpha)},t,\boldsymbol{\pi}\right)dt = 0. \tag{4.34}$$

(ii) *If a flow with k-cycles to the reference* Ξ_0 *is not periodic, then the corresponding condition becomes:*

$$\mathbf{x}_k^{(\alpha)} \neq \mathbf{x}_0^{(\alpha)} \text{ for } t_k \neq t_0 \tag{4.35}$$

and the first integral quantity increment is the difference between the first integrals of the two points $(t_0, \mathbf{x}_0^{(\alpha)})$ *and* $(t_k, \mathbf{x}_k^{(\alpha)})$

$$L^{(\alpha)}(t_0,t_k) = \int_{t_0}^{t_k} \nabla F\left(\mathbf{x}^{(\alpha)},\boldsymbol{\mu}\right)\cdot\mathbf{g}\left(\mathbf{x}^{(\alpha)},t,\boldsymbol{\pi}\right)dt = F\left(\mathbf{x}_k^{(\alpha)},\boldsymbol{\mu}\right) - F\left(\mathbf{x}_0^{(\alpha)},\boldsymbol{\mu}\right). \tag{4.36}$$

In addition, the iteration relation of the first integral quantity from time t_l *to* t_{l+1} *is determined by*

$$E_{l+1} - E_l = \Delta E^{(\alpha)}(t_l,t_{l+1}) \equiv L^{(\alpha)}(t_l,t_{l+1}), \tag{4.37}$$

where for time t_l *to* t_{l+1} *(* $l=0,1,2,\cdots,k-1$ *),*

$$L^{(\alpha)}\left(t_l,t_{l+1}\right)=\int_{t_l}^{t_{l+1}}\nabla F\left(\mathbf{x}^{(\alpha)}\right)\cdot\mathbf{g}\left(\mathbf{x}^{(\alpha)},t,\boldsymbol{\pi}\right)dt=F\left(\mathbf{x}_{l+1}^{(\alpha)},\boldsymbol{\mu}\right)-F\left(\mathbf{x}_l^{(\alpha)},\boldsymbol{\mu}\right). \quad (4.38)$$

PROOF. Using THEOREM 4.1, this theorem can be easily proved. ■

To explain the above two theorems, consider a perturbed flow to a reference surface Ξ_0 in the 3-dimesnional phase space for illustration. A perturbed flow γ ($\cup_\alpha \mathbf{x}^{(\alpha)}(t)$, $\alpha\in\{m_1,m_2,\cdots m_l\}\subseteq\{1,2,\cdots,m\}$) of the non-integrable system is sketched in Fig.4.5 via the dark curves. A perturbed flow close to one cycle to the cross-section surface Ξ_0 from time t_0 to t_1 is presented in Fig.4.5(a) by the dark curve. If $\mathbf{x}_0^{(\alpha)}=\mathbf{x}_1^{(\alpha)}$ and $t_1\neq t_0$, the periodic flow can be formed. For a periodic flow with one closed cycle, the first integral quantity increment is equal to zero. If this flow to the reference Ξ_0 is non-periodic, such an increment is equal to the difference between the two first integral quantities relative to the points $\mathbf{x}_0^{(\alpha)}$ and $\mathbf{x}_1^{(\alpha)}$. The flow to the cross-section surface Ξ_0 with the $(k+1)$ intersected points $\mathbf{x}_l^{(\alpha)}$ ($l=0,1,\cdots,k$) to time from t_0 to t_k is depicted in Fig.4.5(b) by the dark curve. The periodic flow with k-cycles to the reference surface Ξ_0 requires $\mathbf{x}_0^{(\alpha)}=\mathbf{x}_k^{(\alpha)}$ for $t_{l+1}>t_l$ ($l=0,1,\cdots,k-1$), and the first integral quantity increment is equal to zero because the flow is a closed curve to time from t_0 to t_k. The first integral quantity for $t\in[t_l,t_{l+1}]$ can be computed by $L^{(\alpha)}(t_l,t_{l+1})$. So $L^{(\alpha)}(t_0,t_k)=\Sigma_{l=0}^{k-1}L^{(\alpha)}(t_l,t_{l+1})=0$. For the non-periodic flow to with k-cycles to the reference surface Ξ_0, the perturbed flow is an open curve, and the total increment is the summation of the increment during each time interval $t\in[t_l,t_{l+1}]$, i.e., $L^{(\alpha)}(t_0,t_k)=\Sigma_{l=0}^{k-1}L^{(\alpha)}(t_l,t_{l+1})$. For each open cycle, the corresponding iteration relation can be developed. Before doing so, the reference surface should be defined mathematically to determine such an iteration relation.

From the reference surface, it is very crucial to obtain the iterative relation as

$$E_{l+1}-E_l=L^{(\alpha)}\left(t_l,t_{l+1}\right)\ \text{and}\ t_{l+1}=t_l-\Delta T_\alpha\left(t_l,t_{l+1}\right) \quad (4.39)$$

where the time difference $\Delta T_\alpha(t_l,t_{l+1})=t_{l+1}-t_l$ is determined by the dynamical system in Eq.(3.1) with the initial and final constraints on the reference surface Ξ_0, for instance, equation (4.26) gives

$$f\left(\mathbf{x}_l^{(\alpha)}\right)=0\ \text{and}\ f\left(\mathbf{x}_{l+1}^{(\alpha)}\right)=0. \quad (4.40)$$

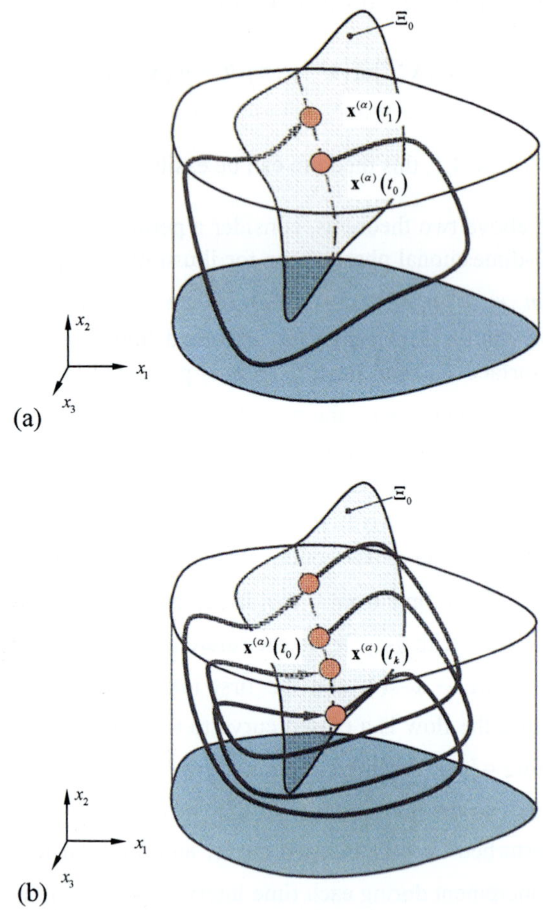

Figure 4.5. A flow γ ($\cup_\alpha \mathbf{x}^{(\alpha)}(t)$, $\alpha \in \{m_1, m_2, \cdots m_l\} \subseteq \{1, 2, \cdots, m\}$) of the non-integrable system being close to: (a) a cycle to the cross-section surface Ξ_0 for time $t \in [t_0, t_1]$; (b) k-cycles to the cross-section surface Ξ_0 for time $t \in [t_0, t_k]$.

The chaotic flows can be investigated through the above mapping form. The 2-dimensional mapping in Eq.(4.39) can be used to investigate chaotic flows in n-dimensional nonlinear dynamical systems. The advantage of the above mapping is to reduce the computation in n-dimensional nonlinear dynamical systems. The mapping relation can be used to the system in Eq.(3.1) with the periodically time-dependent terms. Similarly, the two theorems for the perturbed system in Eq.(3.1) possessing the periodically time-dependent terms will be presented as follows.

THEOREM 4.3. *For a dynamical system in Eq.*(3.1) *possessing periodically time-dependent vector fields with period T, there is a Poincare mapping section* Γ^P, *defined through each period. A perturbed flow* γ *is composed of* $\cup_\alpha \mathbf{x}^{(\alpha)}(t)$ ($\alpha \in \{m_1, m_2, \cdots m_l\} \subseteq \{1,2,\cdots,m\}$)*for* $t \in [t_0, t_1]$ *in all possible domains* Ω_α, *and for one period T, the starting and ending points are* $(t_0, \mathbf{x}_0^{(\alpha)})$ *and* $(t_1, \mathbf{x}_1^{(\alpha)})$ *with* $\mathbf{x}_0^{(\alpha)}$, $\mathbf{x}_1^{(\alpha)} \in \Gamma^P$.

(i) *If the perturbed flow* γ *is periodic from* $(t_0, \mathbf{x}_0^{(\alpha)})$ *to* $(t_1, \mathbf{x}_1^{(\alpha)})$ *with period T, then the periodicity condition is*

$$\mathbf{x}_1^{(\alpha)} = \mathbf{x}_0^{(\alpha)} \text{ for } t_1 = t_0 + T \tag{4.41}$$

and the first integral quantity increment is equal to zero during period T

$$L^{(\alpha)}(t_0, T) = \int_{t_0}^{t_0+T} \nabla F(\mathbf{x}^{(\alpha)}, \boldsymbol{\mu}) \bullet \mathbf{g}(\mathbf{x}^{(\alpha)}, t, \boldsymbol{\pi})\, dt = 0. \tag{4.42}$$

(ii) *If the perturbed flow* γ *is not periodic from* $(t_0, \mathbf{x}_0^{(\alpha)})$ *to* $(t_1, \mathbf{x}_1^{(\alpha)})$ *with period T then the periodicity conditions cannot hold, i.e.,*

$$\mathbf{x}_1^{(\alpha)} \neq \mathbf{x}_0^{(\alpha)} \text{ at } t_1 = t_0 + T \tag{4.43}$$

and the first integral quantity increment is the difference between the first integrals of the two points $(t_0, \mathbf{x}_0^{(\alpha)})$ *to* $(t_1, \mathbf{x}_1^{(\alpha)})$

$$L^{(\alpha)}(t_0, T) = \int_{t_0}^{t_0+T} \nabla F(\mathbf{x}^{(\alpha)}) \bullet \mathbf{g}(\mathbf{x}^{(\alpha)}, t, \boldsymbol{\pi})\, dt = F(\mathbf{x}_1^{(\alpha)}, \boldsymbol{\mu}) - F(\mathbf{x}_0^{(\alpha)}, \boldsymbol{\mu}). \tag{4.44}$$

In the foregoing theorem, the periodic flow with one period is discussed. Once a perturbed flow is not periodic for one period, which does not mean the flow is not periodic. Therefore, the following theorem will be presented for the periodic flow with *k*-periods.

THEOREM 4.4. *For a dynamical system in Eq.*(3.1) *possessing periodically time-dependent terms with period T, there is a Poincare mapping section* Γ^P *which is defined through each period. A perturbed flow* γ *is composed of* $\cup_\alpha \mathbf{x}^{(\alpha)}(t)$ ($\alpha \in \{m_1, m_2, \cdots m_l\} \subseteq \{1,2,\cdots,m\}$) *for* $t \in [t_0, t_k]$ *in all possible domains* Ω_α, *and the starting and ending points on the Poincare surface* Γ^P *are* $\mathbf{x}_l^{(\alpha)} \in \Gamma^P$ ($l = 0,1,2, \cdots, k$) *at time* t_l.

(i) *If the perturbed flow* γ *is periodic from* $(t_0, \mathbf{x}_0^{(\alpha)})$ *to* $(t_k, \mathbf{x}_k^{(\alpha)})$ *with k-periods (kT), the following conditions exist for* $t_l = t_0 + lT$, *i.e.,*

$$\mathbf{x}_l^{(\alpha)} \neq \mathbf{x}_0^{(\alpha)} \text{ for } l = 1, 2, \cdots, k-1 \tag{4.45}$$

$$\mathbf{x}_k^{(\alpha)} = \mathbf{x}_0^{(\alpha)} \text{ for } t_k - t_0 = kT \tag{4.46}$$

and the first integral quantity increment from time t_0 *to* t_l ($l = 1, 2, \cdots, k-1$) *is computed by*

$$L^{(\alpha)}(t_0, lT) = \int_{t_0}^{t_0+lT} \nabla F(\mathbf{x}^{(\alpha)}) \bullet \mathbf{g}(\mathbf{x}^{(\alpha)}, t, \boldsymbol{\pi})\, dt = F(\mathbf{x}_l^{(\alpha)}, \boldsymbol{\mu}) - F(\mathbf{x}_0^{(\alpha)}, \boldsymbol{\mu}), \tag{4.47}$$

and the first integral quantity increment from time t_0 *to* t_k *is zero*

$$L^{(\alpha)}(t_0, kT) = \int_{t_0}^{t_0+kT} \nabla F(\mathbf{x}^{(\alpha)}, \mu) \bullet \mathbf{g}(\mathbf{x}^{(\alpha)}, t, \boldsymbol{\pi})\, dt = 0. \tag{4.48}$$

(ii) *If the perturbed flow* γ *is not periodic from* $(t_0, \mathbf{x}_0^{(\alpha)})$ *to* $(t_k, \mathbf{x}_k^{(\alpha)})$ *with k-periods (kT), then the following conditions exist,*

$$\mathbf{x}_l^{(\alpha)} \neq \mathbf{x}_0^{(\alpha)} \text{ for } l = 1, 2, \cdots, k \tag{4.49}$$

and the first integral quantity increment is the difference between the first integrals of the two points $(t_0, \mathbf{x}_0^{(\alpha)})$ *to* $(t_k, \mathbf{x}_k^{(\alpha)})$

$$L^{(\alpha)}(t_0, kT) = \int_{t_0}^{t_0+kT} \nabla F(\mathbf{x}^{(\alpha)}) \bullet \mathbf{g}(\mathbf{x}^{(\alpha)}, t, \boldsymbol{\pi})\, dt = F(\mathbf{x}_k^{(\alpha)}, \boldsymbol{\mu}) - F(\mathbf{x}_0^{(\alpha)}, \boldsymbol{\mu}), \tag{4.50}$$

and the iteration relation of the first integral quantity from time t_l *to* t_{l+1} *is determined by*

$$\left.\begin{aligned} &E_{l+1} - E_l = \Delta E^{(\alpha)}(t_l, T) \equiv L^{(\alpha)}(t_l, T) \\ &\text{and } t_{l+1} - t_l = T \end{aligned}\right\} \tag{4.51}$$

where for time t_l *to* t_{l+1} ($l = 0, 1, 2, \cdots, k-1$)

$$L^{(\alpha)}(t_l, T) = \int_{t_l}^{t_l+T} \nabla F(\mathbf{x}^{(\alpha)}) \bullet \mathbf{g}(\mathbf{x}^{(\alpha)}, t, \boldsymbol{\pi})\, dt = F(\mathbf{x}_{l+1}^{(\alpha)}, \boldsymbol{\mu}) - F(\mathbf{x}_l^{(\alpha)}, \boldsymbol{\mu}). \tag{4.52}$$

PROOF. The above two theorems can be proved as THEOREM 4.1. ■

Finally, the above theorem is generalized for the periodic flow with k-periods in a dynamical system in Eq.(3.1) with the sth-order, harmonic, time-dependent vector fields. The theorem is stated as follows.

THEOREM 4.5. *For a dynamical system in Eq.*(3.1) *possessing the* s^{th}*-order, har-*

monic, time-dependent vector fields (i.e., period $\frac{1}{s}T$), there is a Poincare mapping section Γ^P defined through each period. A perturbed flow γ is composed of $\cup_\alpha \mathbf{x}^{(\alpha)}(t)$ ($\alpha \in \{m_1, m_2, \cdots m_l\} \subseteq \{1,2,\cdots,m\}$) for $t \in [t_0, t_k]$ in possible domains Ω_α, and the starting and ending points the Poincare surface Γ^P are $\mathbf{x}_l^{(\alpha)} \in \Gamma^P$ ($l = 0,1,2,\cdots,k$) at time t_l.

(i) *If the perturbed flow γ is periodic from $(t_0, \mathbf{x}_0^{(\alpha)})$ to $(t_k, \mathbf{x}_k^{(\alpha)})$ with k-periods of the sth-order harmonic term (i.e., $\frac{k}{s}T$), the following conditions exist for $t_l = t_0 + \frac{l}{s}T$, i.e.,*

$$\mathbf{x}_l^{(\alpha)} \neq \mathbf{x}_0^{(\alpha)} \text{ for } l = 1,2,\cdots,k-1 \tag{4.53}$$

$$\mathbf{x}_k^{(\alpha)} = \mathbf{x}_0^{(\alpha)} \text{ for } t_k - t_0 = \tfrac{k}{s}T \tag{4.54}$$

and the first integral quantity increment from time t_0 to t_l ($l = 1,2,\cdots,k-1$) is computed by

$$L^{(\alpha)}\left(t_0, \tfrac{l}{s}T\right) = \int_{t_0}^{t_0+\frac{l}{s}T} \nabla F\left(\mathbf{x}^{(\alpha)}\right) \bullet \mathbf{g}\left(\mathbf{x}^{(\alpha)}, t, \boldsymbol{\pi}\right) dt = F\left(\mathbf{x}_l^{(\alpha)}, \boldsymbol{\mu}\right) - F\left(\mathbf{x}_0^{(\alpha)}, \boldsymbol{\mu}\right), \tag{4.55}$$

and the first integral quantity increment from time t_0 to t_k is zero

$$L^{(\alpha)}\left(t_0, \tfrac{k}{s}T\right) = \int_{t_0}^{t_0+\frac{k}{s}T} \nabla F\left(\mathbf{x}^{(\alpha)}, \mu\right) \bullet \mathbf{g}\left(\mathbf{x}^{(\alpha)}, t, \boldsymbol{\pi}\right) dt = 0. \tag{4.56}$$

(ii) *If the flow γ is not periodic from $(t_0, \mathbf{x}_0^{(\alpha)})$ to $(t_k, \mathbf{x}_k^{(\alpha)})$ with k-periods of the sth-order harmonic terms (i.e., $\frac{k}{s}T$), then the periodicity cannot hold, i.e.,*

$$\mathbf{x}_l^{(\alpha)} \neq \mathbf{x}_0^{(\alpha)} \text{ for } l = 1,2,\cdots,k \tag{4.57}$$

and the first integral quantity increment is the difference between the first integrals of the two points $(t_0, \mathbf{x}_0^{(\alpha)})$ to $(t_k, \mathbf{x}_k^{(\alpha)})$, i.e.,

$$L^{(\alpha)}\left(t_0, \tfrac{k}{s}T\right) = \int_{t_0}^{t_0+\frac{k}{s}T} \nabla F\left(\mathbf{x}^{(\alpha)}\right) \bullet \mathbf{g}\left(\mathbf{x}^{(\alpha)}, t, \boldsymbol{\pi}\right) dt = F\left(\mathbf{x}_k^{(\alpha)}, \boldsymbol{\mu}\right) - F\left(\mathbf{x}_0^{(\alpha)}, \boldsymbol{\mu}\right), \tag{4.58}$$

and the iteration relation of the first integral quantity from time t_l to t_{l+1} is determined by

$$E_{l+1} - E_l = L^{(\alpha)}\left(t_l, \tfrac{1}{s}T\right) \text{ and } t_{l+1} - t_l = \tfrac{1}{s}T \tag{4.59}$$

where for time t_l to t_{l+1} ($l = 0,1,2,\cdots,k-1$)

$$L^{(\alpha)}\left(t_l,\tfrac{1}{s}T\right)=\int_{t_l}^{t_l+\frac{1}{s}T}\nabla F\left(\mathbf{x}^{(\alpha)}\right)\bullet\mathbf{g}\left(\mathbf{x}^{(\alpha)},t,\boldsymbol{\pi}\right)dt=F\left(\mathbf{x}_{l+1}^{(\alpha)},\boldsymbol{\mu}\right)-F\left(\mathbf{x}_l^{(\alpha)},\boldsymbol{\mu}\right). \quad (4.60)$$

PROOF. This theorem can be proved as Theorem 4.1. ■

In the above theorems, the integrations are based on one period of harmonic terms. To determine the interaction between the integrable and non-integrable vector fields in system in Eq.(3.1), the computation of the first integral quantity increment can be based on the lth-oscillation of the integrable system in Eq.(3.1) on the first integral surface S_{σ_α} from time $t\in[t_k,t_k+T_l]$, i.e.,

$$L^{(\alpha)}(t_k,T_l)=\int_{t_k}^{t_k+T_l}\nabla F(\mathbf{x}^{(\alpha)},\boldsymbol{\mu})\bullet\mathbf{g}(\mathbf{x}^{(\alpha)},t,\boldsymbol{\pi})dt=F(\mathbf{x}_{k+1}^{(\alpha)},\boldsymbol{\mu})-F(\mathbf{x}_k^{(\alpha)},\boldsymbol{\mu}). \quad (4.61)$$

Using a time transformation $t'=t-t_k$, the forgoing equation becomes

$$\begin{aligned}L^{(\alpha)}(t_k,T_l)&=\int_0^{T_l}\nabla F(\mathbf{x}^{(\alpha)}(t-t_k),\boldsymbol{\mu})\bullet\mathbf{g}(\mathbf{x}^{(\alpha)}(t-t_k),t-t_k,\boldsymbol{\pi})dt\\&=F\left(\mathbf{x}_{k+1}^{(\alpha)},\boldsymbol{\mu}\right)-F\left(\mathbf{x}_k^{(\alpha)},\boldsymbol{\mu}\right).\end{aligned} \quad (4.62)$$

If the base surface is selected on the separatrix surface, the period $T_l\to\infty$, so

$$L^{(s)}\left(t_k,T_l\right)=\lim_{T_l\to\infty}\int_0^{T_l}\nabla F(\mathbf{x}^{(\alpha)}(t-t_k))\bullet\mathbf{g}(\mathbf{x}^{(\alpha)}(t-t_k),t-t_k,\boldsymbol{\pi})dt \quad (4.63)$$

From the existing perturbation analysis, if the non-integrable flow $\mathbf{x}^{(\alpha)}(t-t_k)$ is very close to the integrable $\overline{\mathbf{x}}_{\sigma_\alpha}^{(\alpha)}(t-t_k)$ or $\overline{\mathbf{x}}^{(s)}(t-t_k)$ for $t\in[t_k,t_k+T_l]$ with $T_l<K_T$ (const), then, for an arbitrarily small perturbation parameter $\varepsilon>0$, then the following series exist,

$$\mathbf{x}^{(\alpha)}(t-t_k)\approx\overline{\mathbf{x}}_{\sigma_\alpha}^{(\alpha)}(t-t_k)+\Sigma_{j=1}^{\infty}\varepsilon^j\mathbf{x}_{\sigma_\alpha}^{(j,\alpha)}(t-t_k), \quad (4.64)$$

$$\mathbf{x}^{(\alpha)}(t-t_k)\approx\overline{\mathbf{x}}^{(s)}(t-t_k)+\Sigma_{j=1}^{\infty}\varepsilon^j\mathbf{x}_{\sigma_\alpha}^{(j,s)}(t-t_k). \quad (4.65)$$

From Eq.(4.8) in Definition 4.3, for any point $\mathbf{x}^{(\alpha)}$ of the perturbed flow in the dynamical system in Eq.(3.1) on the first integral manifold surface S_{σ_α}, the normal component function of vector fields in the domain Ω_α is rewritten as

$$G_{\sigma_\alpha}(\mathbf{x}^{(\alpha)},t,\mathbf{p})=\mathbf{n}_{\sigma_\alpha}^T\bullet\mathbf{F}(\mathbf{x}^{(\alpha)},t,\mathbf{p})=\nabla F(\mathbf{x}^{(\alpha)},\boldsymbol{\mu})\bullet\mathbf{g}(\mathbf{x}^{(\alpha)},t,\boldsymbol{\pi}), \quad (4.66)$$

Substitution of Eqs.(4.64) and (4.65) into Eq.(4.66), and the Taylor series gives

$$G_{\sigma_\alpha}(\mathbf{x}^{(\alpha)},t,\mathbf{p}) = G_{\sigma_\alpha}(\overline{\mathbf{x}}^{(\alpha)},t,\mathbf{p}) + \sum_{k=1}^{\infty} \varepsilon^k \tfrac{d^k}{d\mathbf{x}^k} G_{\sigma_\alpha}(\mathbf{x}^{(\alpha)},t,\mathbf{p})\Big|_{\overline{\mathbf{x}}^{(\alpha)}} \left(\Sigma_{j=1}^{\infty}\varepsilon^{j-1}\mathbf{x}^{(j,\alpha)}\right)^k, \tag{4.67}$$

$$G_{\sigma_\alpha}(\mathbf{x}^{(\alpha)},t,\mathbf{p}) = G_{s}(\overline{\mathbf{x}}^{(s)},t,\mathbf{p}) + \sum_{k=1}^{\infty} \varepsilon^k \tfrac{d^k}{d\mathbf{x}^k} G_{\sigma_\alpha}(\mathbf{x}^{(\alpha)},t,\mathbf{p})\Big|_{\overline{\mathbf{x}}^{(s)}} \left(\Sigma_{j=1}^{\infty}\varepsilon^{j-1}\mathbf{x}^{(j,s)}\right)^k. \tag{4.68}$$

Using Eqs.(4.67) and (4.68), the first integral quantity increment is computed by

$$L^{(\alpha)}(t_k,T_l) = \int_0^{T_l} \Big[G_s\left(\overline{\mathbf{x}}^{(\alpha)}(t-t_k),t-t_k,\mathbf{p}\right) + \Sigma_{k=1}^{\infty}\varepsilon^k \tfrac{d^k}{d\mathbf{x}^k} G_{\sigma_\alpha}\left(\mathbf{x}^{(\alpha)},t-t_k,\mathbf{p}\right)\Big|_{\overline{\mathbf{x}}^{(s)}(t-t_k)} \times\left(\Sigma_{j=1}^{\infty}\varepsilon^{j-1}\mathbf{x}^{(j,\alpha)}(t-t_k)\right)^k \Big] dt, \tag{4.69}$$

$$L^{(s)}(t_k,T_l) = \lim_{T_l\to\infty} \int_0^{T_l} \Big[G_s\left(\overline{\mathbf{x}}^{(s)}(t-t_k),t-t_k,\mathbf{p}\right) + \Sigma_{k=1}^{\infty}\left(\varepsilon^k \tfrac{d^k}{d\mathbf{x}^k} G_{\sigma_\alpha}(\mathbf{x}^{(\alpha)},t-t_k,\mathbf{p})\right)\Big|_{\overline{\mathbf{x}}^{(s)}(t-t_k)} \times\left(\Sigma_{j=1}^{\infty}\varepsilon^{j-1}\mathbf{x}^{(j,s)}(t-t_k)\right)^k \Big] dt. \tag{4.70}$$

If the following conditions exist for all $k=1,2,\cdots$,

$$\begin{array}{lcl} \| \Sigma_{j=1}^{\infty}\varepsilon^{j-1}\mathbf{x}^{(j,s)}(t-t_k) \| \le K_{\mathbf{x}}, & & \| \Sigma_{j=1}^{\infty}\varepsilon^{j-1}\mathbf{x}^{(j,\alpha)}(t-t_k) \| \le K_{\mathbf{x}}, \\ \| \tfrac{d^k}{d\mathbf{x}^k} G_{\sigma_\alpha}(\overline{\mathbf{x}}^{(s)},t,\mathbf{p}) \| \le K_k, & \text{or} & \| \tfrac{d^k}{d\mathbf{x}^k} G_{\sigma_\alpha}(\overline{\mathbf{x}}^{(\alpha)},t,\mathbf{p}) \| \le K_k, \\ \dfrac{\varepsilon K_{\mathbf{x}} \| \tfrac{d^{k+1}}{d\mathbf{x}^{k+1}} G_{\sigma_\alpha}(\overline{\mathbf{x}}^{(s)},t,\mathbf{p}) \|}{\| \tfrac{d^k}{d\mathbf{x}^k} G_{\sigma_\alpha}(\overline{\mathbf{x}}^{(s)},t,\mathbf{p}) \|} < 1 & & \dfrac{\varepsilon K_{\mathbf{x}} \| \tfrac{d^{k+1}}{d\mathbf{x}^{k+1}} G_{\sigma_\alpha}(\overline{\mathbf{x}}^{(\alpha)},t,\mathbf{p}) \|}{\| \tfrac{d^k}{d\mathbf{x}^k} G_{\sigma_\alpha}(\overline{\mathbf{x}}^{(\alpha)},t,\mathbf{p}) \|} < 1 \end{array} \tag{4.71}$$

when $\varepsilon \to 0$, equations (4.69) and (4.70) give the approximate computations of the increment for the first integral quantity, i.e.,

$$L^{(\alpha)}(t_k,T_l) \approx \int_0^{T_l} G_s(\overline{\mathbf{x}}^{(\alpha)}(t-t_k),t-t_k,\mathbf{p})dt = \int_0^{T_l} \nabla F(\overline{\mathbf{x}}^{(\alpha)}(t-t_k),\boldsymbol{\mu})\bullet\mathbf{g}(\overline{\mathbf{x}}^{(\alpha)}(t-t_k),t-t_k,\boldsymbol{\pi})dt, \tag{4.72}$$

$$L^{(s)}(t_k,T_l) \approx \lim_{T_l\to\infty} \int_0^{T_l} G_s(\overline{\mathbf{x}}^{(s)}(t-t_k),t-t_k,\mathbf{p})dt \approx \lim_{T_l\to\infty} \int_0^{T_l} \nabla F(\overline{\mathbf{x}}^{(s)}(t-t_k),\boldsymbol{\mu})\bullet\mathbf{g}(\overline{\mathbf{x}}^{(s)}(t-t_k),t-t_k,\boldsymbol{\pi})dt. \tag{4.73}$$

Under conditions in Eq.(4.71), equations (4.72) and (4.73) *only* can be used for an approximate estimate of the increment of the first integral quantity. Compared to Eq.(4.61) or (4.62), both of equations in Eqs.(4.72) and (4.73) *cannot* be used for investigation of the chaotic and periodic flows of dynamical systems in Eq.(3.1). For periodic flows, the exact, first integral quantity increment must be equal to zero (i.e., $L^{(\alpha)}(t_k, T_l) = 0$). However, equations (4.72) and (4.73) cannot provide the exact computation. Only when $\varepsilon \to 0$, equations (4.72) and (4.73) may provide an acceptable result. From Eq.(4.72), it is observed that the computation of the first integral quantity increment needs the oscillation period of the reference surface of the integrable systems in Eq.(3.1). However, in many dynamical systems in Eq.(3.1), no frequency ω_l for the integrable systems exists or it is very difficult to determine the frequency ω_l. Such an integral for the increment of the first integral quantity may not be obtained to investigate the chaotic and periodic flows in nonlinear dynamical systems.

From the above five theorems, the increment of the first integral quantity is based on the Poincare mapping section, and the periodic flow can be determined. For a dynamical system in Eq.(3.1), periodic and chaotic flows can be studied by iterative relations, defined by the first integral quantity increment. The periodic and chaotic flows in a dynamical system with periodically time-dependent terms are of great interest. In Theorems 4.1 and 4.2, the reference surface is the cross-section surface of the initial first integral manifold surfaces. In fact, the first integral surface can also be used for the dynamical system in Eq.(3.1) with harmonic, time-dependent vector fields. In a flow band $\Omega_\alpha^\varepsilon$, the first integral manifold surface S_{σ_α} is considered as a reference surface. A flow of the system in Eq.(3.1) passing through the referenced, first integral manifold surface S_{σ_α} with the first integral quantity E_{σ_α} is sketched in Fig.4.6. A flow close to one closed cycle to a reference S_{σ_α} is presented in Fig.4.6(a). For a simple periodic flow, we have $\mathbf{x}_{k+2}^{(\alpha)} = \mathbf{x}_k^{(\alpha)}$. For a time interval $t \in [t_k, t_{k+1}]$, the starting point $(t_k, \mathbf{x}_k^{(\alpha)})$ and ending point $(t_{k+1}, \mathbf{x}_{k+1}^{(\alpha)})$ on the first integral manifold S_{σ_α} give

$$L^{(\alpha)}\left(t_k, t_{k+1}\right) = F\left(\mathbf{x}_{k+1}^{(\alpha)}, \boldsymbol{\mu}\right) - F\left(\mathbf{x}_k^{(\alpha)}, \boldsymbol{\mu}\right) = 0. \tag{4.74}$$

The corresponding flow starting from $(t_k, \mathbf{x}_k^{(\alpha)})$ to $(t_{k+1}, \mathbf{x}_{k+1}^{(\alpha)})$ gives

$$\mathbf{x}_{k+1}^{(\alpha)} = \boldsymbol{\Phi}^{(\alpha)}\left(t_k, \mathbf{x}_k^{(\alpha)}, \mathbf{p}, t_{k+1}\right). \tag{4.75}$$

The two relations in Eqs.(4.74) and (4.75) will determine the flow characteristics on the reference surface S_{σ_α}. The simple relations in Fig.4.6(a) give

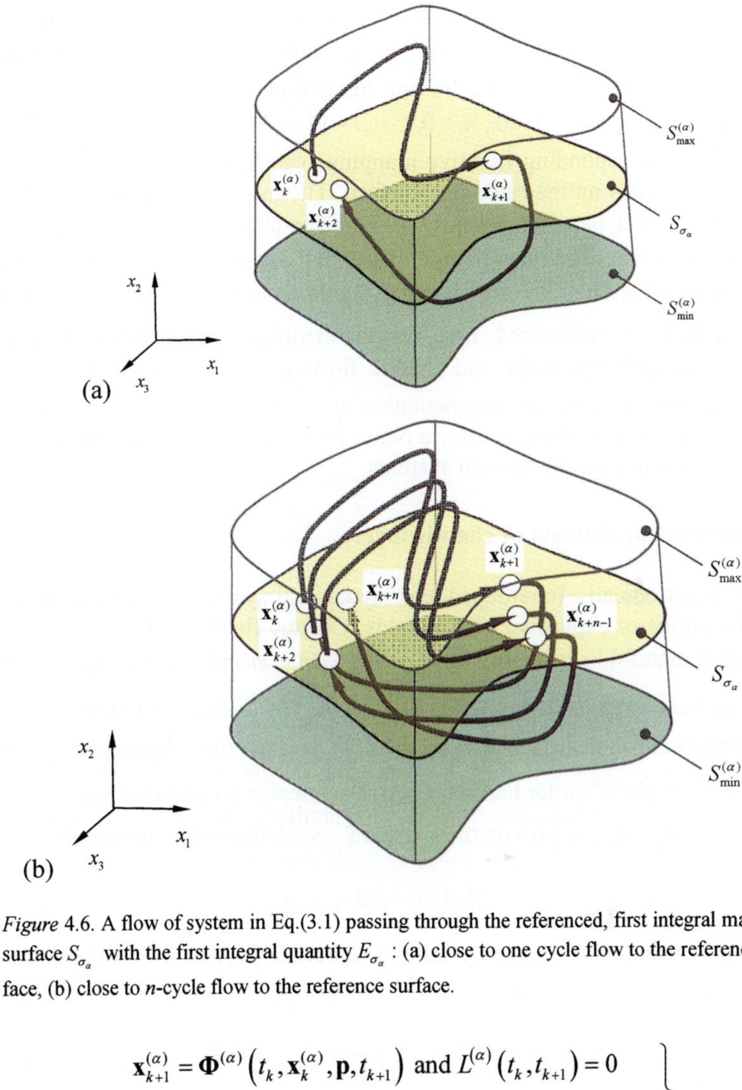

Figure 4.6. A flow of system in Eq.(3.1) passing through the referenced, first integral manifold surface S_{σ_α} with the first integral quantity E_{σ_α} : (a) close to one cycle flow to the reference surface, (b) close to n-cycle flow to the reference surface.

$$\left.\begin{aligned}\mathbf{x}_{k+1}^{(\alpha)} &= \mathbf{\Phi}^{(\alpha)}\left(t_k, \mathbf{x}_k^{(\alpha)}, \mathbf{p}, t_{k+1}\right) \text{ and } L^{(\alpha)}\left(t_k, t_{k+1}\right) = 0 \\ \mathbf{x}_{k+2}^{(\alpha)} &= \mathbf{\Phi}^{(\alpha)}\left(t_{k+1}, \mathbf{x}_{k+1}^{(\alpha)}, \mathbf{p}, t_{k+2}\right) \text{ and } L^{(\alpha)}\left(t_{k+1}, t_{k+2}\right) = 0\end{aligned}\right\} \tag{4.76}$$

The periodicity of the periodic flow is given by

$$\mathbf{x}_{k+2}^{(\alpha)} = \mathbf{x}_k^{(\alpha)},\ t_{k+2} = t_k + T \tag{4.77}$$

where $T = t_{k+2} - t_k$. If $\mathbf{x}_{k+2}^{(\alpha)} \neq \mathbf{x}_k^{(\alpha)}$, the flow will continue, and a flow close to an

l-cycle flow to a reference surface is formed, as also sketched in Fig.4.6(b). The period-1 of the l-cycle flow requires $\mathbf{x}_{k+n}^{(\alpha)} = \mathbf{x}_k^{(\alpha)}$ and $t_{k+n} = t_k + T$ on the reference surface, and all the increments of the first integral quantity for each time period is zero (i.e., $L^{(\alpha)}(t_{k+i}, t_{k+i+1}) = 0$, $i = 0, 1, \cdots, l-1$) and $T = t_{k+l} - t_k = \Sigma_{i=1}^{l} T_i$. As in Eq.(4.76), the corresponding iterative mappings can be developed to determine the switching values on the reference surface. The more discussion for periodic flows in nonlinear dynamical systems will be given later. If $l \to \infty$, a chaotic flow will be obtained from the time-change of the first integral quantity. The strange attractor fragmentation on the first integral manifold surface S_{σ_α} can be discussed in Chapter 8. If the referenced, first integral manifold surface is the separatrix surface S_s, the global periodic and chaotic flows can be discussed in a similar fashion. The strange attractor fragmentation and global transversality on the separatrix surface will be discussed for a better understanding of the complexity in n-dimensional nonlinear dynamical systems.

4.3. Resonance mechanism of chaotic layers

Consider a periodically time-dependent, dynamical system in Eq.(3.1) without either dissipative or expansive vector fields. Chaotic flows in the vicinity of the first integral manifold surface $S_{\sigma_\alpha^0}$ in Ω_α ($\alpha = 1, 2, \cdots, m$) can be investigated, and the chaotic band in Ω_α is defined as in Definition 4.9. In the band, there is a specific integral manifold surface S_{σ_α} ($\sigma_\alpha \in (0,1]$), and the time-dependent dynamical system with the sth-order harmonic terms possesses a resonance with ($(\mathbf{k}:s) = (k_1 : k_2 : \cdots : k_q : s)$, $q \le \frac{1}{2} n$) on the surface S_{σ_α} with the following condition

$$\Sigma_{l=1}^{q} k_l \omega_l - s\Omega = 0 \tag{4.78}$$

where ω_l is oscillation frequency for the separable oscillator and the iterative mapping for a period ($T_l = 2\pi/\omega_l$) is

$$E_{k+1} - E_k = L^{(\alpha)}(t_k, T_l) \text{ and } t_{k+1} - t_k = T_l. \tag{4.79}$$

Let $\varphi_k = \Omega t_k$ be phase, equation (4.79) becomes

$$E_{k+1} - E_k = L^{(\alpha)}(\varphi_k, T_l) \text{ and } \varphi_{k+1} - \varphi_k = \frac{2\pi\Omega}{\omega_l}. \tag{4.80}$$

If the $(\mathbf{k}:s)$-resonance exists on $E_k^{(\mathbf{k}:s)} = E_{k+1}^{(\mathbf{k}:s)}$, we have

$$0 = L^{(\alpha,(\mathbf{k}:s))}\left(\varphi_k, \tfrac{2\pi\Omega}{\omega_l}\right) \text{ and } \varphi_{k+1}^{(\mathbf{k}:s)} - \varphi_k^{(\mathbf{k}:s)} = \tfrac{2\pi}{s}\sum_{p=1}^{q} \tfrac{k_p \omega_p}{\omega_l}. \tag{4.81}$$

Except for $k_l \neq 0$, when the other $k_p = 0$ ($p = 1, 2, \cdots, l-1, l+1, \cdots, q$), the second equation of Eq.(4.81) and the first integral quantity increment become

$$\varphi_{k+1}^{(k_l:s)} - \varphi_k^{(k_l:s)} = \frac{2k_l\pi}{s}, \tag{4.82}$$

$$L^{(\alpha,(k_l:s))}\left(t_k, \tfrac{k_l}{s}T\right) = \int_{t_k}^{t_k + \frac{k_l}{s}T} \nabla F\left(\mathbf{x}^{(\alpha)}\right) \boldsymbol{\cdot} \mathbf{g}\left(\mathbf{x}^{(\alpha)}, t, \boldsymbol{\pi}\right) dt = 0. \tag{4.83}$$

Once the harmonic frequency Ω is given, the resonance condition $\mathbf{k} \cdot \boldsymbol{\omega} - s\Omega = 0$ gives the first integral quantity $E_0^{(\alpha,(\mathbf{k}:s))}$ from which the initial conditions are obtained. If $\mathbf{k} \cdot \boldsymbol{\omega} - s\Omega \approx 0$, the flow of dynamical system in Eq.(3.1) will form a chaotic band in the neighborhood of the first integral manifold surface S_{σ_α} with the $(\mathbf{k}:s)$-resonance. Such a chaotic band is called the $(\mathbf{k}:s)$-*resonant band* in Ω_α. The definition is given as follows.

DEFINITION 4.15: For a dynamical system in Eq.(3.1) without either dissipative or expansive vector fields, if a perturbed flow γ (i.e., $\mathbf{x}^\alpha(t)$) in the neighborhood of the first integral manifold surface S_{σ_α} with the $(\mathbf{k}:s)$-resonance in domain Ω_α is chaotic, then such a neighborhood is termed the $(\mathbf{k}:s)$-*resonant* band in Ω_α.

To investigate the chaotic flow in such a resonant band, equation (4.80) can be employed. The resonant, integral quantities of the chaotic flow should be in the neighborhood of the first integral quantity for the chaotic flow band in Ω_α. This neighborhood of the first integral quantity is determined by Eq.(4.9) or Eqs.(4.11) and (4.12). In other words, the maximum and minimum, integral quantities for such a $(\mathbf{k}:s)$-resonance will bound the first integral quantity, i.e.,

$$E \in \left[E_{\min}^{(\alpha,(\mathbf{k}:s))}, E_{\max}^{(\alpha,(\mathbf{k}:s))}\right]. \tag{4.84}$$

The corresponding maximum and minimum, first integral manifold surfaces are determined by

$$\begin{aligned} S_{\max}^{(\alpha,(\mathbf{k}:s))} &= \left\{\overline{\mathbf{x}}^{(\alpha,(\mathbf{k}:s))} \,\middle|\, E_{\max}^{(\alpha,(\mathbf{k}:s))} = F\left(\overline{\mathbf{x}}^{(\alpha,(\mathbf{k}:s))}, \boldsymbol{\mu}\right)\right\}, \\ S_{\min}^{(\alpha,(\mathbf{k}:s))} &= \left\{\overline{\mathbf{x}}^{(\alpha,(\mathbf{k}:s))} \,\middle|\, E_{\min}^{(\alpha,(\mathbf{k}:s))} = F\left(\overline{\mathbf{x}}^{(\alpha,(\mathbf{k}:s))}, \boldsymbol{\mu}\right)\right\}. \end{aligned} \tag{4.85}$$

The chaotic flow in the resonant band will be bounded by the two surfaces.

THEOREM 4.6. *With assumptions* (A1)-(A4)*, consider a periodically time-dependent, dynamical system in Eq.*(3.1) *without either dissipative or expansive vector fields. For a small* $\varepsilon > 0$*, there is a positive parameter* $\delta > 0$*. If*

$$\left| L^{(\alpha,(\mathbf{k}:s))}\left(\varphi_i, T_l\right) \right| \le \left| E_0^{(\alpha,(\mathbf{k}_1:s))} - E_0^{(\alpha,(\mathbf{k}:s))} \right| < \delta, \tag{4.86}$$

then for all time $t \in [t_0, \infty)$*, at least a point of* $\mathbf{x}^{(\alpha,(\mathbf{k}_1:s))}(t)$ *on the first integral manifold surface* $S_0^{(\alpha,(\mathbf{k}_1:s))}$ *satisfies*

$$\left\| \mathbf{x}^{(\alpha,(\mathbf{k}_1:s))}\left(t\right) - \overline{\mathbf{x}}_0^{(\alpha,(\mathbf{k}:s))} \right\| < \varepsilon, \tag{4.87}$$

i.e., the chaotic flow relative to $\mathbf{x}^{(\alpha,(\mathbf{k}_1:s))}(t)$ *is in the* $(\mathbf{k}:s)$*-resonant band.*

PROOF. For the $(\mathbf{k}:s)$-resonant band, the maximum and minimum, first integral manifold surfaces are $S_{\max}^{(\alpha,(\mathbf{k}:s))}$ and $S_{\min}^{(\alpha,(\mathbf{k}:s))}$, and the corresponding first integral manifolds are $E_{\max}^{(\alpha,(\mathbf{k}:s))}$ and $E_{\min}^{(\alpha,(\mathbf{k}:s))}$. Choosing $\delta = \max\{| E_{\max}^{(\alpha,(\mathbf{k}:s))} - E_0^{(\alpha,(\mathbf{k}:s))} |, | E_0^{(\alpha,(\mathbf{k}:s))} - E_{\min}^{(\alpha,(\mathbf{k}:s))} |\}$, the first integral quantity of a perturbed flow γ ($\mathbf{x}^{(\alpha)}(t)$, $t \in [t_0, \infty)$ in this resonant band satisfies

$$\left| E_{\sigma_\alpha}\left(t\right) - E_0^{(\alpha,(\mathbf{k}:s))} \right| \le \delta.$$

For a given $\varepsilon > 0$, let $\delta = \varepsilon \| F \| > 0$ satisfying

$$\begin{aligned} \left| E_{\sigma_\alpha}\left(t\right) - E_0^{(\alpha,(\mathbf{k}:s))} \right| &= \left| F\left(\mathbf{x}^{(\alpha)}\left(t\right), \boldsymbol{\mu}\right) - F\left(\overline{\mathbf{x}}_0^{(\alpha,(\mathbf{k}:s))}, \boldsymbol{\mu}\right) \right| \\ &\le \left\| F \right\| \left\| \mathbf{x}^{(\alpha)} - \overline{\mathbf{x}}_0^{(\alpha,(\mathbf{k}:s))} \right\| \le \varepsilon \left\| F \right\| = \delta. \end{aligned}$$

Because the first integral manifold $F(\mathbf{x}^{(\alpha)}(t), \boldsymbol{\mu}) = E(t)$ is C^r-continuous ($r \ge 2$),

$$0 < \left\| F \right\| \le \sup\left[\left(\left| F\left(\mathbf{x}^{(\alpha)}\left(t\right), \boldsymbol{\mu}\right) - F\left(\overline{\mathbf{x}}_0^{(\alpha,(\mathbf{k}:s))}, \boldsymbol{\mu}\right) \right| \right) \Big/ \left\| \mathbf{x}^{(\alpha)} - \overline{\mathbf{x}}_0^{(\alpha,(\mathbf{k}:s))} \right\| \right] < \infty$$

for any bounded flow in the resonant band. Therefore, we have

$$\left\| \mathbf{x}^{(\alpha)}\left(t\right) - \overline{\mathbf{x}}_0^{(\alpha,(\mathbf{k}:s))} \right\| \le \delta \big/ \left\| F \right\| = \varepsilon \text{ for all time } t.$$

For the flow γ (i.e., $\mathbf{x}^{(\alpha)}(t)$, $t \in \left[t_k, t_{k+1}\right]$ and $t_{k+1} = T_l + t_k$) starting point on the $S_0^{(\alpha,(\mathbf{k}:s))}$, equation (4.86) gives

$$\left| E_{k+1} - E_0^{(\alpha,(\mathbf{k}:s))} \right| = \left| L^{\alpha,(\mathbf{k}:s)}\left(t_k, T_l\right) \right|$$

If the flow γ ($\mathbf{x}^{(\alpha)}(t)$, $t \in [t_0, \infty)$ possesses at least one point on the $S_0^{(\alpha,(\mathbf{k}_1:s))}$, there is a maximum or minimum integral quantity $E_{\max}^{(\alpha,(\mathbf{k}:s))} = E_0^{(\alpha,(\mathbf{k}_1:s))}$ or $E_{\min}^{(\alpha,(\mathbf{k}:s))}$ $= E_0^{(\alpha,(\mathbf{k}_1:s))}$ and this flow γ is defined as $\mathbf{x}^{(\alpha,(\mathbf{k}_1:s))}(t)$. So for all E_{k+1} for $k = 1, 2, 3, \cdots$, one has

$$\left|L^{\alpha,(\mathbf{k}:s)}\left(t_k, T_l\right)\right| < \left|E_0^{(\alpha,(\mathbf{k}_1:s))} - E_0^{(\alpha,(\mathbf{k}:s))}\right|.$$

From deduction, the farthest distance satisfies the following condition

$$\left\|\mathbf{x}^{(\alpha,(\mathbf{k}_1:s)}\left(t\right) - \overline{\mathbf{x}}_0^{(\alpha,(\mathbf{k}:s))}\right\| < \varepsilon.$$

It is implies that the chaotic flow $\mathbf{x}^{(\alpha,(\mathbf{k}_1:s))}(t)$ relative to the $(\mathbf{k}_1 : s)$ -resonance is in the $(\mathbf{k} : s)$ -resonant band. So this theorem is proved. ■

To explain the above theorem, consider a chaotic flow γ ($\mathbf{x}^{(\alpha)}(t), t \in [t_0, \infty)$) in the $(\mathbf{k} : s)$ - resonant band, as shown in Fig.4.7. For each period T_l of the flow on the referenced first integral manifold surface of the $(\mathbf{k} : s)$ -resonant band, the chaotic flow is represented by a dark curve with circles. In Fig.4.7(a), the chaotic flow is not accessible to the maximum and minimum, first integral manifold surfaces. For achaotic flow in the resonant band, the corresponding width of the chaotic flow in the resonant band can be evaluated by its maximum and minimum, first integral quantities. The corresponding condition for computation of the $(\mathbf{k} : s)$ -resonant width is

$$\max_{\varphi_i \in [0,2\pi]} \left|L^{(\alpha,(\mathbf{k}:s))}\left(\varphi_i, T_l\right)\right| = \max\left\{\left|E_{\max}^{(\alpha,(\mathbf{k}:s))} - E_0^{(\alpha,(\mathbf{k}:s))}\right|, \left|E_{\min}^{(\alpha,(\mathbf{k}:s))} - E_0^{(\alpha,(\mathbf{k}:s))}\right|\right\} \quad (4.88)$$

In Fig.4.7(b), the chaotic flow can be tangential the maximum and minimum, first integral manifold surfaces.

Consider a resonant flow γ ($\mathbf{x}^{(\alpha,(\mathbf{k}:s))}(t), t \in [t_0, \infty)$) just tangential to the resonant flow $S_0^{(\alpha,(\mathbf{k}_1:s))}$ in the $(\mathbf{k} : s)$ -resonant band, as shown in Fig.4.8(a). The dashed curves represent the first integral manifold surface. The thin solid curves are the maximum and minimum first integral manifold surface. The thick solid curves represent the resonant flows. Once the resonant flow or chaotic flow in the vicinity of the resonant flow induced another primary resonant flow, the new resonant flow will be formed. The interaction of the two resonant flows will generate the chaotic motion to be more complex. This mechanism of the resonant flow interaction is sketched in Fig.4.8(b).

If a maximum or minimum first integral quantity surface is the $(\mathbf{k}_1 : s)$ -resonant, referenced, first integral manifold surface, the $(\mathbf{k}_1 : s)$ -resonance will exist

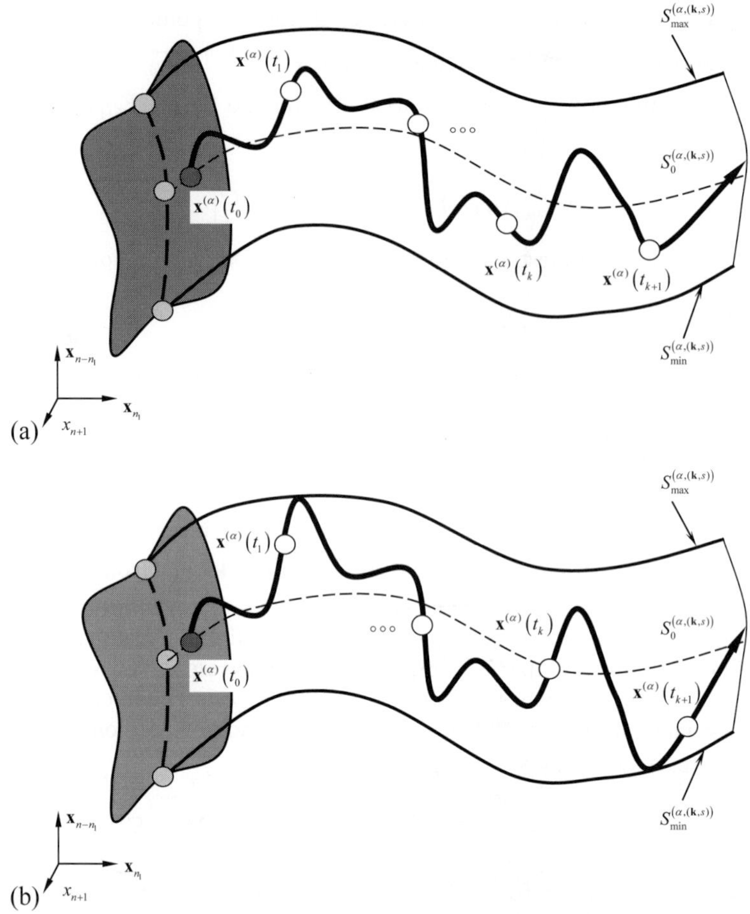

Figure 4.7. A chaotic flow γ ($\mathbf{x}^{(\alpha)}(t)$, $t \in [t_0, \infty)$) in the $(\mathbf{k}:s)$-resonant band: (a) before the resonant band destruction, (b) just the resonant band destruction. The dashed curve represents the first integral manifold surface. The thin solid curves are the maximum and minimum first integral manifold surface. The thick solid curve with circles is a chaotic flow.

in the $(\mathbf{k}:s)$-resonant band, such a $(\mathbf{k}:s)$-resonant band is destroyed by the new $(\mathbf{k}_1:s)$-resonance. The corresponding condition is given by

$$\max_{\varphi_i \in [0,2\pi]} \left| L^{(\alpha,(\mathbf{k}:s))}(\varphi_i, T_l) \right| = \left| E_0^{(\alpha,(\mathbf{k}_1:s))} - E_0^{(\alpha,(\mathbf{k}:s))} \right|. \quad (4.89)$$

However, once such a resonant band is destroyed, the resonant overlap of both the $(\mathbf{k}_1:s)$ and $(\mathbf{k}:s)$-resonances gives a condition as follows.

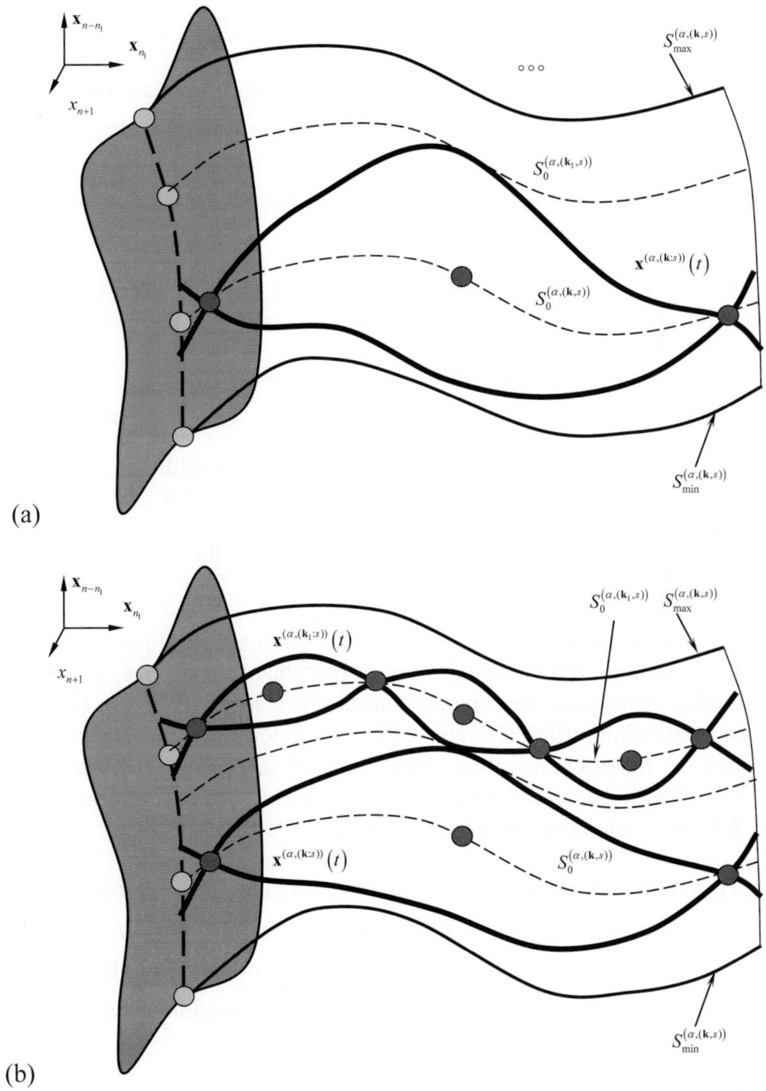

Figure 4.8. Resonant flows γ ($\mathbf{x}^{(\alpha,(\mathbf{k}:s))}(t)$, $t \in [t_0, \infty)$) in the $(\mathbf{k}:s)$-resonant band: (a) just tangential to the resonant flow $S_0^{(\alpha,(\mathbf{k}_1:s))}$, (b) Induced resonant flow $\mathbf{x}^{(\alpha,(\mathbf{k}_1:s))}(t)$. The dashed curve represents the first integral manifold surface. The thin solid curves are the maximum and minimum first integral manifold surface. The thick solid curves represent the resonant flows.

$$\max_{\varphi_i\in[0,2\pi]}\left|L^{(\alpha,(\mathbf{k}:s))}(\varphi_i,T_l)\right| + \max_{\varphi_i\in[0,2\pi]}\left|L^{(\alpha,(\mathbf{k}_1:s))}(\varphi_i,T_l)\right| = \left|E_0^{(\alpha,(\mathbf{k}:s))} - E_0^{(\alpha,(\mathbf{k}_1:s))}\right| \quad (4.90)$$

This criterion does not consider the effects of the sub-resonances of $(\mathbf{k}_1 : s)$ and $(\mathbf{k} : s)$-resonances. But this criterion is similar to the Chirikov overlap criterion in Chirikov (1979). The Chirikov criterion is based on the two-dimensional dynamical systems. However, the new criterion can be used for n-dimensional, nonlinear dynamical systems in Eq.(3.1).

DEFINITION 4.16. For a dynamical system in Eq.(3.1) without the time-periodic vector fields, the resonance is termed the *internal* resonance if

$$\mathbf{k}\bullet\boldsymbol{\omega} = 0 \ \ (\text{or } \textstyle\sum_{l=1}^{q} k_l\omega_l = 0). \quad (4.91)$$

DEFINITION 4.17. For the dynamical system in Eq.(3.1) with the time-periodic vector fields, the resonance is termed the *external* resonance if

$$\mathbf{k}\bullet\boldsymbol{\omega} - s\Omega = 0 \ \ (\text{or } \textstyle\sum_{l=1}^{q} k_l\omega_l - s\Omega = 0). \quad (4.92)$$

Because the separatrix surface is relative to the hyperbolic points, the homoclinic or heteroclinic orbits should exist on the separatrix surface. For 2-dimensional nonlinear systems, the separatrix surface is a separatrix curve (i.e., homoclinic or heteroclinic orbit). For any n ($n > 2$)-dimensional system, such all the homoclinic or heteroclinic orbits cannot occupy the entire separartrix surface (see, Luo, 2006a). Therefore, for an n-dimensional dynamical system, the homoclinic or heteroclinic orbit (or flow) should be defined as follows:

DEFINITION 4.18. For a dynamical system in Eq.(3.1), an unperturbed flow γ (i.e., $\bar{\mathbf{x}}(t) \subset S_s$) on the separatrix surface S_s is called *the homoclinic orbit* if the flow γ is connected by a hyperbolic point $\bar{\mathbf{x}}_0$. A set consisting of all the homoclinic orbits on the separatrix surface S_s is termed *the homoclinic orbit set* $\Gamma_h \subset S_s$ with

$$\Gamma_h = \left\{ \bar{\mathbf{x}}(t) \middle| \gamma : \bar{\mathbf{x}}(t) \subset S_s, \ \lim_{t\to\pm\infty} \bar{\mathbf{x}}(t) = \bar{\mathbf{x}}_0 \text{ and } \mathbf{f}(\bar{\mathbf{x}}_0,\boldsymbol{\mu}) = 0 \right\}. \quad (4.93)$$

DEFINITION 4.19. For a dynamical system in Eq.(3.1), an unperturbed flow γ (i.e., $\bar{\mathbf{x}}(t) \subset S_s$) on the separatrix surface S_s is called *the heteroclinic orbit* if the flow γ is connected by two or more hyperbolic points ${}^1\bar{\mathbf{x}}_0$ and ${}^2\bar{\mathbf{x}}_0$. A set consisting of all the heteroclinic orbits on the separatrix surface S_s is termed *the heteroclinic orbit set* $\Gamma_\gamma \subset S_s$ with

$$\Gamma_h = \left\{ \bar{\mathbf{x}}(t) \left| \begin{array}{l} \gamma : \bar{\mathbf{x}}(t) \subset S_s,\ \lim_{t\to\infty} \bar{\mathbf{x}}(t) = {}^1\bar{\mathbf{x}}_0 \text{ and} \\ \lim_{t\to-\infty} \bar{\mathbf{x}}(t) = {}^2\bar{\mathbf{x}}_0, \text{ and } \mathbf{f}({}^j\bar{\mathbf{x}}_0, \boldsymbol{\mu}) = 0,\ j = 1,2 \end{array} \right. \right\}. \tag{4.94}$$

For a flow $\gamma : \bar{\mathbf{x}}(t) \subset \Gamma_h$ on the separatrix surface, its period is infinity. It indicates that the flow $\gamma : \bar{\mathbf{x}}(t)$ possesses zero frequency. If the resonance is close to the flow $\gamma : \bar{\mathbf{x}}(t)$, one of all frequencies in the dynamical system in Eq.(3.1) should be close to zero. However, in Eq.(4.91) or (4.92), if one of q-frequencies is close to zero (i.e., $\omega_l \to 0$, $l = 1, 2, \cdots, q$), then there is a very large number (i.e., $k_l \to \infty$) to make Eq.(4.78) satisfied. Once the resonant flow is very close to the flow $\gamma : \bar{\mathbf{x}}(t) \subset \Gamma_h$, such a resonant flow has an infinite number of resonant order. So in the neighborhood of the flow $\gamma : \bar{\mathbf{x}}(t) \subset \Gamma_h$, the motion will become very complex or chaotic. Such a neighborhood is called the *stochastic* band in the vicinity of the seperatrix surface. The mathematic definition is given as follows.

DEFINITION 4.20. For a dynamical system in Eq.(3.1), in the neighborhood of the homoclinic or heteroclinic orbit set $\Gamma_h \subset S_s$ with domain Ω_α ($\alpha = 1, 2, \cdots, m$), the flow $\cup_\alpha \mathbf{x}^\alpha(t)$ is chaotic, then such a neighborhood is termed a *stochastic band* of the dynamical system in Eq.(3.1).

Once the resonant flow is close to the homoclinic or heteroclinic orbit set on the separatrix surface (i.e., $\Gamma_h \subset S_s$), the resonant order number will approach the infinity (i.e., $|\mathbf{k}| \to \infty$). In Theorem 4.6, once the $(\mathbf{k} : s)$-resonant, referenced, integral manifold surface is replaced by the separatrix surface. Theorem 4.6 can be used for the resonance in the stochastic band. The corresponding theorem can be stated as follows.

THEOREM 4.7. *With assumptions* (A1)-(A4), *consider a periodically time-dependent, dynamical system in Eq.*(3.1) *without either dissipative or expansive vector fields. In the neighborhood of the separatrix surface* S_s *with domain* Ω_α ($\alpha = 1, 2, \cdots, m$), *for any small* $\varepsilon > 0$, *there is a positive parameter* $\delta > 0$. *If*

$$\left|L^{(\alpha,(\mathbf{k}:s))}\left(\varphi_i,T_l\right)\right| \le \left|E_0^{(\alpha,(\mathbf{k}:s))} - E_s\right| \tag{4.95}$$

then for time $t \in [t_0,\infty)$, *there is a point* $\mathbf{x}^{(\alpha,(\mathbf{k}:s))}(t)$ *on the first integral manifold surface* $S_0^{(\alpha,(\mathbf{k}:s))}$ *satisfies*

$$\left\|\mathbf{x}^{(\alpha,(\mathbf{k}:s))}\left(t\right) - \overline{\mathbf{x}}_s\right\| < \varepsilon, \tag{4.96}$$

i.e., the chaotic flow relative to $\mathbf{x}^{(\alpha,(\mathbf{k}:s))}(t)$ *is in the stochastic band.*

PROOF. The proof is similar to THEOREM 4.6. ■

The stochastic band in a neighborhood of the homoclinic or heteroclinic orbit set on the separatrix surface is sketched in Fig.4.9 through resonant flows. In the vicinity of separatrix surface, there are many different domains Ω_α ($\alpha = 1,2,\cdots,m$). In each sub-domain of such a vicinity of the separatrix surface, the resonance is different from the other domains. In Fig.4.9, two domains Ω_α and Ω_β ($\beta \ne \alpha$, $\beta = 1,2,\cdots,m$) are presented, and the different resonant orbits in the two domains are depicted. The resonant flow γ ($\mathbf{x}^{(\alpha)}(t)$ and $\mathbf{x}^{(\beta)}(t)$, $t \in [t_0,\infty)$) is sketched in the stochastic band. The dashed curves represent the first integral manifold surfaces. The thin solid curves are the maximum and minimum first integral manifold surface. The thick solid curves are the resonant flows, and the thick dashed curve denotes the separatrix surface. Once external periodic vector fields become strong, all resonances in the stochastic band will become a stochastic *n*-dimensional resonant web. If all the resonances will be connected together, the global chaos connecting all the primary resonance will be formed. For a case of internal resonance, the global chaos connecting all primary resonance will be formed on an invariant, first integral manifold surface with strong interaction (also see, Luo, 2006a). For this case, the above presented criterion does not work because the first integral manifold is invariant. The detailed discussion can be found in Luo (2006a). If the exact first integral quantity increment is computed, the aforementioned theorems give the exact prediction of resonances in the chaotic layers.

4.4. Energy increments for perturbed Hamiltonian systems

Consider a $2n$-dimensional perturbed Hamiltonian system in Eq.(3.36) without either dissipative or expansive vector fields. The first integral quantity increment is crucial to determine chaos and periodic motions. Because the first integral quantity is the unperturbed Hamiltonian (i.e., $H_0(\mathbf{x},\mathbf{y},\boldsymbol{\mu})$), the corresponding quantity increment is *the energy increment* (i.e., $L^{(\alpha)}(t_k,t_{k+1}) \equiv \Delta H_0(\mathbf{x}_k,\mathbf{x}_{k+1})$) in

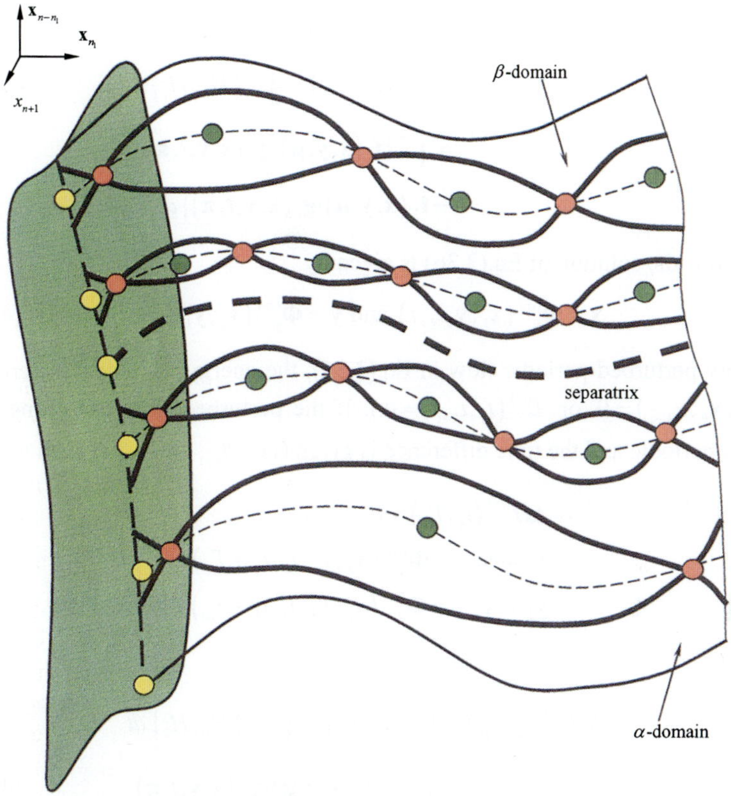

Figure 4.9. Resonant flows γ ($\mathbf{x}^{(\alpha)}(t)$ and $\mathbf{x}^{(\beta)}(t)$, $t\in[t_0,\infty)$, $\alpha,\beta=1,2,\cdots,m$) in the stochastic band. The dashed curves represent the first integral manifold surface. The thin solid curves are the maximum and minimum first integral manifold surface. The thick solid curves are the resonant flow. The thick dash curve represents the separatrix surface.

perturbed Hamiltonian systems. The total energy of the perturbed Hamiltonian system is expressed by the time-independent and time-dependent parts, i.e.,

$$H(\mathbf{x},\mathbf{y},t)=H_0(\mathbf{x},\mathbf{y},\boldsymbol{\mu})+H_1(\mathbf{x},\mathbf{y},t,\boldsymbol{\pi}). \tag{4.97}$$

The energy change rate of the unperturbed Hamiltonian is given by

$$\begin{aligned}\tfrac{d}{dt}H_0(\mathbf{x},\mathbf{y})&=\tfrac{\partial H_0(\mathbf{x},\mathbf{y})}{\partial\mathbf{x}}\bullet\dot{\mathbf{x}}+\tfrac{\partial H_0(\mathbf{x},\mathbf{y})}{\partial\mathbf{y}}\bullet\dot{\mathbf{y}}=\tfrac{\partial H_0(\mathbf{x},\mathbf{y})}{\partial\mathbf{x}}\bullet\tfrac{\partial H_1(\mathbf{x},\mathbf{y},t)}{\partial\mathbf{y}}-\tfrac{\partial H_0(\mathbf{x},\mathbf{y})}{\partial\mathbf{y}}\bullet\tfrac{\partial H_1(\mathbf{x},\mathbf{y},t)}{\partial\mathbf{x}}\\&=\{H_0,H_1\}=\mathbf{f}_1\bullet\mathbf{g}_2-\mathbf{f}_2\bullet\mathbf{g}_1\end{aligned} \tag{4.98}$$

where $\{\bullet,\bullet\}$ is the Poisson bracket. From Eq.(4.25), the energy increment for time interval $[t_k, t_{k+1}]$ is

$$\begin{aligned}\Delta H_0^{(\alpha)}(t_k, t_{k+1}) = L^{(\alpha)}(t_k, t_{k+1}) &= \int_{t_k}^{t_{k+1}} \{H_0, H_1\}\, dt \\ &= \int_{t_k}^{t_{k+1}} \{\mathbf{f}_1(\mathbf{x}, \mathbf{y}, \boldsymbol{\mu}) \boldsymbol{\cdot} \mathbf{g}_2(\mathbf{x}, \mathbf{y}, t, \boldsymbol{\pi}) \\ &\quad - \mathbf{f}_2(\mathbf{x}, \mathbf{y}, \boldsymbol{\mu}) \boldsymbol{\cdot} \mathbf{g}_1(\mathbf{x}, \mathbf{y}, t, \boldsymbol{\pi})\}\, dt,\end{aligned} \tag{4.99}$$

However, the solution of Eq.(3.36) is given by

$$\mathbf{x} = \boldsymbol{\Phi}_{\mathbf{x}}^{(\alpha)}(\mathbf{x}_k, \mathbf{y}_k t_k, t) \text{ and } \mathbf{y} = \boldsymbol{\Phi}_{\mathbf{y}}^{(\alpha)}(\mathbf{x}_k, \mathbf{y}_k t_k, t). \tag{4.100}$$

For any perturbed periodic flow in Eq.(3.36), the energy increment is zero (i.e., $\Delta H_0(\mathbf{x}_k, \mathbf{x}_{k+1}) = 0$ or $L^{(\alpha)}(t_k, t_{k+1}) = 0$). If the perturbed flow from time t_k to t_{k+1} is periodic and the time difference is given (i.e., $t_{k+1} - t_k = T_\alpha$), then we have

$$\begin{aligned}&\Delta H_0^{(\alpha)}(t_k, T_\alpha) = 0, \\ &\mathbf{x}_k = \mathbf{x}_{k+1} = \boldsymbol{\Phi}_{\mathbf{x}}^{(\alpha)}(\mathbf{x}_k, \mathbf{y}_k, t_k, t_k + T_\alpha), \\ &\mathbf{y}_k = \mathbf{y}_{k+1} = \boldsymbol{\Phi}_{\mathbf{y}}^{(\alpha)}(\mathbf{x}_k, \mathbf{y}_k, t_k, t_k + T_\alpha);\end{aligned} \tag{4.101}$$

with

$$\begin{aligned}\Delta H_0^{(\alpha)}(t_k, T_\alpha) = L^{(\alpha)}(t_k, T_\alpha) &= \int_{t_k}^{t_k + T_\alpha} \{H_0, H_1\}\, dt \\ &= \int_{t_k}^{t_k + T_\alpha} \{\mathbf{f}_1(\mathbf{x}, \mathbf{y}, \boldsymbol{\mu}) \boldsymbol{\cdot} \mathbf{g}_2(\mathbf{x}, \mathbf{y}, t, \boldsymbol{\pi}) \\ &\quad - \mathbf{f}_2(\mathbf{x}, \mathbf{y}, \boldsymbol{\mu}) \boldsymbol{\cdot} \mathbf{g}_1(\mathbf{x}, \mathbf{y}, t, \boldsymbol{\pi})\}\, dt,\end{aligned} \tag{4.102}$$

For a chaotic flow in Eq.(3.36), the corresponding iterative relation is given by

$$E_{k+1} - E_k = \Delta H_0^{(\alpha)}(t_k, T_\alpha) \text{ and } t_{k+1} = t_k + T_\alpha. \tag{4.103}$$

The time difference T_α can be the period of a periodic flow on a selected, reference surface S_{σ_α} at $(\bar{\mathbf{x}}_k, \bar{\mathbf{y}}_k) = (\mathbf{x}_k, \mathbf{y}_k)$ for time t_k. Through the iterative equation, the chaotic flow can be investigated. Consider an external frequency Ω and a new phase angle $\varphi = \Omega t$, equation (4.103) becomes

$$E_{k+1} - E_k = \Delta H_0^{(\alpha)}\left(\tfrac{1}{\Omega}\varphi_k, T_\alpha\right) \text{ and } \varphi_{k+1} = \varphi_k + \Omega T_\alpha \tag{4.104}$$

From the above discussion, how to compute the energy increment during the given time interval is very important. Using a transformation of $t = t_k + t'$, equation (4.102) gives

$$\begin{aligned}\Delta H_0^{(\alpha)}(t_k,T_\alpha)=L^{(\alpha)}(t_k,T_\alpha)&=\int_0^{T_\alpha}\{H_0(\mathbf{x},\mathbf{y}),H_1(\mathbf{x},\mathbf{y},t_k+t)\}dt\\&=\int_0^{T_\alpha}\{\mathbf{f}_1(\mathbf{x},\mathbf{y},\boldsymbol{\mu})\bullet\mathbf{g}_2(\mathbf{x},\mathbf{y},t_k+t,\boldsymbol{\pi})\\&\quad-\mathbf{f}_2(\mathbf{x},\mathbf{y},\boldsymbol{\mu})\bullet\mathbf{g}_1(\mathbf{x},\mathbf{y},t_k+t,\boldsymbol{\pi})\}dt.\end{aligned}\tag{4.105}$$

If the separatrix surface is selected as a reference surface, an orbit connecting saddle equilibrium points is used for integration. So, we have $T_\alpha\to\infty$ and

$$\begin{aligned}\Delta H_0^{(s)}(t_k,\infty)=L^{(s)}(t_k,\infty)&=\lim_{T_\alpha\to\infty}\int_0^{T_\alpha}\{H_0(\mathbf{x},\mathbf{y}),H_1(\mathbf{x},\mathbf{y},t_k+t)\}dt\\&=\lim_{T_\alpha\to\infty}\int_0^{T_\alpha}\{\mathbf{f}_1(\mathbf{x},\mathbf{y},\boldsymbol{\mu})\bullet\mathbf{g}_2(\mathbf{x},\mathbf{y},t_k+t,\boldsymbol{\pi})\\&\quad-\mathbf{f}_2(\mathbf{x},\mathbf{y},\boldsymbol{\mu})\bullet\mathbf{g}_1(\mathbf{x},\mathbf{y},t_k+t,\boldsymbol{\pi})\}dt.\end{aligned}\tag{4.106}$$

With application of the perturbation analysis in Section 4.2 to the perturbed Hamiltonian system in Eq.(3.36), equations (4.105) and (4.106) become

$$\begin{aligned}\Delta H_0^{(\alpha)}(t_k,T_\alpha)&=L^{(\alpha)}(t_k,T_\alpha)\\&\approx\int_0^{T_\alpha}\{H_0(\overline{\mathbf{x}}^{(\alpha)},\overline{\mathbf{y}}^{(\alpha)}),H_1(\overline{\mathbf{x}}^{(\alpha)},\overline{\mathbf{y}}^{(\alpha)},t_k+t)\}dt\\&=\int_0^{T_\alpha}\{\mathbf{f}_1(\overline{\mathbf{x}}^{(\alpha)},\overline{\mathbf{y}}^{(\alpha)},\boldsymbol{\mu})\bullet\mathbf{g}_2(\overline{\mathbf{x}}^{(\alpha)},\overline{\mathbf{y}}^{(\alpha)},t_k+t,\boldsymbol{\pi})\\&\quad-\mathbf{f}_2(\overline{\mathbf{x}}^{(\alpha)},\overline{\mathbf{y}}^{(\alpha)},\boldsymbol{\mu})\bullet\mathbf{g}_1(\overline{\mathbf{x}}^{(\alpha)},\overline{\mathbf{y}}^{(\alpha)},t_k+t,\boldsymbol{\pi})\}dt\end{aligned}\tag{4.107}$$

and

$$\begin{aligned}\Delta H_0^{(s)}(t_k,\infty)&=L^{(s)}(t_k,\infty)\\&\approx\lim_{T_\alpha\to\infty}\int_0^{T_\alpha}\{H_0(\overline{\mathbf{x}}^{(\alpha)},\overline{\mathbf{y}}^{(\alpha)}),H_1(\overline{\mathbf{x}}^{(\alpha)},\overline{\mathbf{y}}^{(\alpha)},t_k+t)\}dt\\&=\lim_{T_\alpha\to\infty}\int_0^{T_\alpha}\{\mathbf{f}_1(\overline{\mathbf{x}}^{(\alpha)},\overline{\mathbf{y}}^{(\alpha)},\boldsymbol{\mu})\bullet\mathbf{g}_2(\overline{\mathbf{x}}^{(\alpha)},\overline{\mathbf{y}}^{(\alpha)},t_k+t,\boldsymbol{\pi})\\&\quad-\mathbf{f}_2(\overline{\mathbf{x}}^{(\alpha)},\overline{\mathbf{y}}^{(\alpha)},\boldsymbol{\mu})\bullet\mathbf{g}_1(\overline{\mathbf{x}}^{(\alpha)},\overline{\mathbf{y}}^{(\alpha)},t_k+t,\boldsymbol{\pi})\}dt.\end{aligned}\tag{4.108}$$

As in Section 4.2, equations (4.107) and (4.108) cannot be used to determine the flow complexity characteristics of the perturbed Hamiltonian system in Eq.(3.36) such as chaos and the transversality of a perturbed flow to the separatrix surface. Under conditions in Eq.(4.71), equations (4.107) and (4.108) *only* can be used for an approximate estimate of the energy increment. If the approximate first integral quantity increment is computed, the aforementioned theorems cannot give the exact prediction of chaotic layers in $2n$-dimensional Hamiltonian systems.

Chapter 5

Two-Dimensional Stochastic Layers

In this Chapter, a resonant theory of stochastic layers in two-dimensional nonlinear Hamiltonian systems will be presented for an understanding of the resonant mechanism of chaos in the stochastic layer. The book asserts that the chaos in the stochastic layer is formed by the primary resonance interaction at least in 2-dimensional, nonlinear Hamiltonians systems. In the stochastic layer, the invariant geometrical structure in phase space is generated by the resonance interaction instead of the Smale's horseshoe structure. Based on this idea, a criterion based on an accurate whisker map and resonant conditions will be presented for prediction of the *onset* of resonance in the stochastic layer. The criterion given in this Chapter is applicable to periodically forced, nonlinear Hamiltonian systems with one degree of freedom. The onset of a specific primary in the stochastic layer can be predicted analytically. The improved standard mapping technique will be presented to compare with the existing standard mapping technique. The energy spectrum technique will be presented herein. This energy spectrum method can exactly compute the energy increment (or the first integral quantity increment). However, the approximate predictions are based on approximate computations of the energy increment. Therefore, the energy spectrum method based on the exact energy increment can give an accurate prediction of the primary resonance in the stochastic layer for nonlinear Hamiltonian systems with strong, external perturbations. A forced twin-well Duffing oscillator will be presented to show how to analytically predict a primary resonance in the stochastic layer. Verification of the analytical predictions will be carried out through the energy spectrum method with a symplectic numerical integration scheme.

5.1. Geometric description in phase space

Consider a two-dimensional Hamiltonian system with a time-dependent, periodically perturbed vector field, i.e.,

$$\dot{\mathbf{x}} = \mathbf{f}(\mathbf{x},\boldsymbol{\mu}) + \mu\mathbf{g}(\mathbf{x},t,\boldsymbol{\pi}); \quad \mathbf{x} = (x,y)^T \in \mathbb{R}^2, \tag{5.1}$$

where $\mathbf{f}(\mathbf{x},\boldsymbol{\mu})$ is an unperturbed Hamiltonian vector field on $\mathbb{R}^2$ and $\mathbf{g}(\mathbf{x},t)$ is a periodically perturbed vector field with period $T = 2\pi/\Omega$, and

$$\mathbf{f}(\mathbf{x},\boldsymbol{\mu}) = (f_1(\mathbf{x},\boldsymbol{\mu}), f_2(\mathbf{x},\boldsymbol{\mu}))^T \text{ and } \mathbf{g}(\mathbf{x},t,\boldsymbol{\pi}) = (g_1(\mathbf{x},t,\boldsymbol{\pi}), g_2(\mathbf{x},t,\boldsymbol{\pi}))^T \tag{5.2}$$

are sufficiently smooth $(C^r, r \geq 2)$ and bounded on a bounded set $D \subset \mathbb{R}^2$ in phase space. $f_1 = \partial H_0(x,y)/\partial y$, $f_2 = -\partial H_0(x,y)/\partial x$; $g_1 = \partial H_1(x,y,\Omega t)/\partial y$, $g_2 = -\partial H_1(x,y,\Omega t)/\partial x$. If the perturbation (or forcing term) $\mathbf{g}(\mathbf{x},t)$ vanishes, equation (5.1) reduces to a two-dimensional autonomous system $\dot{\overline{\mathbf{x}}} = \mathbf{f}(\overline{\mathbf{x}},\boldsymbol{\mu})$ corresponding to a one-degree of freedom system in nonlinear Hamiltonian systems. Thus, the total Hamiltonian of equation (5.1) can be expressed by

$$H(x,y,t,\mathbf{p}) = H_0(x,y,\boldsymbol{\mu}) + \mu H_1(x,y,\Omega t,\boldsymbol{\pi}) \tag{5.3}$$

with excitation frequency is Ω and the parameter μ is the strength of the perturbed Hamiltonian $H_1(x,y,t,\boldsymbol{\pi})$. For comparison with the other approximate analysis, such a perturbation parameter is introduced herein. The Hamiltonian of the integrable system in Eq.(5.1) is $H_0(x,y,\boldsymbol{\mu})$. If the initial condition is given, the Hamiltonian $H_0(x,y)$ is invariant (i.e., $H_0(x,y,\boldsymbol{\mu}) = E$), which is the first integral manifold.

To restrict this investigation to a 2-D stochastic layer, four assumptions for Eq.(5.1) are introduced as follows:

(H1) The unperturbed system of Eq.(5.1) possesses a bounded, closed separatrix $q_0(t)$ with at least one hyperbolic point $p_0 : (x_h, y_h)$.

(H2) The neighborhood of $q_0(t)$ for the point $p_0 : (x_h, y_h)$ is filled with at least three families of periodic orbits $q_\sigma(t)$ ($\sigma = \alpha, \beta, \gamma$) with $\alpha, \beta, \gamma \in (0,1]$.

(H3) For the Hamiltonian energy E_σ of $q_\sigma(t)$, its period T_σ is a differentiable function of E_σ.

(H4) The perturbed system of Eq.(5.1) possesses a <u>perturbed</u> orbit $q(t)$ in the neighborhood of the <u>unperturbed</u> separatrix $q_0(t)$.

From the foregoing Hypotheses, the phase portrait of the unperturbed Hamiltonian system in the vicinity of the separatrix is sketched in Fig.5.1. The following point sets and the corresponding Hamiltonian energy are introduced, i.e.,

$$\Gamma_0 \equiv \{(x,y) | (x,y) \in q_0(t), t \in \Re\} \cup \{p_0\} \text{ and } E_0 = H_0(q_0(t)) \tag{5.4}$$

for the separatrix,

$$\Gamma_\sigma \equiv \{(x,y) | (x,y) \in q_\sigma(t), t \in \Re\} \text{ and } E_\sigma = H_0(q_\sigma(t)) \tag{5.5}$$

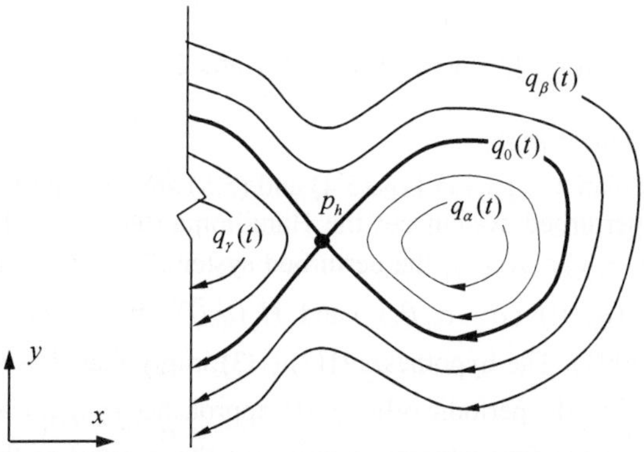

Figure 5.1. The phase portrait of the unperturbed system of Eq. (5.1) near a hyperbolic point p_h. $q_0(t)$ is a separatrix going through the hyperbolic point and splitting the phase into three parts near the hyperbolic point, and the corresponding orbits $q_\sigma(t)$ are termed the σ – orbit ($\sigma = \{\alpha, \beta, \gamma\}$).

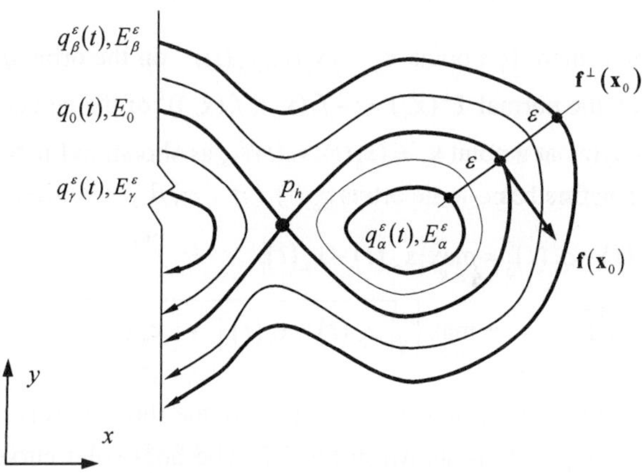

Figure 5.2. The ε -neighborhood of orbit $q_0(t)$. The bold solid curves represent the separatrix $q_0(t)$ and its ε -neighborhood boundaries $q_\sigma^\varepsilon(t)$ determined by $\max_{t\in[0,\infty)} \| q_\sigma^\varepsilon(t) - q_0(t) \| = \varepsilon$ ($\sigma = \alpha, \beta, \gamma$). The solid curves depict all orbits $q_\sigma(t)$ in the ε - neighborhood. The energies on the boundary orbits are given through $E_\sigma^\varepsilon = H_0(q_\sigma^\varepsilon(t))$.

for the unperturbed, σ -periodic orbit and

$$\Gamma = \{(x,y)|(x,y)\in q(t),\, t\in\Re\} \text{ and } E = H_0(q(t)) \tag{5.6}$$

for the perturbed orbit $q(t)$.

The Hamiltonian energies in Eqs.(5.4) and (5.5) are constant for any periodic orbit of the unperturbed system but the Hamiltonian energy in Eq.(5.6) varies with each point $(x,y)\in q(t)$ of the perturbed system. Note that the unperturbed Hamiltonian $H_0(q_\sigma(t))$ ($\sigma=\alpha,\beta,\gamma$) and $H_0(q_0(t))$ are C^r ($r\geq 2$) smooth in Luo and Han (2001). The hypotheses (H2)-(H3) imply that $T_\sigma\to\infty$ monotonically as $\sigma\to 0$ (i.e., the periodic orbit $q_\sigma(t)$ approaches to $q_0(t)$ as $\sigma\to 0$).

The δ -sets of the first integral quantity (or the energy) of the unperturbed system in Eq.(5.1) in Γ_σ ($\sigma=\alpha,\beta,\gamma$), are defined as

$$N_\sigma^\delta(E_0) = \{\, E_\sigma \,\|E_\sigma - E_0| < \delta_\sigma,\ \text{for small } \delta_\sigma > 0\} \tag{5.7}$$

and the union of the three δ -sets with E_0 is

$$N^\delta(E_0) = \cup_\sigma N_\sigma^\delta(E_0)\cup\{E_0\}. \tag{5.8}$$

For some time t, there is a point $\mathbf{x}_\sigma = (x_\sigma(t), y_\sigma(t))^T$ on the orbit $q_\sigma(t)$ and this point is also on the normal $\mathbf{f}^\perp(\mathbf{x}_0) = (-f_2(\mathbf{x}_0), f_1(\mathbf{x}_0))^{\mathrm{T}}$ of the tangential vector of the separatrix $q_0(t)$ at a point $\mathbf{x}_0 = (x_0(t), y_0(t))^T$, as shown in Fig.5.2. Therefore, the distance is defined as

$$\begin{aligned}\|q_\sigma(t) - q_0(t)\| &= \max_{t\in R}|\mathbf{x}_\sigma(t) - \mathbf{x}_0(t)| \\ &= \max_{t\in R}\left\{\sqrt{[x_\sigma(t)-x_0(t)]^2 + [y_\sigma(t)-y_0(t)]^2}\right\}.\end{aligned} \tag{5.9}$$

The ε -neighborhood of orbit $q_0(t)$ is fomed by the three ε -sets of Γ_0 for the σ -orbits ($\sigma=\alpha,\beta,\gamma$), as shown in Fig.5.2. The bold solid curves denote the separatrix $q_0(t)$ and the ε -neighborhood boundaries, $q_\sigma^\varepsilon(t)$ ($\sigma=\alpha,\beta,\gamma$), is determined through $\max_{t\in[0,\infty)}\|q_\sigma^\varepsilon(t)-q_0(t)\|=\varepsilon$ when $E_\sigma^\varepsilon = H_0(q_\sigma^\varepsilon(t))$. The solid curves represent all the σ -orbits $q_\sigma(t)$ in the ε - neighborhood.

The three ε -sets of Γ_0 for the σ -orbits ($\sigma=\alpha,\beta,\gamma$) are defined by

$$\Gamma_\sigma^\varepsilon = \{(x,y)|(x,y)\in q_\sigma(t),\ \|q_\sigma(t)-q_0(t)\| < \varepsilon,\, t\in\mathbb{R}\}. \tag{5.10}$$

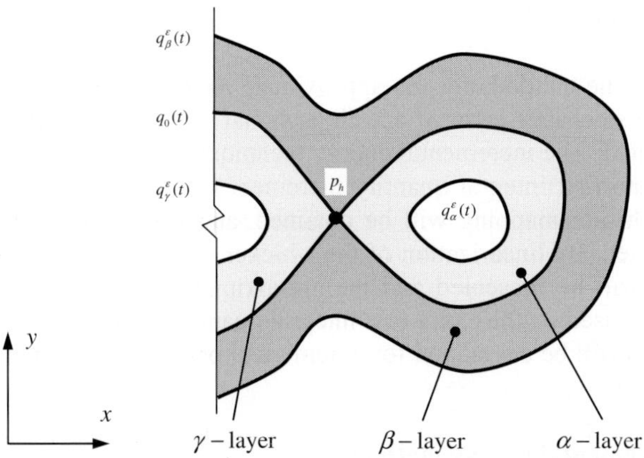

Figure 5.3. A stochastic layer of Eq.(5.1) formed by the Poincare mapping set of $q(t)$ in the ε -neighborhood of $q_0(t)$ for $t \in [0,\infty)$. The separatrix separates the stochastic layer into three sub-stochastic layers (i.e., α – layer and β – layer and γ – layer).

Furthermore, from Eq. (5.8), the unions of the ε -sets with Γ_0 are

$$\Gamma_0^\varepsilon = \bigcup_\sigma \Gamma_\sigma^\varepsilon \cup \Gamma_0, \; \Gamma_{\sigma 0}^\varepsilon = \Gamma_\sigma^\varepsilon \cup \Gamma_0. \tag{5.11}$$

The Poincare map $P:\Gamma^{\mathrm{P}} \to \Gamma^{\mathrm{P}}$, where the corresponding set in phase space is:

$$\Gamma^{\mathrm{P}} \equiv \left\{(x_N, y_N) \middle| (x_N, y_N) \in q(t), \, t_N = \tfrac{2\pi N}{\Omega} + t_0, N = 0,1,\cdots \right\} \subset \Gamma, \tag{5.12}$$

where t_0 is the initial time. Using the above notations, a stochastic layer is defined through the Poincare mapping set with nonzero measure as follows:

DEFINITION 5.1. The Poincaré mapping set Γ^{P} is termed the stochastic layer in the ε -sense if the compact,dense set Γ^{P} belongs to Γ_0^ε (or $\Gamma^{\mathrm{P}} \subset \Gamma_0^\varepsilon$) for $t_N = 2N\pi/\Omega + t_0$ ($N = 0,1,\cdots$). Similarly, the Poincare mapping subset $U_\sigma \subset \Gamma^{\mathrm{P}}$ is the σ -stochastic layer if $U_\sigma \subset \Gamma_{\sigma 0}^\varepsilon$ for $t_N = 2N\pi/\Omega + t_0$.

A stochastic layer of system in Eq.(5.1) is formed via the Poincaré mapping set of $q(t)$ in the ε -neighborhood for time $t \in [0,\infty)$, as illustrated in Fig.5.3. The separation of the stochastic layer by the separatrix gives three sub-stochastic layers shaded in Fig.5.3. The sub-layers relative to the σ -orbits ($\sigma = \alpha, \beta, \gamma$) are termed the σ -stochastic layer.

5.2. Approximate predictions

In this section, the methods for the approximate analytical predictions of the resonance in the stochastic layer of a 2-dimensional nonlinear Hamiltonian system will be presented. The incremental energy technique will be presented first from the approximate first integral quantity increment (or approximate energy increment). The whisker mapping will be obtained, and the corresponding criterion will be presented. By linearization of the whicker mapping, the improved standard mapping will be presented and the approximate prediction of chaos onset will be given. Based on the exact first integral quantity concept, the energy spectrum technique will be developed for a numerical prediction of resonances in the stochastic layers.

5.2.1. An incremental energy method

As in Luo and Han (2001), the incremental energy method will be developed for an understanding of the resonant mechanism of chaos in the stochastic layer. The more discussion is given as follows. Because the conservative energy H_0 is the first integral quantity, for the σ -layer, the map describing the changes of both energy H_0 and phase φ for time transition from t_i to t_i+T_σ in Eq.(5.1) is obtained from Eq.(4.104), i.e.,

$$E_{i+1}=E_i+\Delta H^\sigma(\varphi_i) \text{ and } \varphi_{i+1}=\varphi_i+\Delta\varphi^\sigma(E_{i+1}); \tag{5.13}$$

where $E_i=H(q(t_i)),\varphi_i=\varphi(q(t_i))$. For an external frequency Ω, the initial phase is defined by $\varphi_i=\Omega t_i$. Notice that the energy relationship in the foregoing can be expressed through the action variable. From the energy increment in Chapter 4 (also see, Chirikov, 1979; Lichtenberg and Lieberman,1992), the phase and energy changes $\Delta\varphi^\sigma(E_{i+1})$ and $\Delta H^\sigma(\varphi_i)$ are:

$$\begin{aligned}&\Delta\varphi^\sigma(E_{i+1})\approx\Omega T_\sigma(E_{i+1}), \text{ and}\\&\Delta H^\sigma(\varphi_i)\approx\mu\int_{t_i}^{T_\alpha(E_i)+t_i}\{H_0,H_1\}dt=\mu\int_{t_i}^{T_\sigma(E_i)+t_i}(f_1g_2-f_2g_1)dt.\end{aligned} \tag{5.14}$$

The energy and phase changes in Eq.(5.14) for the system in Eq.(5.1) over one period T_σ of the σ -orbit are sketched in Fig.5.4. If $E_i=E_0$ expresses the energy of the separatrix, equation (5.13) becomes a generalized separatrix map (or a generalized whisker map). If the σ -orbit ($\sigma=\alpha,\beta,\gamma$) is close to the separatrix (i.e., $T_\sigma\to\infty$), the energy increment reduces to the one along the separatrix in Luo and Han (2001):

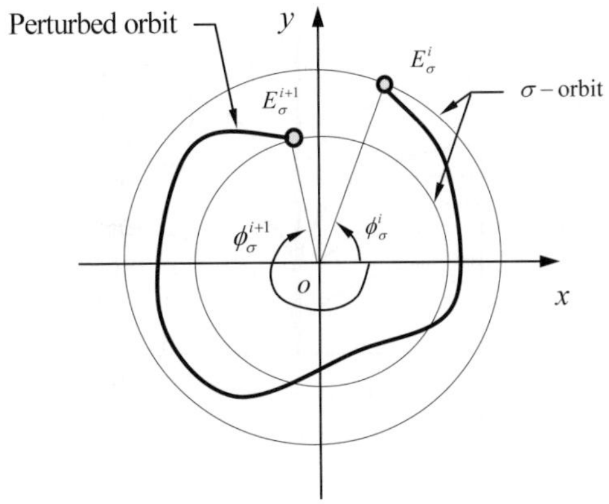

Figure 5.4. The energy and phase changes of a perturbed orbit over one period T_α based on E_σ^i

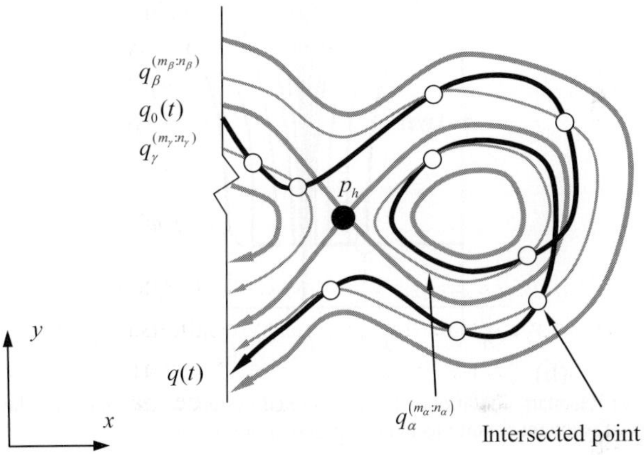

Figure 5.5. A perturbed orbit $q(t)$ for Eq.(5.1) and its resonant interactions with the unperturbed orbits. The hollow circle denotes intersection points between the perturbed and unperturbed orbits. $q_\sigma^{(m_\sigma:n_\sigma)}$ ($\sigma = \alpha, \beta, \gamma$) represents an unperturbed orbit experiencing the $(m_\sigma : n_\sigma)$-resonant interaction with the perturbed orbit.

$$\Delta H^h(\varphi_i) \approx \lim_{T_\sigma \to \infty} \mu \int_{t_i}^{T_\alpha(E_i)+t_i} \{H_0, H_1\} dt = \lim_{T_\sigma \to \infty} \mu \int_{t_i}^{T_\sigma(E_i)+t_i} (f_1 g_2 - f_2 g_1) dt \quad (5.15)$$

which can be obtained through the Melnikov function with a small parameter μ. That is, $\Delta H^h(\varphi_i) = \mu M(t_i)$ (e.g., Rom-Keder, 1990,1994, 1995; Zaslavsky and Abdullaev, 1995; Abdullaev and Zaslavsky, 1995, 1996; Ahn et al, 1996; Iomin and Fishman, 1996).

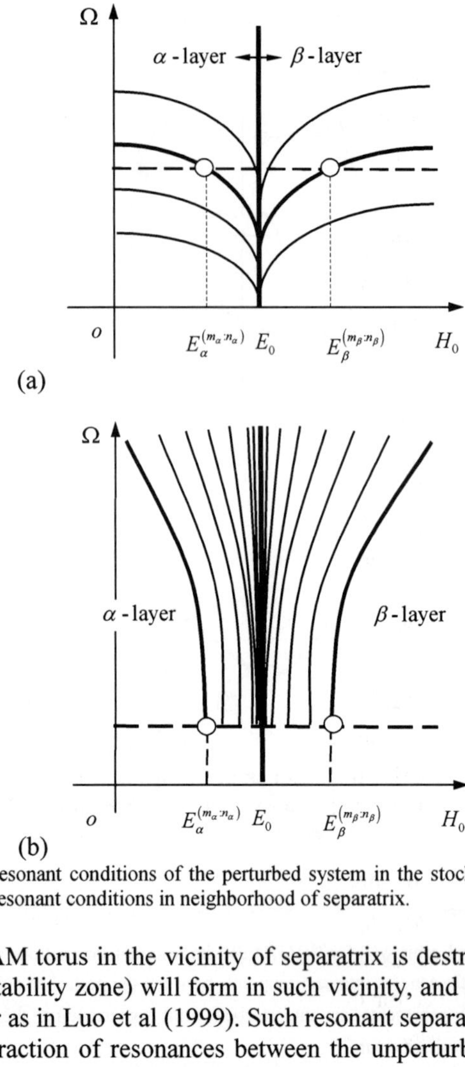

Figure 5.6. (a) Resonant conditions of the perturbed system in the stochastic layer and (b) a zoomed view of resonant conditions in neighborhood of separatrix.

After the KAM torus in the vicinity of separatrix is destroyed, the stochastic layer (or the instability zone) will form in such vicinity, and the resonant separatrix webs appear as in Luo et al (1999). Such resonant separatrix webs are generated by the interaction of resonances between the unperturbed perturbed orbits and the periodic forcing in the stochastic layer. In this section, the prediction of the onset of the primary resonance in the stochastic layer will be of interest. Such resonant interactions in the ε-neighborhood of the separatrix are illustrated in Fig.5.5 qualitatively. The hollow circles depict intersection points of the perturbed and unperturbed orbits. If the perturbed orbit arrives to the $(m_\sigma : n_\sigma)$ primary resonant unperturbed orbit of Eq.(5.1), the resonance between the unperturbed Hamiltonian and perturbation appears, and the resonant condition for a periodically forced system with one degree of freedom is obtained from Eq.(5.3), i.e.,

$$m_\sigma \omega_\sigma = n_\sigma \Omega,\ m_\sigma, n_\sigma \in \mathbb{N} \text{ are irreducible}, \tag{5.16}$$

where $\omega_\sigma = 2\pi / T_\sigma$ is a frequency of the σ-orbit, Ω is the excitation frequency and $\mathbb{N}$ is the natural number set. For the $(i+1)^{\text{th}}$ iteration of Eq.(5.13), if $q(t_{i+1}) = (x_{i+1}, y_{i+1})$ is on the unperturbed, resonant orbit, the phase change in the second of Eq.(5.13) becomes $\varphi_{i+1} - \varphi_i = 2\pi m_\sigma / n_\sigma$ and E_{i+1} is replaced by $E^{(m_\sigma : n_\sigma)}$. The perturbed orbit $q(t)$ relative to the $(m_\sigma : n_\sigma)$-resonance is given by $q^{(m_\sigma : n_\sigma)}(t)$. Since $\omega_\sigma = 2\pi / T_\sigma$ depends on the corresponding energy E_σ, the resonant conditions can be expressed through E_σ and Ω, as shown in Fig.5.6. In Fig.5.6(a), $E_\alpha^{(m_\alpha : n_\alpha)}$ is an unperturbed Hamiltonian energy of the $(m_\alpha : n_\alpha)$-resonant orbit and $E_\beta^{(m_\beta : n_\beta)}$ is the Hamiltonian energy of the $(m_\beta : n_\beta)$-resonant orbit. The resonant number sets in the stochastic layer are introduced for $\sigma = \alpha, \beta, \gamma$:

$$R_\sigma^\varepsilon = \left\{ (m_\sigma : n_\sigma) \left| \begin{array}{l} m_\sigma \omega_\sigma = n_\sigma \Omega,\ m_\sigma, n_\sigma \in \mathbb{N} \text{ are irreducible,} \\ \text{and } \| q_\sigma(t) - q_0(t) \| < \varepsilon \end{array} \right. \right\} \tag{5.17}$$

For a given external frequency Ω with a constant n_σ, the frequency ω_σ relative to E_σ is determined. In Fig.5.6(b), the zoomed view of the resonant relations in the neighborhood of the separatrix with energy $E = E_0$ at an excitation frequency Ω is illustrated through the α- and β-layers. Once the resonance energy is close to the energy of the separatrix, the resonance density of the resonance separatrix increases in the stochastic layer because of the rapid change of elliptic modulus functions. Namely, as $E_\sigma \to E_0$, we have $\omega_\sigma \to 0$. Therefore, from Eq.(5.17), we have $m_\sigma \to \infty$. A resonance with large m_σ will be included in the stochastic layer. For each m_σ, the $(m_\sigma : n_\sigma)$-resonant separatrix possesses m_σ-center points and m_σ-hyperbolic points under the n_σ-external periods. In the stochastic layer, for $E_\sigma \to E_0$, the number for center points and hyperbolic points will approach the infinity. As the energy E_σ approaches E_0, infinite centers and infinite hyperbolic points exists on the resonant separatrix. Such an issue in re-sonant layers will be discussed in detail. As popularly accepted (e.g., Lu, 2007), the chaos is formed by infinite stable and unstable periodic solutions. The total resonant separatrix number is the summation of the resonant orbit relative to the $(m_\sigma : n_\sigma)$-condition with $m_\sigma \in [\min\{m_\sigma\}, \infty) \subset \mathbb{N}$. It implies that the infinity resonances exist in the σ-layer ($\sigma = \alpha, \beta, \gamma$). The stochastic layer complexity will be formed by the resonant-web, which can be refer-red to Luo et al (1999).

When the perturbation becomes stronger, the perturbed orbit will pass over more and more unperturbed, periodic orbits, as shown in Fig.5.5. Therefore, the resonant interaction between the Hamiltonian system and its external perturbation becomes strong with increasing perturbation strength. The resonant interaction leads to a new resonant overlap, and such a resonant overlap produces a new stochastic layer different from the old stochastic layer because a new, resonant separatrix web is involved in the stochastic layer. Thus, for the prediction of the onset of a specific resonant separatrix web in the stochastic layer, Theorem 4.7 gives an approximate condition as in the following theorem since $L^{(\sigma,(\mathbf{k}:s))}(\varphi_i,T_l)$ $=\Delta H^{\sigma}(\varphi_i)$ and $\max_{t\in[0,\infty)}|H(q^{(m_\sigma:n_\sigma)}(t))-E_0|=\delta_\sigma$, i.e.,

THEOREM 5.1. *For a perturbed Hamiltonian Eq.*(5.1) *with* (H1)-(H4), *for an arbitrarily small* $\varepsilon_\sigma>0$, *there exists* $\delta_\sigma>0$, *so that if*

$$\left|\Delta H^{\sigma}(\varphi_i)\right|=\left|E_\sigma^{(m_\sigma:n_\sigma)}-E_0\right|\leq\delta_\sigma \tag{5.18}$$

for $(m_\sigma:n_\sigma)\in R_\sigma^{\varepsilon}$ ($\sigma=\alpha,\beta,\gamma$), *then* $\|q^{(m_\sigma:n_\sigma)}(t_i)-q_0(t_i)\|<\varepsilon_\sigma$ *for all the time* t_i, *i.e., the compact, dense, Poincare mapping set of* $q^{(m_\sigma:n_\sigma)}(t)$ *is in a* σ *-stochastic layer as* $t\in[0,\infty)$.

In Eq.(5.18), the incremental energy $\Delta H^{\sigma}(\varphi_0)$ is a function of the amplitude and frequency of perturbation. The conservative energy $E_\sigma^{(m_\sigma:n_\sigma)}$ of the unperturbed resonant orbit is determined by the resonant condition in Eq.(5.16). With $\max_{t\in[0,\infty)}|H(q^{(m_\sigma^{(1)}:n_\sigma^{(1)})}(t))-E_0|=\delta_\sigma^{(1)}$, the foregoing theorem gives a corollary as.

COROLLARY 5.1. *For a perturbed Hamiltonian in Eq.*(5.1) *with* (H1)-(H4), *for an arbitrarily small* $\varepsilon_\sigma>0$, *there exists* $\delta_\sigma^{(1)}>0$, *so that if*

$$\max_{\sigma\in\{\alpha,\beta,\gamma\}}\left|\Delta H^{\sigma}(\varphi_i)\right|=\max_{R_\sigma^{\varepsilon}}\left|E_\sigma^{(m_\sigma:n_\sigma)}-E_0\right|=\left|E_\sigma^{(m_\sigma^{(1)}:n_\sigma^{(1)})}-E_0\right|\leq\delta_\sigma^{(1)}, \tag{5.19}$$

for $(m_\sigma^{(1)}:n_\sigma^{(1)})\in R_\sigma^{\varepsilon}$ ($\sigma=\alpha,\beta,\gamma$), *then* $\|q^{(m_\sigma^1:n_\sigma^1)}(t)-q_0(t)\|<\varepsilon$ *for all the time t, i.e., the Poincare mapping sets of* $q^{(m_\sigma^{(1)}:n_\sigma^{(1)})}(t)\in\Gamma$ *is the last one absorbed in the* ε *- stochastic layer.*

After a new stochastic layer is formed, the stochastic layer becomes thicker and thicker with increasing the perturbation strength until its destruction occurs. The increase of perturbation strength also leads to the energy increment not satisfying Eq.(5.18), and then, the enlarged stochastic motion domain is termed *the global stochastic layer* in the ε -sense. For an approximate prediction of resonant separatrix webs in the global stochastic layer, we have the following theorem:

THEOREM 5.2. *For a perturbed Hamiltonian in Eq.*(5.1) *with* (H1)-(H4), *for an arbitrarily small* $\varepsilon_\sigma > 0$, *there exists* $\delta_\sigma > 0$, *so that if*

$$\left|\Delta H^\sigma(\varphi_0)\right| = \left|E_\alpha^{(m_\sigma:n_\sigma)} - E_0\right| \geq \delta_\sigma \tag{5.20}$$

for $(m_\sigma : n_\sigma) \notin R_\sigma^\varepsilon$ $(\sigma = \alpha, \beta, \gamma)$, *then* $\| q^{(m_\sigma:n_\sigma)}(t_i) - q_0(t_i) \| > \varepsilon_\sigma$.

The criterion in Eq.(5.18) can be expressed through the action and natural frequency of an unperturbed Hamiltonian system. The onset of a specified resonance in the stochastic layer is predicted by the incremental energy approach. This approach is also applicable for strong excitations when the energy increments can maintain in good accuracy. For a more accurate prediction of the resonance interaction in the stochastic layer, a new computational method for the energy increment should be developed because the sub-resonance is not considered. If the exact energy increment is given, the above theorems can give an exact prediction of the resonance in the stochastic layer.

5.2.2. Accurate standard-mapping technique

Luo (2001a) developed an accurate standard mapping technique to determine the resonant mechanism of the stochastic layer. For linearization of the second equation in Eq.(5.13) at the period-1fixed point on the $(m_\sigma : n_\sigma)$-resonance, for energy $E_{i+1} = E_i = E_\sigma^{(m_\sigma:n_\sigma)}$ and phase $\varphi_{i+1} = \varphi_i + 2\pi m_\sigma/n_\sigma = \varphi^{(m_\sigma:n_\sigma)} + 2\pi m_\sigma/n_\sigma$, equation (5.13) gives:

$$\Delta H_0^\sigma\left(\varphi_\sigma^{(m_\sigma:n_\sigma)}\right) = 0, \text{ and } 2\pi m_\sigma/n_\sigma = \Delta\varphi^\sigma\left(E_\sigma^{(m_\sigma:n_\sigma)}\right), \tag{5.21}$$

from which, $\varphi_\sigma^{(m_\sigma:n_\sigma)}$ and $E_\sigma^{(m_\sigma:n_\sigma)}$ can be obtained. A near energy is

$$E_i = E_\sigma^{(m_\sigma:n_\sigma)} + \Delta E_i, \text{ and } I_i = G_\sigma^{(m_\sigma:n_\sigma)}\Delta E_i; \tag{5.22}$$

where $G_\sigma^{(m_\sigma:n_\sigma)} = \partial(\Delta\varphi(E_{i+1}))/\partial E_{i+1}\big|_{E_{i+1}=E_\sigma^{(m_\sigma:n_\sigma)}}$. With Eq.(5.22), linearization of the second equation of *Eq.*(5.13) leads to

$$I_{i+1} = I_i + G_\sigma^{(m_\sigma:n_\sigma)}\Delta H_0^\sigma(\varphi_i), \text{ and } \varphi_{i+1} \approx \varphi_i + I_{i+1} \tag{5.23}$$

which gives a generalized standard map. In the above derivation, no approximations of the period are required. The criteria for the $(m_\sigma : n_\sigma)$-resonance in the stochastic layer can be obtained through the investigation of the transition from the local stochasticity to global stochasticity in Eq.(5.23). When $G_\sigma^{(m_\sigma : n_\sigma)} \Delta H_0^\sigma(\varphi_i)$ $= K\sin\varphi_i$, such an equation presents a standard map (or the Chirikov-Taylor map), i.e.,

$$I_{i+1} = I_i + K\sin\varphi_i, \text{ and } \varphi_{i+1} \approx \varphi_i + I_{i+1}. \tag{5.24}$$

Greene (1968, 1979) developed a method to numerically determine the strength of the stochasticity parameter $K^* \approx 0.9716\cdots$ when the transition to global stochasticity for Eq.(5.24) occurs. Therefore, for the special case, the perturbation strength of Eq.(5.1) is estimated from

$$\left| G_\sigma^{(m_\sigma : n_\sigma)} \Delta H^\sigma \left(\varphi_{i_i}\right) \right| \approx 0.9716\cdots. \tag{5.25}$$

Other estimates for the strength of stochasticity parameter can be referred to the references (e.g., Chirikov,1979; Lichtenberg and Lieberman,1992). Luo (2001a) developed an accurate standard mapping approach based on the accurate whisker map for such a prediction of the resonance in the stochastic layer. Luo et al (1999) developed an energy spectrum approach (numerical method) for prediction of the onset of resonance in the stochastic layer. A comparison of analytic and numerical predictions is presented in Luo and Han (2000). From numerical results, the previous approaches presented in Luo and Han (2000a) are not very accurate for strong excitations. When the excitation strength is very weak, the accurate and approximate standard-map methods are in good agreement, and the accurate one is applicable to nonlinear systems with strong excitations. Thus, the further improvement should be completed for a more accurate, analytic predicttion of the onset of a new resonant-overlap in stochastic layers of nonlinear Hamiltonian systems with periodical excitations. The energy spectrum is based on the exact energy increments. In next section, the energy spectrum will be discussed and the corresponding layer width will be computed.

5.2.3. Energy spectrum and layer width

For the numerical prediction of resonances in the stochastic layer, an energy spectrum technique was developed in Luo et al (1999), and the resonant characteristics in the stochastic layer are investigated through the energy spectra. This technique computes the maximum and minimum energies of the Poincaré mapping points as in Eq.(5.3). The unperturbed Hamiltonain energy for each Poincare mapping point of Eq.(5.1) is

$$H_0^{(N)} = H_0\left(\mathbf{x}_N, \boldsymbol{\mu}\right) \equiv H_0\left(x_N, y_N, \boldsymbol{\mu}\right), \tag{5.26}$$

and its minimum and maximum energies are determined by

$$E_{\max} = \max_{N\to\infty}\left\{H_0^{(N)}\right\} \text{ and } E_{\min} = \min_{N\to\infty}\left\{H_0^{(N)}\right\}. \tag{5.27}$$

The minimum layer width defined in Luo et al (1999) is

$$w \equiv \min_{t\in[0,\infty)}\left\|\mathbf{x}\left(E^{\max},t\right)-\mathbf{x}\left(E^{\min},t\right)\right\| \equiv \left\|\mathbf{x}^{\max}-\mathbf{x}^{\min}\right\|, \tag{5.28}$$

where $\|\bullet\|$ is a norm. Two points $\mathbf{x}^{\max}$ and $\mathbf{x}^{\min}$ on the normal vector $\mathbf{f}^{\perp}(\mathbf{x}_0)=(-f_2(\mathbf{x}_0),f_1(\mathbf{x}_0))^{\mathrm{T}}$ of the tangential vector of separatrix at point $\mathbf{x}_0$ are the closest between the maximum and minimum energy orbits $\mathbf{x}(E_{\max},t)$ and $\mathbf{x}(E_{\min},t)$ which can be obtained by Eq.(5.27) with $E_{\max}$ and $E_{\min}$, as shown in Fig.5.2.

For a better understanding of the above definition, consider a parametrically excited pendulum in Luo (2001b)

$$\ddot{x}+(\alpha+Q_0\cos\Omega t)\sin x=0, \tag{5.29}$$

where Q_0 and Ω are excitation strength and frequency. The orders of librational and rotational resonances in such a parametric pendulum are $(2M:1)$ and $(M:1)$, respectively. The stochastic layer of the $(4:1)$-librational and $(2:1)$-rotational resonance is used to illustrate resonance interactions, and the width of a stochastic layer is defined in Fig.5.7 for($\alpha=1$, $\Omega=2.5$ and $Q_0=0.05$). The homoclinic orbit and librational and rotational unperturbed orbits relative to the minimum ($E_{\min}$) and maximum ($E_{\max}$) energies are depicted in Fig.5.7, and the width (w) of layer is shown as well. This is located at $x=\pm 2m\pi$ ($m=0,1,2,\cdots$). The librational and rotational stochastic layers are located inside and outside separatrix, respectively. For $Q_0=0.05$ and $\alpha=1.0$ in *Eq.*(5.29), the maximum and minimum energy spectra are computed, as shown in Fig.5.8. The maximum and minimum energies are computed from 10,000 iterations of Poincare map for an excitation strength through the symplectic integrator (e.g., Feng and Qin, 1991; McLachlan and Atela, 1992). The critical excitation frequency $\Omega_l^{(2M:1)}$ (or $\Omega_r^{(M:1)}$) is located at energy jumps in the spectrum when appearance of the $(2M:1)$-librational (or $(M:1)$-rotational) resonance occurs in the stochastic layer. The energy jumps occur at $\Omega_r^{(1:1)}\approx 1.55$ and $\Omega_r^{(2:1)}\approx 2.68$ for the rotational stochastic layer, and the minimum energy jumps at $\Omega_l^{(2:1)}\approx 1.4$ and $\Omega_l^{(4:1)}\approx 2.545$ for the librational layer. If $\Omega>\Omega_r^{(2:1)}\approx 2.68$ for $Q_0=0.05$, then $\Omega>\Omega_l^{(2:1)}$ and $\Omega_l^{(4:1)}$. This implies that the $(1:1)$-rotational resonance and the $(2:1)$- and $(4:1)$-librational resonance will not appear in the stochastic layer. Note that "*r*" and "*l*" denote the rotational

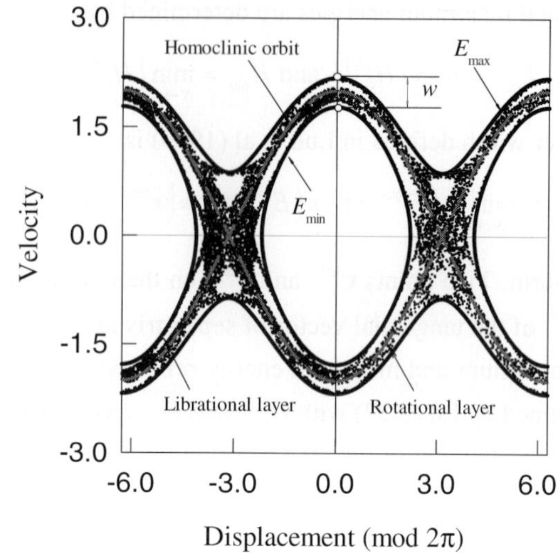

Figure 5.7. Stochastic layer width (w) defined for a parametrically excited pendulum with parameters ($\alpha = 1.0$, $Q_0 = 0.05$, $\Omega = 2.5$).

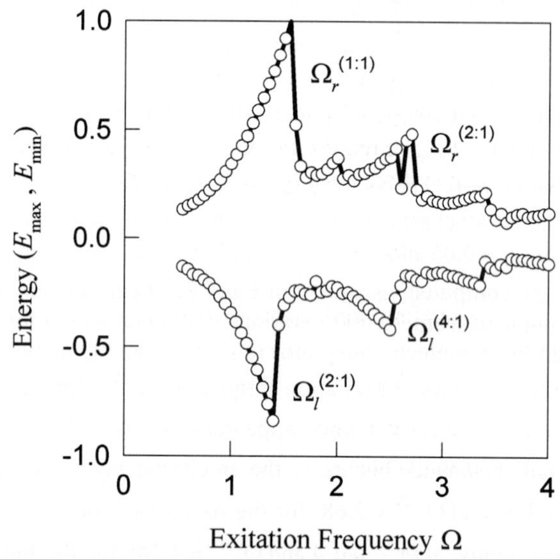

Figure 5.8. Maximum and minimum energy spectra for stochastic layers in a parametrically excited pendulum with parameters ($\alpha = 1.0$, $Q_0 = 0.05$).

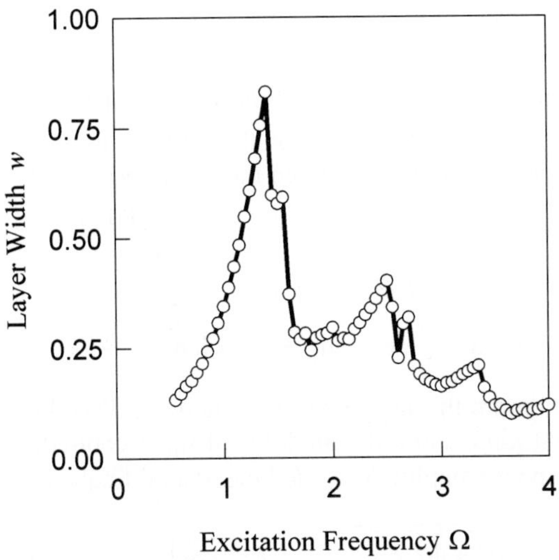

Figure 5.9. The stochastic layer width of a parametrically excited pendulum with parameters ($\alpha = 1.0$, $Q_0 = 0.05$) at $x = \pm 2m\pi$ ($m = 0, 1, 2, \cdots$).

and librational motion, respectively. The minimum and maximum energies in Fig.5.8 are used for computing the width of stochastic layer, as illustrated in Fig.5.9. It is observed that the width of stochastic layer depends on the order of resonance. The layer width decreases with increasing resonance order for an excitation amplitude. Also, the magnitude of the stochastic layer width is not the same order of the magnitude of the excitation strength. Such a conclusion cannot be achieved from the conventional perturbation analysis.

5.3. Stochastic layers in a Duffing oscillator

Consider an undamped Duffing oscillator with a twin-well potential:

$$\ddot{x} - \alpha_1 x + \alpha_2 x^3 = \mu Q_0 \cos(\Omega t), \tag{5.30}$$

where $\alpha_1 > 0$ and $\alpha_2 > 0$ are two system parameters and let $\mu = 1$. Q_0 and Ω are excitation strength and frequency are. The Hamiltonian in Eq.(5.30) consists of two parts $H = H_0 + H_1$ where

$$H_0 = \tfrac{1}{2} y^2 - \tfrac{1}{2}\alpha_1 x^2 + \tfrac{1}{4}\alpha_2 x^4, \tag{5.31}$$

$$H_1 = -x Q_0 \cos(\Omega t), \tag{5.32}$$

and $y = \dot{x}$ is the velocity. The saddle point $(0,0)$ of the unperturbed system of Eq.(5.30) can be easily calculated. The separatrix energy $E_0 = H_0 = 0$ is given. The homoclinic-orbit, α (inner)-orbit and β (outer)-orbit of the unperturbed system in Eq.(5.30) can be determined. For the Duffing oscillator, the sub-layers relative to the α - and β -orbits are termed the inner and outer layers, as sketched in Fig.5.10. For $E_\alpha < E_0$, the solution of the inner-orbit in Luo (1995) is

$$\begin{aligned} &x = \pm e_\alpha \mathrm{dn}[\tfrac{K(k)\omega_\alpha t}{\pi}, k], \text{ and} \\ &y = \dot{x} = \sqrt{\tfrac{\alpha_2}{2}e_\alpha^2 k^2}\mathrm{sn}[\tfrac{K(k)\omega_\alpha t}{\pi}, k]\mathrm{cn}[\tfrac{K(k)\omega_\alpha t}{\pi}, k]; \end{aligned} \tag{5.33}$$

where cn, sn and dn are the Jacobi-elliptic functions, $K(k)$ the complete elliptic integral of the first kind, and k the modulus of the Jacobi-elliptic function. The modulus k, the response amplitude e_α and the natural frequency ω_α are:

$$k = \sqrt{\frac{2\sqrt{\alpha_1^2 + 4\alpha_2 E_\alpha}}{\alpha_1 + \sqrt{\alpha_1^2 + 4\alpha_2 E_\alpha}}}, e_\alpha = \sqrt{\frac{2\alpha_1}{(2-k^2)\alpha_2}}, \omega_\alpha = \frac{\sqrt{\alpha_2} e_\alpha \pi}{\sqrt{2}K(k)}. \tag{5.34}$$

Substitution of Eq.(5.33) into Eqs. (5.31) and (5.32) and expansion of the time-dependent term give:

$$\begin{aligned} H = E_\alpha - \frac{\pi e_\alpha Q_0}{2K}\Big\{\cos\Omega t + \sum_{m=1}^{\infty} \mathrm{sech}\left[\frac{m\pi K'}{K}\right] \\ \times\left[\cos(m\omega_\alpha - \Omega)t + \cos(m\omega_\alpha + \Omega)t\right]\Big\}. \end{aligned} \tag{5.35}$$

From the energy analysis, the resonant condition is

$$m\omega_\alpha = \Omega, \quad \text{for } H_0 = E_\alpha^{(m:1)} < E_0 \tag{5.36}$$

where $E_\alpha^{(m:1)}$ is the energy of the $(m:1)$ resonant orbit.

Because all other terms in H will average to zero over time t except for the term of the primary resonance (i.e., an averaging Hamiltonian in Eq.(5.31)), and the averaging of the Hamiltonian for different $(m:1)$-primary resonance is distinguishing and its magnitude $\bar{H}$ for the inner layer is:

$$\bar{H} \approx \frac{\sqrt{2}\Omega Q_0}{2m\pi\sqrt{\alpha_2}} \mathrm{sech}\left[\frac{m\pi K'\left(k_\alpha^{(m:1)}\right)}{K\left(k_\alpha^{(m:1)}\right)}\right], \tag{5.37}$$

where $K'(k_\alpha^{(m:1)}) = K(k_\alpha'^{(m:1)})$, $k_\alpha'^{(m:1)} = \sqrt{1-(k_\alpha'^{(m:1)})^2}$ and $k_\alpha'^{(m:1)}$ can be obtained from Eq.(5.34) with the resonant condition.

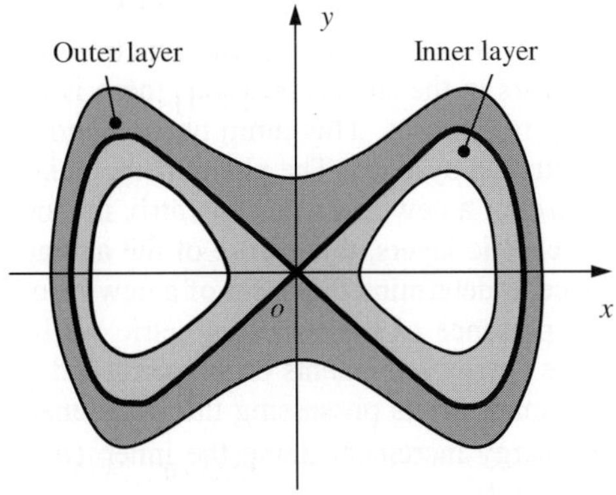

Figure 5.10 Outer and inner stochastic layers for an undamped twin-well Duffing oscillator.

In a same manner, for a given energy $E_\beta > E_0$, the solution of the outer orbit in Luo (1995) is

$$x = e_\beta \text{cn}[\tfrac{2K(k)\omega_\beta t}{\pi}, k], \text{ and}$$
$$y = \dot{x} = \pm\sqrt{\tfrac{\alpha_2}{2}\tfrac{e_\beta^2}{k}}\text{sn}[\tfrac{2K(k)\omega_\beta t}{\pi}, k]\text{dn}[\tfrac{2K(k)\omega_\beta t}{\pi}, k]. \tag{5.38}$$

The modulus k, the response amplitude e_β and the natural frequency ω_β are

$$k = \sqrt{\frac{\alpha_1 + \sqrt{\alpha_1^2 + 4\alpha_2 E_\beta}}{2\sqrt{\alpha_1^2 + 4\alpha_2 E_\beta}}}, e_\beta = \sqrt{\frac{2k^2\alpha_1}{(2k^2-1)\alpha_2}}, \omega_\beta = \frac{\sqrt{\alpha_2}e_\beta\pi}{2\sqrt{2}kK(k)}; \tag{5.39}$$

The corresponding resonant condition given by the energy analysis is

$$(2n-1)\omega_\beta = \Omega, \text{ for } H_0 = E_\beta^{(2n-1:1)} > E_0, \tag{5.40}$$

where $K'(k_\beta^{(2n-1:1)}) = K(k_\beta'^{(2n-1:1)})$ and $k_\beta'^{(2n-1:1)} = \sqrt{1-(k_\beta^{(2n-1:1)})^2}$ are relative to the $(2n-1:1)$-resonance, and $E_\beta^{(2n-1:1)}$ is an energy of the $(2n-1:1)$-resonant orbit. The magnitude $\bar{H}$ for the $(2n-1:1)$-resonance in the outer layer is computed by

$$\bar{H} = \frac{\sqrt{2}\Omega Q_0}{(2n-1)\pi\sqrt{\alpha_1}}\text{sech}\left[(n-\frac{1}{2})\frac{\pi K'(k_\beta^{(2n-1:1)})}{K(k_\beta^{(2n-1:1)})}\right]. \tag{5.41}$$

From Eqs.(5.37) and (5.41), the averaging of Hamiltonian increases with the excitation strength for a specific primary resonance. On the other hand, for given excitation strength, the averaging decreases exponentially with increasing the resonance order because of the hyperbolic "sech" function. Further, when a new primary resonance appears in the stochastic layer, there is a big jump of energy for two closest primary resonances. This jump phenomenon is a very important characteristic of Hamiltonian systems. The phenomena make it possible to numerically observe the onset of a new resonant separatrix in stochastic layers. Thus, in the dynamics of stochastic layers, this nature of the averaging of Hamiltonian is a significant evidence to determine the onset of a new resonance in the stochastic layer. Due to the presence of the two symmetric wells in the unperturbed system in Eq.(5.31), the energy increments of the perturbed orbit should be computed through the two inner orbits possessing the same energy in the stochastic layer. Therefore, the energy increment along the inner $(m:1)$ resonant orbit is computed from Eq. (5.14), i.e.,

$$\begin{aligned}\Delta H^{\alpha}(\varphi_i) &= 2\int_{t_i}^{T_\alpha+t_i}(f_1 g_2 - f_2 g_1)\,dt \\ &\approx 2\sqrt{\frac{2}{\alpha_2}}Q_0\pi\Omega\,\mathrm{sech}\left[\frac{m\pi K'(k_\alpha^{(m:1)})}{K(k_\alpha^{(m:1)})}\right]\sin\varphi_i,\end{aligned} \tag{5.42}$$

where $\varphi_i = \Omega t_i$ and

$$f_1 = y, f_2 = \alpha_1 x - \alpha_2 x^3;\ g_1 = 0, g_2 = Q_0\cos(\Omega t). \tag{5.43}$$

In a same manner, the energy increment along the $(2n-1:1)$-resonant orbit is

$$\begin{aligned}\Delta H^{\beta}(\varphi_i) &= \int_{t_i}^{T_\beta+t_i}(f_1 g_2 - f_2 g_1)\,dt \\ &\approx 2\sqrt{\frac{2}{\alpha_2}}Q_0\pi\Omega\,\mathrm{sech}\left[\frac{(2n-1)\pi K'(k_\beta^{(2n-1:1)})}{2K(k_\beta^{(2n-1:1)})}\right]\sin\varphi_i.\end{aligned} \tag{5.44}$$

From Eq.(5.15) the energy increment along the homoclinic orbit is

$$\begin{aligned}\Delta H^{h}(\varphi_i) &= \lim_{T_\alpha\to\infty}\Delta H^{\alpha} = \lim_{T_\beta\to\infty}\Delta H^{\beta}(\varphi_i) \\ &= \int_{t_i}^{\infty}(f_1 g_2 - f_2 g_1)\,dt \approx 2\sqrt{\frac{2}{\alpha_2}}Q_0\pi\Omega\,\mathrm{sech}\left(\frac{\pi\Omega}{2\sqrt{\alpha_1}}\right]\sin\varphi_i.\end{aligned} \tag{5.45}$$

5.3.1. Incremental energy technique

In this section, the analytical prediction conditions given by the incremental energy approach will be presented. From Eq.(5.18), $E_0 = 0$ and $|\sin\varphi_i| \le 1$, the condition for the onset of the $(m:1)$-resonant separatrix in the inner stochastic layer is

$$\left|\Delta H^{\alpha}(\varphi_i)\right| = \left|E_{\alpha}^{(m:1)}\right|. \tag{5.46}$$

Substitution of Eq.(5.42) into Eq.(5.46) yields the excitation strength for the onset of the $(m:1)$ resonant-separatrix web in the inner stochastic layer:

$$Q_0 = \frac{\left|E_{\alpha}^{(m:1)}\right|}{2\pi\Omega}\sqrt{\frac{\alpha_2}{2}}\cosh\left[\frac{m\pi K'(k_{\alpha}^{(m:1)})}{K(k_{\alpha}^{(m:1)})}\right]. \tag{5.47}$$

If $\Delta H^{\alpha}(\varphi_i)$ is replaced by $\Delta H^{h}(\varphi_0)$ in *Eq.*(5.45), the excitation strength is approximately predicted from

$$Q_0 = \frac{\left|E_{\alpha}^{(m:1)}\right|}{2\pi\Omega}\sqrt{\frac{\alpha_2}{2}}\cosh\left[\frac{\pi\Omega}{2\sqrt{\alpha_1}}\right]. \tag{5.48}$$

From Eq.(5.18) and $E_0 = 0$, the condition for the onset of the $(2n-1:1)$ resonant -separatrix in the outer stochastic layer is

$$\left|\Delta H^{\beta}(\varphi_i)\right| = \left|E_{\beta}^{(2n-1:1)}\right|. \tag{5.49}$$

Substitution of Eq.(5.44) into Eq.(5.49) yields the excitation strength for the onset of the $(2n-1:1)$ resonant-separatrix web, i.e.,

$$Q_0 = \frac{\left|E_{\beta}^{(2n-1:1)}\right|}{2\pi\Omega}\sqrt{\frac{\alpha_2}{2}}\cosh\left[\frac{(2n-1)\pi K'(k_{\beta}^{(2n-1:1)})}{2K(k_{\beta}^{(2n-1:1)})}\right]. \tag{5.50}$$

If $\Delta H^{\beta}(\varphi_i)$ is replaced by $\Delta H^{h}(\varphi_i)$, the excitation strength is given by

$$Q_0 = \frac{\left|E_{\beta}^{(2n-1:1)}\right|}{2\pi\Omega}\sqrt{\frac{\alpha_2}{2}}\cosh\left[\frac{\pi\Omega}{2\sqrt{\alpha_1}}\right]. \tag{5.51}$$

5.3.2. Accurate standard-mapping method

The phase change and the modulus for the α-orbit is from Eq.(5.34):

$$\Delta\varphi^{\alpha}(E_{\alpha}) = \frac{2\pi\Omega}{\omega_{\alpha}} = \frac{2\Omega\sqrt{2-k^2}K(k)}{\sqrt{\alpha_1}}, \text{ and } k = \sqrt{\frac{2\sqrt{\alpha_1^2+4\alpha_2 E_{\alpha}}}{\alpha_1+\sqrt{\alpha_1^2+4\alpha_2 E_{\alpha}}}}. \tag{5.52}$$

Substitution of Eq.(5.52) into Eq.(5.21) for the $(m:1)$ resonance produces

$$\varphi_0 = 0, \pi \text{ and } \Omega\sqrt{2-\left(k_\alpha^{(m:1)}\right)^2}\, K\left(k_\alpha^{(m:1)}\right) = m\pi\sqrt{\alpha_1}. \tag{5.53}$$

Equations (5.52) and (5.53) give $E_\alpha = E_\alpha^m$. From Eqs.(5.25), (5.42) and (5.53) the excitation strength for onset of resonant-orbit in the inner stochastic layer is

$$Q_0 \approx \frac{0.4858}{\pi\Omega G_\alpha^{(m:1)}}\sqrt{\frac{\alpha_2}{2}}\cosh\left[\frac{m\pi K'(k_\alpha^{(m:1)})}{K(k_\alpha^{(m:1)})}\right], \tag{5.54}$$

where

$$G_\alpha^{(m:1)} = -\frac{\Omega\alpha_2}{(k_\alpha^{(m:1)})^4}\left[\frac{2-(k_\alpha^{(m:1)})^2}{\alpha_1}\right]^{\frac{5}{2}}\left[2K(k_\alpha^{(m:1)}) - \frac{2-(k_\alpha^{(m:1)})^2}{1-(k_\alpha^{(m:1)})^2}E(k_\alpha^{(m:1)})\right], \tag{5.55}$$

and $E(k_\alpha^{(m:1)})$ is the second elliptic integral with modulus $k_\alpha^{(m:1)}$.

The phase change and the modulus k for the β-orbit are:

$$\Delta\varphi^\beta\left(E_\beta\right) = \frac{2\pi\Omega}{\omega_\beta} = \frac{4\Omega\sqrt{2k^2-1}\,K(k)}{\sqrt{\alpha_1}} \text{ and } k = \sqrt{\frac{\alpha_1 + \sqrt{\alpha_1^2 + 4\alpha_2 E_\beta}}{2\sqrt{\alpha_1^2 + 4\alpha_2 E_\beta}}}. \tag{5.56}$$

Substitution of Eq.(5.56) into Eq. (5.21) for the $(2n-1:1)$ resonance yields

$$\varphi_0 = 0, \pi \text{ and } 2\sqrt{2\left(k_\beta^{(2n-1:1)}\right)^2 - 1}\, K\left(k_\beta^{(2n-1:1)}\right) = (2n-1)\pi\sqrt{\alpha_1}. \tag{5.57}$$

With Eq.(5.56), the solution to Eq.(5.57) is $E_\beta = E_\beta^{(2n-1:1)}$. Equations (5.25), (5.44) and (5.57) give the excitation strength for the onset of the $(2n-1:1)$-resonant orbit in the outer stochastic layer:

$$Q_0 \approx \frac{0.4858}{\pi\Omega G_\beta^{(2n-1:1)}}\sqrt{\frac{\alpha_2}{2}}\cosh\left[\frac{(2n-1)\pi K'(k_\beta^{(2n-1:1)})}{2K(k_\beta^{(2n-1:1)})}\right], \tag{5.58}$$

where

$$\mathrm{G}_\beta^{(2n-1:1)} = -\frac{2\Omega\alpha_2}{(k_\beta^{(2n-1:1)})^2}\left[\frac{2(k_\beta^{(2n-1:1)})^2-1}{\alpha_1}\right]^{\frac{5}{2}}\left[K(k_\beta^{(2n-1:1)}) - \frac{1-2(k_\beta^{(2n-1:1)})^2}{1-(k_\beta^{(2n-1:1)})^2}E(k_\beta^{(2n-1:1)})\right]. \tag{5.59}$$

The excitation strengths in equation*s* (5.54) and (5.59) distinguish themselves. It indicates that the distinction of the *inner* and *outer* stochastic layers is very significant for a good prediction of the resonant characteristics of the stochastic layer.

5.3.3. Approximate standard-mapping method

The approximate standard map method was developed in Luo (1995) (also see, Luo and Han, 1999). The procedure is also presented herein for comparison of this approximate method with the accurate approach presented in the previous section. Consider the inner layer (or α -layer). As $k \to 1$, the first complete elliptic integral is approximated as in Filonenko et al (1967) and Chirikov (1979)

$$K(k) \approx \frac{1}{2}\log\left(\frac{16}{1-k^2}\right). \tag{5.60}$$

From the foregoing, the energy E_α in Eq. (5.34) and period T_α are given by

$$E_\alpha \approx \frac{(k^2-1)\alpha_1^2}{\alpha_2},\quad T_\alpha = \frac{2\pi}{\omega_\alpha} \approx \frac{1}{\sqrt{\alpha_1}}\log\left(\frac{16\alpha_1^2}{\alpha_2|E_\alpha|}\right). \tag{5.61}$$

The phase change for one period T_α becomes

$$\Delta\varphi(E_\alpha) = \Omega T_\alpha \approx \frac{\Omega}{\sqrt{\alpha_1}}\log\left(\frac{16\alpha_1^2}{\alpha_2|E_\alpha|}\right). \tag{5.62}$$

Using Eqs.(5.45) and (5.62), Eq.(5.13) gives an approximate whisker map

$$\begin{aligned} E_{i+1} &\approx E_i + 2Q_0\pi\Omega\sqrt{\frac{2}{\alpha_2}}\operatorname{sech}\left(\frac{\pi\Omega}{2\sqrt{\alpha_1}}\right)\sin\varphi_i,\\ \varphi_{i+1} &\approx \varphi_i + \frac{\Omega}{\sqrt{\alpha_1}}\log\left(\frac{16\alpha_1^2}{\alpha_2|E_{i+1}|}\right). \end{aligned} \tag{5.63}$$

As in Luo and Han (1999), setting $E_{i+1} = E_i = E_\alpha^{(m:1)}$ and $\varphi_{i+1} = \varphi_i + 2m\pi = \varphi_\alpha^{(m:1)}$ $+2m\pi$ for the $(m:1)$ resonance in Eq.(5.63) gives

$$2Q_0\pi\Omega\operatorname{sech}\left(\frac{\pi\Omega}{2\sqrt{\alpha_1}}\right)\sin\varphi_\alpha^{(m:1)} = 0,\quad \frac{\Omega}{\sqrt{\alpha_1}}\log\left(\frac{16\alpha_1^2}{\alpha_2\left|E_\alpha^{(m:1)}\right|}\right) = 2m\pi. \tag{5.64}$$

Solutions to Eq.(5.64) for $E_\alpha^{(m:1)}$ and $\varphi_\alpha^{(m:1)}$ are

$$\left|E_\alpha^{(m:1)}\right| = \frac{16\alpha_1^2}{\alpha_2}e^{-\left(2m\pi\sqrt{\alpha_1}/\Omega\right)},\quad \varphi_\alpha^{(m:1)} = 0, \pi. \tag{5.65}$$

Perturbation of the energy at $E_\alpha^{(m:1)}$ in Eq. (5.63) ($E_i = E_\alpha^{(m:1)} + \Delta E_i$) and $w_i =$

$-\Omega\Delta E_i/(\sqrt{\alpha_1}\,|\,E_\alpha^{(m:1)}\,|)$ gives a standard map in Eq.(5.24). The strength of stochasticity becomes

$$K = \frac{2Q_0\pi\Omega^2}{\sqrt{\alpha_1}\left|E_\alpha^{(m:1)}\right|}\sqrt{\frac{2}{\alpha_2}}\operatorname{sech}\left(\frac{\pi\Omega}{2\sqrt{\alpha_1}}\right). \tag{5.66}$$

From Eq. (5.24), the excitation strength for the onset of the $(m:1)$ resonant-orbit in the inner stochastic layer is:

$$Q_0 \approx \frac{7.7728}{\pi}\left(\frac{\alpha_1}{\Omega}\right)^2\sqrt{\frac{\alpha_1}{2\alpha_2}}e^{-\left[2m\pi\sqrt{\alpha_1}/\Omega\right]}\cosh\left(\frac{\pi\Omega}{2\sqrt{\alpha_1}}\right). \tag{5.67}$$

For $\alpha_1 = 2, \alpha_1 = 1$ at $\Omega << 1$, equation (5.67) gives one-half of the result given in Reichl and Zheng (1984a,b) and Reichl(1992). This is because only one of the two potential wells in Eq.(5.30) was considered in Reichl and Zheng (1984a,b).

Consider the outer stochastic layer (or β-layer). The energy E_β in Eq.(5.39) and an approximate period T_β, as $k \to 1$, are

$$E_\beta \approx \frac{\alpha_1^2}{\alpha_2}\left(1-k^2\right),\ \ T_\beta \approx \frac{2}{\sqrt{\alpha_2}}\log\left(\frac{16\alpha_1^2}{\alpha_2 E_\beta}\right). \tag{5.68}$$

The phase change over one period T_β is

$$\Delta\varphi\left(E_\beta\right) = \Omega T_\beta \approx \frac{2\Omega}{\sqrt{\alpha_1}}\log\left(\frac{16\alpha_1^2}{\alpha_2 E_\beta}\right). \tag{5.69}$$

Substitution Eqs.(5.45) and (5.69) into Eq.(5.13) yields:

$$\begin{aligned} E_{i+1} &\approx E_i + 2Q_0\pi\Omega\sqrt{\frac{2}{\alpha_2}}\operatorname{sech}\left(\frac{\pi\Omega}{2\sqrt{\alpha_1}}\right)\sin\varphi_i, \\ \varphi_{i+1} &\approx \varphi_i + \frac{2\Omega}{\sqrt{\alpha_1}}\log\left(\frac{16\alpha_1^2}{\alpha_2 E_{i+1}}\right). \end{aligned} \tag{5.70}$$

When $E_{i+1} = E_i = E_\beta^{(2n-1:1)}$ and $\varphi_{i+1} = \varphi_i + 2(2n-1)\pi = \varphi_\alpha^{(2n-1:1)} + 2(2n-1)\pi$ are used, equation (5.70) becomes:

$$2Q_0\pi\Omega\sqrt{\frac{2}{\alpha_2}}\operatorname{sech}\left(\frac{\pi\Omega}{2\sqrt{\alpha_1}}\right)\sin\varphi_\beta^{2n-1} = 0, \tag{5.71a}$$

$$\frac{2\Omega}{\sqrt{\alpha_1}}\log\left(\frac{16\alpha_1^2}{\alpha_2 E_\beta^{2n-1}}\right) = 2(2n-1)\pi. \tag{5.71b}$$

Solving the foregoing gives $E_\beta^{(2n-1:1)}$ and $\varphi_\beta^{(2n-1:1)}$, .i.e.,

$$E_\beta^{(2n-1:1)} = \frac{16\alpha_1^2}{\alpha_2} e^{-(2n-1)\pi\sqrt{\alpha_1}/\Omega}; \ \varphi_\beta^{(2n-1:1)} = 0, \pi. \tag{5.72}$$

In a like fashion, linearization of Eq.(5.70) at $w_i = -2\Omega\Delta E_i / (\sqrt{\alpha_1} E_\beta^{(2n-1:1)})$ for $E_\beta^{(2n-1:1)}$ yields Eq.(5.24). Thus, the stochasticity parameter K is given by

$$K = \frac{4Q_0\pi\Omega^2}{\sqrt{\alpha_1} E_\beta^{2n-1}}\sqrt{\frac{2}{\alpha_2}}\operatorname{sech}\left(\frac{\pi\Omega}{2\sqrt{\alpha_1}}\right). \tag{5.73}$$

The excitation strength for the onset of the $(2n-1:1)$ resonant-overlap in the outer stochastic layer is:

$$Q_0 \approx \frac{3.8864}{\pi}\left(\frac{\alpha_1}{\Omega}\right)^2 \sqrt{\frac{\alpha_1}{2\alpha_2}} e^{-[(2n-1)\pi\sqrt{\alpha_1}/\Omega]} \cosh\left(\frac{\pi\Omega}{2\sqrt{\alpha_1}}\right). \tag{5.74}$$

The excitation strengths in Eqs.(5.58) and (5.74) also distinguish themselves.

5.3.4. Energy spectrum and layer width

Reichl and Zheng (1984a,b) computed the energy varying with time for observation of resonance in the stochastic layer, but those energy responses similar to displacement responses in nonlinear oscillations cannot lead to a better understanding of the dynamic characteristics of stochastic layers. The Poincaré mapping section can provide a useful means to investigate chaotic motion. Herein, the energy for the Poincaré mapping points of the stochastic layer is computed, rather than the energy responses as in Reichl and Zheng (1984a,b). The Poincaré mapping section for the twin-well Duffing oscillator is defined as

$$\Sigma = \left\{ (x(t_N), \dot{x}(t_N)) \left| \begin{array}{l} (x(t_N), \dot{x}(t_N)) \text{ satisfying Eq.(5.30), and} \\ t_N = \dfrac{2N\pi}{\Omega} + t_0, N = 0, 1, \cdots \end{array} \right. \right\} \tag{5.75}$$

where $x(t_N) = x_N, \dot{x}(t_N) = \dot{x}_N$ and $x(t_0) = x_0, \dot{x}(t_0) = \dot{x}_0$ at $t=t_0$ are the initial conditions. The Poincaré map is $P: \Sigma \to \Sigma$. The energy for each Poincaré mapping point of the Duffing oscillator is

$$H_0^{(N)} = \tfrac{1}{2}\dot{x}_N^2 - \tfrac{1}{2}\alpha_1 x_N^2 + \tfrac{1}{4}\alpha_2 x_N^4, \tag{5.76}$$

and its minimum and maximum energy are defined by

$$E_{\max} = \max_{N\to\infty}\left\{H_0^{(N)}\right\} \text{ and } E_{\min} = \min_{N\to\infty}\left\{H_0^{(N)}\right\}. \tag{5.77}$$

The layer width varies with the reference line in the Poincaré plane. Following the definition of the distance between the stable and unstable manifolds in the Melnikov method (also, see Melnikov,1963; Guckenheimer and Holmes, 1983), the minimum layer width is defined as

$$w \equiv \min_{t\in[0,\infty)}\left\|\mathbf{x}\left(E^{\max},t\right)-\mathbf{x}\left(E^{\min},t\right)\right\| \equiv \left\|\mathbf{x}^{\max}-\mathbf{x}^{\min}\right\| \tag{5.78}$$

where $\|\bullet\|$ is a norm. Two points $\mathbf{x}^{\max}$ and $\mathbf{x}^{\min}$ on the normal vector $\mathbf{f}^{\perp}(\mathbf{x}_0) = (-f_2(\mathbf{x}_0), f_1(\mathbf{x}_0))^{\mathrm{T}}$ of the tangential vector of separatrix at point $\mathbf{x}_0$ are the closest between the maximum and minimum energy orbits $\mathbf{x}(E_{\max},t)$ and $\mathbf{x}(E_{\min},t)$ which can be obtained by Eqs.(5.33) and (5.38) with $E_{\max}$ and $E_{\min}$. For a better understanding of the above definitions, a resonant-separatrix web of the 5th-order in the outer stochastic layer for $\alpha_1 = \alpha_2 = 1.0$, $\Omega = 4.0$ and $Q_0 = 0.45$ is shown. The layer width, minimum and maximum energy orbits for that resonant separatrix web are illustrated in Fig.5.11. For the Duffing oscillator, the layer width is $|\mathbf{x}^{\max} - \mathbf{x}^{\min}|$ at $y = 0$ which can be determined numerically.

Using the above definition, the maximum and minimum energy spectra are shown in Fig.5.12 for $Q_0 = 0.2$ and $\alpha_1 = \alpha_2 = 1.0$ in Eq.(5.30). The maximum and minimum energy are computed from 10,000 iterations of Poincaré map for each excitation frequency. $\Omega^{\alpha(m:1)}$ (or $\Omega^{\beta(2n-1:1)}$) denotes a maximum value of excitation frequency when the $(m{:}1)$-order inner (or $(2n-1:1)$-order outer) resonant separatrix disappears in the stochastic layer. To view the energy jump clearly, four specific areas in the spectra are zoomed. The energy jumps occur at $\Omega^{\beta(1:1)} = 1.33$, $\Omega^{\beta(3:1)} = 2.73$, $\Omega^{\beta(5:1)} = 3.51$ for the outer stochastic layer and, at $\Omega^{\alpha(2:1)} = 2.67$ and $\Omega^{\alpha(3:1)} = 3.39$ for the inner stochastic layer. The specific values are critical excitation frequencies for the disappearance of the resonance in the stochastic layer. For instance, if an excitation frequency chosen for $Q_0 = 0.2$ is greater than $\Omega^{\beta(3:1)} = 2.73$, then, the resonant-separatrix of lower than 5th order in the outer-layer and of lower than 3rd order in the inner-layer will not appear. For $Q_0 = 0.2$, the resonant-separatrix of the first order in the inner stochastic layer cannot be observed because the Hamiltonian arrives at the minimum energy E_α

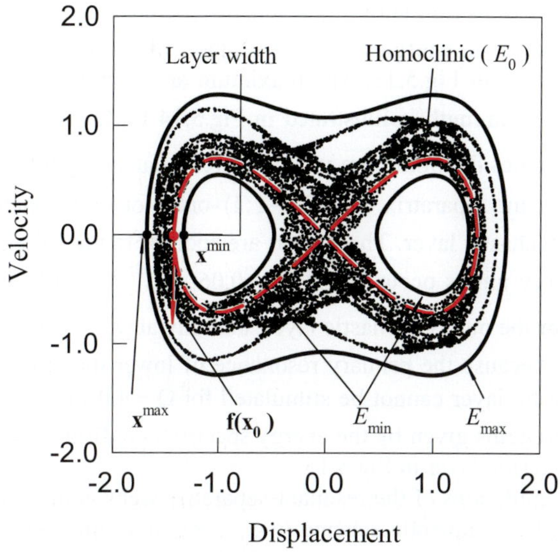

Figure 5.11. A description of layer width, minimum and maximum energy through an outer resonant-separatrix web of the 5th-order in the stochastic layer for the twin-well Duffing oscillator at $Q_0 = 0.45$ and $\Omega = 4.0$.

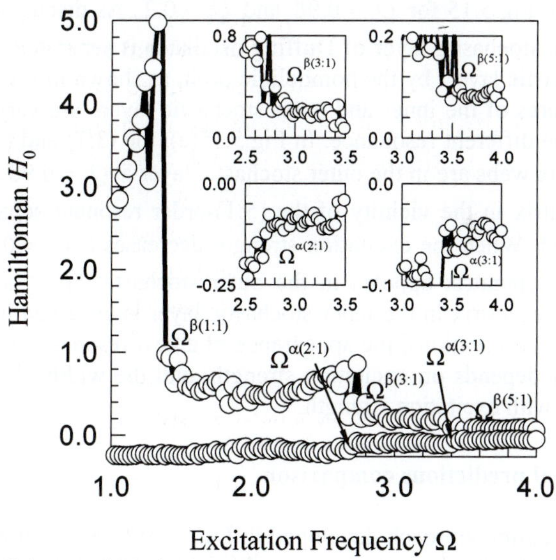

Figure 5.12 Minimum and Maximum energy spectra for the stochastic layer in the twin-well Duffing oscillator at $Q_0 = 0.2$ and $\alpha_1 = \alpha_2 = 1.0$.

$=-0.25$ (for $\alpha_1=\alpha_2=1.0$) until this resonant separatrix appears. Based on the minimum and maximum energies in Fig.5.12, the width of the stochastic layer is computed, as shown in Fig.5.13. The maximum and minimum energy varying with the excitation strength is illustrated in Fig.5.14 for $\Omega=4.0$, where $Q_0^{\alpha(m:1)}$ (or $Q_0^{\beta(2n-1:1)}$) denotes the minimum value of excitation strength when the $(m{:}1)$-order inner resonant separatrix (or $(2n-1:1)$-order outer resonant separatrix) appears in the stochastic layer. Three areas are zoomed to view the energy jumps clearly, the energy jumps occur at $Q_0^{\beta(9:1)}=0.06$, $Q_0^{\beta(7:1)}=0.18$, $Q_0^{\beta(5:1)}=0.44$, $Q_0^{\beta(3:1)}=0.96$ for the outer-stochastic layer, but only at $Q_0^{\alpha(4:1)}=0.11$ for the inner stochastic layer because the primary resonance of lower than the (4:1)-order in the inner stochastic layer cannot be stimulated for $\Omega=4.0$ and $Q_0\leq 1$. The critical excitation strengths given by the energy spectrum confirm the resonant separatrix in the stochastic layer in Fig.5.15.

For numerical simulations of the resonant-separatrix webs in the stochastic layer, we use a 2nd-order symplectic scheme (e.g., Feng and Qin, 1991; McLach-lan and Atela, 1992) with time step $\Delta t=10^{-5}\sim 10^{-7}T$, where $T=2\pi/\Omega$, and a precision of 10^{-6}. For $\alpha_1=\alpha_2=1.0$ and $\Omega=4.0$, the resonant-separatrix webs in the stochastic layers generated by 20,000 Poincaré mapping points of Eq.(5.30) are illustrated in Fig.5.15 for $Q_0=0.98$ and $Q_0=0.2$. As discussed in Luo and Han (2001), the stochastic layer of Duffing oscillator is separated into the inner and outer stochastic layers by the homoclinic orbit, as shown in Fig.5.15, and the resonant structures of the inner and outer stochastic layers are very distinguishing owing to the different resonance. In Fig.5.15(a), the (3:1) and (5:1)-order resonant separatrix webs are in the outer stochastic layer at $Q_0=0.98$, and the sub-resonant separatrix in the vicinity of the (5:1)-order resonant separatrix is also clearly observed. When the excitation strength decreases to $Q_0=0.2$, the (7:1)-order resonant separatrix appears in the outer stochastic layer, and the (4:1)-order resonant separatrix in the inner stochastic layer is observed in Fig.5.15(b). From the above observations, the appearance of a new resonant-separatrix in the stochastic layer depends on excitation strength, and the width of the stochastic layer increases with excitation strength.

5.3.5. Analytical predictions comparison

The energy spectrum approach developed in Luo et al (1999) is used for numerical predictions. Again, the numerical results are generated through a 2nd-order symplectic scheme in (e.g., Feng and Qin, 1991; McLachlan and Atela, 1992) with time step $\Delta t=10^{-5}\sim 10^{-7}T$ (where $T=2\pi/\Omega$ is the perturbation period)

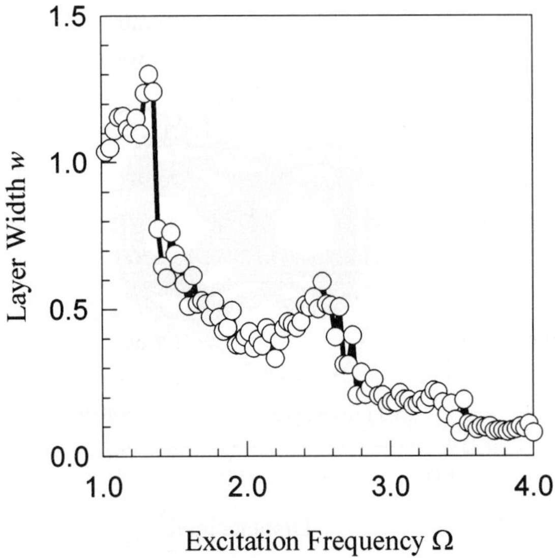

Figure 5.13. The width of the stochastic layer in the twin-well Duffing oscillator for $Q_0 = 0.2$ and $\alpha_1 = \alpha_2 = 1.0$.

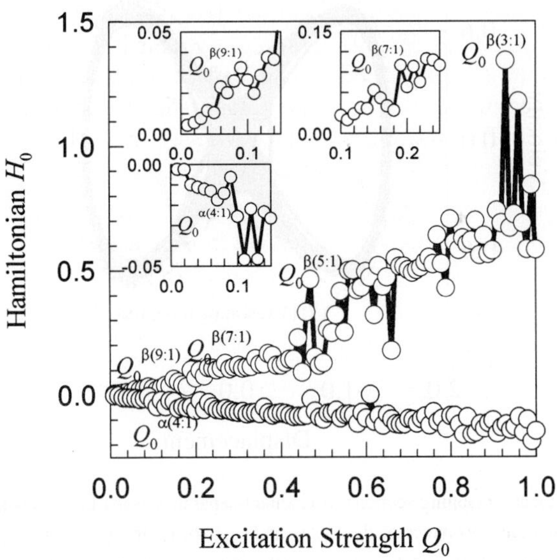

Figure 5.14. Minimum and maximum energy versus excitation strength for the twin-well Duffing oscillator at $\Omega = 4.0$ and $\alpha_1 = \alpha_2 = 1.0$.

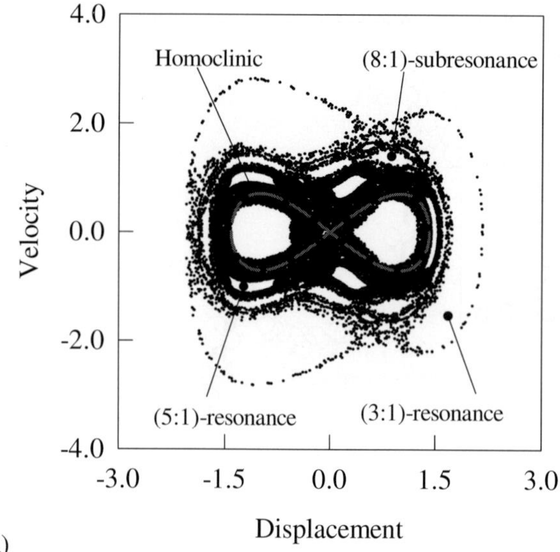

(a)

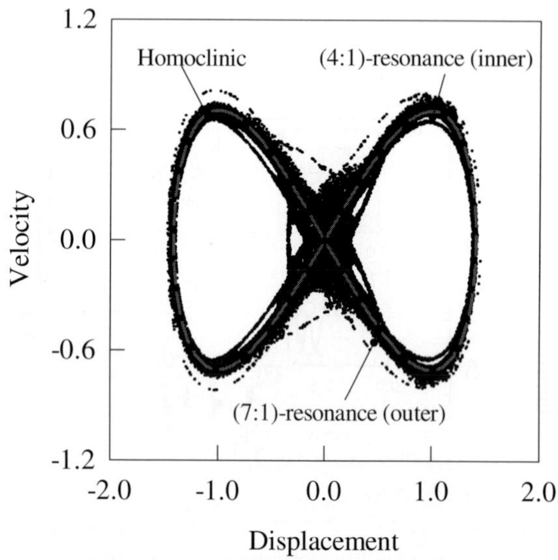

(b)

Figure 5.15. Poincaré mapping sections of resonant-separatrix webs in the stochastic layer at $\Omega = 4.0$. (a) The outer resonance of the (3:1)- and (5:1)-order for $Q_0 = 0.98$ (upper), and (b) the outer resonance of the (7:1)-order and the inner resonance of the (4:1)-order for $Q_0 = 0.2$ (lower).

and a precision of 10^{-6}. When the energy increment is computed, the unperturbed orbit is used to replace the perturbed orbit in the stochastic layer. The computational accuracy of energy increments in Eqs.(5.42), (5.44) and (5.45) at $\alpha_1 = \alpha_2 = 1.0$ is controlled by $\max_{\alpha} | E_{\alpha} - E_0 | = \delta_{\alpha} \approx 0.055$ and $\max_{\beta} | E_{\beta} - E_0 | = \delta_{\beta} \approx 0.076$ pertaining to the modulus $k = 0.9684$ close to $k = 1$ for the separatrix. In Fig.5.16, the solid curves give the predictions of Q_0 through Eqs.(5.47) and (5.50) for the inner and outer layers when $| E_{\alpha} - E_0 | \leq \delta_{\alpha} \approx 0.055$ and $\left| E_{\beta} - E_0 \right| \leq \delta_{\beta} \approx 0.076$. The dashed curves represent the approximate predictions of Q_0 through Eqs.(5.48) and (5.51) for the inner and outer layers. The hollow-circle curves give the numerical predictions. The two predictions of Q_0 through the energy increments based on the resonant orbit and the separatrix are very close, but they are different from the numerical predictions because of the approximation of energy increments. When the above energy constraints $| E_{\alpha} - E_0 | \leq \delta_{\alpha} \approx 0.055$ and $| E_{\beta} - E_0 | \leq \delta_{\beta} \approx 0.076$ are not applied, the approximate predictions of Q_0 are illustrated in Fig.5.17. For the inner layer, the prediction based on resonant orbit is *larger* than the one based on the separatrix, but for the outer layer, the prediction based on resonant orbit is *smaller* than the one based on the separatrix because of $\Delta H_0^{\beta}(\varphi_i) \geq \Delta H_0^{h}(\varphi_i) \geq \Delta H_0^{\alpha}(\varphi_i)$. It is observed that the numerical results for the onset of resonance in the stochastic layer is the same pattern as the analytical prediction based on the resonant orbit through the incremental energy approach whether the excitation strength increases or not. However, this is not the case for the analytical predictions based on the homoclinic orbit. The prediction of Q_0 through an unperturbed, resonant orbit is much close to the numerical prediction, but they are still different from the numerical predictions. In the derivation of analytical conditions, only the primary resonance is used and the incremental energy is computed approximately through the unperturbed orbits instead of the perturbed ones. The above two factors result in the analytical estimates of excitation strength higher than the numerical prediction. Therefore, an accurate computation of the incremental energy needs to be carried out and the self-similar structure of the resonance in the stochastic layer should be developed through the renormalization technique.

The energy spectrum approach developed in Luo et al (1999) is used for numerical predictions. The accurate, analytical predictions of excitation strength Q_0 are given through Eqs.(5.54) and (5.58), and the approximate predictions by Eqs.(5.67) and (5.74). In Fig.5.18, the short-dash curve gives the accurate standard mapping prediction of excitation strength Q_0 at $\alpha_1 = \alpha_2 = 1.0$ for the energy constraints $\max_{\alpha} | E_{\alpha} - E_0 | = 0.055$ and $\max_{\beta} | E_{\beta} - E_0 | = 0.076$ relative to the modulus $k = 0.9684$ close to $k = 1$ for the separatrix. The dash-dot-dot curves

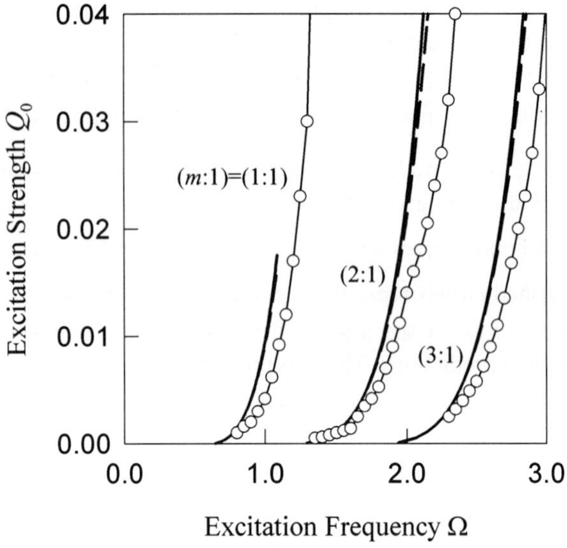

(a)

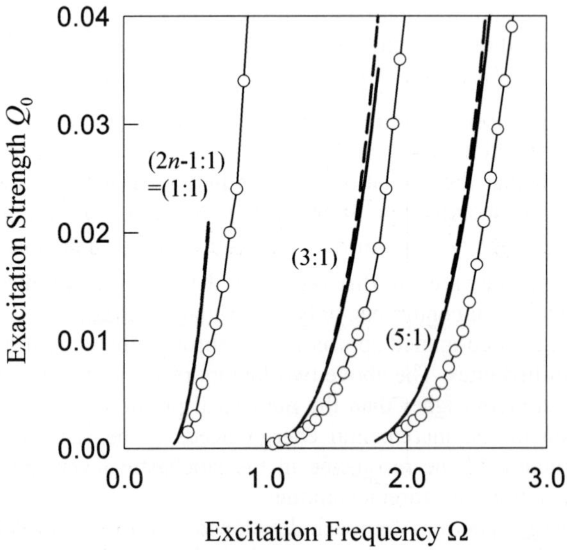

(b)

Figure 5.16. Comparison of excitation strength for (a) the inner layer with $\max_{\alpha} |E_\alpha - E_0| =$ 0.055 and (b) the outer layer (bottom) with $\max_{\beta} |E_\beta - E_0| = 0.076$ in the Duffing oscillator. The solid curve, dashed curve and circular-symbol curve represent the approximate conditions based on the resonant orbit, separatrix and numerical simulations, respectively.

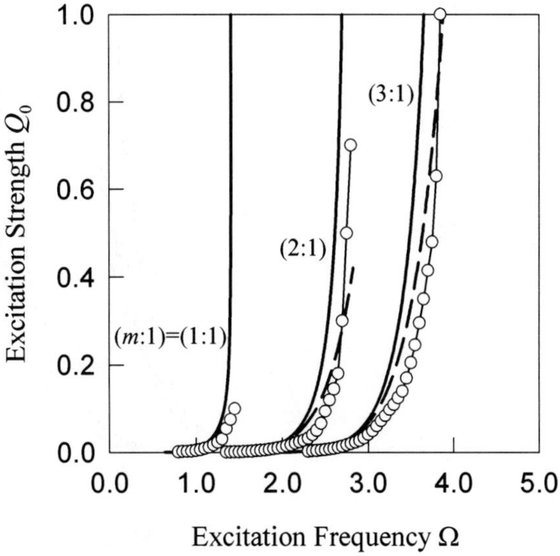

(a)

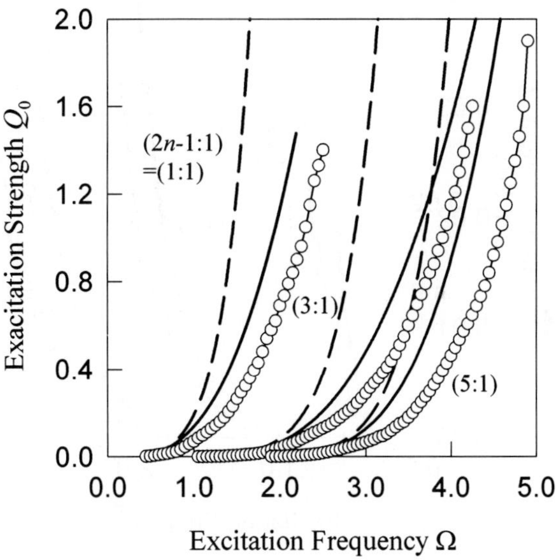

(b)

Figure 5.17. Comparison of excitation strength for (a) the inner layer and (b) the outer layer in the Duffing oscillator. The solid curve, dashed curve and circular-symbol curve represent the approximate conditions based on the resonant orbit, separatrix and numerical simulations, respectively.

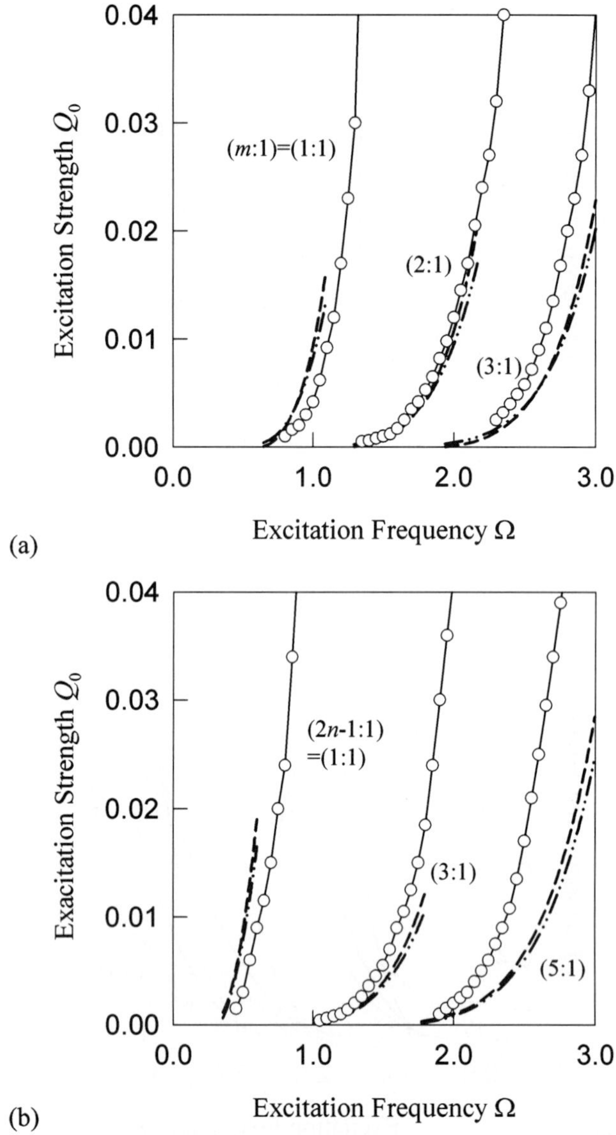

Figure 5.18. Excitation strength for (a) the inner layer with $\max|E_\alpha - E_0| = 0.055$ and (b) the outer layer with $\max|E_\beta - E_0| = 0.076$ in the Duffing oscillator ($\alpha_1 = \alpha_2 = 1.0$). The short-dash curve, dash-dot-dot curve and circular-symbol curve are based on accurate standard mapping, approximate standard mapping and energy spectrum methods, respectively.

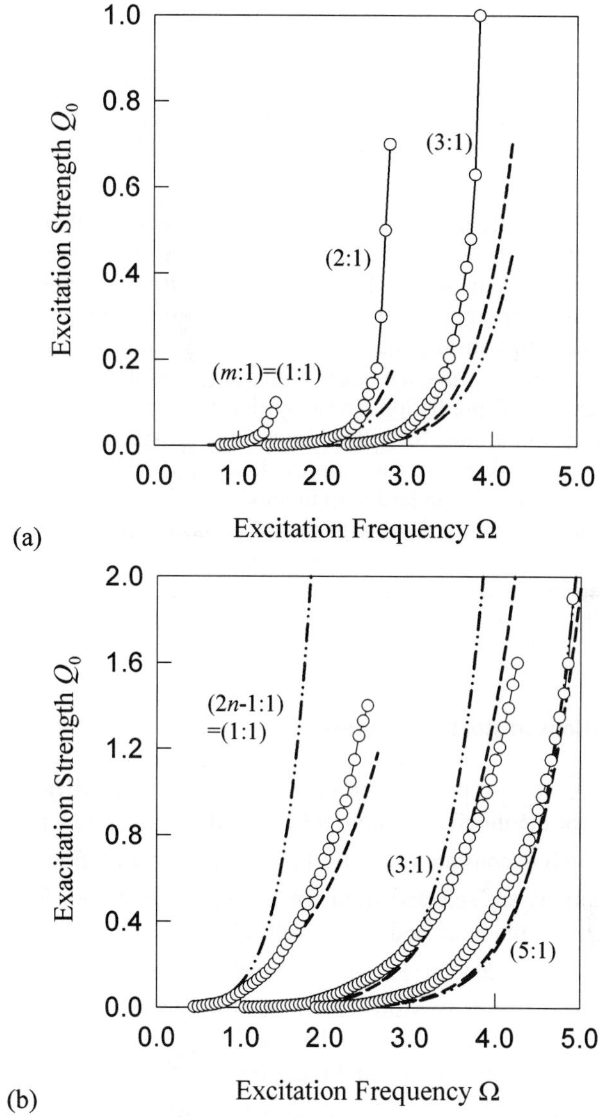

Figure 5.19. Excitation strength for (a) the inner layer and (b) the outer layer in the Duffing oscillator ($\alpha_1 = \alpha_2 = 1.0$).The short-dash curve, dash-dot-dot curve and circular-symbol curve are based on accurate standard mapping, approximate standard mapping and energy spectrum methods, respectively.

represents the approximate standard mapping prediction of Q_0. The hollow-circle lines give the numerical results generated by the energy spectrum method. The two predictions of Q_0 through two standard-map approaches are very close, but both of them are different from the numerical results because of the phase linearization and energy increment approximation in the neighborhood of the resonant separatrix. If $\max_\alpha |E_\alpha - E_0| = 0.055$ and $\max_\beta |E_\beta - E_0| = 0.076$ at $\alpha_1 = \alpha_2 = 1.0$ do not apply to the analytical predictions, the excitation strength versus the excitation frequency is illustrated in Fig.5.19. Compared with the numerical results, the accurate standard-map approach gives a much better prediction than the approximate standard-map approach (or the improved *traditional* standard map). If the excitation becomes strong, the *approximate* standard mapping method gives worse than the *accurate standard map*. With increasing the order number of resonance, the analytical prediction given by that approximate whisker mapping method becomes worse too. Note that the *traditional* standard mapping hinged on the *traditional* whisker map cannot provide any correct predictions. Based on the traditional whisker or standard map method, we cannot compute the numerical results to make a comparison with the proposed accurate approach because the *traditional* standard map in references (e.g., Filonenko et al, 1967 and Chirikov,1979) used two equations instead of six equations of the energy and phase changes for the σ-orbits ($\sigma = \alpha, \beta, \gamma$) in the improved traditional standard mappings.

5.3.6. Illustrations of stochastic layers

From Eq.(5.75), stochastic layers are generated through the differential equation (or Eq.(5.30)) for the initial condition $(0,0)$ instead of the discrete map. When the excitation strength is constant ($Q_0 = 0.01$), several excitation frequencies for numerical simulations are selected from the analytical condition. Fig.5.20(a) and (b) give (Ω, Q_0) for the inner and outer stochastic layers, respectively. At point A ($\Omega = 0.5$), the first order $(m:1) = (1:1)$ resonant overlap is very weak and the resonant-separatrix web is very difficult to be observed in the inner layer. On the other hand, in Fig.5.20(b), it is shown that the first order resonant separatrix (i.e., $(2n-1:1) = (1:1)$) is very strong, and it implies that the resonant-separatrix web will appear in the outer layer. This evidence is clearly shown in Fig.5.21(a). When the excitation frequency increases to $\Omega = 1.0$ (point *B*), a very strong, resonant separatrix of the first-order (i.e., $(m:1) = (1:1)$) appears in the inner layer, but the resonant separatrix of the third-order (i.e., $(2n-1:1) = (3:1)$) becomes very weak in the outer layer. Therefore, the resonant-separatrix of the first order exists in the inner layer, as illustrated in Fig.5.21(b). At point C ($\Omega = 1.5$), the

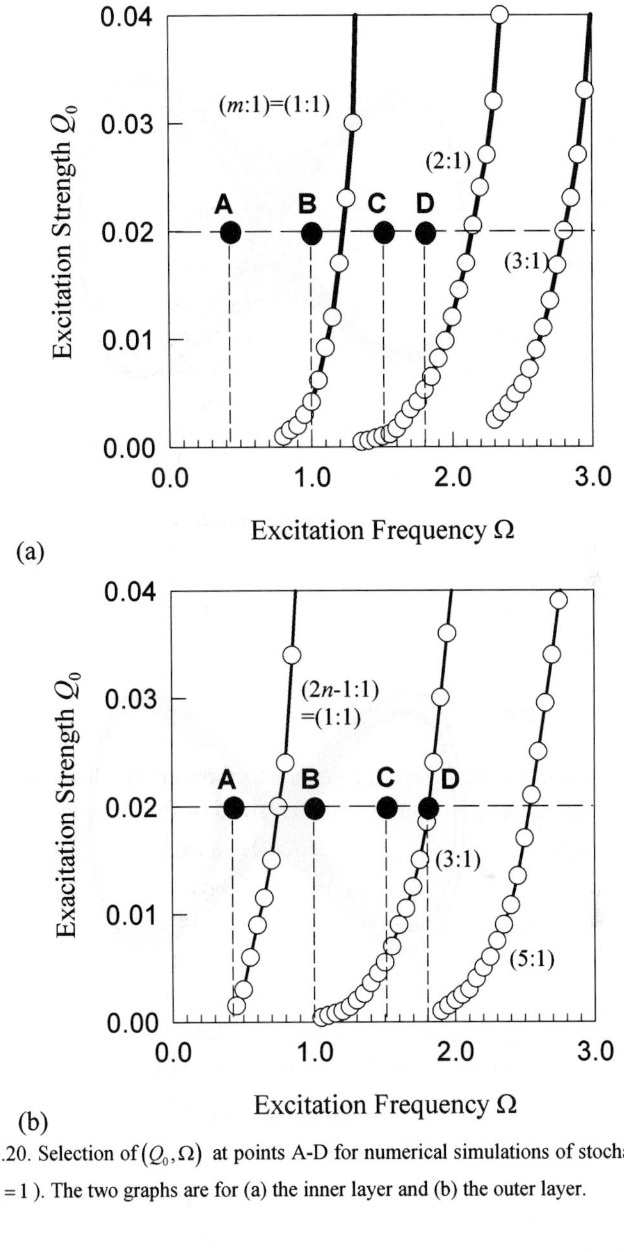

Figure 5.20. Selection of (Q_0, Ω) at points A-D for numerical simulations of stochastic layers. ($\alpha_1 = \alpha_2 = 1$). The two graphs are for (a) the inner layer and (b) the outer layer.

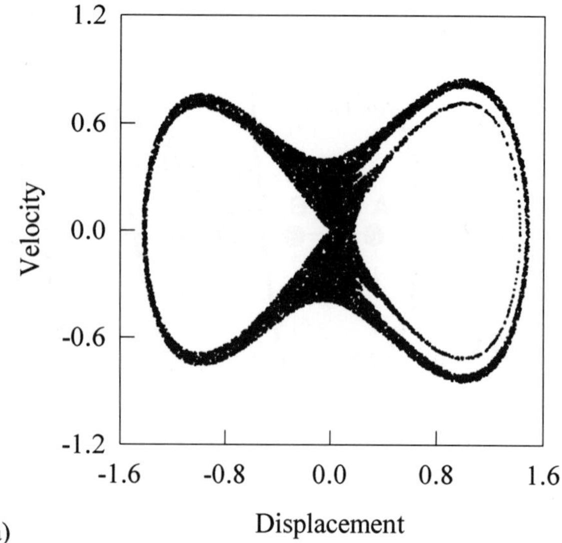

(a)

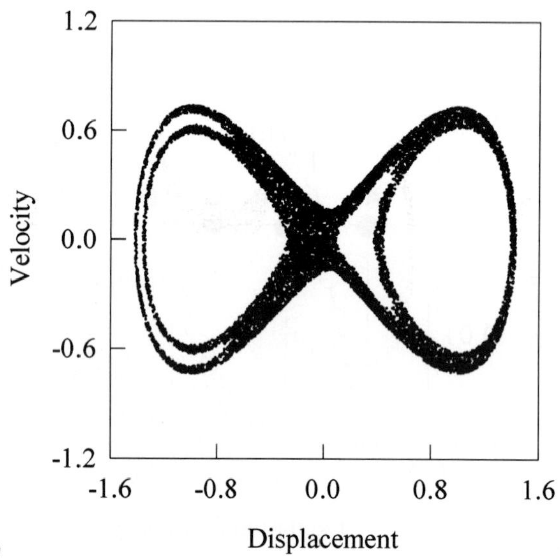

(b)

Figure 5.21. Poincare mapping sections of the stochastic layers in the Duffing oscillator ($\alpha_1 = \alpha_2 = 1.0$) at $Q_0 = 0.01$: (a) Point A $\Omega = 0.5$, (b) Point B $\Omega = 1.0$, (c) Point C $\Omega = 1.5$ and (d) Point D $\Omega = 1.8$.

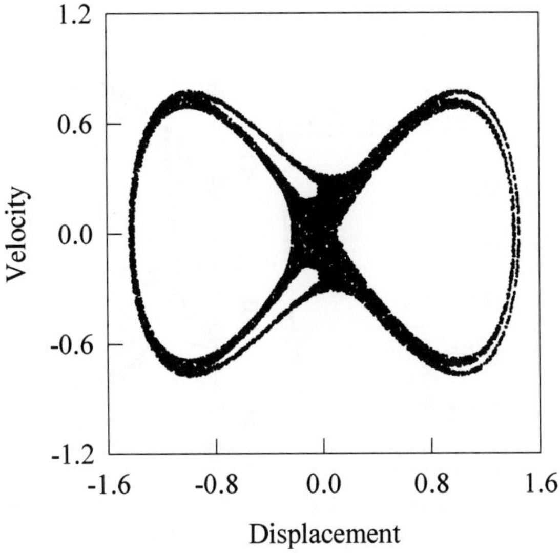

(c)

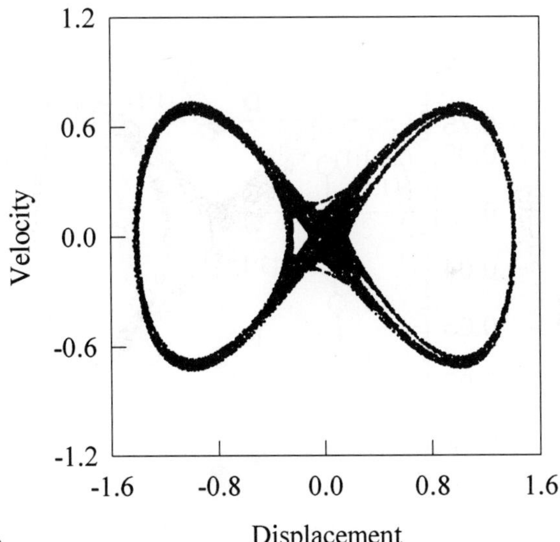

(d)

Figure 5.21 (continued)

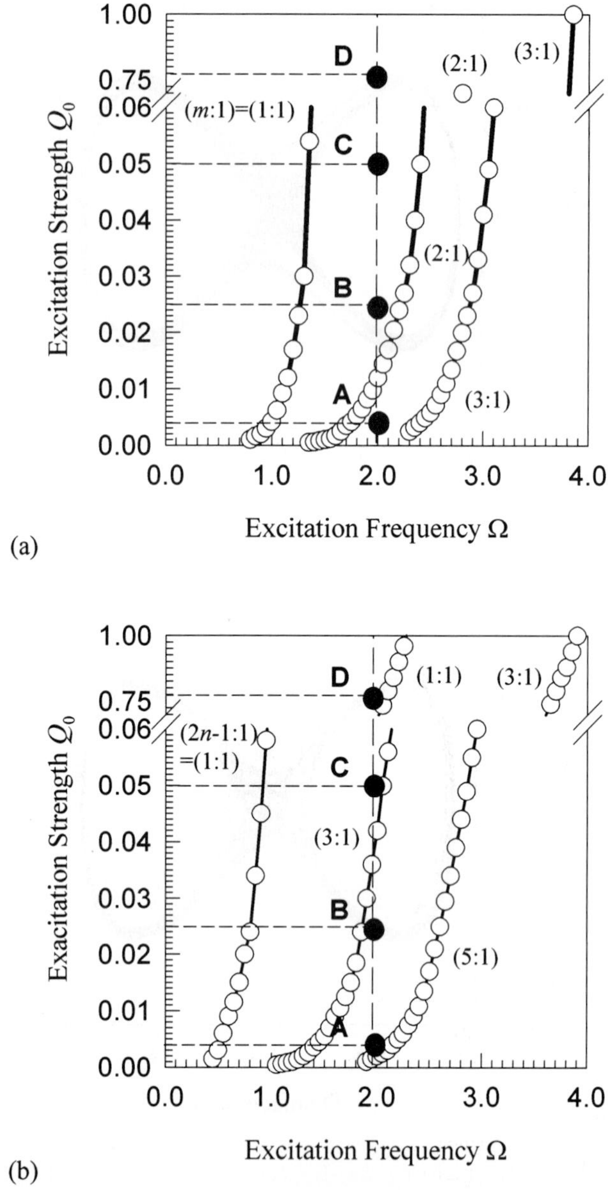

Figure 5.22. Selections of input parameters for numerical simulations of stochastic layers. ($\Omega = 2.0$, $\alpha_1 = \alpha_2 = 1$). The two graphs are for (a) the inner layer and (b) the outer layer.

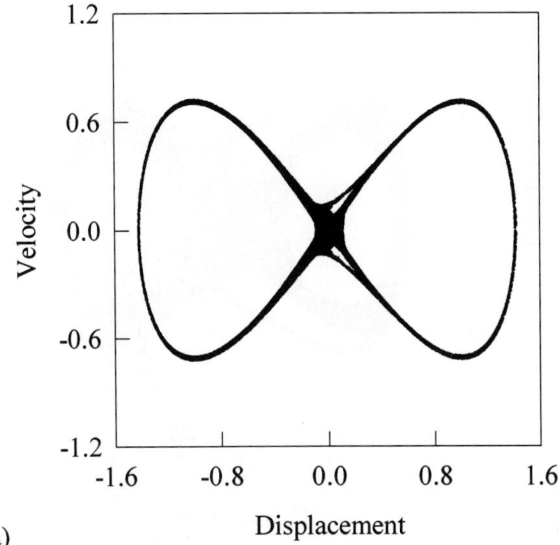

(a)

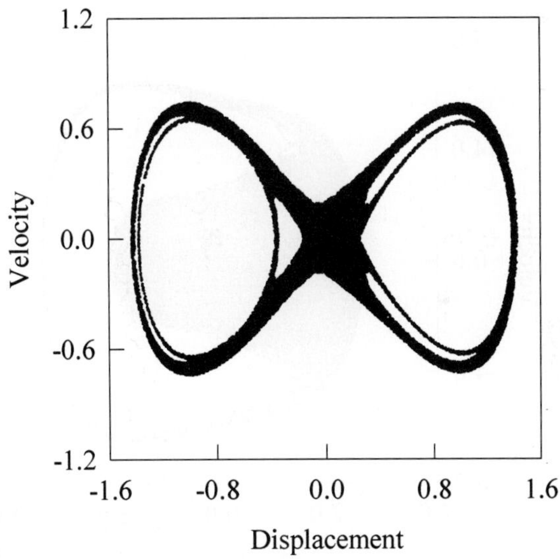

(b)

Figure 5.23. Poincare mapping sections of stochastic layers at $\Omega = 2.0$, (a) Point A ($Q_0 = 0.004$), (b) Point B ($Q_0 = 0.025$), (c) Point C ($Q_0 = 0.05$) and (d) Point D ($Q_0 = 0.75$).

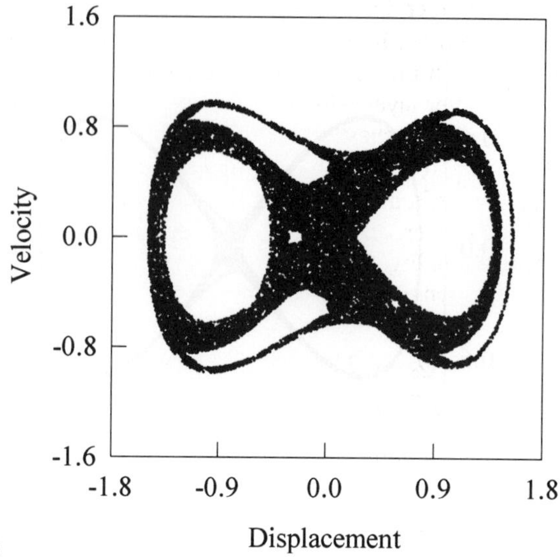

(c)

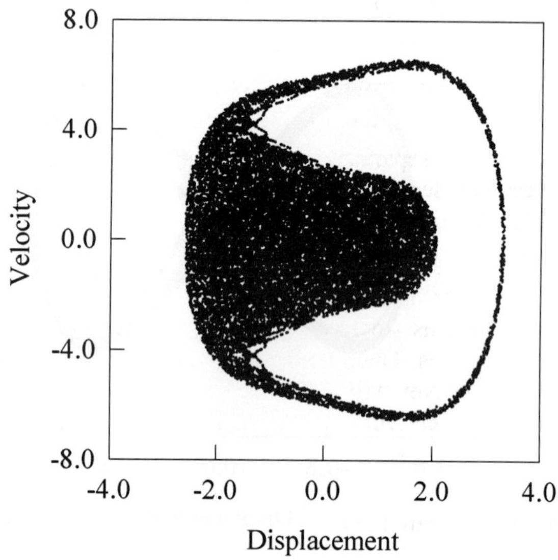

(d)

Figure 5.23 (continued)

excitation frequency further increases to $\Omega = 1.5$, and similarly, the resonant separatrix of the second order in the inner layer is weaker than the resonant separatrix of the third order in the outer layer. Therefore, the resonant separatrix of the third order in the outer layer will be clearer than the one of the second order in the inner layer. Such a stochastic layer characterized by the resonant separatrix is plotted in Fig.5.21(c). Finally, at point $D(\Omega = 1.8)$, the resonant separatrix of the second order in the inner layer is much stronger than the one of the fifth order in the outer layer, and the resonant separatrix of the second order in the inner layer is shown in Fig. 5.21(d).

Several excitation strengths in Fig.5.22 are chosen for numerical simulations. The stochastic layers varying with several excitation strengths are illustrated in Fig.5.23 at $\Omega = 2.0$: (a) $Q_0 = 0.004 > Q_0^{\beta(5:1)} \approx 0.002$, (b) $Q_0 = 0.015 > Q_0^{\alpha(2:1)} \approx 0.012$, (c) $Q_0 = 0.05 > Q_0^{\beta(3:1)} \approx 0.042$ and (d) $Q_0 = 0.75 > Q_0^{\beta(1:1)} \approx 0.69$. Note that $Q_0^{\beta(5:1)}$ denotes the excitation strength for the onset of the resonant separatrix of the fifth order (i.e., $(2n-1:1) = (5:1)$) in the outer layer. If $Q_0 < Q_0^{\beta(5:1)}$, the resonant separatrix of the fifth order will not appear in the stochastic layer. Thus, the resonant separatrix of the fifth order is observed in the outer stochastic layer in Fig.5.23(a) because $Q_0 > Q_0^{\beta(5:1)}$. For $Q_0 > Q_0^{\alpha(2:1)}$, the resonant separatrix of the second order (i.e., $(m:1) = (2:1)$) is absorbed in the stochastic layer, and the width of the stochastic layer becomes thicker, as shown in Fig.5.23(b). With increasing Q_0 and over $Q_0^{\beta(3:1)}$, the width of the stochastic layer becomes thicker and thicker. The resonant separatrix of the second order in the inner layer is destroyed and the resonant separatrix of the third order in the outer is involved in the stochastic layer, as in Fig.5.23(c). For $Q_0 > Q_0^{\beta(1:1)}$, all the possible primary inner resonant separatrices are destroyed, and the resonant separatrix of the first-order in the outer layer is finally in the stochastic layer, as illustrated in Fig.5.23(d). Therefore, this stochastic layer includes all the primary, inner and outer resonant separatrices. Once the resonant separatrix of the first order disappears, the stochastic layer will not exist any longer. The stochastic layers in the pendulum and Mathieu-Duffing oscillators can be in Luo and Han (2000b) and Luo (2001c).

5.4. Conclusions and discussions

In this chapter, the stochastic layer in 2-dimensional perturbed nonlinear Hamiltonian systems was comprehensively presented for a better understanding of the resonant mechanism of chaotic layers in $2n$-dimensional nonlinear Hamiltonian systems. This criterion presented in this chapter is applicable to any periodically forced, 2-dimensional nonlinear Hamiltonian system. Since an exact first integral increment (or energy increment) is very difficult to compute analytically, the approximate expressions of the energy increment is derived for analytical prediction of the stochastic layer. Based on the approximate energy increment, several analytical criteria were developed for the onset of resonance in the stochastic layer. In the energy spectrum method, the minimum and maximum energies are equal to exactly computing minimum and maximum energy increments because the initial energy is given. That is,

$$\begin{aligned} E_{\max} &= \max_{N\in\mathbb{N}} H_0^{(N)}(\mathbf{x}_N,\boldsymbol{\mu}) = \max_{N\in\mathbb{N}}\left\{\Delta H_0(t_0,NT)+H_0(\mathbf{x}_0,\boldsymbol{\mu})\right\} \\ &= \max_{N\in\mathbb{N}}\left\{\Delta H_0(t_0,kT)\right\}+H_0(\mathbf{x}_0,\boldsymbol{\mu}), \end{aligned} \tag{5.79}$$

$$\begin{aligned} E_{\min} &= \min_{N\in\mathbb{N}} H_0^{(N)}(\mathbf{x}_N,\boldsymbol{\mu}) = \min_{N\in\mathbb{N}}\left\{\Delta H_0(t_0,NT)+H_0(\mathbf{x}_0,\boldsymbol{\mu})\right\} \\ &= \min_{N\in\mathbb{N}}\left\{\Delta H_0(t_0,kT)\right\}+H_0(\mathbf{x}_0,\boldsymbol{\mu}). \end{aligned} \tag{5.80}$$

If $H_0(\mathbf{x}_0,\boldsymbol{\mu})=0$, we have

$$E_{\max} = \max_{N\in\mathbb{N}}\left\{\Delta H_0(t_0,kT)\right\} \text{ and } E_{\min} = \min_{N\in\mathbb{N}}\left\{\Delta H_0(t_0,kT)\right\}. \tag{5.81}$$

Based on the maximum and minimum energies, the width of the stochastic layer was computed, which is much better than the estimate of the layer width, given by Melnikov (1963).

The onset of a specified resonance in stochastic layer is exactly predicted by the incremental energy approach if the energy increment is exactly computed. Therefore, such an approach is also applicable for strong excitations when the energy increments still maintain in good accuracy. The accurate standard mapping approach gives a very good prediction compared to the numerical predicttion for the resonant separatrix with the lower order. The approximate standard mapping method was also presented in this chapter. When the excitation is very weak, both of the two standard-map methods are in good agreement. The accurate standard mapping approach is still applicable to one degree of freedom nonlinear Hamiltonian systems with strong, periodical excitations, but the further improvement should be completed for a more accurate, analytical prediction of the onset of a new resonant-separatrix in stochastic layers. For more accurate, analytical predictions of the resonance interaction in the stochastic layer, a new analytical expression for the incremental energy should be developed.

It is observed that the chaotic motion in the stochastic layer is formed by the primary resonance interaction. The sub-resonance in the vicinity of the primary resonance is developed through a renormalization. Therefore, from this chapter, this book asserts that chaotic motions in the stochastic layer of nonlinear Hamiltonian is the resonance inter-action structure instead of the Smale's horseshoe structure. The Smale's horseshoe structure is based on the linearization of the hyperbolic point along the separatrix for one to imagine the topological structure. In fact, chaotic motion in the stochastic layer exists in the vicinity of the separatrix rather than the separatrix only. The difference between the Smale's horseshoe structure and the resonance interaction structure is similar to the one between the resonance and beat phenomena in periodically forced linear oscillation. Such a comparison may not be proper but it provides a physical sense for one to think.

Chapter 6

Stochasticity in Resonant Separatrix Layers

For a periodically forced, nonlinear Hamiltonian system, once the external frequency satisfies a primary resonance condition, the primary resonant separatrix will be formed. As in Chapter 5, for an unperturbed nonlinear Hamiltonian system possessing hyperbolic points, the generic separatrix connecting the hyperbolic points has zero frequency. From the stochastic layer dynamics of the corresponding periodically forced, Hamiltonian system, there are infinite hyperbolic and parabolic points of the resonant separatrix on the generic separatrix. However, for an unperturbed nonlinear Hamiltonian system without any hyperbolic point, the corresponding periodically forced Hamiltonian system only possesses the resonant separatrix layer. If a sub-resonance appears in the vicinity of the primary resonant separatrix, the resonant separatrix webs will be formed in such vicinity. For a better understanding of the dynamics of chaotic motions in the vicinity of the resonant separatrix, in this chapter, a theory of resonant separatrix layers in nonlinear Hamiltonian systems will be presented. A mathematical description of the resonant separatrix layer will be given, and the mechanism for the appearance, formation and destruction of a resonant separatrix layer will be discussed. Analytical techniques for approximate predictions of the onset and destruction of a resonant separatrix layer will be presented. Two examples will be given to show how to predict the resonant separatrix layer in periodically forced nonlinear Hamiltonian systems.

6.1. Two-dimensional resonant separatrix layers

Consider a two-dimensional Hamiltonian system as

$$\dot{\mathbf{x}} = \mathbf{f}(\mathbf{x},\boldsymbol{\mu}) + \mathbf{g}(\mathbf{x},t,\boldsymbol{\pi}); \qquad \mathbf{x} = (x,y)^T \in \mathbb{R}^2, \tag{6.1}$$

where $\mathbf{f}(\mathbf{x},\boldsymbol{\mu})$ is a Hamiltonian vector field defined on $\mathbb{R}^2$ and $\mathbf{g}(\mathbf{x},t,\boldsymbol{\pi})$ is a T $2\pi/\Omega$-periodic (fixed period) Hamiltonian vector field in time t, and Ω denotes excitation frequency. Specifically, they are of the form

$$\mathbf{f}(\mathbf{x},\boldsymbol{\mu}) = (f_1(\mathbf{x},\boldsymbol{\mu}), f_2(\mathbf{x},\boldsymbol{\mu}))^T \text{ and } \mathbf{g}(\mathbf{x},t,\boldsymbol{\pi}) = (g_1(\mathbf{x},t,\boldsymbol{\pi}), g_2(\mathbf{x},t,\boldsymbol{\pi}))^T \tag{6.2}$$

and are assumed to be sufficiently smooth ($C^r, r \geq 2$) and bounded on bounded sets $D \subset \mathbb{R}^2$ in phase space. The total energy of such a system is,

$$H(x,y,t,\mathbf{p}) = H_0(x,y,\boldsymbol{\mu}) + H_1(x,y,t,\boldsymbol{\pi}), \tag{6.3}$$

where $H_0(x,y,\boldsymbol{\mu})$ and $H_1(x,y,t,\boldsymbol{\pi})$ are energy functions of the conservative and perturbed Hamiltonians), respectively.

To restrict the discussion on the two-dimensional resonant layer (or band) in perturbed nonlinear Hamiltonian system, the following hypothesis will be used.

(H1a) There is a bounded open domain $D \subset \mathbb{R}^2$ and in such a domain, there is only one center equilibrium $p_c : (x_c, y_c)$ around which a family of periodic flows $q_\alpha(t) = (x_\alpha(t), y_\alpha(t))$ ($\alpha \in [1,\infty)$) of the unperturbed Hamiltonian exists.

(H1b) There is an open domain $D \subset \mathbb{R}^2$ bounded by a separatrix (i.e., $q_0(t) = (x_0(t), y_0(t)) : \mathbf{x}_0 \in q_0(t)$) with hyperbolic points, and then in such a domain there is a center equilibrium p_c: (x_c, y_c) around which a family of periodic flows (i.e., $\mathbf{x}_\alpha \in q_\alpha(t)$ for $\alpha \in (0,1]$) of unperturbed Hamiltonian with $\limsup_{\alpha\to 0} \sup_{t\in R} \inf_{\mathbf{x}_0 \in q_0(t), \mathbf{x} \in q_\alpha(t)} \| \mathbf{x}_\alpha(t) - \mathbf{x}_0(t) \| = 0$ exists.

(H1c) There is an open domain $D \subset \mathbb{R}^2$ bounded by an internal boundary formed by a separatrix (i.e., $\mathbf{x}_0 \in q_0(t)$) with hyperbolic points. On the outside of the separatrix, a family of periodic flows (i.e., $\mathbf{x}_\alpha \in q_\alpha(t)$ for $\alpha \in (0,\infty)$) of the unperturbed Hamiltonian exists. $\limsup_{\alpha\to 0} \sup_{t\in R} \inf_{\mathbf{x}_0 \in q_0(t), \mathbf{x} \in q(t)} \| \mathbf{x}_\alpha - \mathbf{x}_0 \| = 0$.

(H2) $H_0(q(t)) = E_\alpha$ and T_α is the period of $q_\alpha(t)$ and its frequency ω_α is greater than zero (i.e. $\omega_\alpha > 0$). The frequency is a differentiable function of E_α (i.e., $d\omega_\alpha / dE_\alpha \neq 0$). Namely, $d\omega_\alpha / dE_\alpha > 0$, $d\omega_\alpha / dE_\alpha < 0$ and $d\omega_\alpha / dE_\alpha > 0$ are for cases in (H1a), (H1b) and (H1c), respectively.

From the above Hypotheses, the periodic flows of the unperturbed Hamiltonian system exist in three types of domains in 2-dimensional phase space. The periodic flows in a domain possessing only one center equilibrium are depicted in Fig.6.1. The dashed curve is an open boundary to the infinity for the given bounded domain in Fig.6.1(a). The periodic flow will be around the center point. A family of periodic flows is presented in Fig.6.1(b). The natural frequency at the center point is minimum. With increasing energy, the frequency will increase (i.e., $d\omega_\alpha / dE_\alpha > 0$) and the flow period will decrease (i.e., $dT_\alpha / dE_\alpha < 0$). For linear Hamiltonian systems, the natural frequency is constant. One likes to use the linear frequency for the resonance analysis, which may not be adequate. So, the nonlinear natural frequency is very important. Otherwise, such traditional resonant analysis is just valid for motion in the small neighborhood of center point. For degenerated case, the linear frequency is zero, one tries to investigate dynamical behavior through the nonlinear terms. In fact, the nonlinear frequency can resolve such difficulty directly. The second type of domain is bounded by the separatrix, as shown in Fig.6.2(a). In this domain, there is a center point. The separatrix is sketched by a dashed curve. All the periodic flows in this domain will be formed around the center point in Fig.6.2(b). The natural frequency at the center point is maximum. With increasing energy, the frequency will decrease (i.e., $d\omega_\alpha / dE_\alpha < 0$) or the flow period will increase (i.e., $dT_\alpha / dE_\alpha > 0$). When a periodic flow in the family of periodic flows in such a domain approach the separatrix, the natural frequency will approach zero (i.e., $\lim_{E_\alpha \to E_0} \omega_\alpha = 0$) or the corresponding period approaches infinity (i.e., $\lim_{E_\alpha \to E_0} T_\alpha = \infty$). In the third case, the domain is on the outside of separatrix, and the separatrix will be embedded in the periodic flows, as shown in Fig.6.3. For this case, the natural frequency at the separatrix is minimum ($\omega_\alpha = 0$). In addition, with increasing energy, the frequency will increase (i.e., $d\omega_\alpha / dE_\alpha > 0$) and the flow period will decrease (i.e., $dT_\alpha / dE_\alpha < 0$). For this case, if the separatrix is formed by the heteroclinic separatrices instead of a homoclinic separatrix, the rotation flows will exist. The frequency characteristics are as same as in discussed. The rotation flow in such a domain is presented in Fig.6.4, and frequency feature is the same as in Fig.6.3. The librational flow in the potential well is similar to the second case in Fig.6.2.

6.1.1. Layer dynamics

To investigate chaotic motions in a resonant separatrix layer, for a given energy E_α, we have,

$$H_0(x_\alpha, y_\alpha) = E_\alpha \tag{6.4}$$

from which

$$y_\alpha = y_\alpha(x_\alpha, E_\alpha). \tag{6.5}$$

The action variable is defined by

$$J_\alpha = \oint y_\alpha dx_\alpha. \tag{6.6}$$

So we have

$$H_0(J_\alpha) = E_\alpha. \tag{6.7}$$

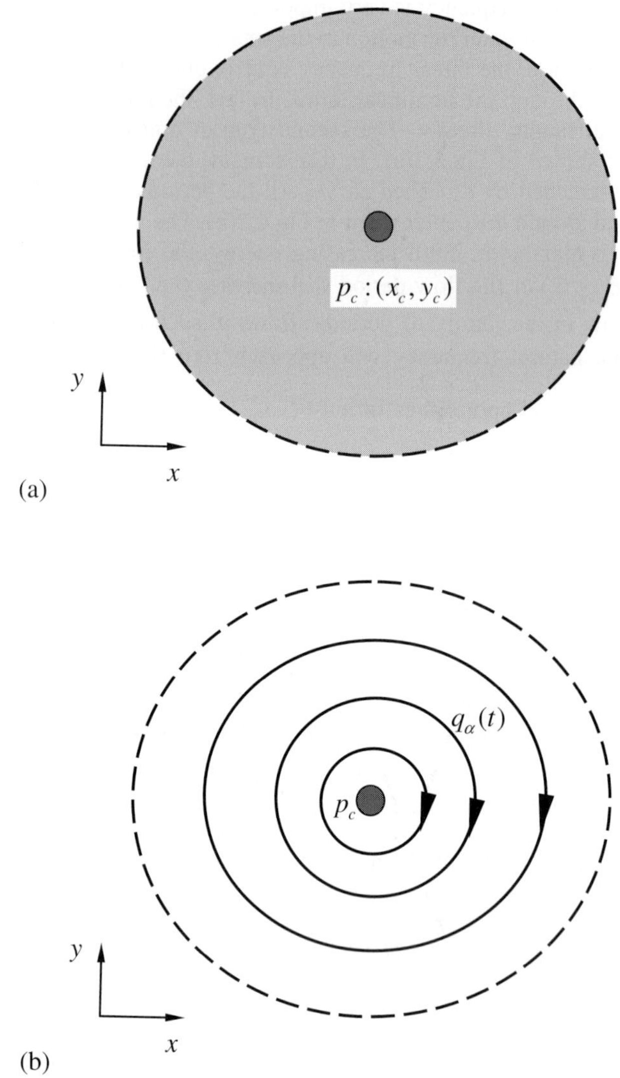

Figure 6.1. (a) An open domain $D \subset \mathbb{R}^2$ with only one center point and (b) periodic flow of the unperturbed system without separatrix. $q_\alpha(t)$ is the periodic flow of the unperturbed system in domain $D \subset \mathbb{R}^2$.

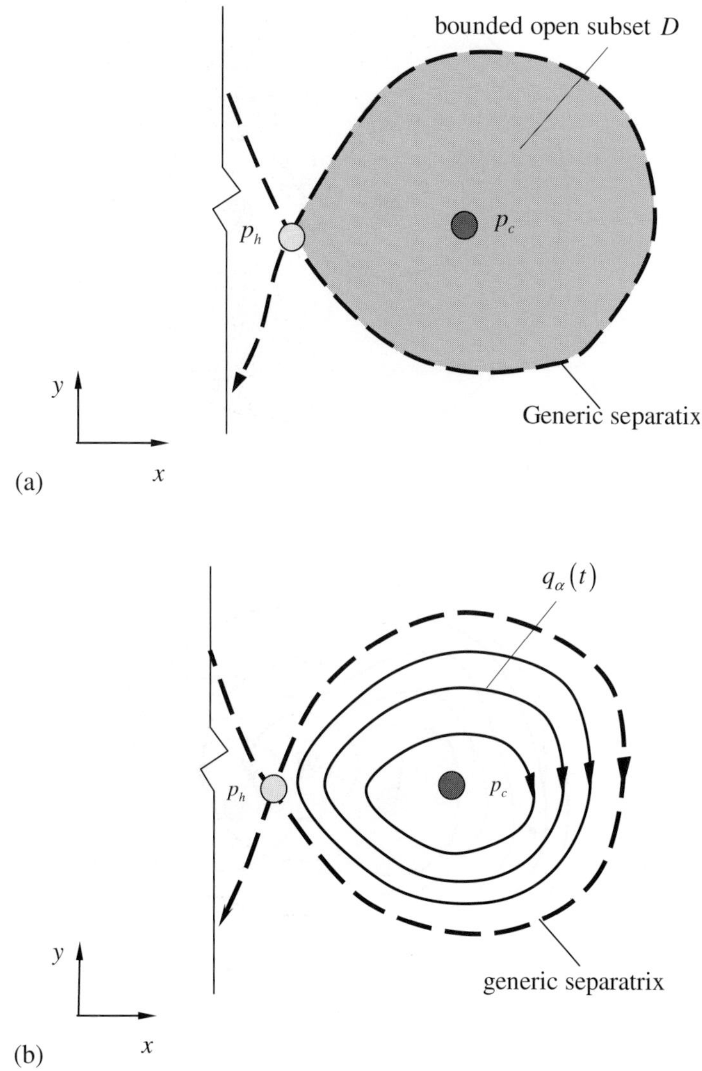

Figure 6.2. (a) An open domain D bounded by the separatrix and (b) periodic flow of the unperturbed system of *Eq.* (6.1). $q_\alpha(t)$ is the periodic flow of the unperturbed system in domain $D \subset \mathbb{R}^2$. This periodic flow existing on the inside of separatrix can be called the librational (or local) periodic flows.

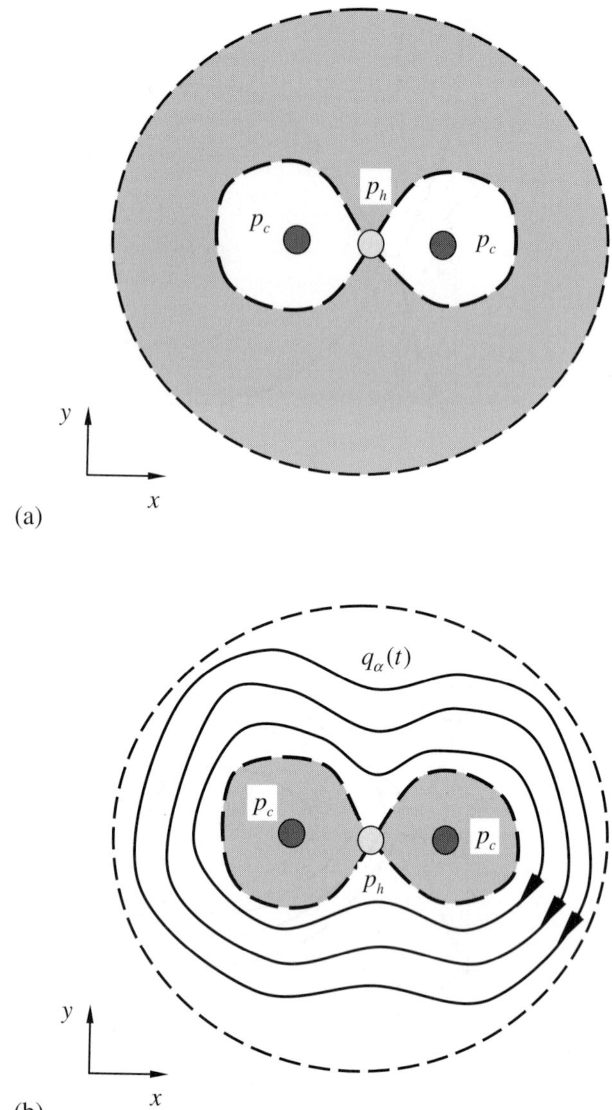

Figure 6.3. (a) An open domain D existing on the outside of the closed separatrix and (b) periodic orbits of the unperturbed system of *Eq.* (6.1) on the outside of the separatrix. $q_\alpha(t)$ is the periodic flow of unperturbed system in domain $D \subset \mathbb{R}^2$. This flow is a kind of rotation (or global) flows.

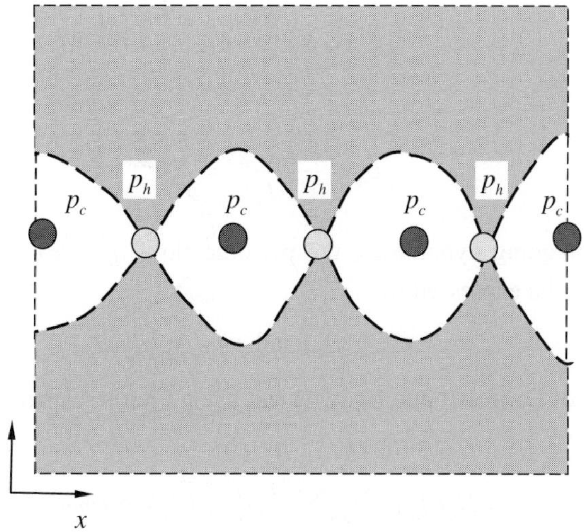

(a)

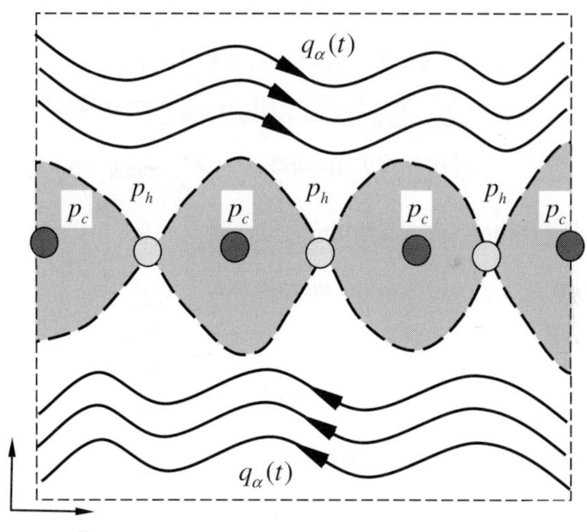

(b)

Figure 6.4. (a) an open domain D divided into two parts by the separatrix and (b) periodic orbits of the unperturbed system of *Eq.*(6.1) on the outside of the separatrix. $q_\alpha(t)$ is the periodic orbit in domain $D \subset \mathbb{R}^2$, which exists in two parts of the domain. This flow is a rotation (or global) flow.

The angle variable is defined by

$$\theta_\alpha = \omega_\alpha t + \theta_{\alpha 0}, \tag{6.8}$$

with

$$\dot{\theta}_\alpha = \frac{\partial H_0(J_\alpha)}{\partial J_\alpha} = \omega_\alpha. \tag{6.9}$$

From the foregoing hypotheses, the periodic flow $q_\alpha(t) = (x_\alpha(t), y_\alpha(t))$ in the domain D can be expressed by

$$x_\alpha = x_\alpha(J_\alpha, \theta_\alpha) \text{ and } y_\alpha = y_\alpha(J_\alpha, \theta_\alpha). \tag{6.10}$$

Substitution of Eq.(6.10) into Eq.(6.2) and using Fourier expansion gives

$$\begin{aligned} H\left(J_\alpha, \theta_\alpha, t\right) &= H_0(J_\alpha) + H_1(J_\alpha, \theta_\alpha, \Omega t) \\ &= H_0(J_\alpha) + \sum_n \sum_m \Big\{ H_{1(m:n)}^{(-)}(J_\alpha)\cos(m\omega_\alpha t - n\Omega t + \psi_{\alpha n}^{(-)}) \\ &\quad + H_{1(m:n)}^{(+)}(J_\alpha)\cos(m\omega_\alpha t + n\Omega t + \psi_{\alpha n}^{(+)}) \Big\}. \end{aligned} \tag{6.11}$$

Letting $\varphi_\alpha = (m\omega_\alpha - n\Omega)t$, we have

$$\begin{aligned} H\left(J_\alpha, \theta_\alpha, t\right) &= H_0(J_\alpha) + H_1(J_\alpha, \theta_\alpha, \Omega t) = H_0(J_\alpha) \\ &\quad + \sum_n \sum_{m_1} \Big\{ H_{1(m_1:n)}^{(-)}(J_\alpha)\cos(\tfrac{m_1}{m}\varphi_\alpha + \tfrac{n(m_1 - m)}{m}\Omega t + \psi_{\alpha n}^{(-)}) \\ &\quad + H_{1(m_1:n)}^{(+)}(J_\alpha)\cos(\tfrac{m_1}{m}\varphi_\alpha + \tfrac{n(m_1 + m)}{m}\Omega t + \psi_{\alpha n}^{(+)}) \Big\}. \end{aligned} \tag{6.12}$$

If the following resonance condition holds as

$$m\omega_\alpha = n\Omega, \tag{6.13}$$

then we have

$$J_\alpha = J_\alpha^{(m:n)} \quad \text{and} \quad E_\alpha = E_\alpha^{(m:n)}. \tag{6.14}$$

To define a generating function, we have

$$\begin{aligned} H_0\left(J_\alpha, \theta_\alpha, t\right) + \tfrac{\partial G}{\partial t} &= H_0(J_\alpha^{(m:n)}) + \left.\frac{\partial H_0(J_\alpha)}{\partial J_\alpha}\right|_{J_\alpha = J_\alpha^{(m:n)}} (J_\alpha - J_\alpha^{(m:n)}) + \frac{\partial G}{\partial t} \\ &\quad + \frac{1}{2}\left.\frac{\partial^2 H_0(J_\alpha)}{\partial J_\alpha^2}\right|_{J_\alpha = J_\alpha^{(m:n)}} (J_\alpha - J_\alpha^{(m:n)})^2 + \cdots + h.o.t. \end{aligned} \tag{6.15a}$$

$$H_1\left(J_\alpha,\theta_\alpha,t\right)=\sum_n\sum_{m_1}\left\{H_{1(m_1:n)}^{(-)}(J_\alpha)\cos\left(\tfrac{m_1}{m}\varphi_\alpha+\tfrac{n(m_1-m)}{m}\Omega t+\psi_{\alpha n}^{(-)}\right)\right.$$
$$\left.+H_{1(m_1:n)}^{(+)}(J_\alpha)\cos\left(\tfrac{m_1}{m}\varphi_\alpha+\tfrac{n(m_1+m)}{m}\Omega t+\psi_{\alpha n}^{(+)}\right)\right\}. \tag{6.15b}$$

Letting

$$\frac{\partial H_0(J_\alpha)}{\partial J_\alpha}(J_\alpha-J_\alpha^{(m:n)})=-\frac{\partial G}{\partial t} \tag{6.16}$$

the generating function can be defined as

$$G=-\omega_\alpha t(J_\alpha-J_\alpha^{(m:n)})=-\frac{(\varphi_\alpha+n\Omega t)}{m}(J_\alpha-J_\alpha^{(m:n)}). \tag{6.17}$$

Furthermore, we have a new coordinate $(\overline{p}_\alpha,\varphi_\alpha)$

$$\overline{p}_\alpha=-\frac{\partial G}{\partial \varphi_\alpha}=\frac{1}{m}(J_\alpha-J_\alpha^{(m:n)})\text{ and }\theta_\alpha=\omega_\alpha t=-\frac{\partial G}{\partial J_\alpha}=\frac{1}{m}(\varphi_\alpha+n\Omega t). \tag{6.18}$$

If $\overline{p}_\alpha=0$, we have $J_\alpha=J_\alpha^{(m:n)}$. The variable $\overline{p}_\alpha$ gives the difference between the instant J_α and the resonance $J_\alpha^{(m:n)}$, which defines the gap of the resonant separatrix layer. Because the natural frequency ω_α is a function of energy E_α or the action variable J_α, the relation between the resonance frequency and energy is illustrated in Fig.6.5 from the resonance relation in Eq.(6.12). The resonant frequency distributions along the conservative energy are different for four resonant layers. Further, their resonant structures are distinguishing from each other. The specific (m:n)-resonant frequency and natural frequency are expressed by $\Omega^{(m:n)}$ and $\omega_\alpha^{(m:n)}$, respectively. The corresponding resonant condition in Eq.(6.12) becomes,

$$m\omega_\alpha^{(m:n)}=n\Omega^{(m:n)}. \tag{6.19}$$

To guarantee the resonant layers in a certain gap, consider a neighborhood of a natural frequency ω_α close to $\omega_\alpha^{(m:n)}$. That is,

$$\left|\omega_\alpha-\omega_\alpha^{(m:n)}\right|=\left|\omega_\alpha-\frac{n}{m}\Omega^{(m:n)}\right|\le\varepsilon\text{ and}$$
$$T\left(E_\alpha^{(m:n)}\right)=\frac{2\pi}{\omega_\alpha^{(m:n)}},\ T\left(E_\alpha\right)=\frac{2\pi}{\omega_\alpha}; \tag{6.20}$$

where $\varepsilon<<1$ and $T\left(E\right)$ is the nonlinear period. From the foregoing condition, equation (6.15) becomes

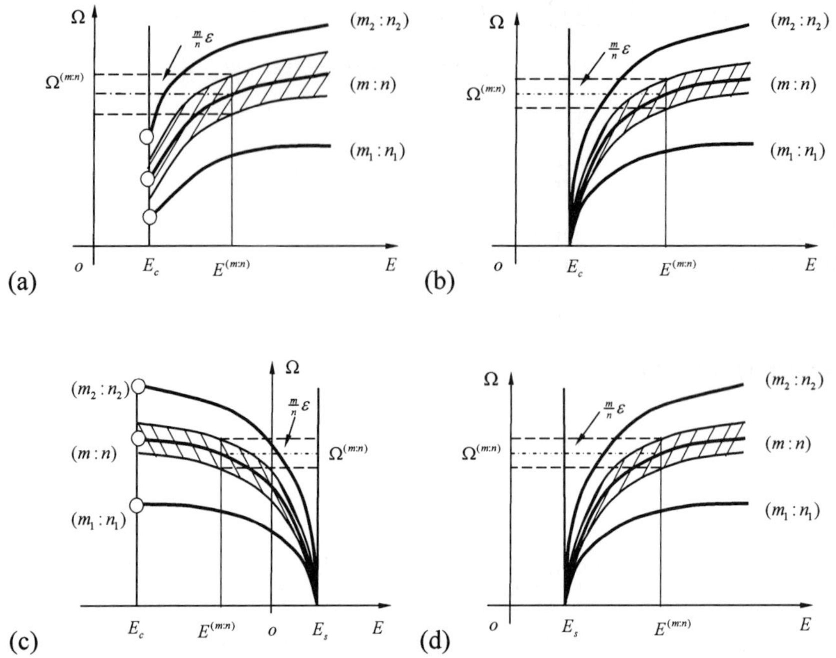

Figure 6.5. The neighborhood of the $(m:n)$-resonant frequency: (a) one center without separatrix, (b) degenerated center without separatrix, (c) inside separatrix, and (d) outside separatrix.

$$
\begin{aligned}
H(J_\alpha,\theta_\alpha,t)+\frac{\partial G}{\partial t} &\approx E_\alpha^{(m:n)}+\frac{1}{2}\frac{\partial^2 H_0(J_\alpha)}{\partial J_\alpha^2}\bigg|_{J_\alpha=J_\alpha^{(m:n)}} m^2\overline{p}_\alpha^2+ \\
&+\sum_n\sum_{m_1}\Big\{H^{(+)}_{1(m_1:n)}(J_\alpha^{(m:n)},\overline{p}_\alpha)\cos(\tfrac{m_1}{m}\varphi_\alpha+\tfrac{n(m_1-m)}{m}\Omega t+\psi^{(+)}_{\alpha n}) \\
&+H^{(-)}_{1(m_1:n)}(J_\alpha^{(m:n)},\overline{p}_\alpha)\cos(\tfrac{m_1}{m}\varphi_\alpha+\tfrac{n(m_1-m)}{m}\Omega t+\psi^{(-)}_{\alpha n})\Big\}
\end{aligned}
\tag{6.21}
$$

Rescaling gives the following variables as

$$
\begin{aligned}
&\overline{H}(J_\alpha,\theta_\alpha,t)=H(J_\alpha,\theta_\alpha,t)+\tfrac{\partial G}{\partial t}-E_\alpha, B=\tfrac{\partial^2 H_0(J_\alpha)}{\partial J_\alpha^2}\Big|_{J_\alpha=J_\alpha^{(m:n)}} \\
&mB\overline{p}_\alpha=p_\alpha,\ H(p_\alpha,\varphi_\alpha,t)=B\overline{H}(J_\alpha,\theta_\alpha,t),\ \Omega_1=2n\Omega \\
&U^{(+)}_{(m:n)}=BH^{(+)}_{1(m:n)}\left(J_\alpha^{(m:n)},p_\alpha\right),U^{(-)}_{(m:n)}=BH^{(-)}_{1(m:n)}\left(J_\alpha^{(m:n)},p_\alpha\right)
\end{aligned}
\tag{6.22}
$$

The new Hamiltonian becomes

$$
\begin{aligned}
H(p_\alpha,\varphi_\alpha,t) \approx & \frac{1}{2}p_\alpha^2 + U_{1(m:n)}^{(-)}\cos(\varphi_\alpha+\psi_{\alpha n}^{(-)}) \\
& + U_{1(m:n)}^{(+)}\cos(\varphi_\alpha+2n\Omega t+\psi_{\alpha n}^{(+)}) \\
& + \sum_{n_1}\sum_{m_1\neq m}\Big\{U_{1(m_1:n_1)}^{(-)}\cos(\tfrac{m_1}{m}\varphi_\alpha+\tfrac{n_1(m_1-m)}{m}\Omega t+\psi_{\alpha n_1}^{(-)}) \\
& + U_{1(m_1:n_1)}^{(+)}\cos(\tfrac{m_1}{m}\varphi_\alpha+\tfrac{n_1(m_1-m)}{m}\Omega t+\psi_{\alpha n_1}^{(+)})\Big\}.
\end{aligned}
\tag{6.23}
$$

Because the primary resonance is relatively isolated, herein the other resonance terms in Eq.(6.23) is ignored except for the $(m:n)$-resonance, the approximate Hamiltonian is expressed by

$$
H(p_\alpha,\varphi_\alpha,t) \approx \frac{1}{2}p_\alpha^2 + U_{1(m:n)}^{(-)}\cos(\varphi_\alpha+\psi_{\alpha n}^{(-)}) + U_{1(m:n)}^{(+)}\cos(\varphi_\alpha+\Omega_1 t+\psi_{\alpha n}^{(+)}). \tag{6.24}
$$

It is assumed that the two parameters $U_{1(m:n)}^{(-)}$ and $U_{1(m:n)}^{(+)}$ are independent of p_α, and the corresponding dynamical system is

$$
\begin{aligned}
\dot{\varphi}_\alpha &= p_\alpha, \\
\dot{p}_\alpha &= -U_{1(m:n)}^{(-)}\sin(\varphi_\alpha+\psi_{\alpha n}^{(-)}) - U_{1(m:n)}^{(+)}\sin(\varphi_\alpha+\Omega_1 t+\psi_{\alpha n}^{(+)}).
\end{aligned}
\tag{6.25}
$$

This equation represents a kind of parametrically excited pendulum. The dynamics in the neighborhood of the $(m:n)$-resonant separatrix can be investigated through Eq.(6.25). The resonance effects of Eq.(6.25) give the sub-resonance for the $(m:n)$-resonance. Without the perturbation, the separatrix of such a parametric pendulum can be illustrated in Fig.6.6(a). The red and yellow circles are the center and hyperbolic points of the unperturbed pendulum in Eq.(6.25), respectively. The two thin dash curves are the inner and outer boundaries of the resonant separatrix layer. Because $U_{1(m:n)}^{(-)}$ is relative to external excitation and energy orbit, the resonant separatrix will be changed with frequency and amplitude. The sub-resonance can be obtained from the perturbed system in Eq.(6.25). On the other hand, from the differential equation, the self-similar structure may not exist. The sub-resonant structure is strongly dependent on the energy analysis of Eq.(6.25). Based on a certain sub-resonance in the $(m:n)$-resonant separatrix layer, the sub-sub-resonance can be obtained by repeating the same procedure to obtain a new equation similar to Eq.(6.25). Such a sub-resonant structure is sketched in Fig6.6(b). This renormalization procedure cannot lead to a self-similar sub-resonance structure. Indeed, the sub-resonant structure in the resonant layer is relative to the corresponding parent-resonance separatrix of Eq.(6.25), but such a sub-resonant structure cannot be generated by simply copying from its parent primary-resonance. From the foregoing discussion, the dynamics of the parametric pendulum is a key to understand the mechanism of stochasticity in a

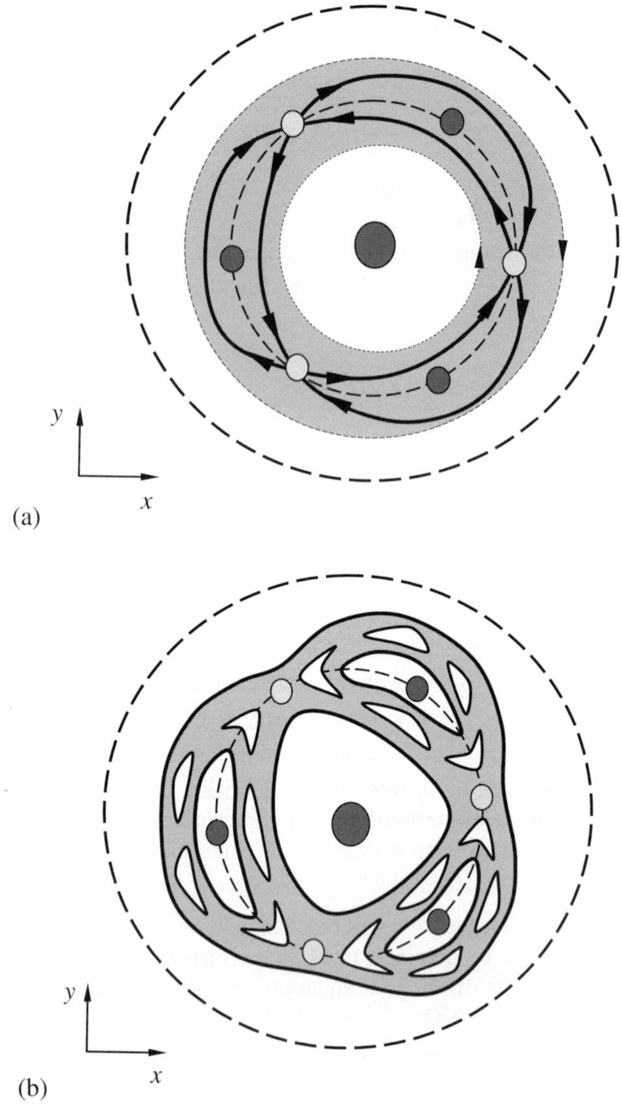

Figure 6.6. Resonant layer without a generic separatrix: (a) resonant separatrix and (b) subresonance. The solid and hollow circles represent the center and hyperbolic points. The irregular small curcles are the sub-resonances in the neighborhood of the primary resoance.

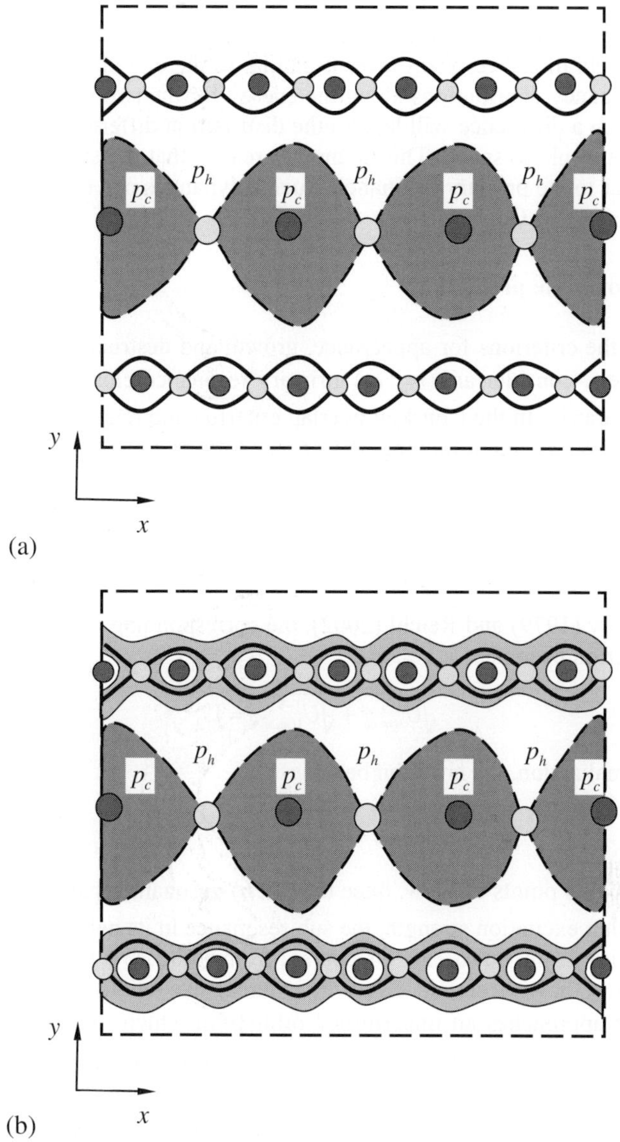

Figure 6.7. (a) resonant separatrix and (b) resonant layer in domain $D \subset \mathbb{R}^2$. The solid and hollow circles represent the center and hyperbolic points. The big circles are for generic equilbirum points, and the small circles are for the resonant equilbirum points based on the dynamics of the resonant layers.

neighborhood of the $(m:n)$-resonant separatrix. The generic separatrix, rotational resonant separatrix and corresponding rotational resonant layer are sketched in Fig.6.7. The librational resonant layer is similar to the one in Fig.6.6, but the resonance frequency distributions along the energy for the two resonant layers are different. Such a difference will lead to the distribution difference of primary resonant layers in phase space. This is another reason that the sub-resonance may not be similar to the primary resonance. Such a dynamics can be discussed by an example later in this Chapter.

6.1.2. Approximate predictions

To develop the criterions for appearance, growth and destruction of primary resonance layers, consider an $(m_1:n_1)$-primary resonance closest to the $(m:n)$-resonant separatrix. In the Chirikov overlap criterion and renormalization technique, from Eq.(6.23), the following new energy form is considered.

$$H(p_\alpha,\varphi_\alpha,t)\approx\frac{1}{2}p_\alpha^2+U_{1(m:n)}^{(-)}\cos(\varphi_\alpha+\psi_{\alpha n}^{(-)})$$
$$+U_{1(m_1:n_1)}^{(-)}\cos(\tfrac{m_1}{m}\varphi_\alpha+\tfrac{n_1(m_1-m)}{m}\Omega t+\psi_{\alpha n_1}^{(-)}). \tag{6.26}$$

As in Chirikov (1979) and Reichl (2002), the corresponding Chirikov resonance overlap criterion becomes

$$\sqrt{U_{1(m:n)}^{(-)}}+\sqrt{U_{1(m_1:n_1)}^{(-)}}=1. \tag{6.27}$$

From renormalization, the criterion becomes

$$\sqrt{U_{1(m:n)}^{(-)}}+\sqrt{U_{1(m_1:n_1)}^{(-)}}\approx 0.7. \tag{6.28}$$

From this book's points of view, once the $(m:n)$-resonant separatrix is form-ed, with increasing excitation strength, the sub-resonance in its neighborhood will be developed first from Eq.(6.25). Of course, from Eq.(6.23), the other pri-mary resonance may have a certain effect as external excitations. Until such a resonant layer approaches an unperturbed orbit from which the $(m_1:n_1)$-primary resonance can be formed, the $(m:n)$-resonant layer almost cannot be destroyed, and the effects of the $(m_1:n_1)$-primary resonance to the $(m:n)$-resonant layer are very small compared to the $U_{1(m:n)}^{(+)}$-term. In other words, before such a resonant layer is destroyed, the $(m_1:n_1)$-primary resonance will not be strongly involved in the $(m:n)$-resonant layer. Therefore, it is very doubtable that the criterions in Eq.(6.27) and (6.28), given by the Chirikov overlap criterion and the renormalization technique, can provide a reasonable prediction of the global stochasticity of

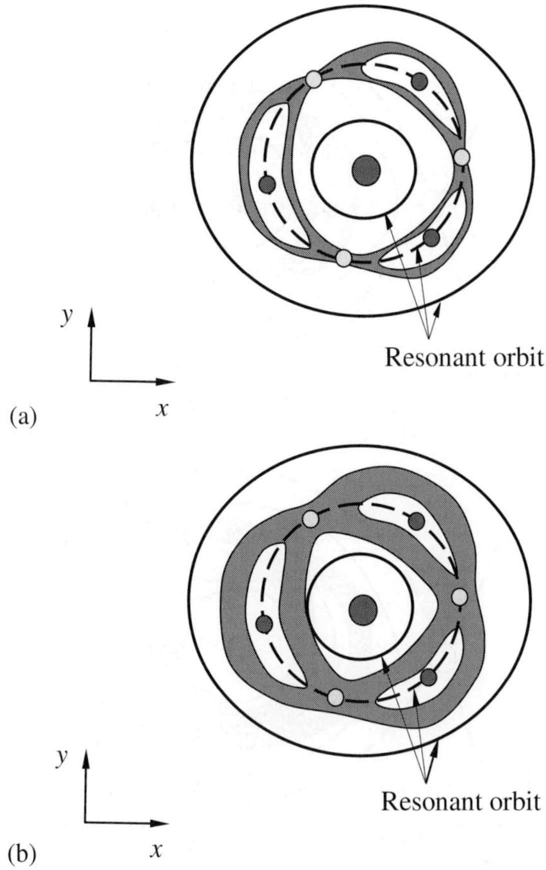

Figure 6.8. Resonant layer in nonlinear Hamiltonian system without separatrix: (a) appearance and (b) disappearance.

the resonant layer. In addition, it is not clear that the two existing criteria can be used for the appearance or disappearance of the resonant layer.

6.1.2a. Onset conditions

To discuss the appearance, growth and destruction of the $(m:n)$-resonant layer, The geometrical intuitions of both just after onset and just before destruction of the resonant layer are sketched. Two kinds of resonant layers are sketched in Figs.6.8 and 6.9. The $(m:n)$-resonant layer is formed in the neighborhood of its primary resonant separatrix. With increasing excitation strength, the width of the resonant layer will increase.

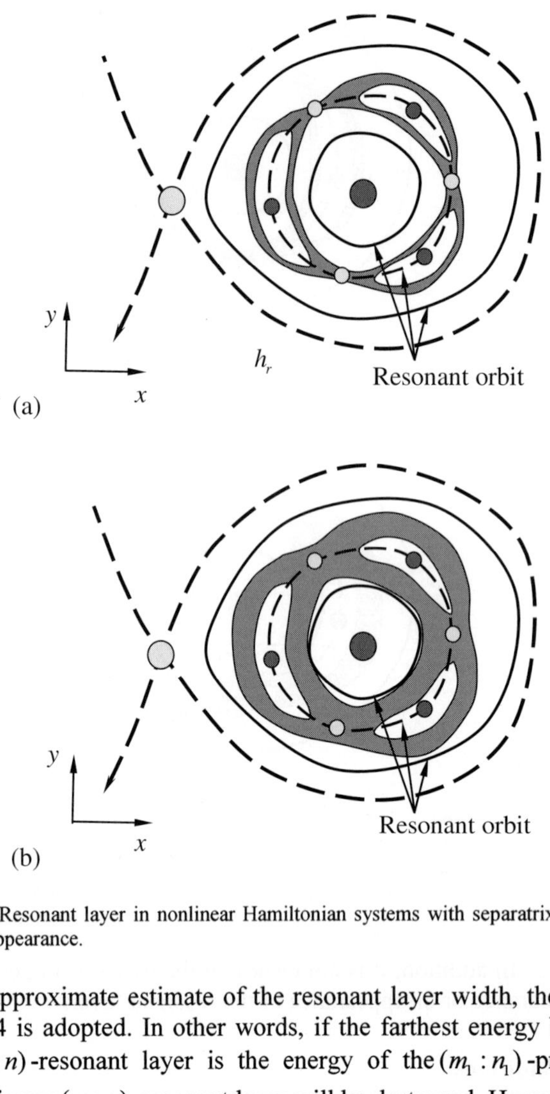

Figure 6.9. Resonant layer in nonlinear Hamiltonian systems with separatrix: (a) appearance and (b) disappearance.

For an approximate estimate of the resonant layer width, the theorem given in Chapter 4 is adopted. In other words, if the farthest energy boundary of the primary $(m:n)$-resonant layer is the energy of the $(m_1:n_1)$-primary resonant orbit, the primary $(m:n)$-resonant layer will be destroyed. However, the onset of the primary resonant layer will be estimated through the standard mapping technique. To understand the formation mechanism of the primary resonant layer, the onset of the primary $(m:n)$-resonant layer will be discussed first as follows.

From Eq.(4.105), the energy increment along the $(m:n)$-resonant orbit of the perturbed conservative system is approximated by

$$\begin{aligned}\Delta H_0 &= \int_{t_i}^{T_\alpha(E_i)+t_i} \tfrac{dH(x,y,t)}{dt} dt = \int_{t_i}^{T_\alpha(E_i)+t_i} \{H_0, H_1\}_{\text{Possion}}\, dt \\ &= \int_{t_i}^{T_\alpha(E_i)+t_i} (f_1 g_2 - f_2 g_1) dt = U_0 f(\varphi_i),\end{aligned} \tag{6.29}$$

where t_i is the initial time and $f(\phi_i)$ is a bounded and periodic function. Without loss generality, in Eq.(6.29), consider the following form,

$$\Delta H_0 = U_0 \sin\phi_i, \tag{6.30}$$

in which U_0 is a system parameter function excluding the initial phase angle $\phi_i = \Omega t_i$. For a perturbed orbit in the neighborhood of the prescribed resonant orbit, the change of phase angle over one period is

$$\Delta\varphi = \varphi_{i+1} - \varphi_i = \Omega T(E_{i+1}) = V_0(E_{i+1}), \tag{6.31}$$

where V_0 is a function associated with energy E_i. To calculate this new energy iteratively, we introduce the following notation: $E_{i+1} = w_{i+1}$ at the $(i+1)^{\text{th}}$ period and the corresponding phase angle is φ_{i+1}. Equations (6.30) and (6.31) can now be written as,

$$w_{i+1} = w_i + U_0 \sin\varphi_i \text{ and } \varphi_{i+1} = \varphi_i + V_0(w_{i+1}). \tag{6.32}$$

It is observed that Eq.(6.32) is now expressed in a form that permits the resonant separatrix layer to be investigated iteratively. Although this mapping is not based on the accurate energy increment, it is enough good as an approximate, analytical expression because the exact computation of the energy increment only can be done numerically. In the neighborhood of the resonant separatrix layer, equation (6.31) can be linearized about a fixed point and the standard mapping can be obtained. Considering a period-1 of the iterative map, its fixed point can be easily determined by $w_{i+1} = w_i = w_0$ and $\varphi_{i+1} = \varphi_i + 2m\pi/n = \varphi_0 + 2m\pi/n$. This implies,

$$U_0 \sin\varphi_0 = 0,\ V_0(w_0) = 2\pi \tfrac{m}{n}. \tag{6.33}$$

Defining a new dimensionless energy

$$I_i = \left.\frac{\partial V_0(w_{i+1})}{\partial w_{i+1}}\right|_{w_{i+1}=w_0} (w_i - w_0), \tag{6.34}$$

and linearization of Eq.(6.31) about the fixed point yields

$$I_{i+1} = I_i + \Xi \sin\varphi_i \text{ and } \varphi_{i+1} = \varphi_i + I_{i+1}, \tag{6.35}$$

where $\Xi = U_0\,\partial V_0/\partial w_{i+1}\big|_{w_{i+1}=w_0}$. From Eq.(6.35), the mechanism involved in the transition to the global stochasticity in a nonlinear Hamiltonian system is very clear. The coefficient Ξ is the only control parameter for the characterization of the KAM tori. For the standard map, a critical value of Ξ is attained when $\Xi = \Xi^* = 0.9716354\cdots$. At this value, the last remaining KAM torus is broken. When this happens, we have

$$U_0\frac{\partial V_0}{\partial w_0} = \Xi^*. \tag{6.36}$$

The transition from the local stochasticity to global stochasticity implies chaos appears in such a standard mapping. This appearance condition is as a condition for the appearance of resonant layers in the neighborhood of the $(m:n)$-resonant separatrix. For a generalized case in Eq.(6.29), it can be discussed in a similar fashion. The corresponding criteria can be developed for the global sto-chasticity of motion in the primary resonant layer.

6.1.2b. Vanishing conditions

Once this resonant separatrix layer is formed, with increasing the excitation, the other sub-resonant separatrix layers will merge in the resonant layer until they come into contact the closest resonant orbit. When this case occurs, the resonant layer will be destroyed, and a new stochastic motion near that resonant orbit will be involved in, and suddenly, the width of the resonant layer will become large. The two primary resonant layers will be overlapped each other. Based on this reason, the Chirikov resonant overlap criterion and the renormalized criterion may be used as a condition for the destruction of a certain, primary resonant layer. Such a mechanism is qualitatively sketched in Figs.6.8 and 6.9. Again it is postulated that when the resonant layer is destroyed, the energy increment in Eq.(6.29) is given by the energy difference between the two closest resonant orbits, one of which is associated with the destroyed resonant layer. From Eq.(4.86), we have

$$\min\left(\left|E_\alpha^{(m_2:n_2)} - E_\alpha^{(m:n)}\right|,\ \left|E_\alpha^{(m_1:n_1)} - E_\alpha^{(m:n)}\right|\right) = \left|\Delta H_0^{(m:n)}\right| \approx U_0\left|f(\phi_0)\right|. \tag{6.37}$$

Equation (6.37) constitutes the critical condition for the disappearance of the $(m:n)$-resonant separatrix layer. From the foregoing equation, the excitation strength for *disappearance* of the resonant layer can be computed. To determine the excitation strength for *appearance* of the resonant layer, equation (6.36) will be used. For a better prediction of resonant layers, the effects of the secondary resonances should be considered in the vicinity of the primary resonant layer. Because the energy increment is computed by an approximate expression, with increasing excitation strength, such a prediction is not accurate. To verify the valid of the approximate prediction, the numerical prediction should be completed through the energy increment for nonlinear Hamiltonian system.

6.1.2c. Energy increment spectrum

The exact energy increment can be computed numerically. Luo et al (1999) developed the energy spectrum approach, which was used for the numerical prediction of the onset of resonance in the stochastic layer. In the energy spectrum, the maximum and minimum conservative energies are computed through the Poincaré mapping section. As discussed in Chapter 5, this energy spectrum is equal to the energy increment spectrum. Using the Poincaré map section in Eq.(6.1), the Poincaré map is defined by $P:\Sigma\to\Sigma$. Such a technique computes the maximum and minimum energies of the Poincare mapping points as in Eq.(6.3). The unperturbed Hamiltonian energy for each Poincare mapping point of Eq.(6.1) is

$$H_0^{(N)}=H_0\left(\mathbf{x}_N,\boldsymbol{\mu}\right)\equiv H_0\left(x_N,y_N,\boldsymbol{\mu}\right). \tag{6.38}$$

However, in computation of conservative energy for a specified resonant layer, the energy changes in energy spectrums cannot be observed clearly. To observe the energy changes caused by the sub-resonance for the specified resonant layer, the minimum and maximum energy increments with respect to the unperturbed resonant orbit are introduced herein. On the other hand, the initial condition is chosen from the specific resonant orbit. The initial energy is $H_0(\mathbf{x}_0,\boldsymbol{\mu})=E_\alpha^{(m:n)}$, so the energy increment (or the first integral quantity increment) should be computed by $L(t_0,kT)=\Delta H_0(t_0,kT)=H_0^{(N)}(\mathbf{x}_N,\boldsymbol{\mu})-H_0(\mathbf{x}_0,\boldsymbol{\mu})$. That is,

$$\Delta E_{\max}=\max_{N\to\infty}\left\{H_0^{(N)}-E_\alpha^{(m:n)}\right\}\ \text{ and }\ \Delta E_{\min}=\min_{N\to\infty}\left\{H_0^{(N)}-E_\alpha^{(m:n)}\right\}. \tag{6.39}$$

For the appearance and disappearance of the resonant layer for a specific resonance, the maximum or minimum energy increment will have a big jump between the two closest resonant separatrices. From the minimum and maximum energy increment spectra, The width of a resonant separatrix layer is computed like the one of a stochastic layer in Luo et al (1999), i.e.,

$$w\equiv\min_{t\in[0,\infty)}\|\mathbf{x}(E_\alpha^{\max},t)-\mathbf{x}(E_\alpha^{\min},t)\|\equiv\|\mathbf{x}_\alpha^{\max}-\mathbf{x}_\alpha^{\min}\| \tag{6.40}$$

where $\|\bullet\|$ is a norm and the minimum and maximum energies are computed by $E_\alpha^{\max}=\Delta E_{\max}+E_\alpha^{(m:n)}$, $E_\alpha^{\min}=\Delta E_{\min}-E_\alpha^{(m:n)}$. Two points $\mathbf{x}_\alpha^{\max}$ and $\mathbf{x}_\alpha^{\min}$ on the normal vector $\mathbf{f}^{\perp}(\mathbf{x}_\alpha^{(m:n)})=(-f_2(\mathbf{x}_\alpha^{(m:n)}),f_1(\mathbf{x}_\alpha^{(m:n)}))^{\mathrm{T}}$ of the tangential vector of unperturbed resonant orbit at point $\mathbf{x}_\alpha^{(m:n)}$ are the closest between the maximum and minimum energy orbits $\mathbf{x}(E_\alpha^{\max},t)$ and $\mathbf{x}(E_\alpha^{\min},t)$, which can be obtained from Eq.(6.3) with $E_\alpha^{\max}$ and $E_\alpha^{\min}$. The detailed discussion on the energy increment spectrum can be referred to Luo (2002).

6.2. 2*n*-dimensional resonant separatrix layers

This section will extend the ideas in Section 6.1 to the 2*n*-dimensional nonlinear Hamiltonian systems. Consider a 2*n*-dimensional nonlinear Hamiltonian system with periodic excitations

$$\left.\begin{aligned}\dot{\mathbf{x}} = \mathbf{F}_1(\mathbf{x},\mathbf{y},t,\mathbf{p}) \equiv {}^s\mathbf{f}_1(\mathbf{x},\mathbf{y},\boldsymbol{\mu}_1) + {}^c\mathbf{f}_1(\mathbf{x},\mathbf{y},\boldsymbol{\mu}_2) + \mathbf{g}_1(\mathbf{x},\mathbf{y},t,\boldsymbol{\pi})\\ \dot{\mathbf{y}} = \mathbf{F}_2(\mathbf{x},\mathbf{y},t,\mathbf{p}) \equiv {}^s\mathbf{f}_2(\mathbf{x},\mathbf{y},\boldsymbol{\mu}_1) + {}^c\mathbf{f}_2(\mathbf{x},\mathbf{y},\boldsymbol{\mu}_2) + \mathbf{g}_2(\mathbf{x},\mathbf{y},t,\boldsymbol{\pi})\end{aligned}\right\} \tag{6.41}$$

where $\mathbf{x} = (x_1,\cdots,x_n)^T$ and $\mathbf{y} = (y_1,\cdots,y_n)^T$. ${}^s\mathbf{f}_1(\mathbf{x},\mathbf{y},\boldsymbol{\mu}_1)$ and ${}^s\mathbf{f}_2(\mathbf{x},\mathbf{y},\boldsymbol{\mu}_1)$ are the <u>s</u>eparable Hamiltonian vector fields on $\mathbb{R}^{2n}$, and ${}^c\mathbf{f}_1(\mathbf{x},\mathbf{y},\boldsymbol{\mu}_1)$ and ${}^c\mathbf{f}_2(\mathbf{x},\mathbf{y},\boldsymbol{\mu}_1)$ are the <u>c</u>oupled Hamiltonian vector field on $\mathbb{R}^{2n}$. In addition, the external vector fields $\mathbf{g}_1(\mathbf{x},\mathbf{y},t,\boldsymbol{\pi})$ and $\mathbf{g}_2(\mathbf{x},\mathbf{y},t,\boldsymbol{\pi})$ are periodic in time with period $T_i = \frac{2\pi}{\Omega_i}$ ($i \in \{1,2,\cdots,m\}$). All the vector fields are sufficiently smooth ($C^r, r \geq 2$) and bounded on a bounded set $D \subset \mathbb{R}^{2n}$ in phase space. If the vector field functions are for Hamiltonian systems as

$$\begin{aligned}{}^s\mathbf{f}_1 &= \frac{\partial^c H_0(\mathbf{x},\mathbf{y},\boldsymbol{\mu}_1)}{\partial \mathbf{x}}, \quad {}^s\mathbf{f}_2 = -\frac{\partial^c H_0(\mathbf{x},\mathbf{y},\boldsymbol{\mu}_1)}{\partial \mathbf{y}};\\ {}^c\mathbf{f}_1 &= \frac{\partial^c H_0(\mathbf{x},\mathbf{y},\boldsymbol{\mu}_2)}{\partial \mathbf{x}}, \quad {}^c\mathbf{f}_2 = -\frac{\partial^c H_0(\mathbf{x},\mathbf{y},\boldsymbol{\mu}_2)}{\partial \mathbf{y}};\\ \mathbf{g}_1 &= \frac{\partial H_1(\mathbf{x},\mathbf{y},t,\boldsymbol{\pi})}{\partial \mathbf{x}}, \quad \mathbf{g}_2 = -\frac{\partial H_1(\mathbf{x},\mathbf{y},t,\boldsymbol{\pi})}{\partial \mathbf{y}};\end{aligned} \tag{6.42}$$

then Eq.(6.41) has the Hamiltonian as

$$H(\mathbf{x},\mathbf{y},t) = {}^sH_0(\mathbf{x},\mathbf{y},\boldsymbol{\mu}_1) + {}^cH_0(\mathbf{x},\mathbf{y},\boldsymbol{\mu}_2) + H_1(\mathbf{x},\mathbf{y},\boldsymbol{\Omega}t,\boldsymbol{\pi}). \tag{6.43}$$

For a Hamiltonian system, x_j ($j = 1,2,\cdots,n$) represents the generalized coordinates and y_j are the corresponding generalized velocities. Because a separable Hamiltonian system consist on *n*-independent parts (oscillators), the corresponding Hamiltonian can be expressed by

$${}^sH_0(\mathbf{x},\mathbf{y},\boldsymbol{\mu}_1) = \sum\nolimits_{j=1}^{n} {}^sH_{0(j)}(x_j,y_j,\boldsymbol{\mu}_1). \tag{6.44}$$

Consider a point $(x_j, y_j)^T \in D_j \subset \mathbb{R}^2$ ($j = 1, 2, \cdots, n$) in the jth-separable Hamiltonian. In the domain D_j, the jth-separable Hamiltonian system satisfies the Hypotheses in (H1a-c) and (H2). For a given energy $E_j^{(\alpha_j)}$, we have

$$ {}^{s}H_{0(j)}(x_j^{(\alpha_j)}, y_j^{(\alpha_j)}) = E^{(\alpha_j)} \tag{6.45}$$

from which

$$ y_j^{(\alpha_j)} = y_j^{(\alpha_j)}(x_j^{(\alpha_j)}, E_j^{(\alpha_j)}). \tag{6.46}$$

The action variable is defined by

$$ J^{(\alpha_j)} = \oint y_j^{(\alpha_j)} dx_j^{(\alpha_j)}, \tag{6.47}$$

So we have

$$ {}^{s}H_{0(j)}(J_j^{(\alpha_j)}) = E^{(\alpha_j)}. \tag{6.48}$$

The angle variable is defined by

$$ \theta^{(\alpha_j)} = \omega_j^{(\alpha_j)} t + \theta_0^{(\alpha_j)} \tag{6.49}$$

with

$$ \dot{\theta}^{(\alpha_j)} = \frac{\partial\, {}^{s}H_{0(j)}(J^{(\alpha_j)})}{\partial J^{(\alpha_j)}} = \omega^{(\alpha_j)}. \tag{6.50}$$

From the foregoing hypotheses, in the domain $D_j \subset \mathbb{R}^2$, a periodic flow $q_j^{(\alpha_j)}(t)$ $= (x_j^{(\alpha_j)}(t), y_j^{(\alpha_j)}(t))$ can be expressed by

$$ x_j^{(\alpha_j)} = x_j^{(\alpha_j)}(J^{(\alpha_j)}, \theta^{(\alpha_j)}) \text{ and } y_j^{(\alpha_j)} = y_j^{(\alpha_j)}(J^{(\alpha_j)}, \theta^{(\alpha_j)}). \tag{6.51}$$

Substitution of Eq.(6.51) into Eq.(6.43) and using Fourier expansion gives

$$ \begin{aligned} H\left(\mathbf{J}^{\alpha}, \boldsymbol{\theta}^{\alpha}, t\right) &= H_0(\mathbf{J}^{\alpha}) + {}^{c}H_0(\mathbf{J}^{\alpha}, \boldsymbol{\theta}^{\alpha}) + H_1(\mathbf{J}^{\alpha}, \boldsymbol{\theta}^{\alpha}, t, \boldsymbol{\pi}) \\ &= \sum_{j=1}^{n} {}^{s}H_{0(j)}(J_j^{(\alpha_i)}) + \sum_{j=1}^{n}\sum_{\mathbf{s}^{(j)}} {}^{c}H_{0(\mathbf{s}^{(j)})}(\mathbf{J}^{\alpha}) \exp(i\mathbf{s}^{(j)} \cdot \boldsymbol{\omega}^{\alpha} t) \\ &\quad + \sum_{\mathbf{t}}\sum_{j=1}^{n}\sum_{\mathbf{r}^{(j)}} H_{1(\mathbf{r}^{(j)})}(\mathbf{J}^{\alpha}) \exp[i(\mathbf{r}^{(j)} \cdot \boldsymbol{\omega}^{\alpha} + \mathbf{t} \cdot \boldsymbol{\Omega})t]. \end{aligned} \tag{6.52}$$

In the foregoing equation, the notations are defined as

$$i=\sqrt{-1},\mathbf{J}^{\boldsymbol{\alpha}}=(J^{(\alpha_1)},\cdots,J^{(\alpha_n)})^T,\boldsymbol{\theta}^{\boldsymbol{\alpha}}=(\theta^{(\alpha_1)},\cdots,\theta^{(\alpha_n)})^T;$$
$$\boldsymbol{\omega}^{\boldsymbol{\alpha}}=(\omega^{(\alpha_1)},\cdots,\omega^{(\alpha_n)})^T,\mathbf{r}^{(i)}=(r_1^{(i)},\cdots,r_n^{(i)})^T, \tag{6.53}$$
$$\mathbf{s}^{(i)}=(s_1^{(i)},\cdots,s_n^{(i)})^T,\mathbf{t}=(t_1,\cdots t_m)^T.$$

6.2.1. Internal resonant layer dynamics

To consider the dynamics of a internal resonant layer, the internal resonant conditions are very important. Without the external Hamiltonian vector field in Eq.(6.41), the dynamical system is an unperturbed Hamiltonian system. The nonlinear dynamics on the equi-energy surface will be discussed in Chapter 7. From Eq.(6.52), the internal resonance condition is

$$\mathbf{s}^{(j)}\cdot\boldsymbol{\omega}^{\boldsymbol{\alpha}}=0 \quad \text{for } j\in\{1,2,\cdots,n\}. \tag{6.54}$$

Let

$$\mathbf{s}^{(k)}\cdot\boldsymbol{\theta}^{\boldsymbol{\alpha}}=\varphi^{(k)} \quad \text{for } k=1,2,\cdots,n \tag{6.55}$$

and the corresponding generating function are

$$G=\sum_{k=1}^{n}G_k=-\sum_{k=1}^{n}\varphi^{(k)}I_k=-\sum_{k=1}^{n}(\mathbf{s}^{(k)}\cdot\boldsymbol{\theta}^{\boldsymbol{\alpha}})I_k, \tag{6.56}$$

such that

$$J^{(\alpha_j)}=-\frac{\partial G_j}{\partial\theta^{(\alpha_j)}}=\sum_{k=1}^{n}s_j^{(k)}I_k \text{ and } \overline{p}_{\alpha_j}=-\frac{\partial G_k}{\partial\varphi^{(k)}}=I_k \tag{6.57}$$

for $j=1,2,\cdots,n$. From Eq.(6.55), we have

$$\begin{aligned}&\boldsymbol{\theta}^{\boldsymbol{\alpha}}=S^{-1}\boldsymbol{\varphi} \text{ with matrix } S=(s_i^{(k)})_{n\times n}\\&\boldsymbol{\varphi}=\left(\varphi^{(1)},\cdots,\varphi^{(n)}\right).\end{aligned} \tag{6.58}$$

Further,

$$\begin{aligned}H(\mathbf{I},\boldsymbol{\varphi},t)&={}^sH_0(\mathbf{I})+{}^cH_0(\mathbf{I},\boldsymbol{\varphi})+H_1(\mathbf{I},\boldsymbol{\varphi},t,\boldsymbol{\pi})\\&=\sum_{k=1}^{n}{}^sH_{0(k)}(I_k)+\sum_{k=1}^{n}{}^cH_{0(k)}(\mathbf{I})\exp(i\varphi^{(k)})\\&\quad+\sum_{\mathbf{t}}\sum_{k=1}^{n}\sum_{\mathbf{r}^{(k)}}H_{1(\mathbf{r}^{(k)})}(\mathbf{I})\exp[i(\mathbf{r}^{(k)}\cdot S^{-1}\boldsymbol{\varphi}+\mathbf{t}\cdot\boldsymbol{\Omega}t)].\end{aligned} \tag{6.59}$$

A new Hamiltonian is

$$
\begin{aligned}
H(\mathbf{I},\boldsymbol{\varphi},t)+\tfrac{\partial G_j}{\partial t} &= {}^sH_0(\mathbf{I})+{}^cH_0(\mathbf{I},\boldsymbol{\varphi})+H_1(\mathbf{I},\boldsymbol{\varphi},t,\boldsymbol{\pi})+\tfrac{\partial G_j}{\partial t}\\
&\approx {}^sH_0(I_j)+\frac{1}{2}\left.\frac{\partial^2\,{}^sH_0(I_j)}{\partial I_j^2}\right|_{I_j=0}(\Delta I_j)^2+\cdots+h.o.t\\
&+{}^cH_{0(j)}(\mathbf{I})\exp(i\varphi^{(j)})\\
&+\sum_{k=1,k\neq j}^{n}{}^sH_{0(k)}(I_k)+\sum_{k=1,k\neq j}^{n}{}^cH_{0(k)}(\mathbf{I})\exp(i\varphi^{(k)})\\
&+\sum_{\mathbf{t}}\sum_{k=1}^{n}\sum_{\mathbf{r}^{(k)}}H_{1(\mathbf{r}^{(k)})}(\mathbf{I})\exp[i(\mathbf{r}^{(k)}\cdot S^{-1}\boldsymbol{\varphi}+\mathbf{t}\cdot\boldsymbol{\Omega}t)].
\end{aligned}
\tag{6.60}
$$

Introduce the new parameters

$$
\begin{aligned}
&B_j=\left.\frac{\partial^2\,{}^cH_0(I_j)}{\partial I_j^2}\right|_{I_j=0},\ p_j=B_j\Delta I_j,\\
&\bar{H}(\mathbf{I},\boldsymbol{\varphi},t)=B_j(H(\mathbf{I},\boldsymbol{\varphi},t)+\tfrac{\partial G_j}{\partial t}-{}^sH_0(I_j)),\\
&{}^c\bar{H}_{0(j)}(p_j)=B_j\,{}^cH_{0(j)}(p_j),\ {}^s\bar{H}_{0(k)}(I_k)=B_j\,{}^sH_{0(k)}(I_k),\\
&\bar{H}_{1(\mathbf{r}^{(k)})}(\mathbf{I})=B_j\,{}^sH_{0(k)}(I_k).
\end{aligned}
\tag{6.61}
$$

Rescaling the new Hamiltonian with $I_j=0$ gives

$$
\begin{aligned}
\bar{H}(\mathbf{I},\boldsymbol{\varphi},t)&\approx\frac{1}{2}p_j^2+{}^c\bar{H}_{0(j)}(\mathbf{I})\exp(i\varphi^{(j)})\\
&+\sum_{k=1,k\neq j}^{n}{}^s\bar{H}_{0(k)}(I_k)+\sum_{k=1,k\neq j}^{n}{}^c\bar{H}_{0(k)}(\mathbf{I})\exp(i\varphi^{(k)})\\
&+\sum_{\mathbf{t}}\sum_{k=1}^{n}\sum_{\mathbf{r}^{(k)}}\bar{H}_{1(\mathbf{r}^{(k)})}(\mathbf{I})\exp[i(\mathbf{r}^{(k)}\cdot S^{-1}\boldsymbol{\varphi}+\mathbf{t}\cdot\boldsymbol{\Omega}t)]
\end{aligned}
\tag{6.62}
$$

As in Eq.(6.25), the corresponding dynamics of the internal $\mathbf{s}^{(j)}$-resonant layer can be investigated through the nonlinear Hamiltonian system in Eq.(6.62). The internal $\mathbf{s}^{(j)}$-resonant separatrix requires $\varphi^{(j)}=(2n-1)\pi$ ($n=1,2,\cdots$) and $I_j=0$. Without the external excitation, equation (6.62) becomes

$$
\begin{aligned}
\bar{H}(\mathbf{I},\boldsymbol{\varphi},t)&\approx\frac{1}{2}p_j^2+{}^c\bar{H}_{0(j)}(\mathbf{I})\exp(i\varphi^{(j)})\\
&+\sum_{k=1,k\neq j}^{n}{}^s\bar{H}_{0(k)}(I_k)+\sum_{k=1,k\neq j}^{n}{}^c\bar{H}_{0(k)}(\mathbf{I})\exp(i\varphi^{(k)}).
\end{aligned}
\tag{6.63}
$$

From the new Hamiltonian, dynamics of the resonant separatrix on the energy surface can be analyzed approximately. Such an analysis is only valid for the very weak coupling. So far, no criterion is developed for such an internal resonant separatrix. For the strong coupling, the above analysis may not be very good.

6.2.2. External resonant layer dynamics

In a similar fashion, dynamics of the external resonant layer can be discussed. From Eq.(6.52), the external resonance condition is

$$\mathbf{r}^{(j)} \cdot \boldsymbol{\omega}^{\alpha} + \mathbf{t} \cdot \boldsymbol{\Omega} = 0 \quad \text{for } j \in \{1, 2, \cdots, n\} . \tag{6.64}$$

Let

$$\mathbf{r}^{(k)} \cdot \boldsymbol{\theta}^{\alpha} + \mathbf{t} \cdot \boldsymbol{\Omega} t = \varphi^{(k)} \quad \text{for } k = 1, 2, \cdots, n \tag{6.65}$$

and the corresponding generating function is

$$G = \sum_{k=1}^{n} G_k = -\sum_{k=1}^{n} \varphi^{(k)} I_k = -\sum_{k=1}^{n} (\mathbf{r}^{(k)} \cdot \boldsymbol{\theta}^{\alpha} - \mathbf{t} \cdot \boldsymbol{\Omega} t) I_k , \tag{6.66}$$

such that

$$J^{(\alpha_j)} = -\frac{\partial G_j}{\partial \theta^{(\alpha_j)}} = \sum_{k=1}^{n} r_j^{(k)} I_k \quad \text{for } j = 1, 2, \cdots, n. \tag{6.67}$$

From Eq.(6.65), we have

$$\begin{aligned} \boldsymbol{\theta}^{\alpha} &= R^{-1} \boldsymbol{\varphi} \text{ with matrix } R = (r_i^{(k)})_{n \times n} , \\ \bar{\boldsymbol{\varphi}} &= \left(\varphi^{(1)}, \cdots, \varphi^{(j)} - \mathbf{t} \cdot \boldsymbol{\Omega} t, \cdots, \varphi^{(n)} \right). \end{aligned} \tag{6.68}$$

Further,

$$\begin{aligned} H(\mathbf{I}, \boldsymbol{\varphi}, t) &= {}^{s}H_0(\mathbf{I}) + {}^{c}H_0(\mathbf{I}, \boldsymbol{\varphi}) + H_1(\mathbf{I}, \boldsymbol{\varphi}, t, \boldsymbol{\pi}) \\ &= \sum_{k=1}^{n} {}^{s}H_{0(k)}(I_k) + \sum_{k=1}^{n} \sum_{\mathbf{s}^{(k)}} {}^{c}H_{0(k)}(\mathbf{I}) \exp[i(\mathbf{s}^{(k)} \cdot R^{-1} \bar{\boldsymbol{\varphi}})] \\ &\quad + H_{1(j)}(\mathbf{I}) \exp(i\varphi^{(j)}) + \sum_{\mathbf{t}} \sum_{k=1, k \neq j}^{n} H_{1(k)}(\mathbf{I}) \exp(i\varphi^{(k)} + \mathbf{t} \cdot \boldsymbol{\Omega} t). \end{aligned} \tag{6.69}$$

A new Hamiltonian for the $(\mathbf{s}^{(j)} : \mathbf{t})$-external resonant layer is

$$H(\mathbf{I}, \boldsymbol{\varphi}, t) + \tfrac{\partial G_j}{\partial t} = {}^{s}H_0(\mathbf{I}, \boldsymbol{\varphi}, t) + \tfrac{\partial G_j}{\partial t} + {}^{c}H_0(\mathbf{I}, \boldsymbol{\varphi}) + H_1(\mathbf{I}, \boldsymbol{\varphi}, t, \boldsymbol{\pi}) \tag{6.70a}$$

$$
\begin{aligned}
H(\mathbf{I},\boldsymbol{\varphi},t)+\tfrac{\partial G_j}{\partial t} &\approx {}^sH_0(I_j)+\frac{1}{2}\frac{\partial^2\, {}^sH_0(I_j)}{\partial I_j^2}\bigg|_{I_j=0}\left(\Delta I_j\right)^2+\cdots+h.o.t\\
&+\sum_{k=1,k\neq j}^{n} {}^sH_{0(k)}(I_k)+H_{1(j)}(\mathbf{I})\exp(i\varphi^{(j)})\\
&+\sum_{k=1}^{n}\sum_{\mathbf{s}^{(k)}} {}^cH_{0(k)}(\mathbf{I})\exp[i(\mathbf{s}^{(k)}\cdot R^{-1}\overline{\boldsymbol{\varphi}})]\\
&+\sum_{\mathbf{t}}\sum_{k=1,k\neq j}^{n} H_{1(k)}(\mathbf{I})\exp(i\varphi^{(k)}+\mathbf{t}\cdot\boldsymbol{\Omega}t).
\end{aligned}
\tag{6.70b}
$$

Introducing the new parameters as in Eq.(6.61) and rescaling the new Hamiltonian with $I_j=0$ gives

$$
\begin{aligned}
\overline{H}(\mathbf{I},\boldsymbol{\varphi},t) &\approx \frac{1}{2}p_j^2+\overline{H}_{1(j)}(\mathbf{I})\exp(i\varphi^{(j)})\\
&+\sum_{k=1,k\neq j}^{n} {}^s\overline{H}_{0(k)}(I_k)+\sum_{k=1}^{n}\sum_{\mathbf{s}^{(k)}} {}^cH_{0(k)}(\mathbf{I})\exp[i(\mathbf{s}^{(k)}\cdot R^{-1}\overline{\boldsymbol{\varphi}})]\\
&+\sum_{\mathbf{t}}\sum_{k=1,k\neq j}^{n} \overline{H}_{1(k)}(\mathbf{I})\exp(i\varphi^{(k)}+\mathbf{t}\cdot\boldsymbol{\Omega}t).
\end{aligned}
\tag{6.71}
$$

From the new Hamiltonian, dynamics of the $(\mathbf{s}^{(j)}:\mathbf{t})$-external resonant layer can be investigated.

6.2.3. Approximate criteria

For developing approximate criterion, based on the $(\mathbf{s}:\mathbf{t})$-resonant surface, the energy increment for the jth separable oscillator is

$$
\begin{aligned}
&\Delta\, {}^sH_0^{(j)}(\theta_i^{(\alpha_1)},\cdots,\theta_i^{(\alpha_{j-1})},\theta_i^{(\alpha_{j+1})},\cdots,\theta_i^{(\alpha_n)},\boldsymbol{\Omega}t_i)\\
&=\int_{t_j}^{T_j^{(\alpha_j)}}[{}^sH_0^{(j)},{}^cH_0^{(j)}]dt+\int_{t_j}^{T_j^{(\alpha_j)}}[{}^sH_0^{(j)},H_1^{(j)}]dt\\
&=\int_{t_j}^{T_j^{(\alpha_j)}}({}^sf_1^{(j)}\,{}^cf_2^{(j)}-{}^sf_2^{(j)}\,{}^cf_1^{(j)})dt+\int_{t_j}^{T_j^{(\alpha_j)}}({}^sf_1^{(j)}g_2^{(j)}-{}^sf_2^{(j)}g_1^{(j)})dt
\end{aligned}
\tag{6.72}
$$

The iterative mappings are for ($j=1,2,\cdots,n$)

$$
\begin{aligned}
E_{i+1}^{(\alpha_j)}-E_i^{(\alpha_j)} &= \Delta\, {}^sH_0^{(j)}(\theta_i^{(\alpha_1)},\cdots,\theta_i^{(\alpha_{j-1})},\theta_i^{(\alpha_{j+1})},\cdots\theta_i^{(\alpha_n)},\boldsymbol{\Omega}t_i)\\
\theta_{i+1}^{(\alpha_j)}-\theta_i^{(\alpha_j)} &= \nu^{(\alpha_j)}(E_{i+1}^{(\alpha_j)})
\end{aligned}
\tag{6.73}
$$

From the foregoing mappings, the condition from the local stochasticity to the global stochasticity can be obtained. Such a condition is the condition for onset of the $(\mathbf{s}:\mathbf{t})$-resonant layer. Setting $E_{i+1}^{(\alpha_j)} = E_i^{(\alpha_j)} = E_0^{(\alpha_j),(\mathbf{s}:\mathbf{t})}$, $\theta_{i+1}^{(\alpha_j)} = 2s^j\pi + \theta_0^{(\alpha_j)}$ and $\theta_i^{(\alpha_j)} = \theta_0^{(\alpha_j)}$, the fixed points for mapping (6.73) can be determined. The standard mapping procedure can be developed for the global stochasticity.

Consider a primary $(\mathbf{s}_1:\mathbf{t}_1)$-resonance closest to the primary $(\mathbf{s}:\mathbf{t})$-resonant separatrix. As in (6.37), the approximate estimate for the destruction of the primary $(\mathbf{s}:\mathbf{t})$-resonant separatrix can be given for ($j = 1,2,\cdots,n$) by

$$\left|\Delta^s H_0^{(j)}\right| = \left|E^{(\alpha_j),(\mathbf{s},\mathbf{t})} - E^{(\alpha_j),(\mathbf{s}_1,\mathbf{t}_1)}\right|. \tag{6.74}$$

To verify the above analysis, the energy spectrum can be developed for each separable system in nonlinear Hamiltonian systems. The unperturbed Hamiltonian energy for each Poincare mapping point of Eq.(6.41) is

$${}^sH_0^{(\alpha_j),(N)} = {}^sH_0^{(\alpha_j)}\left(\mathbf{x}_N^{(\alpha_j)}, \boldsymbol{\mu}_1\right) \equiv H_0\left(x_N^{(\alpha_j)}, y_N^{(\alpha_j)}, \boldsymbol{\mu}_1\right). \tag{6.75}$$

To observe the energy changes caused by the sub-resonance for the specified resonant layer, the minimum and maximum energy increments with respect to the unperturbed resonant orbit are introduced herein. That is, for ($j = 1,2,\cdots,n$)

$$\begin{aligned} \Delta E_{\max}^{(\alpha_j)} &= \max_{N\to\infty}\left\{{}^sH_0^{(\alpha_j),(N)} - E^{(\alpha_j),(\mathbf{s}:\mathbf{t})}\right\}, \\ \Delta E_{\min}^{(\alpha_j)} &= \min_{N\to\infty}\left\{{}^sH_0^{(\alpha_j),(N)} - E^{(\alpha_j),(\mathbf{s}:\mathbf{t})}\right\}. \end{aligned} \tag{6.76}$$

As described before, equation (6.76) computed the exact first integral quantity increment for each separable Hamiltonian system.

6.3. Resonant layers in a Duffing oscillator

Consider a hardening spring Duffing oscillator as

$$\begin{aligned} \dot{x} &= y, \\ \dot{y} &= -\alpha_1 x - \alpha_2 x^3 + Q_0 \cos\Omega t. \end{aligned} \tag{6.77}$$

This unperturbed Hamiltonian system has only one center equilibrium at point $(0,0)$. Therefore, the resonant layer will exist in the entire domain. For this system, the Hamiltonian is expressed by,

$$H = \tfrac{1}{2}y^2 + \tfrac{1}{2}\alpha_1 x^2 + \tfrac{1}{4}\alpha_2 x^4 - xQ_0\cos\Omega t. \tag{6.78}$$

Furthermore, the Hamiltonian in Eq.(6.78) can be separated into the time-independent part H_0 and the time-dependent part H_1, i.e., $H = H_0 + H_1$ where

$$H_0 = \tfrac{1}{2}y^2 + \tfrac{1}{2}\alpha_1 x^2 + \tfrac{1}{4}\alpha_2 x^4 \text{ and } H_1 = -xQ_0 \cos \Omega t. \tag{6.79}$$

6.3.1. Resonant layer dynamics

For given values of H_0, the energy contour in phase plane is sketched in Fig.6.10 for $\alpha_1 = \alpha_2 = 1.0$. Three initial energies are $H_0 = E_0 < E_1 < E_2$. For an initial energy $H_0 = E_0$, the solution of the unperturbed system in Eq.(6.77) is

$$\begin{aligned} x &= h\mathrm{cn}\left[\tfrac{2K(k)\theta}{\pi}, k\right], \text{ and} \\ y &= \pm\sqrt{\frac{\alpha_2}{2}}\frac{h^2}{k}\mathrm{sn}\left[\tfrac{2K(k)\theta}{\pi}, k\right]\mathrm{dn}\left[\tfrac{2K(k)\theta}{\pi}, k\right]; \end{aligned} \tag{6.80}$$

in which cn, sn, dn are the Jacobi-elliptic functions, $K(k)$ is the complete elliptic integral of the first kind and k is its modulus. The other parameters h, θ of the solutions are defined by

$$h^2 = \frac{2k^2\alpha_1}{(1-2k^2)\alpha_2} \quad \text{and} \quad \theta = \omega t; \tag{6.81}$$

where ω is the nonlinear natural frequency of the hardening spring system. The nonlinear natural frequency and the corresponding period are computed by

$$\omega = \frac{1}{2}\sqrt{\frac{\alpha_2}{2}}\frac{h\pi}{kK(k)} \quad \text{and } T = \frac{2\pi}{\omega}. \tag{6.82}$$

The particular value E_0 is expressed by

$$E_0 = \frac{k^2(1-k^2)\alpha_1^2}{(1-2k^2)^2\alpha_2}; \tag{6.83}$$

and the action variable J for an orbit is

$$J = \frac{\sqrt{2\alpha_2 h^3}}{3\pi k^3}\left[(1-k^2)K(k) + (2k^2-1)E(k)\right]. \tag{6.84}$$

Note that $E(k)$ in Eq.(6.84) denotes the complete elliptic integral of the second kind. Substitution of Eq.(6.80) into Eq.(6.78) and expanding the time-dependent term, the complete Hamiltonian now becomes

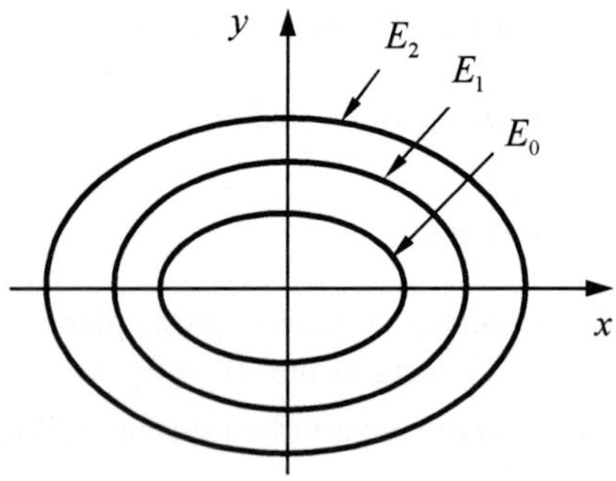

Figure 6.10. Conservative trajectories for the given energy of a hardening spring oscillator.

$$H = H_0(J) - Q_0 \sum_{n=1}^{\infty} 2Q_{2n-1} \cos\left[(2n-1)\omega t\right] \cos \Omega t \quad \text{or}$$
$$H = H_0(J) - Q_0 \sum_{n=1}^{\infty} Q_{2n-1} \left\{ \cos\left[(2n-1)\omega - \Omega\right] t + \cos\left[(2n-1)\omega + \Omega\right] t \right\}. \tag{6.85}$$

in which

$$Q_{2n-1} = \frac{\pi h}{2kK \cosh\left[\pi\left(n - \frac{1}{2}\right)\frac{K'}{K}\right]}. \tag{6.86}$$

Notice that $K' = K(k')$, and $k' = \sqrt{1-k^2}$. Except for the $(2n-1:1)$-primary resonance, all other terms in H will average to zero over time t. Thus, from Eqs.(6.81) and (6.82), the $(2n-1:1)$-primary resonance conditions are

$$(2n-1)\omega = \Omega, \text{ i.e., } \Omega = \frac{(2n-1)\pi}{2K(k)}\sqrt{\frac{\alpha_1}{1-2k^2}}, \tag{6.87}$$

from which the modulus k at the $(2n-1:1)$-primary resonance for a given n. (i.e., $k^{(2n-1:1)}$). Likewise, $J^{(2n-1:1)}$ is a value for action variable J at the $(2n-1:1)$-primary resonance. We are interested in the region between the $(2n-1:1)$- and the $(2n+1:1)$-primary resonances. Assuming the effects of other resonances to be negligible, the Hamiltonian can be expanded through a Taylor series near the $(2n-1:1)$ primary resonance.

$$H \approx H_0(J^{(2n-1:1)}) + \left.\tfrac{dH_0}{dJ}\right|_{J^{(2n-1:1)}} (J - J^{(2n-1:1)}) + \tfrac{1}{2}\left.\tfrac{d^2H_0}{dJ^2}\right|_{J^{(2n-1:1)}} (J - J^{(2n-1:1)})^2 + \cdots + \text{h.o.t.}$$
$$-Q_0\left\{Q_{2n-1}(J^{(2n-1:1)})\cos[(2n-1)\omega-\Omega]t + Q_{2n-1}(J^{(2n-1:1)})\cos[(2n-1)\omega+\Omega]t\right\}$$
$$-Q_0\left\{Q_{2n+1}(J^{(2n+1:1)})\cos[(2n+1)\omega-\Omega]t + Q_{2n+1}(J^{(2n+1:1)})\cos[(2n+1)\omega+\Omega]t\right\}. \quad (6.88)$$

Observing that $H_0 = E_0$, the following parameter is introduced as

$$B_0 = \left(\tfrac{d^2E_0}{dJ^2}\right)_{J^{(2n-1:1)}} = \frac{\pi^2}{4h^2\left[K(k^{(2n-1:1)})\right]^3}\left[K\left(k^{(2n-1:1)}\right) - \frac{1-2(k^{(2n-1:1)})^2}{1-(k^{(2n-1:1)})^2}E\left(k^{(2n-1:1)}\right)\right]. \quad (6.89)$$

A new canonical coordinate $(\bar{p},\bar{\phi})$ is introduced at $\bar{p}=0$ with $J = J^{(2n-1:1)}$. The generating function is

$$G(J,\bar{\phi}) = -\left(J - J^{(2n-1:1)}\right)\left(\frac{\bar{\phi}+\Omega t}{2n-1}\right), \quad (6.90)$$

and then we have

$$\bar{p} = -\frac{\partial G}{\partial \bar{\phi}} = \frac{J - J^{(2n-1:1)}}{2n-1} \text{ and } \theta = \omega t = -\frac{\partial G}{\partial J} = \frac{\bar{\phi}+\Omega t}{2n-1}. \quad (6.91)$$

The new Hamiltonian with two resonances is now given by

$$\bar{H} = H + \tfrac{\partial G}{\partial t} \approx H_0\left(J^{(2n-1:1)}\right) - \frac{1}{2}B_0(2n-1)^2\bar{p}^2$$
$$-Q_0Q_{2n-1}\left(J^{(2n-1:1)}\right)\left\{\cos\bar{\phi} + Q_{2n-1}\cos(\bar{\phi}+2\Omega t)\right\} \quad (6.92)$$
$$-Q_0Q_{2n+1}\left(J^{(2n+1:1)}\right)\left\{\cos\left[\frac{2n+1}{2n-1}\bar{\phi}+\Omega_1 t\right] + \cos\left[\frac{2n+1}{2n-1}\bar{\phi}+\Omega_2 t\right]\right\};$$

in which $\Omega_1 = 2\Omega/(2n-1)$ and $\Omega_2 = 2n\Omega/(2n-1)$. To re-scale Eq.(6.92), the following quantities are introduced by

$$\phi = \bar{\phi},\ p = \frac{(2n-1)^2(2n+1)B_0}{\Omega}\bar{p},$$
$$H = \frac{(2n-1)^2(2n+1)^2B_0}{\Omega^2}\left[\bar{H} - E_0\left(J^{(2n-1:1)}\right)\right]; \quad (6.93)$$

and thus, the re-scaled Hamiltonian is

$$H = \tfrac{1}{2}p^2 - U_0\left[\cos\phi + \cos(\phi + 2\Omega t)\right]$$
$$-V_0\left[\cos\left(\frac{2n-1}{2n+1}\phi + \Omega_1 t\right) + \cos\left(\frac{2n-1}{2n+1}\phi\phi + \Omega_2 t\right)\right]; \tag{6.94}$$

where

$$U_0 = \frac{(2n-1)^2(2n+1)^2 B_0}{\Omega^2} Q_0 Q_{2n-1},$$
$$V_0 = \frac{(2n-1)^2(2n+1)^2 B_0}{\Omega^2} Q_0 Q_{2n+1}; \tag{6.95}$$

$$Q_{2n-1} = \frac{\sqrt{2}\Omega}{\sqrt{\alpha_2}(2n-1)\cosh\left[\pi\left(n-\frac{1}{2}\right)\frac{K'^{(2n-1:1)}}{K^{(2n-1:1)}}\right]},$$
$$Q_{2n+1} = \frac{\sqrt{2}\Omega}{\sqrt{\alpha_2}(2n+1)\cosh\left[\pi\left(n+\frac{1}{2}\right)\frac{K'^{(2n+1:1)}}{K^{(2n+1:1)}}\right]}. \tag{6.96}$$

Due to $Q_{2n+1} \ll Q_{2n-1}$, we have $U_0 \gg V_0$. For an approximate analysis, the following new Hamiltonian for only one resonance in Eq.(6.88) is

$$H \approx \tfrac{1}{2}p^2 - U_0\left[\cos\phi + \cos(\phi + 2\Omega t)\right]. \tag{6.97}$$

The dynamical system for the $(2n-1:1)$-resonant layer is

$$\dot{\phi} = p$$
$$\dot{p} = -U_0\left[\sin\phi + \sin(\phi + 2\Omega t)\right] \tag{6.98}$$

from which the equilibrium points of the unperturbed are at

$$p = 0,\ \phi = m\pi \ \text{ for } m = 0,1,2\cdots,2(2n-1). \tag{6.99}$$

The points for $m = 0,2,\cdots,2(2n-1)$ are the center points for the $(2n-1:1)$-primary resonant separatrix, and the points for $m = 1,3,\cdots,2(2n-1)-1$ are the hyperbolic points for the $(2n-1:1)$-primary resonant separatrix. From Eq.(6.91), we have

$$J = J^{(2n-1:1)} \text{ and } \theta = \frac{m\pi + \Omega t}{2n-1} \quad \text{for } (m = 0,1,2\cdots,2(2n-1)). \tag{6.100}$$

From Eq.(6.84), the resonant elliptic modulus value (i.e., $k^{(2n-1:1)}$) can be computed, and the initial energy $E_0^{(2n-1:1)}$ can be computed. By using Eq.(6.100), equation (6.80) gives the location of the equilibrium points on the $(2n-1:1)$-primary resonant orbit of the unperturbed Hamiltonian system in Eq.(6.77). Herein, we will not continue to use Eq.(6.98) to compute the sub-resonant separatrix. To avoid repeating, the similar results will be presented later in this chapter.

6.3.2. Analytical conditions

Substitution of Eq.(6.81) into Eq.(6.82) gives

$$\omega = \frac{\pi}{2K(k)}\sqrt{\frac{\alpha_1}{1-2k^2}} \tag{6.101}$$

Let the amplitude $h \to 0$ in Eq.(6.80), the minimum period orbit will be obtained. At $h = 0$, we have $k = 0$ in Eq.(6.81) and $K(k) = \pi/2$. So the minimum nonlinear natural frequency in Eq.(6.101) is $\omega = \sqrt{\alpha_1}$. As $h \to \infty$, we have $k \to 1/\sqrt{2}$. $K(k) \approx 1.8541$, $\omega \to \infty$. From the resonant condition, the resonant excitation frequency versus the conservative energy is presented in Fig. 6.11 for $\alpha_1 = \alpha_2 = 1$. So we have $\omega = 1$ at $E_0 = 0$ (i.e. at the center). It is observed that we have $\Omega^{(2n-1:1)} = (2n-1)\omega = 2n-1$ at $E_0 = 0$. With increasing conservative energy E_0, the nonlinear frequency ω and the resonant excitation frequency $\Omega^{(2n-1:1)}$ will increase. The (1:1) and (3:1)-resonant layers are sketched in Fig.6.12 to help one understand the resonant separatrix layer network in such a Hamiltonian system in Eq.(6.77). The red and yellow circular symbols represent the center and hyperbolic points for resonant layer. The large read symbol is the generic center of Eq.(6.77). On the inside of the (3:1)-resonant layer, the $(2n-1:1)$-primary resonant layers ($n = 3,4,5\cdots$) can be included. Due to the external excitation, for each period, the flow in Eq.(6.77) cannot return the initial point. The period based on ω related to the energy E_1 is computed by

$$T(E_1) = 2\pi/\omega. \tag{6.102}$$

The change of phase angle over one period $T(E_1)$ is

$$\Delta\phi = \Omega T(E_1) = V_1(E_1) \equiv 4\Omega K(k)\sqrt{\frac{2}{\alpha_2}\frac{k}{h}} \tag{6.103}$$

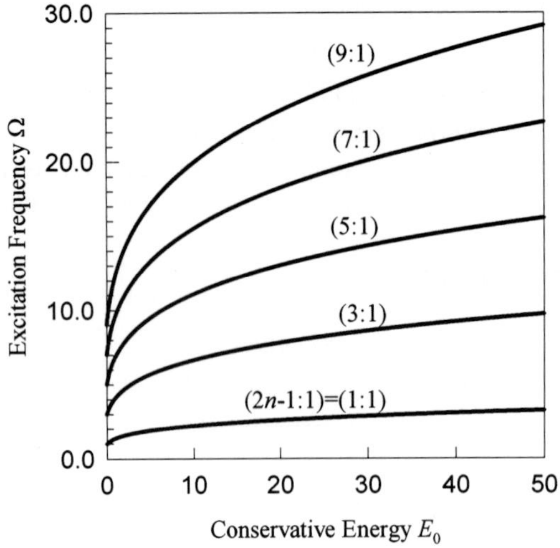

Figure 6.11. Resonant excitation frequency versus conservative energy ($\alpha_1 = \alpha_2 = 1$).

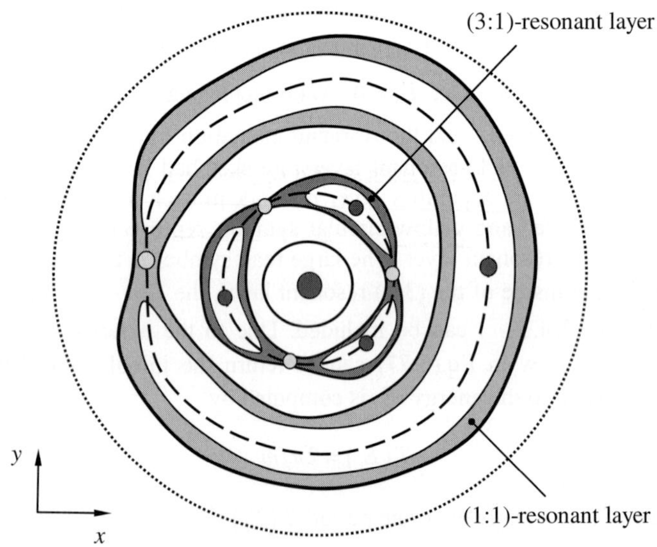

Figure 6.12. A sketch of the (1:1) and (3:1)- resonant layers in the hardening spring system.

From Eq.(6.29), the energy increment over one period $T(E_1)$ is approximated by

$$\Delta H_0(\varphi_i) \approx \int_{t_i}^{T(E_i)+t_i} yQ_0 \cos(\Omega t)\,dt = U_1 \sin(\Omega t_i), \qquad (6.104)$$

$$U_1 = \frac{2\sqrt{2}\pi\Omega Q_0}{\sqrt{\alpha_2}} \operatorname{sech}\left[\left(n-\tfrac{1}{2}\right)\pi \frac{K'(k^{(2n-1:1)})}{K(k^{(2n-1:1)})}\right]. \qquad (6.105)$$

Equations (6.103) and (6.104) are re-written as

$$E_{i+1} = E_i + U_1 \sin(\varphi_i) \text{ and } \varphi_{i+1} = \varphi_i + V_1(w_{i+1}). \qquad (6.106)$$

To obtain the standard map, linearization of Eq. (6.106) in the neighborhood of the resonant orbits about the period-1 fixed point with $E_{i+1} = E_i = E_0^{(2n-1:1)}$ and $\varphi_{i+1} - \varphi_i = 2(2n-1)\pi$. The resonant energy is obtained from

$$2(2n-1)\pi = V_1\left(E_0^{(2n-1:1)}\right). \qquad (6.107)$$

A new energy $I_i = G_1 \Delta E_i$ at $G_1 = \partial V_1 / \partial E_{i+1}$ and $E_{i+1} = E_i = E_0^{(2n-1:1)}$ is given by

$$G_1 = -\frac{2\Omega\alpha_2}{(k^{(2n-1:1)})^2}\left[\frac{1-2(k^{(2n-1:1)})^2}{\alpha_1}\right]^{\frac{5}{2}}\left[K(k^{(2n-1:1)}) - \frac{1-2(k^{(2n-1:1)})^2}{1-(k^{(2n-1:1)})^2} E(k^{(2n-1:1)})\right]. \qquad (6.108)$$

Linearization of Eq.(6.106) about the fixed point gives

$$I_{i+1} = I_i \pm \Xi \sin\varphi_i \text{ and } \varphi_{i+1} = \varphi_i + I_{i+1} \qquad (6.109)$$

where $\Xi = U_0 |G_1|$. Equation (6.109) is now a standard map. Employing the universal constant of the standard mapping, i.e. $\Xi = \Xi^* = 0.9716354\cdots$, the condition for the *appearance* of the $(2n-1:1)$-resonant layer is,

$$Q_0 \approx \frac{0.9716354\sqrt{\alpha_2}}{2\sqrt{2}\pi\Omega|G_1|} \cosh\left[\left(n-\tfrac{1}{2}\right)\pi \frac{K'(k^{(2n-1:1)})}{K(k^{(2n-1:1)})}\right]. \qquad (6.110)$$

From Eq.(5:37), the condition for the *disappearance* of the $(2n-1:1)$-resonant layer is given by

$$\begin{aligned} &2Q_0\pi\Omega\sqrt{\tfrac{2}{\alpha_2}} \operatorname{sech}\left[\left(n-\tfrac{1}{2}\right)\pi \frac{K'(k^{(2n-1:1)})}{K(k^{(2n-1:1)})}\right] \\ &\approx \min\left\{\left|E_0^{(2n+1:1)} - E_0^{(2n-1:1)}\right| \text{ and } \left|E_0^{(2n-1:1)} - E_0^{(2n-3:1)}\right|\right\}. \end{aligned} \qquad (6.111)$$

Before the maximum energy of the $(2n-1:1)$-resonant layer arrives to $E_0^c = 0$ at the center point without other primary resonance, the *maximum* critical condition for the accumulated disappearance strength:

$$2Q_0\pi\Omega\sqrt{\frac{2}{\alpha_2}}\text{sech}\left[\left(n-\tfrac{1}{2}\right)\pi\frac{K'(k^{(2n-1:1)})}{K(k^{(2n-1:1)})}\right] = E_0^{(2n-1:1)}. \tag{6.112}$$

REMARK: $E_0^c \neq 0$ at the center point, the approximate prediction for the $(m:n)$-resonant layer in 2-dimensional nonlinear Hamiltonian systems is given by

$$\left|\Delta H_0^{(m:n)}(\varphi_i)\right| = \left|E_0^{(2n-1:1)} - E_0^c\right|. \tag{6.113}$$

If the sub-resonant separatrix is very strong, the criterion for the disappearance of the $(m:n)$-parimary resonant layer should consider the effects of the sub-resonant separatrix. Suppose the sub-resonance of the $(m:n)$-primary resonance is of the $(m_1:n_1)$-order, and the sub-resonance of the $(m_i:n_i)$ sub-resonance is the $(m_{i+1}:n_{i+1})$-sub-resonance, all the energy increment of the sub-resonance series is expressed by $\Delta H_0^{(m:n),(m_1:n_1),\cdots,(m_i,n_i)}$. Further, the criterion can be postulated as

$$\left|\Delta H_0^{(m:n)}\right| + \sum\nolimits_{i=1}^{\infty}\left|\Delta H_0^{(m:n),(m_1:n_1),\cdots,(m_i,n_i)}\right| = \left|E_0^{(m:n)} - E_0^c\right|. \tag{6.114}$$

To obtain $\Delta H_0^{(m:n),(m_1:n_1),\cdots,(m_i:n_i)}$, the renormalization procedure of the nonlinear Hamiltonian system should be developed. If the sub-resonance is very weak, the energy increment $\Sigma_{i=1}^{\infty} \mid \Delta H_0^{(m:n),(m_1:n_1),\cdots,(m_i:n_i)} \mid$ can be ignored. Further, Equation (6.114) can reduce to Eq.(6.113). The condition for approximate estimates of an $(M:N)$-primary resonance induced in the $(m:n)$-resonant layer is

$$\left|\Delta H_0^{(m:n)}\right| + \sum_{i=1}^{\infty}\left|\Delta H_0^{(m:n),(m_1:n_1),\cdots,(m_i:n_i)}\right| = \left|E_0^{(m:n)} - E_0^{(M:N)}\right|. \tag{6.115}$$

Again, if the sub-resonance effect is very weak, the approximate condition is

$$\left|\Delta H_0^{(m:n)}\right| \approx \left|E_0^{(m:n)} - E_0^{(M:N)}\right|. \tag{6.116}$$

6.3.3. Numerical predictions

If we can obtain an exact expression of $\Delta H_0^{(m:n)}$ for the primary $(m:n)$-resonance, including sub-resonant effects, equations (6.113) and (6.116) can give an exact prediction. In fact, the exact expression of the energy increment is very difficult.

However, the exact energy increment can be computed numerically. Therefore, to predict the resonant layer, the energy increment spectrum should be used to verify the approximate analytical prediction. The Poincaré mapping section for the harden spring oscillator is defined as

$$\Sigma=\left\{\left(x(t_N),y(t_N)\right)\left|\begin{array}{l}\left(x(t_N),y(t_N)\right)\text{ satisfying Eq.(6.77), and}\\ t_N=\dfrac{2N\pi}{\Omega}+t_0, N=0,1,\cdots\end{array}\right.\right\} \quad (6.117)$$

where $x(t_N)=x_N, \dot{x}(t_N)=\dot{x}_N$ and $x(t_0)=x_0, \dot{x}(t_0)=\dot{x}_0$ at $t=t_0$ are the initial conditions. The Poincaré map is $P:\Sigma\rightarrow\Sigma$. The energy for each Poincaré mapping point of the Duffing oscillator is

$$H_0^{(N)}=\tfrac{1}{2}\dot{x}_N^2+\tfrac{1}{2}\alpha_1 x_N^2+\tfrac{1}{4}\alpha_2 x_N^4, \quad (6.118)$$

and as in Eq.(6.39), its minimum and maximum energy increment based on the resonant energy $E^{(2n-1:1)}$ are computed for the $(2n-1:1)$-resonant layer as

$$\Delta E_{\max}=\max_{N\rightarrow\infty}\left\{H_0^{(N)}-E_0^{(2n-1:1)}\right\} \text{ and } \Delta E_{\min}=\min_{N\rightarrow\infty}\left\{H_0^{(N)}-E_0^{(2n-1:1)}\right\}. \quad (6.119)$$

Consider parameters $\alpha_1=\alpha_2=1$ and $Q_0=10$ in the hardening spring system in Eq.(6.77). The conservative energy increment spectrum is illustrated in Fig.6.13. This energy increment spectrum is illustrated via the maximum and minimum conservative energy increments versus excitation frequency. In Fig.6.13(a), the range for the (3:1)-resonant layer is shaded. The range for the sub-resonant web in the (3:1)-resonant layer is presented. The acronym "RL" denotes the primary "resonant layer". The onset of the (3:1)-primary resonant layer is formed at excitation frequency of $\Omega^{\text{app}(3:1)}\approx 6.374556$. With decreasing excitation frequency, the sub-resonant effects on the (3:1) -primary resonance become more effective. It is assumed that such a primary resonant layer is destroyed once the sub-resonance becomes significant. The critical value for the destruction of the (3:1)-resonant layer are $\Omega^{\text{des}(3:1)}\approx 5.566478$. If a sub-resonant network in the (3:1)-primary resonant layer is developed, the (3:1)-primary resonant layer with the sub-resonant network will be formed. Once the (1:1)-primary resonant is involved in the (3:1)-resonant layer, the resonant web are formed by two primary resonance overlap. Therefore, in Fig.6.13(b), the excitation range for the (1:1)-resonant layer induced in the (3:1) resonant layer are presented.

The approximate analytical predictions of the resonant layer in such a system in Eq.(6.77) are given by Eqs.(6.110) and (6.111) or (6.112). Such approximate analytical predictions are presented in Fig.6.14. The solid and chain curves represent the analytical predictions for the destruction and onset of the resonant layer by Eqs.(6.110) and (6.111) or (6.112), respectively. The hollow and solid circul-

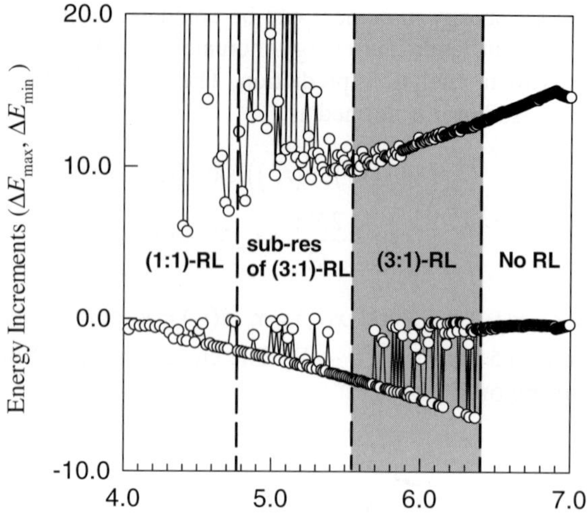

(a)

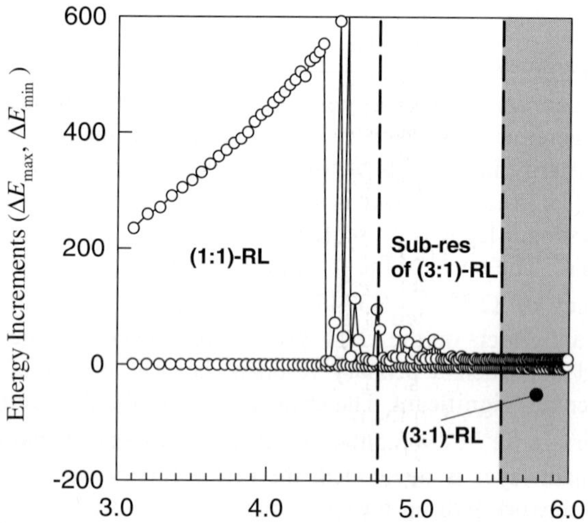

(b)

Figure 6.13. Minimum and maximum energy increment spectrums for the hardening spring system ($\alpha_1 = \alpha_2 = 1$, $Q_0 = 10.0$). (a) (3:1)-resonant layer, (b) (1:1)-resonant layer induced in (3:1) resonant layer. The range for the sub-resonant web in the (3:1)-resonant layer is presented. The shaded area is the (3:1)-primary resonant layer. The acronym "RL" denotes "resonant layer". "Sub-res" represents "sub-resonance".

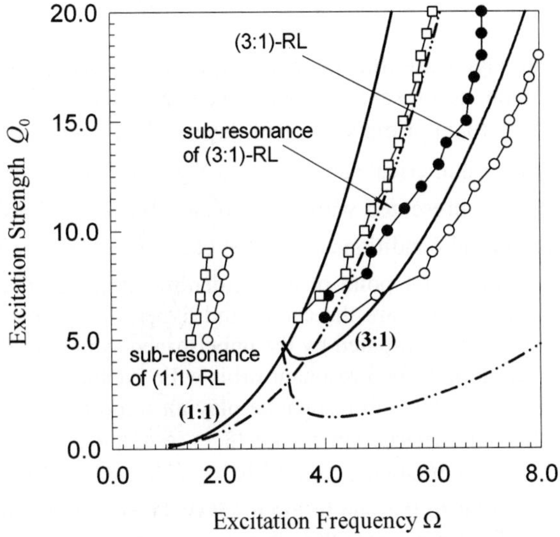

Figure 6.14. Comparison between the approximate analytical prediction and numerical prediction ($\alpha_1 = \alpha_2 = 1$). The solid and chain curves represent the analytical predictions for the destruction and onset of the resonant layer. The hollow and solid circular symbol curves represent the numerical predictions for the destruction and onset of the resonant layer. The square symbol curve represents a new primary resonant layer involved. The region between the solid circular symbol and squared curves are for the sub-resonant web is involved in the primary resonant layer. The acronym "RL" denotes resonant layer.

by Eqs.(6.110) and (6.111) or (6.112), respectively. The hollow and solid circular symbol curves represent the numerical predictions for the destruction and onset of the resonant layer, which is given by the energy increment spectrum technique. The square symbol curve represents a new primary resonant layer involved. The region between the solid circular symbol and squared curves are for the sub-resonant web involved in the primary resonant layer. Because the sub-resonance is effective, in the analytical conditions, the energy increment for the sub-resonance layer is not computed. So the analytical and numerical predictions of such a resonant layer have a big difference. So the resonant layer dynamics should be investigated through the renormalization procedure of the nonlinear Hamiltonian system in Eq.(6.77). For the (1:1)-primary resonant layer, the sub-resonant web will play a very important role to form the resonant network web. The dynamics of the corresponding sub-resonant layer should be discussed. Surprisingly, the numerical prediction for the (1:1)-primary resonant layer induced in the (3:1)-primary resonant layer agrees very well with the analytical condition for the onset of the (1:1)-primary resonant layer.

6.3.4. Illustrations of resonant layers

The (3:1)-resonant layers are illustrated in Fig.6.15 for a better understanding of the onset, destruction, sub-resonant network web and the (1:1)-primary resonant layer induced in the (3:1)-primary resonant layer. The parameters ($\alpha_1 = \alpha_2 = 1$, $Q_0 = 10$) are used for numerical simulation. In Fig. 6.15(a), the near onset of the (3:1)-resonant layer is presented with the resonant frequency $\Omega \approx 6.374556$, and the corresponding initial conditions ($x_0 \approx 0.990831$, $y_0 \approx 3.89054$). The big red circular symbol is the center point of the hardening spring oscillator. The dark and yellow circular symbols are the center and hyperbolic points for the (3:1)-resonant layer, which are computed by the unperturbed system in Eq.(6.98). The dashed curve is the unperturbed resonant orbit with excitation frequency of $\Omega \approx 6.374556$. Just before the sub-resonance play an important role in the (3:1)-resonant layer. Just the destruction of the (3:1)-resonant layer is presented in Fig.6.15(b) with ($\Omega \approx 5.566487$, $x_0 \approx 0.833013$, $y_0 \approx 2.839953$). The profile of resonant layer is similar to the one in Fig.6.15(a). The circular symbols keep of the same meaning. To demonstrate the sub-resonant network web in the resonant layer, the excitation frequency of $\Omega \approx 5.143333$ with the corresponding initial conditions ($x_0 \approx 0.746687$, $y_0 \approx 2.342441$) are used in Fig.6.15(c). The different circular symbols represent the center and hyperbolic points for the different levels of the resonant layer. The original center is the largest one. The center and hyperbolic points for the third generation of the primary resonance (or sub-sub-resonance) are the smallest points. The sub-resonant network web in the (3:1)-resonant layer is presented. It is very difficult to find the self-similarity. In Fig.6.15(d), the (1:1)-primary resonant layer induced in the (3:1)-resonant layer is presented with $\Omega = 4.587102$, $x_0 \approx 0.626703$ and $y_0 \approx 1.740723$. For a clear illustration, only the center and hyperbolic points for the (1:1) and (3:1)-primary resonant layers are marked. It is obviously seen that the (1:1)-primary resonant layer is induced in the (3:1)-resonant layer. For the better prediction of such a resonant layer, analytical and numerical techniques should be further developed. The primary resonant layer in the soften-spring Duffing oscillator can be found in Luo (1995) (or Han and Luo, 1998). For the Duffing oscillator with the negative linear stiffness, the resonant layer can be found in Luo (1995), Luo and Gu (2003). The resonant layer in the forced pendulum was presented in Luo and Han (2000b). The resonant layer in the MEMS was presented in Luo and Wang (2002, 2004). For parametric Duffing and pendulum oscillators, the resonant layers were investigated in Luo (2001b, 2002, 2004b). Luo (2005b) applied such a theory to investigate chaotic motion in traveling plates. In Eq.(6.98), it is observed that the rescaled new dynamical system is a kind of parametric pendulum oscillator. Therefore, in the next section, the resonant layers in the parametric pendulum will be discussed.

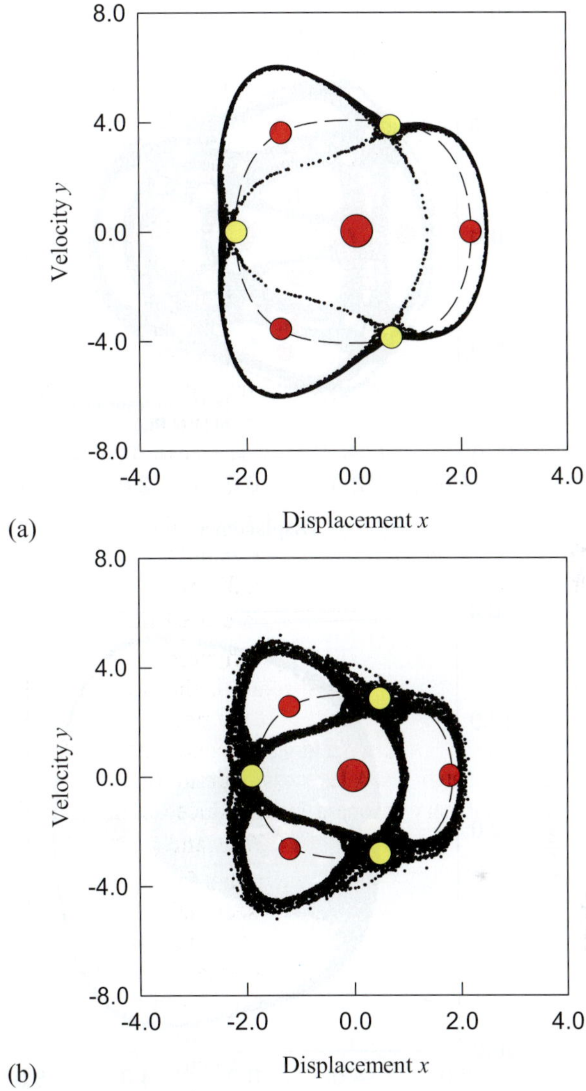

Figure 6.15. Numerical illustrations for a (3:1)-resonant layer ($\alpha_1 = \alpha_2 = 1$, $Q_0 = 10$): (a) near onset ($\Omega \approx 6.374556$, $x_0 \approx 0.990831$, $y_0 \approx 3.89054$), (b) near destruction ($\Omega \approx 5.566487$, $x_0 \approx 0.833013$, $y_0 \approx 2.839953$), (c) sub-resonant web ($\Omega \approx 5.143333$, $x_0 \approx 0.746687$, $y_0 \approx 2.342441$), (d) (1:1)-resonant layer induced ($\Omega = 4.587102$, $x_0 \approx 0.626703$, $y_0 \approx 1.740723$). The circular symbols represent the center and hyperbolic points for resonant layers and sub-resonant network webs.

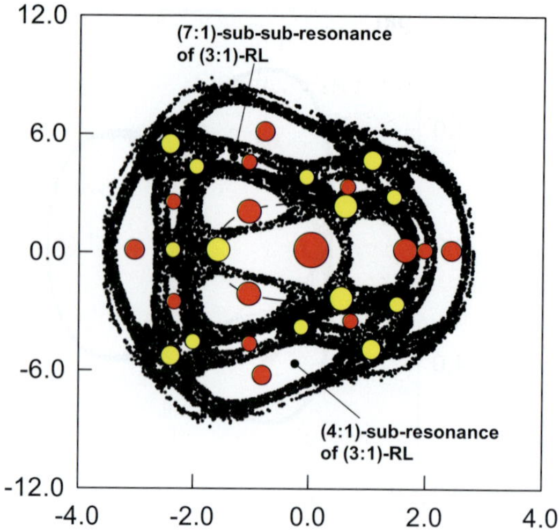

(c)

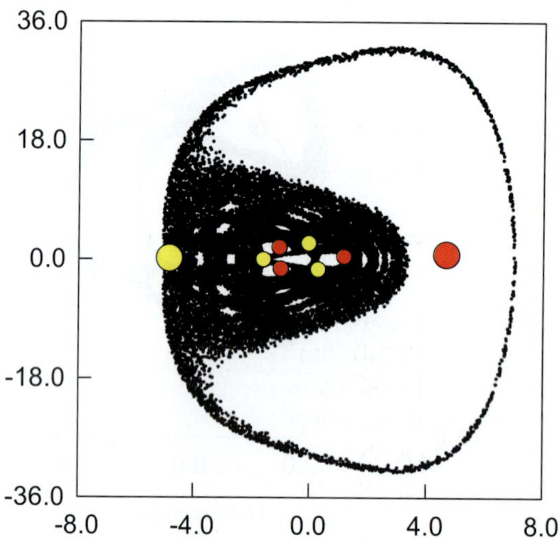

(d)

Figure 6.15. (continued)

6.4. Resonant layers in a parametric pendulum

The pendulum oscillator possesses qualitative, dynamic characteristics that exist in other nonlinear dynamic systems. As in Eqs.(6.26) and (6.98), it is necessary to investigate dynamics in parametric pendulum for a better understanding of dynamics in the resonant layer. Luo and Han (2000b) investigated the stochastic and resonant layers in the periodically driven pendulum. The dynamic behaviors in a parametrically excited pendulum are different from the ones in the periodically forced one. Therefore, consider a parametrically excited pendulum in order to further understand the mechanism of the resonant separatrix in 2-dimesnional, periodically forced, nonlinear Hamiltonian systems. The parametric pendulum is given by

$$\ddot{x}+(\alpha+Q_0\cos\Omega t)\sin x=0, \tag{6.120}$$

where Q_0 and Ω are excitation strength and frequency. The Hamiltonian of Eq. (6.120) is

$$\begin{aligned} H(x,y,t)&=\tfrac{1}{2}y^2-\alpha\cos x-Q_0\cos\Omega t\cos x\\ &=H_0(x,y)+H_1(x,y,t) \end{aligned} \tag{6.121}$$

where

$$\begin{aligned} &H_0=\tfrac{1}{2}y^2-\alpha\cos x \quad \text{and}\\ &H_1=-Q_0\cos\Omega t\cos x=-\tfrac{1}{2}Q_0\left[\cos(x+\Omega t)+\cos(x-\Omega t)\right]; \end{aligned} \tag{6.122}$$

with $y=\dot{x}$. In the conservative system of Eq.(6.120), elliptic points $(\pm 2j\pi,0)$ and hyperbolic points $(\pm(2j+1)\pi,\ 0)$ for ($j=0,1,\cdots$) exist. Two homoclinic orbits connecting all the hyperbolic points separates the phase space in the conservative system of Eq.(6.120) into the librational and rotational motions. The resonant layers located in regions associated with the librational and rotational motion regions are termed *the librational and rotational resonant layers* in the parametrically excited pendulum. The resonant layers for the parametrically excited pendulum are presented through the Poincaré mapping section given by

$$\Sigma=\left\{\left(x(t_N),\dot{x}(t_N)\right)\,\middle|\,\begin{array}{l}\text{satisfying Eq.(6.120)},\ x(t_N)\bmod 2\pi\\ \text{and } t_N=\dfrac{2N\pi}{\Omega}+t_0, N=0,1,\cdots\end{array}\right\} \tag{6.123}$$

where $x(t_N)=x_N$, $\dot{x}(t_N)=\dot{x}_N$ and $x(t_0)=x_0$, $\dot{x}(t_0)=\dot{x}_0$ at $t=t_0$ are initial conditions. The Poincaré map is $P:\Sigma\rightarrow\Sigma$.

To understand the mechanism of the resonant separatrix layer of nonlinear

Hamiltonian systems, it is necessary to show the dynamic characteristics of the stocha-stic layer in Eq.(6.120). With the parameter ($\alpha = 1.0$), the stochastic layer of the parametric pendulum is shown in Fig.6.16. The thin stochastic layer with ($Q_0 = 0.01, \Omega = 1.9$) is presented in Fig.6.16(a). The initial condition for the stochastic layer is chosen at hyperbolic points $(\pm(2j+1)\pi, 0)$. The stochastic layer includes the (4:1) and higher order resonant separatrices in the librational stochastic layer and the (3:1) and higher order resonant separatrices in the rotational stochastic layer. With a similar resonance structure, the thick (3:1)-*rotational* stochastic layer is illustrated in Fig.6.16(b) with ($Q_0 = 0.05, \Omega = 2.5$). It is observed that the stochastic layer with resonant separatrices become much thicker than in Fig.6.16(a). The investigation on dynamics in the stochastic layer of the parametric pendulum can be found in Luo (2001b). In this Chapter, dynamics in the resonant layer will be focused on for a better understanding of the sub-resonant separatrix effects in the resonant layer.

Because the primary resonant layer is formed in the vicinity of the primary resonant separatrix in Eq.(6.77). With varying parameters, the corresponding sub-resonance separatrices will exist in such a primary resonant layer, as shown in Fig.6.15(c). With increasing effects of the sub-resonant separatrix, the primary resonant layer will overlap with the other closest primary resonant separatrix as in Fig.6.15(d). To understand the sub-resonance mechanism of chaotic motions in the primary resonant layer of 2-dimensional, periodically forced, nonlinear Hamiltonian systems, the resonant separatrix layers in Eq.(6.120) should be investigated. Before presenting the analytical prediction, the numerical simulations are presented first for an intuitive impression of the primary resonant layer in the parametric pendulum. In Fig.6.17(a), the (2:1)- and (4:1)-*librational* resonant layers inside the separatrix (or heteroclinic orbit) are presented with parameters ($\Omega \approx 1.5721$, $Q_0 = 0.06$, $x_0 = 0$, $y_0 \approx 1.603122$) and ($\Omega \approx 2.328332$, $Q_0 = 0.02$, $x_0 = 0$, $\dot{x}_0 \approx 1.920937$), respectively. Fig. 6.17(b) gives the (2:1)- primary resonant layers outside the separatrix with excitation strength ($Q_0 = 0.1$) and resonant excitation frequency and initial conditions ($\Omega \approx 3.222314$, $x_0 \approx 1.938710$ and $y_0 \approx 1.555847$) and ($\Omega \approx 3.987793$, $x_0 \approx 1.526633$ and $y_0 \approx 1.615023$), respectively. The following subsection gives approximate prediction of the resonant layers in the parametrically excited pendulum.

6.4.1. Librational resonant layer

From a conservative energy $H_0 = E_l$ $(0 \le E_l < \alpha)$, a time-dependent Hamiltonian H_1 in Eq.(6.121) is approximated by

$$H_1^l \approx -Q_0 \cos(\Omega t) \cos x_l^0. \tag{6.124}$$

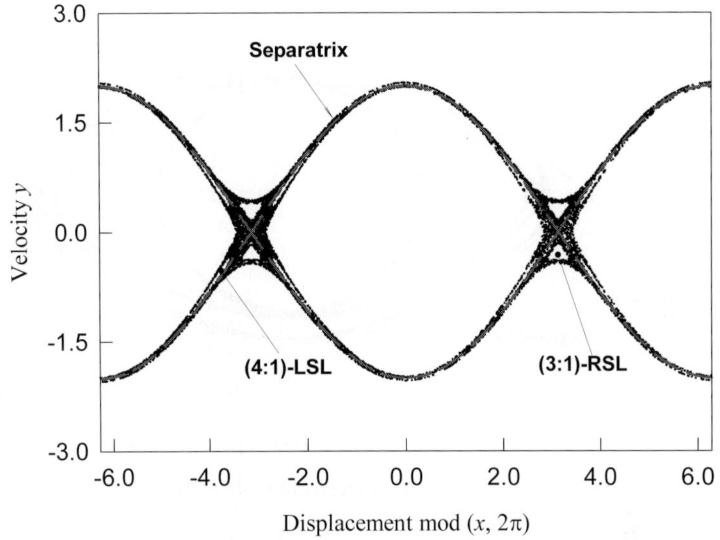

(a)

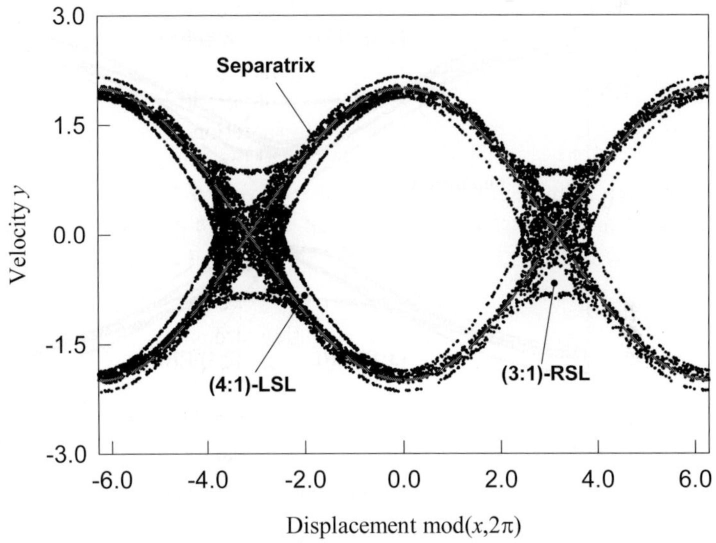

(b)

Figure 6.16. Stochastic layers with resonance near separatrix ($\alpha = 1.0$): (a) thin layer (Q_0 = 0.01, $\Omega = 1.9$) and (b) thick layer (Q_0 = 0.05, $\Omega = 2.5$). Acronyms "LSL" and "RSL" represents librational and rotational stochastic layers, respectively.

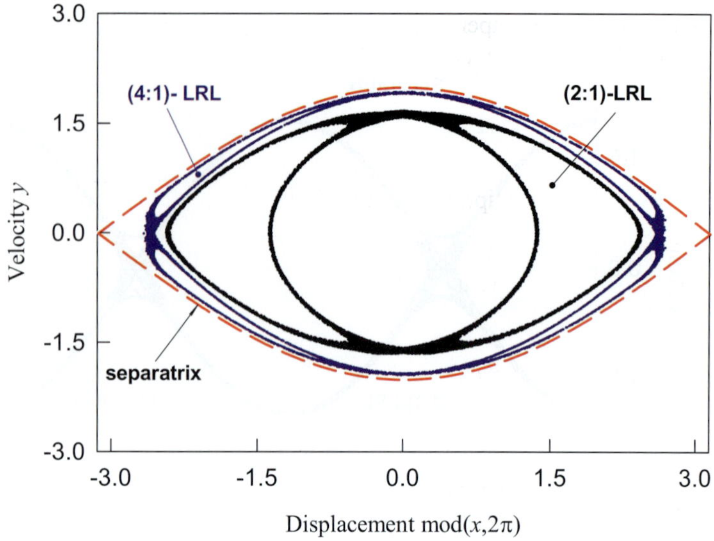

(a)

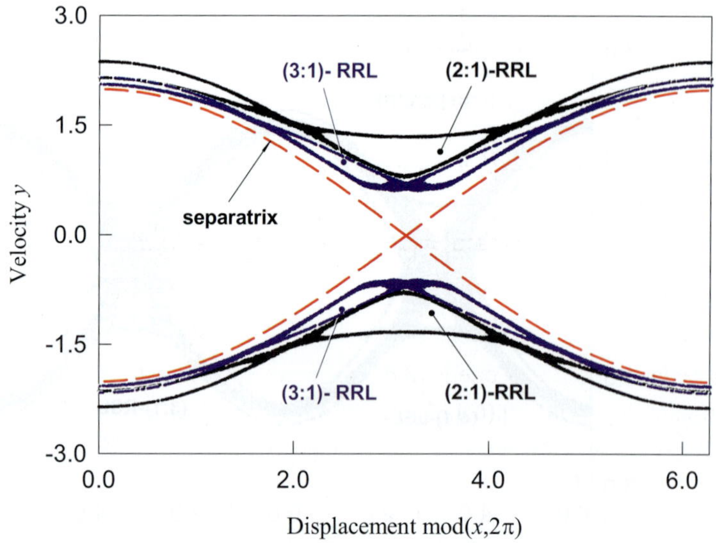

(b)

Figure 6.17. (a) Two *librational* resonant layers (**LRL**) ((2:1)-layer: $\Omega \approx 1.572100$, $Q_0 = 0.06$, $x_0 = 0$, $y_0 \approx 1.603122$, and (4:1)-layer: $\Omega \approx 2.328332$, $Q_0 = 0.02$, $x_0 = 0$, $y_0 \approx 1.920937$); (b) the *rotational* resonant layers (**RRL**) ((2:1)-layer: $\Omega \approx 3.222313$, $Q_0 = 0.1$, $x_0 \approx 1.93871$, $y_0 \approx 1.555847$, and (3:1)-layer: $\Omega \approx 3.987793$, $Q_0 = 0.1$, $x_0 \approx 1.526633$, $y_0 \approx 1.615023$).

where the subscript (or superscript) l denotes the libration. x_l^0 is the librational solution of the unperturbed pendulum,

$$x_l^0 = \pm 2j\pi \pm 2\arcsin\left[k_l \mathrm{sn}\left(\tfrac{2K(k_l)\theta_l}{\pi}, k_l\right)\right], \tag{6.125}$$

where sn is the Jacobi elliptic functions, and $k_l = 1/\sqrt{2}\sqrt{1+E_l/\alpha}$ is the elliptic modulus. $K(k_l)$ is the elliptic integral of the first kind. The phase angle φ^l, natural frequency ω_l (or period T_l) are

$$\varphi^l = \omega_l t + \varphi_0^l \quad \text{and} \quad \omega_l = \frac{2\pi}{T_l} = \frac{\pi\sqrt{\alpha}}{2K(k_l)}, \tag{6.126}$$

where φ_0^l is the initial phase angle.

For the librational potential well, substitution of the Fourier expansion of x_l^0 into Eq.(6.124) leads to

$$\begin{aligned} H_1^l \approx -Q_0 \cos\Omega t + \frac{2\pi^2 Q_0}{K_l^2} \sum_{M=1}^{\infty}\sum_{m=1}^{\infty}\sum_{n=1}^{\infty}\left(p_{2m-1}^l p_{2n-1}^l \delta_{m+n-1}^M - p_{2m-1}^l p_{2n-1}^l \delta_{|m-n|}^M\right) \\ \times\left[\sin(2M\omega_l + \Omega)t + \sin(2M\omega_l - \Omega)t\right], \end{aligned} \tag{6.127}$$

where the Delta function $\delta_i^j = 1$ $(i=j)$ or 0 $(i \neq j)$, $|\bullet|$ is the absolute value and

$$p_{2i-1}^l = \mathrm{csch}\left[\pi\left(i-\tfrac{1}{2}\right)\frac{K_{2i-1}^{\prime l}}{K_{2i-1}^l}\right] \quad \text{for } i \in \{m,n\}. \tag{6.128}$$

where $K_{2i-1}^{\prime l} = K(k_l')$ and $k_l' = \sqrt{1-k_l^2}$. All the other time-dependent terms in H_1^l will average to zero over one period $2\pi/\Omega$ except for the $(2M\omega_l - \Omega)$- harmonic terms. Thus, from Eq.(6.127), the librational resonant condition and the corresponding modulus are

$$2M\omega_l = \Omega \quad \text{and} \quad k_l^{(2M:1)} = \frac{\sqrt{\alpha + E_l^{(2M:1)}}}{\sqrt{2\alpha}}, \quad \text{at } H_0 < \alpha, \tag{6.129}$$

where $E_l^{(2M:1)}$ denotes the conservative energy related to the $(2M:1)$-librational resonance.

Since the resonant separatrix can be approximated by the corresponding two unperturbed orbits in Eq.(6.125), from two unperturb-ed orbits relative to the $(2M:1)$-librational resonant separatrix, the energy increment isapproximated by

$$\Delta H_0^l = 2\int_{t_0}^{T_l+t_0} \{H_0, H_1\}\,dt = 2\int_{t_0}^{T_l+t_0} (f_1 g_2 - f_2 g_1)dt \approx 2Q_0 Q_l^{(2M:1)} \sin \Omega t_0, \quad (6.130)$$

with

$$Q_l^{(2M:1)} \approx \frac{4\pi^2}{\sqrt{\alpha}K_l\Omega}\sum_{m=1}^{\infty}\sum_{n=1}^{\infty}\left(p_{2m-1}^l p_{2n-1}^l \delta_{m+n-1}^M - p_{2m-1}^l p_{2n-1}^l \delta_{|m-n|}^M\right), \quad (6.131)$$

$$p_{2i-1}^l = \operatorname{csch}\left[\pi\left(i-\tfrac{1}{2}\right)\frac{K_{2i-1}'^l}{K_{2i-1}^l}\right] \quad \text{for } i \in \{m,n\}. \quad (6.132)$$

where $k_l' = \sqrt{1-k_l^2}$, $\{\bullet,\bullet\}$ represents the Poisson bracket and

$$\begin{aligned} &f_1 = y \approx y_l^0,\; g_1 = 0;\;\; f_2 = \alpha \sin x \approx \alpha \sin x_l^0, \\ &g_2 = Q_0 \cos \Omega t \sin x \approx Q_0 \cos \Omega t \sin x_l^0. \end{aligned} \quad (6.133)$$

As presented in references (i.e., Luo and Han, 1999; Luo et al, 1999), the energy increments (ΔH_0^l) and phase changes ($\Delta\varphi_i^l = \Omega T_l$) give an whisker maps for the $(2M:1)$ - librational resonant layer:

$$E_{i+1}^l \approx E_i^l + 2Q_0 Q_l^{(2M:1)} \sin \varphi_{i+1}^l \;\text{ and }\; \varphi_{i+1}^l = \varphi_i^l + \frac{4\Omega K(k_l)}{\sqrt{\alpha}}, \quad (6.134)$$

where $\varphi_i^l = \Omega t_i + \varphi_0^l$, E_i^l is the conservative energy and $\Delta H_0^l = E_{i+1}^l - E_i^l$, $\Delta\varphi_i^l = \varphi_{i+1}^l - \varphi_i^l$. In the second one of Eq.(6.122), $\varphi_{i+1}^l - \varphi_i^l = 2(2M)\pi$ gives the resonant energy $E_l^{(2M:1)}$. From the accurate standard mapping approach (e.g., Luo, 1995; Luo and Han,1999), an approximate, critical condition for the onset of the librational resonant layer is from Eq.(6.36)

$$\begin{aligned} &Q_0 \approx \frac{0.9716354}{2Q_l^{(2M:1)}\left|G_l^{(2M:1)}\right|}, \text{ and} \\ &G_l^{(2M:1)} = \frac{\Omega}{(\sqrt{\alpha})^3 (k_l^{(2M:1)})^2}\left[\frac{E(k_l^{(2M:1)})}{1-(k_l^{(2M:1)})^2} - K(k_l^{(2M:1)})\right], \end{aligned} \quad (6.135)$$

where

$$k_l^{(2M:1)} = \sqrt{\alpha + E_l^{(2M:1)}}\Big/\sqrt{2\alpha}. \quad (6.136)$$

Let $E_i^l = E_l^{(2M:1)}$ and $E_{i+1}^l = E_l^{(2M+2:1)}$ and the excitation strength Q_0 for the $(2M:1)$

-librational resonant layer destroyed by the $(2M+2:1)$ resonance is approximately predicted from Eq.(6.37) (also see, Luo, 1995; Han and Luo, 1998) by

$$Q_0 = \frac{\left|E_l^{(2M+2:1)} - E_l^{(2M:1)}\right|}{2Q_l^{(2M:1)}} \text{ and } \Omega = \frac{2M\pi\sqrt{\alpha}}{K(k_l^{(2M:1)})} \tag{6.137}$$

Setting $E_i^l = E_l^{(2M:1)}$ and $E_{i+1}^l = \alpha$, the excitation strength Q_0 for the $(2M:1)$-order librational resonant layer involved in the stochastic layer is roughly estimated by

$$Q_0 = \frac{\left|\alpha - E_l^{(2M:1)}\right|}{2Q_l^{(2M:1)}} \text{ and } \Omega = \frac{2M\pi\sqrt{\alpha}}{K(k_l^{(2M:1)})}. \tag{6.138}$$

6.4.2. Rotational resonant layer

In a like fashion, for a conservative energy $H_0 = E_r > \alpha$, the time-dependent Hamiltonian H_1 in Eq. (6.122) for the rotation is approximated by

$$H_1^r \approx -Q_0 \cos(\Omega t) \cos x_r^0, \tag{6.139}$$

where the subscript (or superscript) r denotes the rotation, and

$$x_r^0 = \pm 2\mathrm{am}\left[\frac{K(k_r)\varphi_r}{\pi}, k_r\right]; \tag{6.140}$$

where am is the amplitude of Jacobi elliptic function. The modulus $k_r = \frac{\sqrt{2\alpha}}{\sqrt{E_r+\alpha}}$ and phase angle $\varphi_r = \omega_r t + \varphi_{r0}$. The natural frequency ω_r (or period T_r) is

$$\omega_r = \frac{2\pi}{T_r} = \frac{\pi\sqrt{\alpha}}{k_r K(k_r)}. \tag{6.141}$$

Substitution of Eq.(6.140) into H_1 in Eq.(6.139) and the Fourier expansion of the elliptic function (*am*) gives

$$\begin{aligned} H_1^r \approx& -Q_0 \cos\Omega t + \frac{\pi^2 Q_0}{k_r^2 K_r^2} \sum_{M=1}^{\infty}\sum_{m=1}^{\infty}\sum_{n=1}^{\infty} \left(p_{2m-1}^r p_{2n-1}^r \delta_{m+n-1}^M - p_{2m-1}^r p_{2n-1}^r \delta_{|m-n|}^M\right) \\ &\times\left[\sin(M\omega_r + \Omega)t + \sin(M\omega_r - \Omega)t\right]. \end{aligned} \tag{6.142}$$

Therefore, in the vicinity of the $(M:1)$-primary resonance, all other terms in H_1^r in Eq.(6.142) will average to zero over a period $2\pi/\Omega$ except for the

$(M\omega_r - \Omega)$ - term. Thus, the rotational resonant condition and the corresponding modulus are

$$M\omega_r = \Omega \text{ and } k_r^{(M:1)} = \frac{\sqrt{2\alpha}}{\sqrt{\alpha + E_r^{(M:1)}}}, \quad \text{for } H_0 > \alpha, \tag{6.143}$$

where $E_r^{(M:1)}$ denote the conservative energy associated with the $(M:1)$-rotational resonance. The energy increment, based on two unperturbed orbits relative to $(M:1)$-rotational resonant separatrix, is approximately computed, i.e.,

$$\Delta H_0^r = 2\int_{t_i}^{T_r+t_i} \{H_0, H_1\}\, dt = 2\int_{t_i}^{T_r+t_i} (f_1 g_2 - f_2 g_1) dt \approx 2Q_0 Q_r^{(M:1)} \sin(\Omega t_i), \tag{6.144}$$

where

$$Q_r^{(M:1)} \approx \frac{2\pi^2}{\sqrt{\alpha}k_r K_r \Omega} \sum_{m=0}^{\infty}\sum_{n=1}^{\infty} \left(p_{2m-1}^r p_{2n-1}^r \delta_{m+n-1}^M - p_{2m-1}^r p_{2n-1}^r \delta_{|m-n|}^M \right). \tag{6.145}$$

Similarly, the energy increment (ΔH_0^r) and phase change ($\Delta\varphi_i^r = \Omega T_r$) generate a whisker map for the $(M:1)$-rotational resonant layer. So we have,

$$E_{i+1}^r \approx E_i^r + 2Q_0 Q_r^{(M:1)} \sin\varphi_i^r \text{ and } \varphi_{i+1}^r = \varphi_i^r + \frac{2\Omega k_r K(k_r)}{\sqrt{\alpha}}, \tag{6.146}$$

From the accurate standard mapping approach in Luo (2001a), the condition for the appearance of the (M:1)-rotational resonant layer is obtained from Eq.(6.36), i.e.,

$$Q_0 \approx \frac{0.9716354}{2Q_r^{(M:1)}\left|G_r^{(M:1)}\right|} \text{ and } G_r^{(M:1)} = -\frac{\Omega (k_r^{(M:1)})^3}{2(\sqrt{\alpha})^3[1-(k_r^{(M:1)})^2]} E(k_r^{(M:1)}). \tag{6.147}$$

$$k_r^{(M:1)} = \sqrt{2\alpha}\Big/\sqrt{\alpha + E_r^{(M:1)}}. \tag{6.148}$$

When $E_i^r = E_l^{(M:1)}$ and $E_{i+1}^r = E_l^{(M+1:1)}$ are used, the excitation strength Q_0 for the $(M:1)$-rotational resonant layer destroyed by the $(M+1:1)$-resonance is approximately computed from Eq.(6.37) (also see Luo, 1995; Han and Luo,1998), i.e.,

$$Q_0 = \frac{\left|E_r^{(M+1:1)} - E_r^{(M:1)}\right|}{2Q_r^{(M:1)}} \text{ and } \Omega = \frac{M\pi\sqrt{\alpha}}{k_r^{(M:1)} K(k_r^{(M:1)})}. \tag{6.149}$$

Setting $E_i^r = E_r^{(M:1)}$ and $E_{i+1}^r = \alpha$, the excitation strength Q_0 for the $(M:1)$-rotational resonant layer involved in the stochastic layer is estimated by

$$Q_0 = \frac{\left|\alpha - E_r^{(M:1)}\right|}{2Q_r^{(M:1)}} \text{ and } \Omega = \frac{M\pi\sqrt{\alpha}}{k_r^{(M:1)}K(k_r^{(M:1)})}. \tag{6.150}$$

6.4.3. Energy increment spectrum

In the previous analytical predictions, the energy increment is computed approximately. As stated before, the exact energy increment can be numerically computed. Therefore, in 1999, Luo et al developed the energy spectrum method, which is used to determine the onset of resonance in the stochastic layer. In the energy spectrum, the maximum and minimum conservative energies are computed by the Poincaré mapping section. Using the Poincaré map section in Eq.(6.123), the Poincaré map is defined by $P:\Sigma\to\Sigma$. The conservative energy for each Poincaré mapping point of the parametrically excited pendulum is

$$H_0^{(N)} = \tfrac{1}{2}\dot{x}_N^2 - \alpha\cos x_N. \tag{6.151}$$

However, in the computation of conservative energy for specified resonant layer, the energy changes in energy spectrums cannot be observed clearly. To detect the energy changes due to the sub-resonance in the specified resonant layer, the minimum and maximum energy increments with respect to the unperturbed resonant orbit are computed by

$$\Delta E_{\max} = \max_{N\to\infty}\left\{H_0^{(N)} - E_l^{(2M:1)}\right\} \text{ and } \Delta E_{\min} = \min_{N\to\infty}\left\{H_0^{(N)} - E_l^{(2M:1)}\right\} \tag{6.152}$$

for the $(2M:1)$-librational resonant layer, and

$$\Delta E_{\max} = \max_{N\to\infty}\left\{H_0^{(N)} - E_r^{(M:1)}\right\} \text{ and } \Delta E_{\min} = \min_{N\to\infty}\left\{H_0^{(N)} - E_r^{(M:1)}\right\} \tag{6.153}$$

for the $(M:1)$-rotational resonant layer. Such an energy increment versus excitation frequency is termed the energy increment spectrum as in Section 6.1.2c. For $Q_0 = 0.05$ and $\alpha = 1.0$ in Eq.(6.120), the maximum and minimum energy increment spectra are computed. The two plots of Fig.6.18 give the energy increment spectrum for the $(4:1)$-librational and $(3:1)$-rotational resonant layers. We obtain $\Omega^{\text{app}} \approx 2.7695$ and $\Omega^{\text{dis}} \approx 2.6632$ for the $(4:1)$-librational resonant layers, and $\Omega^{\text{app}} \approx 4.0381$ and $\Omega^{\text{dis}} \approx 3.6230$ are for the $(3:1)$-rotational resonant layers. The maximum and minimum energy increments are computed from 10,000 iterations of Poincaré map for each specified excitation frequency and strength.

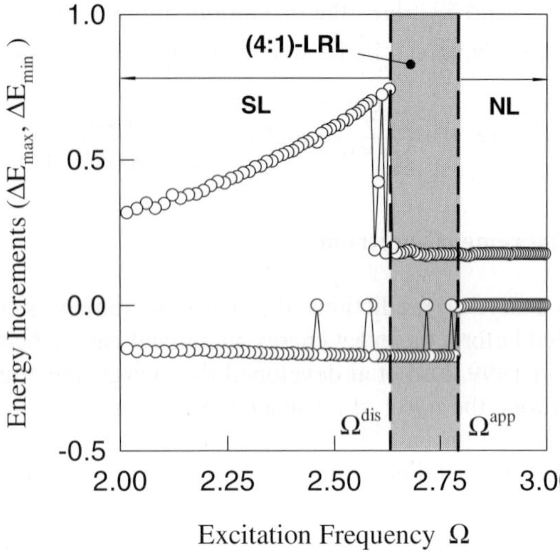

(a)

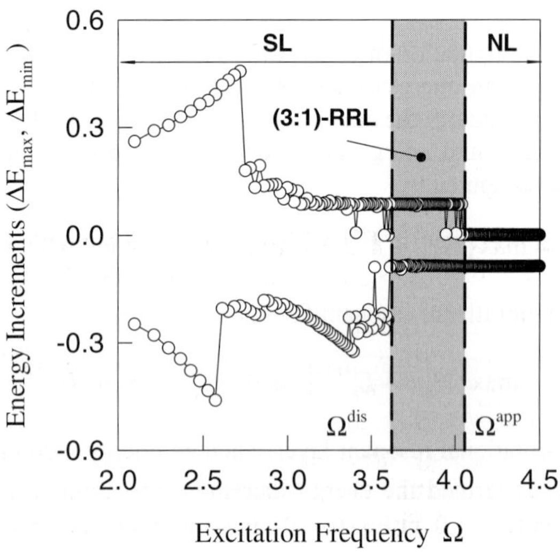

(b)

Figure 6.18. The energy increment spectrum for (a) the (4 : 1) -librational and (b) (3 : 1) -rotational resonant layers at $Q_0 = 0.05$ and $\alpha = 1.0$. The acronyms "NL", "SL","LRL" and "RRL" represent "No Layer", "Stochastic Layer", "Librational Resonant Layer" and "Rotational Resonnant Layer", respectively.

The width of a resonant layer can be computed as in references (e.g., Luo et al, 1999; Luo and Han, 2001). From the minimum and maximum energy increment spectra, the width of the librational resonant separatrix layer is computed by

$$w \equiv \min_{t\in[0,\infty)} \| \mathbf{x}(E_l^{\max},t) - \mathbf{x}(E_l^{\min},t) \| \equiv \| \mathbf{x}_l^{\max} - \mathbf{x}_l^{\min} \| \tag{6.154}$$

where $\|\bullet\|$ is a norm and the minimum and maximum energies are $E_l^{\max} = \Delta E^{\max} + E_l^{(2M:1)}$, $E_l^{\min} = \Delta E^{\min} + E_l^{(2M:1)}$. Two points $\mathbf{x}_l^{\min}$ and $\mathbf{x}_l^{\min}$ on the normal vector $\mathbf{f}^{\perp}(\mathbf{x}_l^0) = (-f_2(\mathbf{x}_l^0), f_1(\mathbf{x}_l^0))^{\mathrm{T}}$ of the tangential vector of unperturbed librational, resonant orbit at point $\mathbf{x}_l^0$ are the closest between the maximum and minimum energy orbits $\mathbf{x}(E_l^{\max},t)$ and $\mathbf{x}(E_l^{\min},t)$, which can be obtained by Eq.(6.122) with $E_l^{\max}$ and $E_l^{\min}$. Similarly, the width of the rotational resonant separatrix layer is computed by

$$w \equiv \min_{t\in[0,\infty)} \| \mathbf{x}(E_r^{\max},t) - \mathbf{x}(E_r^{\min},t) \| \equiv \| \mathbf{x}_r^{\max} - \mathbf{x}_r^{\min} \| \tag{6.155}$$

where the minimum and maximum energies are $E_r^{\max} = \Delta E^{\max} + E_r^{(M:1)}$, $E_r^{\min} = \Delta E^{\min} + E_r^{(M:1)}$. Illustration of the $(4:1)$-librational and $(3:1)$-rotational resonant layers is given in Fig.6.19 for a better understanding of the resonant layer width. Using parameters ($x_0 = 0, \dot{x}_0 \approx 1.8221$, $\Omega \approx 2.6950$, $Q_0 = 0.05$ and $\alpha = 1.0 \quad 1.0$), the $(4:1)$-librational resonant layer is placed in 6.19(a). The $(3:1)$-rotational resonant layer is presented in Fig.6.19(b), which are simulated with parameters ($x_0 \approx 1.6153$, $\dot{x}_0 \approx 1.4970$, $\Omega \approx 3.6230$, $Q_0 = 0.05$ and $\alpha = 1.0$). The homoclinic orbit (or separatrix) and the librational and rotational unperturbed orbits relative to the minimum and maximum energies ($E^{\min}$ and $E^{\max}$) are also depicted in Fig.6.19, and the layer width (w) is sketched as well. For resonant layer widths, the corresponding locations are selected at $x = \pm 2m\pi$ ($m = 0,1,2, \cdots$). The minimum and maximum energy increments with respect to the unperturbed resonant energy are used for computing the width of stochastic layer. The width of resonant layers is illustrated in Fig.6.20. The ranges for non-layer and the stochastic layers are marked as well. The width of the resonant layer is almost constant instead of the exponential decaying with excitation frequency. Such a result is different from the asymptotic analysis (e.g., Melnikov, 1963).

6.4.4. Analytical predictions with comparison

The analytical conditions for the appearance and destruction of the resonant layer have been presented and the energy spectrum approach has been given for the

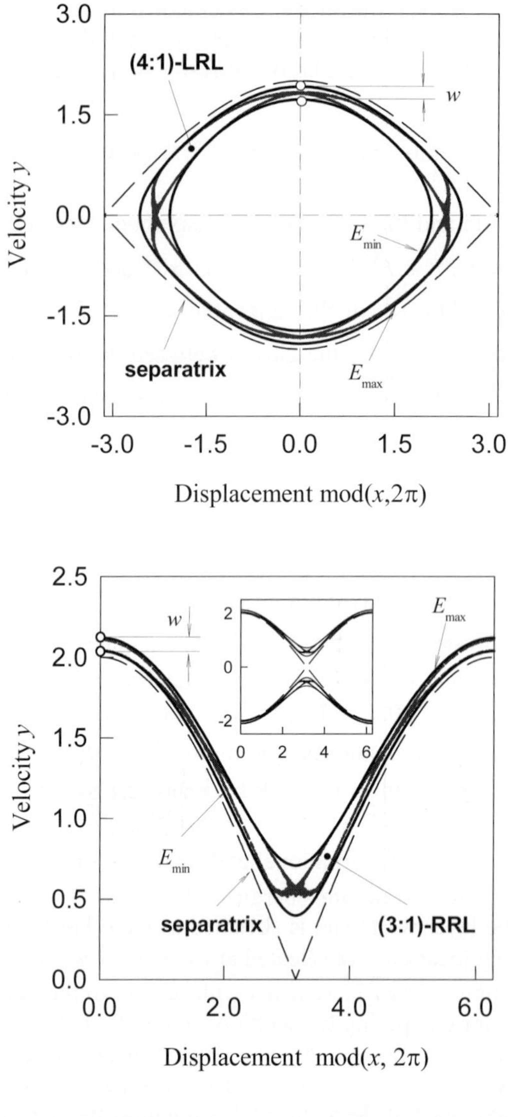

(a)

(b)

Figure 6.19. The resonant layer width ($Q_0 = 0.05$ and $\alpha = 1.0$): (a) the $(4:1)$ -librational resonant separatrix layer ($\Omega \approx 2.6950$, $x_0 = 0$, $\dot{x}_0 \approx 1.8221$) and (b) the $(3:1)$ -rotational resonant separatrix layer ($\Omega \approx 3.6230$, $x_0 \approx 1.6153$, $\dot{x}_0 \approx 1.4970$) . The acronyms "LRL" and "RRL" represent "Librational Resonant Layer" and "Rotational Resonant Layer", respectively.

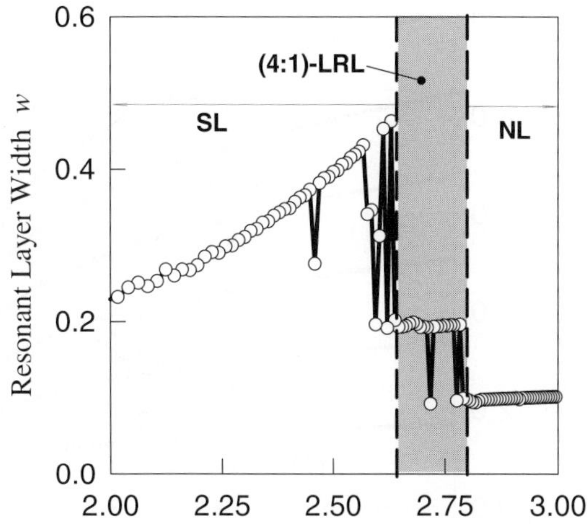

(a)

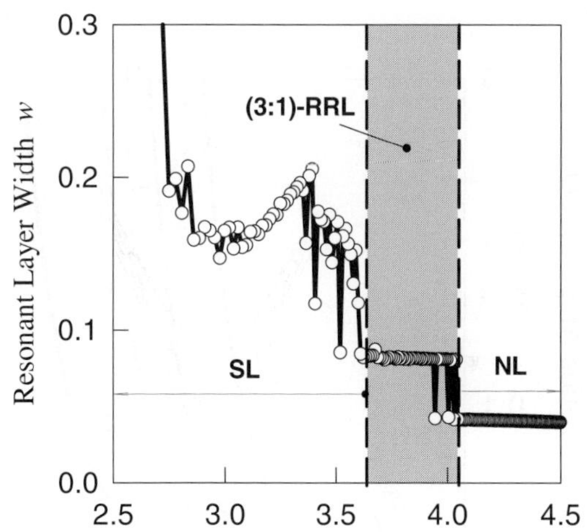

(b)

Figure 6.20. The resonant layer widths for (a) the (4:1) -librational and (b) (3:1) -rotational (lower) resonant layers at $Q_0 = 0.05$ and $\alpha = 1.0$. The acronyms "NL", "SL","LRL" and "RRL" represent "No Layer", "Stochastic Layer", "Librational Resonant Layer" and "Rotational Resonant Layer", respectively.

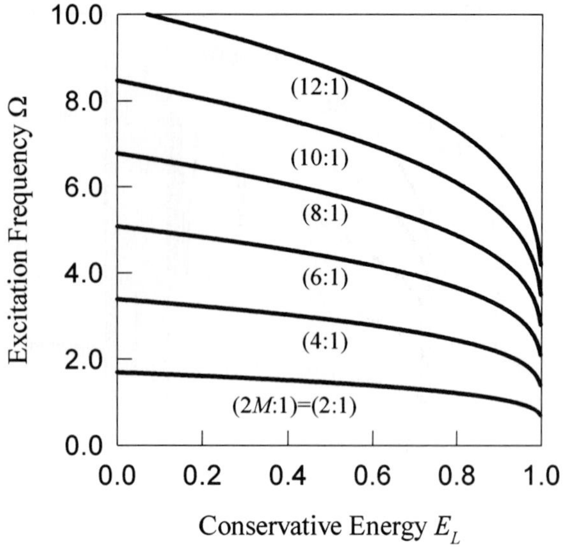

(a)

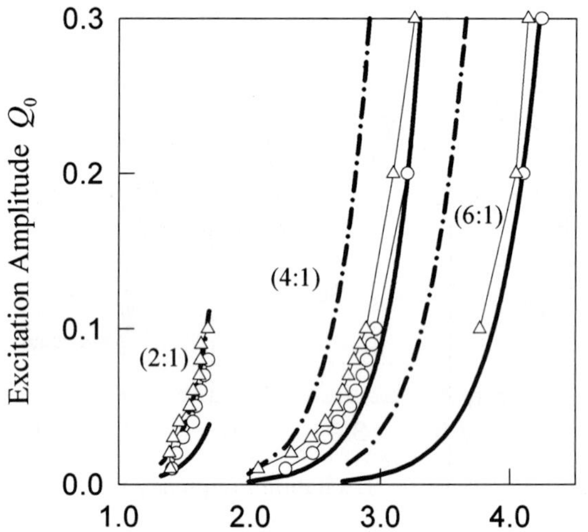

(b)

Figure 6.21. (a) The (2M:1)-librational resonant conditions and (b) the excitation strength conditions for the appearance (solid line) and disappearance (dash-dot) of the librational resonant layers at $\alpha = 1$. The circular and triangle symbol curves give the numerical predictions of the appearance and disappearance of the resonant layer.

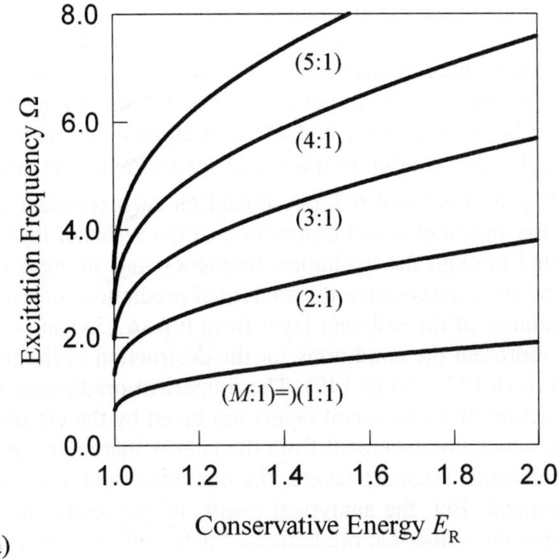

(a)

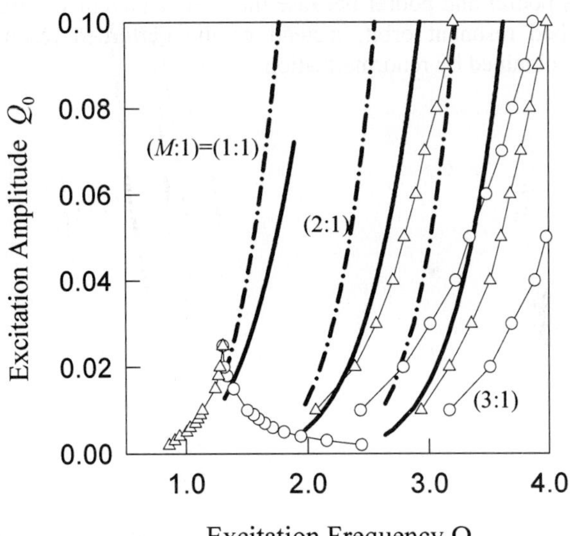

(b)

Figure 6.22.(a) The (M:1)-rotational resonant conditions, and(b) the excitation strength conditions for the appearance (solid) and disappearance (dash-dot) of the rotational resonant layers at $\alpha = 1$. The circular and triangle symbol curves give the numerical predictions of the appearance and disappearance of the resonant layers.

numerical prediction of the resonant layer. To make a comparison of two predictions, the excitation frequency versus the excitation strength for a specific resonant layer will be presented in this section. To determine a specific resonant layer, the resonant condition is very important. Thus, for the parametrically excited pendulum, the (2*M*:1)-librational and (*M*:1)-rotational resonant conditions are presented through the excitation frequency Ω versus the conservative energy H_0, as illustrated Figs.6.21 (a) and 6.22(a). Based on such resonant conditions, the conditions for the appearance and destruction of the resonant layers are computed and illustrated through the excitation frequency and strength in Figs.6.21(b) and 6.22(b). The solid curves give the analytical predictions of excitation strength for the appearance of the resonant layer from Eqs.(6.135) and (6.147), and the dashed curves represent the conditions for the destruction of the resonant layers, computed by Eqs.(6.137) and (6.149). The numerical predictions of the appearance and destruction of the resonant layers are given by the circular and triangle symbol-curves, which are generated from the energy increment spectrum technique. For the librational resonant layers, the numerical and analytical results are in a good agreement. But, the analytical results of the rotational resonant layer are different from the numerical predictions. Such a phenomenon may be caused by the sub-resonance. For the higher order resonant layer, the analytical prediction becomes poorer and poorer because the energy increments are computed by the unperturbed resonant orbit, instead of the perturbed resonant separatrix which can be obtained by renormalization.

Chapter 7

Nonlinear Dynamics on an Equi-energy Surface

In Chapter 6, the resonant layer dynamics of nonlinear Hamiltonian systems with time-periodic perturbation have been discussed. Without external perturbations, the mechanism of chaotic motion on a (2n-1)-dimensional equi-energy surface in 2n-dimensional nonlinear Hamiltonian systems will be presented in this Chapter. The internal resonances and the corresponding chaotic motions will be predicted analytically for weak interactions. To exactly compute the energy exchange, the energy spectrum will be determined from each separable system in a 2n-dimensional nonlinear Hamiltonian system. The corresponding numerical method based on the energy spectrum will be introduced for prediction of quasi-periodic and chaotic motions in nonlinear Hamiltonian systems. A two-DOF integrable, nonlinear Hamiltonian system will be discussed for demonstration of the procedure to determine the chaotic motion on the equi-energy surface in nonlinear Hamiltonian systems. The Poincare mapping surfaces of chaotic motions in such nonlinear Hamiltonian systems will be illustrated. The phase planes, displacement surfaces (or potential domains), and the velocity surfaces (or kinetic energy domains) for the chaotic and quasi-periodic motions will be presented. The analytical estimates of regular and chaotic motions on the equi-energy surface in nonlinear Hamiltonian systems need to be further investigated. The mathematical theory should be further developed for a better prediction of chaotic and quasi-periodic motions in nonlinear Hamiltonian systems with many degrees of freedom. Such an idea presented in this chapter is applicable to extendable to the nonlinear dynamics on the equi-first integral quantity surface in the n-dimesional, conserved, nonlinear dynamical systems.

7.1. Hamiltonian systems

The chaotic motion on the separatrix surface in 4-dimesnional nonlinear Hamiltonian system was presented in Luo (2006c). The materials in this chapter are mainly based on the results presented in Luo (2006a). The theory on dynamics on the (2n-1)-dimensional equi-energy surface for nonlinear Hamiltonian systems was initially discussed. From a physical point of view, consider a Hamiltonian (or the total energy) with *n-degrees of freedom* as

$$H(q_1,\cdots,q_n;p_1,\cdots,p_n)=T(p_1,\cdots,p_n)+V(q_1,\cdots,q_n) \tag{7.1}$$

with the momentum $p_i = m_i \, dq_i / dt$ and position q_i ($i = 1, \cdots, n$). The velocity is expressed by $\dot{q}_i = dq_i / dt$. Without loss generality, the kinetic and total potential energy can accordingly be expressed as

$$T = \sum_{i=1}^{n} \frac{1}{2m_i} p_i^2 \text{ and } V = \sum_{i=1}^{n} V_i(q_i) + H_1(q_1, \cdots, q_n). \tag{7.2}$$

Note that m_i ($i = 1, \cdots, n$) represents positive parameters including the mass and moment of inertia. $H_1(q_1, \cdots, q_n)$ is the interaction potential energy. Therefore, equation (7.1) is rewritten as

$$H(q_1, \cdots, q_n; p_1, \cdots, p_n) = H_0(q_1, \cdots, q_n; p_1, \cdots, p_n) + H_1(q_1, \cdots, q_n) \tag{7.3}$$

and the separable Hamiltonian expresses as

$$H_0(q_1, \cdots, q_n; p_1, \cdots, p_n) = \sum_{i=1}^{n} H_0^{(i)}(p_i, q_i). \tag{7.4}$$

Consider a single Hamiltonian without coupling in the form of

$$H_0^{(i)}(p_i, q_i) = \frac{p_i^2}{2m_i} + V_i(q_i) \qquad (i = 1, 2, \cdots, n). \tag{7.5}$$

The total energy expression becomes

$$H(q_1, \cdots, q_n; p_1, \cdots, p_n) = \sum_{i=1}^{n} \frac{p_i^2}{2m_i} + \sum_{i=1}^{n} V_i(q_i) + H_1(q_1, \cdots, q_n). \tag{7.6}$$

For any given energy $H(p_1, \cdots, p_n; q_1, \cdots, q_n) = h$, the Hamiltonian becomes

$$\sum_{i=1}^{n} \frac{p_i^2}{2m_i} + \sum_{i=1}^{n} V_i(q_i) + H_1(q_1, \cdots, q_n) = h. \tag{7.7}$$

Owing to $p_i^2 \geq 0$ and $m_i > 0$, so

$$V \equiv \sum_{i=1}^{n} V_i(q_i) + H_1(q_1, \cdots, q_n) \leq h \tag{7.8}$$

from which the maximum equi-potential energy $V_{\max} = h$. Namely, the maximum (n-1)-dimensional potential surface is

$$\sum_{i=1}^{n} V_i(q_i) + H_1(q_1, \cdots, q_n) = h \tag{7.9}$$

The positions given by Eq.(7.9) possess the maximum potential energy for a given total energy *h*. On the other hand, on the maximum equi-potential energy surface, the minimum kinetic energy is zero. However, if there is a globally minimum, finite, potential energy $V = V_{\min}$, the maximum kinetic energy is computed by

$$T_{\max} \equiv \sum_{i=1}^{n} \frac{p_i^2}{2m_i} = h - V_{\min} \tag{7.10}$$

The kinetic energy surface changes in the following range.

$$0 \leq \sum_{i=1}^{n} \frac{p_i^2}{2m_i} \leq h - V_{\min} \tag{7.11}$$

from which $h \geq V_{\min}$ should hold. Furthermore, for given total energy *h*, the potential energy surface is bounded by

$$V_{\min} \leq \sum_{i=1}^{n} V_i(q_i) + H_1(q_1, \cdots, q_n) \leq h\,. \tag{7.12}$$

From the foregoing equations, the momentums and positions are bounded. For the above Hamiltonian system, the Hamilton's equation for motion is

$$\dot{q}_i \equiv \frac{dq_i}{dt} = \frac{\partial H}{\partial p_i} \text{ and } \dot{p}_i \equiv \frac{dp_i}{dt} = -\frac{\partial H}{\partial q_i} \quad (i = 1, 2, \cdots, n). \tag{7.13}$$

7.2. Nonlinear resonance

In Chapter 6, the internal resonance has been discussed. Herein, nonlinear resonance will be discussed from a different aspect. Consider a Hamiltonian system in Eq.(7.1) at least possessing a potential well with the local minimum potential energy $V_{\min}^{loc}$ with the corresponding position $(q_1^*, \cdots, q_n^*)$, which are determined by the stable equilibrium of Eq.(7.13). For any given time *t*, the energy distribution for each decoupled Hamiltonian is

$$H_0^{(i)}(p_i, q_i) = \frac{p_i^2}{2m_i} + V_i(q_i) = E_i \quad (i = 1, 2, \cdots, n) \tag{7.14}$$

where E_i is an given energy for time *t*.

The action-angle coordinates (I_i, θ_i) are defined as in references (e.g., Arnold, 1989; Goldstein et al, 2002). From the foregoing equation, the *instantaneous* action is defined as

$$I_i = \frac{1}{2\pi}\oint_{\gamma_{E_i}} p_i dq_i \quad (i = 1, 2, \cdots, n) \tag{7.15}$$

$$\theta_i - \theta_i^0 = \int_{\theta_i^0}^{\theta_i} d\theta = \frac{\partial}{\partial I}\int_{q_i^0}^{q_i} p_i dq_i \tag{7.16}$$

where γ_{E_i} is the close path based on the given energy E_i. For each given E_i, the action I_i is constant.

$$\theta_i - \theta_i^0 = \omega_i t \tag{7.17}$$

and

$$\omega_i = \frac{\partial H_0^{(i)}(I_i)}{\partial I_i}. \tag{7.18}$$

Based on the new coordinates, the Hamiltonian in Eq.(7.6) becomes

$$H(I_1, \cdots, I_n; \theta_1, \cdots, \theta_n) = \sum_{i=1}^{n} H_0^{(i)}(I_i) + H_1(I_1, \cdots, I_n; \theta_1, \cdots, \theta_n) \tag{7.19}$$

Using the Fourier series of the interaction energy $H_1(I_1, \cdots, I_n; \theta_1, \cdots, \theta_n)$ gives

$$H_1(I_1, \cdots, I_n; \theta_1, \cdots, \theta_n) = \sum_{m_{i_1}, \cdots, m_{i_n}}^{\infty} H_{m_{i_1} \cdots m_{i_n}}^{(1)}(I_1, \cdots, I_n) e^{\left[\left(m_{i_1}\theta_1 + \cdots + m_{i_n}\theta_n\right)\mathrm{i}\right]} \tag{7.20}$$

where $\mathrm{i} = \sqrt{-1}$ and the constants m_{i_α} $(\alpha = 1, 2, \cdots, n)$ are uniquely determined by the interaction potential energy. Without the interaction potential energy, the Hamiltonian is separable with the *n*-independent Hamiltonians. Once the initial conditions are given, the *n*-independent Hamiltonians possess the periodic orbit with equi-energy surface. However, once the interaction potential energy exists, the *n*-independent Hamiltonians will have energy change through the interaction potential energy. Consider the averaging of Hamiltonian $H(I_1, \cdots, I_n; \theta_1, \cdots, \theta_n)$ for all the phase angles, i.e.,

$$\langle H(I_1, \cdots, I_n; \theta_1, \cdots, \theta_n)\rangle = \sum_{i=1}^{n}\langle H_0^{(i)}(I_i)\rangle + \langle H_1(I_1, \cdots, I_n; \theta_1, \cdots, \theta_n)\rangle \tag{7.21}$$

The angled parentheses represent the averaging of the energy as

$$\langle\bullet\rangle = \frac{1}{(2\pi)^n}\int_{\theta_i^0}^{\theta_i^0 + 2\pi} d\theta_1 \cdots \int_{\theta_n^0}^{\theta_n^0 + 2\pi} (\bullet) d\theta_n. \tag{7.22}$$

Because $H_0^{(i)}$ is independent of the phase angle, the fluctuating energy based on the averaged Hamiltonian is written like

$$H-\langle H\rangle = H_1-\langle H_1\rangle \tag{7.23}$$

For the averaged interaction potential energy,

$$\langle H_1\rangle = \frac{-\mathrm{i}}{(2\pi)^n}\sum_{m_{i_1},\cdots,m_{i_n}}^{\infty}\frac{H^{(1)}_{m_{i_1}\cdots m_{i_n}}(I_1,\cdots,I_n)}{m_{i_1}\omega_1+\cdots+m_{i_n}\omega_n}e^{\left[(m_{i_1}\theta_1+\cdots+m_{i_n}\theta_n)\mathrm{i}\right]}\Bigg|_{(\theta_1^0,\cdots,\theta_n^0)}^{(\theta_1^0+2\pi,\cdots,\theta_n^0+2\pi)} \tag{7.24}$$

From the foregoing equation, if there is a set of integers $(m_{i_1}:\cdots:m_{i_n})$ to make the following relation hold

$$m_{i_1}\omega_1+\cdots+m_{i_n}\omega_n=0 \tag{7.25}$$

and then the foregoing condition is termed the *resonance* condition for Hamiltonian in Eq.(7.3). To label this resonance condition, the integer sets is expressed by the notation $(m_{i_1}:\cdots:m_{i_n})$. Therefore the averaged potential energy is constant, i.e.,

$$\langle H_1\rangle = C \tag{7.26}$$

for non-resonance $(m_{i_1}\omega_1+\cdots+m_{i_n}\omega_n)\neq 0$ and

$$\langle H_1\rangle_{(m_{i_1}\cdots m_{i_n})} = C-\mathrm{i}\sum_{m_{i_1},\cdots,m_{i_n}}^{\infty} H^{(1)}_{m_{i_1}\cdots m_{i_n}}(I_1,\cdots,I_n) \tag{7.27}$$

for the resonance condition $(m_{i_1}\omega_1+\cdots+m_{i_n}\omega_n)=0$. Note that C is a constant. From Eq.(7.27), once the resonance occurs, the averaged energy will be jumped. The averaged potential energy decreases with increasing $\Sigma=|m_{i_1}|+\cdots+|m_{i_n}|$. Therefore, when the resonance of the Hamiltonian exists, the individual Hamiltonians have an energy jump because the averaged potential energy has a sudden change. Chaotic regions occur when the isolating integrals of motion are destroyed locally by nonlinear resonances. As in Chirikov (1979), when the isolateing integrals of motion are destroyed locally by nonlinear resonances, at least two nonlinear resonances are overlapped. Once the two resonances are overlapped, the energy exchanges of decoupled Hamiltonians will occur between the two resonances. Because two resonances have different averaging interaction energies, the maximum and minimum amplitudes for the interaction energy and all the individual energy will be discontinuous when the initial condition is varied for the constant total energy. Only one resonance exists, before the KAM torus is destroyed by the sub-resonance, the averaged potential energy will change smoothly with varying the initial conditions for the initial total energy. Once the sub-resonance appears, such smoothness of the energy changes will be destroyed. Before this resonance overlaps with the other primary resonance, the small variation of the averaged interaction potential energy can be observed. The resonance bands or layers in such a Hamiltonian system will be discussed in sequel. To predict the resonance overlap, the energy spectrum will be presented in the following section.

7.3. Energy spectrum

In Chapter 6, based on the approximate energy exchange of each separable oscillator, the analytical prediction was discussed. However, for the exact energy exchange, it is very difficult to compute analytically. The numerical computation of the energy exchange can be carried out. Further, the energy spectrum can be obtained, from which the chaos and quasi-periodic motions can be determined. Before introduction of the Poincare surface, the $(2n-1)$-dimensional reference surface in the $2n$-dimensional phase space is introduced through the following set

$$\Xi = \left\{ (q_1,\cdots,q_n;p_1,\cdots,p_n) \left| \begin{array}{l} \sum_{i=1}^{n}(a_i p_i + b_i q_i) - c = 0 \\ \text{with arbitrarily constants } a_i, b_i \text{ and } c \end{array} \right. \right\} \tag{7.28}$$

which is decomposed into the 2-dimensional sub-phase planes as

$$\Xi = {}^1\Xi \oplus \cdots \oplus {}^n\Xi \tag{7.29}$$

where $\oplus$ is the direct summation.

The energy surface in $2n$-dimensional phase space is defined by

$$\Pi_h = \{(q_1,\cdots,q_n;p_1,\cdots,p_n) | H(q_1,\cdots,q_n;p_1,\cdots,p_n) = h\}. \tag{7.30}$$

Similarly, the equi-energy sets in the $2n$-dimensional phase space are projected on the 2-dimensional sub-phase planes

$$\Pi_h = {}^1\Pi_h \oplus \cdots \oplus {}^n\Pi_h. \tag{7.31}$$

To conveniently express the motion of the Hamiltonian, the equi-energy surface can be decomposed into the momentum and position components

$$\Pi_h = {}^M\Pi_h \oplus {}^P\Pi_h \tag{7.32}$$

where ${}^M\Pi$ and ${}^P\Pi$ are the momentum space and the position space. Further, the kinetic and potential energies surfaces can be easily defined on the foregoing sub-spaces. By use of this domain, equi-kinetic energy and equi-potential energy are expressed.

For a given energy $H(q_1,\cdots,q_n;p_1,\cdots,p_n)=h$ in Eq.(7.3), the time change rate of the total energy is given as

$$\begin{aligned}&\frac{dH(q_1,\cdots,q_n;p_1,\cdots,p_n)}{dt}\\&=\frac{dH_0(q_1,\cdots,q_n;p_1,\cdots,p_n)}{dt}+\frac{dH_1(q_1,\cdots,q_n)}{dt}=0\end{aligned} \tag{7.33}$$

which implies

$$\frac{dH_0(q_1,\cdots,q_n;p_1,\cdots,p_n)}{dt}=-\frac{dH_1(q_1,\cdots,q_n)}{dt} \tag{7.34}$$

where

$$\frac{dH_0(q_1,\cdots,q_n;p_1,\cdots,p_n)}{dt}=\sum_{i=1}^{n}\frac{dH_0^{(i)}(q_i,p_i)}{dt}. \tag{7.35}$$

Because the total energy is conservative, the time-change rate of the first integral quantity cannot be used, which cause more difficulty to investigate the dynamics on the equi-energy surface. However, the time-change rates of energy for each decoupled Hamiltonian $H_0^{(i)}(p_i,q_i)$ and the interaction Hamiltonian $H_1(q_1,\cdots,q_n)$ can be computed by

$$\frac{dH_0^{(i)}(q_i,p_i)}{dt}=\frac{\partial H_0^{(i)}(q_i,p_i)}{\partial t}+\left\{H_0^{(i)},H_1\right\}_{\text{Poisson}}=\left\{H_0^{(i)},H_1\right\}_{\text{Poisson}}, \tag{7.36}$$

$$\begin{aligned}\frac{dH_1(q_1,\cdots,q_n)}{dt}&=\frac{\partial H_1(q_1,\cdots,q_n)}{\partial t}+\sum_{i=1}^{n}\frac{\partial H_1(q_1,\cdots,q_n)}{\partial q_i}\dot{q}_i\\&=\sum_{i=1}^{n}\frac{\partial H_1(q_1,\cdots,q_n)}{\partial q_i}\dot{q}_i=-\sum_{i=1}^{n}\left\{H_0^{(i)},H_1\right\}_{\text{Poisson}}.\end{aligned} \tag{7.37}$$

The time change rate of the interaction potential energy is equal to the total time-change rates of all the decoupled Hamiltonians. Therefore, owing to the energy conservation, it is clear how the interaction potential energy is distributed to all the decoupled Hamiltonians.

As in Chapter 4, the first integral quantity increment (or energy increment in nonlinear Hamiltonian systems) can be defined from two points $(p_i^{(k)},q_i^{(k)},t_k)$ to

$(p_i^{(k+1)}, q_i^{(k+1)}, t_{k+1})$ as

$$L^{(i)}(t_k, t_{k+1}) = \Delta H_0^{(i)}(p_i^{(k)}, q_i^{(k)}, p_i^{(k+1)}, q_i^{(k+1)}) = \int_{t_k}^{t_{k+1}} \left\{H_0^{(i)}, H_1\right\}_{\text{Poisson}} dt \quad (7.38)$$

for ($i = 1, 2, \cdots, n$).

However, it is very difficult to analytically compute the energy increment from the points $(p_i^{(k)}, q_i^{(k)}, t_k)$ to $(p_i^{(k+1)}, q_i^{(k+1)}, t_{k+1})$. Therefore the numerical computation should be completed. Once the two points are on the same reference surface as the Poincare mapping surface, the energy increment can be computed. As in the precious section, the approximate increment can be computed through a specific orbit obtained from a separable oscillator without any interaction with the other separable oscillators. The incremental energy technique and the accurate standard mapping approach can be used for approximate analytical prediction of the onset and destruction of the chaotic motions, which was already discussed in Chapter 6. Herein, we will not repeat such materials.

To predict chaotic motions in Hamiltonian system of Eq.(7.3), the Poincare mapping surface is introduced as

$$\Sigma_h^{\Xi} = \left\{\left(q_1^{(N)}, \cdots, q_n^{(N)}; p_1^{(N)}, \cdots, p_n^{(N)}\right) \middle| N \in \mathbb{N}\right\} \subset \Xi \cap \Pi_h \quad (7.39)$$

where the sub-Poincare surfaces are decomposed into the n-sub-planes for each separable oscillator.

$$\Sigma_h^{\Xi} = {}^1\Sigma_h^{\Xi} \oplus \cdots \oplus {}^n\Sigma_h^{\Xi} \quad (7.40)$$

$${}^i\Sigma_h^{\Xi} = \left\{\left(q_i^{(N)}, p_i^{(N)}\right) \middle| N \in \mathbb{N}\right\} \subset {}^i\Xi \cap {}^i\Pi_h \quad (7.41)$$

On each sub-Poinacre surface, ${}^i\Sigma_h^{\Xi}$, the corresponding energy for decoupled Hamiltonians is computed as

$$H_0^{(i)}\left(q_i^{(N)}, p_i^{(N)}\right) = \frac{\left(p_i^{(N)}\right)^2}{2m_i} + V_i\left(q_i^{(N)}\right) \quad (7.42)$$

and the maximum and minimum energies for the decoupled Hamiltonians are computed

$${}^{\max}H_0^{(i)} = \max_{N\in\mathbb{N}} H_0^{(i)}\left(q_i^{(N)}, p_i^{(N)}\right) \text{ and } {}^{\min}H_0^{(i)} = \min_{N\in\mathbb{N}} H_0^{(i)}\left(q_i^{(N)}, p_i^{(N)}\right) \quad (7.43)$$

To determine chaos in vicinity of the internal resonant separatrix, the resonant analysis should be completed and the corresponding energy $E^{(i),(s_1:s_2:\cdots:s_n)}$ is determined as in 2-dimensional periodically forced nonlinear Hamiltonian systems.

$$\Delta E_{\max}^{(i)} = \max_{N\in\mathbb{N}}\left\{H_0^{(i)}\left(q_i^{(N)}, p_i^{(N)}\right) - E^{(i),(s_1:s_2:\cdots:s_n)}\right\},$$
$$\Delta E_{\min}^{(i)} = \min_{N\in\mathbb{N}}\left\{H_0^{(i)}\left(q_i^{(N)}, p_i^{(N)}\right) - E^{(i),(s_1:s_2:\cdots:s_n)}\right\}. \tag{7.44}$$

From the relation between the resonance and chaotic motion, once the maximum or minimum energy spectrum has a big jump, an old resonance structure of motion in Hamiltonian will disappear and a new resonance will disappear. In a domain of the special initial conditions of Hamiltonian, the maximum or minimum energy spectrum possesses random variation. It implies that the chaotic motion exists because of the primary resonance overlapping. Once one of the individual Hamiltonians has resonance or chaos, the entire Hamiltonian will be resonance and chaos. Therefore, the more possibility for motion complexity exists in many degrees of freedom systems.

7.4. Chaotic motions on equi-energy surfaces

To demonstrate the complexity in Hamiltonian systems with many degrees of freedom, consider a simple Hamiltonian system in Luo (2006c)

$$H = \tfrac{1}{2}y_1^2 + \tfrac{1}{2}y_2^2 - \tfrac{1}{2}\alpha x_1^2 + \tfrac{1}{4}\beta x_1^4 + \tfrac{1}{2}\gamma x_2^2 - \varepsilon x_1 x_2, \tag{7.45}$$

where $m_i = 1$ and $y_i = dx_i/dt$ ($i = 1,2$). The coefficients $\{\alpha,\beta,\gamma,\varepsilon\}$ are positive. The corresponding Hamilton's equation for motion is

$$\dot{x}_1 = y_1,\ \ \dot{y}_1 = \alpha x_1 - \beta x_1^3 + \varepsilon x_2;$$
$$\dot{x}_2 = y_2,\ \ \dot{y}_2 = -\gamma x_2 + \varepsilon x_1. \tag{7.46}$$

The equilibriums in Eq.(7.46) are given by three points $(0,0,0,0)$, $(\sqrt{\frac{\alpha}{\beta}+\frac{\varepsilon^2}{\gamma\beta}},0,$ $\frac{\varepsilon}{\gamma}\sqrt{\frac{\alpha}{\beta}+\frac{\varepsilon^2}{\gamma\beta}},0)$ and $(-\sqrt{\frac{\alpha}{\beta}+\frac{\varepsilon^2}{\gamma\beta}},0,-\frac{\varepsilon}{\gamma}\sqrt{\frac{\alpha}{\beta}+\frac{\varepsilon^2}{\gamma\beta}},0)$. By the local stability analysis, the equilibrium $(0,0,0,0)$ is saddle and the other equilibriums are center. Substitution of equilibrium $(0,0,0,0)$ into Eq.(7.45), the total energies are

$$H(x_1,x_2;y_1,y_2) = \tfrac{1}{2}y_1^2 + \tfrac{1}{2}y_2^2 - \tfrac{1}{2}\alpha x_1^2 + \tfrac{1}{4}\beta x_1^4 + \tfrac{1}{2}\gamma x_2^2 - \varepsilon x_1 x_2 = 0 \tag{7.47}$$

The equi-energy surface relative to the saddle in Eq.(7.47) is termed the homoclinic separatrix surface. On such a surface, it is observed that the total energy is always zero. Since the kinetic energy is non-negative, the maximum potential energy surface relative to the homoclinic separatrix surface is

$$V(x_1,x_2) = -\tfrac{1}{2}\alpha x_1^2 + \tfrac{1}{4}\beta x_1^4 + \tfrac{1}{2}\gamma x_2^2 - \varepsilon x_1 x_2 = 0 \tag{7.48}$$

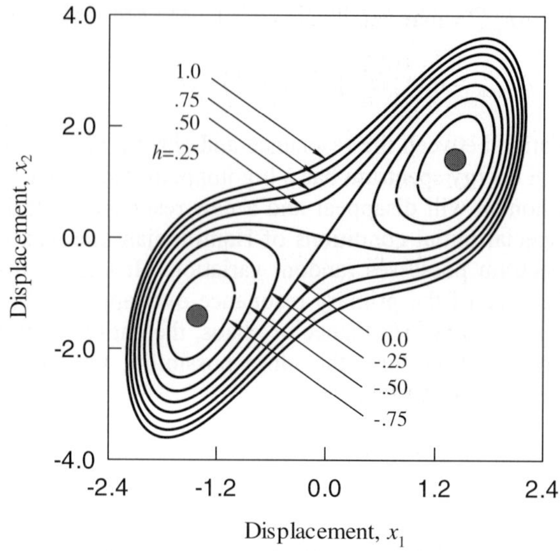

Figure 7.1. The maximum equi-potential curves for given energies $h = \{-0.75, -0.5, \cdots, 1.0\}$. The equilibrium points $(\pm\sqrt{2}, \pm\sqrt{2})$ with the minimum potential energy $V_{\min} = -1.0$ are labeled by the filled circular symbols. The equilibrium point $(0,0)$ is connected by the curve with maximum equi-potential energy with $h = 0$.

on which curve the maximum equi-potential energy with a zero value. On the homoclinic surface, only the orbits passing through the hyperbolic point has zero frequency. As in Chapter 4, the set of homoclinic orbits on the homoclinic surface is its subset. In addition to homoclinic orbits, the quasi-periodic and chaotic motions exist on the homoclinic surface. From the other equilibriums, the total energies

$$\begin{aligned} H(x_1, x_2; y_1, y_2) &= \tfrac{1}{2}y_1^2 + \tfrac{1}{2}y_2^2 - \tfrac{1}{2}\alpha x_1^2 + \tfrac{1}{2}\beta x_1^4 + \tfrac{1}{2}\gamma x_2^2 - \varepsilon x_1 x_2 \\ &= -\tfrac{1}{4\gamma^2\beta}(\alpha\gamma + \varepsilon^2)^2. \end{aligned} \tag{7.49}$$

Because the minimum potential energy is $V_{\min} = -\frac{1}{4\gamma^2\beta}(\alpha\gamma + \varepsilon^2)^2$ and the kinetic energy is positive, the maximum equi-potential energy equals to the minimum value. For any given energy $H(x_1, x_2; y_1, y_2) = h$, the maximum boundary in the displacement space is determined by the equi-potential curve with $V(x_1, x_2) = h$ and the motion of the Hamiltonian will be bounded in a domain computed by $V(x_1, x_2) \le h$. Consider a set of parameters ($\alpha = \beta = \gamma = \varepsilon = 1$), the equi-potential energy curve in displacement space is shown in Fig.7.1 for the total energy

$h = \{-0.75, -0.5, \cdots, 1.0\}$. The equilibrium points $(\pm\sqrt{2}, \pm\sqrt{2})$ in displacement with the minimum potential energy $V_{\min} = -1.0$ are labeled by the filled circular symbols. The equilibrium $(0,0)$ in displacement space is connected by the maximum equi-potential energy of $h = 0$. The potential wells are symmetry for $x_1 = x_2$ which is demonstrated through the equi-potential curves in Fig.7.1.

7.4.1. Resonance and averaging of Hamiltonians

In Eq.(7.45), for $H(y_1, y_2, x_1, x_2) = h$, the instantaneous energy distributions are

$$\begin{aligned} &H_0^{(1)}(x_1, y_1) = \tfrac{1}{2} y_1^2 - \tfrac{1}{2}\alpha x_1^2 + \tfrac{1}{4}\beta x_1^4 = E^{(1)}, \\ &H_0^{(2)}(x_2, y_2) = \tfrac{1}{2} y_2^2 + \tfrac{1}{2}\gamma x_2^2 = E^{(2)}, \\ &H_1(x_1, x_2) = -\varepsilon x_1 x_2 = E^{(3)} \text{ and } \textstyle\sum_{i=1}^{3} E^{(i)} = h. \end{aligned} \tag{7.50}$$

The energy $H_0^{(1)} = E_0^{(1)} = 0$ at the saddle point $(0,0)$ of the decoupled Hamiltonian $H_0^{(1)}$ possesses a homoclinic orbit. For $H_0^{(1)} = E_S^{(1)} < E_0^{(1)}$, two symmetric small periodic orbits exist in the two potential wells. However, $H_0^{(1)} = E_L^{(1)} > E_0^{(1)}$, the large periodic orbits exist outside of the homoclinic orbit. For the given energy $E_S^{(1)}$ satisfying $H_0^{(1)} = E_S^{(1)} \dot{<} E_0^{(1)}$, the solution of the small-orbit is

$$\begin{aligned} x_1 &= \pm e_S \mathrm{dn}\left[\frac{K(k_S^{(1)})\omega_S^{(1)} t}{\pi}, k_S^{(1)}\right], \\ y_1 &= \dot{x}_1 = \sqrt{\frac{\beta}{2}} e_S^2 (k_S^{(1)})^2 \mathrm{sn}\left[\frac{K(k_S^{(1)})\omega_S^{(1)} t}{\pi}, k_{1S}\right] \mathrm{cn}\left[\frac{K(k_S^{(1)})\omega_S^{(1)} t}{\pi}, k_S^{(1)}\right]; \end{aligned} \tag{7.51}$$

where cn, sn and dn are the Jacobi-elliptic functions, $K(k_S^{(1)})$ the complete elliptic integral of the first kind, and $k_S^{(1)}$ the modulus of the Jacobi-elliptic function. The modulus $k_S^{(1)}$, the response amplitude $e_S^{(1)}$ and the natural frequency $\omega_S^{(1)}$ are:

$$k_S^{(1)} = \sqrt{\frac{2\sqrt{\alpha^2 + 4\beta E_S^{(1)}}}{\alpha^2 + \sqrt{\alpha^2 + 4\beta E_S^{(1)}}}}, \quad e_S = \sqrt{\frac{2\alpha}{\left[2 - (k_S^{(1)})^2\right]\beta}}, \quad \omega_S^{(1)} = \frac{\sqrt{\alpha}\, e_S \pi}{\sqrt{2} K(k_S^{(1)})}. \tag{7.52}$$

The homoclinic orbit of the first oscillator may not be the homoclinic orbit for the entire system. So such a homoclinic orbits may not be on the homoclinic surface. Herein, we adopt such separable system in order to do analytical predicttion. However, the solution to the second decoupled Hamiltonian $H_0^{(2)} = E^{(2)}$ is

$$x_2 = a\cos\left(\omega^{(2)}t+\varphi_0\right), y_2 = \dot{x}_2 = -a\omega^{(2)}\sin\left(\omega^{(2)}t+\varphi^{(2)}\right), \tag{7.53}$$

$$\omega^{(2)} = \sqrt{\gamma}, a = \sqrt{E_2/\gamma}, \varphi^{(2)} = \arctan\left(-\tfrac{y_2(0)}{x_2(0)}\right). \tag{7.54}$$

With Eq.(7.51) and (7.53), the interaction energy in Eq.(7.45) is approximately estimated through the Fourier expansion, i.e.,

$$\begin{aligned} H_1(x_1,x_2) = -\frac{\pi e_S a}{2K}\Bigg(\cos\left(\omega^{(2)}t+\varphi^{(2)}\right)+\sum_{m=1}^{\infty}\operatorname{sech}\left(\frac{m\pi K'}{K}\right) \\ \times\left\{\cos\left[\left(m\omega_S^{(1)}-\omega^{(2)}\right)t-\varphi^{(2)}\right]+\cos\left[\left(m\omega_S^{(1)}+\omega^{(2)}\right)t-\varphi^{(2)}\right]\right\}\Bigg) \end{aligned} \tag{7.55}$$

From which, the resonance condition is

$$m\omega_S^{(1)} - \omega^{(2)} = 0 \tag{7.56}$$

which resonance condition is called the $(m:1)$-resonant condition, and the non-resonant condition is

$$m\omega_S^{(1)} - \omega^{(2)} \neq 0 \tag{7.57}$$

For the non-resonant case, the averaged interaction energy is zero. However, for resonance case, all other terms in $H_1(x_1,x_2)$ will average to zero over time t except for the term of the primary resonance (i.e., an averaging Hamiltonian in Eq.(7.55)). The averaging of the Hamiltonian for the different primary resonance is distinguishing and its magnitude $\langle H_1\rangle$ along the small orbit or the periodic orbit of the second Hamiltonian $H_0^{(2)}$ is :

$$\langle H_1\rangle^{(m:1)} \approx \frac{\sqrt{2}\sqrt{E^{(2)}}}{2m\sqrt{\beta}}\operatorname{sech}\left[\frac{m\pi K'\left(k_S^{(1)}\right)}{K\left(k_S^{(1)}\right)}\right], \tag{7.58}$$

where $K'(k_S^{(1)}) = K(k_S'^{(1)})$, $k_A'^{(1)} = \sqrt{1-(k_S^{(1)})^2}$.

For $H_0^{(1)} = E_L^{(1)} > E_0^{(1)}$, a large orbit for the decoupled Hamiltonian $H_0^{(1)}(x_1,y_1)$ is

$$x_1 = e_L \operatorname{cn}\left[\frac{2K(k_L^{(1)})\omega_L^{(1)}t}{\pi}, k_L^{(1)}\right], \tag{7.59a}$$

$$y_1 = \dot{x}_1 = \pm\sqrt{\frac{\beta}{2}\frac{(e_L^{(1)})^2}{k_L^{(1)}}}\,\mathrm{sn}\left[\frac{2K(k_L^{(1)})\omega_L^{(1)}t}{\pi},k\right]\mathrm{dn}\left[\frac{2K(k_L^{(1)})\omega_L^{(1)}t}{\pi},k_L^{(1)}\right]. \quad (7.59\text{b})$$

The modulus $k_L^{(1)}$, the response amplitude $e_L^{(1)}$ and the natural frequency $\omega_L^{(1)}$ are

$$k_L^{(1)} = \sqrt{\frac{\alpha+\sqrt{\alpha^2+4\beta E_L^{(1)}}}{2\sqrt{\alpha^2+4\beta E_L^{(1)}}}}, e_L^{(1)} = \sqrt{\frac{2(k_L^{(1)})^2\alpha}{[2(k_L^{(1)})^2-1]\beta}}, \omega_L^{(1)} = \frac{\sqrt{\beta}e_L^{(1)}\pi}{2\sqrt{2}k_L^{(1)}K(k_L^{(1)})}. \quad (7.60)$$

The resonance condition for the large orbit of the decoupled Hamiltonian $H_0^{(1)}(x_1, y_1)$ is given from the interaction energy $H_1(x_1, x_2)$, i.e.,

$$(2n-1)\omega_L^{(1)} - \omega^{(2)} = 0. \quad (7.61)$$

With such a resonant condition, the averaged interaction energy $\langle H_1\rangle$ is

$$\langle H_1\rangle^{(2n-1:1)} = \frac{\sqrt{2}\sqrt{E^{(2)}}}{(2n-1)\sqrt{\beta}}\,\mathrm{sech}\left[\left(n-\frac{1}{2}\right)\frac{\pi K'(k_L^{(1)})}{K(k_L^{(1)})}\right]. \quad (7.62)$$

From Eqs.(7.58) and (7.62), the averaging of the interaction energy decreases exponentially with increasing the resonance order due to the hyperbolic "sech" function. Furthermore, when a new primary resonance appears, there is a big jump of energy for two closest primary resonances. This jump phenomenon is a very important characteristic of Hamiltonian. Because of the energy conservation of Hamiltonian, the interaction energy has a jump as the resonance occurs, so the decoupled Hamiltonian will have the corresponding resonance jump to keep the energy be conservative. Therefore, the jump phenomena make it possible to numerically detect the onset of a new resonant-separatrix. As discussed before, the two resonances overlap, the averaging of decoupled Hamiltonians will switch from one resonance to another one. The energy spectrum will not be smooth, which is a significant evidence to determine the chaotic motion. Of course, the sub-resonance will cause the non-smoothness. But the energy jump caused by the primary resonance will be larger than the one caused by the sub-resonance.

7.4.2. Energy exchanges and criterion

From the resonance conditions in Eq.(7.56) and (7.61), the relationship between energy $E^{(1)}$ and $E^{(2)}$ can be computed. Namely, the resonance relationships can be expressed by

$$f^{(m:1)}\left(E_S^{(1)}, E^{(2)}\right) = 0 \text{ and } f^{(2n-1:1)}\left(E_L^{(1)}, E^{(2)}\right) = 0. \tag{7.63}$$

The decoupled Hamiltonian $H^{(1)}(x_1, y_1)$ possesses two potential wells, the energy exchange of the decoupled Hamiltonian $H_0^{(1)}(x_1, y_1)$ along its periodic orbit of the $(m:1)$-order resonance is computed by

$$\begin{aligned}\Delta H_{0(S)}^{(1),(m:1)}\left(\varphi^{(2)}\right) &= \int_0^{T_S^{(1)}} \left\{H_0^{(1)}, H_1\right\}_{\text{Poisson}} dt \approx \varepsilon \int_0^{T_S^{(1)}} y_1 x_2 dt \\ &= \sqrt{\frac{2}{\beta}} \varepsilon\pi\sqrt{E^{(2)}} \operatorname{sech}\left(\frac{m\pi K'\left(k_S^{(1)}\right)}{K\left(k_S^{(1)}\right)}\right) \sin\varphi^{(2)},\end{aligned} \tag{7.64}$$

and the energy exchange of the decoupled Hamiltonian $H_0^{(2)}(x_2, y_2)$ along its periodic orbit relative to the $(m:1)$-order resonance is computed by

$$\begin{aligned}\Delta H_{0(S)}^{(2),(m:1)}\left(\varphi^{(2)}\right) &= \int_0^{T^{(2)}} \left\{H_0^{(2)}, H_1\right\}_{\text{Poisson}} dt \approx \varepsilon \int_0^{T^{(2)}} y_2 x_1 dt \\ &= \sqrt{\frac{2}{\beta}} \varepsilon\pi\sqrt{E^{(2)}} \operatorname{sech}\left(\frac{m\pi K'\left(k_S^{(1)}\right)}{K\left(k_S^{(1)}\right)}\right) \sin\varphi^{(2)}.\end{aligned} \tag{7.65}$$

From the foregoing two equations, the interaction energy gives the same energy exchange during their corresponding periods for this particular Hamiltonian. In Eqs.(7.64) and (7.65), the different period are used. When the same period of $T_S^{(1)}$ is considered, the energy exchange distribution have the following relation

$$\Delta H_{0(S)}^{(2),(m:1)}\left(\varphi^{(2)}\right) = m\Delta H_{0(S)}^{(1),(m:1)}\left(\varphi^{(2)}\right) \tag{7.66}$$

From Eq.(7.39), the two energy exchange for $H_1\left(x_1, x_2\right)$ are

$$\Delta H_{1(S)}^{(1)(m:1)}(\varphi^{(2)}) = -\Delta H_{0(S)}^{(1),(m:1)}(\varphi^{(2)}) \text{ and } \Delta H_{1(S)}^{(2),(m:1)}(\varphi^{(2)}) = -\Delta H_{0(S)}^{(2),(m:1)}(\varphi^{(2)}) \tag{7.67}$$

and the energy exchange of the interaction energy $H_1\left(x_1, x_2\right)$ is

$$\Delta H_{1(S)}^{(m:1)}\left(\varphi^{(2)}\right) = -\Delta H_{0(S)}^{(1),(m:1)}\left(\varphi^{(2)}\right) - \Delta H_{0(S)}^{(2),(m:1)}\left(\varphi^{(2)}\right). \tag{7.68}$$

The approximate exchange of the interaction energy for one period $T_S^{(1)}$ is

$$\Delta H_{1(S)}^{(m:1)}\left(\varphi^{(2)}\right) \approx -(m+1)\sqrt{\frac{2}{\beta}} \varepsilon\pi\sqrt{E^{(2)}} \operatorname{sech}\left(\frac{m\pi K'\left(k_S^{(1)}\right)}{K\left(k_S^{(1)}\right)}\right) \sin\varphi^{(2)}, \tag{7.69}$$

The energy exchanges on the space of (x_1, y_1) make the energy of the orbit of the coupled system vary from $E_S^{(1),(m:1)}$ to $E_0^{(1)} = 0$

$$\sqrt{\frac{2}{\beta}}\varepsilon\pi\sqrt{E^{(2),(m:1)}}\operatorname{sech}\left(\frac{m\pi K'\left(k_S^{(1)}\right)}{K\left(k_S^{(1)}\right)}\right)\sin\varphi^{(2)} = \left|E_S^{(1),(m:1)}\right| \tag{7.70}$$

which implies the resonant orbit will arrive to the homoclinic orbit in (x_1, y_1). Because many resonances exist, a motion near the homoclinic orbit will become very complex. From the resonant overlap theory, such a condition indicates the chaotic motion will exist. In fact, two resonances overlapping will cause the chaotic motions. Therefore, the approximate, minimum condition for resonant overlap chaos in the coupled Hamiltonian is given by

$$\sqrt{\frac{2}{\beta}}\varepsilon\pi\sqrt{E^{(2),(m:1)}}\operatorname{sech}\left(\frac{m\pi K'\left(k_S^{(1)}\right)}{K\left(k_S^{(1)}\right)}\right)\sin\varphi^{(2)} = \left|E_S^{(1),(m:1)} - E_S^{(1),(m+1:1)}\right|. \tag{7.71}$$

The foregoing condition can be given through the second decoupled Hamiltonian.

$$\sqrt{\frac{2}{\beta}}\varepsilon\pi\sqrt{E^{(2),(m:1)}}\operatorname{sech}\left(\frac{m\pi K'\left(k_S^{(1)}\right)}{K\left(k_S^{(1)}\right)}\right)\sin\varphi^{(2)} = \left|E^{(2),(m:1)} - E^{(2),(m+1:1)}\right|. \tag{7.72}$$

The energy exchange for the decoupled Hamiltonian $H_0^{(1)}(x_1, y_1)$ along the resonant orbit of the $(2n-1:1)$-order is

$$\begin{aligned}\Delta H_{0(L)}^{(1),(2n-1:1)}\left(\varphi^{(2)}\right) &= \int_0^{T_L^{(1)}} \left\{H_0^{(1)}, H_1\right\}_{\text{Poisson}} dt \approx \varepsilon\int_0^{T_L^{(1)}} y_1 x_2 dt \\ &\approx 2\sqrt{\frac{2}{\beta}}\varepsilon\pi\sqrt{E^{(2)}}\operatorname{sech}\left(\frac{(2n-1)\pi K'\left(k_L^{(1)}\right)}{2K\left(k_L^{(1)}\right)}\right)\sin\varphi^{(2)}.\end{aligned} \tag{7.73}$$

The energy exchanges on the space of (x_1, y_1) make the energy of the orbit of the coupled system vary from $E_S^{(1),(m:1)}$ to $E_0^{(1)} = 0$

$$2\sqrt{\frac{2}{\beta}}\varepsilon\pi\sqrt{E^{(2),(2n-1:1)}}\operatorname{sech}\left(\frac{(2n-1)\pi K'\left(k_L^{(1)}\right)}{2K\left(k_L^{(1)}\right)}\right)\sin\varphi^{(2)} = \left|E_L^{(1),(2n-1:1)}\right|. \tag{7.74}$$

The approximate, minimum condition for chaos in the coupled Hamiltonian is given by

$$2\sqrt{\frac{2}{\beta}}\varepsilon\pi\sqrt{E^{(2),(2n-1:1)}}\operatorname{sech}\left(\frac{m\pi K'\left(k_L^{(1)}\right)}{K\left(k_L^{(1)}\right)}\right)\sin\varphi^{(2)} = \left|E_L^{(1),(2n-1:1)} - E_L^{(1),(2n+1:1)}\right|. \quad (7.75)$$

The above conditions are acceptable for small ε from the KAM theory. With increasing ε, the conditions for chaotic motions cannot give a good prediction, but such criteria can be as a rough estimate to guide the numerical investigation, which should be very significant. Although the analytical prediction cannot be accurate for large ε, the primary resonance should exist no matter how the parameter ε is small or large. However, with increasing ε, the sub-resonance will become the significant importance. For accurate predictions of the chaotic flows on the equi-energy surface, the exact energy increment should be computed. However, it is very difficult to obtain the analytical prediction. So the numerical prediction should be conducted.

7.4.3. Energy spectrum computation

Since the analytical prediction is strongly dependent on the small coupling parameter ε, the numerical prediction of the chaotic motion will become important. The exact energy exchange can be computed through this numerical method. If we consider the Poincaré mapping point, such a numerical method is called the energy spectrum technique as in Chapter 6. To consider the Poincaré mapping section as

$$\begin{aligned}\Xi_1 &= \left\{\left(x_1, y_1, x_2, y_2\right)\middle| x_2 = c\right\} = {}^1\Xi_1 \oplus {}^2\Xi_1, \text{ or}\\ \Xi_2 &= \left\{\left(x_1, y_1, x_2, y_2\right)\middle| y_2 = c\right\} = {}^1\Xi_2 \oplus {}^2\Xi_2;\end{aligned} \quad (7.76)$$

where

$$\begin{aligned}{}^1\Xi_1 &= \left\{\left(x_1, y_1\right)\middle| x_2 = c\right\}, {}^2\Xi_1 = \left\{\left(c, y_2\right)\middle| x_2 = c\right\},\\ {}^1\Xi_2 &= \left\{\left(x_1, y_1\right)\middle| y_2 = c\right\}, {}^2\Xi_2 = \left\{\left(x_2, c\right)\middle| y_2 = c\right\}.\end{aligned} \quad (7.77)$$

From Eqs.(7.30) and (7.45), the energy surface is defined as

$$\Pi_h = \left\{\left(x_1, x_2, y_1, y_2\right)\middle| H\left(x_1, x_2, y_1, y_2\right) = h\right\}, \quad (7.78)$$

$$\Sigma_h^{\Xi_j} = \left\{\left(x_1^{(N)}, x_2^{(N)}; y_1^{(N)}, y_2^{(N)}\right)\middle| N \in \mathrm{N}\right\} \subset \Xi_j \cap \Pi_h \quad (7.79)$$

where $j \in \{1, 2\}$, and the sub-Poincare surfaces are decomposed into the n-sub-planes.

$$\Sigma_h^{\Xi_j} = {}^1\Sigma_h^{\Xi_j} \oplus {}^2\Sigma_h^{\Xi_j}, \tag{7.80}$$

$${}^i\Sigma_h^{\Xi_j} = \left\{ \left(x_i^{(N)}, y_i^{(N)} \right) \middle| N \in \mathbb{N} \right\} \subset {}^i\Xi_j \cap {}^i\Pi \;\; (i \in \{1,2\}). \tag{7.81}$$

On each sub-Poincare surface, ${}^i\Sigma_h^{\Xi_j}$, the corresponding energies for decoupled Hamiltonians in Eq.(7.50) are

$$H_0^{(1)}\left(x_1^{(N)}, y_1^{(N)}\right) = \tfrac{1}{2}\left(y_1^{(N)}\right)^2 - \tfrac{1}{2}\alpha\left(x_1^{(N)}\right)^2 + \tfrac{1}{4}\beta\left(x_1^{(N)}\right)^4, \tag{7.82}$$

$$H_0^{(2)}\left(x_2^{(N)}, y_2^{(N)}\right) = \tfrac{1}{2}\left(y_2^{(N)}\right)^2 + \tfrac{1}{2}\gamma\left(x_2^{(N)}\right)^2. \tag{7.83}$$

In addition, on the Poincare section, the interaction energy is computed as

$$\begin{aligned} &H_1\left(x_1^{(N)}, x_2^{(N)}\right) = -\varepsilon x_2^{(N)} x_1^{(N)} \text{ and} \\ &\sum\nolimits_{i=1}^{2} H_0^{(i)}\left(x_i^{(N)}, y_i^{(N)}\right) + H_1\left(x_1^{(N)}, x_2^{(N)}\right) = h. \end{aligned} \tag{7.84}$$

The maximum and minimum energies of the decoupled Hamiltonian are computed for $i \in \{1,2\}$

$${}^{\max}H_0^{(i)} = \max_{N\in\mathbb{N}} H_0^{(i)}\left(x_i^{(N)}, y_i^{(N)}\right) \text{ and } {}^{\min}H_0^{(i)} = \min_{N\in\mathbb{N}} H_0^{(i)}\left(x_i^{(N)}, y_i^{(N)}\right) \tag{7.85}$$

In energy spectrum illustrations, the white-filled, shaded and grey-filled areas are quasi-periodic, resonance layer and chaotic motions, respectively. As in Chapter 6 (also see, Luo, 2002), the chaotic motion near the resonant separatrix is termed *the resonant layer* on the equi-energy surface. The acronyms "RL", "QP" and "NM" represent "resonant motion in the resonant layer", "Quasi-periodic motion" and "non-motion", respectively. The dashed vertical lines are the boundaries for the aforementioned three motions. The dark solid vertical lines are the non-motion boundary. For the maximum and minimum energy spectrum illustrations, a set of parameters ($\alpha = \beta = \gamma = \varepsilon = 1$) are used. Both the energy spectrums are very smooth for the quasi-periodic motion. When one of the two energy spectrums becomes rough, the motion switches from the quasi-periodic motion into the resonance layer motion. Once at least one of two energy spectrums becomes random, the chaotic motion exists.

Owing to the symmetry of the two potential wells, for a given total energy, there are two separated regions in potential wells. The value of x_1 is symmetric for $x_2 = 0$. Thus, the energy spectrums are symmetric. For given y_2, the velocity y_1 can be computed from $H(x_1, y_1, x_2, y_2) = h$ for the varied displacement x_1 on ${}^1\Sigma_h^{\Xi_1}$. For the motions in two potential wells, the maximum and minimum energy

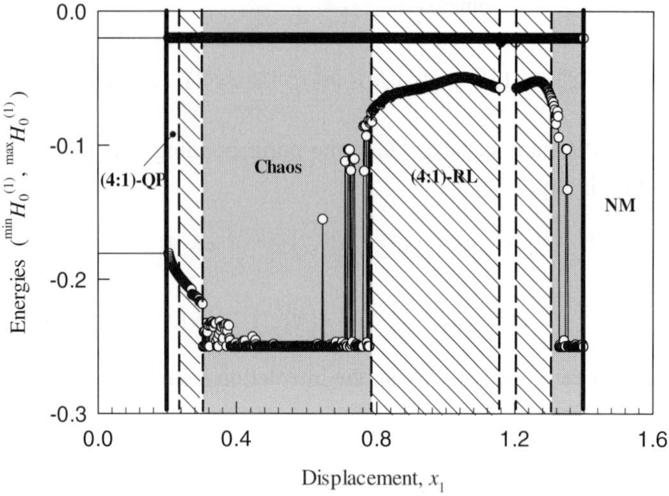

(a)

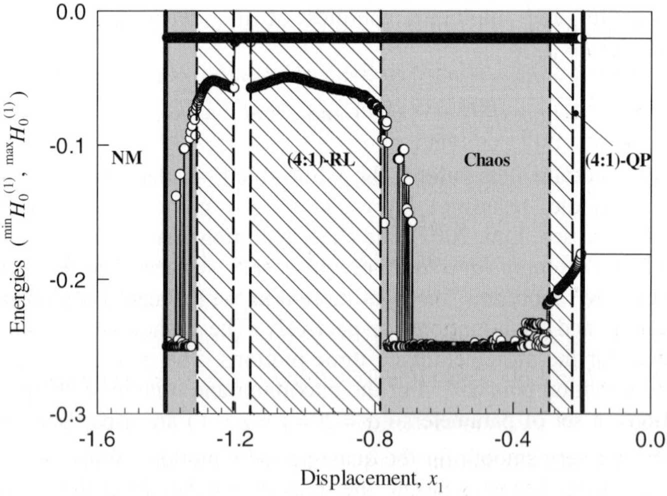

(b)

Figure 7.2. Maximum and minimum energy spectrums varying with displacement x_1 ($h = -0.02$) on the sub-Poincare mapping section ${}^{1}\Sigma_{-0.02}^{\Xi_1}$ in: (a) the right potential well and (b) the left-potential well. ($\alpha = \beta = \gamma = \varepsilon = 1$, $c = 0$). The white-filled, shaded and grey-filled areas are quasi-periodic, resonant layer and chaotic motion motions, respectively. The dashed vertical lines are the boundaries separated from one motion to another one. The dark solid vertical line is the boundary for non-motion.

spectrums on the sub-Poincaré mapping section ${}^1\Sigma_{-0.02}^{\Xi_1}$ are illustrated in Fig.7.2 for $h=-0.02$ and $c=0$. From the energy spectrum in the left potential, the chaotic motion exists in the region $x_1 \in \{(-1.398,-1.312),\ (-0.704,-0.294)\}$. The (4:1)-resonance layer exists in the following regions $x_1 \in \{(-1.312,-1.204),$ $(-1.156,-0.704),\ (-0.294,-0.220)\}$. The quasi-periodic motions pertaining to the $(4:1)$-resonance exist in the region $x_1 \in \{(-1.204,-1.156),(-0.224,-0.204)\}$. Due to the symmetry, the regions are $x_1 \in \{(1.156,1.204),(0.204,0.224)\}$ for the $(4:1)$-quasi-periodic motion in the right potential well; $x_1 \in \{(0.220,0.294),$ $(0.704,1.156)\ (1.204,-1.312)\}$ for the $(4:1)$-resonant layer; and for the chaotic motion $x_1 \in \{(1.312,1.398),\ (0.294,0.704)\}$, respectively. No any motion exists for $x_1 \in (-0.202,0.202)$, $(-\infty,-1.398)$ and $(1.398,\infty)$ with $x_2=0$, which is determined by the maximum potential energy domain. Through the quasi-periodic motion, it is possible that the corresponding periodic motion be numerically discovered. In addition, for small ε, the motion relative to the $(2:1)$-resonance will be observed. For $\varepsilon \in (0.4,1.2)$, the motion relative to the $(2:1)$-resonance disappears, but the motion relative to the $(4:1)$-resonance is observed. Such a resonance switching need to be investigated further.

The maximum and minimum energy spectrums for $h=0.0$ are presented in Fig.7.3 on the Poincare mapping section ${}^1\Sigma_0^{\Xi_1}$. For this case, the energy surface is a homoclinic surface. The ranges of the displacement x_1 on the corresponding Poincaré mapping section are: $(0,0.768)$, $(1.282,1.312)$, $(-1.312,-1.282)$ and $(-0.768,0)$ for chaotic motion; $(1.312,1.338)$, $(0.768,0.816)$, $(-0.816,-0.768)$, and $(-1.338,-1.312)$ for the $(4:1)$-resonant separatrix layer; $(0.862,1.282)$ and $(-1.282,-0.862)$ for the quasi-periodic motion relative to the $(4:1)$- resonance. At $x_i=0$ ($i=1,2$), no motion exists. For $|x_1|>\sqrt{2}$ and $x_2=0$, no motion can be observed. The chaotic motion is distributed near the maximum potential energy boundary and the periodic motion is distributed on the central area of the potential well domain. It is observed that the quasi-periodic motion jumps to the chaotic motion without the corresponding resonance.

Consider the maximum and minimum energy spectrums with $h=0.2$ for the large orbits beyond the homoclinic surface, as shown in Fig.7.4. The energy spectrums are based on the Poincare mapping sections ${}^1\Sigma_{0.2}^{\Xi_1}$ with $c=\{0,1.0\}$. For the case of $c=0$, the energy spectrums on the Poincare mapping section shows the chaotic motion exists in the entire range on ${}^1\Sigma_{0.2}^{\Xi_1}$, i.e., $x_1 \in (-1.530,1.530)$ and

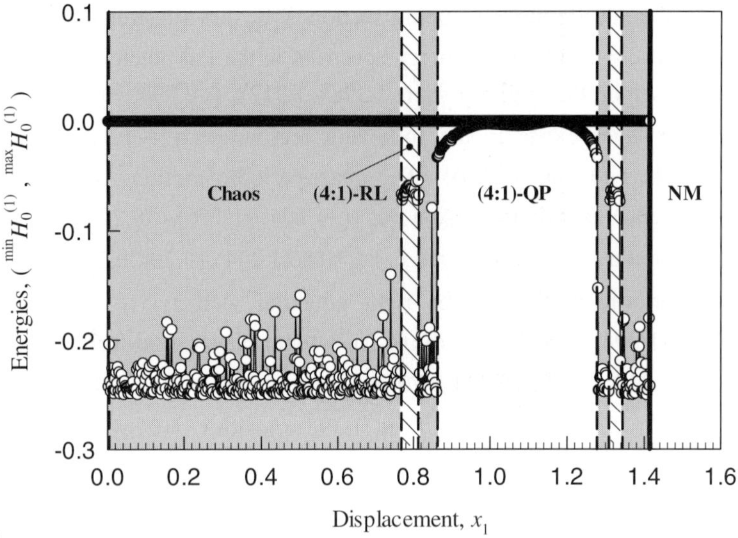

(a)

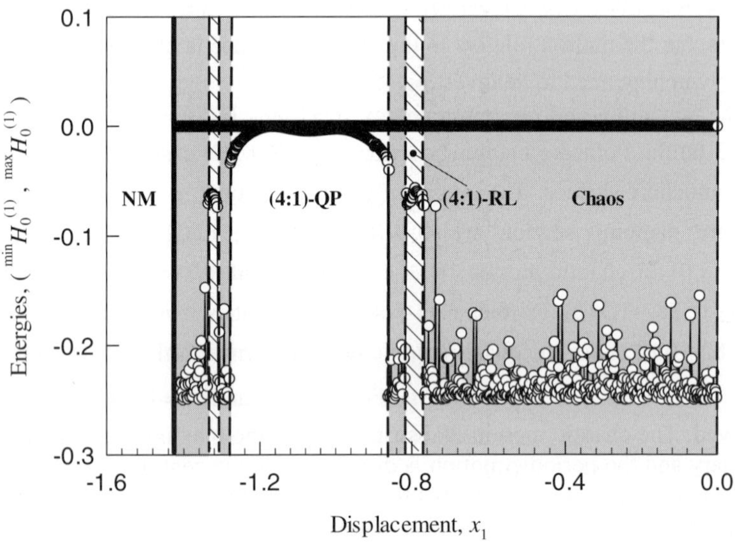

(b)

Figure 7.3. Maximum and minimum energy spectrums varying with displacement x_1 ($h = 0$) on the sub-Poincare mapping section ${}^1\Sigma_0^{\Xi_1}$ in: (a) the right potential well and (b) the left-potential well. ($\alpha = \beta = \gamma = \varepsilon = 1$, $c = 0$). The white-filled, shaded and grey-filled areas are quasi-periodic, resonant layer and chaotic motion motions, respectively. The dashed vertical lines are the boundaries separated from one motion to another one. The dark solid vertical line is the boundary for non-motion.

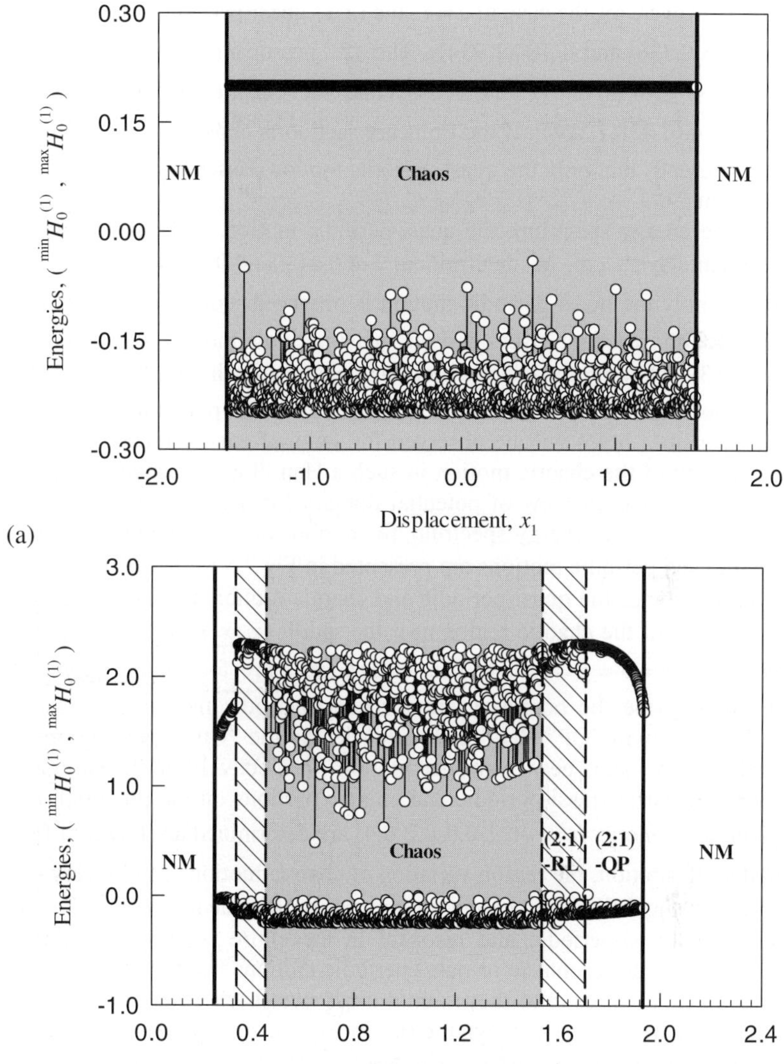

Figure 7.4. Maximum and minimum energy spectrums varying with displacement x_1 ($h = 0.2$) on the sub-Poincare mapping section ${}^1\Sigma_{0.2}^{\Xi_1}$ in: (a) ($c = 0$) and (b) ($c = 1$). ($\alpha = \beta = \gamma = \varepsilon = 1$). The white-filled, shaded and grey-filled areas are quasi-periodic, resonant layer and chaotic motion motions, respectively. The dashed vertical lines are the boundaries separated from one motion to another one. The dark solid vertical line is the boundary for non-motion.

$x_2 = 0$. However, for the case of $c = 1$, the (2:1)-quasi-periodic motion exists for $x_1 \in (0.266, 0.336)$ and $(1.704, 1.934)$. The (2:1)-resonant layer motion appears for $x_1 \in (0.336, 0.438)$ and $(1.524, 1.704)$; and the chaotic motion exists in the range of $x_1 \in (0.438, 1.524)$. If the Poincare section is chosen for $c = 2$, the energy spectrum tells that only the quasi-periodic motion passes through such a Poincaré section.

From the energy spectrum, the quasi-periodic motion, and chaotic motion in the resonant layers can be determined. For $h \in [-1, -0.09)$, no chaotic motion exists and only the quasi-periodic motion is observed. For $h \in (-0.09, 0.38)$, the chaotic motion, resonant layers and the quasi-periodic motion will be observed. For $h \in (0.39, \infty)$, only the chaotic motion is observed, which implies the chaotic motion approaches the minimum potential energy in the potential well. Of course, the resonance holes will be observed for some of energy levels. For a better understanding of the chaotic motion in such a Hamiltonian system, it is necessary to have an overall view of potential domain for quasi-periodic and chaotic motions. Through the energy spectrum, the regions on the potential well domain for chaotic and periodic motions are presented in Fig.7.5 for $h = 0$. In Fig.7.5(a), the overall view of the quasi-periodic and chaotic motion are illustrated and the de-tailed view of the chaotic region near the saddle $(x_1, x_2) = (0, 0)$ are given for parameters ($\alpha = \beta = \gamma = \varepsilon = 1$). The shaded areas are for chaotic motion. The white areas inside the boundary are for quasi-periodic motion and resonance layer. The acronym "QP","RL" and "NR" denote the "Quasi-periodic motion", "resonant layers" and "new resonance motion", respectively. In the similar manner, the regions for quasi-periodic and chaotic motions on the potential domain for given total energy $h = \{-0.2, 0.0, 0.2, 0.4\}$ are determined as shown in Fig.7.6. From this illustration, the region variation of chaotic motions can be very clearly observed. In Fig.7.6(a), no chaotic motion can be obtained for $h = -0.2$. All the motions are quasi-periodic and resonant in the entire solution domain. Each curve has the same resonance or quasi-periodic motion. The motion is trapped in the two potential wells, and they cannot cross over each other. The motions in two potential wells are skew-symmetric. In Fig.7.6(b), chaotic motion appears on the homoclinic energy surface. The resonant layer and quasi-periodic motion also exists. The motions in two potential well will not merge together because the equilibrium point $(0, 0, 0, 0)$ is saddle. Once an energy level is greater than the energy $h = 0$, motions in the two potential wells will merge together. Thus, for an energy $h = 0.2$, in Fig.7.6(c), the chaotic motions near the saddle point of the Hamiltonian merges together. The regions of chaotic motion for this energy level are enlarged, which is compared to the one for $h = 0$. For $h > h_{cr}$, the chaotic motions almost occupy the entire region on the potential well domain. In Fig.7.6(d),

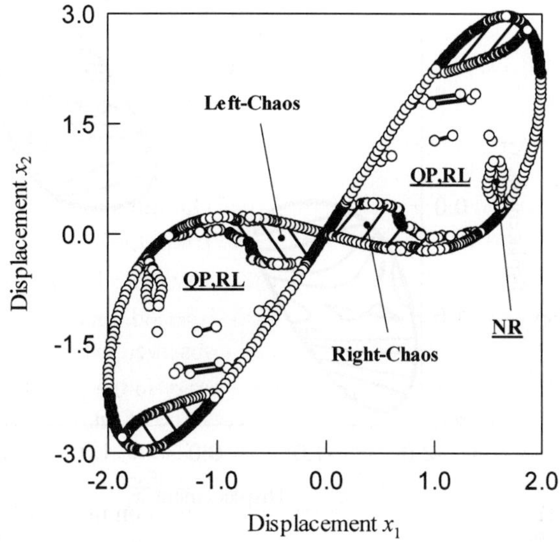

(a)

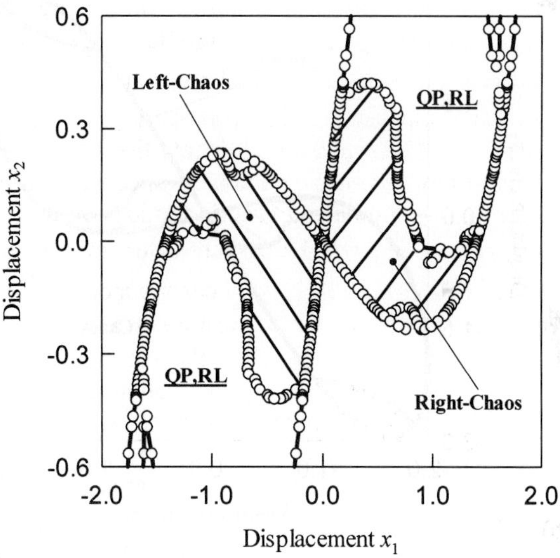

(b)

Figure 7.5. The domains on the potential well for chaotic and periodic motions ($h = 0$): (a) overall view, (b) detailed view. ($\alpha = \beta = \gamma = \varepsilon = 1$). The shaded areas are for chaotic motion. The white areas inside the boundary are for quasi-periodic motion and resonance layer. The acronym "QP","RL" and "NR" represent the "Quasi-periodic motion", "resonant layers" and "new resonance motion", respectively.

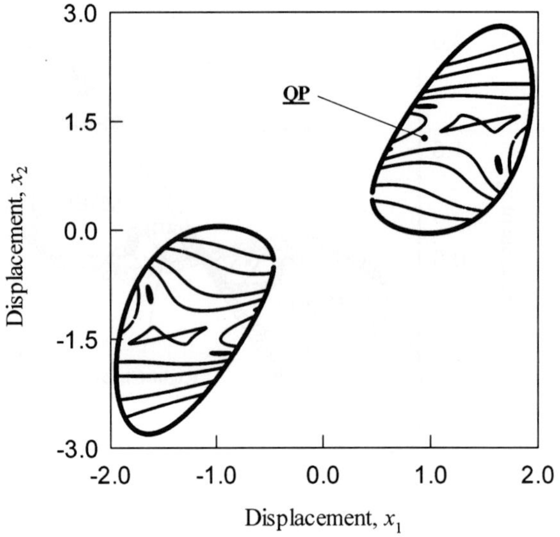

(a)

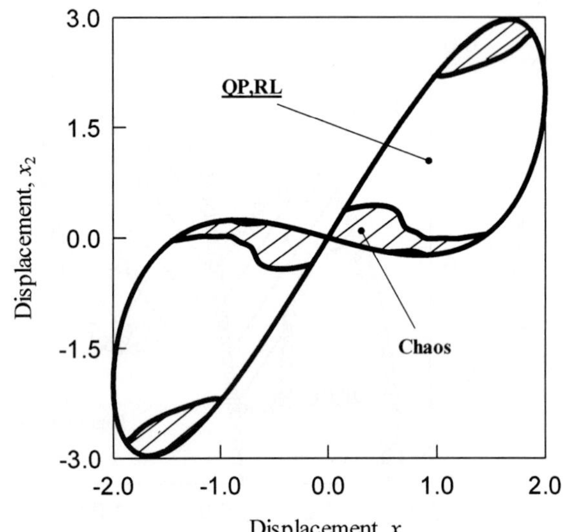

(b)

Figure 7.6. The domains on the potential well for chaotic and periodic motions: (a) $h=-0.2$, (b) $h=0$, (c) $h=0.2$ and (d) $h=0.4$. ($\alpha=\beta=\gamma=\varepsilon=1$). The shaded areas are for chaotic motion. The white areas inside the boundary are for quasi-periodic motion and resonance layer. The acronym "QP" and "RL" represent the "Quasi-periodic motion" and "resonant layers", respectively.

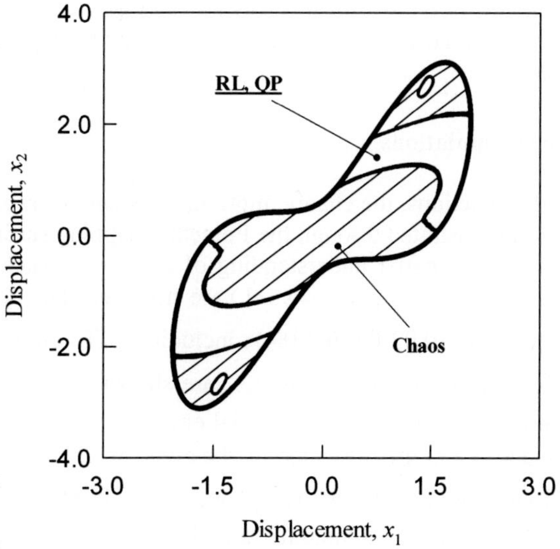

(c)

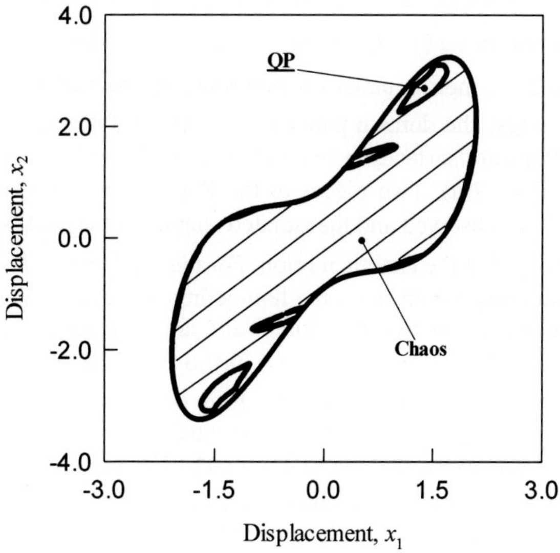

(d)

Figure 7.6. (continue)

the region for $h = 0.4$ are illustrated. It is observed that the chaotic motion takes the almost entire potential well domain except a few spots of resonant and quasi-periodic motions.

7.4.4. Numerical simulations

After numerical predictions, the chaotic motion, resonant layers and quasi-periodic motions can be illustrated through the Poincare mapping surfaces and decoupled phase planes for a better understanding of motions in such a Hamiltonian system. The parameters ($\alpha = \beta = \gamma = \varepsilon = 1$) are still used. The Poincare surfaces for given energies $h = \{-0.02, 0.0, 0.2, 0.4\}$, including ${}^{P}\Pi_h$ on the position plane (x_1, x_2) and $\Sigma_h^{\Xi_2}$ on the surface (x_1, y_1, x_2), are shown in Figs.7.7-7.8. The red boundary represents maximum equi-potential energy domain. The Poincare surfaces for $y_1 > 0$ and $y_1 < 0$ are presented by the green and blue colors, respectively. Four initial conditions of $(x_1, x_2; y_1, y_2)$ are $(\mp 0.6, 0.0; \pm 0.5052, 0.0)$, $(\pm 0.86, 0.0; \pm 0.6827, 0.0)$, $(1.56, 2.8; 0.8766, 0.0)$ and $(1.3, 2.8; 0.7085, 0.0)$ for $h = \{-0.02, 0, 0.2, 0.4\}$, accordingly. On the Poincare mapping surface, $y_2 = 0$ holds from the surface definition in Eq.(7.79). In Fig.7.7, $\Sigma_h^{\Xi_2}$ on the surface (x_1, y_1, x_2) are illustrated. The chaotic motion pattern is presented on the surface $\Sigma_h^{\Xi_2}$. Hence, for a given total energy, the domain pattern of the chaotic domain is uniquely determined. The Poincare surfaces in two potential wells for $h = -0.02$ are separated, as shown in Fig.7.7(a). Two pieces of the Poincare mapping surfaces for each potential well are observed and the isolated islands of the $(4:1)$-and $(2:1)$-resonances are involved in the chaotic motion. For the total energy $h = 0.0$, the motion of the Hamiltonian system is on the homoclinic surface. In Fig.7.7(b), it seems that the motions in the two potential wells are connected together. However, because the equilibrium $(x_1, x_2; y_1, y_2) = (0, 0; 0, 0)$ is saddle, the motion of the Hamiltonian system cannot pass through from one potential well to the other one because of the energy conservation. The surface area for the chaotic motion becomes large compared the one of $h = -0.02$. The motion relative to the (4:1)-resonance becomes small in the chaotic motion of $h = 0.0$. The most of area on the displacement (or position) plane of the Poincare mapping is quasi-periodic motion relative to certain resonances, as shown in Fig.7.8. The quasi-periodic motion relative to the $(2:1)$-resonance islands are embedded in chaotic motion. The Poincare mapping surface for $h = 0.2$ is illustrated in Fig.7.7(c). The chaotic motions in the vicinity of the equilibrium $(x_1, x_2; y_1, y_2) = (0, 0; 0, 0)$ are merged together. The areas on surface for quasi-periodic motions are further shrunken. With further increasing the total energy of the Hamiltonian system over a special

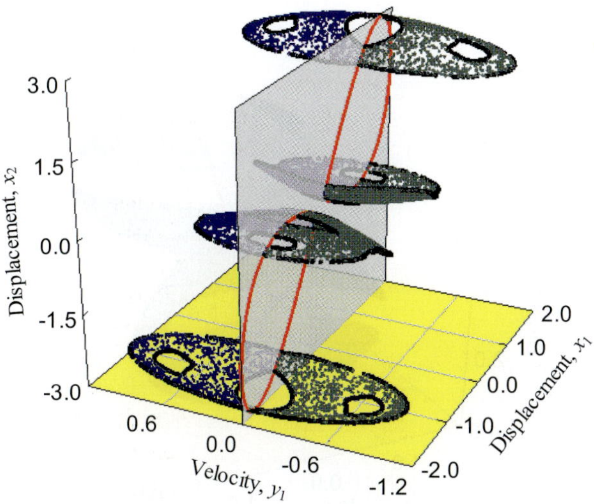

(a)

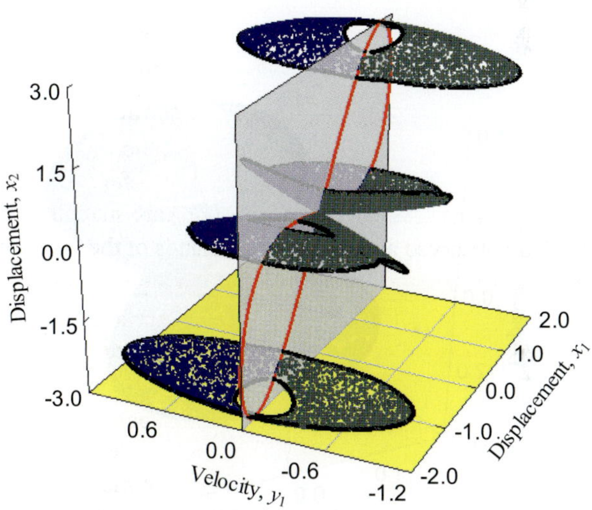

(b)

Figure 7.8. Poincaré surface $\Sigma_h^{\Xi_2}$ on the plane (x_1, y_1, x_2) with initial conditions $(x_1, x_2; y_1, y_2)$: (a) $(\mp 0.6,0.0;\pm 0.5052,0.0)$ for $h=-0.02$, (b) $(\pm 0.86,0.0;\pm 0.6827,0.0)$ for $h=0$, (c) (1.56,2.8; 0.8766,0.0) for $h=0.2$ and (d) (1.3,2.8;0.7085,0.0) for $h=0.4$. ($\alpha=\beta=\gamma=\varepsilon=1,\ c=0$). The red boundary is the maximum equi-potential energy domain. The Poincaré surfaces for $y_1>0$ and $y_1<0$ are presented by the green and blue colors, respectively.

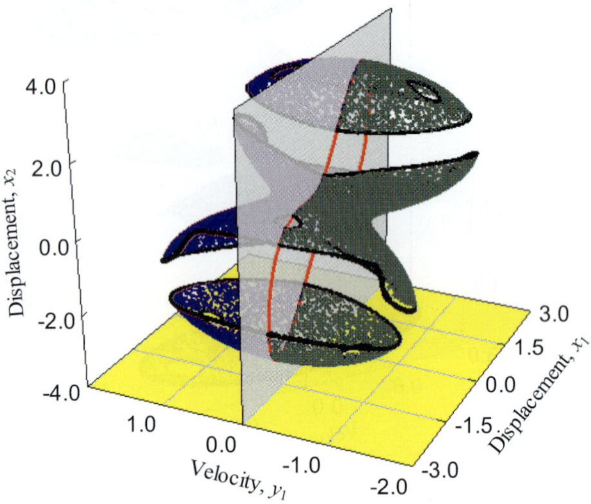

(c)

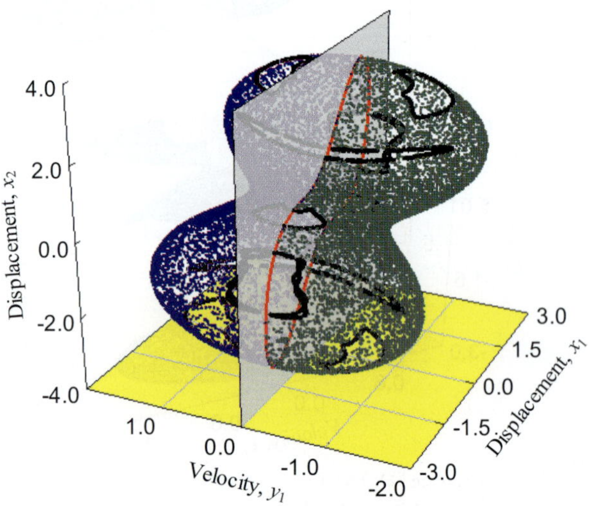

(d)

Figure 7.7. (continued)

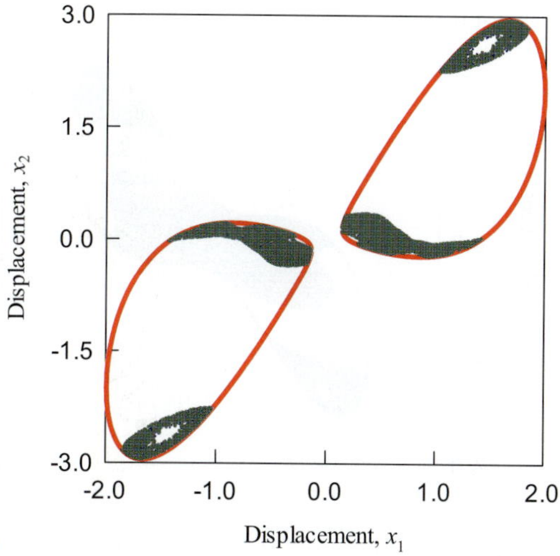

(a)

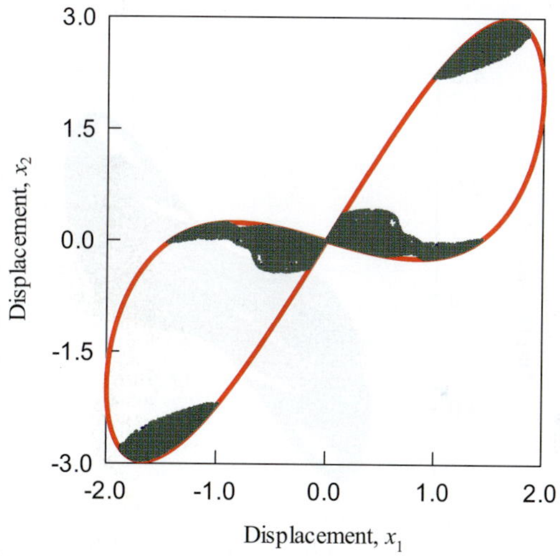

(b)

Figure 7.8. Poincare surface ${}^{P}\Pi_h$ on the plane (x_1, x_2) with initial conditions $(x_1, x_2; y_1, y_2)$: (a) $(\mp 0.6, 0.0; \pm 0.5052, 0.0)$ for $h = -0.02$, (b) $(\pm 0.86, 0.0; \pm 0.6827, 0.0)$ for $h = 0$,(c) (1.56,2.8; 0.8766,0.0) for $h = 0.2$ and (d) (1.3,2.8;0.7085,0.0) for $h = 0.4$. ($\alpha = \beta = \gamma = \varepsilon = 1$, $c = 0$). The red boundary is the maximum equi-potential energy domain. The Poincare surfaces for $y_1 > 0$ and $y_1 < 0$ are presented by the green and blue colors, respectively.

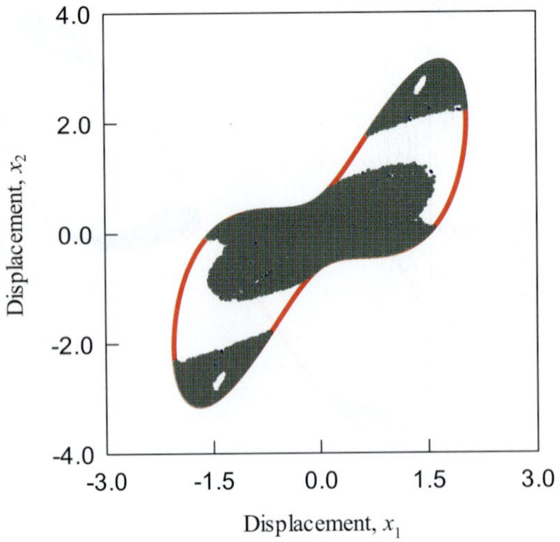

(c)

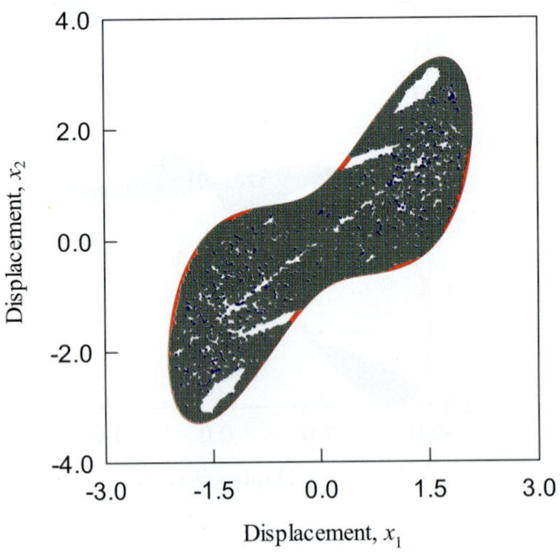

(d)

Figure 7.8. (continued)

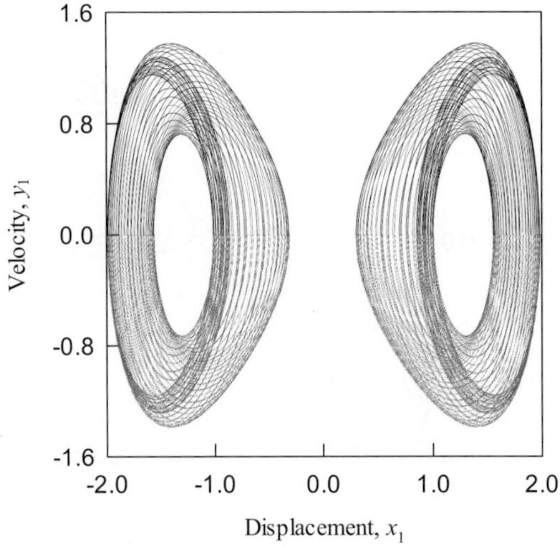

(a)

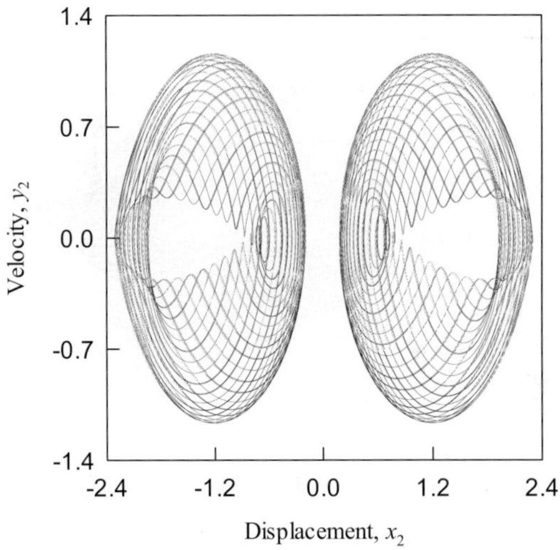

(b)

Figure 7.9. Quasi-periodic motion for $h = -0.02$: (a) phase plane (x_1, y_1) , (b) phase plane (x_2, y_2) , (c) the displacement plane (or potential domain) (x_1, x_2) and (d) velocity plane (or kinetic energy domain) (y_1, y_2) . ($\alpha = \beta = \gamma = \varepsilon = 1$) . $(x_1, x_2; y_1, y_2) \approx (1.5, 0.2; 0.4886, 0.0)$ is used as the initial condition .

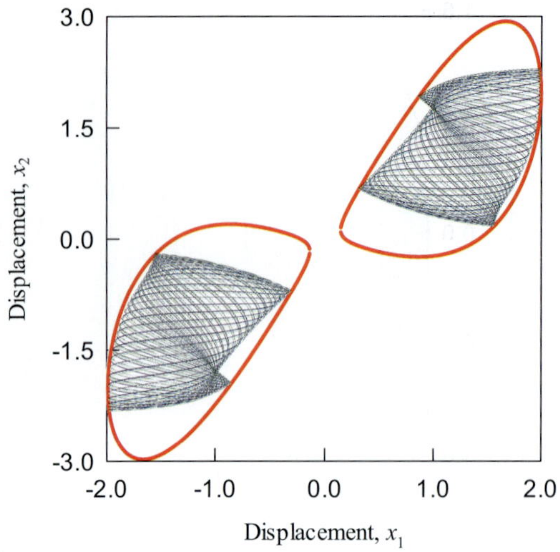

(c)

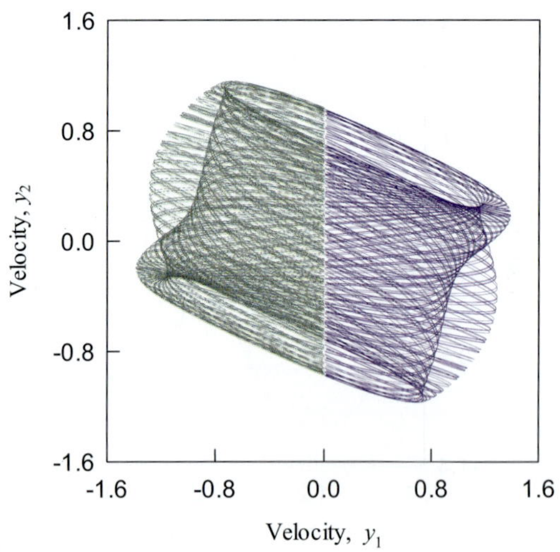

(d)

Figure 7.9. (continued)

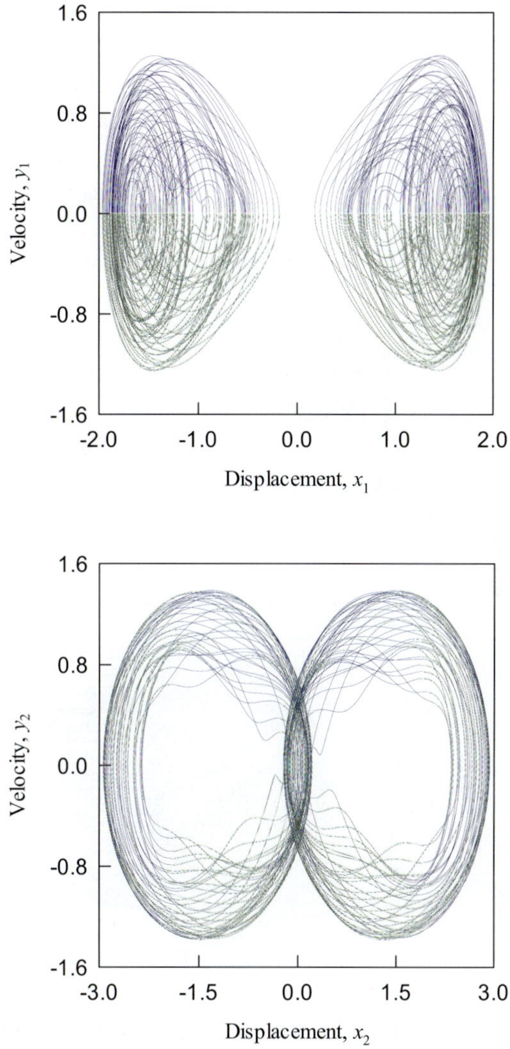

(a)

(b)

Figure 7.10. Chaotic motion of $h = -0.02$: (a) phase plane (x_1, y_1), (b) phase plane (x_2, y_2), (c) displacement plane (potential domain) (x_1, x_2) and (d) velocity plane (or kinetic energy domain) (y_1, y_2) .($\alpha = \beta = \gamma = \varepsilon = 1$). $(x_1, x_2; y_1, y_2) \approx (0.4, 0.2; 0.4767, 0.0)$ is the initial condition.

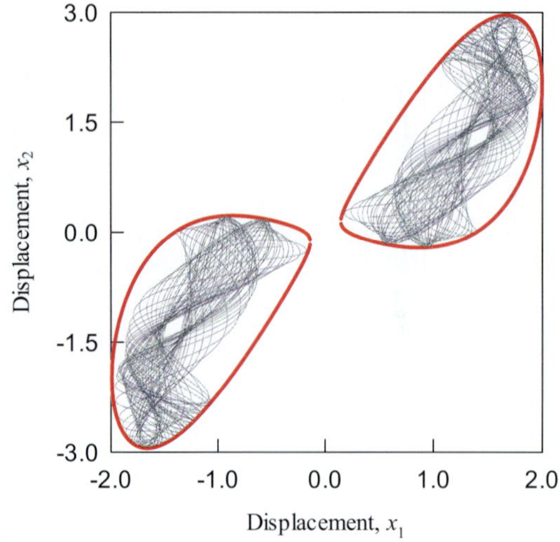

(c)

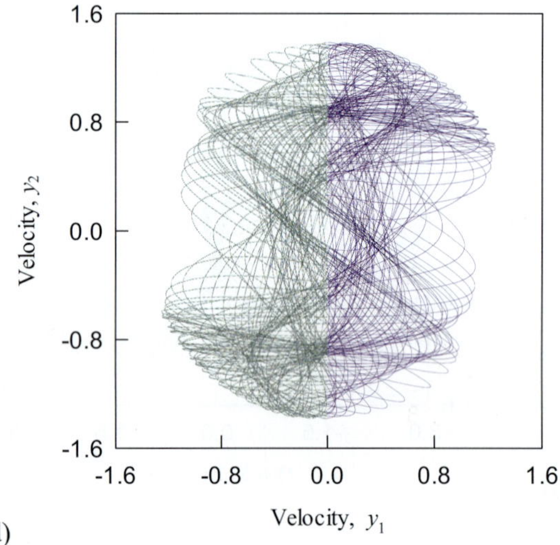

(d)

Figure 7.10. (continued)

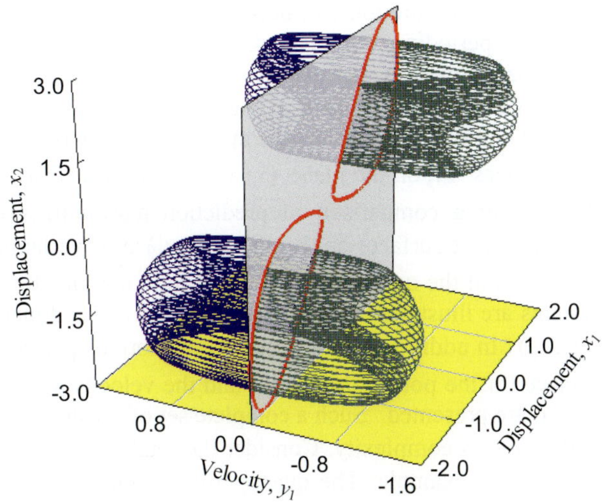

(a)

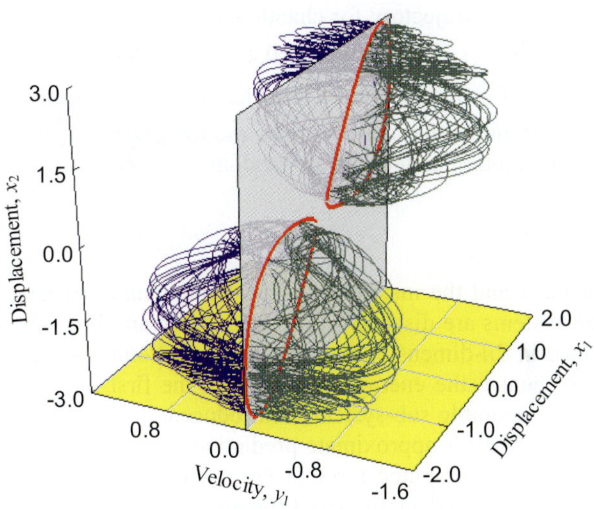

(b)

Figure 7.11. Motions on the surface (x_1, y_1, x_2) for $h = -0.02$: (a) quasi-periodic motion with the initial condition $(x_1, x_2; y_1, y_2) \approx (1.5, 0.2, 0.4886, 0.0)$ and (b) chaos with the initial condition $(x_1, x_2; y_1, y_2) \approx (0.4, 0.2; 0.4767, 0.0)$. ($\alpha = \beta = \gamma = \varepsilon = 1$).

critical value, the chaotic motion will occupy the entire Poincaré surface. Only a few isolated, quasi-periodic motion islands relative to certain resonances are embedded on the chaotic motion. For instance, the Poincaré surface of the Hamiltonian system with $h = 0.4$ is presented in Fig.7.7(d). From such illustrations, the resonance tunnel embedded in the chaotic motion is clearly observed. In Fig.7.8, the Poincaré surfaces ${}^P\Pi_h$ on the plane (x_1, x_2) are illustrated which is similar to those in Fig.7.6 for a comparison of prediction and simulation. Because of symmetry, the Poincaré surfaces on the plane (x_1, x_2) are overlapped.

The trajectories of the quasi-periodic and chaotic motions for two-degrees of freedom systems are illustrated through phase planes and 3-dimensional illustration of (x_1, y_1, x_2). In addition, the trajectories of motions projecting on the displacement space (or the potential domain) and the velocity space (or the kinetic energy domain) are presented. Such a complete set of illustrations will help ones understand the motion complexity. Consider the total energy $h = -0.02$ for ($\alpha = \beta = \gamma = \varepsilon = 1$) as an example. The quasi-periodic motion relative to the (4:1)-resonance are arranged in Fig.7.9 with the initial condition $(x_1, x_2; y_1, y_2) \approx (1.5, 0.2; 0.4886, 0.0)$. The trajectory for chaotic motion is plotted in Fig.7.10 with an initial condition $(x_1, x_2; y_1, y_2) \approx (0.4, 0.2; 0.4767, 0.0)$. The 3-dimensional illustrations of (x_1, y_1, x_2) for quasi-periodic and chaotic motions are given in Fig.7.11. From such illustrations, the quasi-periodic motion gives a regular pattern. However, the chaotic motion pattern is very random.

7.5. Conclusions

The boundedness and the maximum equi-potential super-surfaces for nonlinear Hamiltonian systems are discussed for dynamics on an (2*n*-1)-dimensional equi-energy surface in 2*n*-dimensional, nonlinear Hamiltonian systems. To compute the energy exchange, the energy increment (or the first integral quantity increment) for each separable sub-system is introduced. The approximate energy increment can be used for approximate predictions of chaotic motions. The exact energy increment is computed through the numerical method. In this Chapter, the internal resonance and energy spectrum method are presented for a numerical prediction of chaotic motion in nonlinear Hamiltonian systems. This method is applied to an integrable, nonlinear Hamiltonian system with two degrees of freedom as an example. The analytical estimates of regular and chaotic motions in nonlinear Hamiltonian systems need to be further investigated. The mathematical theory should be developed. The theory presented in this Chapter may be extendable to the nonlinear dynamics on the equi-first integral quantity surface in the *n*-dimensional, conserved, nonlinear dynamical systems.

Chapter 8

Stability and Grazing in Dissipative Systems

For Chapters 4-7, the chaotic motions in non-dissipative systems with periodic excitations were discussed. It was discovered that the mechanism of chaotic motions in such non-dissipative systems is the resonant interaction rather than the Smale's horseshoe structure because such resonant interaction generates the resonant separatrix and resonance overlap. The analytical conditions for such chaotic motions were presented from the first integral quantity increments. Once the dissipation exists in dynamical system, the characteristics of flow in such a system are different from non-dissipative system. Therefore, in this Chapter, the stability for equilibrium and periodic flows in n-dimensional dissipative systems will be discussed through the first integral quantity increments. The expansiveness, conservativeness and dissipativeness of dynamical systems will be introduced. The strong stability condition based on the G-function will be presented, which is equivalent to the traditional Lyapunov stability condition. The weak stability condition based on the L-function (i.e., the first integral quantity increment) will be presented. Using the L-function, the stability of periodic flows in n-dimensional systems will be discussed. To investigate the complexity of periodic flows, the local grazing bifurcation to a selected, first integral manifold will be presented. This bifurcation is different from the periodic doubling bifurcation. The global grazing bifurcation to the separatrix will be discussed. The mapping cluster of the post-grazing will be introduced to discuss the complexity of a flow, and the global chaotic flows will be discussed though the mapping structure. The strange attractor fragmentation of chaotic flows will be presented. However, the further development of mathematical theory on such a strange attractor fragmentation needs to be carried out.

8.1. Equilibrium stability

The flow and vector field characteristics of a dynamical system in Eq.(3.1) are discussed through both the normal component of the vector fields to the first integral manifold surface and the first integral quantity increment. Further, the corresponding theorems for the stability of equilibrium in Eq.(3.1) will be developed herein. Based on the normal component of vector fields to the first integral manifold surface, the conditions for equilibrium stability is similar to the ones in the Lyapunov sense of stability. However, based on such conditions, the stability condition for equilibriums is very strong. Thus, the stability conditions based on the first integral quantity increment will be developed as well in this section.

8.1.1. System dissipativeness and expansiveness

Before the stability theory for equilibrium is discussed, the following concepts are defined first.

DEFINITION 8.1. For a dynamical system in Eq.(3.1), a flow γ (i.e., $\mathbf{x}^{\alpha}(t)$) from $\mathbf{x}_k$ to $\mathbf{x}_{k+1}$ for $t \in [t_k, t_{k+1}]$ in a sub-domain Ω_α is:

(i) uniformly dissipative to the first integral quantity in Ω_α if all points $\mathbf{x}^{\alpha}(t)$ for $t \in [t_k, t_{k+1}]$ on the flow γ satisfy the following condition

$$G_{\sigma_\alpha}(\mathbf{x}^{(\alpha)}, t, \mathbf{p}) = \nabla F\left(\mathbf{x}^{(\alpha)}, \boldsymbol{\mu}\right) \cdot \mathbf{g}\left(\mathbf{x}^{(\alpha)}, t, \boldsymbol{\pi}\right) < 0; \tag{8.1}$$

(ii) uniformly expansive to the first integral quantity in Ω_α if all points $\mathbf{x}^{\alpha}(t)$ for $t \in [t_k, t_{k+1}]$ on the flow γ satisfy the following condition

$$G_{\sigma_\alpha}(\mathbf{x}^{(\alpha)}, t, \mathbf{p}) = \nabla F\left(\mathbf{x}^{(\alpha)}, \boldsymbol{\mu}\right) \cdot \mathbf{g}\left(\mathbf{x}^{(\alpha)}, t, \boldsymbol{\pi}\right) > 0; \tag{8.2}$$

(iii) uniformly conservative to the first integral quantity in Ω_α if all points $\mathbf{x}^{\alpha}(t)$ for $t \in [t_k, t_{k+1}]$ on the flow γ satisfy the following condition

$$G_{\sigma_\alpha}(\mathbf{x}^{(\alpha)}, t, \mathbf{p}) = \nabla F\left(\mathbf{x}^{(\alpha)}, \boldsymbol{\mu}\right) \cdot \mathbf{g}\left(\mathbf{x}^{(\alpha)}, t, \boldsymbol{\pi}\right) = 0. \tag{8.3}$$

DEFINITION 8.2. For a dynamical system in Eq.(3.1), a flow $\mathbf{x}^{\alpha}(t)$ from $\mathbf{x}_k$ to $\mathbf{x}_{k+1}$ with $t \in [t_k, t_{k+1}]$ in a sub-domain Ω_α is:

(i) weakly dissipative to the first integral quantity in Ω_α if the first integral quantity increment for $t \in [t_k, t_{k+1}]$ is less than zero, i.e.,

$$L^{(\alpha)}\left(t_k, t_{k+1}\right) \equiv \int_{t_k}^{t_{k+1}} \nabla F\left(\mathbf{x}^{(\alpha)}, \boldsymbol{\mu}\right) \cdot \mathbf{g}\left(\mathbf{x}^{(\alpha)}, t, \boldsymbol{\pi}\right) dt < 0; \tag{8.4}$$

(ii) weakly expansive to the first integral quantity in Ω_α if the first integral quantity increment for $t \in [t_k, t_{k+1}]$ is greater than zero, i.e.,

$$L^{(\alpha)}\left(t_k, t_{k+1}\right) \equiv \int_{t_k}^{t_{k+1}} \nabla F\left(\mathbf{x}^{(\alpha)}, \boldsymbol{\mu}\right) \cdot \mathbf{g}\left(\mathbf{x}^{(\alpha)}, t, \boldsymbol{\pi}\right) dt > 0; \tag{8.5}$$

(iii) conservative to the first integral quantity in Ω_α if the first integral quantity increment for $t \in [t_k, t_{k+1}]$ equal to zero, i.e.,

$$L^{(\alpha)}(t_k, t_{k+1}) \equiv \int_{t_k}^{t_{k+1}} \nabla F(\mathbf{x}^{(\alpha)}, \boldsymbol{\mu}) \cdot \mathbf{g}(\mathbf{x}^{(\alpha)}, t, \boldsymbol{\pi})\, dt = 0. \tag{8.6}$$

DEFINITION 8.3. For a dynamical system in Eq.(3.1), a flow γ (i.e., $\mathbf{x}^{\alpha}(t)$) starting at any non-singular point $\mathbf{x}_k^{(\alpha)}$ with $t \in [t_k, \infty)$ in a sub-domain Ω_α is:

(i) absolutely and uniformly *dissipative* to the first integral quantity in Ω_α if all points $\mathbf{x}^{\alpha}(t)$ ($t \in [t_k, \infty)$) on the flow γ satisfy Eq.(8.1);

(ii) absolutely and uniformly *expansive* to the first integral quantity in Ω_α if all points $\mathbf{x}^{\alpha}(t)$ ($t \in [t_k, \infty)$) on the flow γ satisfy Eq.(8.2);

(iii) absolutely and uniformly *conservative* to the first integral quantity in Ω_α if all points $\mathbf{x}^{\alpha}(t)$ ($t \in [t_k, \infty)$) on the flow satisfy Eq.(8.3).

DEFINITION 8.4. For a dynamical system in Eq.(3.1), the vector field $\mathbf{g}(\mathbf{x}^{(\alpha)}, t, \boldsymbol{\pi})$ in a sub-domain Ω_α is:

(i) *dissipative* to the first integral quantity in Ω_α if all the flows γ (i.e., $\mathbf{x}^{\alpha}(t)$) starting at any non-singular point $\mathbf{x}_k^{(\alpha)}$ with $t \in [t_k, \infty)$ is absolutely and uniformly *dissipative*;

(ii) *expansive* to the first integral quantity in Ω_α if all the flows γ (i.e., $\mathbf{x}^{\alpha}(t)$) starting at any non-singular point $\mathbf{x}_k^{(\alpha)}$ with $t \in [t_k, \infty)$ is absolutely and uniformly *expansive*;

(iii) *conservative* to the first integral quantity in Ω_α if all the flows γ (i.e., $\mathbf{x}^{\alpha}(t)$) starting at any non-singular point $\mathbf{x}_k^{(\alpha)}$ with $t \in [t_k, \infty)$ is absolutely and uniformly *conservative*.

REMARK 8.1. If the vector field $\mathbf{g}(\mathbf{x}, t, \boldsymbol{\pi})$ in phase space is dissipative, or expansive or conservative, then the vector field $\mathbf{F}(\mathbf{x}, t, \mathbf{p})$ is also dissipative or, expansive or conservative, accordingly. After introduction of the foregoing concepts, the local and global minimum and maximum of the first integral quantity are defined as follows.

DEFINITION 8.5. For a dynamical system in Eq.(3.1), there is a point $\mathbf{x}_0$ with $E_0 = F(\mathbf{x}_0, \boldsymbol{\mu})$ and $|E_0| \le K(\text{const}) < \infty$. For any given $\varepsilon > 0$, there is a neighborhood of $\mathbf{x}_0$, i.e.,

$$U_{\mathbf{x}_0}^{\varepsilon} = \{\mathbf{x} \| \|\mathbf{x} - \mathbf{x}_0\| < \varepsilon\}. \tag{8.7}$$

(i) The first integral quantity $F(\mathbf{x}_0, \boldsymbol{\mu}) = E_0$ in $U_{\mathbf{x}_0}^{\varepsilon}$ is *the local minimum* if there is a positive $\delta > 0$ to make the following relation hold

$$F(\mathbf{x}, \boldsymbol{\mu}) = E \in (E_0, E_0 + \delta] \tag{8.8}$$

for all $\mathbf{x} \in U_{\mathbf{x}_0}^{\varepsilon} \subset \mathbb{R}^n$ and $\mathbf{x} \neq \mathbf{x}_0$, and the corresponding first integral quantity set is defined as

$$N_{E_0}^{\delta+} = \left\{ E \left| \begin{array}{l} F(\mathbf{x}, \boldsymbol{\mu}) = E \in [E_0, E_0 + \delta] \\ \text{for } \mathbf{x} \in U_{\mathbf{x}_0}^{\varepsilon} \text{ and only } F(\mathbf{x}_0, \boldsymbol{\mu}) = E_0 \end{array} \right. \right\}. \tag{8.9}$$

(ii) The first integral quantity $F(\mathbf{x}_0, \boldsymbol{\mu}) = E_0$ in $U_{\mathbf{x}_0}^{\varepsilon}$ is *the local maximum* if there is a positive $\delta > 0$ to make the following relation hold

$$F(\mathbf{x}, \boldsymbol{\mu}) = E \in [E_0 - \delta, E_0) \tag{8.10}$$

for all $\mathbf{x} \in U_{\mathbf{x}_0}^{\varepsilon} \subset \mathbb{R}^n$ and $\mathbf{x} \neq \mathbf{x}_0$, and the corresponding first integral quantity set is defined as

$$N_{E_0}^{\delta-} = \left\{ E \left| \begin{array}{l} F(\mathbf{x}, \boldsymbol{\mu}) = E \in [E_0 - \delta, E_0] \\ \text{for } \mathbf{x} \in U_{\mathbf{x}_0}^{\varepsilon} \text{ and only } F(\mathbf{x}_0, \boldsymbol{\mu}) = E_0 \end{array} \right. \right\}. \tag{8.11}$$

DEFINITION 8.6. For the dynamical system in Eq.(3.1), a contour set is defined as

$${}^{E}U = \{\mathbf{x} | F(\mathbf{x}) = E\}, \tag{8.12}$$

and an intersection set with $U_{\mathbf{x}_0}^{\varepsilon}$ and the corresponding complement set are:

$${}^{E}U_{\mathbf{x}_0}^{\varepsilon} = {}^{E}U \cap U_{\mathbf{x}_0}^{\varepsilon} \text{ and } {}^{E}V_{\mathbf{x}_0}^{\varepsilon} = {}^{E}U / {}^{E}U_{\mathbf{x}_0}^{\varepsilon} \tag{8.13}$$

DEFINITION 8.7. For a dynamical system in Eq.(3.1), there is a point $\mathbf{x}_0$ with $\mathbf{x} \in U_{\mathbf{x}_0}^{\varepsilon} \subset \mathbb{R}^n$ and $|E_0| < \infty$. The first integral quantity $F(\mathbf{x}, \boldsymbol{\mu})$ is bounded with the infimum if

$$F(\mathbf{x}, \boldsymbol{\mu}) \in [E_0, \infty) \quad \text{for all } \mathbf{x} \in U_x \subset \mathbb{R}^n. \tag{8.14}$$

The first integral quantity $F(\mathbf{x},\boldsymbol{\mu})$ is bounded with the supremum if

$$F(\mathbf{x},\boldsymbol{\mu}) \in (-\infty, E_0] \quad \text{for all } \mathbf{x} \in U_x \subset \mathbb{R}^n. \tag{8.15}$$

DEFINITION 8.8. For a dynamical system in Eq.(3.1), there are two points $\mathbf{x}_i$ ($i = 1,2$) with $F(\mathbf{x}_i,\boldsymbol{\mu}) = E_i$, $|E_i| < \infty$ and $E_1 < E_2$. The first integral quantity $F(\mathbf{x},\boldsymbol{\mu})$ is absolutely bounded if

$$F(\mathbf{x},\boldsymbol{\mu}) \in [E_1, E_2] \quad \text{for all } \mathbf{x} \in U_x \subset \mathbb{R}^n. \tag{8.16}$$

The infimum, supremum, and local minimum and maximum of the first integral quantity of a dynamical system in Eq.(3.1) are shown in Fig.8.1. The first integral quantity possesses an infimum, as shown in Fig.8.1(a). The corresponding local minimum and maximum of the first integral quantity with infimum are presented. The local domain and the interval of the first integral quantity are presented. The dark, gray and white circles represent the global minimum and maximum (i.e., infimum and supremum), local minimum and local maximum of the first integral quantity, respectively. The corresponding domains and the integral quantity interval are represented through solid lines. Similarly, the first integral quantity possessing a supremum is sketched in Fig.8.1(b). For such a first integral quantity, the corresponding local minimum and maximum is also presented.

8.1.2. Stability definitions

The definition for the flow stability of dynamical system in Eq.(3.1) is given as follows.

DEFINITION 8.9. For a dynamical system in Eq.(3.1), there is a solution $\mathbf{x}^*(t)$. For any given $\varepsilon > 0$, there is a positive parameter $\delta(t_0,\varepsilon) > 0$.

(i) The solution $\mathbf{x}^*(t)$ for $t \in [t_0,\infty)$ is stable if

$$\left\|\mathbf{x}(t_0) - \mathbf{x}^*(t_0)\right\| < \delta \tag{8.17}$$

from which the following relation holds for $t \in [t_0,\infty)$

$$\left\|\mathbf{x}(t) - \mathbf{x}^*(t)\right\| < \varepsilon. \tag{8.18}$$

(ii) Otherwise, the solution $\mathbf{x}^*(t)$ for $t \in [t_0,\infty)$ is unstable.

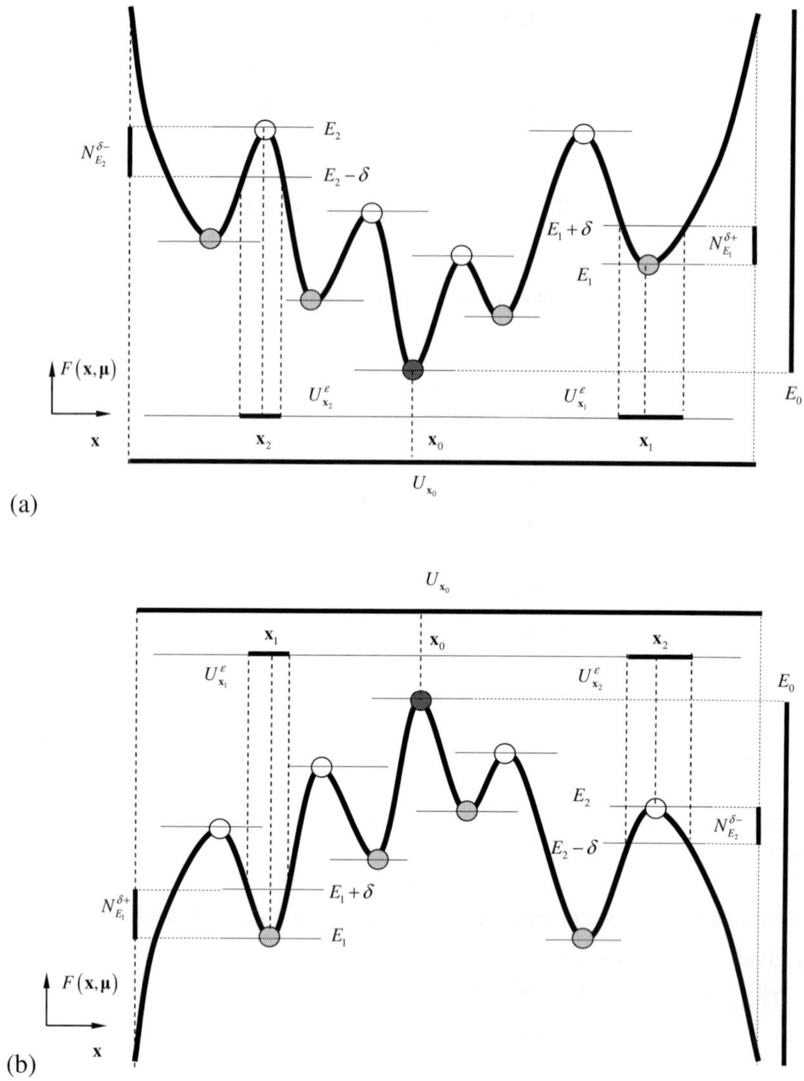

Figure 8.1. The local maximum and minimum first integral quantity with: (a) infimum and (b) supremum. The dark, gray and white circles represent the global minimum or maximum (i.e., infimum or supremum), local minimum and local maximum of the first integral quantity, respectively. The corresponding domains and the integral quantity interval are represented through solid lines.

DEFINITION 8.10. For a dynamical system in Eq.(3.1), there is a solution $\mathbf{x}^*(t)$. For any given $\varepsilon > 0$, there is a positive parameter $\delta(t_0, \varepsilon) > 0$.
The solution $\mathbf{x}^*(t)$ for $t \in [t_0, \infty)$ is asymptotically stable if, under the condition in Eq.(8.14), the following relations hold for $t \in [t_0, \infty)$

$$\left\|\mathbf{x}(t) - \mathbf{x}^*(t)\right\| < \varepsilon \text{ and } \lim_{t\to\infty}\left\|\mathbf{x}(t) - \mathbf{x}^*(t)\right\| = 0. \tag{8.19}$$

DEFINITION 8.11. For a dynamical system in Eq.(3.1), there is a solution $\mathbf{x}^*(t)$. For any given $\varepsilon > 0$, there is a positive parameter $\delta(\varepsilon) > 0$ independent of initial time t_0.

(i) The solution $\mathbf{x}^*(t)$ for $t \in [t_0, \infty)$ is uniformly stable if, under the condition in Eq.(8.17), equation (8.18) is satisfied
(ii) The solution $\mathbf{x}^*(t)$ for $t \in [t_0, \infty)$ is uniformly and asymptotically stable if Eq. (8.19) is satisfied under the condition in Eq.(8.16).

The three definitions are applicable to equilibrium and periodic motion stability in n-dimensional dynamical systems. Next, the equilibrium stability is discussed.

8.1.3. Stability conditions

From the stability definition, the stability conditions for equilibriums are developed as follows:

THEOREM 8.1. *For a dynamical system in Eq.(3.1), there is an equilibrium point* $\mathbf{x}^*(t) = \mathbf{x}_0$ *for* $t \in [t_0, \infty)$. *For any* $\varepsilon > 0$, *there is a positive parameter* $\delta > 0$. *For* $\mathbf{x}(t) \in U^{\varepsilon}_{\mathbf{x}_0}$, *the first integral quantity is given by*

$$\begin{aligned} &F(\mathbf{x}, \boldsymbol{\mu}) = E \in N^{\delta+}_{E_0} \ (\text{or } F(\mathbf{x}, \boldsymbol{\mu}) = E \in N^{\delta-}_{E_0}) \\ &\text{and } {}^{E}V^{\varepsilon}_{\mathbf{x}_0} \cap {}^{E}U^{\varepsilon}_{\mathbf{x}_0} = \{\varnothing\}. \end{aligned} \tag{8.20}$$

(i) *The equilibrium* $\mathbf{x}_0$ *is stable for* $t \in [t_0, \infty)$ *in* $U^{\varepsilon}_{\mathbf{x}_0}$ *if*

$$G(t, \mathbf{x}) \equiv \nabla F(\mathbf{x}, \boldsymbol{\mu}) \cdot \mathbf{g}(\mathbf{x}, t, \boldsymbol{\pi}) \le 0 \ \ (\text{or } G(t, \mathbf{x}) \ge 0). \tag{8.21}$$

(ii) *The equilibrium* $\mathbf{x}_0$ *is unstable for* $t \in [t_0, \infty)$ *in* $U^{\varepsilon}_{\mathbf{x}_0}$ *if*

$$G(t, \mathbf{x}) \equiv \nabla F(\mathbf{x}, \boldsymbol{\mu}) \cdot \mathbf{g}(\mathbf{x}, t, \boldsymbol{\pi}) > 0 \ \ (\text{or } G(t, \mathbf{x}) < 0). \tag{8.22}$$

PROOF. (i) Because the first integral quantity $F(\mathbf{x},\boldsymbol{\mu}) = E$ is determined by the C^r-integrable vector field ($r > 1$) in Eq.(3.1). So $F(\mathbf{x},\boldsymbol{\mu})$ and its derivative are continuous. Consider an equilibrium point $\mathbf{x}^*(t) = \mathbf{x}_0$ for $t \in [t_0, \infty)$ with $F(\mathbf{x}_0,\boldsymbol{\mu}) = E_0$. For any $\varepsilon > 0$, there is a neighborhood $U^{\varepsilon}_{\mathbf{x}_0}$ of point $\mathbf{x}_0$. For $\mathbf{x}(t) \in U^{\varepsilon}_{\mathbf{x}_0}$, the first integral quantity is expressed by $F(\mathbf{x},\boldsymbol{\mu}) = E \in N^{\delta+}_{E_0}$,which implies $F(\mathbf{x},\boldsymbol{\mu}) = E \geq E_0$ with $F(\mathbf{x}_0,\boldsymbol{\mu}) = E_0$ .If $G(t,\mathbf{x}) \equiv \nabla F(\mathbf{x},\boldsymbol{\mu}) \cdot \mathbf{g}(\mathbf{x},t,\boldsymbol{\pi}) \leq -M < 0$ (i.e. $M > 0$), so we have

$$\begin{aligned} F(\mathbf{x}(t),\boldsymbol{\mu}) - F(\mathbf{x}(t_0),\boldsymbol{\mu}) &= \int_{t_0}^{t} G(t,\mathbf{x})\,dt \\ &\equiv \int_{t_0}^{t} \nabla F(\mathbf{x},\boldsymbol{\mu}) \cdot \mathbf{g}(\mathbf{x},t,\boldsymbol{\pi})dt \leq -M \times (t - t_0) \end{aligned}$$

from which $F(\mathbf{x}(t),\boldsymbol{\mu}) \leq F(\mathbf{x}(t_0),\boldsymbol{\mu}) - M \times (t - t_0)$ for $t \in [t_0, \infty)$. $F(\mathbf{x}(t),\boldsymbol{\mu}) \to -\infty$ as $t \to \infty$. Because $F(\mathbf{x}_0,\boldsymbol{\mu}) = E_0$ is *only* the minimum for $\mathbf{x}(t) \in U^{\varepsilon}_{\mathbf{x}_0}$ owing to ${}^{E}V^{\varepsilon}_{\mathbf{x}_0} \cap {}^{E}U^{\varepsilon}_{\mathbf{x}_0} = \{\varnothing\}$, $F(\mathbf{x}(t),\boldsymbol{\mu})$ must arrive to the minimum at $\mathbf{x}(t) = \mathbf{x}_0$ before time $t = t_0 + \frac{1}{M}[F(\mathbf{x}(t_0),\boldsymbol{\mu}) - E_0]$. Therefore, the equilibrium $\mathbf{x}_0$ is stable for $t \in [t_0, \infty)$ in $U^{\varepsilon}_{\mathbf{x}_0}$. Similarly, for $F(\mathbf{x},\boldsymbol{\mu}) = E \in N^{\delta-}_{E_0}$, if $G(t,\mathbf{x}) > 0$, for $t \in [t_0, \infty)$ $F(\mathbf{x}(t),\boldsymbol{\mu}) \geq F(\mathbf{x}(t_0),\boldsymbol{\mu}) + M \times (t - t_0)$. As $t \to \infty$, $F(\mathbf{x}(t),\boldsymbol{\mu}) \to \infty$. Because $E_0 = F(\mathbf{x}_0,\boldsymbol{\mu})$ is *only* the maximum for $\mathbf{x}(t) \in U^{\varepsilon}_{\mathbf{x}_0}$. Due to ${}^{E}V^{\varepsilon}_{\mathbf{x}_0} \cap {}^{E}U^{\varepsilon}_{\mathbf{x}_0} = \{\varnothing\}$, at $\mathbf{x}(t) = \mathbf{x}_0$, $F(\mathbf{x}(t),\boldsymbol{\mu})$ must arrive to the maximum before $t = \frac{1}{M}[F(\mathbf{x}(t_0),\boldsymbol{\mu}) - E_0]$ $+t_0$. Therefore, the equilibrium $\mathbf{x}_0$ is stable for $t \in [t_0, \infty)$ in $U^{\varepsilon}_{\mathbf{x}_0}$.

For any $\varepsilon > 0$,if $\| \mathbf{x}(t_0) - \mathbf{x}_0 \| \leq \varepsilon$, let $\delta = \| F \| \varepsilon$, $\| F(\mathbf{x}(t_0),\boldsymbol{\mu}) - F(\mathbf{x}_0,\boldsymbol{\mu}) \| \leq \delta$. For $F(\mathbf{x},\boldsymbol{\mu}) = E \in N^{\delta+}_{E_0}$, if $G(t,\mathbf{x}) = 0$, the foregoing equation gives

$$\begin{aligned} 0 &= \left\| F(\mathbf{x}(t),\boldsymbol{\mu}) - F(\mathbf{x}(t_0),\boldsymbol{\mu}) \right\| \\ &\geq \left\| F(\mathbf{x}(t),\boldsymbol{\mu}) - F(\mathbf{x}_0,\boldsymbol{\mu}) \right\| - \left\| F(\mathbf{x}(t_0),\boldsymbol{\mu}) - F(\mathbf{x}_0,\boldsymbol{\mu}) \right\| \\ &\geq \left\| F(\mathbf{x}(t),\boldsymbol{\mu}) - F(\mathbf{x}_0,\boldsymbol{\mu}) \right\| - \delta \end{aligned}$$

from which one has

$$\left\| F(\mathbf{x}(t),\boldsymbol{\mu}) - F(\mathbf{x}_0,\boldsymbol{\mu}) \right\| \leq \delta.$$

Since $\varepsilon > 0$ is chosen arbitrarily, as $\varepsilon \to 0$, $\delta = \| F \| \varepsilon \to 0$ because $\| F \|$ is bounded. Therefore, from Definition 8.9, the equilibrium $\mathbf{x}_0$ is stable. Similarly, for

$\varepsilon > 0$, if $\| \mathbf{x}(t_0) - \mathbf{x}_0 \| \le \varepsilon$, let $\delta = \| F \| \varepsilon$, and then $\| F(\mathbf{x}(t_0), \boldsymbol{\mu}) - F(\mathbf{x}_0, \boldsymbol{\mu}) \| \le \delta$. For $F(\mathbf{x}, \boldsymbol{\mu}) = E \in N_{E_0}^{\delta-}$, if $G(t, \mathbf{x}) = 0$, $\| F(\mathbf{x}(t), \boldsymbol{\mu}) - F(\mathbf{x}_0, \boldsymbol{\mu}) \| \le \delta$. Since $\varepsilon > 0$ is chosen arbitrarily, as $\varepsilon \to 0$, $\delta = \| F \| \varepsilon \to 0$ because $\| F \|$ is bounded. Therefore, from Definition 8.9, the equilibrium $\mathbf{x}_0$ is stable.

(ii) In a similar fashion, for $\mathbf{x}(t) \in U_{\mathbf{x}_0}^{\varepsilon}$, the first integral quantity $F(\mathbf{x}, \boldsymbol{\mu}) = E \in N_{E_0}^{\delta+}$, which implies $F(\mathbf{x}, \boldsymbol{\mu}) = E \le E_0$ with $F(\mathbf{x}_0, \boldsymbol{\mu}) = E_0$. Consider $G(t, \mathbf{x}) \equiv \nabla F(\mathbf{x}, \boldsymbol{\mu}) \cdot \mathbf{g}(\mathbf{x}, t, \boldsymbol{\pi}) \ge M > 0$, so we have

$$\begin{aligned} F(\mathbf{x}(t), \boldsymbol{\mu}) - F(\mathbf{x}(t_0), \boldsymbol{\mu}) &= \int_{t_0}^{t} G(t, \mathbf{x})\, dt \\ &\equiv \int_{t_0}^{t} \nabla F(\mathbf{x}, \boldsymbol{\mu}) \cdot \mathbf{g}(\mathbf{x}, t, \boldsymbol{\pi}) dt \\ &\ge M \times (t - t_0). \end{aligned}$$

$F(\mathbf{x}, \boldsymbol{\mu}) = E_0 + \delta$ is the maximum for $\mathbf{x}(t) \in U_{\mathbf{x}_0}^{\varepsilon}$. As $t \to \infty$, $F(\mathbf{x}(t)) \to +\infty$. Thus $F(\mathbf{x}, \boldsymbol{\mu})$ is out of the boundary. Therefore, the equilibrium $\mathbf{x}_0$ is unstable for $t \in [t_0, \infty)$ in $U_{\mathbf{x}_0}^{\varepsilon}$. Similarly, for $F(\mathbf{x}, \boldsymbol{\mu}) = E \in N_{E_0}^{\delta-}$, if $G(t, \mathbf{x}) < 0$, $F(\mathbf{x}(t), \boldsymbol{\mu}) \ge F(\mathbf{x}(t_0), \boldsymbol{\mu}) - M \times (t - t_0)$ (i.e., $M > 0$) for $t \in [t_0, \infty)$. $F(\mathbf{x}, \boldsymbol{\mu}) = E_0 - \delta$ is the minimum for $\mathbf{x}(t) \in U_{\mathbf{x}_0}^{\varepsilon}$. As $t \to \infty$, $F(\mathbf{x}(t), \boldsymbol{\mu}) \to -\infty$. Thus $F(\mathbf{x}, \boldsymbol{\mu})$ is out of the boundary. Therefore, the equilibrium $\mathbf{x}_0$ is unstable for $t \in [t_0, \infty)$ in $U_{\mathbf{x}_0}^{\varepsilon}$. This theorem is proved. ■

REMARK 8.2. For convenience, $E(\mathbf{x}(t)) \equiv F(\mathbf{x}(t), \boldsymbol{\mu})$ and $E(\mathbf{x}(t_0)) = F(\mathbf{x}(t_0), \boldsymbol{\mu})$ from now on. This theorem is similar to the Lyaponuv stability theorem. The first integral quantity of the integrable system of Eq.(3.1) is employed as a candidate of the Lyaponuv function. The first integral quantity is determined by the C^r-integrable vector field ($r > 1$) in Eq.(3.1). So $F(\mathbf{x}, \boldsymbol{\mu})$ and its derivative are continuous. ${}^{E}V_{\mathbf{x}_0}^{\varepsilon} \cap {}^{E}U_{\mathbf{x}_0}^{\varepsilon} = \{\varnothing\}$ implies the set ${}^{E}U_{\mathbf{x}_0}^{\varepsilon}$ is isolated with only one local minimum at $F(\mathbf{x}_0) = E_0$. The stability conditions of equilibrium in the foregoing theorem are very strict. They are valid only for a uniformly dissipative, or expansive or conservative system. However, the theorem for a weak stability condition of equilibrium is given as follows.

THEOREM 8.2. *For a dynamical system in Eq.*(3.1), *there is an equilibrium point* $\mathbf{x}^*(t) = \mathbf{x}_0$ *for* $t \in [t_0, \infty)$. *For any* $\varepsilon > 0$, *there is a positive parameter* $\delta > 0$.

For $\mathbf{x}(t)\in U^{\varepsilon}_{\mathbf{x}_0}$, *the first integral quantity possesses the property in Eq.*(8.20).

(i) *The equilibrium* $\mathbf{x}_0$ *is stable in* $U^{\varepsilon}_{\mathbf{x}_0}$ *if for* $\mathbf{x}(t_0)\in U^{\varepsilon}_{\mathbf{x}_0}$ *and* $t\in[t_1,\infty)\subset[t_0,\infty)$

$$\begin{gathered}L(t_0,t)\equiv\int_{t_0}^{t}\nabla F(\mathbf{x},\boldsymbol{\mu})\cdot\mathbf{g}(\mathbf{x},t,\boldsymbol{\pi})dt\le E_0-E(\mathbf{x}(t_0))\\(\text{or } L(t_0,t)\ge E_0-E(\mathbf{x}(t_0))).\end{gathered}\tag{8.23}$$

(ii) *The equilibrium* $\mathbf{x}_0$ *is unstable in* $U^{\varepsilon}_{\mathbf{x}_0}$ *if for* $\mathbf{x}(t_0)\in U^{\varepsilon}_{\mathbf{x}_0}$ *and* $t\in[t_1,\infty)\subset[t_0,\infty)$

$$\begin{gathered}L(t_0,t)\equiv\int_{t_0}^{t}F(\mathbf{x},\boldsymbol{\mu})\cdot\mathbf{g}(\mathbf{x},t,\boldsymbol{\pi})dt> E_0+\delta-E(\mathbf{x}(t_0))\\(\text{or } L(t_0,t)< E_0-\delta-E(\mathbf{x}(t_0))).\end{gathered}\tag{8.24}$$

PROOF. Consider an equilibrium point $\mathbf{x}^*(t)=\mathbf{x}_0$ for $t\in[t_0,\infty)$ with $F(\mathbf{x}_0,\boldsymbol{\mu})=E_0$. For any $\varepsilon>0$, there is a neighborhood $U^{\varepsilon}_{\mathbf{x}_0}$ of point $\mathbf{x}_0$. For $\mathbf{x}(t)\in U^{\varepsilon}_{\mathbf{x}_0}$, the first integral quantity $F(\mathbf{x},\boldsymbol{\mu})=E\in N^{\delta+}_{E_0}$, i.e., $F(\mathbf{x},\boldsymbol{\mu})=E\ge E_0$ with $F(\mathbf{x}_0,\boldsymbol{\mu})=E_0$. Because of

$$\int_{t_0}^{t}\nabla F(\mathbf{x},\boldsymbol{\mu})\cdot\mathbf{g}(\mathbf{x},t,\boldsymbol{\pi})dt=F(\mathbf{x}(t),\boldsymbol{\mu})-F(\mathbf{x}(t_0),\boldsymbol{\mu})\le E_0-F(\mathbf{x}(t_0),\boldsymbol{\mu}),$$

for $t\in[t_1,\infty)\subset[t_0,\infty)$, from which one has

$$E_0\ge F(\mathbf{x}(t),\boldsymbol{\mu}).$$

However, $F(\mathbf{x}_0,\boldsymbol{\mu})=E_0$ is *only* the minimum for $\mathbf{x}(t)\in U^{\varepsilon}_{\mathbf{x}_0}$ because ${}^{E}V^{\varepsilon}_{\mathbf{x}_0}\cap{}^{E}U^{\varepsilon}_{\mathbf{x}_0}=\{\varnothing\}$, thus $F(\mathbf{x}(t),\boldsymbol{\mu})\ge E_0$. Therefore $E_0=F(\mathbf{x}(t))$ from time t, which implies the equilibrium $\mathbf{x}_0$ is stable for $t\in[t_0,\infty)$ in $U^{\varepsilon}_{\mathbf{x}_0}$. Similarly, for $\mathbf{x}(t)\in U^{\varepsilon}_{\mathbf{x}_0}$, the first integral quantity $F(\mathbf{x},\boldsymbol{\mu})=E\in N^{\delta-}_{E_0}$, i.e., $F(\mathbf{x},\boldsymbol{\mu})=E\le E_0$ with $F(\mathbf{x}_0,\boldsymbol{\mu})=E_0$. Similarly, if $L(t,t_0)\ge E_0-F(\mathbf{x}(t_0),\boldsymbol{\mu})$, then the equilibrium $\mathbf{x}_0$ is stable for $t\in[t_0,\infty)$ in $U^{\varepsilon}_{\mathbf{x}_0}$.

(ii) In the similar fashion, for any $\varepsilon>0$ and $\mathbf{x}(t)\in U^{\varepsilon}_{\mathbf{x}_0}$, the first integral quantity $F(\mathbf{x})=E\in N^{\delta+}_{E_0}$, which implies $F(\mathbf{x},\boldsymbol{\mu})=E\ge E_0$ with $F(\mathbf{x}_0,\boldsymbol{\mu})=E_0$. Because

$$\int_{t_0}^{t}\nabla F(\mathbf{x},\boldsymbol{\mu})\cdot\mathbf{g}(\mathbf{x},t,\boldsymbol{\pi})dt=F(\mathbf{x}(t),\boldsymbol{\mu})-F(\mathbf{x}(t_0),\boldsymbol{\mu})> E_0+\delta-F(\mathbf{x}(t_0),\boldsymbol{\mu}),$$

for $t \in [t_1, \infty) \subset [t_0, \infty)$, from which one has

$$F(\mathbf{x}(t), \boldsymbol{\mu}) > \delta + E_0.$$

However, $F(\mathbf{x}, \boldsymbol{\mu}) = E_0 + \delta$ is the maximum for $\mathbf{x}(t) \in U^{\varepsilon}_{\mathbf{x}_0}$. Thus, the flow passes over the boundary of $N^{\delta+}_{E_0}$, which indicates the equilibrium $\mathbf{x}_0$ is unstable for $t \in [t_0, \infty)$ in $U^{\varepsilon}_{\mathbf{x}_0}$. For $\mathbf{x}(t) \in U^{\varepsilon}_{\mathbf{x}_0}$, the first integral quantity $F(\mathbf{x}, \boldsymbol{\mu}) = E \in N^{\delta-}_{E_0}$, i.e., $F(\mathbf{x}, \boldsymbol{\mu}) = E \le E_0$ with $F(\mathbf{x}_0, \boldsymbol{\mu}) = E_0$ .Similarly, if $L(t, t_0) < E_0 + \delta - F(\mathbf{x}(t_0), \boldsymbol{\mu})$, we have $F(\mathbf{x}(t), \boldsymbol{\mu}) < \delta + E_0$, so the flow passes over the boundary of $N^{\delta-}_{E_0}$. So, the equilibrium $\mathbf{x}_0$ is unstable for $t \in [t_0, \infty)$ in $U^{\varepsilon}_{\mathbf{x}_0}$. This theorem is proved. ■

REMARK 8.3. In the foregoing theorem, if Eq.(8.23) or (8.24) holds only for $t_1 \to \infty$, an equilibrium $\mathbf{x}_0$ is asymptotically stable or unstable in $U^{\varepsilon}_{\mathbf{x}_0}$ i.e., if for $\mathbf{x}(t_0) \in U^{\varepsilon}_{\mathbf{x}_0}$ and $t \in [t_1, \infty) \subset [t_0, \infty)$, (i) the equilibrium $\mathbf{x}_0$ is stable in $U^{\varepsilon}_{\mathbf{x}_0}$

$$\begin{aligned} &L(t_0, \infty) \equiv \int_{t_0}^{\infty} \nabla F(\mathbf{x}, \boldsymbol{\mu}) \cdot \mathbf{g}(\mathbf{x}, t, \boldsymbol{\pi}) dt \le E_0 - E(\mathbf{x}(t_0)) \\ &(\text{or } L(t_0, \infty) \ge E_0 - E(\mathbf{x}(t_0))), \end{aligned} \tag{8.25}$$

and (ii) the equilibrium $\mathbf{x}_0$ is unstable in $U^{\varepsilon}_{\mathbf{x}_0}$

$$\begin{aligned} &L(t_0, \infty) \equiv \int_{t_0}^{\infty} \nabla F(\mathbf{x}, \boldsymbol{\mu}) \cdot \mathbf{g}(\mathbf{x}, t, \boldsymbol{\pi}) dt > E_0 + \delta - E(\mathbf{x}(t_0)) \\ &(\text{or } L(t_0, \infty) < E_0 - \delta - E(\mathbf{x}(t_0))). \end{aligned} \tag{8.26}$$

In the above theorem, the first integral quantity increment is computed from the time period of the flow. The following theorem is given through the integral quantity increment based on the normal surface to the first integral manifold surface (or Poincaré cross section).

THEOREM 8.3. *For a dynamical system in Eq.(3.1), there is an equilibrium point* $\mathbf{x}^*(t) = \mathbf{x}_0$ *for* $t \in [t_0, \infty)$. *For any* $\varepsilon > 0$, *there is a positive parameter* $\delta > 0$. *For* $\mathbf{x}(t) \in U^{\varepsilon}_{\mathbf{x}_0}$, *the first integral quantity possesses the property in Eq.*(8.20). *For* $\mathbf{x}(t_0) \in U^{\varepsilon}_{\mathbf{x}_0}$, *an initial normal surface to the first integral manifold surface is defined by*

$$\Xi_0 = \left\{ \mathbf{x} \middle| \left(\mathbf{x} - \mathbf{x}_0^* \right) \bullet \dot{\mathbf{x}}^* (t_0) = 0 \right\} \tag{8.27}$$

and there are k-cycles with points $(t_l, \mathbf{x}(t_l))$ ($l = 1, 2, \cdots, k$) *on a normal reference surface.*

(i) *The equilibrium* $\mathbf{x}_0$ *is stable for* $t \in [t_0, \infty)$ *in* $U_{\mathbf{x}_0}^{\varepsilon}$ *if for* $\mathbf{x}(t_0) \in U_{\mathbf{x}_0}^{\varepsilon}$

$$L^{(l)}\left(t_{l-1}, t_l\right) \equiv \int_{t_{l-1}}^{t_l} \nabla F(\mathbf{x}, \boldsymbol{\mu}) \cdot \mathbf{g}(\mathbf{x}, t, \boldsymbol{\pi}) dt \leq 0 \ \text{ for } (l = 1, 2, \cdots, k). \tag{8.28}$$

(ii) *The equilibrium* $\mathbf{x}_0$ *is unstable for* $t \in [t_0, \infty)$ *in* $U_{\mathbf{x}_0}^{\varepsilon}$ *if* $\mathbf{x}(t_0) \in U_{\mathbf{x}_0}^{\varepsilon}$

$$L^{(l)}\left(t_{l-1}, t_l\right) \equiv \int_{t_{l-1}}^{t_l} \nabla F(\mathbf{x}, \boldsymbol{\mu}) \cdot \mathbf{g}(\mathbf{x}, t, \boldsymbol{\pi}) dt > 0 \ \text{ for } (l = 1, 2, \cdots, k). \tag{8.29}$$

PROOF. Consider an equilibrium point $\mathbf{x}^*(t) = \mathbf{x}_0$ for $t \in [t_0, \infty)$ with $F(\mathbf{x}_0, \boldsymbol{\mu}) = E_0$. For any $\varepsilon > 0$, there is a neighborhood $U_{\mathbf{x}_0}^{\varepsilon}$ of point $\mathbf{x}_0$. For $\mathbf{x}(t) \in U_{\mathbf{x}_0}^{\varepsilon}$, the first integral quantity $F(\mathbf{x}, \boldsymbol{\mu}) = E \in N_{E_0}^{\delta+}$, which implies $F(\mathbf{x}, \boldsymbol{\mu}) = E \geq E_0$ with $F(\mathbf{x}_0, \boldsymbol{\mu}) = E_0$. Consider $L(t_{l-1}, t_l) < 0$ in Eq.(8.28), on the reference surface Ξ_0,

$$F\left(\mathbf{x}(t_k), \boldsymbol{\mu}\right) - F\left(\mathbf{x}(t_0), \boldsymbol{\mu}\right) = \sum_{l=0}^{k} L^{(l)}\left(t_{l-1}, t_l\right) < 0$$

With increasing k, $F(\mathbf{x}, \boldsymbol{\mu})$ decreases more and more. Once $F(\mathbf{x}, \boldsymbol{\mu}) \to E_0$, the orbit $\mathbf{x}(t) \to \mathbf{x}_0$, which implies that the equilibrium $\mathbf{x}_0$ is stable for $t \in [t_0, \infty)$ in $U_{\mathbf{x}_0}^{\varepsilon}$. If $L(t_{l-1}, t_l) = 0$, then

$$F\left(\mathbf{x}(t_k), \boldsymbol{\mu}\right) = F\left(\mathbf{x}(t_0), \boldsymbol{\mu}\right) \ \text{ for } l = 1, 2, \cdots, k.$$

Because the first integral quantity $F(\mathbf{x}, \boldsymbol{\mu}) = E$ is determined by the C^r-integrable vector field ($r > 1$) in Eq.(3.1). So $F(\mathbf{x})$ and its derivative are continuous. For any small $\varepsilon > 0$, if $\| \mathbf{x}(t_0) - \mathbf{x}_0 \| \leq \varepsilon$, let $\delta = \| F \| \varepsilon$, and then the relation $\| F(\mathbf{x}(t_0), \boldsymbol{\mu}) - F(\mathbf{x}_0, \boldsymbol{\mu}) \| \leq \delta$ holds. Hence, the foregoing equation gives

$$\begin{aligned} 0 &= \left\| F\left(\mathbf{x}(t_1), \boldsymbol{\mu}\right) - F\left(\mathbf{x}(t_0), \boldsymbol{\mu}\right) \right\| \\ &\geq \left\| F\left(\mathbf{x}(t_1), \boldsymbol{\mu}\right) - F\left(\mathbf{x}_0, \boldsymbol{\mu}\right) \right\| - \left\| F\left(\mathbf{x}(t_0), \boldsymbol{\mu}\right) - F\left(\mathbf{x}_0, \boldsymbol{\mu}\right) \right\| \\ &\geq \left\| F\left(\mathbf{x}(t_1), \boldsymbol{\mu}\right) - F\left(\mathbf{x}_0, \boldsymbol{\mu}\right) \right\| - \delta \end{aligned}$$

from which one has

$$\left\| F\left(\mathbf{x}(t_1),\boldsymbol{\mu}\right) - F\left(\mathbf{x}_0,\boldsymbol{\mu}\right)\right\| \le \delta.$$

Since $\varepsilon > 0$ is selected arbitrarily, when $\varepsilon \to 0$ and $\delta \to 0$. Therefore, from Definition 8.9, the equilibrium $\mathbf{x}_0$ is stable.

(ii) Consider $L(t_{l-1},t_l) > 0$ in Eq.(8.29), so on the reference surface Ξ_0,

$$F\left(\mathbf{x}(t_k),\boldsymbol{\mu}\right) - F\left(\mathbf{x}(t_0),\boldsymbol{\mu}\right) = \sum_{l=0}^{k} L^{(l)}\left(t_{l-1},t_l\right) > 0.$$

With increasing *k*, $F(\mathbf{x},\boldsymbol{\mu})$ increases more and more. Once $F(\mathbf{x},\boldsymbol{\mu}) > E_0 + \delta$, the orbit $\mathbf{x}(t) \notin U_{\mathbf{x}_0}^{\varepsilon}$, which implies that the equilibrium $\mathbf{x}_0$ is unstable for $t \in [t_0,\infty)$ in $U_{\mathbf{x}_0}^{\varepsilon}$. This theorem is proved ■

From the definition of the equilibrium stability, the forgoing theorem can be generalized as follows.

THEOREM 8.4. *For a dynamical system in Eq.*(3.1), *there is an equilibrium point* $\mathbf{x}^*(t) = \mathbf{x}_0$ *for* $t \in [t_0,\infty)$. *For any* $\varepsilon > 0$, *there is a positive parameter* $\delta > 0$. *For* $\mathbf{x}(t) \in U_{\mathbf{x}_0}^{\varepsilon}$, *the first integral quantity possessing the property in Eq.*(8.20) *and the initial condition satisfies*

$$\left\|\mathbf{x}(t_0) - \mathbf{x}_0\right\| \le \varepsilon/2. \tag{8.30}$$

The equilibrium $\mathbf{x}_0$ *is stable for* $t \in [t_0,\infty)$ *in* $U_{\mathbf{x}_0}^{\varepsilon}$ *if for* $\mathbf{x}(t_0) \in U_{\mathbf{x}_0}^{\varepsilon}$

$$\left|L\left(t_0,t\right)\right| \equiv \left|\int_{t_0}^{t} \nabla F\left(\mathbf{x},\boldsymbol{\mu}\right)\cdot \mathbf{g}\left(\mathbf{x},t,\boldsymbol{\pi}\right) dt\right| \le \delta/2 \quad \text{for} \quad t \in \left[t_0,\infty\right). \tag{8.31}$$

PROOF. Consider an equilibrium point $\mathbf{x}^*(t) = \mathbf{x}_0$ for $t \in [t_0,\infty)$ with $F(\mathbf{x}_0,\boldsymbol{\mu}) = E_0$. For any $\varepsilon > 0$, there is a neighborhood $U_{\mathbf{x}_0}^{\varepsilon}$ of point $\mathbf{x}_0$. For $\mathbf{x}(t) \in U_{\mathbf{x}_0}^{\varepsilon}$, the first integral quantity $F(\mathbf{x}) = E \in N_{E_0}^{\delta+}$, which implies $F(\mathbf{x},\boldsymbol{\mu}) = E \ge E_0$ with $E_0 = F(\mathbf{x}_0,\boldsymbol{\mu})$. Since the first integral quantity is C^r ($r \ge 2$) smooth,

$$0 < \left\|F\right\| \equiv \sup[\left|F\left(\mathbf{x}(t),\boldsymbol{\mu}\right) - F\left(\mathbf{x}_0,\boldsymbol{\mu}\right)\right| / \left\|\mathbf{x}(t) - \mathbf{x}_0\right\|] < \infty.$$

If $\| \mathbf{x}(t_0) - \mathbf{x}_0 \| < \varepsilon/2$, let $\delta = \| F \| \varepsilon$, we have

$$\left|F\left(\mathbf{x}\left(t_0\right),\boldsymbol{\mu}\right) - F\left(\mathbf{x}_0,\boldsymbol{\mu}\right)\right| \le \delta/2$$

Because

$$\left|\int_{t_0}^{t} \nabla F(\mathbf{x},\boldsymbol{\mu})\cdot \mathbf{g}(\mathbf{x},t,\pi)\,dt\right| = \left|F(\mathbf{x}(t),\boldsymbol{\mu}) - F(\mathbf{x}(t_0),\boldsymbol{\mu})\right|$$
$$\geq \left|F(\mathbf{x}(t),\boldsymbol{\mu}) - F(\mathbf{x}_0,\boldsymbol{\mu})\right| - \left|F(\mathbf{x}(t_0),\boldsymbol{\mu}) - F(\mathbf{x}_0,\boldsymbol{\mu})\right|,$$

equations (8.30) and (8.31) give

$$\delta/2 \geq \left|F(\mathbf{x}(t),\boldsymbol{\mu}) - F(\mathbf{x}_0,\boldsymbol{\mu})\right| - \left|F(\mathbf{x}(t_0),\boldsymbol{\mu}) - F(\mathbf{x}_0,\boldsymbol{\mu})\right|$$
$$\geq \left|F(\mathbf{x}(t),\boldsymbol{\mu}) - F(\mathbf{x}_0,\boldsymbol{\mu})\right| - \delta/2.$$

From the foregoing equation, we have

$$\left|F(\mathbf{x}(t),\boldsymbol{\mu}) - F(\mathbf{x}_0,\boldsymbol{\mu})\right| \leq \delta$$

Therefore, $\| \mathbf{x}(t) - \mathbf{x}_0 \| < \delta / \| F \| = \varepsilon$. This theorem is proved. ■

The above theorems cannot be applied to the hyperbolic equilibrium. The flow passing through the hyperbolic equilibriums possess the global behavior. Such a flow will be discussed later.

8.2. Periodic flow stability

In this section, the stability of periodic flows in a dynamical system of Eq.(3.1) will be discussed. Before discussing the stability of periodic flow, the flow trajectory and its neighborhood in phase space are defined. The corresponding stability will be discussed through the first integral quantity increment on the Poincaré mapping section.

DEFINITION 8.12. For a dynamical system in Eq.(3.1), the positive and negative flows with the initial condition $(t_0, \mathbf{x}_0)$ are respectively defined by

$$\Gamma^{+}_{\mathbf{x}^*} = \left\{ \mathbf{x}^*(t) \in U_{\mathbf{x}} \,\middle|\, \mathbf{x}^*(t) = \boldsymbol{\Phi}(\mathbf{x}_0, t_0, t),\ \mathbf{x}_0 \in U_{\mathbf{x}}, t \in [t_0, \infty) \right\}, \tag{8.32}$$

and

$$\Gamma^{-}_{\mathbf{x}^*} = \left\{ \mathbf{x}^*(t) \in U_{\mathbf{x}} \,\middle|\, \mathbf{x}^*(t) = \boldsymbol{\Phi}(\mathbf{x}_0, t_0, t),\ \mathbf{x}_0 \in U_{\mathbf{x}}, t \in (-\infty, t_0] \right\}. \tag{8.33}$$

Therefore, a complete flow $\Gamma_{\mathbf{x}^*}$ with the initial condition $(t_0, \mathbf{x}_0)$ is defined by

$$\Gamma_{\mathbf{x}^*} = \Gamma^{-}_{\mathbf{x}^*} \cup \Gamma^{+}_{\mathbf{x}^*} = \left\{ \mathbf{x}^*(t) \in U_{\mathbf{x}} \,\middle|\, \mathbf{x}^*(t) = \boldsymbol{\Phi}(\mathbf{x}_0, t_0, t),\ \mathbf{x}_0 \in U_{\mathbf{x}}, t \in \mathbb{R} \right\}. \tag{8.34}$$

DEFINITION 8.13. For the dynamical system in Eq.(3.1), there is a flow $\Gamma_{\mathbf{x}^*} \subset U_{\mathbf{x}}$. The normal surface of the flow $\Gamma_{\mathbf{x}^*}$ to the first integral manifold surface at time t is defined by

$$\Xi_{\Gamma_{\mathbf{x}^*}}(t) = \left\{ \mathbf{x}(t) \left| \begin{array}{l} \left(\mathbf{x}(t) - \mathbf{x}^*(t)\right) \bullet \dot{\mathbf{x}}^*(t) = 0 \\ \text{for } \mathbf{x}^*(t) \in \Gamma_{\mathbf{x}^*}, t \in [t_0, \infty) \end{array} \right. \right\} \subset \mathbb{R}^{n-1} \otimes \mathbb{R}. \tag{8.35}$$

For any given $\varepsilon > 0$, a neighborhood of $\Gamma_{\mathbf{x}^*}$ is defined by

$$U_{\Gamma_{\mathbf{x}^*}}^{\varepsilon} = \left\{ \mathbf{x}(t) \in U_{\mathbf{x}} \left| \begin{array}{l} \left\| \mathbf{x}(t) - \mathbf{x}^*(t) \right\| < \varepsilon, \\ \text{for } \mathbf{x}^*(t) \in \Gamma_{\mathbf{x}^*} \text{ and } t \in [t_0, \infty) \\ \mathbf{x}(t_0) \in \Xi_{\Gamma_{\mathbf{x}^*}}(t_0), \mathbf{x}^*(t_0) \in \Xi_{\Gamma_{\mathbf{x}^*}}(t_0) \end{array} \right. \right\} \subset \mathbb{R}^n \otimes \mathbb{R}. \tag{8.36}$$

The first integral surface of the flow $\Gamma_{\mathbf{x}^*}$ is defined as

$$S_{\mathbf{x}^*}(t) = \left\{ \mathbf{x}(t) \left| \begin{array}{l} F(\mathbf{x}(t), \boldsymbol{\mu}) = F\left(\mathbf{x}^*(t), \boldsymbol{\mu}\right), \\ \mathbf{x}(t) \in \Xi_{\Gamma_{\mathbf{x}^*}}(t) \\ \text{for } \mathbf{x}^*(t) \in \Gamma_{\mathbf{x}^*},\ t \in [t_0, \infty) \end{array} \right. \right\} \subset \mathbb{R}^{n-1} \otimes \mathbb{R}. \tag{8.37}$$

8.2.1. Stability conditions

THEOREM 8.5. *For a dynamical system in Eq.*(3.1)*, there is a steady-state flow* $\Gamma_{\mathbf{x}^*}$ *(i.e.,* $\mathbf{x}^*(t), t \in [t_0, \infty]$*). For any given* $\varepsilon > 0$*, there is a flow* $\mathbf{x}(t) \in U_{\Gamma_{\mathbf{x}^*}}^{\varepsilon}$ *with the initial integral quantity satisfying* $|F(\mathbf{x}(t_0), \boldsymbol{\mu}) - F(\mathbf{x}^*(t_0), \boldsymbol{\mu})| < \delta$ *for a finite positive* $\delta > 0$.

(i) *The manifold* $S_{\mathbf{x}^*}(t) \cap U_{\Gamma_{\mathbf{x}^*}}^{\varepsilon}$ *of the flow* $\Gamma_{\mathbf{x}^*}$ *is bounded if* $\mathbf{x}(t_i') \in \Xi_{\Gamma_{\mathbf{x}^*}}(t_i) \cap U_{\Gamma_{\mathbf{x}^*}}^{\varepsilon} \cap U_{\Gamma_{\mathbf{x}^*}}^{\varepsilon}$ *at* $t_i \in [t_0, \infty)$ *and* $t_i' \in [t_0, \infty)$,

$$\left. \begin{array}{l} \left[F\left(\mathbf{x}(t_i'), \boldsymbol{\mu}\right) - F\left(\mathbf{x}^*(t_i), \boldsymbol{\mu}\right)\right] \times \left[F\left(\mathbf{x}(t'), \boldsymbol{\mu}\right) - F\left(\mathbf{x}^*(t), \boldsymbol{\mu}\right)\right] > 0, \\ \left[F\left(\mathbf{x}(t_i'), \boldsymbol{\mu}\right) - F\left(\mathbf{x}^*(t_i), \boldsymbol{\mu}\right)\right] \times \left[G\left(\mathbf{x}, t_i', \mathbf{p}\right) - G\left(\mathbf{x}^*, t_i, \mathbf{p}\right)\right] \leq 0. \end{array} \right\} \tag{8.38}$$

From the foregoing condition, the steady-state flow $\Gamma_{\mathbf{x}^*}$ *for the 2-dimensional system in Eq.(3.1) is stable.*

(ii) *The manifold* $S_{\mathbf{x}^*}(t)\cap U_{\Gamma_{\mathbf{x}^*}}^{\varepsilon_i}$ *of the steady-state flow* $\Gamma_{\mathbf{x}^*}$ *is unbounded if* $\mathbf{x}(t_i')$ $\in \Xi_{\Gamma_{\mathbf{x}^*}}\cap U_{\Gamma_{\mathbf{x}^*}}^{\varepsilon}$ $\mathbf{x}(t_i')\in \Xi_{\Gamma_{\mathbf{x}^*}}\cap U_{\Gamma_{\mathbf{x}^*}}^{\varepsilon}$ *at* $t_i\in[t_0,\infty)$ *and* $t_i'\in[t_0,\infty)$

$$\left.\begin{aligned}&\left[F\left(\mathbf{x}(t_i'),\boldsymbol{\mu}\right)-F\left(\mathbf{x}^*(t_i),\boldsymbol{\mu}\right)\right]\times\left[F\left(\mathbf{x}(t'),,\boldsymbol{\mu}\right)-F\left(\mathbf{x}^*(t),\boldsymbol{\mu}\right)\right]>0,\\&\left[F\left(\mathbf{x}(t_i'),\boldsymbol{\mu}\right)-F\left(\mathbf{x}^*(t_i),\boldsymbol{\mu}\right)\right]\times\left[G\left(\mathbf{x},t_i',\mathbf{p}\right)-G\left(\mathbf{x}^*,t_i,\mathbf{p}\right)\right]>0.\end{aligned}\right\}\tag{8.39}$$

and the steady-state flow $\Gamma_{\mathbf{x}^*}$ *is unstable.*

PROOF. Consider a periodic flow $\Gamma_{\mathbf{x}^*}$ (i.e., $\mathbf{x}^*(t), t\in[t_0,\infty)$) in an n-dimensional, dynamical system in Eq.(3.1) at time $t_i\in[t_0,\infty)$. Consider a flow $\mathbf{x}(t)\in U_{\Gamma_{\mathbf{x}^*}}^{\varepsilon_i}$ (t $\in[t_0,\infty)$). For any point $\mathbf{x}(t_i')\in\Xi(t_i)\cap U_{\Gamma_{\mathbf{x}^*}}^{\varepsilon}$ for an enough small $\varepsilon>0$, let $t=t_i$ $+\delta_t$ and $t'=t_i'+\delta_{t'}$ with any small increment $\delta_t>0$, under $\|DG(\mathbf{x},t_i',\mathbf{p})\|\le K'$ $<\infty$ and $\|DG(\mathbf{x}^*,t_i,\mathbf{p})\|\le K<\infty$, the difference of the first integral quantity of $\mathbf{x}(t')\in\Xi(t)\cap U_{\Gamma_{\mathbf{x}^*}}^{\varepsilon}$

$$\begin{aligned}F\left(\mathbf{x}(t'),\boldsymbol{\mu}\right)-F\left(\mathbf{x}^*(t),\boldsymbol{\mu}\right)=&\left[F\left(\mathbf{x}(t_i'),\boldsymbol{\mu}\right)-F\left(\mathbf{x}^*(t_i),\boldsymbol{\mu}\right)\right]\\&+\left[G\left(\mathbf{x}(t_i')\right)\delta_{t'}-G\left(\mathbf{x}^*(t_i)\right)\delta_t\right]+o(\delta_t^2)+o(\delta_{t'}^2)\end{aligned}$$

Further, the following equation is achieved,

$$\begin{aligned}&\left[F\left(\mathbf{x}(t_i'),\boldsymbol{\mu}\right)-F\left(\mathbf{x}^*(t_i),\boldsymbol{\mu}\right)\right]\times\left[F\left(\mathbf{x}(t'),\boldsymbol{\mu}\right)-F\left(\mathbf{x}^*(t),\boldsymbol{\mu}\right)\right]\\&=\left[F\left(\mathbf{x}(t_i'),\boldsymbol{\mu}\right)-F\left(\mathbf{x}^*(t_i),\boldsymbol{\mu}\right)\right]^2\\&+\left[F\left(\mathbf{x}(t_i'),\boldsymbol{\mu}\right)-F\left(\mathbf{x}^*(t_i),\boldsymbol{\mu}\right)\right]\times\left[G\left(\mathbf{x},t_i',\mathbf{p}\right)\delta_{t'}-G\left(\mathbf{x}^*,t_i,\mathbf{p}\right)\delta_t\right].\end{aligned}$$

(i) Due to the infinitesimal increment $\delta_t>0$ and $\delta_{t'}>0$, we have $\delta_t\approx\delta_{t'}$. With Eq.(8.38), the foregoing equation gives

$$\begin{aligned}&\left[F\left(\mathbf{x}(t_i),\boldsymbol{\mu}\right)-F\left(\mathbf{x}^*(t_i),\boldsymbol{\mu}\right)\right]\times\left[F\left(\mathbf{x}(t'),\boldsymbol{\mu}\right)-F\left(\mathbf{x}^*(t),\boldsymbol{\mu}\right)\right]\\&\le\left[F\left(\mathbf{x}(t_i),\boldsymbol{\mu}\right)-F\left(\mathbf{x}^*(t_i),\boldsymbol{\mu}\right)\right]^2\end{aligned}$$

For $[F(\mathbf{x}(t_i),\boldsymbol{\mu})-F(\mathbf{x}^*(t_i),\boldsymbol{\mu})]>0$ and $[F(\mathbf{x}(t'),\boldsymbol{\mu})-F(\mathbf{x}^*(t),\boldsymbol{\mu})]>0$, we have

$$0<\left[F\left(\mathbf{x}(t'),\boldsymbol{\mu}\right)-F\left(\mathbf{x}^*(t),\boldsymbol{\mu}\right)\right]\le\left[F\left(\mathbf{x}(t_i'),\boldsymbol{\mu}\right)-F\left(\mathbf{x}^*(t_i),\boldsymbol{\mu}\right)\right].$$

For $[F(\mathbf{x}(t_i),\boldsymbol{\mu})-F(\mathbf{x}^*(t_i),\boldsymbol{\mu})]<0$ and $[F(\mathbf{x}(t'),\boldsymbol{\mu})-F(\mathbf{x}^*(t),\boldsymbol{\mu})]<0$, we have

$$0>\left[F\left(\mathbf{x}(t'),\boldsymbol{\mu}\right)-F\left(\mathbf{x}^*(t),\boldsymbol{\mu}\right)\right]>\left[F\left(\mathbf{x}(t_i'),\boldsymbol{\mu}\right)-F\left(\mathbf{x}^*(t_i),\boldsymbol{\mu}\right)\right].$$

Because of $[F(\mathbf{x}(t_i),\boldsymbol{\mu})-F(\mathbf{x}^*(t_i),\boldsymbol{\mu})]\neq 0$, we have

$$\left|F\left(\mathbf{x}(t'),\boldsymbol{\mu}\right)-F\left(\mathbf{x}^*(t),\boldsymbol{\mu}\right)\right|\leq\left|F\left(\mathbf{x}(t_i'),\boldsymbol{\mu}\right)-F\left(\mathbf{x}^*(t_i),\boldsymbol{\mu}\right)\right| \text{ for } t\in[t_i,t_i+\delta_t]$$

For the case of inequality, with increasing time to $t=t_j>t_i$, we have

$$\left|F\left(\mathbf{x}(t_j'),\boldsymbol{\mu}\right)-F\left(\mathbf{x}^*(t_j),\boldsymbol{\mu}\right)\right|=0.$$

For $t\in[t_j,\infty)\subset[t_i,\infty)$, we have $F(\mathbf{x}(t_j'),\boldsymbol{\mu})=F(\mathbf{x}^*(t_j),\boldsymbol{\mu})$. It implies that the first integral surface of the flow $\Gamma_{\mathbf{x}^*}$ is the same as the one of $\mathbf{x}(t)\in U^{\varepsilon}_{\Gamma_{\mathbf{x}^*}}$ for $t\in[t_j,\infty)\ \subset[t_0,\infty)$. That is, the first integral surface of the steady state flow $\Gamma_{\mathbf{x}^*}$ for $t\in\ [t_0,\infty)$ in a neighborhood of flow $\Gamma_{\mathbf{x}^*}$ is bounded.

For the case of the equality, for all time $t_i\in[t_0,\infty)$ in a neighborhood of flow $\Gamma_{\mathbf{x}^*}$, the following equation exists.

$$\left|F\left(\mathbf{x}(t'),\boldsymbol{\mu}\right)-F\left(\mathbf{x}^*(t),\boldsymbol{\mu}\right)\right|=\left|F\left(\mathbf{x}(t_i'),\boldsymbol{\mu}\right)-F\left(\mathbf{x}^*(t_i),\boldsymbol{\mu}\right)\right|<\delta.$$

So the first integral surface of the steady state flow $\Gamma_{\mathbf{x}^*}$ for $t\in[t_0,\infty)$ in a neighborhood of flow $\Gamma_{\mathbf{x}^*}$ is bounded. For a two-dimensional system in Eq.(3.1), the intersection between the first integral manifold surface and cross-section is a line. The first integral surface of the steady state flow is a curve of the steady state flow. For $t=t'>t_j>t_0$, $\mathbf{x}(t')=\mathbf{x}^*(t)$ because $|\,F(\mathbf{x}(t_j'),\boldsymbol{\mu})-F(\mathbf{x}^*(t_j),\boldsymbol{\mu})\,|$ $=0$. It implies that the steady-state flow is stable.

(ii) Due to the infinitesimal increment $\delta_t>0$ and $\delta_{t'}>0$, we have $\delta_t\approx\delta_{t'}$. With Eq.(8.39), we have

$$\left[F\left(\mathbf{x}(t_i'),\boldsymbol{\mu}\right)-F\left(\mathbf{x}^*(t_i),\boldsymbol{\mu}\right)\right]\times\left[F\left(\mathbf{x}(t'),\boldsymbol{\mu}\right)-F\left(\mathbf{x}^*(t),\boldsymbol{\mu}\right)\right]$$
$$>\left[F\left(\mathbf{x}(t_i'),\boldsymbol{\mu}\right)-F\left(\mathbf{x}^*(t_i),\boldsymbol{\mu}\right)\right]^2.$$

Because of $[F(\mathbf{x}(t_i),\boldsymbol{\mu})-F(\mathbf{x}^*(t_i),\boldsymbol{\mu})]\neq 0$, we have

$$\left|F\left(\mathbf{x}(t'),\boldsymbol{\mu}\right)-F\left(\mathbf{x}^*(t),\boldsymbol{\mu}\right)\right|>\left|F\left(\mathbf{x}(t_i'),\boldsymbol{\mu}\right)-F\left(\mathbf{x}^*(t_i),\boldsymbol{\mu}\right)\right|$$

for $t \in [t_i, t_i + \delta_t] \subset [t_0, \infty)$ .For the case of the inequality, with increasing time to $t_j > t_i$, on the cross-section surface $\Xi_{\Gamma_{\mathbf{x}^*}}(t_j)$, we have the difference of the two first integral quantities with time $t'_j \neq t_j$ as

$$\left| F\left(\mathbf{x}(t'_j), \boldsymbol{\mu}\right) - F\left(\mathbf{x}^*(t_j), \boldsymbol{\mu}\right) \right| > \delta.$$

It means that $F(\mathbf{x}(t'_j), \boldsymbol{\mu})$ is out of boundary, from which the first integral manifold of a flow $\Gamma_{\mathbf{x}^*}$ is divergent. Therefore, the steady-state flow $\Gamma_{\mathbf{x}^*}$ is unstable. This theorem is proved. This theorem is proved. ■

REMARK 8.4. From Theorem 8.5, the first integral quantities of a steady-state flow and a flow in its neighborhood are compared on the same cross-section surface $\Xi_{\Gamma_{\mathbf{x}^*}}(t_j)$. For the first case, there is a bounded, first integral surface of the flow $\Gamma_{\mathbf{x}^*}$. On the surface, so far, we cannot prove whether the steady-state flow is stable or not. On the first integral surface, there are many orbits which may be chaotic. However, for the second case, the first integral surface of the steady-state flow is divergent. It indicates that all the flows on such a surface will be unstable. Therefore, the steady-state flow is unstable.

From Theorem 8.5, the boundedness of the first integral manifold surface of the steady-state flow $\Gamma_{\mathbf{x}^*}$ can be investigated. This criterion does not need to know the solution of the steady state flow. It can be easily to embed such a criterion in numerical computations. If the flow $\Gamma_{\mathbf{x}^*}$ is a periodic flow with period *T*, the foregoing theorem can be applied to the boundedness of the first integral surface of the periodic motion can be developed. Based on the Poincaré section ideas, the following theorem for the stability of periodic flows in Eq.(3.1) is given through the increment of the first integral quantity, i.e.,

THEOREM 8.6. *For a dynamical system in Eq.*(3.1), *there is a periodic flow* $\Gamma_{\mathbf{x}^*}$ *(i.e.,* $\mathbf{x}^*(t)$, $t \in [t_0, t_0 + T^*]$*) with period* T^* *with* $\mathbf{x}^*(t_0 + T^*) = \mathbf{x}^*(t_0)$, *and only the initial and final states* $\mathbf{x}^*(t_0)$ *and* $\mathbf{x}^*(t_0 + T^*)$ *are on the initial normal surface (i.e.,* $\Xi_{\Gamma_{\mathbf{x}^*}}(t_0)$*) to the first integral manifold surface. The total first integral increment during one period* T^* *for this periodic flow is zero, i.e.,*

$$L^{(*)}\left(t_0, T^*\right) \equiv \int_{t_0}^{t_0+T^*} \nabla F\left(\mathbf{x}^*, \boldsymbol{\mu}\right) \cdot \mathbf{g}\left(\mathbf{x}^*, t, \boldsymbol{\pi}\right) dt = 0. \tag{8.40}$$

For a flow $\mathbf{x}(t) \in U^{\varepsilon}_{\Gamma_{\mathbf{x}^*}}$ *with the initial state* $\mathbf{x}(t_0) \in \Xi_{\Gamma_{\mathbf{x}^*}}(t_0) \cap U^{\varepsilon}_{\Gamma_{\mathbf{x}^*}}$, *the points in-*

tersected with the reference surface are $\mathbf{x}(t_l) = \Xi_{\Gamma_{\mathbf{x}^*}}(t_0) \cap U^{\varepsilon}_{\Gamma_{\mathbf{x}^*}}$ *with period* T_l ($l = 1, 2, \cdots, k$).

(i) *The first integral manifold surface of the periodic flow* $\Gamma_{\mathbf{x}^*}$ *is bounded if*

$$\left.\begin{array}{l}\left[F(\mathbf{x}_{l-1},\boldsymbol{\mu}) - F(\mathbf{x}_0^*,\boldsymbol{\mu})\right] \times \left[F(\mathbf{x}_l,\boldsymbol{\mu}) - F(\mathbf{x}_0^*,\boldsymbol{\mu})\right] \geq 0 \\ \left[F(\mathbf{x}_{l-1},\boldsymbol{\mu}) - F(\mathbf{x}_0^*,\boldsymbol{\mu})\right] \times L^{(l)}(t_{l-1},t_l) \leq 0\end{array}\right\} \tag{8.41}$$

From the foregoing condition, the corresponding periodic flow $\Gamma_{\mathbf{x}^*}$ *in a 2-D dynamical system in Eq.(3.1) is stable.*

(ii) *The first integral manifold surface of the periodic flow* $\Gamma_{\mathbf{x}^*}$ *is unbounded if*

$$\left.\begin{array}{l}\left[F(\mathbf{x}_{l-1},\boldsymbol{\mu}) - F(\mathbf{x}_0^*,\boldsymbol{\mu})\right] \times \left[F(\mathbf{x}_l,\boldsymbol{\mu}) - F(\mathbf{x}_0^*,\boldsymbol{\mu})\right] > 0 \\ \left[F(\mathbf{x}_{l-1},\boldsymbol{\mu}) - F(\mathbf{x}_0^*,\boldsymbol{\mu})\right] \times L^{(l)}(t_{l-1},t_l) > 0\end{array}\right\} \tag{8.42}$$

and the periodic flow $\Gamma_{\mathbf{x}^*}$ *is unstable.*

PROOF. For a dynamical system in Eq.(3.1), because a periodic flow $\Gamma_{\mathbf{x}^*}$ (i.e., $\mathbf{x}^*(t)$, $t \in [t_0, t_0 + T^*]$) with period T^* with $\mathbf{x}^*(t_0 + T^*) = \mathbf{x}^*(t_0)$ is a closed curve in $U_{\mathbf{x}} \subset \Re^n$. From Theorem 4.1, equation (8.40) should be satisfied. For a flow $\mathbf{x}(t) \in U^{\varepsilon}_{\Gamma_{\mathbf{x}^*}}$ with the initial state $\mathbf{x}(t_0) \in \Xi_{\Gamma_{\mathbf{x}^*}}(t_0) \cap U^{\varepsilon}_{\Gamma_{\mathbf{x}^*}}$, the intersected points on the reference surface is $\mathbf{x}(t_l) = \Xi_{\Gamma_{\mathbf{x}^*}}(t_0) \cap U^{\varepsilon}_{\Gamma_{\mathbf{x}^*}}$ with period $T_l = t_l - t_{l-1} \neq T^*$ ($l = 1, 2, \cdots$).

$$\begin{aligned}L(t_{l-1},t_l) &= \int_{t_{l-1}}^{t_l} G(\mathbf{x},t,\mathbf{p})\,dt = F(\mathbf{x}_l,\boldsymbol{\mu}) - F(\mathbf{x}_{l-1},\boldsymbol{\mu}) \\ &= F(\mathbf{x}_l,\boldsymbol{\mu}) - F(\mathbf{x}_0^*,\boldsymbol{\mu}) - F(\mathbf{x}_{l-1},\boldsymbol{\mu}) + F(\mathbf{x}_0^*,\boldsymbol{\mu})\end{aligned}$$

Multiplying $[F(\mathbf{x}_{l-1},\boldsymbol{\mu}) - F(\mathbf{x}_0^*,\boldsymbol{\mu})]$ on both sides of the above equation yields

$$\begin{aligned}&\left[F(\mathbf{x}_{l-1},\boldsymbol{\mu}) - F(\mathbf{x}_0^*,\boldsymbol{\mu})\right] \times L(t_{l-1},t_l) \\ &= \left[F(\mathbf{x}_{l-1},\boldsymbol{\mu}) - F(\mathbf{x}_0^*,\boldsymbol{\mu})\right] \times \left[F(\mathbf{x}_l,\boldsymbol{\mu}) - F(\mathbf{x}_0^*,\boldsymbol{\mu})\right] \\ &\quad - \left[F(\mathbf{x}_{l-1},\boldsymbol{\mu}) - F(\mathbf{x}_0^*,\boldsymbol{\mu})\right]^2\end{aligned}$$

(i) Using the second equation of Eq.(8.41) gives

$$\left[F(\mathbf{x}_{l-1},\boldsymbol{\mu})-F(\mathbf{x}_0^*,\boldsymbol{\mu})\right]\times\left[F(\mathbf{x}_l,\boldsymbol{\mu})-F(\mathbf{x}_0^*,\boldsymbol{\mu})\right]\le\left[F(\mathbf{x}_{l-1},\boldsymbol{\mu})-F(\mathbf{x}_0^*,\boldsymbol{\mu})\right]^2$$

For $[F(\mathbf{x}_{l-1},\boldsymbol{\mu})-F(\mathbf{x}_0^*,\boldsymbol{\mu})]>0$, we have $[F(\mathbf{x}_l,\boldsymbol{\mu})-F(\mathbf{x}_0^*,\boldsymbol{\mu})]>0$, and the foregoing inequality gives

$$0<\left[F(\mathbf{x}_l,\boldsymbol{\mu})-F(\mathbf{x}_0^*,\boldsymbol{\mu})\right]\le\left[F(\mathbf{x}_{l-1},\boldsymbol{\mu})-F(\mathbf{x}_0^*,\boldsymbol{\mu})\right].$$

For $[F(\mathbf{x}_{l-1},\boldsymbol{\mu})-F(\mathbf{x}_0^*,\boldsymbol{\mu})]<0$, we have $[F(\mathbf{x}_l,\boldsymbol{\mu})-F(\mathbf{x}_0^*,\boldsymbol{\mu})]<0$, and

$$0>\left[F(\mathbf{x}_l,\boldsymbol{\mu})-F(\mathbf{x}_0^*,\boldsymbol{\mu})\right]\ge\left[F(\mathbf{x}_{l-1},\boldsymbol{\mu})-F(\mathbf{x}_0^*,\boldsymbol{\mu})\right].$$

Therefore, for $l=1,2,\cdots$,

$$\left|F(\mathbf{x}_l,\boldsymbol{\mu})-F(\mathbf{x}_0^*,\boldsymbol{\mu})\right|\le\left|F(\mathbf{x}_{l-1},\boldsymbol{\mu})-F(\mathbf{x}_0^*,\boldsymbol{\mu})\right|.$$

For the inequality case, we have

$$\left|F(\mathbf{x}_l,\boldsymbol{\mu})-F(\mathbf{x}_0^*,\boldsymbol{\mu})\right|<\left|F(\mathbf{x}_{l-1},\boldsymbol{\mu})-F(\mathbf{x}_0^*,\boldsymbol{\mu})\right|<$$
$$\cdots<\left|F(\mathbf{x}_0,\boldsymbol{\mu})-F(\mathbf{x}_0^*,\boldsymbol{\mu})\right|<\delta.$$

For $l=k$, we obtain

$$\left|F(\mathbf{x}_k,\boldsymbol{\mu})-F(\mathbf{x}_0^*,\boldsymbol{\mu})\right|=0.$$

The flow $\mathbf{x}(t)\in U^{\varepsilon}_{\Gamma_{\mathbf{x}^*}}$ with the initial condition is convergent to the first integral manifold surface of $\Gamma_{\mathbf{x}^*}$. So the first integral surface of $\Gamma_{\mathbf{x}^*}$ is bounded.

For the equality case, we have for $l=1,2,\cdots$,

$$\left|F(\mathbf{x}_l,\boldsymbol{\mu})-F(\mathbf{x}_0^*,\boldsymbol{\mu})\right|=\cdots=\left|F(\mathbf{x}_0,\boldsymbol{\mu})-F(\mathbf{x}_0^*,\boldsymbol{\mu})\right|<\delta$$

Because $\delta>0$ is selected arbitrarily, once $\delta\to 0$, we have

$$\left|F(\mathbf{x}_l,\boldsymbol{\mu})-F(\mathbf{x}_0^*,\boldsymbol{\mu})\right|\to 0,\ \text{i.e., } F(\mathbf{x}_l,\boldsymbol{\mu})\to F(\mathbf{x}_0^*,\boldsymbol{\mu})$$

So the first integral surface of $\Gamma_{\mathbf{x}^*}$ is bounded.

For a two-dimensional system in Eq.(3.1), the initial normal surface (i.e., $\Xi_{\Gamma_{\mathbf{x}^*}}(t_0)$) is a line. The intersections between the first integral manifold surface and the initial normal surface (i.e., $\Xi_{\Gamma_{\mathbf{x}^*}}(t_0)$) are points $\mathbf{x}_l$ ($l=0,1,2\cdots$) such a line. So the relation $|F(\mathbf{x}_k,\boldsymbol{\mu})-F(\mathbf{x}_0^*,\boldsymbol{\mu})|=0$ gives $\mathbf{x}_k=\mathbf{x}_0^*$. The series $\mathbf{x}_l$ ($l=$

$0,1,2\cdots$) is convergent to $\mathbf{x}_0^*$ at $l=k$. It implies that the steady-state flow is stable. The relation $|F(\mathbf{x}_l,\boldsymbol{\mu})-F(\mathbf{x}_0^*,\boldsymbol{\mu})|=|F(\mathbf{x}_0,\boldsymbol{\mu})-F(\mathbf{x}_0^*,\boldsymbol{\mu})|<\delta$ gives two periodic flow are parallel, once $\delta\to 0$, $F(\mathbf{x}_l,\boldsymbol{\mu})\to F(\mathbf{x}_0^*,\boldsymbol{\mu})$ implies $\mathbf{x}_l\to\mathbf{x}_0^*$. So the steady-state flow is stable.

(ii) Using the second equation of Eq.(8.42) gives

$$\left[F\left(\mathbf{x}_{l-1},\boldsymbol{\mu}\right)-F\left(\mathbf{x}_0^*,\boldsymbol{\mu}\right)\right]\times\left[F\left(\mathbf{x}_l,\boldsymbol{\mu}\right)-F\left(\mathbf{x}_0^*,\boldsymbol{\mu}\right)\right]>\left[F\left(\mathbf{x}_{l-1},\boldsymbol{\mu}\right)-F\left(\mathbf{x}_0^*,\boldsymbol{\mu}\right)\right]^2$$

For $[F(\mathbf{x}_{l-1},\boldsymbol{\mu})-F(\mathbf{x}_0^*,\boldsymbol{\mu})]>0$, we have $[F(\mathbf{x}_l,\boldsymbol{\mu})-F(\mathbf{x}_0^*,\boldsymbol{\mu})]>0$, and the foregoing inequality gives

$$\left[F\left(\mathbf{x}_l,\boldsymbol{\mu}\right)-F\left(\mathbf{x}_0^*,\boldsymbol{\mu}\right)\right]>\left[F\left(\mathbf{x}_{l-1},\boldsymbol{\mu}\right)-F\left(\mathbf{x}_0^*,\boldsymbol{\mu}\right)\right]>0$$

For $[F(\mathbf{x}_{l-1},\boldsymbol{\mu})-F(\mathbf{x}_0^*,\boldsymbol{\mu})]<0$, we have $[F(\mathbf{x}_l,\boldsymbol{\mu})-F(\mathbf{x}_0^*,\boldsymbol{\mu})]<0$, and the foregoing inequality gives

$$\left[F\left(\mathbf{x}_l,\boldsymbol{\mu}\right)-F\left(\mathbf{x}_0^*,\boldsymbol{\mu}\right)\right]<\left[F\left(\mathbf{x}_{l-1},\boldsymbol{\mu}\right)-F\left(\mathbf{x}_0^*,\boldsymbol{\mu}\right)\right]<0$$

Therefore for $l=1,2,\cdots$,

$$\left|F\left(\mathbf{x}_l,\boldsymbol{\mu}\right)-F\left(\mathbf{x}_0^*,\boldsymbol{\mu}\right)\right|>\left|F\left(\mathbf{x}_{l-1},\boldsymbol{\mu}\right)-F\left(\mathbf{x}_0^*,\boldsymbol{\mu}\right)\right|$$

i.e.,

$$\begin{aligned}\left|F\left(\mathbf{x}_l,\boldsymbol{\mu}\right)-F\left(\mathbf{x}_0^*,\boldsymbol{\mu}\right)\right|&>\left|F\left(\mathbf{x}_{l-1},\boldsymbol{\mu}\right)-F\left(\mathbf{x}_0^*,\boldsymbol{\mu}\right)\right|\\&>\cdots>\left|F\left(\mathbf{x}_l,\boldsymbol{\mu}\right)-F\left(\mathbf{x}_0^*,\boldsymbol{\mu}\right)\right|\end{aligned}$$

For $l=k$, we have

$$\left|F\left(\mathbf{x}_k,\boldsymbol{\mu}\right)-F\left(\mathbf{x}_0^*,\boldsymbol{\mu}\right)\right|>\delta \text{ and } \left\|\mathbf{x}_k-\mathbf{x}_0^*\right\|>\varepsilon.$$

So the first integral surface of $\Gamma_{\mathbf{x}^*}$ is unbounded, and the periodic motion on such an unbounded integral surface is also unstable. This theorem is proved. ■

The foregoing theorem discussed the stability of the periodic flow with the single cycle. The period-doubling bifurcation can be observed with the symmetry break of the periodic flow. The corresponding periodic flow will have two cycles to the reference surface, and the first integral quantity increment will become two positive and negative parts. For the stability of a periodic flow with *k*-cycles to the reference surface, the above theorem is generalized as follows.

THEOREM 8.7. *For a dynamical system in Eq.(3.1), there is a periodic flow* $\Gamma_{\mathbf{x}^*}$ *(i.e.,* $\mathbf{x}^*(t)$, $t\in[t_0,t_0+T^*]$*) with period* T^* *with* $\mathbf{x}^*(t_0+T^*)=\mathbf{x}^*(t_0)$*, and only the initial and final states* $\mathbf{x}^*(t_0)$ *and* $\mathbf{x}^*(t_0+T^*)$ *are on the normal surface to the first integral manifold surface* $\Xi_{\Gamma_{\mathbf{x}^*}}(t_0)$*. For the k-cycle, periodic flow, the total first integral increment during one period* $T^*=\Sigma_{l=1}^{k}T_l^*$ *to the reference surface* Ξ_0 *is zero, i.e.,*

$$\left.\begin{aligned} L(t_0,T^*) &= \sum\nolimits_{l=1}^{k} L^{(l)}(t_{l-1},T_l^*)\\ &= \sum\nolimits_{l=1}^{k}\int_{t_{l-1}}^{t_{l-1}+T_l^*}\nabla F(\mathbf{x}^*,\boldsymbol{\mu})\cdot\mathbf{g}(\mathbf{x}^*,t,\boldsymbol{\pi})\,dt=0\\ t_l &= t_0+\sum\nolimits_{I=1}^{l}T_I^*,\ \text{and } L^{(l)}(t_{l-1},T_l^*)\neq 0 \end{aligned}\right\}\tag{8.43}$$

For a flow $\mathbf{x}(t)\in U^{\varepsilon}_{\Gamma_{\mathbf{x}^*}}$ *with the initial state* $\mathbf{x}(t_0)\in\Xi_{\Gamma_{\mathbf{x}^*}}(t_0)\cap U^{\varepsilon}_{\Gamma_{\mathbf{x}^*}}$*, the points intersected with the reference surface is* $\mathbf{x}(t_l^{(j)})=\Xi_{\Gamma_{\mathbf{x}^*}}(t_0)\cap U^{\varepsilon}_{\Gamma_{\mathbf{x}^*}}$ *with period* $T_l^{(j)}$ *(*$l=1,2,\cdots,k$ *and* $j=1,2,\cdots$*).* $T_j=\Sigma_{l=1}^{k}T_l^{(j)}$, $t_l^{(j)}=t_0+\Sigma_{J=1}^{j-1}\Sigma_{J=1}^{k}T_I^{(J)}+\Sigma_{I=1}^{l}T_I^{(j)}$ *and* $t_j=t_k^{(j)}$.

(i) *The first integral manifold surface of the periodic flow* $\Gamma_{\mathbf{x}^*}$ *is unbounded if*

$$\left.\begin{aligned} &[F(\mathbf{x}_{j-1},\boldsymbol{\mu})-F(\mathbf{x}_0^*,\boldsymbol{\mu})]\times[F(\mathbf{x}_j,\boldsymbol{\mu})-F(\mathbf{x}_0^*,\boldsymbol{\mu})]\geq 0\\ &[F(\mathbf{x}_{j-1},\boldsymbol{\mu})-F(\mathbf{x}_0^*,\boldsymbol{\mu})]\times L(t_{j-1},T_j)\leq 0 \end{aligned}\right\}\tag{8.44}$$

From the foregoing condition, the periodic flow $\Gamma_{\mathbf{x}^*}$ *for the two-dimensional system in Eq.(3.1) is stable.*

(ii) *The first integral manifold of the periodic flow* $\Gamma_{\mathbf{x}^*}$ *is unbounded if*

$$\left.\begin{aligned} &[F(\mathbf{x}_{j-1},\boldsymbol{\mu})-F(\mathbf{x}_0^*,\boldsymbol{\mu})]\times[F(\mathbf{x}_j,\boldsymbol{\mu})-F(\mathbf{x}_0^*,\boldsymbol{\mu})]> 0\\ &[F(\mathbf{x}_{j-1},\boldsymbol{\mu})-F(\mathbf{x}_0^*,\boldsymbol{\mu})]\times L(t_{j-1},T_j)> 0 \end{aligned}\right\}\tag{8.45}$$

and the periodic flow $\Gamma_{\mathbf{x}^*}$ *is unstable.*

PROOF. As proved in Theorem 8.6, this theorem can be proved. ■

In the above theorems, the Poincaré section is selected from a normal surface of the initial first integral manifold surface. In application, the Poincaré mapping section is defined from the external period. For a dynamical system in Eq.(3.1) possessing periodically time-dependent terms with period T, a periodic flow γ (i.e., $\mathbf{x}^*(t)$) for $t \in [t_0, t_0 + kT]$ intersected with a Poincaré mapping section are $\mathbf{x}_l^* \in \Gamma_{\mathbf{x}^*}^P$ ($l = 0,1,2,\cdots,k$) at time t_l, and the first integral quantity increment from time t_0 to t_k is zero, i.e.,

$$\begin{aligned} L^*(t_0, kT) &= \sum_{l=1}^{k} L^{(l)}(t_{l-1}, T) \\ &= \sum_{l=1}^{k} \int_{t_{l-1}}^{t_{l-1}+T} \nabla F(\mathbf{x}^*, \boldsymbol{\mu}) \cdot \mathbf{g}(\mathbf{x}^*, t, \boldsymbol{\pi})\, dt = 0, \end{aligned} \tag{8.46}$$

with

$$L^{(l)}(t_{l-1}, T) \neq 0 \quad \text{for } l = 0,1,2,\cdots,k \tag{8.47}$$

The Poincaré section is defined as

$$\Xi_T = \left\{ \mathbf{x}(t_N) \middle| \begin{array}{l} \mathbf{x}(t_N) = \boldsymbol{\Phi}(\mathbf{x}_0, t_N, t_0), \\ t_N = t_0 + NT, N = 1,2,\cdots \end{array} \right\} \subset \Re^n. \tag{8.48}$$

For a flow $\mathbf{x}(t) \in U_{\Gamma_{\mathbf{x}^*}}^{\varepsilon}$ with the initial state $\mathbf{x}(t_0) \in \Xi_T \cap U_{\Gamma_{\mathbf{x}^*}}^{\varepsilon}$, the intersected points on the reference surface are $\mathbf{x}(t_l) \in \Xi_T \cap U_{\Gamma_{\mathbf{x}^*}}^{\varepsilon}$ with $t_l^{(j)} = t_0 + [(j-1)k + l]T$ for ($l = 0,1,2,\cdots,k$ and $j = 1,2,\cdots$) and $t_j = t_k^{(j)}$.

$$L(t_{j-1}, T_j) \neq 0 \text{ and } T_j = kT. \tag{8.49}$$

Therefore, the stability conditions in Theorem 8.7 can be used for this case. If $k \to \infty$, all the flows in the neighborhood of the periodical flow approach such a steady-state flow. This periodic flow is asymptotically stable. Next, we discuss the existence of limit cycle in dissipative systems. The discussion will be given through a 2-dimensional nonlinear dynamical system.

8.2.2. Limit cycles

Limit cycles in two-dimensional, nonlinear dynamical systems can be discussed through the first integral quantity increment (or the energy increment). In the existing theory, the proof of the existence of a limit cycle in 2-dimensional nonlinear dynamical system is tried to find a zero increment of energy. The stability of the periodic flows can also be determined from the above theorems. Consider a typical system in the following theorem.

THEOREM 8.8. *Consider a dynamical system given by*

$$\ddot{x} + f(x,y)y + g(x) = 0 \text{ and } y = \dot{x} \tag{8.50}$$

where f *and* g *are continuous and the first integral quantity is defined by*

$$F(x,y) = \frac{1}{2}y^2 + \int_{x_0}^{x} g(x)\,dx, \tag{8.51}$$

which possesses an infimum E_0 *with* $|E_0| < \infty$ *at* $x = x_0$, *and there is a domain*

$$\mho = \{(x,y) | F(x,y) \geq E_0\}. \tag{8.52}$$

(i) *If there are* k*-closed orbits* $\gamma_j \in \mho$ $(j = 1,2,\cdots,k)$ *for the dynamical system in Eq.(8.50), then* $f(x,y) = 0$ *at least has* $2k$*-solutions on the* k*-closed orbits.*
(ii) *If* $f(x,y) = 0$ *has* $2k$*-non-repeated roots, then the dynamical system in Eq.(8.50) has at most* k*-closed orbits* $\gamma_j \in \mho$ $(j = 1,2,\cdots,k)$.

PROOF. The system in Eq.(8.50) is expressed by

$$\dot{x} = y \text{ and } \dot{y} = -g(x) - f(x,y)y.$$

The increment of the first integral quantity in Eq.(8.50) is computed for each closed orbit $\gamma_j \in \mho$ $(j = 1,2,\cdots,k)$ by

$$L(t_i, T_{\gamma_j}) = \int_{t_i}^{t_i+T_{\gamma_j}} f(x(t),y(t))[y(t)]^2\,dt$$
$$\underbrace{=}_{\text{Mean Value theorem}} f(x(t_m),y(t_m)) \int_{t_i}^{t_i+T_{\gamma_j}} [y(t)]^2\,dt.$$

Because $[y(t)]^2 \geq 0$ for $t \in [t_i, t_i + T_{\gamma_j}]$, but $[y(t)]^2 = 0$ cannot hold for the entire time interval, we have

$$\int_{t_k}^{t_k+T_{\gamma_i}} [y(t)]^2\,dt > 0.$$

(i) The closed orbit $\gamma_j \in \mho$ with period T_{γ_j} requires $L(t_i, T_{\gamma_j}) = 0$. So there is a point $(x(t_m), y(t_m))$ on the closed orbit $\gamma_j \in \Omega$ for $t_m \in [t_i, t_i + T_{\gamma_j}]$ with

$$f(x(t_m), y(t_m)) = 0$$

to make $L(t_i, T_{\gamma_j}) = 0$ with $G^{(1)}(x(t_m), y(t_m)) \neq 0$. However,

$$G\left(x\left(t_m\right), y\left(t_m\right)\right) = f\left(x\left(t_m\right), y\left(t_m\right)\right)\left[y\left(t_m\right)\right]^2 = 0$$

and

$$\frac{1}{2}y^2 + \int_{x_0}^{x} g\left(x\right)dx = \frac{1}{2}\left[y\left(t_m\right)\right]^2 + \int_{x_0}^{x_m} g\left(x\right)dx$$

So we have

$$\frac{1}{2}y^2 + \int_{x_m}^{x} g\left(x\right)dx = \frac{1}{2}\left[y\left(t_m\right)\right]^2$$

Choosing $(x(t_m), y(t_m))$ on the closed orbit γ_j to be a starting point, we have

$$\begin{aligned} L\left(t_m, T_{\gamma_j}\right) &= \int_{t_m}^{t_m+T_{\gamma_j}} f\left(x(t), y(t)\right)\left[y(t)\right]^2 dt \\ &\underset{\text{Mean Value theorem}}{=} f\left(x\left(t_n\right), y\left(t_n\right)\right) \int_{t_m}^{t_m+T_{\gamma_j}} \left[y(t)\right]^2 dt = 0. \end{aligned}$$

In a similar fashion, we have

$$f\left(x\left(t_n\right), y\left(t_n\right)\right) = 0 \text{ at } t_n \in \left(t_m, t_m + T_{\gamma_j}\right).$$

So the point $(x(t_n), y(t_n))$ is the second point on the closed curve to make the first integral increment $L(t_i, T_{\gamma_j}) = 0$. This point is the second point $(x(t_n), y(t_n)) \neq (x(t_m), y(t_m))$ of the closed orbit and the first integral curves. Of course, there are more than two points on this closed curve. Because the system in Eq.(8.50) have k-closed orbits, $f(x, y) = 0$ will have at least $2k$-solutions on the k-closed orbits to make $L(t_i, T_{\gamma_j}) = 0$ ($j = 1, 2, \cdots, k$).

(ii) Because of $f(x(t_m), y(t_m)) = 0$, there is $L(t_i, T_{\gamma_j}) = 0$. From (i), to form a closed orbit need at least two points of $f(x(t_m), y(t_m)) = 0$ with $G(x(t_m), y(t_m)) = 0$. If all the $2k$-non-repeated roots of $f(x(t_m), y(t_m)) = 0$ can be at most k-different closed orbits, Thus, the system in Eq.(8.50) has at most k-closed orbits $\gamma_j \in \Omega$ ($j = 1, 2, \cdots, k$). This theorem is proved. ∎

To explain the above theorem on limit cycles, consider a limit cycle for a two-dimensional system with two points as shown in Fig.8.2. The gray curves represent the first integral manifolds and the dark solid, closed curve represents a limit cycle. Two hollow circles may denote the grazing or transversal points

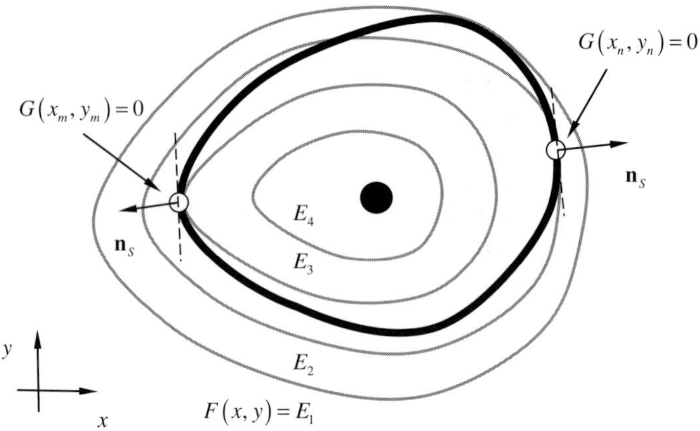

Figure 8.2. A limit cycle for a two-dimensional system with two points having $G = 0$. The gray curves represent the first integral manifolds and the dark solid, closed curve represents a limit cycle. The two hollow circles may be the grazing or transversal points with $G = 0$. The filled circle is equilibrium point.

with $G = 0$. The filled circle is equilibrium point. The two points with $G = 0$ will change the first integral quantity increment of the closed orbit. From point $(x(t_m), y(t_m))$ to $(x(t_n), y(t_n))$, if the first integral quantity increment increases, and from $(x(t_n), y(t_n))$ to $(x(t_m), y(t_m))$, the first integral quantity increment decreases. The increased and decreased first integral quantity increments are of the same. So $L(t_i, T_{\gamma_j}) = 0$. On this closed orbit, there can be $2l$-points to form l-pairs of points ($l = 1, 2, 3\cdots$), and the closed curve can be knotted together to make the motion more complex. For the general case, it is periodic motions. Once an external and periodical forcing is considered, the limit cycle will be more interesting. Because the periodic flow requires the L-function to be zero, the stability of the periodic flow can be easily determined through the L-function. In other words, If an initial point is arbitrarily selected on the outside of periodic orbit, we have $L(t_i, T_{\gamma_j}) < 0$, and if an initial point is arbitrarily selected on the inside of periodic orbit, $L(t_i, T_{\gamma_j}) > 0$ exists. Therefore, the periodic flow is stable. Otherwise, the periodic flow is unstable.

8.3. Local grazing bifurcation

In Section 8.2, the stability of periodic flows in nonlinear dynamical systems was discussed. Once the stable periodic flow loses its stability, the new periodic or chaotic flow will be formed. The traditional stability analysis is completed through the eigenvalue analysis. The eiganvalue analysis needs the analytical solutions of periodic flows. However, many nonlinear dynamical systems cannot find analytical expressions of periodic flows. Thus, one has used the Poincaré mapping to determine the stability of periodic flows. From the previous sections, the criteria with the first integral quantity increment based on the Poincaré mapping were presented to determine the stability of the periodic flow. The Poincaré mapping section is often based on the two traditional selections. In this section, the first integral manifold surface is chosen as the Poincaré mapping section to investigate periodic flow bifurcations and the strange attractor fragmentation of chaos, and the transversality of a flow on the prescribed, first integral manifold surface will be discussed.

8.3.1. Local grazing

Before discussing the grazing bifurcation of a periodic motion, the flow grazing to a referenced, first integral surface is defined as follows:

DEFINITION 8.14. For a dynamical system in Eq.(3.1), consider the normal vector direction of the first integral surface to be in the direction of increasing the first integral quantity. A flow $\mathbf{x}^{(\alpha)} = \mathbf{\Phi}^{(\alpha)}(t_0, \mathbf{x}_0^{(\alpha)}, \mathbf{p}, t)$ of Eq.(3.1) in a domain Ω_α ($\alpha = 1, 2, \cdots, m$) for $t \in [t_0, \infty)$ is of the $(2l-1)^{th}$-order grazing ($l = 1, 2, \cdots$) to a referenced, first integral surface S_{σ_α} at point $(t_k, \mathbf{x}_k)$ with $t_k \in [t_0, \infty)$ if the following condition exists

$$\left.\begin{array}{ll} & G_{\sigma_\alpha}^{(r)}(\mathbf{x}_k^{(\alpha)}, t_k, \mathbf{p}) = 0 \text{ for } r = 0, 1, 2, \cdots, 2l-2 \\ \text{either} & G_{\sigma_\alpha}^{(2l-1)}(\mathbf{x}_k^{(\alpha)}, t_k, \mathbf{p}) < 0 \;\; \text{for } F\left(\mathbf{x}^{(\alpha)}, \boldsymbol{\mu}\right) < E_{\sigma_\alpha} \\ \text{or} & G_{\sigma_\alpha}^{(2l-1)}(\mathbf{x}_k^{(\alpha)}, t_k, \mathbf{p}) > 0 \;\; \text{for } F\left(\mathbf{x}^{(\alpha)}, \boldsymbol{\mu}\right) > E_{\sigma_\alpha}. \end{array}\right\} \tag{8.53}$$

The flow grazing to a referenced, first integral surface S_{σ_α} with the $(2l-1)^{th}$-order is termed *the local grazing bifurcation of* the flow in domain Ω_α ($\alpha = 1, 2, \cdots, m$) with the $(2l-1)^{th}$-order.

After such a local grazing bifurcation, the flow still stays in a domain Ω_α. To investigate the local grazing bifurcation of a flow to a specific, first integral surface, consider the Poincare mapping section defined on the referenced, first integral surface S_{σ_α}, instead of the traditional, Poincare mapping section. To construct the Poincaré mapping, the mapping sets should be introduced first. Based on the mapping sets, the mapping can be defined to describe the motion on both sides of the referenced, first integral surface. Hence, the mapping sets on the referenced, first integral surface S_{σ_α} with $F(\overline{\mathbf{x}}^{(\alpha)}_{\sigma_\alpha},\boldsymbol{\mu}) = E_{\sigma_\alpha}$ are defined as follows:

$$\Sigma^1_{\sigma_\sigma} = \left\{ \left(t_k, \mathbf{x}^{(\alpha)}_k\right) \left| \begin{array}{l} \mathbf{x}^{(\alpha)}_l = \overline{\mathbf{x}}^{(\alpha)}_{\sigma_\alpha}(t_l) \in S_{\sigma_\alpha}, l \in \{k, k+1\} \\ \text{and } F\left(\mathbf{x}^{(\alpha)},\boldsymbol{\mu}\right) > F\left(\overline{\mathbf{x}}^{(\alpha)}_{\sigma_\alpha},\boldsymbol{\mu}\right), \\ \mathbf{x}^{(\alpha)} = \boldsymbol{\Phi}\left(t_k, \mathbf{x}^{(\alpha)}_k, \mathbf{p}, t\right),\ t \in (t_k, t_{k+1}) \end{array} \right. \right\} \tag{8.54}$$

$$\Sigma^2_{\sigma_\sigma} = \left\{ \left(t_k, \mathbf{x}^{(\alpha)}_k\right) \left| \begin{array}{l} \mathbf{x}^{(\alpha)}_l = \overline{\mathbf{x}}^{(\alpha)}_{\sigma_\alpha}(t_l) \in S_{\sigma_\alpha}, l \in \{k, k+1\} \\ \text{and } F\left(\mathbf{x}^{(\alpha)},\boldsymbol{\mu}\right) < F\left(\overline{\mathbf{x}}^{(\alpha)}_{\sigma_\alpha},\boldsymbol{\mu}\right), \\ \mathbf{x}^{(\alpha)} = \boldsymbol{\Phi}\left(t_k, \mathbf{x}^{(\alpha)}_k, \mathbf{p}, t\right),\ t \in (t_k, t_{k+1}) \end{array} \right. \right\}. \tag{8.55}$$

8.3.2. Grazing mapping

From the above mapping sets, the two mappings are defined as

$$P_1 : \Sigma^1_{\sigma_\sigma} \to \Sigma^2_{\sigma_\sigma} \text{ and } P_2 : \Sigma^2_{\sigma_\sigma} \to \Sigma^1_{\sigma_\sigma}. \tag{8.56}$$

Because the referenced surface arbitrarily selected in the domain Ω_α is non-singular, the two mappings P_1 and P_2 defined on such a referenced surface are called the *local mapping*. Let $\mathbf{y}_k = (t_k, \mathbf{x}^{(\alpha)}_k)^T$, the mapping relation can be given by

$$\mathbf{y}_{k+1} = P_\lambda \mathbf{y}_k \ (\lambda=1,2). \tag{8.57}$$

The mapping structure consisting of two mappings is on the reference surface S_{σ_α}, i.e.,

$$P = P_2 \circ P_1 : \Sigma^1_{\sigma_\sigma} \to \Sigma^1_{\sigma_\sigma} \tag{8.58}$$

with the corresponding relation $\mathbf{y}_{k+2} = P\mathbf{y}_k$. As in Eq.(4.76), the governing relations is given by

$$\left.\begin{array}{l} \mathbf{x}^{(\alpha)}_{k+1} = \boldsymbol{\Phi}^{(\alpha)}\left(t_k, \mathbf{x}^{(\alpha)}_k, \mathbf{p}, t_{k+1}\right) \text{ and } L^{(\alpha)}\left(t_k, t_{k+1}\right) = 0, \\ \mathbf{x}^{(\alpha)}_{k+2} = \boldsymbol{\Phi}^{(\alpha)}\left(t_{k+1}, \mathbf{x}^{(\alpha)}_{k+1}, \mathbf{p}, t_{k+2}\right) \text{ and } L^{(\alpha)}\left(t_{k+1}, t_{k+2}\right) = 0. \end{array}\right\} \tag{8.59}$$

The two mappings and mapping sets on the referenced, first integral surface S_{σ_α} are sketched in Fig.8.3 through the dark curves and two local areas on the

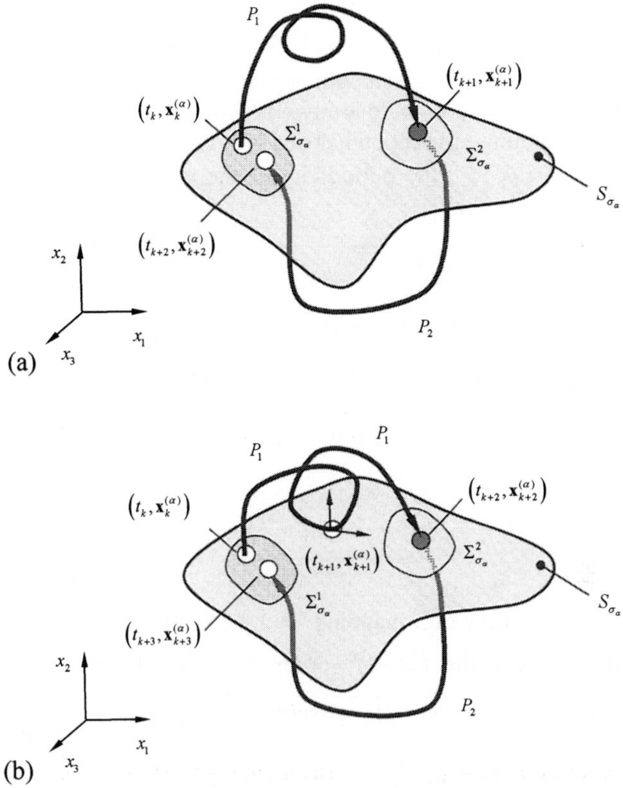

Figure 8.3. A flow mapping defined on the referenced, first integral manifold surface S_{σ_α} with the first integral quantity E_{σ_α} : (a) two mappings to the reference surface and (b) three mappings with a flow tangential to the reference surface.

surface. The mappings P_1 and P_2 are determined by the flows to satisfy $\mathbf{x}^{(\alpha)}$ relative to $F(\mathbf{x}^{(\alpha)}) > F(\overline{\mathbf{x}}_{\sigma_\alpha}^{(\alpha)})$ and $F(\mathbf{x}^{(\alpha)}) < F(\overline{\mathbf{x}}_{\sigma_\alpha}^{(\alpha)})$, respectively. In Fig.8.3(a), the flows relative to two mappings are regular without grazing. However, the flow pertaining to the mapping with a grazing on the referenced surface S_{σ_α} is shown Fig.8.3(b). At that tangential point, the conditions in Eq.(8.53) should be satisfied. After grazing, the mapping becomes $P_1 \to P_2 \circ P_1$. The old mapping should become $P = P_2 \circ P_1 \circ P_2 \circ P_1$. From this mapping structure, it seems a period-doubling bifurcation. In fact, this mapping structure is caused by a grazing bifurcation, and the period may not be doubled. For the period-doubling bifurcation, the mapping points should start from the old mapping sets. However, for the grazing mapping, the mapping sets can start anywhere on the referenced surface S_{σ_α} if possible, and two new mapping sets will appear for this mapping structure. For this simple mapping structure, there are two possibilities of the grazing bifurcation. The period-doubling bifurcation has one bifurcation possibility. Before grazing of mapping $P = P_2 \circ P_1$, the periodicity of the periodic flow based on the mapping structure $P = P_2 \circ P_1$ is

$$\mathbf{x}_{k+2}^{(\alpha)} = \mathbf{x}_k^{(\alpha)}, t_{k+2} = t_k + T_\alpha. \tag{8.60}$$

where the period T_α is constant in domain Ω_α. If $T_\alpha = NT$ and T is a period of the external periodic vector field, then this flow is a periodic flow in domain Ω_α under N-periods of external vector fields.

Consider a generalized periodic flow with a mapping structure before grazing on the reference surface S_{σ_α}

$$P = P_{(21)^s} = \underbrace{(P_2 \circ P_1) \circ \cdots \circ (P_2 \circ P_1)}_{s-pairs} \quad (s = 1, 2, \cdots) \tag{8.61}$$

with a period T_α. If a flow of a mapping P_λ ($\lambda \in \{1,2\}$) is tangential to the reference surface S_{σ_α} with the $(2l-1)^{th}$-order at $(t_\eta, \mathbf{x}_\eta)$ for the location index $\eta \in \{k + 2(s-j) + 1, k + 2(s-j) + 2\}$, governed by

$$\left.\begin{aligned} & G_{\sigma_\alpha}^{(r)}(\mathbf{x}_\eta^{(\alpha)}, t_\eta, \mathbf{p}) = 0 \text{ for } r = 0, 1, \cdots, l-2; \\ & \text{either } \; G_{\sigma_\alpha}^{(2k-1)}(\mathbf{x}_\eta^{(\alpha)}, t_\eta, \mathbf{p}) < 0 \;\text{ for } F\left(\mathbf{x}^{(\alpha)}, \boldsymbol{\mu}\right) < E_{\sigma_\alpha} \\ & \text{or } \quad\; G_{\sigma_\alpha}^{(2k-1)}(\mathbf{x}_\eta^{(\alpha)}, t_\eta, \mathbf{p}) > 0 \;\text{ for } F\left(\mathbf{x}^{(\alpha)}, \boldsymbol{\mu}\right) > E_{\sigma_\alpha}, \end{aligned}\right\} \tag{8.62}$$

then the corresponding mapping structures is

$$P = P_{(21)^k \lambda (21)^{s-k}} = \underbrace{(P_2 \circ P_1) \circ \cdots \circ (P_2 \circ P_1)}_{k-pairs} \circ P_\lambda \circ \underbrace{(P_2 \circ P_1) \circ \cdots \circ (P_2 \circ P_1)}_{(s-k)-pairs}. \tag{8.63}$$

After grazing to the reference surface S_{σ_α}, the mapping structure for a new periodic flow with period T_α^* becomes

$$P = P_{(21)^{k+1}} = \underbrace{(P_2 \circ P_1) \circ \cdots \circ (P_2 \circ P_1)}_{(k+1)-pairs}. \tag{8.64}$$

Consider two groups of mapping subsets Ξ_i^1 and Ξ_i^2 ($i = 1, 2, \cdots, s$), so the mapping sets can be expressed as

$$\Sigma^1_{\sigma_\sigma} = \cup_i \Xi^1_i \text{ and } \Sigma^2_{\sigma_\sigma} = \cup_i \Xi^2_i . \tag{8.65}$$

For the mapping structures $P_{(21)^s}$, the mapping subsets and mapping relations before and after grazing are sketched in Fig.8.4. The mapping sets are gray and hollow cycled areas. The new subsets are filled with diagonal lines. The mapping subsets for the flow relative to the mapping $P_{(21)^s}$ are depicted in Fig.8.4(a). For a simple periodic flow of $\mathbf{y}_{k+2s} = P_{(21)^s}\mathbf{y}_k$, all the $2s$-mapping subsets are $2s$-points. Further, the points in the mapping subsets for all the periodic flow of $\mathbf{y}_{k+2s} = P_{(21)^s}\mathbf{y}_k$ are finite and countable. For a chaotic flow relative to $\mathbf{y}_{k+2s} = P_{(21)^s}\mathbf{y}_k$, all the $2s$- mapping subsets are dense. After one of the flows relative of the jth mapping pair in the mapping structure has a grazing bifurcation, two mapping subsets are added, as shown in Fig.8.4(b). After grazing, the mapping structure changes from $P_{(21)^s}$ to $P_{(21)^{s+1}}$. For a mapping structure $P_{(21)^s}$, at least there are $2s$-possibilities of grazing bifurcation. For a periodic flow relative to $P_{(21)^s}$, after grazing, only two more points in two new subsets are added on the referenced surface. For the chaotic flow of $P_{(21)^s}$, the two more mapping subsets are generated. Such a phenomenon is termed the strange attractor fragmentation, as in Luo (2006b), and the comprehensive discussion will be given later.

For a periodic flow relative to $\mathbf{y}_{k+2s} = P_{(21)^s}\mathbf{y}_k$ with $\mathbf{y}_{k+2s} = \mathbf{y}_k$, before the grazing bifurcation occurs, if the period-doubling bifurcation occurs to this periodic flow, then the corresponding mapping structure is given by

$$P = P_{(21)^s} \circ P_{(21)^s} = P_{(21)^{2s}} . \tag{8.66}$$

After period-doubling bifurcation, the mapping points should still be in the $2s$-mapping subsets Ξ^1_i and Ξ^2_i ($i = 1, 2, \cdots, s$) and the corresponding period is doubled (i.e., $2T_\alpha$). Before the grazing bifurcation happens, the period-doubling bifurcation continues to take place and the chaotic flow is induced by the period-doubling bifurcation. The corresponding mapping structure can be ex-pressed by

$$P = P_{(21)^{2^m s}} \circ P_{(21)^{2^m s}} = P_{(21)^{2^{m+1} s}} \text{ for } s = 1, 2, \cdots \text{ and } m = 1, 2, \cdots . \tag{8.67}$$

As $m \to \infty$, the chaotic flow is induced by the period-doubling bifurcation. It is interesting that the number of mapping subsets still is $2s$. However, the subsets

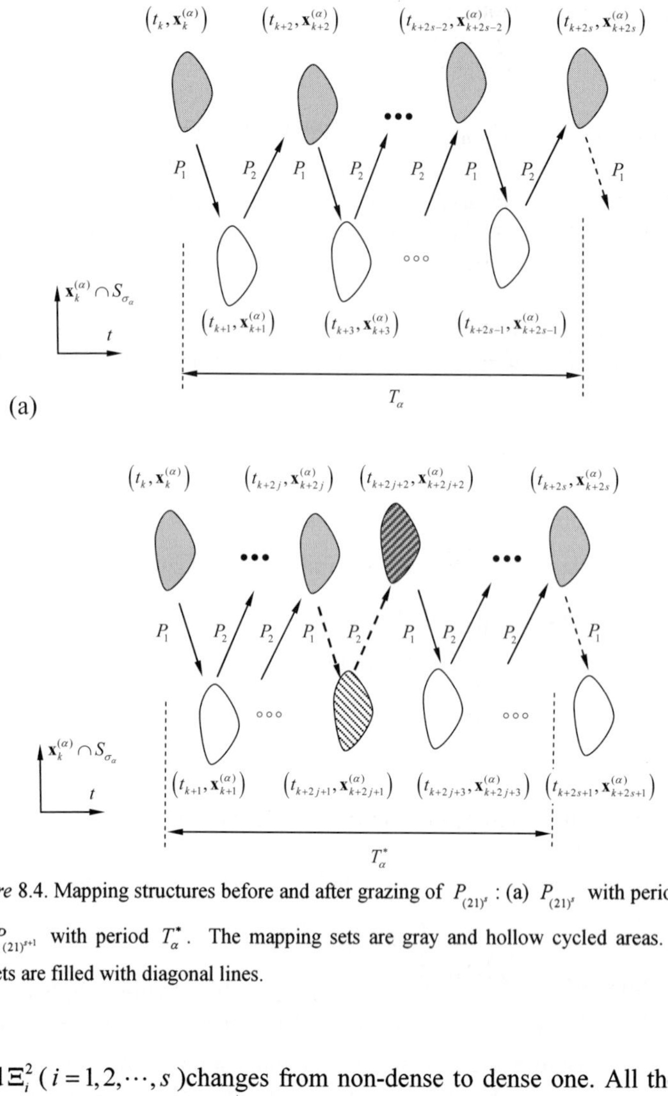

Figure 8.4. Mapping structures before and after grazing of $P_{(21)^s}$: (a) $P_{(21)^s}$ with period T_α and (b) $P_{(21)^{s+1}}$ with period T_α^*. The mapping sets are gray and hollow cycled areas. The new subsets are filled with diagonal lines.

Ξ_i^1 and Ξ_i^2 ($i = 1, 2, \cdots, s$)changes from non-dense to dense one. All the periodic and chaotic flows possess the same number of the mapping subsets. This class of flows is called *the flow of mapping* $P_{(21)^s}$ on the referenced surface. For the grazing bifurcation, the mapping structure of the post-grazing changes to $P_{(21)^{s+1}}$, and the number of subsets Ξ_i^1 and Ξ_i^2 ($i = 1, 2, \cdots, s+1$) becomes $2(s+1)$. To discuss the grazing bifurcation, the initial grazing set should be introduced. As discussed above, the periodic or chaotic flow can be investigated through the appropriate mapping structure. The key is how to determine the grazing bifurcation because the mapping sets have been changed.

8.3.3. Grazing determination

To determine the grazing bifurcation, as in Luo (2006b), the initial and final grazing sets of the mappings to the referenced, first integral surface should be defined from the grazing conditions in Definition 8.14.

DEFINITION 8.15. For a dynamical system in Eq.(3.1), consider the normal vector direction of the first integral surface to be in the direction of increasing the first integral quantity. If $\Sigma_{\sigma_\alpha}^{\lambda} \subset S_{\sigma_\alpha}$ ($\lambda = 1,2$), a subset ${}^{(\mathrm{i})}\Gamma_\lambda$ for mapping $P_\lambda : \Sigma_{\sigma_\alpha}^{\lambda} \to \Sigma_{\sigma_\alpha}^{\mathrm{mod}(\lambda,2)+1}$ is called *the initial set of grazing mapping* on the referenced surface S_{σ_α},

$$
{}^{(\mathrm{i})}\Gamma_\lambda \equiv \left\{ \left(\mathrm{mod}\left(t_i, T_\sigma\right), \mathbf{x}_i\right) \left| \begin{array}{l} P_\lambda \mathbf{y}_i = \mathbf{y}_{i+1} \text{ where } \mathbf{y}_i = (t_i, \mathbf{x}_i^{(\alpha)})^{\mathrm{T}}, \\ G_{\sigma_\alpha}^{(r)}\left(\mathbf{x}_{i+1}^{(\alpha)}, t_{i+1}, \mathbf{p}\right) = 0 \text{ for } r=0,1,\cdots 2l-2, \\ G_{\sigma_\alpha}^{(2l-1)}(\mathbf{x}_{i+1}^{(\alpha)}, t_{i+1}, \mathbf{p}) < 0 \text{ for } F\left(\mathbf{x}^{(\alpha)}, \boldsymbol{\mu}\right) < E_{\sigma_\alpha} \text{ or} \\ G_{\sigma_\alpha}^{(2l-1)}(\mathbf{x}_{i+1}^{(\alpha)}, t_{i+1}, \mathbf{p}) > 0 \text{ for } F\left(\mathbf{x}^{(\alpha)}, \boldsymbol{\mu}\right) > E_{\sigma_\alpha}. \end{array} \right. \right\} \subset \Sigma_{\sigma_\alpha}^{\lambda} \quad (8.68)
$$

The grazing set on the referenced surface S_{σ_α} is defined as

$$
{}^{(\mathrm{f})}\mathrm{G}_\lambda \equiv \left\{ \left(\mathrm{mod}\left(t_{i+1}, T_\sigma\right), \mathbf{x}_{i+1}\right) \left| \begin{array}{l} P_\lambda \mathbf{y}_i = \mathbf{y}_{i+1} \text{ where } \mathbf{y}_i = (t_i, \mathbf{x}_i^{(\alpha)})^{\mathrm{T}}, \\ G_{\sigma_\alpha}^{(r)}\left(\mathbf{x}_{i+1}^{(\alpha)}, t_{i+1}, \mathbf{p}\right) = 0 \text{ for } r=0,1,\cdots 2l-2, \\ G_{\sigma_\alpha}^{(2l-1)}(\mathbf{x}_{i+1}^{(\alpha)}, t_{i+1}, \mathbf{p}) < 0 \text{ for } F\left(\mathbf{x}^{(\alpha)}, \boldsymbol{\mu}\right) < E_{\sigma_\alpha} \text{ or} \\ G_{\sigma_\alpha}^{(2l-1)}(\mathbf{x}_{i+1}^{(\alpha)}, t_{i+1}, \mathbf{p}) > 0 \text{ for } F\left(\mathbf{x}^{(\alpha)}, \boldsymbol{\mu}\right) > E_{\sigma_\alpha}. \end{array} \right. \right\} \subset \Sigma_{\sigma_\alpha}^{\mathrm{mod}(\lambda,2)+1}. \quad (8.69)
$$

DEFINITION 8.16. For a dynamical system in Eq.(3.1), consider the normal vector direction of the first integral surface to be in the direction of increasing the first integral quantity. If $\Sigma_{\sigma_\alpha}^{\lambda} \subset S_{\sigma_\alpha}$ ($\lambda = 1,2$), a subset ${}^{(\mathrm{i})}\Gamma_\lambda$ for mapping $P_\lambda : \Sigma_{\sigma_\alpha}^{\lambda} \to \Sigma_{\sigma_\alpha}^{\mathrm{mod}(\lambda,2)+1}$ is called *the final set of grazing post-mapping* on the referenced surface S_{σ_α}:

$$
{}^{(\mathrm{f})}\Gamma_\lambda \equiv \left\{ \left(\mathrm{mod}\left(t_{i+1}, T_\sigma\right), \mathbf{x}_{i+1}\right) \middle| \begin{array}{l} P_\lambda \mathbf{y}_i = \mathbf{y}_{i+1}, \text{ where } \mathbf{y}_i = (t_i, \mathbf{x}_i^{(\alpha)})^{\mathrm{T}}, \\ G_{\sigma_\alpha}^{(r)}\left(\mathbf{x}_i^{(\alpha)}, t_i, \mathbf{p}\right) = 0 \text{ for } r=0,1,\cdots 2l-2, \\ G_{\sigma_\alpha}^{(2l-1)}(\mathbf{x}_i^{(\alpha)}, t_i, \mathbf{p}) < 0 \text{ for } F\left(\mathbf{x}^{(\alpha)}, \boldsymbol{\mu}\right) < E_{\sigma_\alpha} \text{ or} \\ G_{\sigma_\alpha}^{(2l-1)}(\mathbf{x}_i^{(\alpha)}, t_i, \mathbf{p}) > 0 \text{ for } F\left(\mathbf{x}^{(\alpha)}, \boldsymbol{\mu}\right) > E_{\sigma_\alpha}. \end{array} \right\} \tag{8.70}
$$
$$
\subset \Sigma_{\sigma_\alpha}^{\mathrm{mod}(\lambda,2)+1}
$$

The grazing set of *the post-grazing mapping* on the referenced surface S_{σ_α} is defined as

$$
{}^{(\mathrm{i})}G_\lambda \equiv \left\{ \left(\mathrm{mod}\left(t_i, T_\sigma\right), \mathbf{x}_i\right) \middle| \begin{array}{l} P_\lambda \mathbf{y}_i = \mathbf{y}_{i+1}, \text{ where } \mathbf{y}_i = (t_i, \mathbf{x}_i^{(\alpha)})^{\mathrm{T}}, \\ G^{(r)}\left(\mathbf{x}_i^{(\alpha)}, t_i, \mathbf{p}\right) = 0 \text{ for } r = 0,1,2,\cdots,2l-2 \\ G_{\sigma_\alpha}^{(2l-1)}(\mathbf{x}_i^{(\alpha)}, t_i, \mathbf{p}) < 0 \text{ for } F\left(\mathbf{x}^{(\alpha)}, \boldsymbol{\mu}\right) < E_{\sigma_\alpha} \text{ or} \\ G_{\sigma_\alpha}^{(2l-1)}(\mathbf{x}_i^{(\alpha)}, t_i, \mathbf{p}) > 0 \text{ for } F\left(\mathbf{x}^{(\alpha)}, \boldsymbol{\mu}\right) > E_{\sigma_\alpha} \end{array} \right\} \tag{8.71}
$$
$$
\subset \Sigma_{\sigma_\alpha}^{\lambda}.
$$

The grazing and post-grazing mappings of the flow are presented through a trajectory in phase space, as shown in Fig.8.5. The initial set of the grazing mapping and the final set of post-grazing mapping on the referenced surface S_{σ_α} are sketched. The grazing sets for the grazing and post-grazing mappings are also presented. At the grazing set, the trajectory grazing to the referenced surface S_{σ_α} is given. For a further discussion of the grazing characteristics, the relation between the subsets and the initial sets of grazing mapping should be presented. From the above definition, the initial set of the grazing mapping and the final sets of the post-grazing mapping can be determined. To explain this concept, the non-grazing sets and strange attractor fragmentation are sketched in Fig.8.6. The dashed and dashed-dotted curves represent the initial sets of grazing mapping and the final sets of the post-grazing mapping, respectively. The subsets are presented by cycled domains. If $\Xi_i^\lambda \cap {}^{(\mathrm{i})}\Gamma_\lambda = \{\varnothing\}$, the flow relative to all mapping subsets Ξ_i^λ will not have any grazing bifurcation, as in Fig.8.6(a). That is, the periodic flow with finite and countable points in Ξ_i^λ does not have any grazing bifurcation, and the chaotic flow with dense points in Ξ_i^λ does not have any gra-

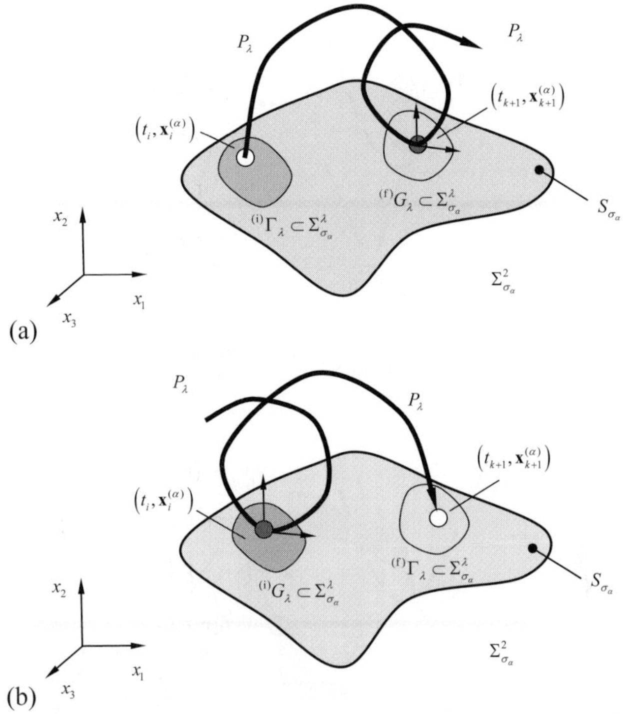

Figure 8.5. (a) The initial set of the grazing mapping and (b) the final set of post-grazing mapping on the referenced surface S_{σ_α}. The corresponding grazing sets for the grazing and post-grazing mappings are presented.

zing bifurcation. However, if $\Xi_i^\lambda \cap {}^{(\mathrm{i})}\Gamma_\lambda \neq \{\varnothing\}$, the flow relative to the mapping subset Ξ_i^λ will have grazing bifurcation. The periodic flow with finite and countable points in Ξ_i^λ has at least a grazing bifurcation. Such a phenomenon is called the *grazing bifurcation* of the periodic flow. Because the grazing bifurcation of the periodic flow occurs on the non-separatrix surface, only two more mapping subsets with finite and countable points are added. However, for the grazing bifurcation of a chaotic flow, two more dense subsets are generated. This phenomenon is called *the strange attractor fragmentation*, as shown in Fig. 8.6(b). If the strange attractor has $2s$-dense subsets, then there are at least $2s$-chances to cause the strange attractor fragmentation. In a single domain, the grazing bifurcation to the first integral surface will cause the flow switching from one periodic flow to another periodic flow or a chaotic flow. If a system in Eq. (3.1) has at least a separatrix, with varying system parameters, the flow may be tangential to the separatrix surface. Hence, the global grazing bifurcation of sys-

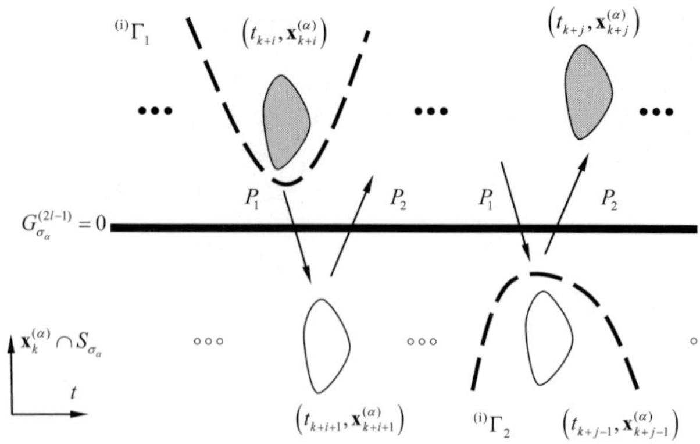

(a)

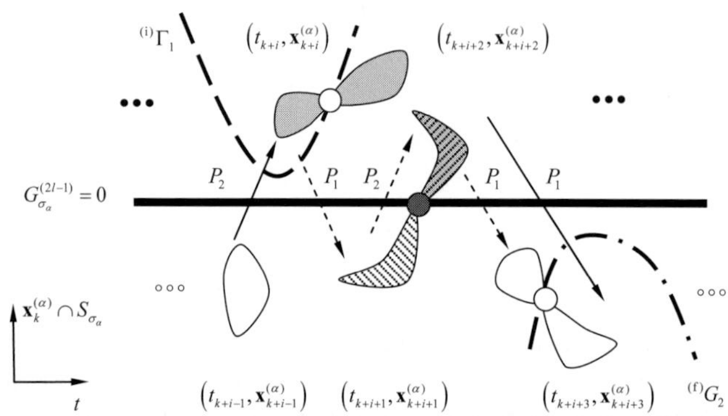

(b)

Figure 8.6. Non-grazing and grazing mapping subsets: (a) non-grazing sets and (b) strange attractor fragmentation. The dashed and dashed-dotted curves represented the initial sets of grazing mapping and the final sets of the post-grazing mapping, respectively. The mapping sets are gray and hollow cycled areas. The fragmentized subsets are filled with diagonal lines.

tem in Eq.(3.1) will occur, furthermore, the flow will become more complex. In next section, the global grazing bifurcation to separatrix will be discussed. This local grazing fragmentation can be used for chaos independent of the generic separatrix in dynamical systems.

8.4. Global grazing bifurcation

In a single domain Ω_α, the solution characteristic will not be changed, which is independent of selection of the first integral manifold surface as a reference surface. However, if a flow in Eq.(3.1) exists in different domains separated by the separatrix surface, the characteristics of a flow in different domains are different. Hence, the flow involving with the separatrix should be investigated by different mappings with the different solution properties. The solution characteristics will lead to the flow complexity existing in the vicinity of the separatrix surface. The grazing bifurcation to the separatrix surface becomes very crucial. This grazing bifurcation is called *the global grazing bifurcation.* A flow tangential or grazing to the separeatrix is also called the global tangency of the flow as in Chapter 3.

8.4.1. Global grazing and singular sets

Before the global grazing bifurcation of the periodic and chaotic flows on the separatrix surface is discussed, the following concept is introduced.

DEFINITION 8.17. For a dynamical system in Eq.(3.1), consider the normal vector direction of the first integral manifold surface to be in the direction of increasing the first integral quantity. A flow $\mathbf{x}^{(\alpha)} = \mathbf{\Phi}^{(\alpha)}(t_0, \mathbf{x}_0^{(\alpha)}, \mathbf{p}, t)$ in domain Ω_α ($\alpha = \{i, j\} \in 1, 2, \cdots, m$) for $t \in [t_0, \infty)$ is of $(2l-1)^{th}$-order grazing to the separatrix surface $\partial\Omega_{ij} \subset S_s$ at point $\mathbf{x}_k^{(\alpha)} = \bar{\mathbf{x}}_k^{(s)}$ at time $t_k \in [t_0, \infty)$ if

$$\left.\begin{array}{ll} G_{\partial\Omega_{ij}}^{(r)}(\mathbf{x}_k^{(\alpha)}, t_k, \mathbf{p}) = 0 \text{ for } r = 0,1,2,\cdots,2l-2; & \\ \text{either } \; G_{\partial\Omega_{ij}}^{(2l-1)}(\mathbf{x}_k^{(\alpha)}, t_k, \mathbf{p}) < 0 \;\text{ for } F\left(\mathbf{x}^{(\alpha)}, \boldsymbol{\mu}\right) < E_s & \\ \text{or } \quad G_{\partial\Omega_{ij}}^{(2l-1)}(\mathbf{x}_k^{(\alpha)}, t_k, \mathbf{p}) > 0 \;\text{ for } F\left(\mathbf{x}^{(\alpha)}, \boldsymbol{\mu}\right) > E_s. & \end{array}\right\} \tag{8.72}$$

The $(2l-1)^{th}$-order grazing to the separatrix surface S_s is termed *the global grazing bifurcation* of the flow in the domain Ω_α ($\alpha = 1, 2, \cdots, m$).

For convenience, without loss generality, the switching sets are from the separatrix boundary $\partial\Omega_{p_1p_2} \subseteq S_s$, and p_1 and p_2 are used to represent the domains, and the "i" and "j" will be used for mapping indices. So

$$\Sigma_s^{(p_1p_2)} = \left\{ (t_i, \mathbf{x}_i) \left| \begin{array}{l} F_{p_1p_2}(t_i, \mathbf{x}_i) = E_s \\ p_1, p_2 \in \{0, 1, 2, \cdots, m\} \end{array} \right. \right\} \subseteq \partial\Omega_{p_1p_2}. \tag{8.73}$$

Further, singular sets on the separatrix surface are introduced herein.

DEFINITION 8.18. For a dynamical system in Eq.(3.1), a set on the separatrix boundary $\partial\Omega_{p_1p_2} \subseteq S_s$ for $p_1, p_2 \in \{1, 2, \cdots, m\}$ is termed:

(i) the singular set if

$$\Gamma_{p_1p_2}=\left\{\left(t_i,\mathbf{x}_i^{(0)}\right)\middle|\begin{array}{l}\mathbf{x}_i^{(0)}\in\partial\Omega_{p_1p_2},\mathbf{x}\in\Omega_\alpha,\ \lim\limits_{\mathbf{x}\to\mathbf{x}_i^{(0)}}\left(\mathbf{n}_{\partial\Omega_{p_1p_2}}^T\bullet\dot{\mathbf{x}}\right)=0,\\ \alpha\in\{p_1,p_2\},i\in\mathbb{N}\end{array}\right\}; \tag{8.74}$$

(ii) the semi-singular sets in domain Ω_α if

$$\Gamma_{p_1p_2}^{(\alpha)}=\left\{\left(t_i,\mathbf{x}_i^{(0)}\right)\middle|\begin{array}{l}\mathbf{x}_i^{(0)}\in\partial\Omega_{p_1p_2},\mathbf{x}\in\Omega_\alpha,\ \lim\limits_{\mathbf{x}\to\mathbf{x}_i^{(0)}}\left(\mathbf{n}_{\partial\Omega_{p_1p_2}}^T\bullet\dot{\mathbf{x}}\right)=0,\\ \alpha=\{p_1\text{ or }p_2\},\ i\in\mathbb{N}\end{array}\right\}\subset\Gamma_{p_1p_2}; \tag{8.75}$$

(iii) the full-singular set if

$$\Gamma_{\overline{p_1p_2}}=\left\{\left(t_i,\mathbf{x}_i^{(0)}\right)\middle|\begin{array}{l}\mathbf{x}_i^{(0)}\in\partial\Omega_{p_1p_2},\mathbf{x}\in\Omega_\alpha,\ \lim\limits_{\mathbf{x}\to\mathbf{x}_i^{(0)}}\left(\mathbf{n}_{\partial\Omega_{p_1p_2}}^T\bullet\dot{\mathbf{x}}\right)=0,\\ \alpha=\{p_1\text{ and }p_2\},\ i\in\mathbb{N}\end{array}\right\}\subset\Gamma_{p_1p_2}. \tag{8.76}$$

8.4.2. Global grazing conditions

A switching set on $\partial\Omega_{p_1p_2}\subseteq S_s$ is formed by two neighbored boundaries $\Xi_{q_1}\subseteq\overrightarrow{\partial\Omega}_{p_1p_2}$ and $\Xi_{q_2}\subseteq\overleftarrow{\partial\Omega}_{p_1p_2}$ connected by a singular point $\Gamma_{p_1p_2}$, i.e.,

$$\Sigma_s^{(p_1p_2)}=\Xi_{q_1}\cup\Xi_{q_2}\cup\Gamma_{p_1p_2}\subseteq\partial\Omega_{p_1p_2} \tag{8.77}$$

for $q_1,q_2\in\{1,2,\cdots,M\}$. The local mappings near the switching sets $\Xi_{q_1q_2}$ are defined as

$$P_{J_1}:\Xi_{q_1}\to\Xi_{q_2},P_{J_2}:\Xi_{q_2}\to\Xi_{q_1} \tag{8.78}$$

For specific $J_1,J_2\in\{1,2,\cdots,N\}$. If $\Xi_{q_1}\subseteq\partial\Omega_{p_1p_2}$ and $\Xi_{q_3}\not\subset\partial\Omega_{p_1p_2}$, the global mapping in a domain Ω_α ($\alpha\in\{p_1,p_2\}$) is defined as

$$P_{J_3}:\Xi_{q_1}\to\Xi_{q_3} \tag{8.79}$$

for specific $J_3\in\{1,2,\cdots,N\}$. Such generic mappings are illustrated in Fig.8.7.

After the global grazing bifurcation to the separatrix surface, the new flow of the post grazing will exist in different domains, and many new mappings will be added into the old mapping structures. Consider a post-grazing mapping cluster of a specific local or global mapping P_J ($J\in\{1,2,\cdots,\ N\}$). For a local mapping P_{J_1} on the boundary $\partial\Omega_{p_1p_2}$, the pre-grazing, grazing and post-grazing flows are

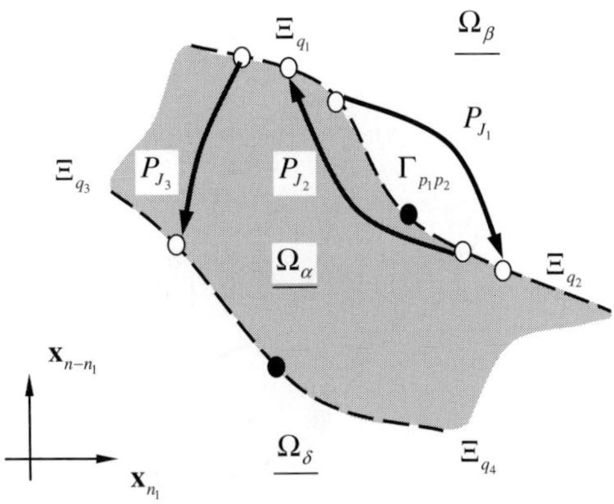

Figure 8.7. Switching sets, local and global mappings in the shaded domain Ω_α.

illustrated in Fig.8.8(a)-(d). There are many clusters of post-grazing mappings that determine the property of the post-grazing flow. Two clusters of the post-grazing mappings for mapping P_{J_1} are sketched. After grazing, the relation between the pre-grazing and post-grazing is given by

$$P_{J_1} \underset{\text{pre-grazing}}{\overset{\text{post-grazing}}{\rightleftharpoons}} P_{J_1} \circ \underbrace{P_{J_{n_2}} \circ \cdots \circ P_{J_3} \circ P_{J_2}}_{\text{local mapping cluster}} = P_{J_1(J_{n_2} \cdots J_3 J_2)} \tag{8.80}$$

For ($J_1, J_2, \cdots J_{n_2} \in \{1,2,\cdots,N\}$). The index J_2 can be J_1, but the other index J_i ($J_i = J_3, J_4, \cdots J_{n_2}$) should not be J_1. P_{J_2} ($\neq P_{J_1}$) can be any mapping on the *same* part of the separatrix boundary. Similarly, consider a global mapping P_{J_1} to map the flow on the boundary $\partial\Omega_{p_1p_2}$ to another boundary $\partial\Omega_{p_2p_3}$ in domain Ω_{p_2}. The pre-grazing, grazing and post-grazing flows for the global mapping P_{J_1} are illustrated in Fig.8.9(a)-(d). The relation between the pre-grazing and post-grazing is given by

$$P_{J_1} \underset{\text{pre-grazing}}{\overset{\text{post-grazing}}{\rightleftharpoons}} \underbrace{P_{J_{n_2}} \circ \cdots \circ P_{J_3}}_{\text{grazing mapping cluster}} \circ P_{J_2} = P_{(J_{n_2} \cdots J_3) J_2} \tag{8.81}$$

for($J_1, J_2, \cdots J_{n_2} \in \{1,2,\cdots,N\}$). The index J_2 can be J_1, but the other index J_i ($J_i = J_3, J_4, \cdots J_{n_2}$) is not be J_1. In post-grazing mapping clusters, the mappings can be local mappings on the boundary $\partial\Omega_{p_2p_3}$. Two clusters of post-grazing mapping structure for such a mapping grazing are sketched in Fig.8.9(c) and (d).

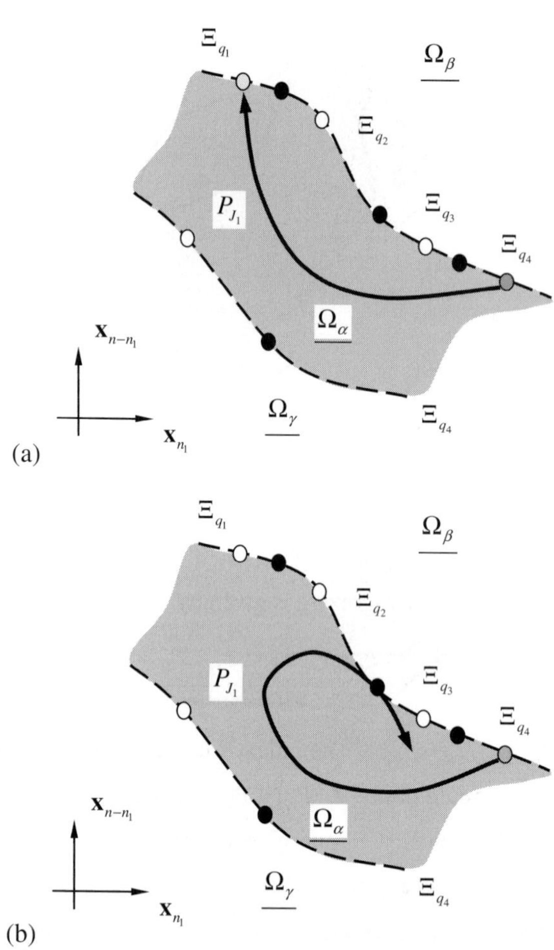

Figure 8.8. Local mapping grazing switching P_{J_1} ($J_1 \in \{1, 2, \cdots, N\}$): (a) pre-grazing mapping, (b) grazing mapping, (c, d) two possible post-grazing mappings. The black solid circular symbols are grazing points. The rest circular points are switching points.

For the grazing occurrence of the mapping P_{J_1} on the boundary $\partial\Omega_{p_1 p_2}$, the post-grazing mapping structure is the same as in Eq.(8.80). However, the mapping P_{J_1} is global and the index J_2 cannot be J_1 because P_{J_2} is any mapping rather than P_{J_1} or another global mapping. This mapping can be a local mapping. Without the mapping J_1, this post-grazing mapping structure is a local mapping grazing structure, which is considered as a case in Eq.(8.80). This concept is ex-

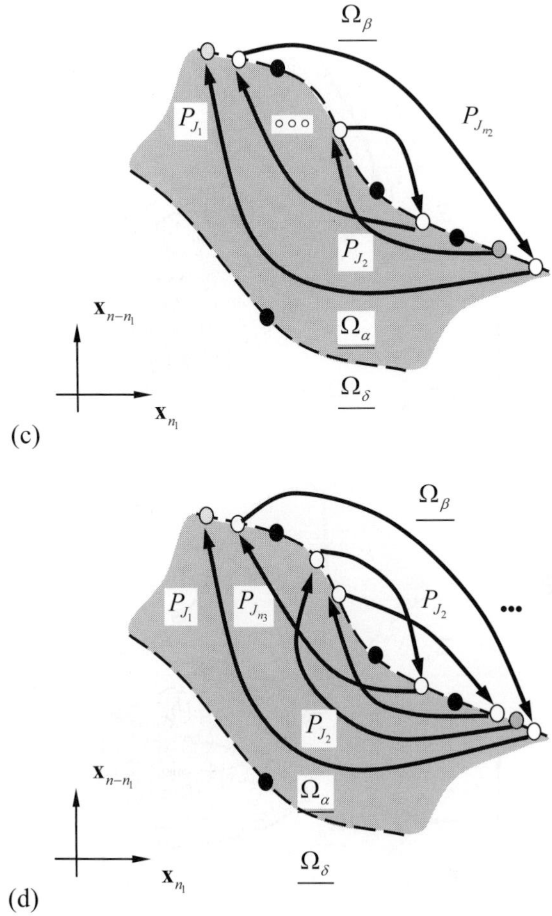

Figure 8.8. (Continued)

tended to the more generalized case. The post-mapping cluster can include any possible mappings rather than the local mappings only.

DEFINITION 8.19. For a dynamical system in Eq.(3.1), consider the normal vector direction of the first integral surface to be in the direction of increasing the first integral quantity. If $\Xi_{q_1} \subset \Sigma_s^{(p_1 p_2)}$ and $\Xi_{q_2} \subset \Sigma_s^{(p_2 \alpha)}$ with ($\alpha \in \{p_1, p_3\}$ and p_2, $\alpha \in \{1, 2, \cdots, m\}$), a subset ${}^{(i)}\Gamma_J$ for mapping $P_J : \Xi_{q_1} \to \Xi_{q_2}$ is called *the initial set of grazing mapping* on the separatrix surface $\partial\Omega_{p_2\alpha}$ for a specific period T_s,

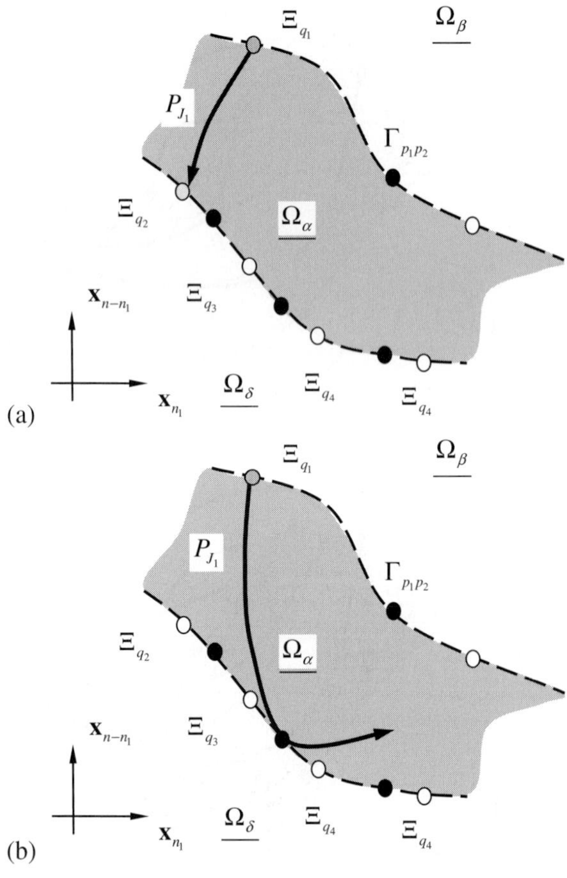

Figure 8.9. Global mapping grazing switching: (a) pre-grazing mapping, (b) grazing mapping, (c, d) two possible post grazing mappings. The black solid circular symbols are grazing points. The rest circular points are switching points.

$$
{}^{(\mathrm{i})}\Gamma_J \equiv \left\{ \left(\operatorname{mod}(t_i, T_s), \mathbf{x}_i\right) \middle| \begin{array}{l} P_J \mathbf{y}_i = \mathbf{y}_{i+1} \text{ where } \mathbf{y}_i = (t_i, \mathbf{x}_i)^{\mathrm{T}}, \\ G^{(r)}_{\partial\Omega_{p_2\alpha}}(\mathbf{x}^{(p_2)}_{i+1}, t_{i+1}, \mathbf{p}) = 0 \text{ for } r = 0,1,2,\cdots,2l-2, \\ G^{(2l-1)}_{\partial\Omega_{p_2\alpha}}(\mathbf{x}^{(p_2)}_{i+1}, t_{i+1}, \mathbf{p}) < 0 \text{ for } F\left(\mathbf{x}^{(p_2)}, \boldsymbol{\mu}\right) < E_s \text{ or} \\ G^{(2l-1)}_{\partial\Omega_{p_2\alpha}}(\mathbf{x}^{(p_2)}_{i+1}, t_{i+1}, \mathbf{p}) > 0 \text{ for } F\left(\mathbf{x}^{(p_2)}, \boldsymbol{\mu}\right) > E_s. \end{array} \right\} \subset \Xi_{q_1}
\tag{8.82}
$$

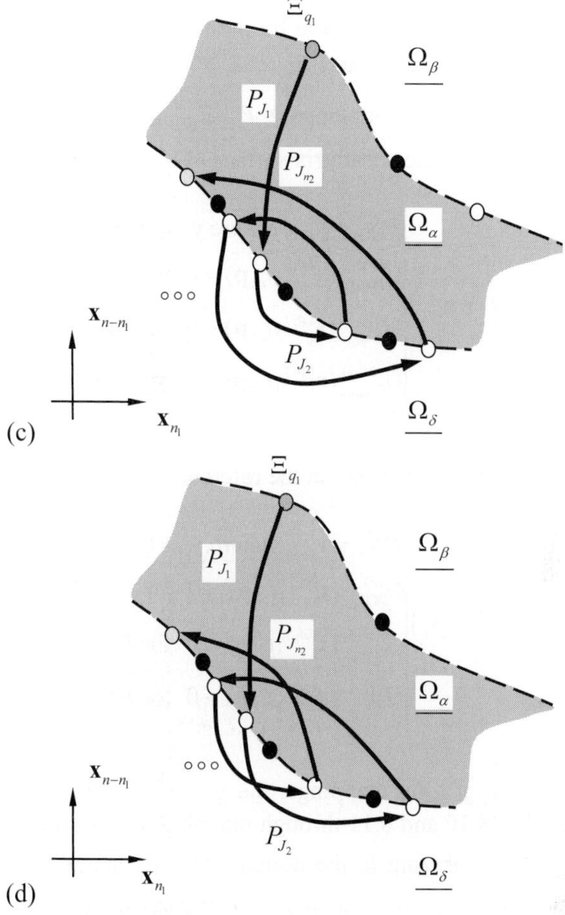

Figure 8.9. (Continued)

The corresponding grazing set on the separatrix surface $\partial\Omega_{p_2\alpha}$ is defined as

$$
{}^{(\mathrm{f})}G_J \equiv \left\{ \left(\operatorname{mod}\left(t_{i+1}, T_s\right), \mathbf{x}_{i+1}\right) \left| \begin{array}{l} P_J \mathbf{y}_i = \mathbf{y}_{i+1}, \text{ where } \mathbf{y}_i = (t_i, \mathbf{x}_i)^{\mathrm{T}} \\ G^{(r)}_{S_{p2\alpha}}(\mathbf{x}^{(p_2)}_{i+1}, t_{i+1}, \mathbf{p}) = 0 \;\; \text{for } r = 0,1,2,\cdots,2l-2; \\ G^{(2l-1)}_{S_{p2\alpha}}(\mathbf{x}^{(p_2)}_{i+1}, t_{i+1}, \mathbf{p}) < 0 \text{ for } F\left(\mathbf{x}^{(p_2)}, \boldsymbol{\mu}\right) < E_s \text{ or} \\ G^{(2l-1)}_{S_{p2\alpha}}(\mathbf{x}^{(p_2)}_{i+1}, t_{i+1}, \mathbf{p}) > 0 \text{ for } F\left(\mathbf{x}^{(p_2)}, \boldsymbol{\mu}\right) > E_s. \end{array} \right. \right\} \subset \Xi_{q_2}
\tag{8.83}
$$

DEFINITION 8.20. For a dynamical system in Eq.(3.1), consider the normal vector direction of the first integral surface to be in the direction of increasing the first integral quantity. If $\Xi_{q_1} \subset \Sigma_s^{(p_1 p_2)}$ and $\Xi_{q_2} \subset \Sigma_s^{(p_2\alpha)}$ with ($\alpha \in \{p_1, p_3\}$ and $p_1, \alpha \in$ $\in \{1,2,\cdots,m\}$), a subset ${}^{(\mathrm{i})}\Gamma_J$ for mapping $P_J : \Xi_{q_1} \to \Xi_{q_2}$ is called *the final set of grazing post-mapping* on the separatrix surface $\partial\Omega_{p_1 p_2}$ for a specific period T_s,

$$
{}^{(\mathrm{f})}\Gamma_J \equiv \left\{ \left(\mathrm{mod}\left(t_{i+1}, T_s\right), \mathbf{x}_{i+1}\right) \middle| \begin{array}{l} P_J \mathbf{y}_i = \mathbf{y}_{i+1}, \text{ where } \mathbf{y}_i = (t_i, \mathbf{x}_i)^{\mathrm{T}} \\ G^{(r)}_{\partial\Omega_{p_1 p_2}}(\mathbf{x}_i^{(p_2)}, t_i, \mathbf{p}) = 0 \text{ for } r = 0,1,2,\cdots,2l-2 \\ G^{(2l-1)}_{\partial\Omega_{p_1 p_2}}(\mathbf{x}_i^{(p_2)}, t_i, \mathbf{p}) < 0 \text{ for } F\left(\mathbf{x}^{(p_2)}, \boldsymbol{\mu}\right) < E_s \text{ or} \\ G^{(2l-1)}_{\partial\Omega_{p_1 p_2}}(\mathbf{x}_i^{(p_2)}, t_i, \mathbf{p}) > 0 \text{ for } F\left(\mathbf{x}^{(p_2)}, \boldsymbol{\mu}\right) > E_s. \end{array} \right\} \subset \Xi_{q_2} \tag{8.84}
$$

The *grazing set of the post-grazing* on the referenced surface $\partial\Omega_{p_2\alpha}$ is defined as

$$
{}^{(\mathrm{i})}G_J \equiv \left\{ \left(\mathrm{mod}\left(t_i, T_s\right), \mathbf{x}_i\right) \middle| \begin{array}{l} P_J \mathbf{y}_i = \mathbf{y}_{i+1}, \text{ where } \mathbf{y}_i = (t_i, \mathbf{x}_i)^{\mathrm{T}} \\ G^{(r)}_{\partial\Omega_{p_2\alpha}}(\mathbf{x}_i^{(p_2)}, t_i, \mathbf{p}) = 0 \text{ for } r = 0,1,2,\cdots,2l-2 \\ G^{(2l-1)}_{\partial\Omega_{p_2\alpha}}(\mathbf{x}_i^{(p_2)}, t_i, \mathbf{p}) < 0 \text{ for } F\left(\mathbf{x}^{(p_2)}, \boldsymbol{\mu}\right) < E_s \text{ or} \\ G^{(2l-1)}_{\partial\Omega_{p_2\alpha}}(\mathbf{x}_i^{(p_2)}, t_i, \mathbf{p}) > 0 \text{ for } F\left(\mathbf{x}^{(p_2)}, \boldsymbol{\mu}\right) > E_s. \end{array} \right\} \subset \Xi_{q_1} \tag{8.85}
$$

For global and local grazing mappings, the grazing and post-grazing mapping are sketched in Figs.8.10 and 8.11 through mapping P_J ($J \in \{J_1, \cdots, J_m\}$) on the boundary $\partial\Omega_{p_1 p_2}$. The grazing in the domain Ω_{p_2} occurs at the final points of the grazing mapping P_J on the boundary $\partial\Omega_{p_2\alpha}$ ($\alpha \in p_2, p_3$). The above definitions for both the initial grazing sets of grazing mapping and the final sets of post-grazing mapping are illustrated. The hollow symbols are either the initial point for grazing mapping or final point of grazing post-mapping. The circular symbols are the grazing points of the grazing and post grazing mappings. The governing equations of mapping P_J with the final point on the boundary $\partial\Omega_{p_2\alpha}$ in Ω_α ($\alpha \in \{p_1, p_3\}$) is expressed by

$$
\left.\begin{array}{l} \boldsymbol{\Phi}_J^{(\alpha)}\left(t_i, \mathbf{x}_i, t_{i+1}\right) - \mathbf{x}_{i+1} = 0 \\ F_{p_1 p_2}\left(\mathbf{x}_i, \boldsymbol{\mu}\right) = E_s, \ F_{p_2\alpha}\left(\mathbf{x}_{i+1}, \boldsymbol{\mu}\right) = E_s. \end{array}\right\} \tag{8.86}
$$

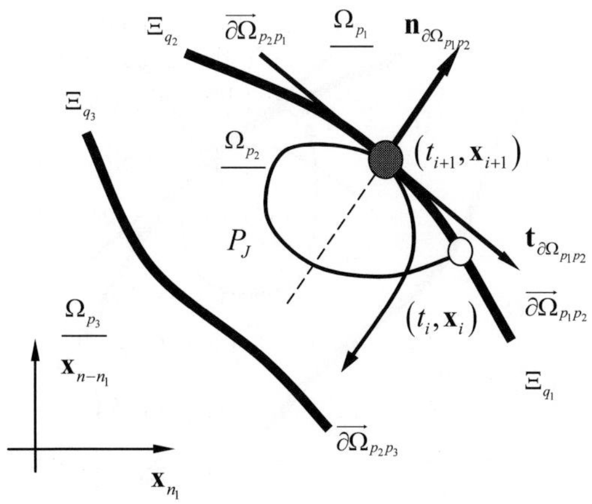

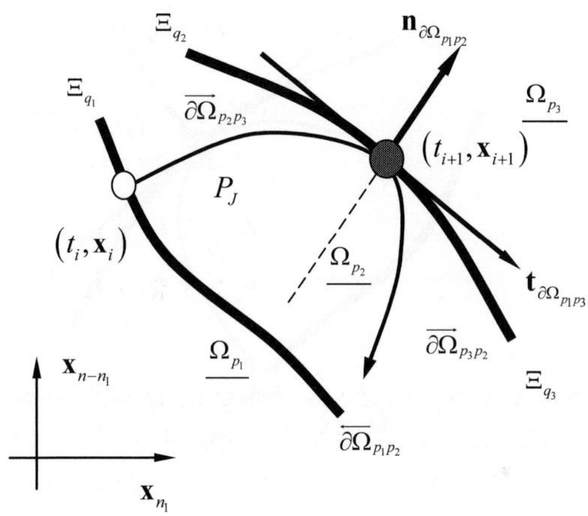

Figure 8.10 (a) Local and (b) global grazing mappings. The filled solid circular symbols are grazing points. The hollow circular symbols are initial switching points.

The grazing necessary conditions for mapping P_J at the singular set $\Gamma^{(0)}_{p_2\alpha}$ in the sub-domain Ω_α ($\alpha,\beta=\{p_1,p_2\},\alpha\neq\beta$) is:

$$G^{(2l-1)}_{\partial\Omega_{p_2\alpha}}(\mathbf{x}^{(p_2)}_{i+1},t_{i+1},\mathbf{p})=0; \tag{8.87}$$

To guarantee the occurrence of the grazing flow at the singular points, the sufficient conditions should be considered as follows.

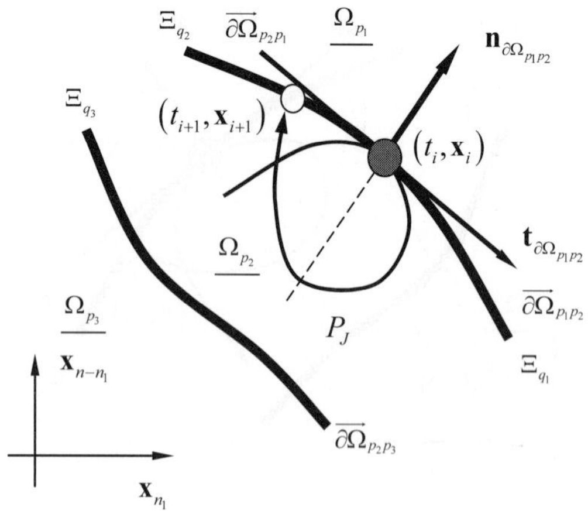

(a)

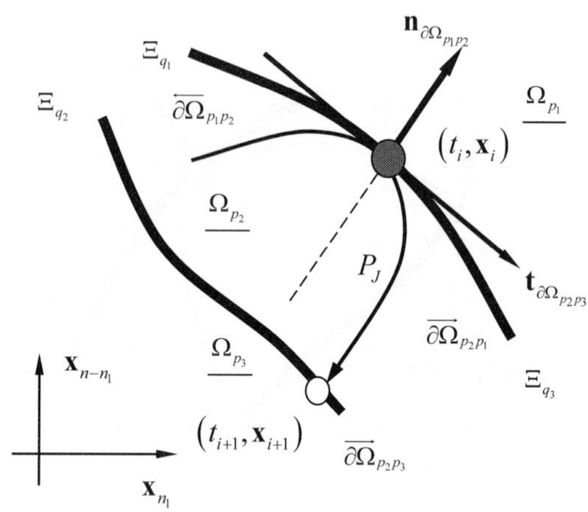

(b)

Figure 8.11 (a) Local and (b) global grazing post-mappings mappings. The filled solid circular symbols are grazing points. The hollow circular symbols are initial switching points.

$$\left.\begin{array}{l} G^{(2l-1)}_{\partial\Omega_{p_{2\alpha}}}(\mathbf{x}^{(p_2)}_{i+1},t_{i+1},\mathbf{p})<0 \;\text{ for } F\left(\mathbf{x}^{(p_1)},\boldsymbol{\mu}\right)<E_s \;\text{ or}\\ G^{(2l-1)}_{\partial\Omega_{p_{2\alpha}}}(\mathbf{x}^{(p_2)}_{i+1},t_{i+1},\mathbf{p})>0 \;\text{ for } F\left(\mathbf{x}^{(p_1)},\boldsymbol{\mu}\right)>E_s. \end{array}\right\} \tag{8.88}$$

From Eqs.(8.86) and (8.87), the initial set of grazing mapping is on an $(n-1)$ surface because of the $(n+3)$-equations with the $2(n+1)$-unknowns. Equation (8.88) is the sufficient condition for the initial set. This $(n-1)$ surface in phase

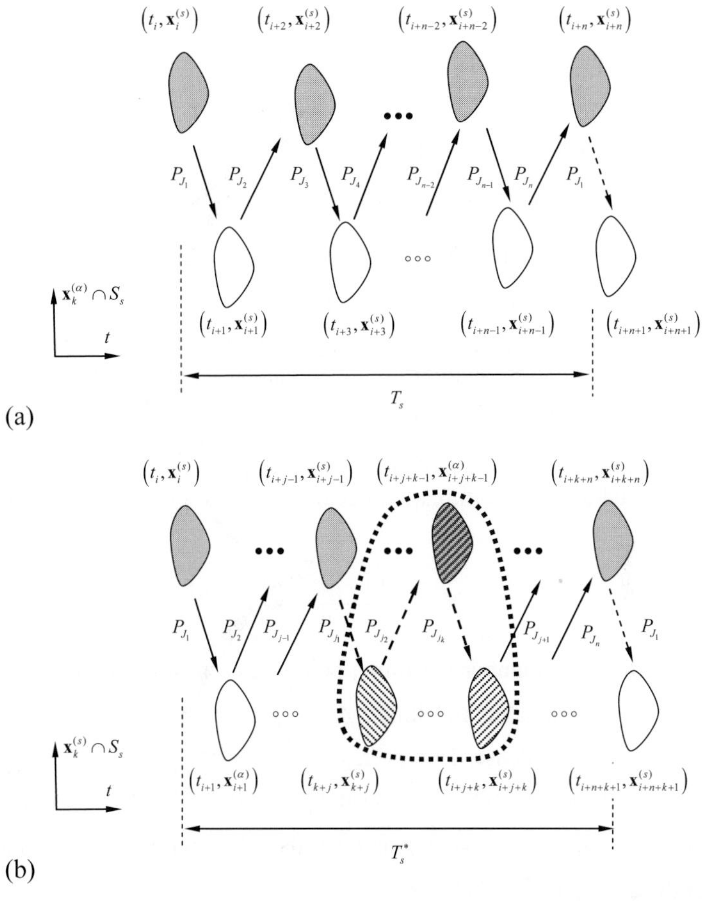

Figure 8.12. Mapping structures before and after grazing of $P_{J_k \cdots J_2 J_1}$: (a) $P_{J_k \cdots J_2 J_1}$ with period T_s and (b) $P_{J_k \cdots J_{j+1}(J_{j1} \cdots J_{j2} J_{j1}) J_j \cdots J_2 J_1}$ with period T_s^*. The mapping sets are gray and hollow cycled areas. The new subsets after grazing are filled with diagonal lines.

space $(\text{mod}(t_i, T_s), \mathbf{x}_i \cap \partial\Omega_{p_1 p_2})$ is called the *initial grazing manifold* which will be used in discussion of the strange attractor fragmentation. The switching time conditions $t_{i+1} > t_i$ should be inserted. The boundary $\partial\Omega_{p_1\alpha}$ is determined by $F_{p_2\alpha}(\mathbf{x}_i, \boldsymbol{\mu}) = E_s$. Similarly, when the switching times t_i and t_{i+1} in Eqs.(8.86) and (8.87) are exchanged, *the final grazing manifold* can be determined through the final set of grazing post-mapping.

Consider a global periodic flow with a mapping structure $P_{J_k \cdots J_2 J_1} = P_{J_k} \circ \cdots \circ P_{J_2} \circ P_{J_1}$ to the separatrix surface for $J_1, J_2, \cdots J_k \in \{1,2,\cdots,N\}$. For $\mathbf{y}_i = (t_i, \mathbf{x}_i)^{\mathrm{T}}$, the periodic flow during NT_s can be described by $\mathbf{y}_{i+k} = P_{J_k \cdots J_2 J_1} \mathbf{y}_i$. The periodicity of the periodic flow requires

$$\mathbf{x}_{i+k} = \mathbf{x}_i \text{ and } \mathrm{mod}(t_{i+k}, T_s) = t_i. \tag{8.89}$$

Based on the mapping structure, there are k-mapping subsets $\Xi_s^{(j)} \subset \Xi_{q_j}$ with $\cap_{j=1}^{k} \Xi_s^{(j)} = \{\varnothing\}$ ($q_j \in \{1,2,\cdots,M\}$ and $j = 1,2,\cdots,k$) on the separatrix surface. If the mapping relation can be expressed analytically, the stability and bifurcation of the periodic flow can be analyzed by the eigenvalue analysis. In addition, the stability of the periodic flow can be done through the first integral increment quantity based on the Poincare section. In this section, the grazing bifurcation of the periodic and chaotic flows is of great interest. Before grazing of this periodic flow, if its period-doubling bifurcation occurs, the corresponding mapping structure is

$$P_{(J_k \cdots J_2 J_1)^{2^l}} = \underbrace{P_{J_k \cdots J_2 J_1} \circ \cdots \circ P_{J_k \cdots J_2 J_1}}_{\text{Period-}l\text{ doubling}}. \tag{8.90}$$

If the period-doubling number l is finite, points in each of k-subsets $\Xi_s^{(j)} \subset \Xi_{q_j}$ ($j = 1,2,\cdots,l$) are countable and finite. As $l \to \infty$, points in each of k-subsets $\Xi_s^{(j)}$ ($j = 1,2,\cdots,k$) are dense. However, both the periodic and chaotic flows of $P_{J_k \cdots J_2 J_1}$ have the same number of invariant switching subsets on the separatrix surface. If a flow of $P_{J_n \cdots J_2 J_1}$ has a grazing at the jth mapping P_{J_j}, the mapping cluster of the post-grazing of P_{J_j} is

$${}^{C}P_{J_j} = P_{J_{Jm}} \circ \cdots \circ P_{J_{J1}}. \tag{8.91}$$

The mapping structure of the post-grazing becomes

$$\begin{aligned} P_{J_k \cdots J_{j+2}} \circ {}^{C}P_{J_j} \circ P_{J_{j-1} \cdots J_2 J_1} = P_{J_k} \circ \cdots \circ P_{J_{j+2}} \circ P_{J_{Jm+1}} \circ \cdots \\ \circ P_{J_{J1}} \circ P_{J_{j-1}} \circ \cdots \circ P_{J_2} \circ P_{J_1} \end{aligned} \tag{8.92}$$

from which the number of the invariant subsets increases from k to $(m+k)$. To interpret the above concept, the mapping structures before and after grazing of $P_{J_k \cdots J_2 J_1}$ are sketched in Fig.8.12. The mapping sets are gray and hollow cycled areas. The new subsets after grazing are filled with diagonal lines, which are caused by the grazing. In Fig.8.12(a), the k-invariant subsets (i.e., $\Xi_s^{(j)} \subset \Xi_{q_j}$ and $\cap_{j=1}^{k} \Xi_s^{(j)} = \{\varnothing\}$) and the corresponding mappings are presented. After grazing, the post-grazing mapping structure is presented in Fig.8.12(b), and the mapping cluster of the post-grazing is circled by a dotted cycle and the new mappings caused by the grazing are presented through the dark, dashed arrows. The mapping subset number increases after grazing, and the flow relative to the old mapping structure suddenly become complex, which may lead to a chaotic flow. Compared with the local grazing in the single domain, the global grazing to the separatrix surface will generate more complex flows of the post-grazing. If after the grazing, the post-flow of the periodic flow is chaotic, this grazing will be an onset of the chaotic flow. Therefore, the global grazing may imply the global chaos on the separatrix surface. If the flow of $P_{J_k \cdots J_2 J_1}$, before grazing, is chaotic, then there are k-invariant subsets to form a strange attractor relative to mapping $P_{J_k \cdots J_2 J_1}$. Once the grazing bifurcation occurs, the strange attractor with the k-invariant subsets will be fragmentized. Thus, the fragmentation of the strange attractor will be discussed as follows.

8.4.3. Global strange attractor fragmentation

Once the intersection exists between the invariant subsets of the strange attractor and one of the *initial* grazing manifolds of the generic mappings, the fragmentation of the strange attractor will occur. For a subset $\Xi_s^{(j)}$ and an initial, the final subset $\Xi_s^{(j+1)}$ of the mapping P_{J_j} will be fragmentized. If one of the initial, grazing manifolds is tangential to one of the strange attractor subsets, the strange attractor fragmentation may appear or disappear. Therefore, a grazing manifold ${}^{(\mathrm{i})}\Gamma_{J_j}$ relative to a mapping P_{J_j}, if $\Xi_s^{(j)} \cap {}^{(\mathrm{i})}\Gamma_{J_j} \neq \{\varnothing\}$, then mathematical definition of strange attractor fragmentation is given as follows.

DEFINITION 8.21. For a dynamical system in Eq.(3.1), there is a chaotic flow of mapping $P_{J_k \cdots J_2 J_1}$ with k-invariant subsets $\Xi_s^{(j)}$ ($j = 1, 2, \cdots, k$). For a mapping P_{J_j} : $\Xi_s^{(j)} \to \Xi_s^{(j+1)}$ ($J_j \in \{1, 2, \cdots, N\}$), if ${}^{(i)}\Pi_s^{(j)} \equiv \Xi_s^{(j)} \cap {}^{(\mathrm{i})}\Gamma_{J_j} \neq \{\varnothing\}$, the invariant set $\Xi_s^{(j+1)}$ is fragmentized. Further, there is a mapping cluster $P_{J_{j_{m+1}} \cdots J_{j_2} J_{j_1}}$ with $(m+1)$ mappings $P_{J_{j_\kappa}}$ ($\kappa = 1, 2, \cdots, m+1$) and m invariant subsets $\Xi_s^{(j_\kappa)}$. For the post-grazing mapping cluster $P_{J_{j_{m+1}} \cdots J_{j_2} J_{j_1}}$, the corresponding map-ping chain is given by

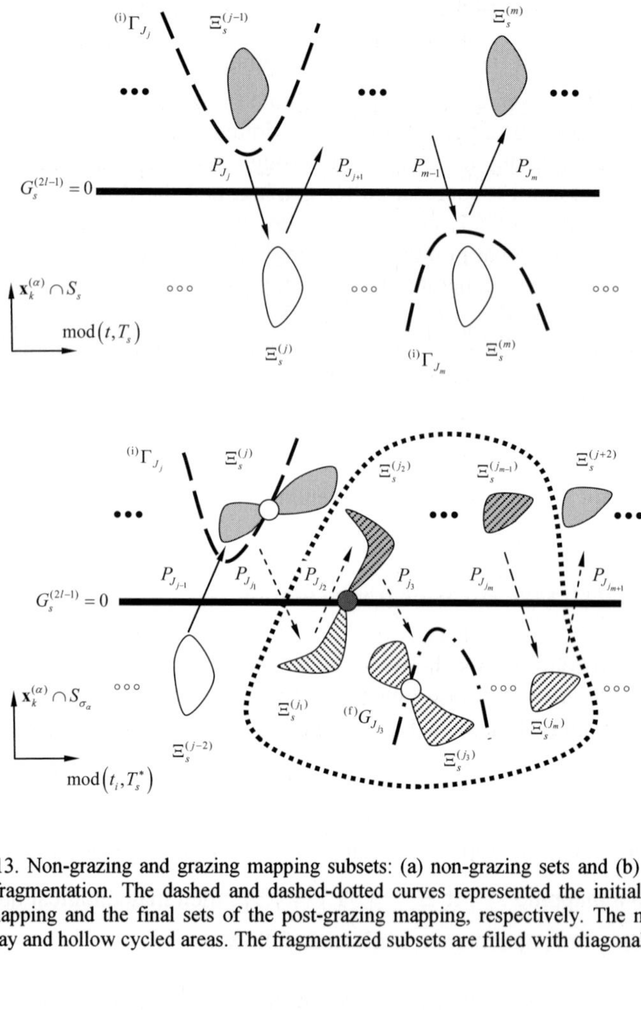

(a)

(b)

Figure 8.13. Non-grazing and grazing mapping subsets: (a) non-grazing sets and (b) strange attractor fragmentation. The dashed and dashed-dotted curves represented the initial sets of grazing mapping and the final sets of the post-grazing mapping, respectively. The mapping sets are gray and hollow cycled areas. The fragmentized subsets are filled with diagonal lines.

$$\left.\begin{array}{l}
\Xi_s^{(1)} \xrightarrow{P_{J_1}} \Xi_s^{(2)} \cdots \Xi_s^{(j-1)} \xrightarrow{P_{J_{j-1}}} \Xi_s^{(j)} \\
\Xi_s^{(j)} \xrightarrow{P_{J_{j_1}}} \underbrace{\Xi_s^{(j_1)} \cdots \Xi_s^{(j_{m-1})} \xrightarrow{P_{J_{j_m}}} \Xi_s^{(j_m)}}_{m-\text{mappings}} \xrightarrow{P_{J_{j_{m+1}}}} \Xi_s^{(j+2)} \\
\Xi_s^{(j+2)} \xrightarrow{P_{J_{j+2}}} \Xi_s^{(j+2)} \cdots \Xi_s^{(k-1)} \xrightarrow{P_{J_k}} \Xi_s^{(k)}
\end{array}\right\} \tag{8.93}$$

The union of all the switching sets $\cup_{\kappa=1}^{m} \Xi_s^{(j_\kappa)}$ is termed *a fragmentation set* of the invariant set $\Xi_s^{(j+1)}$ under the mapping cluster $P_{J_{j_{m+1}} \cdots J_{j_2} J_{j_1}}$.

Non-grazing and grazing mapping subsets on the separatrix manifold surface $\mathbf{x}_i^{(\alpha)} \cap S_s$, relative to mapping structure $P_{J_k \cdots J_2 J_1}$, are illustrated in Fig.8.13 to explain the fragmentation mechanism of strange attractors in nonlinear dynamic systems. The dashed and dashed-dotted curves represented the initial sets of grazing mapping and the final sets of the post-grazing mapping, respectively. The mapping sets are gray and hollow cycled areas. The fragmentized subsets are filled with diagonal lines. For $\Xi_s^{(j)} \cap {}^{(\mathrm{i})}\Gamma_{J_j} = \{\varnothing\}$ ($j = 1, 2, \cdots, k$), no fragmentation of the strange attractor $\cup_{i=1}^{k} \Xi_s^{(j)}$ occurs for the mapping structure $P_{J_k \cdots J_2 J_1}$, as shown in Fig.8.13(a). The invariant sets and the initial sets of grazing mappings are independent each other. If the intersection between one invariant set $\Xi_s^{(j)}$ and the initial set ${}^{(\mathrm{i})}\Gamma_{J_j}$ is non-empty (i.e., $\Xi_s^{(j)} \cap {}^{(\mathrm{i})}\Gamma_{J_j} \neq \{\varnothing\}$), the fragmentation of the invariant set $\Xi_s^{(j-1)}$ will occur. The fragmentation of the strange attractor on S_s is illustrated in Fig.8.13(b). The fragmentized set is $\cup_{\kappa=1}^{m} \Xi_s^{(j_k)}$ and the corresponding mapping cluster is a mapping $P_{J_{j_{m+1}} \cdots J_{j_2} J_{j_1}}$ to make Eq.(8.93) hold, and the mappings in the cluster of grazing mappings are depicted through the dashed arrows. The fragmentized invariant sets are cycled by the dotted curves. Finally, the grazing-induced fragmentation of strange attractors of chaotic flows in dynamical systems is discussed. The mathematical theory for such a fragmentation of strange attractors should be further developed. The invariant sets on the separatrix surface also show the transversality of the flow on the separatrix, which is very important for one to understand the complexity of the global flow.

Chapter 9

Global Dynamics in Two-dimensional Dynamical Systems

In this Chapter, the first integral quantity increment (i.e., the energy increment) for a certain time interval will be derived for periodic flows and chaos in the 2-dimensional nonlinear dynamical systems. Under perturbation assumptions and convergent conditions, the Melnikov function will be derived from the first integral quantity increment. Further, a periodically forced, damped Duffing oscillator with a separatrix will be investigated as a sample problem. The switching planes and the local and global mappings will be defined on the separatrix. The mapping structures for local and global periodic flows passing through the separatrix will be discussed from Chapter 8. The mapping structures of chaos in the damped Duffing oscillator will be presented as well. Bifurcation scenarios of the damped Duffing oscillator will be given numerically through the Poincaré mapping section and the switching planes. The first integral quantity increment (i.e., L-function) will be presented to observe the periodicity of flows. The global tangency and transversality of periodic flows in such an oscillator will be measured by the G-function and $G^{(1)}$-function. Numerical simulations of chaos in such a Duffing oscillator will be carried out through the Poincare mapping sections. The conservative energy distribution, G-function and L-function along the displacement of Poincaré mapping points will be presented to observe the complexity of chaos. The switching planes of chaos will be presented on the separatrix for a better understanding of the global transversality of a perturbed flow to the separatrix. The switching point distribution on the separatrix will be presented and the switching G-function on the separatrix will be given to show the global transversality of chaos on the separatrix.

9.1. Tangency and transversality

The global, tangential flows to the separatrix of a perturbed Hamiltonian system in Eq.(3.52) were discussed. Based on this separatrix surface, the corresponding boundary $\partial\Omega_{ij}$ for domains Ω_i and Ω_j is defined. For the nonlinear Hamiltonian system in Eq.(3.52), the normal vector direction of the conservative energy surface (the first integral manifold surface) is abreast of the direction of increaseing conservative Hamiltonian of Eq.(3.52). Therefore, consider a flow $\mathbf{x}^{(\alpha)}(t) = (x^{(\alpha)}(t), y^{(\alpha)}(t))^T$ arriving at a specific integral manifold $S_0^{(\alpha)}$ with $H_0 = E_0$ at point $(x_m^{(\alpha)}, y_m^{(\alpha)})$ with time t_m. If the following conditions are satisfied,

$$\left.\begin{array}{l} H_0(x_m^{(\alpha)}, y_m^{(\alpha)}, \boldsymbol{\mu}) = E_0 \text{ and} \\ G_\alpha\left(x_m^{(\alpha)}, y_m^{(\alpha)}, t_m, \mathbf{p}\right) > 0 \quad \text{for } E_\alpha > E_0 \text{ or} \\ G_\alpha\left(x_m^{(\alpha)}, y_m^{(\alpha)}, t_m, \mathbf{p}\right) < 0 \quad \text{for } E_\alpha < E_0, \end{array}\right\} \tag{9.1}$$

then a flow $\mathbf{x}^{(\alpha)}(t) = (x^{(\alpha)}(t), y^{(\alpha)}(t))^T$ will pass through the specific manifold $S_0^{(\alpha)}$ from one side to another side of the manifold $S_0^{(\alpha)}$. For $H_0 = E_s$, the necessary and sufficient conditions for the *global* transversal flow to the separatrix boundary $\partial\Omega_{ij}$ for the perturbed Hamiltonian system in Eq.(3.52) is given by Eq.(3.62). Consider a flow $\mathbf{x}^{(\alpha)}(t) = (x^{(\alpha)}(t), y^{(\alpha)}(t))^T$ arriving at a specific integral manifold $S_0^{(\alpha)}$ with $H_0 = E_0$ at point $(x_m^{(\alpha)}, y_m^{(\alpha)})$ with time t_m. If the following conditions are satisfied,

$$\left.\begin{array}{l} H_0(x_m^{(\alpha)}, y_m^{(\alpha)}, \boldsymbol{\mu}) = E_0, \; G\left(x_m^{(\alpha)}, y_m^{(\alpha)}, t_m, \mathbf{p}\right) = 0 \text{ and} \\ G_\alpha^{(1)}\left(x_m^{(\alpha)}, y_m^{(\alpha)}, t_m, \mathbf{p}\right) < 0 \quad \text{for } E_\alpha < E_0 \text{ or} \\ G_\alpha^{(1)}\left(x_m^{(\alpha)}, y_m^{(\alpha)}, t_m, \mathbf{p}\right) > 0 \quad \text{for } E_\alpha > E_0, \end{array}\right\} \tag{9.2}$$

then the flow $\mathbf{x}^{(\alpha)}(t) = (x^{(\alpha)}(t), y^{(\alpha)}(t))^T$ will be tangential to the specific manifold $S_0^{(\alpha)}$ in one of the two sides of the manifold $S_0^{(\alpha)}$. Once a flow $\mathbf{x}^{(\alpha)}(t) = (x^{(\alpha)}(t), y^{(\alpha)}(t))^T$ ($\alpha \in \{i, j\}$) arrives to the separatrix $\partial\Omega_{ij}$, if conditions in Eq.(9.2) are satisfied, there is a global flow from the domain Ω_α into Ω_β passing through the separatrix boundary $\partial\Omega_{ij}$. The necessary and sufficient conditions for the *tangential* flow of the perturbed Hamiltonian system in Eq.(3.52) are given in Eq.(3.65). Further, the above conditions can be generalized. Consider a flow $\mathbf{x}^{(\alpha)}(t) = (x^{(\alpha)}(t), y^{(\alpha)}(t))^T$ to arrive to a specific integral manifold $S_0^{(\alpha)}$ with $H_0 = E_0$ at point $(x_m^{(\alpha)}, y_m^{(\alpha)})$ with time t_m. If the following conditions are satisfied,

$$\left.\begin{array}{l} H_0(x_m^{(\alpha)}, y_m^{(\alpha)}, \boldsymbol{\mu}) = E_0, \\ G_\alpha^{(r)}\left(x_m^{(\alpha)}, y_m^{(\alpha)}, t_m, \mathbf{p}\right) = 0 \text{ for } r = 1, 2, \cdots, 2l-1 \text{ and} \\ G_\alpha^{(2l)}\left(x_m^{(\alpha)}, y_m^{(\alpha)}, t_m, \mathbf{p}\right) > 0 \text{ for } E_\alpha < E_0 \text{ or} \\ G_\alpha^{(2l)}\left(x_m^{(\alpha)}, y_m^{(\alpha)}, t_m, \mathbf{p}\right) < 0 \text{ for } E_\alpha > E_0, \end{array}\right\} \tag{9.3}$$

then a flow $\mathbf{x}^{(\alpha)}(t)=(x^{(\alpha)}(t),y^{(\alpha)}(t))^T$ will be the $(2l)^{th}$-order, transversal flow to the specific manifold $S_0^{(\alpha)}$ from one of the two sides of the manifold $S_0^{(\alpha)}$. Again, if $H_0=E_s$, the necessary and sufficient conditions for the $(2l)^{th}$-order, *global* transversal flow to the separatrix boundary $\partial\Omega_{ij}$ for the perturbed Hamiltonian system in Eq.(3.52) is obtained. Consider a flow $\mathbf{x}^{(\alpha)}(t)=(x^{(\alpha)}(t),y^{(\alpha)}(t))^T$ arriving to a specific integral manifold $S_0^{(\alpha)}$ with $H_0=E_0$ at point $(x_m^{(\alpha)},y_m^{(\alpha)})$ with time t_m. If the following conditions are satisfied,

$$\left.\begin{aligned}&H_0(x_m^{(\alpha)},y_m^{(\alpha)},\boldsymbol{\mu})=E_0,\\&G_\alpha^{(r)}\left(x_m^{(\alpha)},y_m^{(\alpha)},t_m,\mathbf{p}\right)=0\text{ for }r=0,1,2,\cdots 2l-2\text{ and}\\&G_\alpha^{(2l-1)}\left(x_m^{(\alpha)},y_m^{(\alpha)},t_m,\mathbf{p}\right)<0\quad\text{for }E_\alpha<E_0\text{ or}\\&G_\alpha^{(2l-1)}\left(x_m^{(\alpha)},y_m^{(\alpha)},t_m,\mathbf{p}\right)>0\quad\text{for }E_\alpha>E_0,\end{aligned}\right\}\tag{9.4}$$

then a flow $\mathbf{x}^{(\alpha)}(t)=\left(x^{(\alpha)}(t),y^{(\alpha)}(t)\right)^T$ will be the $(2l-1)^{th}$-order tangential flow to the specific manifold $S_0^{(\alpha)}$ from one of the two sides of the manifold $S_0^{(\alpha)}$. If $H_0=E_s$, the necessary and sufficient conditions for the $(2l-1)^{th}$-order, global tangential flow $\mathbf{x}^{(\alpha)}(t)$ to the separatrix for a perturbed Hamiltonian system in Eq.(3.52) are achieved. In Chapter 3, the global transversality and tangency to the separatrix are discussed in detail.

9.2. Energy increment and Melnikov function

To investigate the periodic and chaotic flows in perturbed Hamiltonian system, the first integral quantity increment is used. Because the first integral quantity is the unperturbed Hamiltonian (i.e., $H_0(x,y,\boldsymbol{\mu})$), the first integral quantity increment is termed *an energy increment* (i.e., $L^{(\alpha)}(t_k,t_{k+1})\equiv\Delta H_0(\mathbf{x}_k^{(\alpha)},\mathbf{x}_{k+1}^{(\alpha)})$) in the perturbed Hamiltonian system where $\mathbf{x}_k^{(\alpha)}=(x_k^{(\alpha)},y_k^{(\alpha)})^{\mathrm{T}}$. The time-change rate of the unperturbed Hamiltonian energy is given by

$$\begin{aligned}\frac{d}{dt}H_0(x,y)&=\frac{\partial H_0(x,y)}{\partial x}\dot{x}+\frac{\partial H_0(x,y)}{\partial y}\dot{y}\\&=f_1g_2-f_2g_1=G(x,y,t,\mathbf{p}).\end{aligned}\tag{9.5}$$

As in Luo (2007b), the energy increment for time interval $[t_k,t_{k+1}]$ is

$$\begin{aligned}L^{(\alpha)}\left(t_k,t_{k+1}\right)&=\Delta H_0^{(\alpha)}\left(\mathbf{x}_k^{(\alpha)},\mathbf{x}_{k+1}^{(\alpha)}\right)\\&=\int_{t_k}^{t_{k+1}}G_\alpha\left(\mathbf{x}^{(\alpha)},t,\mathbf{p}\right)dt=\int_{t_k}^{t_{k+1}}\left(f_1\bullet g_2-f_2\bullet g_1\right)dt.\end{aligned}\tag{9.6}$$

However, the solution to Eq.(3.52) is given by

$$\left.\begin{aligned}&x^{(\alpha)}=\mathbf{\Phi}_x^{(\alpha)}\left(\mathbf{x}_k^{(\alpha)},t_k,t\right)\text{ and }y^{(\alpha)}=\mathbf{\Phi}_y^{(\alpha)}\left(\mathbf{x}_k^{(\alpha)},t_k,t\right);\\&\text{or }\mathbf{x}^{(\alpha)}=\mathbf{\Phi}^{(\alpha)}\left(\mathbf{x}_k^{(\alpha)},t_k,t\right).\end{aligned}\right\}\tag{9.7}$$

For a periodic flow of the dynamical system in Eq.(3.52), the energy increment should be zero (i.e., $\Delta H_0\left(\mathbf{x}_k,\mathbf{x}_{k+1}\right)=0$ or $L^{(\alpha)}\left(t_k,t_{k+1}\right)=0$). If the flow from time t_k to t_{k+1} is of period-1 and the time difference is given (i.e., $t_{k+1}-t_k=T$ $=\frac{2\pi}{\Omega}$), then the foregoing equation becomes

$$\left.\begin{aligned}&L^{(\alpha)}\left(t_k,T\right)=\Delta H_0^{(\alpha)}\left(\mathbf{x}_k^{(\alpha)},\mathbf{x}_{k+1}^{(\alpha)}\right)=0,\\&t_{k+1}=t_k+T,\\&\mathbf{x}_k^{(\alpha)}=\mathbf{x}_{k+1}^{(\alpha)}=\mathbf{\Phi}^{(\alpha)}\left(\mathbf{x}_k^{(\alpha)},t_k,t_k+T\right).\end{aligned}\right\}\tag{9.8}$$

with

$$\begin{aligned}\Delta H_0^{(\alpha)}\left(\mathbf{x}_k^{(\alpha)},\mathbf{x}_{k+1}^{(\alpha)}\right)=L^{(\alpha)}\left(t_k,T\right)&=\int_{t_k}^{t_k+T}G\left(x^{(\alpha)},y^{(\alpha)},t,\mathbf{p}\right)dt\\&=\int_{t_k}^{t_k+T}\left(f_1\bullet g_2-f_2\bullet g_1\right)dt.\end{aligned}\tag{9.9}$$

For a chaotic flow in Eq.(3.52), the corresponding iterative relation is given by

$$\left.\begin{aligned}&E_{k+1}-E_k=\Delta H_0^{(\alpha)}\left(\mathbf{x}_k^{(\alpha)},\mathbf{x}_{k+1}^{(\alpha)}\right)=L^{(\alpha)}\left(t_k,T\right),\\&t_{k+1}=t_k+T,\\&\mathbf{x}_{k+1}^{(\alpha)}=\mathbf{\Phi}^{(\alpha)}\left(\mathbf{x}_k^{(\alpha)},t_k,t_k+T\right).\end{aligned}\right\}\tag{9.10}$$

For a local period-n flow, the corresponding relations are given by

$$\left.\begin{aligned}&L^{(\alpha)}(t_k,t_k+iT)=\Delta H_0^{(\alpha)}\left(\mathbf{x}_k^{(\alpha)},\mathbf{x}_{k+i}^{(\alpha)}\right)\neq 0,i=0,1,\cdots n-1\\&\mathbf{x}_{k+i}^{(\alpha)}=\mathbf{\Phi}^{(\alpha)}\left(\mathbf{x}_k^{(\alpha)},t_k,t_k+iT\right);\\&L^{(\alpha)}(t_k,t_k+nT)=\Delta H_0^{(\alpha)}\left(\mathbf{x}_k^{(\alpha)},\mathbf{x}_{k+n}^{(\alpha)}\right)=0,t_{k+n}=t_k+nT,\\&\mathbf{x}_k^{(\alpha)}=\mathbf{x}_{k+n}^{(\alpha)}=\mathbf{\Phi}^{(\alpha)}\left(\mathbf{x}_k^{(\alpha)},t_k,t_k+nT\right).\end{aligned}\right\}\tag{9.11}$$

From the foregoing equations, the L-function of the period-n local flow has (n-1) -values at $t=t_k+iT$ ($i=1,2,\cdots,n-1$) before the L-function becomes zero at $t=t_k+nT$. If the L-function is non-zero for $n\to\infty$, then a local chaotic flow can be observed, which can be determined by the iterative relations in Eq.(9.10). If a global periodic flow in Eq.(3.52) consists of M-segments of orbit in N-domains (i.e., $\cup_{j=1}^{M}\mathbf{x}^{(\alpha_j)}(t)$) for $t\in[t_k,t_{k+1}]$ with M-sub-intervals $[t_{k_j},t_{k_j+1}]$ ($j=1,2,\cdots,M$), the total energy increment computed by the Poincaré section should be zero. In the other words, with $t_{k+1}=t_{k_M+1}$ and $t_k=t_{k_1}$, the following relations exist

$$\left.\begin{aligned}
&L^{(s)}(t_k,T)=\sum\nolimits_{j=1}^{M}L^{(\alpha_j)}\left(t_{k_j+1},t_{k_j}\right)=0,\\
&T=t_{k+1}-t_k=\sum\nolimits_{j=1}^{M}\left(t_{k_j+1}-t_{k_j+1}\right),\mathbf{x}_{k_M+1}^{(\alpha_1)}=\mathbf{x}_{k_1}^{(\alpha_1)}\\
&\mathbf{x}_{k_j+1}^{(\alpha_j)}=\mathbf{\Phi}^{(\alpha_j)}\left(\mathbf{x}_{k_j}^{(\alpha_j)},t_{k_j},t_{k_j+1}\right)
\end{aligned}\right\}\tag{9.12}$$

for $\alpha_j,\alpha_1\in\{1,2,\cdots,N\}$ and $j=1,2,\cdots,M$. If an intersected point (x_m,y_m) of the separatrix and a global flow in Eq.(3.52) satisfies the necessary and sufficient conditions as in Eq.(9.4) with $H_0=E_s$, then the $(2l-1)^{th}$-order, global grazing bifurcation of the global periodic flow occurs. For a global chaotic motion, we have the following relationship

$$\left.\begin{aligned}
&E_{k+1}^{(\alpha_l)}-E_k^{(\alpha_1)}=\Delta H_0(\mathbf{x}_{k_{M_k}+1}^{(\alpha_l)},\mathbf{x}_{k_1}^{(\alpha_1)})=L^{(s)}(t_k,t_k+T),\\
&T=t_{k+1}-t_k=\sum\nolimits_{j=1}^{M_k}\left(t_{k_j+1}-t_{k_j}\right)\\
&\mathbf{x}_{k_j+1}^{(\alpha_j)}=\mathbf{\Phi}^{(\alpha_j)}\left(\mathbf{x}_{k_j}^{(\alpha_j)},t_{k_j+1},t_{k_j}\right)
\end{aligned}\right\}\tag{9.13}$$

where M_k-piece segments of the global flow will form a complete flow for $t\in[t_k,t_{k+1}]$ and $k=0,1,2,\cdots$. For a global period-1 flow, $M_k=\text{const}\triangleq M$. Similarly, for a global period-n flow, the L-function satisfies the following relation as

$$\left.\begin{aligned}
&L^{(s)}(t_k,t_k+iT)=\Delta H_0(\mathbf{x}_{(k+i)_{M(k+i)}+1}^{(\alpha_{M(k+i)})},\mathbf{x}_{k_1}^{(\alpha_1)})\neq0,\ i=1,2,\cdots,n-1\\
&\mathbf{x}_{(k+i)_j+1}^{(\alpha_j)}=\mathbf{\Phi}^{(\alpha_j)}\left(\mathbf{x}_{(k+i)_j}^{(\alpha_j)},t_{(k+i)_j+1},t_{(k+i)_j}\right),j=1,2,\cdots M_{(k+i)}\\
&L^{(s)}(t_k,t_k+nT)=\Delta H_0(\mathbf{x}_{(k+n-1)_{M(k+n-1)}+1}^{(\alpha_l)},\mathbf{x}_{k_1}^{(\alpha_1)})=0\\
&t_{k+n}-t_k=nT=\sum\nolimits_{i=0}^{n-1}\sum\nolimits_{j=1}^{M_{(k+i)}}\left(t_{(k+i)_j+1}-t_{(k+i)_j}\right)\\
&\mathbf{x}_{(k+n-1)_j+1}^{(\alpha_j)}=\mathbf{\Phi}^{(\alpha_j)}\left(\mathbf{x}_{(k+n-1)_j}^{(\alpha_j)},t_{(k+n-1)_j+1},t_{(k+n-1)_j}\right),j=1,2,\cdots M_{(k+n-1)}
\end{aligned}\right\}\tag{9.14}$$

The time difference T_α can be the period of a periodic flow on a chosen reference surface $S_0^{(\alpha)}$ at $(\bar{x}_k, \bar{y}_k) = (x_k, y_k)$ for time t_k. The period T for the relations in Eqs.(9.6)-(9.12) will be changed to the period of the flow based on the selected reference surface. Consider a relation for a local, chaotic flow as an example. Using an external frequency Ω and a new phase angle $\varphi = \Omega t$, equation (9.10) becomes

$$\left.\begin{aligned} E_{k+1} - E_k &= \Delta H_0^{(\alpha)}\left(\mathbf{x}_k^{(\alpha)}, \mathbf{x}_{k+1}^{(\alpha)}\right) = L^{(\alpha)}\left(t_k, T_\alpha\right), \\ \varphi_{k+1} &= \varphi_k + \Omega T_\alpha, \\ \mathbf{x}_{k+1}^{(\alpha)} &= \mathbf{\Phi}^{(\alpha)}\left(\mathbf{x}_k^{(\alpha)}, \tfrac{\varphi_k}{\Omega}, \tfrac{\varphi_{k+1}}{\Omega}\right). \end{aligned}\right\} \tag{9.15}$$

From the above discussion, how to compute the energy increment (or the L-function) during the given time interval is very crucial. Using a transformation of $t = t_k + t'$ and $t_{k+1} = t_k + T_\alpha$, equation (9.9) gives

$$\begin{aligned} \Delta H_0^{(\alpha)}\left(t_k, T_\alpha\right) = L^{(\alpha)}\left(t_k, T_\alpha\right) &= \int_0^{T_\alpha} G\left(x, y, t_k + t, \mathbf{p}\right) dt \\ &= \int_0^{T_\alpha} \left[f_1\left(x, y, \mu\right) \cdot g_2\left(x, y, t_k + t, \pi\right) \right. \\ &\quad \left. - f_2\left(x, y, \mu\right) \cdot g_1\left(x, y, t_k + t, \pi\right) \right] dt. \end{aligned} \tag{9.16}$$

If the separatrix is selected as a reference curve, such an orbit connecting saddle points is used for integration. Therefore, we have $T_\alpha \to \infty$, and

$$\begin{aligned} \Delta H_0^{(s)}\left(t_k, \infty\right) = L^{(s)}\left(t_k, \infty\right) &= \lim_{T_\alpha \to \infty} \int_0^{T_\alpha} G\left(x, y, t_k + t, \mathbf{p}\right) dt \\ &= \lim_{T_\alpha \to \infty} \int_0^{T_\alpha} \left[f_1\left(x, y, \mu\right) \cdot g_2\left(x, y, t_k + t, \pi\right) \right. \\ &\quad \left. - f_2\left(x, y, \mu\right) \cdot g_1\left(x, y, t_k + t, \pi\right) \right] dt \end{aligned} \tag{9.17}$$

Consider $g_\sigma(x, y, t, \boldsymbol{\pi}) \equiv \varepsilon g_\sigma(x, y, t, \boldsymbol{\pi})$ ($\sigma = 1, 2$) with small $\varepsilon > 0$. From the perturbbation analysis, if a non-integrable flow $(x^{(\alpha)}(t - t_k), y^{(\alpha)}(t - t_k))$ is very close to the integrable $(\bar{x}_{\sigma_\alpha}^{(\alpha)}(t - t_k), \bar{y}_{\sigma_\alpha}^{(\alpha)}(t - t_k))$ or $(\bar{x}_{\sigma_\alpha}^{(s)}(t - t_k), \bar{y}_{\sigma_\alpha}^{(s)}(t - t_k))$ for $t \in [t_k, t_k + T_\alpha]$ with $T_\alpha < \infty$, then the following series exist,

$$\left.\begin{aligned} x^{(\alpha)}(t - t_k) &\approx \bar{x}_{\sigma_\alpha}^{(\alpha)}(t - t_k) + \sum_{l=1}^{\infty} \varepsilon^l x_{\sigma_\alpha}^{(l,\alpha)}(t - t_k), \\ y^{(\alpha)}(t - t_k) &\approx \bar{y}_{\sigma_\alpha}^{(\alpha)}(t - t_k) + \sum_{l=1}^{\infty} \varepsilon^l y_{\sigma_\alpha}^{(l,\alpha)}(t - t_k); \end{aligned}\right\} \tag{9.18}$$

or

$$\left.\begin{aligned} x^{(\alpha)}(t-t_k) &\approx \overline{x}^{(s)}(t-t_k)+\sum\nolimits_{l=1}^{\infty}\varepsilon^l x_{\sigma_\alpha}^{(l,s)}(t-t_k),\\ x^{(\alpha)}(t-t_k) &\approx \overline{x}^{(s)}(t-t_k)+\sum\nolimits_{l=1}^{\infty}\varepsilon^l x_{\sigma_\alpha}^{(l,s)}(t-t_k).\end{aligned}\right\} \tag{9.19}$$

From the above Taylor series expression, we have

$$\begin{aligned} G_{\sigma_\alpha}(x^{(\alpha)},y^{(\alpha)},t,\mathbf{p}) &= G_{\sigma_\alpha}(\overline{x}^{(\alpha)},\overline{y}^{(\alpha)},t,\mathbf{p})\\ &+\sum\nolimits_{k=1}^{\infty}\varepsilon^k \tfrac{\partial^k}{\partial x^k}G_{\sigma_\alpha}(x^{(\alpha)},y^{(\alpha)},t,\mathbf{p})\Big|_{(\overline{x}^{(\alpha)},\overline{y}^{(\alpha)})}\left(\Sigma_{j=1}^{\infty}\varepsilon^{j-1}x^{(j,\alpha)}\right)^k\\ &+\sum\nolimits_{k=1}^{\infty}\varepsilon^k \tfrac{\partial^k}{\partial y^k}G_{\sigma_\alpha}(x^{(\alpha)},y^{(\alpha)},t,\mathbf{p})\Big|_{(\overline{x}^{(\alpha)},\overline{y}^{(\alpha)})}\left(\Sigma_{j=1}^{\infty}\varepsilon^{j-1}y^{(j,\alpha)}\right)^k, \end{aligned} \tag{9.20}$$

$$\begin{aligned} G_{\sigma_\alpha}(x^{(\alpha)},y^{(\alpha)},t,\mathbf{p}) &= G_{s}(\overline{x}^{(s)},\overline{y}^{(s)},t,\mathbf{p})\\ &+\sum\nolimits_{k=1}^{\infty}\varepsilon^k \tfrac{\partial^k}{\partial x^k}G_{\sigma_\alpha}(x^{(\alpha)},y^{(\alpha)},t,\mathbf{p})\Big|_{(\overline{x}^{(s)},\overline{y}^{(s)})}\left(\Sigma_{j=1}^{\infty}\varepsilon^{j-1}x^{(j,s)}\right)^k\\ &+\sum\nolimits_{k=1}^{\infty}\varepsilon^k \tfrac{\partial^k}{\partial y^k}G_{\sigma_\alpha}(x^{(\alpha)},y^{(\alpha)},t,\mathbf{p})\Big|_{(\overline{x}^{(s)},\overline{y}^{(s)})}\left(\Sigma_{j=1}^{\infty}\varepsilon^{j-1}y^{(j,s)}\right)^k. \end{aligned} \tag{9.21}$$

The convergent conditions of the above expressions are

$$\left.\begin{aligned} &\left|\tfrac{\partial^k}{\partial x^k}G_{\sigma_\alpha}(x^{(\alpha)},y^{(\alpha)},t,\mathbf{p})\right|+\left|\tfrac{\partial^k}{\partial y^k}G_{\sigma_\alpha}(x^{(\alpha)},y^{(\alpha)},t,\mathbf{p})\right| < K_k\\ &\frac{\varepsilon\left[\left|\tfrac{\partial^{k+1}}{\partial x^{k+1}}G_{\sigma_\alpha}(x^{(\alpha)},y^{(\alpha)},t,\mathbf{p})\right|+\left|\tfrac{\partial^{k+1}}{\partial y^{k+1}}G_{\sigma_\alpha}(x^{(\alpha)},y^{(\alpha)},t,\mathbf{p})\right|\right]}{\left|\tfrac{\partial^k}{\partial x^k}G_{\sigma_\alpha}(x^{(\alpha)},y^{(\alpha)},t,\mathbf{p})\right|+\left|\tfrac{\partial^k}{\partial y^k}G_{\sigma_\alpha}(x^{(\alpha)},y^{(\alpha)},t,\mathbf{p})\right|}<1 \end{aligned}\right\} \tag{9.22}$$

and

$$\left.\begin{aligned} &\left|\tfrac{\partial^k}{\partial x^k}G_{\sigma_\alpha}(x^{(s)},y^{(s)},t,\mathbf{p})\right|+\left|\tfrac{\partial^k}{\partial y^k}G_{\sigma_\alpha}(x^{(s)},y^{(s)},t,\mathbf{p})\right| < K_k,\\ &\frac{\varepsilon\left[\left|\tfrac{\partial^{k+1}}{\partial x^{k+1}}G_{\sigma_\alpha}(x^{(s)},y^{(s)},t,\mathbf{p})\right|+\left|\tfrac{\partial^{k+1}}{\partial y^{k+1}}G_{\sigma_\alpha}(x^{(s)},y^{(s)},t,\mathbf{p})\right|\right]}{\left|\tfrac{\partial^k}{\partial x^k}G_{\sigma_\alpha}(x^{(s)},y^{(s)},t,\mathbf{p})\right|+\left|\tfrac{\partial^k}{\partial y^k}G_{\sigma_\alpha}(x^{(s)},y^{(s)},t,\mathbf{p})\right|}<1. \end{aligned}\right\} \tag{9.23}$$

all $k=1,2,\cdots$. As $\varepsilon\to 0$, an approximate first integral quantity increment is

$$\begin{aligned} L^{(\alpha)}(t_k,T_\alpha) \approx \int_0^{T_\alpha}\Big\{ & f_1\left(\overline{x}^{(\alpha)},\overline{y}^{(\alpha)},\boldsymbol{\mu}\right)\boldsymbol{\cdot} g_2\left(\overline{x}^{(\alpha)},\overline{y}^{(\alpha)},t_k+t,\boldsymbol{\pi}\right)\\ & -f_2\left(\overline{x}^{(\alpha)},\overline{y}^{(\alpha)},\boldsymbol{\mu}\right)\boldsymbol{\cdot} g_1\left(\overline{x}^{(\alpha)},\overline{y}^{(\alpha)},t_k+t,\boldsymbol{\pi}\right)\Big\}\,dt \end{aligned} \tag{9.24}$$

and

$$\begin{aligned} L^{(s)}\left(t_k,\infty\right) \approx \lim_{T_\alpha \to \infty} \int_0^{T_\alpha} \Big\{ & f_1\left(\overline{x}^{(\alpha)},\overline{y}^{(\alpha)},\boldsymbol{\mu}\right)\bullet g_2\left(\overline{x}^{(\alpha)},\overline{y}^{(\alpha)},t_k+t,\boldsymbol{\pi}\right) \\ & -f_2\left(\overline{x}^{(\alpha)},\overline{y}^{(\alpha)},\boldsymbol{\mu}\right)\bullet g_1\left(\overline{x}^{(\alpha)},\overline{y}^{(\alpha)},t_k+t,\boldsymbol{\pi}\right)\Big\}\, dt. \end{aligned} \tag{9.25}$$

If the period of an unperturbed periodic orbit in domain Ω_α satisfies $T_\alpha = mT/n$ with m and n relatively prime, where $T = 2\pi/\Omega$ is the period of the perturbation, then equation (9.24) gives

$$\begin{aligned} L^{(\alpha)}\left(t_k,\tfrac{m}{n}T\right) \approx \int_0^{\frac{m}{n}T} \Big\{ & f_1\left(\overline{x}^{(m_\alpha:n_\alpha)},\overline{y}^{(m_\alpha:n_\alpha)},\boldsymbol{\mu}\right)\bullet g_2\left(\overline{x}^{(m_\alpha:n_\alpha)},\overline{y}^{(m_\alpha:n_\alpha)},t_k+t,\boldsymbol{\pi}\right) \\ & -f_2\left(\overline{x}^{(m_\alpha:n_\alpha)},\overline{y}^{(m_\alpha:n_\alpha)},\boldsymbol{\mu}\right)\bullet g_1\left(\overline{x}^{(m_\alpha:n_\alpha)},\overline{y}^{(m_\alpha:n_\alpha)},t_k+t,\boldsymbol{\pi}\right)\Big\}\, dt. \end{aligned} \tag{9.26}$$

Equations (9.25) and (9.26) give the Melnikov functions for the separatrix (i.e., homoclinic or heteroclinic orbit) and subharmonic orbits in the current existing theory, respectively. Under conditions in Eqs.(9.22) and (9.23), equations (9.25)-(9.26) cannot be used to determine the flow complexity characteristics of the perturbed Hamiltonian system in Eq.(3.52), such as chaos and the transversality of flow to the separatrix surface. Equations (9.25)-(9.26) *only* can be used for an approximate estimate of the energy increment (e.g., see, Luo,1995; Luo and Han, 2001).

9.3. Mapping structures

In this section, consider a periodically forced, damped Duffing oscillator with a twin-well potential in Eq.(3.66) as an example to show the methodology. The global periodic flows and chaos in the damped Duffing oscillator will be discuss-ed. The traditional Poincare mapping section and the Poincace mapping are introduced. The switching planes and the corresponding mappings are defined on the separatrix. The periodic motion is described through the mappings defined on the switching phase. The functions $L^{(\alpha)}(t_k,T_\alpha)$ $(\alpha=1,2,3)$ and $L^{(s)}(t_k,T_s)$ are important for local and global periodic motions in Eq.(3.66), respectively. Consider a local flow $\gamma_\alpha : (x^{(\alpha)}(t), y^{(\alpha)}(t))$ for $t\in[t_k,t_{k+1}]$, so the corresponding L-functions are computed by

$$\begin{aligned} L^{(\alpha)}\left(t_k,t_{k+1}\right) = \Delta H^{(\alpha)}\left(t_k,t_{k+1}\right) &= \int_{t_k}^{t_{k+1}} G\left(x^{(\alpha)},y^{(\alpha)},t,\mathbf{p}\right)dt \\ &= \int_{t_k}^{t_{k+1}} y^{(\alpha)}\left(Q_0\cos\Omega t-\delta\, y^{(\alpha)}\right)dt \end{aligned} \tag{9.27}$$

for the local flow and

$$
\begin{aligned}
L^{(s)}\left(t_k, t_{k+1}\right) &= \sum_{j_k=1}^{M_k} \Delta H^{\left(\alpha_{jk}\right)}\left(t_{k_{j_k}}, t_{k_{j_k}+1}\right) \\
&= \sum_{j_k=1}^{M_k} \int_{t_{k_{jk}}}^{t_{k_{jk}+1}} G\left(x^{\left(\alpha_{jk}\right)}, y^{\left(\alpha_{jk}\right)}, t, \mathbf{p}\right) dt \\
&= \sum_{j_k=1}^{M_k} \int_{t_{k_{jk}}}^{t_{k_{jk}+1}} y^{\left(\alpha_{jk}\right)}\left(Q_0 \cos \Omega t - \delta y^{\left(\alpha_{jk}\right)}\right) dt
\end{aligned}
\tag{9.28}
$$

for the global flow, and $\alpha_{jk} \in \{1,2,3\}$.

Introduce the traditional Poincare mapping section as follows

$$
\Sigma = \left\{ (x_k, y_k) \middle| t_k = t_0 + kT, k = 0,1,2,\cdots \right\}. \tag{9.29}
$$

The Poincare mapping is defined as

$$
P : \Sigma \to \Sigma \tag{9.30}
$$

from which we have

$$
\mathbf{x}_{k+1} = P\mathbf{x}_k. \tag{9.31}
$$

For a period-1 flow of system in Eq.(3.66), the periodicity conditions $\mathbf{x}_{k+1} = \mathbf{x}_k$ and $t_{k+1} = t_k + T$ are required. Therefore, from the traditional Poincare section, the time t_{k+1} of the L-function in Eq.(9.27) or (9.28) is $t_{k+1} = t_k + T$. For a period-n flow, the mapping relation is given by

$$
\mathbf{x}_{k+n} = P^{(n)}\mathbf{x}_k = P \circ P^{(n-1)}\mathbf{x}_k \tag{9.32}
$$

where $P^{(n)} = P \circ P^{(n-1)}$ and $P^{(0)} = 1$. If the corresponding periodicity condition $\mathbf{x}_{k+n} = \mathbf{x}_k$ and $t_{k+n} = t_k + nT$ is satisfied, the upper integration limit t_{k+1} of the L-function is changed to t_{k+n}, and $L(t_k, t_{k+n})$ can be computed. The L-functions (i.e., $L(t_k, t_{k+i})$ with $i = 1,2,\cdots,n-1$) are computed from Eq.(9.27) or Eq.(9.28). Once the L-function is computed, and if the relations in Eqs.(9.8)-(9.14) are satisfied, the periodic and chaotic flows will be achieved. In addition, the property of the Poincare points (i.e., (x_k, y_k) for $k = 0,1,2,\cdots$) to the corresponding first integral invariant manifold of Eq.(3.66) is very important for one to better understand the periodic and chaotic flows. However, the first integral invariant manifold of the Poincare point (x_k, y_k) can be computed by

$$
H_0(x_k, y_k, \boldsymbol{\mu}) = E_0. \tag{9.33}
$$

To this invariant manifold, the G-function at the Poincare mapping point can be given, which is very important to understand the flow property at the Poincare mapping point. Further, for the local and global flows in Eq.(3.66), the corresponding transversal and tangential conditions of the Poincare points to the first invariant integral manifolds are given by Eqs.(9.1), (3.62), (9.2) and (3.65).

To further investigate the transversality and tangency of the flow to a specific, first integral manifold $S_0^{(\alpha)}$, such a manifold will be considered as a Poincare section, i.e.,

$$\Xi_{E_0^{(\alpha)}} = \left\{ (\text{mod}(\Omega t_k, 2\pi), x_k, y_k) \middle| H_0(x_k, y_k, \boldsymbol{\mu}) = E_0^{(\alpha)} \right\} \subset S_0^{(\alpha)}. \tag{9.34}$$

Once the first integral manifold $S_0^{(\alpha)}$ is chosen as a reference surface, the domain Ω_α will be divided into the two parts, i.e., $\Omega_\alpha = \Omega_\alpha^{(+)} \cup \Omega_\alpha^{(-)}$. Note that $\Omega_\alpha^{(+)}$ is the domain pointed by the normal vector of the manifold $S_0^{(\alpha)}$, the rest domain is $\Omega_\alpha^{(-)}$. In the domain $\Omega_\alpha^{(+)}$, the relation $E_\alpha > E_0^{(\alpha)}$ holds, and in the domain $\Omega_\alpha^{(-)}$, $E_\alpha < E_0^{(\alpha)}$. Again, the transversality and tangency of the local flow to such a reference manifold are given in Eqs.(9.1) and (9.2). For the local periodic flow, the detailed discussion can be referred to the Chapter 8. Herein, a simple local periodic flow to the invariant integral manifold is discussed for a better understanding. We have to introduce the two subsets

$$\Xi_{E_0^{(\alpha)}}^{(+)} = \left\{ (\text{mod}(\Omega t_k, 2\pi), x_k, y_k) \middle| \begin{array}{l} H_0(x_k, y_k, \boldsymbol{\mu}) = E_0^{(\alpha)}, \\ H_0(x, y, \boldsymbol{\mu}) > E_0^{(\alpha)} \\ \text{for } t \in (t_k, t_{k+1}) \end{array} \right\} \subset S_0^{(\alpha)}, \tag{9.35}$$

$$\Xi_{E_0^{(\alpha)}}^{(-)} = \left\{ (\text{mod}(\Omega t_k, 2\pi), x_k, y_k) \middle| \begin{array}{l} H_0(x_k, y_k, \boldsymbol{\mu}) = E_0^{(\alpha)}, \\ H_0(x, y, \boldsymbol{\mu}) < E_0^{(\alpha)} \\ \text{for } t \in (t_k, t_{k+1}) \end{array} \right\} \subset S_0^{(\alpha)}. \tag{9.36}$$

The two mappings are introduced as

$${}^\alpha P_1 : \Xi_{E_0^{(\alpha)}}^{(+)} \to \Xi_{E_0^{(\alpha)}}^{(-)} \text{ and } {}^\alpha P_2 : \Xi_{E_0^{(\alpha)}}^{(-)} \to \Xi_{E_0^{(\alpha)}}^{(+)}. \tag{9.37}$$

Define a new vector

$$\mathbf{y}_k = (\text{mod}(\Omega t_k, 2\pi), x_k, y_k)^{\mathrm{T}}. \tag{9.38}$$

From the above mappings, the Poincare mapping of the local flow is:

$$P = {}^\alpha P_2 \circ {}^\alpha P_1 : \Xi_{E_0^{(\alpha)}}^{(+)} \to \Xi_{E_0^{(\alpha)}}^{(-)}. \tag{9.39}$$

So the mapping relations are

$$\mathbf{y}_{k+2} = P\mathbf{y}_k \text{ and } H_0(x_{k+i}, y_{k+i}, \boldsymbol{\mu}) = E_0^{(\alpha)} \text{ for } i = 0,1,2. \tag{9.40}$$

For a period-1 flow with N- external periods, the periodicity conditions are

$$\mathbf{x}_{k+2} = \mathbf{x}_k \text{ and } t_{k+2} = t_k + NT. \tag{9.41}$$

Introduce the following mapping notation

$$P^{(n)} = \underbrace{{}^{\alpha}P_2 \circ {}^{\alpha}P_1 \circ \cdots \circ {}^{\alpha}P_2 \circ {}^{\alpha}P_1}_{n-sets} \equiv {}^{\alpha}P_{(21)^n}. \tag{9.42}$$

For a period-n flow with N-external periods, its mapping structure is

$$P^{(n)} = {}^{\alpha}P_{(21)^n} : \Xi^{(+)}_{E_0^{(\alpha)}} \to \Xi^{(-)}_{E_0^{(\alpha)}} \tag{9.43}$$

with the periodicity is

$$\mathbf{x}_{k+2n} = \mathbf{x}_k \text{ and } t_{k+2n} = t_k + NT. \tag{9.44}$$

As $n \to \infty$, the local chaotic flow in domain Ω_α can be investigated. The detailed discussion can be referred to Chapter 8.

To investigate the transversality and tangency of a global flow to the separatrix S, such a manifold is considered as a Poincare section, i.e.,

$$\Xi_s = \left\{(\mathrm{mod}(\Omega t_k, 2\pi), x_k, y_k) \middle| H_0(x_k, y_k, \boldsymbol{\mu}) = E_s = 0\right\} \subset S. \tag{9.45}$$

The separatrix S of the dynamical system in Eq.(3.66) divides the phase domain into three sub-domains Ω_α ($\alpha = 1,2,3$). Therefore the subsets are introduced as

$$\Xi_{\alpha\beta} = \left\{ (\mathrm{mod}(\Omega t_k, 2\pi), x_k, y_k) \middle| \begin{array}{l} H_0(x_k, y_k, \boldsymbol{\mu}) = E_s = 0, \\ \boldsymbol{\Phi}(x_k, y_k, t_k, t) \subset \Omega_\beta \\ \text{for } t \in (t_k, t_{k+1}) \end{array} \right\} \subset S \tag{9.46}$$

for $\alpha, \beta = \{1,2,3\}$. Because Ω_1 and Ω_2 do not have the common boundary on the separatrix, only four subsets are expressed by Ξ_{13}, Ξ_{31}, Ξ_{23} and Ξ_{32}. Four local mappings relative to the separatrix are:

$$\begin{array}{l} {}^{s}P_1 : \Xi_{31} \to \Xi_{13}, \ {}^{s}P_2 : \Xi_{13} \to \Xi_{31}, \\ {}^{s}P_4 : \Xi_{32} \to \Xi_{23}, \ {}^{s}P_5 : \Xi_{23} \to \Xi_{32}; \end{array} \tag{9.47}$$

and two global mappings relative to the separatrix are

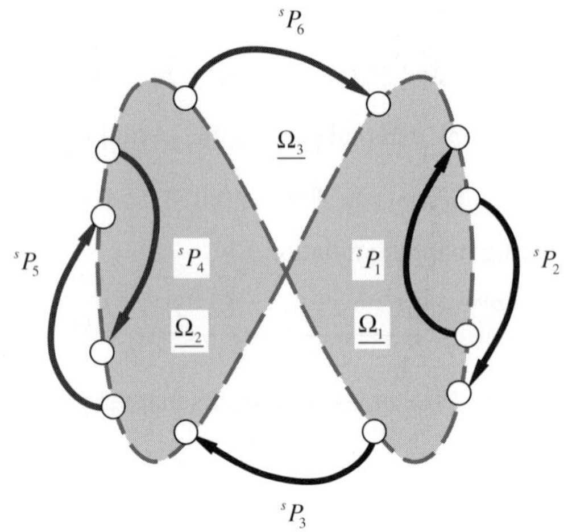

Figure 9.1. Generic mappings relative to the separatrix.

$$ {}^sP_3 : \Xi_{13} \to \Xi_{32},\ {}^sP_6 : \Xi_{23} \to \Xi_{31}. \tag{9.48}$$

The basic local and global mappings are illustrated in Fig.9.1. For a global flow to the separatrix, the separatrix is selected as a reference surface. The periodic flow is based on the two local mappings, and the Poincaré mapping for the simplest periodic flow based on the two local mappings is defined as follows:

$$\left.\begin{array}{l} {}^sP_{21} = {}^sP_2 \circ {}^sP_1 : \Xi_{31} \to \Xi_{31} \text{ or } {}^sP_{12} = {}^sP_1 \circ {}^sP_2 : \Xi_{13} \to \Xi_{13} \\ {}^sP_{54} = {}^sP_5 \circ {}^sP_4 : \Xi_{32} \to \Xi_{32} \text{ or } {}^sP_{45} = {}^sP_4 \circ {}^sP_5 : \Xi_{23} \to \Xi_{23} \end{array}\right\} \tag{9.49}$$

For mappings ${}^sP_{12}$ and ${}^sP_{21}$, they present the same periodic flow with different switching sets. Similarly, the asymmetric periodic flow of mappings ${}^sP_{12}$ and ${}^sP_{21}$ are given by mappings ${}^sP_{45}$ and ${}^sP_{54}$. So one of the two mapping structures is chosen to investigated the periodic flow. The mapping relations for the two asymmetric periodic flows are

$$\left.\begin{array}{l} \mathbf{y}_{k+2} = {}^sP_{12}\mathbf{y}_k \text{ and } H_0(x_{k+i}, y_{k+i}, \boldsymbol{\mu}) = E_s \text{ for } i = 0,1,2 \\ \mathbf{y}_{k+2} = {}^sP_{45}\mathbf{y}_k \text{ and } H_0(x_{k+i}, y_{k+i}, \boldsymbol{\mu}) = E_s \text{ for } i = 0,1,2 \end{array}\right\} \tag{9.50}$$

and the periodicity of a periodic flow with N-external periods requires the conditions in Eq.(9.41). Using the alike fashion in Eq.(9.42), the mapping structure for a period-n flow with N-external periods is

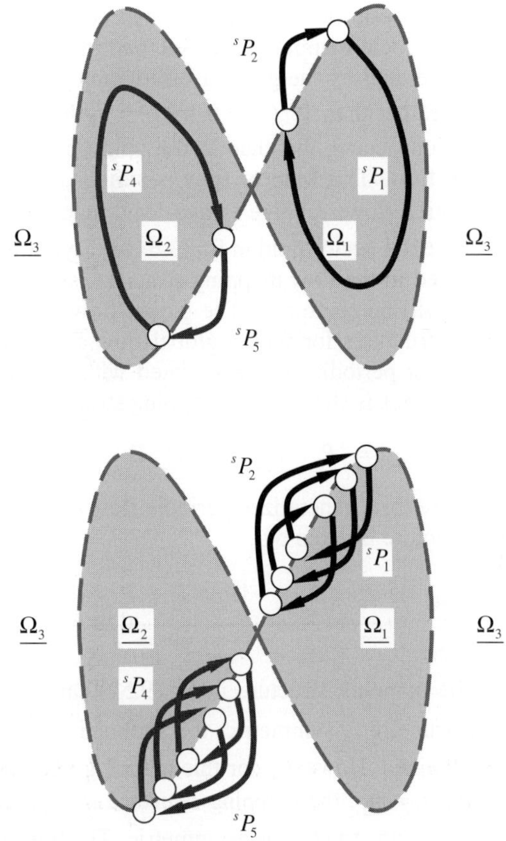

(a) (b)

Figure 9.2. Global periodic flows with: (a) two local mappings, (b) n-pairs local mappings.

$$\begin{aligned}
{}^sP_{(12)^n} &= \underbrace{{}^sP_1 \circ {}^sP_2 \circ \cdots \circ {}^sP_1 \circ {}^sP_2}_{n-\text{pairs}} : \Xi_{13} \to \Xi_{13}, \\
{}^sP_{(45)^n} &= \underbrace{{}^sP_4 \circ {}^sP_5 \circ \cdots \circ {}^sP_4 \circ {}^sP_5}_{n-\text{pairs}} : \Xi_{23} \to \Xi_{23}.
\end{aligned} \tag{9.51}$$

The periodicity for such a periodic flow is similar to Eq.(9.44). As $n \to \infty$, the chaotic flow on the separatrix is obtained. The strange attractors of the two chaotic flows are asymmetric. Such flows relative to local mapping on the separatrix are illustrated in Fig.9.2. In Fig.9.2(a), the simplest periodic flow with two local mappings is presented. The periodic flow is relative to two sub-domains. If the period-doubling of the periodic flow pertaining to the aforementioned mapping structure occurs, the chaotic flow can be obtained. Thus, the mapping structure for n-pairs of the two local mappings are sketched in Fig.9.2(b). This mapping structure may be generated by the grazing bifurcation. So the number n can be any natural number rather than 2^m ($m = 0, 1, 2, \cdots$) in the period-doubling. Once the grazing bifurcation occurs, the chaotic flow may be formed because of the discontinuity of the flow. This kind of flow is similar to the flow in the single domain (i.e., Ω_α). Such flows relative to two local mappings are asymmetric to the separatrix because the vector field in Eq.(3.64) is asymmetric. Once the global mapping is involved in the flow mapping structure, such two flows relative to the local mappings will be combined. The global flow will become more complex. The mapping structures for such a global flow are sketched in Fig.9.3. In Fig.9.3(a), the simplest periodic flow associated with the mapping structure of local and global mappings is shown. The mapping structure is expressed by

$$ {}^sP_{6431} = {}^sP_6 \circ {}^sP_4 \circ {}^sP_3 \circ {}^sP_1 : \Xi_{31} \to \Xi_{31}. \tag{9.52} $$

The mapping structure for a generalized periodic flow is illustrated in Fig.9.3(b) and the mapping structure is given by

$$ {}^sP_{6(45)^{m_2}43(12)^{m_1}1} = \underbrace{{}^sP_6 \circ {}^sP_{(45)^{m_2}} \circ {}^sP_4 \circ {}^sP_3 \circ {}^sP_{(12)^{m_1}} \circ {}^sP_1}_{(m_2+m_1+4)-\text{mappings}} : \Xi_{31} \to \Xi_{31}. \tag{9.53} $$

For $m_1 = m_2 = 0$, the mapping structure in Eq.(9.52) can be recovered. If $m_1 = m_2$, the mapping structure is symmetric, the symmetric flow for the system in Eq.(3.66) can be obtained. However, for this mapping structure, the asymmetric flows also exist. If $m_1 \neq m_2$, the mapping structure is asymmetric and the flow relative to such a mapping structure is asymmetric. The transition from a periodic flow pertaining to a specific mapping structure to another one is completed through a global grazing bifurcation. The global grazing bifurcation may cause a chaotic flow because flows in different domains possess different properties. After period doubling, the corresponding mapping structure may be invariant. Namely, after the period-doubling of a flow relative to ${}^sP_{6(45)^{m_2}43(12)^{m_1}1}$ gives the mapping structures as

$$ \left.\begin{aligned} {}^sP_{(6(45)^{m_2}43(12)^{m_1}1)^2} &= {}^sP_{6(45)^{m_2}43(12)^{m_1}1} \circ {}^sP_{6(45)^{m_2}43(12)^{m_1}1}, \\ {}^sP_{(6(45)^{m_2}43(12)^{m_1}1)^4} &= {}^sP_{(6(45)^{m_2}43(12)^{m_1}1)^2} \circ {}^sP_{(6(45)^{m_2}43(12)^{m_1}1)^2}, \\ &\vdots \\ {}^sP_{(6(45)^{m_2}43(12)^{m_1}1)^{2^m}} &= {}^sP_{(6(45)^{m_2}43(12)^{m_1}1)^{2^{m-1}}} \circ {}^sP_{(6(45)^{m_2}43(12)^{m_1}1)^{2^{m-1}}}. \end{aligned}\right\} \tag{9.54} $$

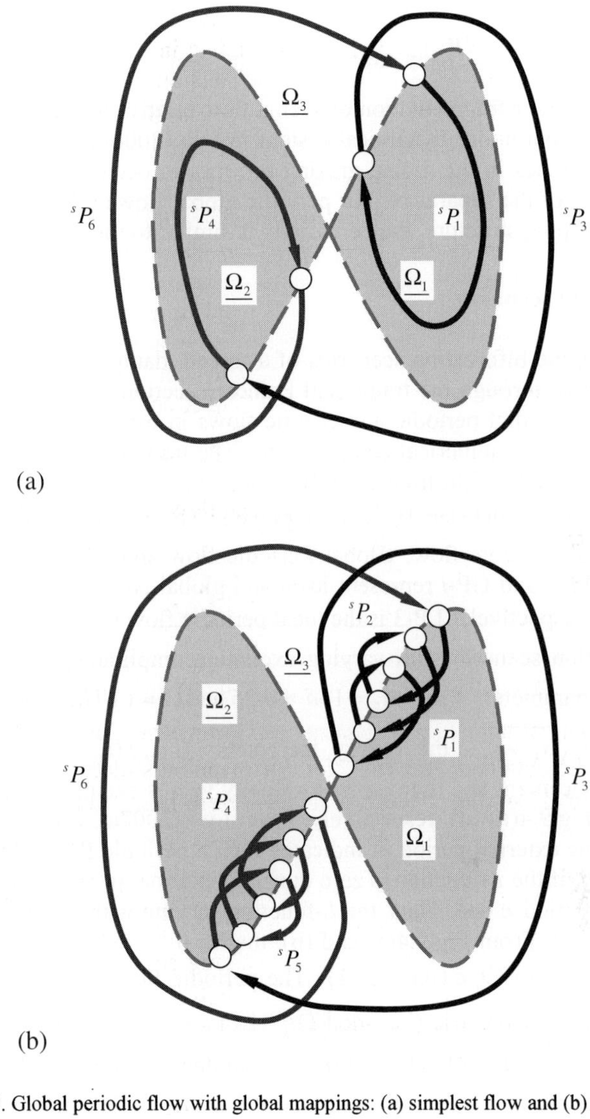

Figure 9.3. Global periodic flow with global mappings: (a) simplest flow and (b) complex flow.

However, the grazing bifurcation of the flow relative to ${}^{s}P_{6(45)^{m_2}43(12)^{m_1}1}$ gives the mapping structures as

$$\begin{aligned} {}^{S}P_{6(45)^{m_2}43(12)^{m_1}1} &\to {}^{S}P_{6(45)^{m_2}43(12)^{m_1+1}1}, \\ {}^{S}P_{6(45)^{m_2}43(12)^{m_1}1} &\to {}^{S}P_{6(45)^{m_2+1}43(12)^{m_1}1}. \end{aligned} \tag{9.55}$$

The strange attractor fragmentation of such a flow relative to the separatrix is similar to the discontinuous dynamical system in Luo (2006a). The complexity of the chaotic flow lies in the global grazing bifurcation occurring at the non-zero measure subset of the separatrix. For periodic global flows relative to the aforementioned mapping structure, the periodicity is similar to the ones in Eq.(9.44).

9.4. Bifurcation scenario

In this section, the bifurcation scenarios of a forced, damped, Duffing oscillator will be presented through the traditional Poincaré section and switching planes. To numerically predict periodic and chaotic flows in Eq. (3.66), the symplectic algorithm is used for numerical computations. The traditional Poincaré mapping section will be used to illustrate the bifurcation scenarios. Consider parameters $\alpha_1 = \alpha_2 = 1$ and the time step $10^{-4}T$ is employed. The acronyms LP, GP and GC represent *L*ocal *P*eriodic flow, *G*lobal *P*eriodic flow and *G*lobal *C*haotic flow, respectively. LP-*i* and GP-*i* represent local and global periodic flows in domain Ω_i ($i = 1,2$), respectively. LP-3 is the local periodic flow in domain Ω_3.

A bifurcation scenario with varying excitation amplitude Q_0 is presented in Fig.9.4 with parameters ($\alpha_1 = \alpha_2 = 1$, $\delta = 0.25$ and $\Omega = 1$).The response vector amplitude $A_k = \sqrt{x_k^2 + y_k^2}$, the energy increments (i.e., $L(t_0, kT)$-function), the response displacement and the response velocity versus excitation amplitude are illustrated in Fig.9.4(a)-(d), respectively. From Luo (2007a), if the L-function is zero during one external period, it indicates such a periodic flow is the period-1 one. Generally, if the L-function is zero during n-external periods, such a periodic flow is the period-n one. Thus, the L-function varying with external amplitude is very important. From Fig.9.4(a) and (b), the local period-1 flows in domains Ω_i ($i = 1,2$) exist for $Q_0 \in (0.0, 0.264)$. The periodic flows in the two domains Ω_i ($i = 1,2$) are asymmetric (i.e., $\mathrm{mod}(\Omega t_k^{(1)}, 2\pi) = \mathrm{mod}(\Omega t_k^{(2)}, 2\pi) + \pi$, and $\mathbf{x}_k^{(1)} = -\mathbf{x}_k^{(2)}$). However, for $Q_0 \in (0.159, 0.167)$, another local period-1 flow with a large orbit in domain Ω_i ($i = 1,2$) is observed. For $Q_0 \approx 0.167$, the grazing bifurcation occurs. The global flow in both the domains Ω_i ($i = 1,2$) and domain Ω_3 appears for $Q_0 \in (0.167, 0.191)$. Such a global flow relative to the domains Ω_i ($i = 1,2$) is denoted by the acronym "GP-*i*". The global period-1 flow relative to the two domains Ω_1 (or Ω_2) and Ω_3 exists for $Q_0 \in (0.167, 0.182)$. Due to period-doubling bifurcation at $Q_0 \approx 0.182$, the global period-2 flow for the two domains lies in $Q_0 \in (0.182, 0.187)$. The period-doubling bifurcation at $Q_0 \approx 0.187$ leads to the global period-4 flow in the range of $Q_0 \in (0.187, 0.189)$, and the chaotic flow relative to the local mappings lies in $Q_0 \in (0.190, 0.191)$. Once the local period-1 flow disappears, the grazing bifurcation occurs, and the global chaotic flows are in the range of $Q_0 \in (0.264, 0.320)$ and $(0.334, 0.453)$. In the range of $Q_0 \in (0.321, 0.332)$, the global period-3 flow with its associated periodic flows caused by period-doubling bifurcation exists. In the neighborhood of $Q_0 \approx 0.349$, there is a period-5 flow. For $Q_0 \in (0.453, 0.5)$, the local period-1 flow in domain Ω_3 is observed. Of course, with increasing excitation amplitude, the chaotic flow in the local domain Ω_3 can be observed. Such a phenomenon will not be considered. The displacement and velocity of Poincare mapping responses of damped Duffing oscillator in Eq.(3.66) are given in Fig.9.4(c) and (d). The asymmetry of the responses can be observed. However, illustrations will be given later.

The global transversality versus excitation strength is presented in Fig.9.5 via switching phase, switching distribution on the separatrix, switching displacement and velocity. Such illustrations will give a clear picture of the global transversality. The switching phase (i.e., mod$(\Omega t_k, 2\pi)$) varying with the excitation strength is illustrated in Fig.9.5(a). It is clearly observed that the switching phase changes with excitation strength. In Fig.9.5(b), the switching displacement and velocity distribution on the separatrix is presented, from which the global transversality location can be determined. Furthermore, the density or probability of the global transversality can be computed. In Fig.9.5(c) and (d), the switching displacement and velocity varying with excitation amplitude are presented. With switching phase together, the global transversality of a flow to the separatrix is determined completely. For local flows, no numerical results relative to the separatrix can be presented.

In Figs.9.4 and 9.5, the bifurcation scenario and global transversality against excitation amplitude are illustrated under the prescribed other parameters, and the ranges for local and global periodic flows and global chaotic flows are given. To understand the periodic and chaotic flows varying with damping coefficient, the bifurcation scenario with damping coefficients are illustrated in Fig.9.6 with the parameters ($\alpha_1 = \alpha_2 = 1, Q_0 = 0.4$ and $\Omega = 1$). For $\delta = 0$, the stochastic or resonant chaotic layers exists, which is dependent on the excitation strength and frequency plus the initial conditions. Such chaotic flows were extensively investigated (e.g., Luo, 1995; Han and Luo,1998; Luo et al,1999; Luo and Han, 2001). For $\delta \in (0.0, 0.201)$, the periodic flow in domain Ω_3 exists. It implies that no any periodic flow intersects with the separatrix. For $\delta \in (0.201, 0.648)$, the global flow exists. The period-1 global flow lies in about $\delta \in (0.584, 0.648)$. The range

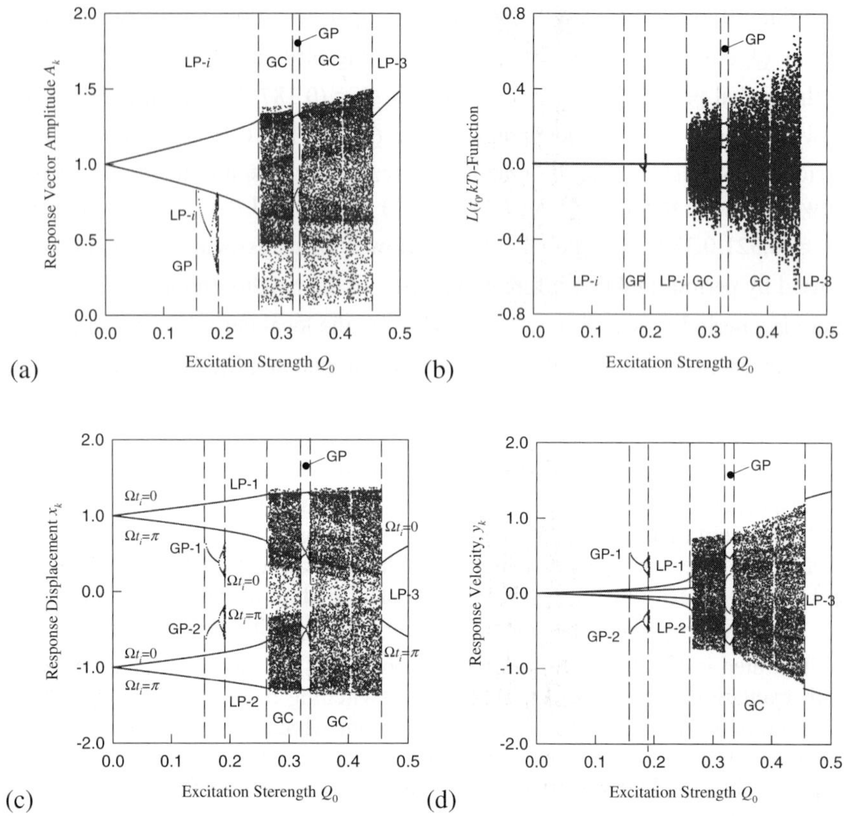

Figure 9.4. Bifurcation scenario varying with excitation strength Q_0 : (a) response vector amplitude A_k, (b) energy increment $L(t_0,kT)$ -function, (c) response displacement and (d) response velocity. ($\alpha_1=\alpha_2=1$, $\delta=0.25$ and $\Omega=1$) . The acronyms LP, GP and GC represent *L*ocal *P*eriodic flow, *G*lobal *P*eriodic flow and *G*lobal *C*haotic flow, respectively. LP-*i* and GP-*i* respectively, represent local and global periodic flows in domain Ω_i ($i=1,2$). LP-3 is the local periodic flow in domain Ω_3.

of the damping coefficient for the period-2 global flow is $\delta\in(0.244,0.248)$ and $(0.564,0.584)$. The period-2 flow for $\delta\in(0.564,0.584)$ is induced by period-doubling bifurcation at approximate $\delta\approx 0.654$. However, for $\delta\in(0.244,0.248)$, the period-2 global flow is obtained by the grazing bifurcation. The period-3 global flow lies in about $\delta\in(0.328,0.342)$. The period-3 global flow lies in about $\delta\in(0.514,0.518)$. The parameter ranges for the period-4 global flow are

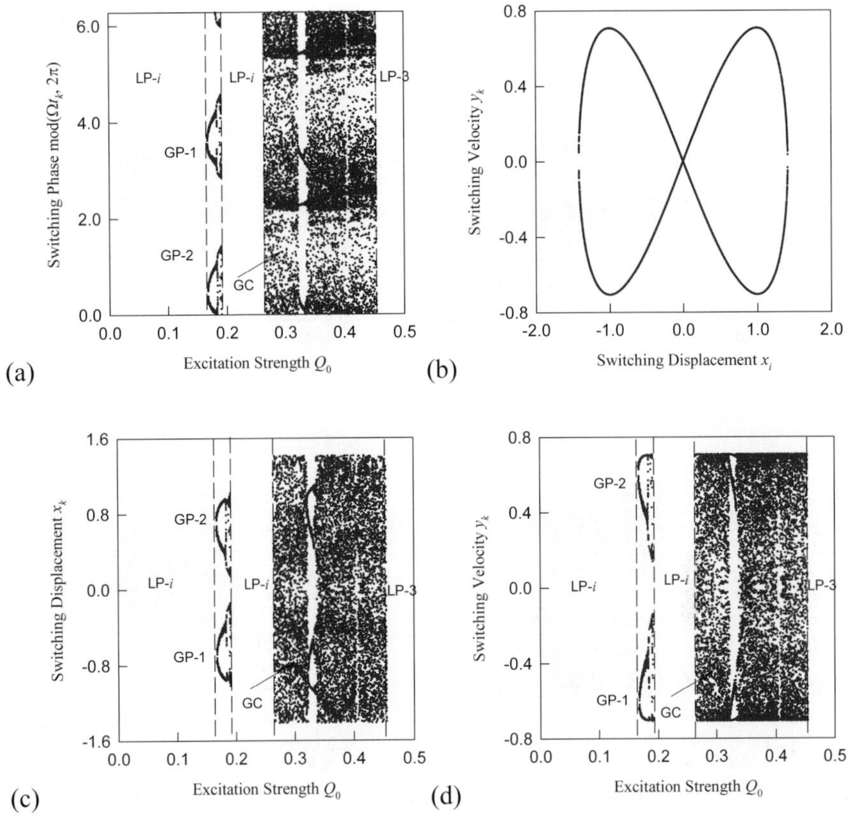

Figure 9.5. Global transversality varying with excitation strength Q_0 : (a) switching phase $\mathrm{mod}(\Omega t_k, 2\pi)$, (b) switching point distribution on the separatrix, (c) response displacement and (d) response velocity. ($\alpha_1 = \alpha_2 = 1$, $\delta = 0.25$ and $\Omega = 1$) . The acronyms LP, GP and GC represent *L*ocal *P*eriodic flow, *G*lobal *P*eriodic flow and *G*lobal *C*haotic flow, respectively. LP-*i* and GP-*i* respectively, represent local and global periodic flows in domain Ω_i ($i = 1, 2$). LP-3 is the local periodic flow in domain Ω_3 .

$\delta \in (0.378, 0.382)$ and $(0.562, 0.564)$. The period-5 global flow lies in about $\delta \in (0.418, 0.426)$ and $(0.502, 0.506)$. The period-7 global flow exists in about $\delta \in (0.440, 0.444)$ and $(0.468, 0.490)$. Without period-doubling bifurcation, the appearance and disappearance of global periodic flows are caused by the grazing bifurcation to the separatrix. For the rest range of δ , the global flow is called the global chaotic flow. Of course, there are some higher periodic global flows in

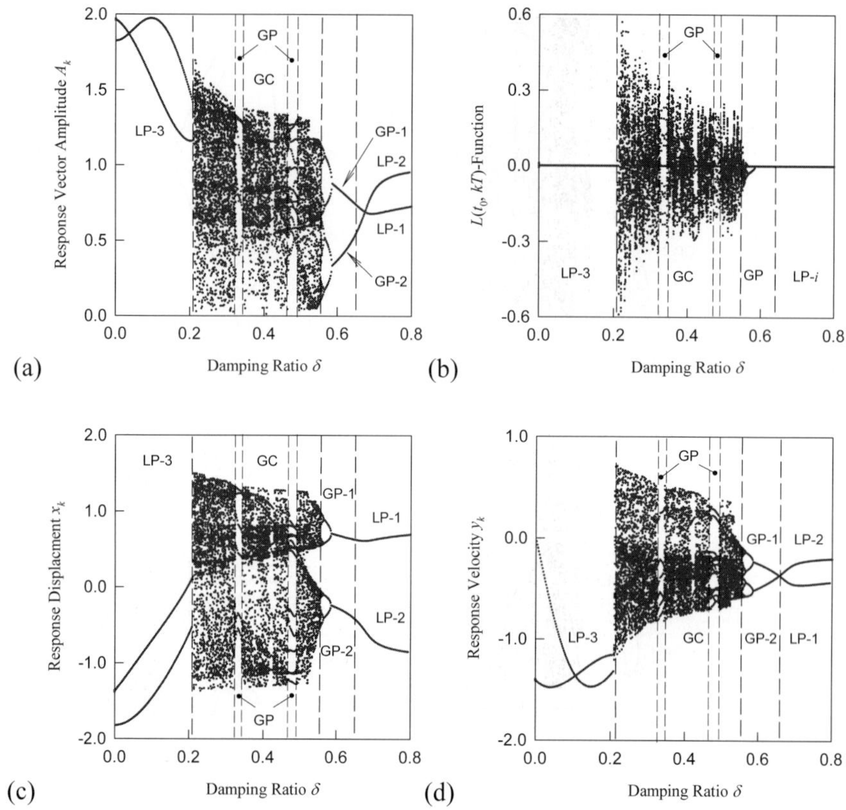

Figure 9.6. Bifurcation scenario varying with damping ratio δ : (a) response vector amplitude A_k, (b) energy increment $L(t_0,kT)$-function, (c) response displacement and (d) response velocity. ($\alpha_1 = \alpha_2 = 1$, $Q_0 = 0.4$ and $\Omega = 1$). The acronyms LP, GP and GC represent *L*ocal *P*eriodic flow, *G*lobal *P*eriodic flow and *G*lobal *C*haotic flow, respectively. LP-*i* and GP-*i* respectively, represent local and global periodic flows in domain Ω_i ($i = 1,2$). LP-3 is the local periodic flow in domain Ω_3.

that range, but such ranges for global periodic flows are too small to list herein. The grazing bifurcation from a local to global periodic flow occurs at $\delta \approx 0.648$. The local period-1 flow in domain Ω_1 or Ω_2 exists for $\delta \in (0.648, 0.80)$. The global transversality versus damping coefficient is presented in Fig.9.7 through the switching phase, switching distribution on the separatrix, switching displacement and velocity. Such illustrations give a clear picture of the global transversality. In Fig.9.7(a), the switching phase (i.e., $\mathrm{mod}(\Omega t_k, 2\pi)$) varying with damp-

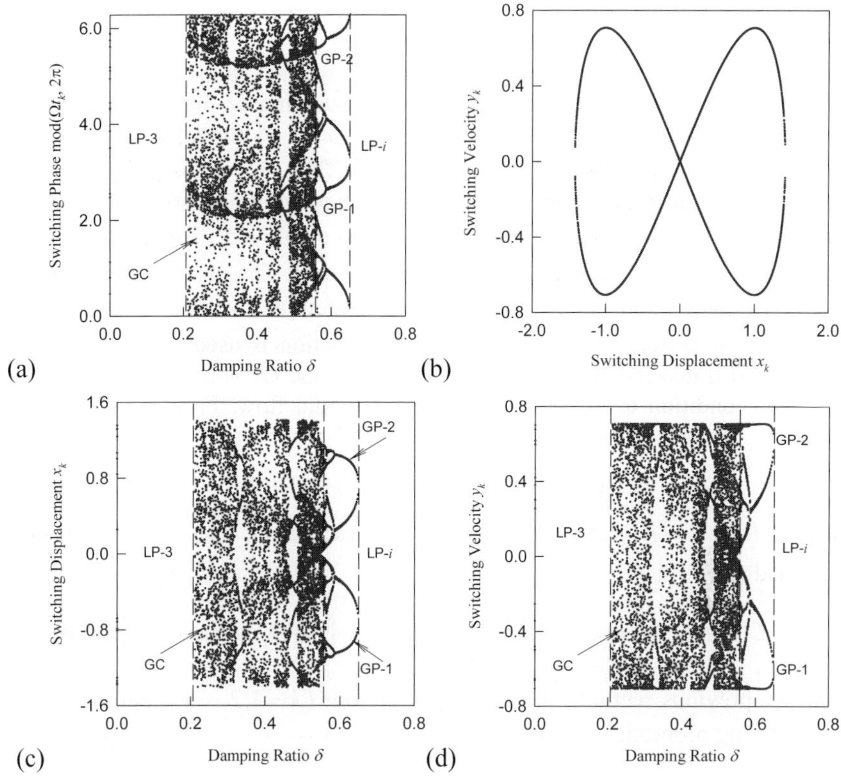

Figure 9.7. Global transversality varying with damping ratio δ : (a) switching phase $\mathrm{mod}(\Omega t_k, 2\pi)$, (b) distribution on the separatrix, (c) switching displacement and (d) switching velocity. ($\alpha_1 = \alpha_2 = 1$, $Q_0 = 0.4$ and $\Omega = 1$). The acronyms LP, GP and GC represent *L*ocal *P*eriodic flow, *G*lobal *P*eriodic flow and *G*lobal *C*haotic flow, respectively. LP-*i* and GP-*i* respectively, represent local and global periodic flows in domain Ω_i ($i = 1, 2$). LP-3 is the local periodic flow in domain Ω_3 .

ing coefficient is illustrated, and the switching phase distribution is given for a specific excitation strength. For $\delta = 0$, the switching phase for the stochastic and chaotic layers of the undamped Duffing oscillator is clearly observed. For non-zero and very small damping, the chaotic flow suddenly disappears. It indicates that the non-damped and damped systems possess different flow characteristics. In Fig.9.7(b), the transversality distribution on the separatrix is presented via switching displacement and velocity. The switching displacement and velocity varying with damping coefficients are presented In Fig.9.7(c) and (d).

9.5. Numerical illustrations

In this section, the global chaotic and periodic responses in the damped Duffing oscillator will be illustrated for a further understanding of the global tangency and transversality in nonlinear dynamical systems. In addition, the L-function is used to determine the periodicity of the periodic flow. The G-function and $G^{(1)}$-function are used to determine the global tangency and transversality. Consider parameters $\alpha_1 = \alpha_2 = 1$ and the time step $10^{-4}T$ is employed. The periodic flows and grazing flows to the separatrix will be presented. The phase plane, G-function, $G^{(1)}$-function and $L(t_0,t)$-function, displacement and velocity responses are illustrated for periodic and grazing flows. G-function is used to determine the grazing and transversality of periodic flows, and the $G^{(1)}$-function is used as a sufficient condition to observe the grazing periodic flow. From Eq.(3.33), for simplicity, the following notation is used as

$$G^{(k)}(t) \equiv G^{(k)}(x,y,t,\mathbf{p}) \quad \text{for } k = 0,1,2,\cdots \tag{9.56}$$

where $G(t) \equiv G^{(0)}(x,y,t,\mathbf{p})$ and $G^{(1)}(t) \equiv G^{(1)}(x,y,t,\mathbf{p})$. The $L(t_0,t)$-function is defined as

$$L(t_0,t) = \Delta H(t_0,t) = \int_{t_0}^{t} G(x,y,t,\mathbf{p})\,dt = \int_{t_0}^{t} y(Q_0 \cos\Omega t - \delta y)\,dt. \tag{9.57}$$

From the previous sections, if a flow for k-periods ($k = 1,2,\cdots$) is periodic, then the following properties exists

$$\begin{aligned} L(t_0,kT) = \Delta H(t_0,kT) &= \int_{t_0}^{t_0+kT} G(x,y,t,\mathbf{p})\,dt \\ &= \int_{t_0}^{t_0+kT} y(Q_0 \cos\Omega t - \delta y)\,dt = 0. \end{aligned} \tag{9.58}$$

For a grazing point (x_m, y_m) of a periodic flow at time t_m, the following relations can be observed

$$G(t_m) \equiv G^{(0)}(x_m,y_m,t_m,\mathbf{p}) = 0 \quad \text{and} \quad H_0(x_m,y_m) = E_s = 0, \tag{9.59}$$

$$\begin{aligned} &G^{(1)}(t_m) \equiv G^{(1)}(x_m,y_m,t_m,\mathbf{p}) < 0 \ \text{ in } \Omega_i \text{ for } i = 1,2 \text{ or} \\ &G^{(1)}(t_m) \equiv G^{(1)}(x_m,y_m,t_m,\mathbf{p}) > 0 \ \text{ in } \Omega_3. \end{aligned} \tag{9.60}$$

9.5.1 Periodic flows

Consider local periodic flows in Ω_i ($i = 1,2,3$) without any grazing to the separatrix. The parameters ($Q_0 = 0.160$, $\Omega = 1.0$, $\delta = 0.25$, $\alpha_1 = \alpha_2 = 1.0$) are used for illustrations in Fig.9.8. For this set of parameters, two local periodic flows coexist. Again, the solid and dashed curves are relative to the domains Ω_1 (right potential well) and Ω_2 (left potential well), respectively. Thus, two sets of initial conditions are considered. In domain Ω_1, the initial condition for $\mathrm{mod}(\Omega t_0, 2\pi) = 0.0$ is employed, but in domain Ω_2, the initial conditions for $\mathrm{mod}(\Omega t_0, 2\pi) = \pi$ is adopted. For local periodic flows, the phase plane and $L(t_0, t)$-function are very important, which shows the periodicity. The phase plane and $L(t_0, t)$-function for the *large* orbits of two periodic flows in Ω_i ($i = 1, 2$) are presented in Fig.9.8(a) and (b). The function $L(t_0, t_k) = 0$ for $t_k = t_0 + kT$ ($k = 1, 2, \cdots$). However, the phase plane and $L(t_0, t)$-function for the *small* orbits of two periodic flows in Ω_i ($i = 1, 2$) are shown in Fig.9.8(c) and (d). For $\mathrm{mod}(\Omega t_0, 2\pi) = 0.0$, the solid curves are for the initial conditions: ($x_0 = 0.6220842$, y_0=0.5307933) (large) and ($x_0 = 1.1545742$, y_0=0.0561600) (small). For $\mathrm{mod}(\Omega t_0, 2\pi) = \pi$, the dashed curves are for initial conditions:($x_0 = -0.6220842$, $y_0 = -0.5307933$) (large) and ($x_0 = -1.1545742$, $y_0 = -0.0561600$) (small). It is observed that the two periodic flows are asymmetric. Local periodic flows in Ω_3 are demonstrated through parameters ($Q_0 = 0.454$, $\Omega = 1.0$, $\delta = 0.25$, $\alpha_1 = \alpha_2 = 1.0$). The phase plane and the $L(t_0, t)$-function for the symmetric local periodic flow in Ω_3 are presented in Fig.9.9(a) and (b). The local periodic flow relative to the initial condition (i.e., $\mathrm{mod}(\Omega t_0, 2\pi) = 0.0$, $x_0 = 0.3646916$, $y_0 = 1.2598329$) is denoted by the solid curve, and the local periodic flow pertaining to the initial condition (i.e., $\mathrm{mod}(\Omega t_0, 2\pi) = \pi$, $x_0 = -0.3646916$, $y_0 = -1.2598329$) is represented by the circle symbol curve. The two periodic flows are overlapped owing to the symmetry of two periodic flows. It is observed that the function $L(t_0, t_k) = 0$ for $t_k = t_0 + kT$ ($k = 1, 2, \cdots$).

To illustrate global periodic flows, a symmetric global periodic flow in three domains is shown in Fig.9.10 through a set of parameters ($Q_0 = 0.325$, $\Omega = 1.0$, $\delta = 0.25$, $\alpha_1 = \alpha_2 = 1.0$). The global periodic flow possesses the global transversality to the separatrix. Hence, the *G*-function versus the displacement is illustrated. An important characteristic of the periodic flow requires $L(t_0, t_k) = 0$ with $t_k = t_0 + kT$ ($k = 3m$ for $m = 1, 2, \cdots$). The solid curve represents the responses of mapping structure ${}^{s}P_{3164}$ with the initial condition: $\mathrm{mod}(\Omega t_0, 2\pi) = 0.0$, $x_0 = -0.4751176$ and $y_0 = 0.6524778$. The circular symbol curve denotes responses for mapping structure ${}^{s}P_{6431}$ with the initial condition: $\mathrm{mod}(\Omega t_0, 2\pi) = \pi$, $x_0 =$

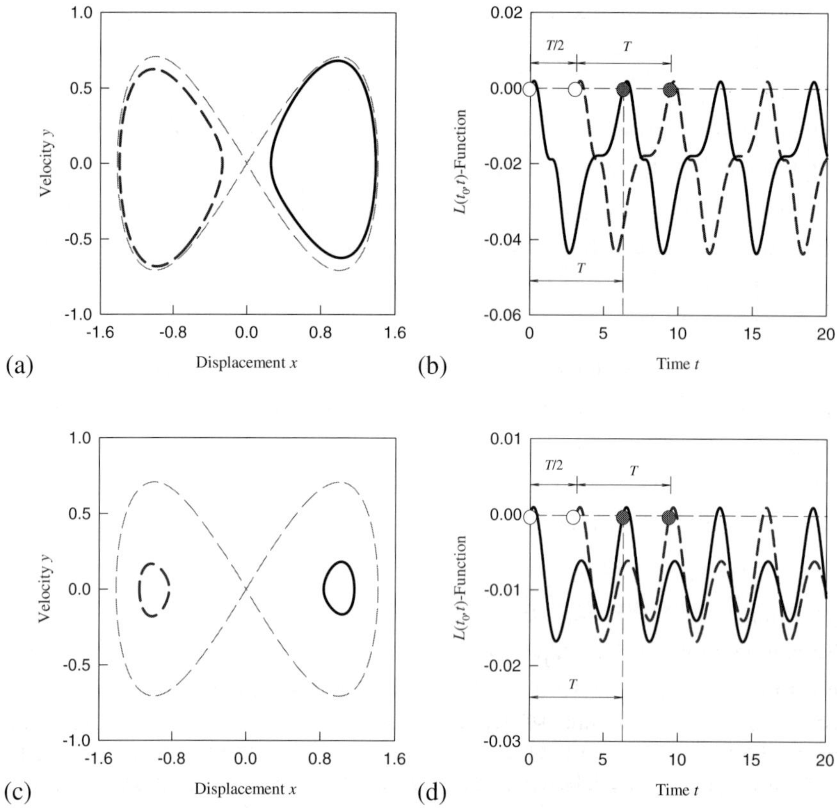

Figure 9.8. Two possible periodic flows co-existence in Ω_i ($i = 1,2$): (a) phase plane and (b) $L(t_0, t)$ -function for the large orbit; (c) phase plane and (d) $L(t_0, t)$ -function for the small orbit; ($Q_0 = 0.160$, $\Omega = 1.0$, $\delta = 0.25$, $\alpha_1 = \alpha_2 = 1.0$). The solid curve is for the initial condition: $\text{mod}(\Omega t_0, 2\pi) = 0.0$; $x_0 = 0.6220842$, y_0=0.5307933 (large) and $x_0 = 1.1545742$, y_0=0.0561600 (small). The dashed curve is for the initial condition: $\text{mod}(\Omega t_0, 2\pi) = \pi$, $x_0 = -0.6220842$, $y_0 = -0.5307933$ (large) and $x_0 = -1.1545742$, $y_0 = -0.0561600$ (small).

0.4751176, and $y_0 = -0.6524778$. From phase plane in Fig.9.10(a), the global flow appears in the three domains and the two periodic flow responses are identical because of the symmetry in phase plane. In Fig.9.10(b), the global transversality of a periodic flow to the separatrix is clearly presented. The flow from Ω_3 to Ω_i ($i = 1,2$) requires $G(t_m) < 0$ and the flow from Ω_i ($i = 1,2$) to Ω_3 requires $G(t_m) > 0$. Such a property is confirmed in Fig.9.10(b). Due to a skew symmetry

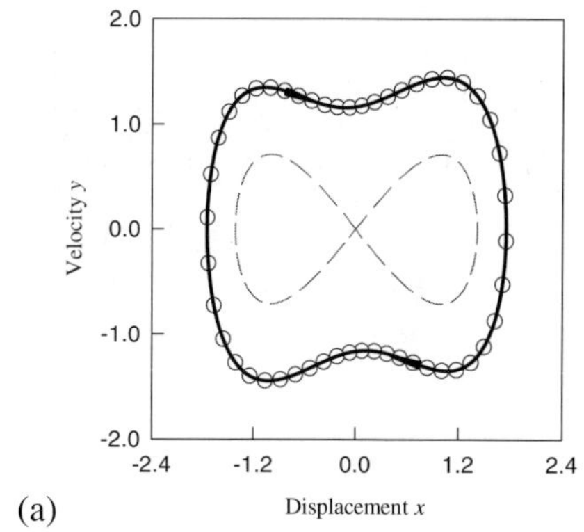

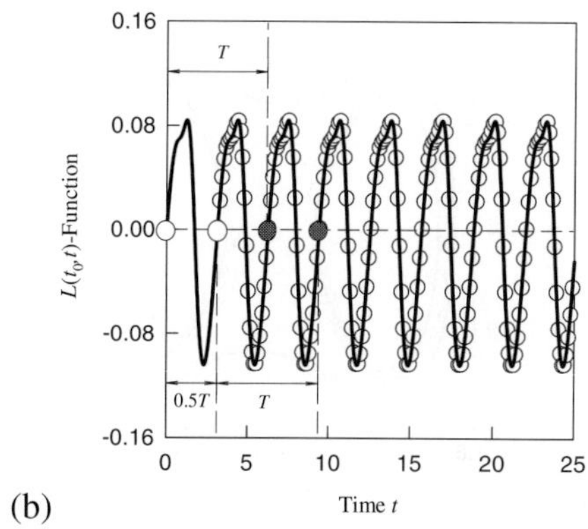

Figure 9.9. Symmetric local periodic flow in Ω_3: (a) phase plane and (b) $L(t_0,t)$-function for the large orbit; ($Q_0=0.454,\ \Omega=1.0,\ \delta=0.25,\ \alpha_1=\alpha_2=1.0$). The solid curve is for the initial condition: $\text{mod}(\Omega t_0,2\pi)=0.0, x_0=0.3646916, y_0=1.2598329$. The dotted curve is for the initial condition: $\text{mod}(\Omega t_0,2\pi)=\pi, x_0=-0.3646916, y_0=-1.2598329$.

of periodic flows, the displacement responses of two periodic flows with the different initial conditions are different, as shown in Fig.9.10(c). The periodicity of displacement responses with period-3 is presented. In addition, the $L(t_0,t_k)$-function is presented in Fig.9.10(d) to further show the periodicity of periodic flows. The two periodic flows are formed during three periods. The starting points are white circular symbol, and the ending points are the dark circular symbol. The intermediate points are labeled by the gray circular symbols. It is observed that $L(t_0,t_k)=0$ for $\text{mod}(k,3)=0$ and $L(t_0,t_k)\neq 0$ for $\text{mod}(k,3)\neq 0$.

In addition to the pre-described global periodic flow, there is another global periodic flow in the two domains Ω_i ($i=1$ or 2) and Ω_3. Consider a set of parameters ($Q_0=0.180,\ \Omega=1.0,\ \delta=0.25,\ \alpha_1=\alpha_2=1.0$). The phase plane, the *G*-function versus displacement, displacement response and the $L(t_0,t)$-function for a period-1 global flow are presented in Fig.9.11(a)-(d), respectively. The solid curve is the periodic flow relative to mapping structure ${}^sP_{12}={}^sP_1\circ{}^sP_2$ with the initial condition (i.e., $\text{mod}(\Omega t_0,2\pi)=0.0,\ x_0=0.3892045,\ y_0=0.3720676$). The dashed curve is a periodic flow of mapping structure ${}^sP_{45}={}^sP_4\circ{}^sP_5$ with the initial condition (i.e., $\text{mod}(\Omega t_0,2\pi)=\pi,\ x_0=-0.3892045,\ y_0=-0.3720676$). The two periodic flows asymmetric to the displacement axis are formed through the grazing bifurcation of a local periodic flow in domain Ω_3. Such a symmetry is illustrated in Fig.9.11(a). Again, the *G*- function versus displacement demonstrates the global transversality of the flow to the separatrix in Fig.9.11(b). It is

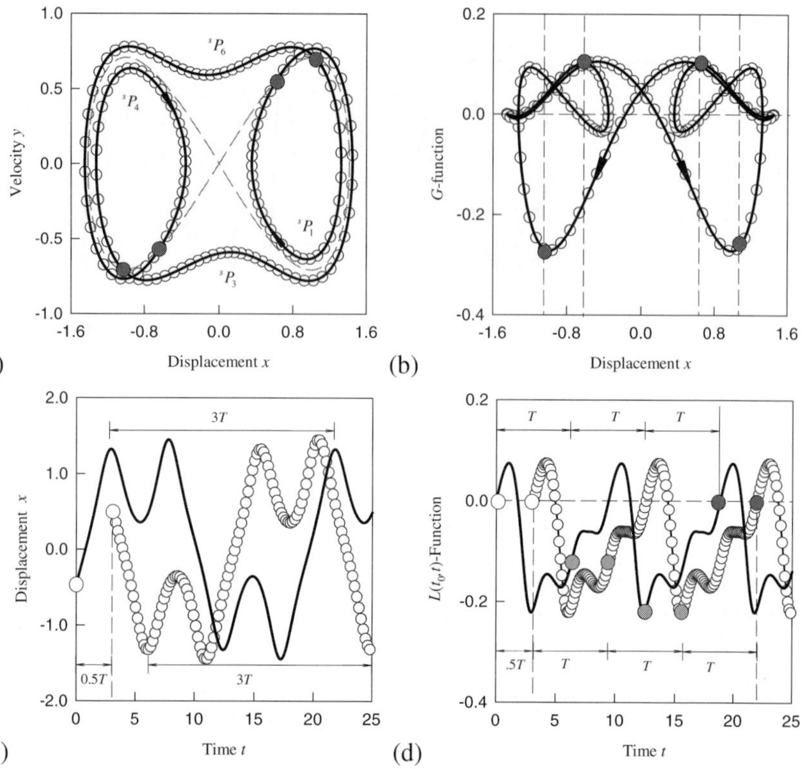

(a) (b) (c) (d)

Figure 9.10. Symmetric global periodic flow in all domains: (a) phase plane and (b) G-function versus displacement, (c) displacement response and (d) $L(t_0,t)$-function. ($Q_0 = 0.325$, $\Omega = 1.0$, $\delta = 0.25$, $\alpha_1 = \alpha_2 = 1.0$). The solid curve is for mapping structure $^sP_{3164}$ with the initial condition: $\mathrm{mod}(\Omega t_0, 2\pi) = 0.0$, $x_0 = -0.4751176$, $y_0 = 0.6524778$. The dashed curve is for mapping structure $^sP_{6431}$ with the initial condition: $\mathrm{mod}(\Omega t_0, 2\pi) = \pi$, $x_0 = 0.4751176$, $y_0 = -0.6524778$.

observed that $G(t_k) > 0$ from Ω_i ($i = 1,2$) to Ω_3 and $G(t_k) < 0$ from Ω_3 to Ω_i ($i = 1,2$). The periodicity of the periodic flow is shown in Fig.9.11(c) by use of the displacement response, and the $L(t_0,t)$-function with $L(t_0,t_k) = 0$ with $t_k = t_0 + kT$ ($k = 1,2,\cdots$) is given in Fig.9.11(d). It is numerically proved that the periodic flow possesses $L(t_0,t_k) = 0$ for $t_k = t_0 + kT$.

After period doubling, the global period-2 flow will appear, and parameters ($Q_0 = 0.18$, $\Omega = 1.0$, $\delta = 0.25$, $\alpha_1 = \alpha_2 = 1.0$)are used for illustration in Fig.9.12.

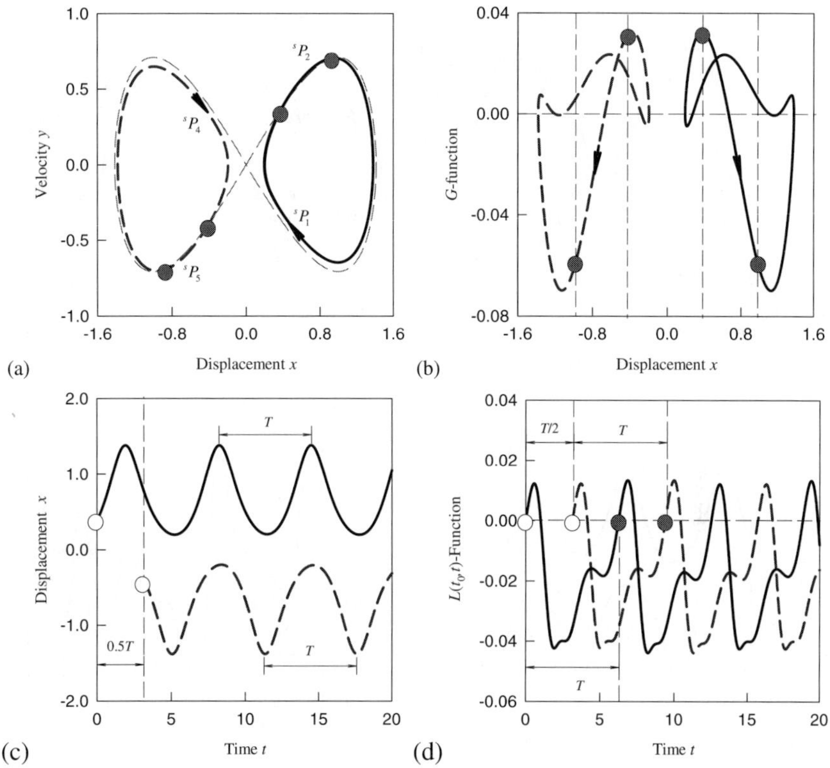

Figure 9.11. Global periodic flow in the two domains: (a) phase plane and (b) G-function versus displacement, (c) phase plane and (d) $L(t_0,t)$ -function. ($Q_0 = 0.180$, $\Omega = 1.0$, $\delta = 0.25$, $\alpha_1 = \alpha_2 = 1.0$). The solid curve is for mapping structure $^sP_{21}$ with the initial condition: $\text{mod}(\Omega t_0, 2\pi) = 0.0$, $x_0 = 0.3892045$, $y_0 = 0.3720676$. The dashed curve is for mapping structure $^sP_{54}$ with the initial condition: $\text{mod}(\Omega t_0, 2\pi) = \pi$, $x_0 = -0.3892045$, $y_0 = -0.3720676$.

The solid curve is the period-2 flow for a mapping structure $^sP_{(12)^2} = {}^sP_{12} \circ {}^sP_{12}$ with the initial conditions (i.e., $\text{mod}(\Omega t_0, 2\pi) = 0.0$, $x_0 = 0.3130951$ and $y_0 =$ 0.3196078), and the dashed curve is the period-2 flow for mapping structure $^sP_{(45)^2} = {}^sP_{45} \circ {}^sP_{45}$ with the initial condition: $\text{mod}(\Omega t_0, 2\pi) = \pi$, $x_0 = -0.3130951$ and $y_0 = -0.3196078$. The trajectories of the period-2 flows in phase plane are presented with two cycles during two-periods in Fig.9.12(a). The two period-doubled periodic flows are symmetric. The G-function versus displacement is

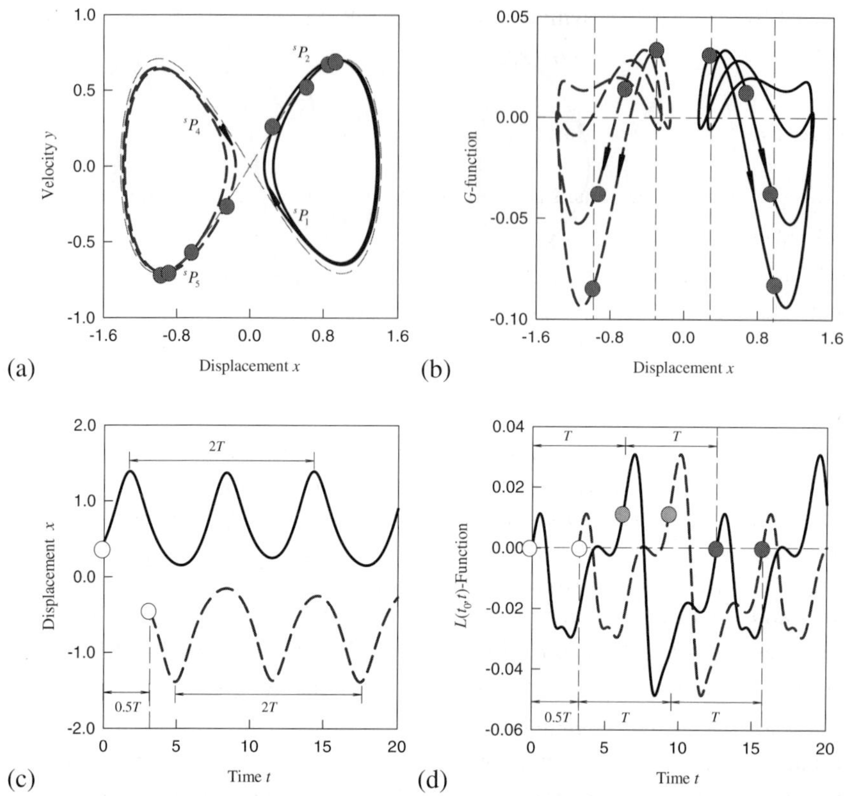

Figure 9.12. Global period-2 flow in the two domains: (a) phase plane and (b) G-function versus displacement, (c) displacement responses and (d) $L(t_0,t)$-function. ($Q_0 = 0.183$, $\Omega = 1.0$, $\delta = 0.25$, $\alpha_1 = \alpha_2 = 1.0$). The solid curve is for mapping structure ${}^sP_{(21)^2}$ with the initial condition: $\text{mod}(\Omega t_0, 2\pi) = 0.0$, $x_0 = 0.3130951$, y_0=0.3196078. The dashed curve is for mapping structure ${}^sP_{(54)^2}$ with the initial condition: $\text{mod}(\Omega t_0, 2\pi) = \pi$, $x_0 = -0.3130951$, $y_0 = -0.3196078$.

shown in Fig.9.12(b) to determine the transversality of the period-2 flow. $G(t_m) > 0$ from Ω_i ($i = 1,2$) to Ω_3 and $G(t_k) < 0$ from Ω_3 to Ω_i ($i = 1,2$) are verified again. The two displacement responses asymmetric to the displacement axis in phase plane are illustrated in Fig.9.12(c), and the two periodic flows are formed with two periods. For the further illustration, the L-function is presented for two period-doubled flows in Fig.9.12(d). The starting point is white circular symbol, and the ending point is the dark circular symbol. The intermediate points are labeled by gray circular symbols. It is observed that $L(t_0,t_k)=0$ for $\mathrm{mod}(k,2)=0$ and $L(t_0,t_k)\neq 0$ for $\mathrm{mod}(k,2)\neq 0$.

9.5.2. Grazing periodic flows

After previous discussions, there are two periodic flows in domain Ω_i ($i=1,2$). In Fig.9.13, a periodic flow grazing to the separatrix in the two domain Ω_i ($i=1,2$) is presented for parameters ($Q_0=0.167$, $\Omega=1.0$, $\delta=0.25$, $\alpha_1=\alpha_2=1.0$) first. The solid curve gives the periodic grazing flow with the initial condition (i.e., $\mathrm{mod}(\Omega t_0,2\pi)=0.0$, $x_0=0.4788839$ and $y_0=$ 0.4375262), and the dashed curve gives the asymmetric periodic grazing flow relative to the initial condition (i.e., $\mathrm{mod}(\Omega t_0,2\pi)=\pi$, $x_0=-0.4788839$ and $y_0=-0.4375262$). The corresponding phase plane, the G-function versus displacement, the $G^{(1)}$-function versus displacement are presented in Fig.9.13(a)-(c), respectively. In the phase plane, the two periodic flows in domains Ω_1 and Ω_2 are observed, and the two periodic flows are tangential to the separatrix, accordingly. The two periodic flows are asymmetric to the displacement axis in phase plane, and the grazing points are marked with circular symbols, and labeled with "grazing". The G- and $G^{(1)}$-functions in Fig.9.13(b) and (c) show $G(t_m)=0$ and $G^{(1)}(t_m)<0$ at the grazing point (x_m,y_m) on the separatrix for time t_m, and the two conditions confirm the necessary and sufficient conditions for grazing flow (i.e., Eqs. (9.59) and (9.60)). In addition, the $L(t_0,t)$-function time response, displacement-time response and velocity-time response are also presented for showing the periodicity of the two periodic flows. The function $L(t_0,t)$ for points relative to $t=t_0+kT$ ($k=1,2,\cdots$) is zero, which is shown in Fig.9.13(d). This is because the periodic flow is of period-1. The starting points are marked with the white circular symbols, and the ending points for one period are marked with the dark circular symbols. The illustrations of the displacement and velocity responses are given in Fig.9.13(e) and (f), respectively. The solid and dashed curves are relative to the domains Ω_1 (right potential well) and Ω_2 (left potential well), respectively.

Once the small periodic flow becomes asymmetric to the displacement axis in phase plane, the grazing may occur. The small periodic flow with grazing in domain Ω_i ($i=1,2$) is shown in Fig.9.14 for parameters ($Q_0=0.4$, $\Omega=0.695$, $\delta=0.25$, $\alpha_1=\alpha_2=1.0$). The solid curve gives a grazing flow for the initial condition (i.e., $\mathrm{mod}(\Omega t_0,2\pi)=0.0$, $x_0=-0.8594988$ and $y_0=0.1724927$) and the dashed curve gives an asymmetric grazing flow for the corresponding initial condition (i.e., $\mathrm{mod}(\Omega t_0,2\pi)=\pi$, $x_0=0.8594988$, $y_0=-0.1724927$). The phase

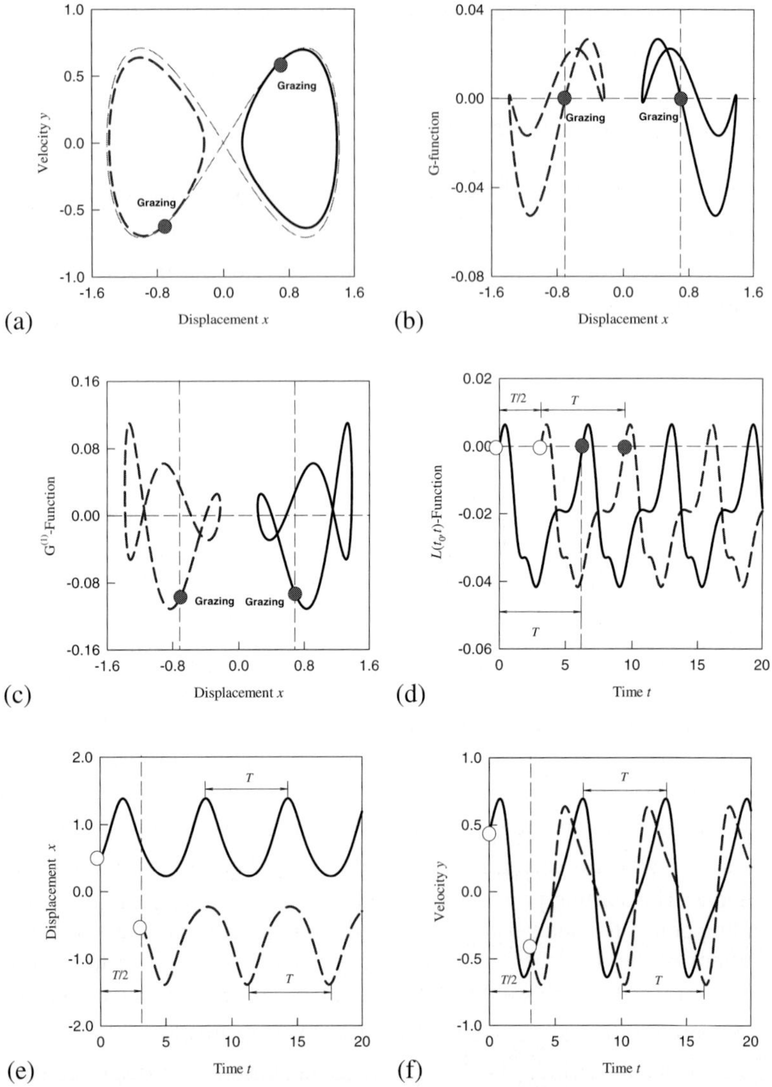

Figure 9.13. Periodic flow grazing to the separatrix: (a) phase plane, (b) G-function versus displacement, (c) $G^{(1)}$-function versus displacement, (d) $L(t_0,t)$-function response, (e) displacement response and (f) velocity response. ($Q_0 = 0.167,\ \Omega = 1.0,\ \delta = 0.25,\ \alpha_1 = \alpha_2 = 1.0$). The solid curve is for initial condition: $\text{mod}(\Omega t_0, 2\pi) = 0.0,\ x_0 = 0.4788839,\ y_0 = 0.4375262$. The dashed curve is for initial condition: $\text{mod}(\Omega t_0, 2\pi) = \pi,\ x_0 = -0.4788839,\ y_0 = -0.4375262$.

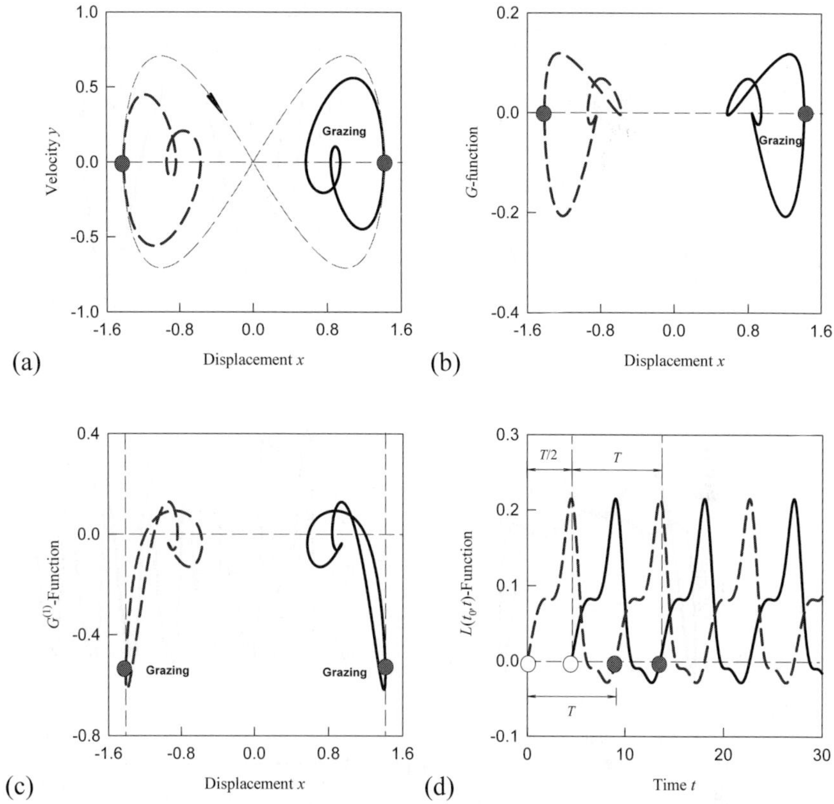

Figure 9.14. Period-1 flow with infinity-order grazing in domain Ω_i ($i=1,2$): (a) phase plane and (b) G-function versus displacement, (c) $G^{(1)}$–function and (d) $L(t_0,t)$ -function. ($Q_0=0.4$, $\Omega=0.695$, $\delta=0.25$, $\alpha_1=\alpha_2=1.0$). The dashed curve is for the initial condition: $\text{mod}(\Omega t_0, 2\pi)=0.0$, $x_0=-0.8594988$, $y_0=0.1724927$. The solid curve is for the initial condition: $\text{mod}(\Omega t_0, 2\pi)=\pi$, $x_0=0.8594988$, $y_0=-0.1724927$.

plane, the G-function versus displacement, the $G^{(1)}$–function and the L-function are presented in Fig.9.14(a)-(d), respectively. The grazing of the two small periodic flows occurs at $y_m=0$. The $G(t_m)=0$ and $G^{(1)}(t_m)<0$ at the grazing points are observed in Fig.9.14(b) and (c). $G(t_m)=0$ and $G^{(1)}(t_m)<0$ at the grazing point (x_m, y_m) on the separatrix for time t_m satisfy the conditions in Eqs.(9.59) and (9.60). The L-function in Fig.9.14(d) shows the periodic flow has $L(t_0,t_k)=0$ for $t_k=t_0+kT$ ($k=1,2,\cdots$).

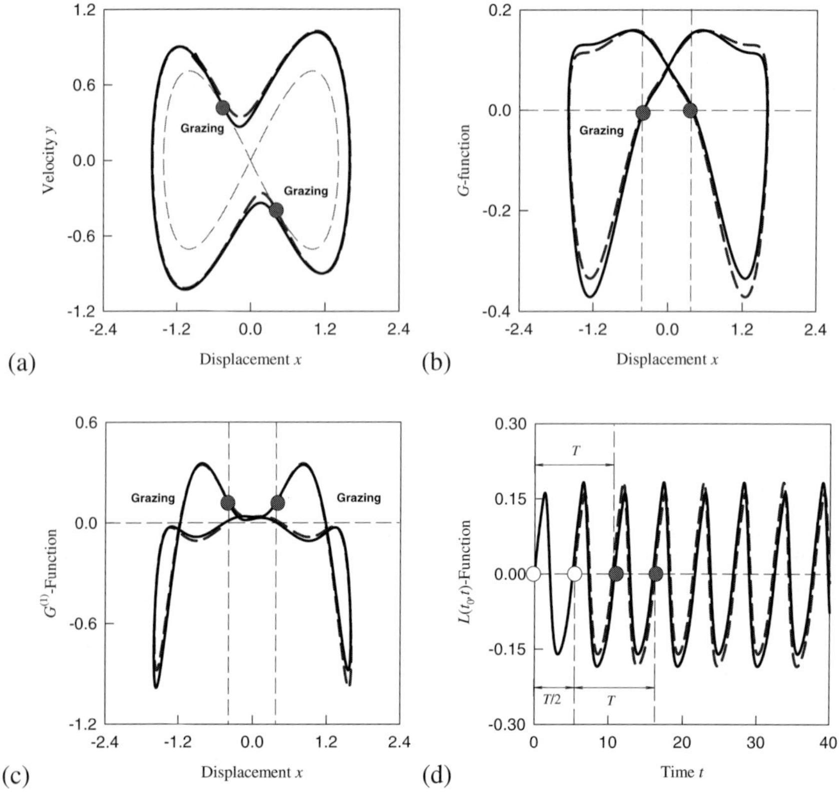

Figure 9.15. Period-1 grazing flow in domain Ω_3: (a) phase plane and (b) G-function versus displacement, (c) $G^{(1)}$–function and (d) $L(t_0,t)$ -function. ($Q_0 = 0.4$, $\Omega = 0.581$, $\delta = 0.25$, $\alpha_1 = \alpha_2 = 1.0$). The solid curve is for the initial condition: $\mathrm{mod}(\Omega t_0, 2\pi) = 0.0$, $x_0 = 0.5302920$, y_0=0.7836607. The dashed curve is for initial condition: $\mathrm{mod}(\Omega t_0, 2\pi) = \pi$, $x_0 = -0.5302920$, $y_0 = -0.7836607$.

With parameter variation, the symmetric periodic flow in domain Ω_3 will become asymmetric. The grazing of an asymmetric period-1 flow in domain Ω_3 is shown in Fig.9.15 with parameters ($Q_0 = 0.4$, $\Omega = 0.581$, $\delta = 0.25$, $\alpha_1 = \alpha_2 = 1.0$). The phase plane, the G-function versus displacement, the $G^{(1)}$–function and $L(t_0,t)$ -function are also presented in Fig.9.15(a)-(d), respectively. The solid curve gives the asymmetric periodic flow with the initial condition ($\mathrm{mod}(\Omega t_0, 2\pi) = 1.0$, $x_0 = 0.5302920$, and y_0=0.7836607). The dashed curve is the asymmetric periodic

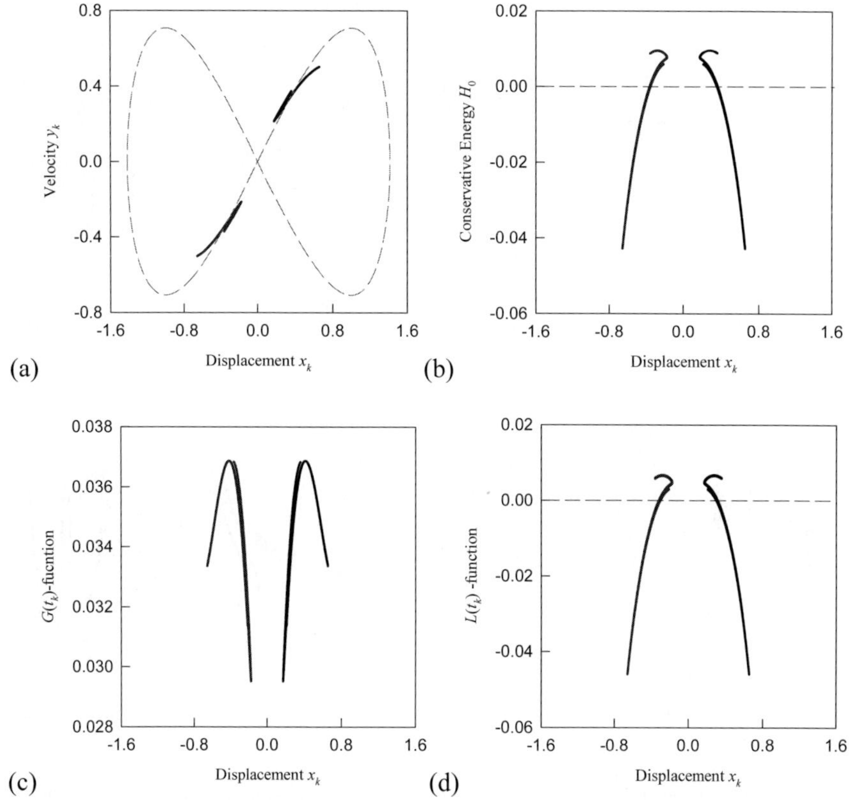

Figure 9.16. Local chaotic flow in domain Ω_i $i = 1$ or 2 and Ω_3: (a) Poincare mapping section, (b) conservative energy $H_0(x_k, y_k)$, (c) $G(x_k, y_k)$–function and (d) $L(t_k)$ -function. ($Q_0 = 0.192$, $\Omega = 1.0$, $\delta = 0.25$, $\alpha_1 = \alpha_2 = 1.0$). The initial condition for right-hand side is $\mathrm{mod}(\Omega t_0, 2\pi) = 0.0$, $x_0 = 0.3199100, y_0 = 0.3206936$. The initial condition for the left-hand side is $\mathrm{mod}(\Omega t_0, 2\pi) = \pi$, $x_0 = -0.3199100, y_0 = -0.3206936$.

flow with the initial condition (i.e., $\mathrm{mod}(\Omega t_0, 2\pi) = \pi$, $x_0 = -0.5302920$ and $y_0 = -0.7836607$). In phase plane, two periodic flows are grazing to the separatrix. The grazing points are marked with circular symbols, and labeled with"grazing". The G-and $G^{(1)}$-functions in Fig.9.15(b) and (c) give $G(t_m) = 0$ and $G^{(1)}(t_m) > 0$ at the grazing point (x_m, y_m) on the separatrix for time t_m, which satisfy the conditions in Eqs.(9.59) and (9.60), respectively. In Fig.9.15(d), it is observed that the function $L(t_0, t)$ for the point relative to $t_k = t_0 + kT$ ($k = 1, 2, \cdots$) is zero, which indicates that the two grazing periodic flows are of period-1.

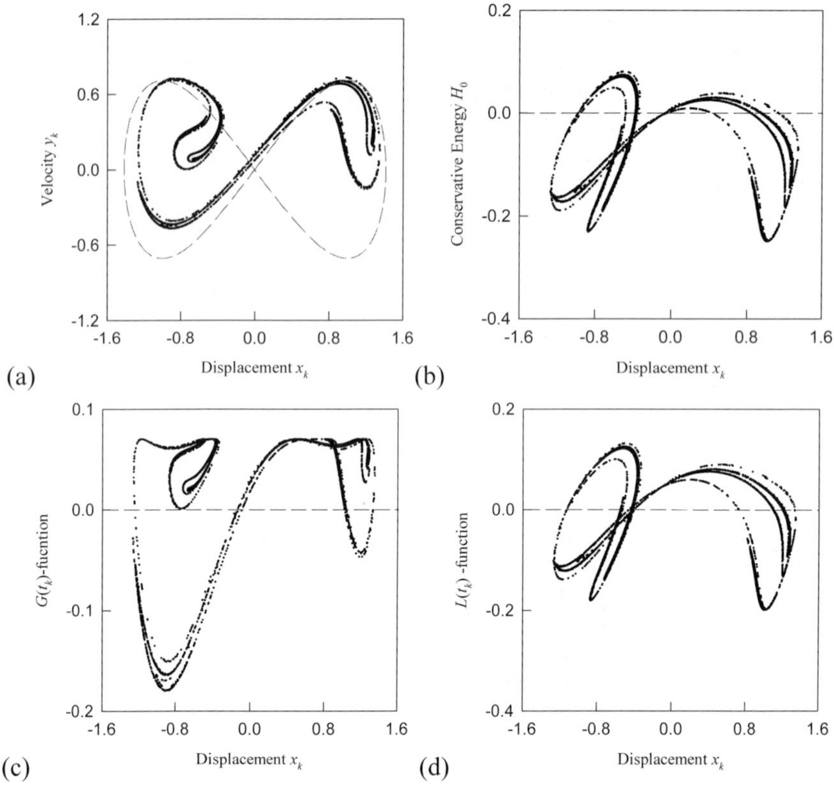

Figure 9.17. Global chaotic flow in domains Ω_i ($i=1,2,3$): (a) Poincare mapping section, (b) conservative energy $H_0(x_k, y_k)$, (c) $G(x_k, y_k)$ –function and (d) $L(t_k)$ -function. ($Q_0 = 0.265$, $\Omega = 1.0$, $\delta = 0.25$, $\alpha_1 = \alpha_2 = 1.0$). The initial condition for right-hand side is $\mathrm{mod}(\Omega t_0, 2\pi) = 0.0$, $x_0 = -0.4180597, y_0$=0.2394332.

9.5.3. Poincaré mapping sections of chaos

After illustration of periodic flows and periodic grazing flows, the characteristics of chaos in the damped Duffing oscillator will be presented. The Poincaré section of chaos, and the corresponding conservative energy, G-function and L-function varying with displacement are illustrated. The local chaos existing in two domains with a common boundary is presented first. In Fig.9.16, system parameters (i.e., $Q_0 = 0.192$, $\Omega = 1.0$, $\delta = 0.25$, $\alpha_1 = \alpha_2 = 1.0$) are used. The initial condition for the right-hand side of the potential well (i.e., $\mathrm{mod}(\Omega t_0, 2\pi) = 0.0$, x_0 = 0.3199100 and $y_0 = 0.3206936$) are employed for illustration, and another initial condition (i.e., $\mathrm{mod}(\Omega t_0, 2\pi) = \pi$, $x_0 = -0.3199100$ and $y_0 = -0.3206936$) for the left-hand side of the potential well are used. The strange attractor in Poincaré section is formed by two pieces in Fig.9.16(a). The conservative energy of the Poincaré mapping points varying along displacement is presented in Fig.9.18(b). As in showing periodic motion, the G-function varying with displacement for the Poincaré mapping points is shown in Fig.9.16(c). The L-function varying with displacement for the Poincaré points is illustrated in Fig.9.16(d). $L(t_k)$ for $t_k = t_0 + kT$ ($k = 1, 2, \cdots$) with $k = 10,000$ is non-zero and chaotic. The $L_k(t_k)$-function are identical to the conservative energy $H_0(x_k, y_k)$ for time t_k because of $L(t_0, t_k) = \Delta H_0(x_k, y_k) = H_0(x_k, y_k) - H_0(x_0, y_0)$.

The global chaos existing in the three domains Ω_i ($i = 1, 2, 3$) is of interest. Consider the system parameters ($Q_0 = 0.265$, $\Omega = 1.0$, $\delta = 0.25$, $\alpha_1 = \alpha_2 = 1.0$) for illustration of the global chaos. The initial condition (i.e., $\mathrm{mod}(\Omega t_0, 2\pi) = 0.$, $x_0 = -0.4180597$ and $y_0 = 0.2394332$) is chosen. With such parameters and initial conditions, the Poincaré mapping section of chaos is shown in Fig.9.17(a). To observe the conservative energy change of global chaos, the conservative energy varying with displacement are plotted in Fig.9.17(b) for the Poincaré mapping points. The G-function and L-function of the Poincaré points are pre-sented via the distribution along the displacement, as shown in Fig.9.17(c) and (d). The G-function on the switching points on the separatrix gives the global transversality of chaos to the separatrix. From the L-function in Fig.9.17(d), non-zero L-function of the Poincaré points can be obtained, which implies the flow is chaotic.

9.5.4. Global transversality of chaos

The previous illustrations of chaos are based on the Poincare mapping section. To understand the global transversality of chaos to the separatrix, chaos can be demonstrated through the switching displacement, switching velocity versus the switching phase. The distribution of the switching points on separatrix can be given. The switching G-function versus the switching displacement will be presented, from which the global transversality and tangency to the separatrix can be understood better. Consider the parameters ($Q_0 = 0.191$, $\Omega = 1.0$, $\delta = 0.25$, $\delta = 0.25$, $\alpha_1 = \alpha_2 = 1.0$). The initial conditions ($\mathrm{mod}(\Omega t_0, 2\pi) = 6.0399156$, $x_0 = 0.1632408$ and $y_0 = 0.1621496$) for the right-hand side and ($\mathrm{mod}(\Omega t_0, 2\pi) = 2.8783230$, $x_0 = -0.1632408$, and $y_0 = -0.1621496$) for the left-hand side of the potential well are employed. The initial conditions satisfy $H_0(x_0, y_0) = E_s = 0$ on the separatrix, and all the switching points are on the separatrix in order to look into the global transversality. From such parameters and initial conditions,

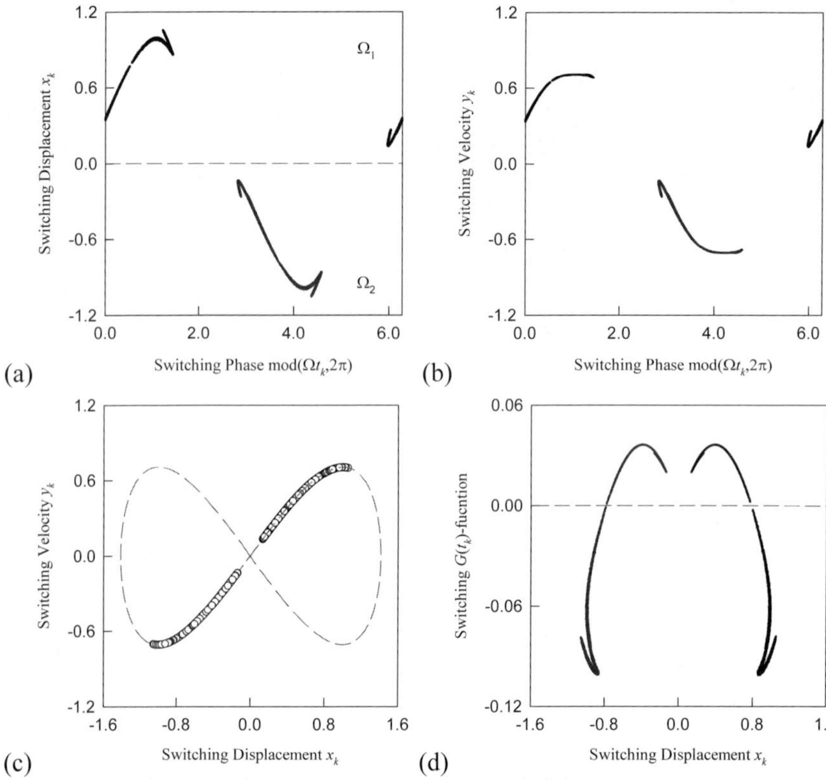

Figure 9.18.Local chaotic flow in domain Ω_i ($i=1$ or 2) and Ω_3 : (a) switching displacement, (b) switching displacement, (c) switching points on separatrix, (d) switching $G(t_k)$ -function versus switching displacement.($Q_0=0.191$, $\Omega=1.0$, $\delta=0.25$, $\alpha_1=\alpha_2=1.0$). The initial conditions are mod($\Omega t_0, 2\pi$) = 6.0399156, $x_0=0.1632408$, $y_0=0.1621496$; and mod($\Omega t_0, 2\pi$) = 2.8783230, $x_0=-0.1632408$, $y_0=-0.1621496$.

the switching displacement and velocity versus the switching phase are shown in Fig. 9.18(a) and (b). The two braches of the switching points on the separatrix for chaos are clearly observed. The two braches are asymmetric, which also can be observed from Fig.9.18(c). All the switching points are on the separatrix, but only the partial separatrix is occupied by the switching points. Therefore, this chaotic flow is a local one. For the global transversality of the chaotic flow, the switching $G(t_k)>0$ implies the flow is globally transversal from domain Ω_i (i = 1,2) to domain Ω_3. On the other hand, the switching $G(t_k)<0$ implies the flow is globally transversal from domain Ω_3 to domain Ω_i ($i=1,2$). In Fig.9.18(b),

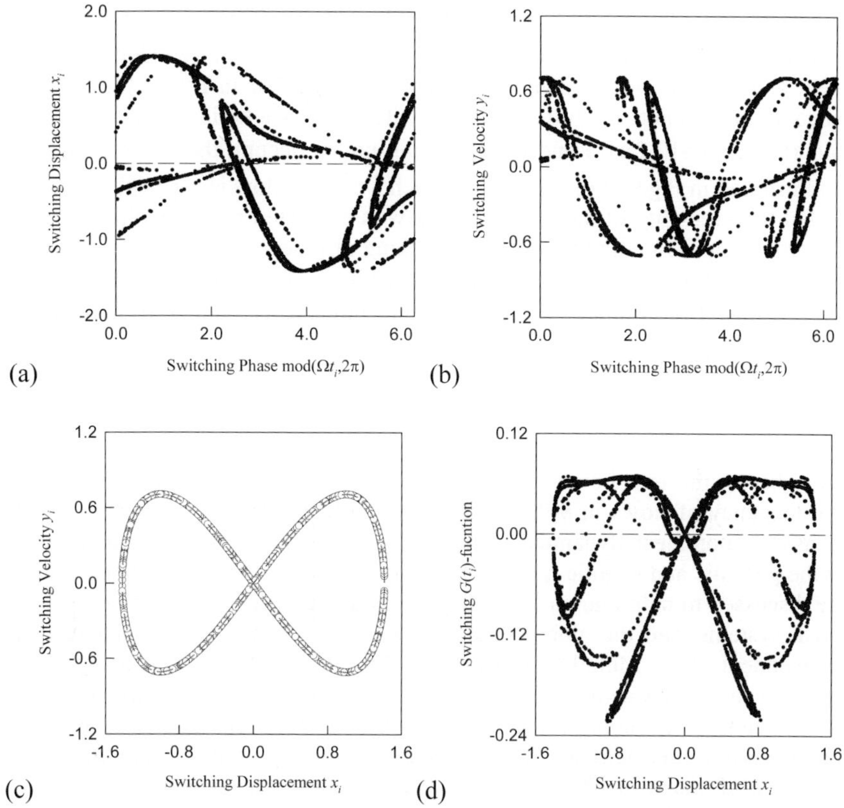

Figure 9.19. Global chaotic flow in domains Ω_i ($i=1,2,3$): (a) switching displacement, (b) switching displacement, (c) switching points on separatrix, (d) switching $G(t_k)$ -function versus switching displacement. ($Q_0=0.265$, $\Omega=1.0$, $\delta=0.25$, $\alpha_1=\alpha_2=1.0$). The initial condition is $\text{mod}(\Omega t_0, 2\pi)=0.0924066$, $x_0=1.0401499, y_0=0.7047306$.

such a property is clearly presented. The non-zero switching G-function (i.e., $G(t_k)\neq 0$) is observed. The global flow from domain Ω_i ($i=1,2$) into domain Ω_3 on the separatrix is close to the hyperbolic point, but the global flow from domain Ω_3 to domain Ω_i ($i=1,2$) on the separatrix is far away from the hyperbolic point. If $G(t_m)=0$, the grazing bifurcation can be observed. The corresponding strange attractor fragmentation can be discussed as in Luo (2006b,d).

For the switching points almost occupying the separatrix, the system parameters ($Q_0=0.265$, $\Omega=1.0$, $\delta=0.25$, $\alpha_1=\alpha_2=1.0$) and the initial condition (i.e.,

$\mathrm{mod}(\Omega t_0, 2\pi) = 0.02924066$, $x_0 = 1.0401499$ and $y_0 = 0.7047306$) are used for another illustration of the global transversality of chaos. The switching displacement and velocity versus switching phase are presented in Fig.9.19(a) and (b). The switching points lie in the range of switching phase lie in the entire range of $\mathrm{mod}(\Omega t_k, 2\pi) \in [0, 2\pi]$) and the switching points occupy the entire separatrix, as illustrated in Fig.9.19(c). Therefore, such a chaotic flow is a global chaos. The global transversality of such a chaotic flow is presented in Fig.9.19(d) through the switching G-function versus the switching displacement. Again the non-zero switching G-function is observed. The switching G-function distribution for the chaotic flow from domain Ω_i ($i = 1, 2$) into domain Ω_3 and from domain Ω_3 into domain Ω_i ($i = 1, 2$) are chaotic.

9.6. Conclusions

This chapter presents how to apply a general theory for the global transversality and tangency of flows in n-dimensional nonlinear dynamical systems to a 2-dimensional nonlinear dynamical system. The global tangency and transversality of the periodic and chaotic motions to the separatrix for such a nonlinear system are discussed to help one understand the complexity of chaos in nonlinear dynamical systems, and the global transversality and tangency to the separatrix are independent of the Melnikov function (or the energy increment). Chaos in nonlinear dynamical systems makes the exact energy increment quantity be chaotic no matter how the nonlinear dynamical systems have separatrices or not. The simple zero of the Melnikov function cannot be simply used for determining the existence of chaos in nonlinear dynamical systems. The first integral quantity increment (i.e., the energy increment) for a certain time interval is achieved for periodic flows and chaos in 2-dimensional nonlinear dynamical systems. Under perturbation assumptions and convergent conditions, the Melnikov function is recovered from the first integral quantity increment. The periodic motions and chaos for the periodically forced, damped Duffing oscillator with a separatrix are presented. The first integral quantity increment of any periodic flow is zero. The first integral quantity increment (i.e., L-function) of chaotic flows at the Poincare mapping points is non-zero and chaotic. The L-function is an exact expression instead of the other approximate expression such as the Melnikov function. The analytical conditions for the global transversality and tangency are obtained from the new theory. The new conditions are exact and independent of the small parameters. The global tangency (or grazing) of a certain motion to the separatrix may induce chaos in nonlinear dynamical systems.

Chapter 10

Flow Switchability in Discontinuous Dynamical Systems

In previous Chapters, the global transversality of a flow to the separatrix in continuous dynamical systems was discussed. Such a global transversality is an origin of chaotic motions in nonlinear dynamic systems, and the resonance mechanism of chaos is also based on the resonant separatrix. On the other hand, for discontinuous dynamical systems, in addition to the generic separatrix, the separation boundaries exist as generalized separatrices. To investigate the transversality of a flow to such separation boundaries, the *G*-functions for discontinuous dynamical systems will be introduced in this Chapter. Based on the new *G*-function, the theory of Luo (2005a, 2006b) for discontinuous dynamical systems can be extended. The switchability of a flow from a domain to an adjacent one will be discussed as a global transversality. The non-passable and passable flows to the separation boundary in discontinuous dynamical systems will be discussed. However, in continuous dynamical systems, the flow is always passable. Thus, the switching bifurcations between the passable and non-passable flows in discontinuous dynamical systems will be presented. Finally, the first integral quantity increment for discontinuous dynamical systems will be presented instead of the Melnikov function to develop the iterative mapping relations.

10.1 Discontinuous dynamic systems

As in Luo (2005a, 2006b), consider a dynamic system consisting of N sub-dynamic systems in a universal domain $\mho \subset \mathbb{R}^n$. The universal domain is divided into N accessible sub-domains Ω_i plus the inaccessible domain Ω_0. The union of the accessible sub-domains is $\cup_{i=1}^{N}\Omega_i$, and the universal domain is $\mho = \cup_{i=1}^{N}\Omega_i \cup \Omega_0$, which can be expressed by an n_1-dimensional sub-vector $\mathbf{x}_{n_1}$ and an $(n-n_1)$-dimensional sub-vector $\mathbf{x}_{n-n_1}$. The complement of the union of the accessible sub-domains $\Omega_0 = \mho \setminus \cup_{i=1}^{m}\Omega_i$ is the union of the inaccessible domains. On the ith-open sub-domain Ω_i, there is a C^r-continuous system ($r \ge 1$) in a form of

$$\dot{\mathbf{x}}^{(i)} \equiv \mathbf{F}^{(i)}(\mathbf{x}^{(i)}, t, \mathbf{p}_i) \in \mathbb{R}^n,\ \ \mathbf{x}^{(i)} = (x_1^{(i)}, x_2^{(i)}, \cdots, x_n^{(i)})^{\mathrm{T}} \in \Omega_i. \tag{10.1}$$

The time is t and $\dot{\mathbf{x}}^{(i)} = d\mathbf{x}^{(i)}/dt$. In an accessible sub-domain Ω_i, the vector field

$\mathbf{F}^{(i)}(\mathbf{x},t,\mathbf{p}_i)$ with parameter vectors $\mathbf{p}=(p_i^{(1)},p_i^{(2)},\cdots,p_i^{(l)})^{\mathrm{T}}\in\mathbb{R}^l$ is C^r-continuous ($r\geq 1$) in a state vector $\mathbf{x}$ and for all time t; and a continuous flow $\mathbf{x}^{(i)}(t)=\mathbf{\Phi}^{(i)}(\mathbf{x}^{(i)}(t_0),t,\mathbf{p}_i)$ with $\mathbf{x}^{(i)}(t_0)=\mathbf{\Phi}^{(i)}(\mathbf{x}^{(i)}(t_0),t_0,\mathbf{p}_i)$ in Eq.(10.1) is C^{r+1}-continuous for time t. The discontinuous dynamics theory developed in this Chapter holds for the following hypothesis.

H1: The switching between two adjacent sub-systems possesses time-continuity.
H2: For an unbounded, accessible sub-domain Ω_i, there is a bounded domain $D_i\subset\Omega_i$ and the corresponding vector field and its flow are bounded, i.e.,

$$\|\mathbf{F}^{(i)}\|\leq K_1(\text{const}) \text{ and } \|\mathbf{\Phi}^{(i)}\|\leq K_2(\text{const}) \text{ on } D_i \text{ for } t\in[0,\infty). \quad (10.2)$$

H3: For a bounded, accessible domain Ω_i, there is a bounded domain $D_i\subset\Omega_i$ and the corresponding vector field is bounded, but the flow may be unbounded, i.e.,

$$\|\mathbf{F}^{(i)}\|\leq K_1(\text{const}) \text{ and } \|\mathbf{\Phi}^{(i)}\|<\infty \text{ on } D_i \text{ for } t\in[0,\infty). \quad (10.3)$$

Because dynamical systems on different accessible sub-domains are distinguishing, the relation between flows in two sub-domains should be developed for the flow continuity. For a sub-domain Ω_i, there are k_i-pieces of boundaries ($k_i\leq n-1$). Consider a boundary of any two adjacent sub-domains, formed by the intersection of the closed sub-domains, i.e., $\partial\Omega_{ij}=\bar{\Omega}_i\cap\bar{\Omega}_j$ ($i,j\in\{1,2,\cdots N\}$, $j\neq i$), as shown in Fig.10.1.

DEFINITION 10.1. The boundary in n-dimensional phase space is defined as

$$\begin{aligned} S_{ij}&\equiv\partial\Omega_{ij}=\bar{\Omega}_i\cap\bar{\Omega}_j\\ &=\left\{\mathbf{x}\middle|\varphi_{ij}(\mathbf{x},t,\boldsymbol{\lambda})=0,\ \varphi_{ij}\text{ is }C^r\text{-continuous }(r\geq1)\right\}\subset\mathbb{R}^{n-1}. \end{aligned} \quad (10.4)$$

The boundary values $\mathbf{x}^{(\alpha)}=(x_1^{(\alpha)},x_2^{(\alpha)},\cdots,x_n^{(\alpha)})^{\mathrm{T}}$ with $\alpha\in\{i,j\}$ pertain to the two open domains Ω_i and Ω_j, respectively. Based on the boundary definition, we have $\partial\Omega_{ij}=\partial\Omega_{ji}$. The separation boundary $\partial\Omega_{ij}$ with $\varphi_{ij}(\mathbf{x},t,\boldsymbol{\lambda})=0$ can be determined by

$$\dot{\mathbf{x}}^{(0)}=\mathbf{F}^{(0)}(\mathbf{x}^{(0)},t,\boldsymbol{\lambda}) \quad (10.5)$$

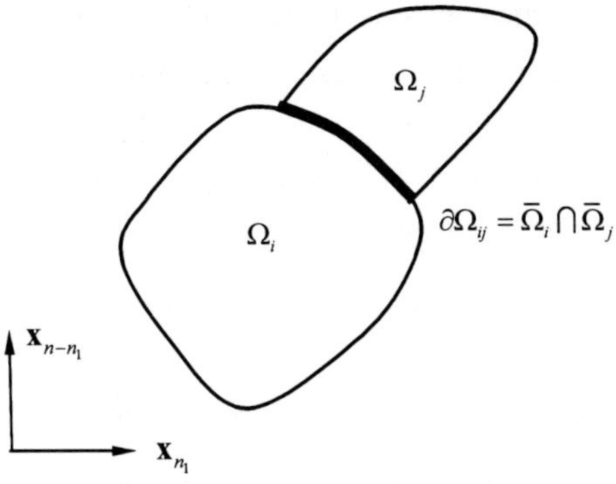

Figure 10.1. Sub-domains Ω_i and Ω_j, the corresponding boundary $\partial\Omega_{ij}$.

where $\mathbf{x}^{(0)} = (x_1^{(0)}, x_1^{(0)}, \cdots, x_1^{(0)})^{\mathrm{T}}$. The flow is given by $\mathbf{x}^{(0)}(t) = \mathbf{\Phi}^{(0)}(\mathbf{x}^{(0)}(t_0), t, \boldsymbol{\lambda})$ with $\mathbf{x}^{(0)}(t_0) = \mathbf{\Phi}^{(0)}(\mathbf{x}^{(0)}(t_0), t_0, \boldsymbol{\lambda})$ is C^{r+1}-continuous for time t.

Because a flow $\mathbf{x}^{(\alpha)}(t)$ approaches the separation boundary with the zero-order $\mathbf{x}^{(\alpha)}(t_{m-}) = \mathbf{x}_m = \mathbf{x}^{(0)}(t_m)$, as in Chapter 2, the G-function is defined as

$$\begin{aligned} G_{\partial\Omega_{ij}}^{(\alpha)}(\mathbf{x}_m, t_m, \mathbf{p}_\alpha, \boldsymbol{\lambda}) &\equiv \mathbf{n}_{\partial\Omega_{ij}}^T(\mathbf{x}^{(0)}, t, \boldsymbol{\lambda}) \bullet [\dot{\mathbf{x}}^{(\alpha)}(t) - \dot{\mathbf{x}}^{(0)}(t)]\Big|_{(\mathbf{x}_m^{(0)}, \mathbf{x}_m^{(\alpha)}, t_{m\pm})} \\ &= \mathbf{n}_{\partial\Omega_{ij}}^T(\mathbf{x}^{(0)}, t, \boldsymbol{\lambda}) \bullet \dot{\mathbf{x}}^{(\alpha)}(t) + \frac{\partial\varphi_{ij}(\mathbf{x}^{(\alpha)}, t, \boldsymbol{\lambda})}{\partial t}\Big|_{(\mathbf{x}_m^{(0)}, \mathbf{x}_m^{(\alpha)}, t_{m\pm})} \\ &= \nabla\varphi_{ij}(\mathbf{x}^{(0)}, t, \boldsymbol{\lambda}) \bullet \dot{\mathbf{x}}^{(\alpha)}(t) + \frac{\partial\varphi_{ij}(\mathbf{x}^{(\alpha)}, t, \boldsymbol{\lambda})}{\partial t}\Big|_{(\mathbf{x}_m^{(0)}, \mathbf{x}_m^{(\alpha)}, t_{m\pm})} \end{aligned} \tag{10.6}$$

where the normal vector of the boundary $\partial\Omega_{ij}$ is

$$\mathbf{n}_{\partial\Omega_{ij}}(\mathbf{x}_m^{(0)}, t_m, \boldsymbol{\lambda}) = \nabla\varphi_{ij}(\mathbf{x}_m^{(0)}, t_m, \boldsymbol{\lambda}) = \left(\frac{\partial\varphi_{ij}}{\partial x_1}, \frac{\partial\varphi_{ij}}{\partial x_2}, \cdots, \frac{\partial\varphi_{ij}}{\partial x_n}\right)^{\mathrm{T}}_{(t_m, \mathbf{x}_m^{(0)})}. \tag{10.7}$$

With Eqs.(10.1) and (10.5), equation (10.6) can be rewritten as

$$\begin{aligned} G_{\partial\Omega_{ij}}^{(\alpha)}(\mathbf{x}_m, t_m, \mathbf{p}_\alpha, \boldsymbol{\lambda}) &\equiv \mathbf{n}_{\partial\Omega_{ij}}^T(\mathbf{x}^{(0)}, t, \boldsymbol{\lambda}) \bullet [\mathbf{F}(\mathbf{x}^{(\alpha)}, t, \mathbf{p}_\alpha) - \mathbf{F}^{(0)}(\mathbf{x}^{(0)}, t, \boldsymbol{\lambda})]\Big|_{(\mathbf{x}_m^{(0)}, \mathbf{x}_m^{(\alpha)}, t_{m\pm})} \\ &= \mathbf{n}_{\partial\Omega_{ij}}^T(\mathbf{x}^{(0)}, t, \boldsymbol{\lambda}) \bullet \mathbf{F}(\mathbf{x}^{(\alpha)}, t, \mathbf{p}_\alpha) + \frac{\partial\varphi_{ij}(\mathbf{x}^{(0)}, t, \boldsymbol{\lambda})}{\partial t}\Big|_{(\mathbf{x}_m^{(0)}, \mathbf{x}_m^{(\alpha)}, t_{m\pm})} \\ &= \nabla\varphi_{ij}(\mathbf{x}^{(0)}, t, \boldsymbol{\lambda}) \bullet \mathbf{F}(\mathbf{x}^{(\alpha)}, t, \mathbf{p}_\alpha) + \frac{\partial\varphi_{ij}(\mathbf{x}^{(0)}, t, \boldsymbol{\lambda})}{\partial t}\Big|_{(\mathbf{x}_m^{(0)}, \mathbf{x}_m^{(\alpha)}, t_{m\pm})}. \end{aligned} \tag{10.8}$$

The G-function $G^{(\alpha)}_{\partial\Omega_{ij}}$ is the component of the vector field in the normal direction of the boundary. The higher derivatives of the G-function is defined as

$$\begin{aligned} G^{(k,\alpha)}_{\partial\Omega_{ij}}(\mathbf{x}_m,t_m,\mathbf{p}_\alpha,\boldsymbol{\lambda}) &= \sum\nolimits_{r=1}^{k+1} C^r_{k+1} D^{k+1-r}_{\mathbf{x}^{(0)}} \mathbf{n}^T_{\partial\Omega_{ij}}(\mathbf{x}^{(0)},t,\boldsymbol{\lambda})\bullet\Big[\tfrac{d^r\mathbf{x}}{dt^r}-\tfrac{d^r\mathbf{x}^{(0)}}{dt^r}\Big]\Big|_{(\bar{\mathbf{x}}^{(0)}_m,\mathbf{x}^{(\alpha)}_m,t_m)} \\ &= \sum\nolimits_{r=1}^{k+1} C^r_{k+1} D^{k+1-r}_{\bar{\mathbf{x}}} \mathbf{n}^T_{\partial\Omega_{ij}}(\mathbf{x}^{(0)},t,\boldsymbol{\lambda})\bullet\Big[D^{r-1}_{\mathbf{x}}\mathbf{F}(\mathbf{x},t,\mathbf{p}_\alpha) \\ &\quad -D^{r-1}_{\mathbf{x}^{(0)}}\mathbf{F}^{(0)}(\mathbf{x}^{(0)},t,\boldsymbol{\lambda})\Big]\Big|_{(\mathbf{x}^{(0)}_m,\mathbf{x}^{(\alpha)}_m,t_m)}. \end{aligned} \tag{10.9}$$

For $k=0$, $G^{(k,\alpha)}_{\partial\Omega_{ij}}(\mathbf{x}_m,t_m,\mathbf{p}_\alpha,\boldsymbol{\lambda}) = G^{(\alpha)}_{\partial\Omega_{ij}}(\mathbf{x}_m,t_m,\mathbf{p}_\alpha,\boldsymbol{\lambda})$. From now on, $\mathbf{n}_{\partial\Omega_{ij}}(\mathbf{x}^{(0)}) \equiv \mathbf{n}_{\partial\Omega_{ij}}(\mathbf{x}^{(0)},t,\boldsymbol{\lambda})$. The *G*-function $G^{(k,\alpha)}_{\partial\Omega_{ij}}$ is the time change rate of $G^{(k-1,\alpha)}_{\partial\Omega_{ij}}$.

10.2. Passable flows

Before we discuss the flow passability in discontinuous dynamical system, as in Luo (2005c, 2006b), the imagery flow concept is introduced first.

DEFINITION 10.2. The C^{r+1} ($r \geq 1$) -continuous flow $\mathbf{x}^{(j)}_i(t)$ is termed the jth -imaginary flow in the ith -open sub-domain Ω_i if the flow $\mathbf{x}^{(j)}_i(t)$ is determined by the C^r -continuous vector field in the jth -open sub-domain Ω_j, to the ith -open sub-domain Ω_i, i.e.,

$$\dot{\mathbf{x}}^{(j)}_i \equiv \mathbf{F}^{(j)}\big(\mathbf{x}^{(j)}_i,t,\boldsymbol{\mu}_j\big) \in \mathbb{R}^n,\ \mathbf{x}^{(j)}_i = \big(x^{(j)}_{1i},x^{(j)}_{2i},\cdots,x^{(j)}_{ni}\big)^T \in \Omega_i, \tag{10.10}$$

with the initial conditions

$$\mathbf{x}^{(j)}_i(t_0) = \mathbf{\Phi}^{(j)}(\mathbf{x}^{(j)}_i(t_0),t_0,\mathbf{p}_j). \tag{10.11}$$

For discontinuous dynamical systems, the passability of a flow to the separation boundary has many cases, compared to the continuous dynamical systems. The passable flow to the separation boundary is discussed first. A passable flow is sketched in Fig.10.2. $\mathbf{x}^{(i)}(t)$ and $\mathbf{x}^{(j)}(t)$ represent the *real* flows in domains Ω_i and Ω_j, respectively. They are depicted by the thin solid curves. $\mathbf{x}^{(j)}_i(t)$ and $\mathbf{x}^{(i)}_j(t)$ are the *imaginary* flows in domains Ω_i and Ω_j, respectively controlled by the vector fields in Ω_j and Ω_i. Such imaginary flows are depicted by the dashed curves. The hollow circles are the switching points and the shaded circles are the starting points. The detail discussion of the real and imaginary flows can be seen

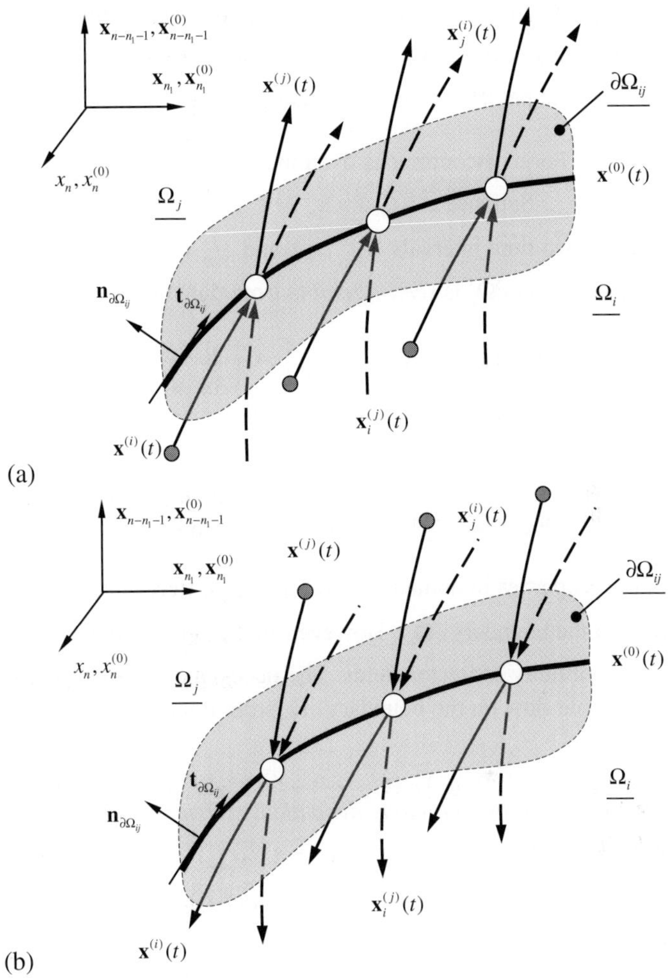

Figure 10.2. Passable flows: (a) from Ω_i to Ω_j with $(2k_i : 2k_j)$-order and (b) from Ω_j to Ω_i with $(2k_j : 2k_i)$-order. $\mathbf{x}^{(i)}(t)$ and $\mathbf{x}^{(j)}(t)$ represent the *real* flows in domains Ω_i and Ω_j, respectively, which are depicted by the thin solid curves. $\mathbf{x}_i^{(j)}(t)$ and $\mathbf{x}_j^{(i)}(t)$ represent the *imaginary* flows in domains Ω_i and Ω_j, respectively controlled by the vector fields in Ω_j and Ω_i, which are depicted by the dashed curves. The flow on the boundary is described by $\mathbf{x}^{(0)}(t)$. The normal and tangential vectors $\mathbf{n}_{\partial\Omega_{ij}}$ and $\mathbf{t}_{\partial\Omega_{ij}}$ on the boundary are depicted. The hollow circles are the switching points and the shaded circles are the starting points.

from Luo (2005c, 2006b). The flow on the boundary is described by $\mathbf{x}^{(0)}(t)$. The normal and tangential vectors $\mathbf{n}_{\partial\Omega_{ij}}$ and $\mathbf{t}_{\partial\Omega_{ij}}$ on the boundary are depicted as well. The definition of the passable flow is given as follows.

DEFINITION 10.3. For a discontinuous dynamical system in Eq.(10.1), $\mathbf{x}(t_m) \equiv \mathbf{x}_m \in \partial\Omega_{ij}$ at time t_m. Suppose $\mathbf{x}^{(i)}(t_{m-}) = \mathbf{x}_m = \mathbf{x}^{(j)}(t_{m+})$. For an arbitrarily small $\varepsilon > 0$, there are two time intervals $[t_{m-\varepsilon}, t_m)$ and $(t_m, t_{m+\varepsilon}]$. The flow $\mathbf{x}^{(i)}(t)$ and $\mathbf{x}^{(j)}(t)$ to the boundary $\partial\Omega_{ij}$ is *semi-passable* from domain Ω_i to Ω_j if

$$\left.\begin{array}{ll}\text{either} & \left.\begin{array}{l}\mathbf{n}^T_{\partial\Omega_{ij}}(\mathbf{x}^{(0)}_{m-\varepsilon})\bullet\left[\mathbf{x}^{(0)}_{m-\varepsilon} - \mathbf{x}^{(i)}_{m-\varepsilon}\right] > 0\\ \mathbf{n}^T_{\partial\Omega_{ij}}(\mathbf{x}^{(0)}_{m+\varepsilon})\bullet\left[\mathbf{x}^{(j)}_{m+\varepsilon} - \mathbf{x}^{(0)}_{m+\varepsilon}\right] > 0\end{array}\right\}\text{for } \mathbf{n}_{\partial\Omega_{ij}} \to \Omega_j\\ \text{or} & \left.\begin{array}{l}\mathbf{n}^T_{\partial\Omega_{ij}}(\mathbf{x}^{(0)}_{m-\varepsilon})\bullet\left[\mathbf{x}^{(0)}_{m-\varepsilon} - \mathbf{x}^{(i)}_{m-\varepsilon}\right] < 0\\ \mathbf{n}^T_{\partial\Omega_{ij}}(\mathbf{x}^{(0)}_{m+\varepsilon})\bullet\left[\mathbf{x}^{(j)}_{m+\varepsilon} - \mathbf{x}^{(0)}_{m+\varepsilon}\right] < 0\end{array}\right\}\text{for } \mathbf{n}_{\partial\Omega_{ij}} \to \Omega_i.\end{array}\right\} \tag{10.12}$$

Since flow properties in domains Ω_i and Ω_j are different, so at point $(t_m, \mathbf{x}_m)$, $G^{(i)}_{\partial\Omega_{ij}} \neq G^{(j)}_{\partial\Omega_{ij}}$ on the boundary $\partial\Omega_{ij}$. However, in Chapter 3, we have $G^{(i)}_{\partial\Omega_{ij}} = G^{(j)}_{\partial\Omega_{ij}}$ owing to the continuity of vector fields. The necessary and sufficient conditions for such a passable flow on the boundary $\partial\Omega_{ij}$ from domain Ω_i to Ω_j are given as follows.

THEOREM 10.1. *For a discontinuous dynamical system in Eq.*(10.1), $\mathbf{x}(t_m) = \mathbf{x}_m \in \partial\Omega_{ij}$ *at time* t_m. *Suppose* $\mathbf{x}^{(i)}(t_{m-}) = \mathbf{x}_m = \mathbf{x}^{(j)}(t_{m+})$. *For an arbitrarily small* $\varepsilon > 0$, *there are two time intervals* $[t_{m-\varepsilon}, t_m)$ *and* $(t_m, t_{m+\varepsilon}]$. *Both flows* $\mathbf{x}^{(i)}(t)$ *and* $\mathbf{x}^{(j)}(t)$ *are* $C^r_{[t_{m-\varepsilon}, t_m)}$ *and* $C^r_{(t_m, t_{m+\varepsilon}]}$*-continuous (*$r \geq 1$*) for time t, respectively, and* $\| d^{r+1}\mathbf{x}^{(\alpha)}/dt^{r+1} \| < \infty$ ($\alpha \in \{i, j\}$). *The flow* $\mathbf{x}^{(i)}(t)$ *and* $\mathbf{x}^{(j)}(t)$ *to the boundary* $\partial\Omega_{ij}$ *is semi-passable from domain* Ω_i *to* Ω_j *iff*

$$\left.\begin{array}{ll}\text{either} & \left.\begin{array}{l}G^{(i)}_{\partial\Omega_{ij}}(\mathbf{x}_m, t_m, \mathbf{p}_i, \boldsymbol{\lambda}) > 0\\ G^{(j)}_{\partial\Omega_{ij}}(\mathbf{x}_m, t_m, \mathbf{p}_j, \boldsymbol{\lambda}) > 0\end{array}\right\}\text{for } \mathbf{n}_{\partial\Omega_{ij}} \to \Omega_j,\\ \text{or} & \left.\begin{array}{l}G^{(i)}_{\partial\Omega_{ij}}(\mathbf{x}_m, t_m, \mathbf{p}_i, \boldsymbol{\lambda}) < 0\\ G^{(j)}_{\partial\Omega_{ij}}(\mathbf{x}_m, t_m, \mathbf{p}_j, \boldsymbol{\lambda}) < 0\end{array}\right\}\text{for } \mathbf{n}_{\partial\Omega_{ij}} \to \Omega_i.\end{array}\right\} \tag{10.13}$$

PROOF. For a point $\mathbf{x}_m \in \partial\Omega_{ij}$, suppose $\mathbf{x}^{(i)}(t_{m-}) = \mathbf{x}_m = \mathbf{x}^{(j)}(t_{m+})$. $\mathbf{x}^{(i)}(t)$ and $\mathbf{x}^{(j)}(t)$ are $C^r_{[t_{m-\varepsilon}, t_m)}$ and $C^r_{(t_m, t_{m+\varepsilon}]}$-continuous ($r \geq 1$) for time *t*. $\|\ddot{\mathbf{x}}^{(\alpha)}(t)\| < \infty$ ($\alpha \in \{i, j\}$) for $0 < \varepsilon << 1$. For $a \in [t_{m-\varepsilon}, t_{m-})$ or $a \in (t_{m+}, t_{m+\varepsilon}]$, the Taylor series expansions of $\mathbf{x}^{(\alpha)}(t_{m\pm\varepsilon})$ with $t_{m\pm\varepsilon} = t_m \pm \varepsilon$ ($\alpha \in \{i, j\}$) to $\mathbf{x}^{(\alpha)}(a)$ give

$$\mathbf{x}^{(\alpha)}_{m\pm\varepsilon} \equiv \mathbf{x}^{(\alpha)}\left(t_{m\pm} \pm \varepsilon\right) = \mathbf{x}^{(\alpha)}(a) + \dot{\mathbf{x}}^{(\alpha)}(a)\left(t_{m\pm} \pm \varepsilon - a\right) + o(\left(t_{m\pm} \pm \varepsilon - a\right)),$$

As $a \to t_{m\pm}$, taking the limit of the foregoing equation leads to

$$\mathbf{x}^{(\alpha)}_{m\pm\varepsilon} \equiv \mathbf{x}^{(\alpha)}\left(t_{m\pm} \pm \varepsilon\right) = \mathbf{x}^{(\alpha)}(t_{m\pm}) + \dot{\mathbf{x}}^{(\alpha)}(t_{m\pm})\left(\pm\varepsilon\right) + o(\left(\pm\varepsilon\right)),$$

In a similar fashion, we have

$$\mathbf{x}^{(0)}_{m\pm\varepsilon} \equiv \mathbf{x}^{(0)}\left(t_m \pm \varepsilon\right) = \mathbf{x}^{(0)}_{m\pm} + \dot{\mathbf{x}}^{(0)}_{m\pm}(\pm\varepsilon) + o\left(\varepsilon\right),$$

$$\mathbf{n}_{\partial\Omega_{ij}}\left(\mathbf{x}^{(0)}_m\right) = \mathbf{n}_{\partial\Omega_{ij}}\left(\mathbf{x}^{(0)}_m\right) + D_{\mathbf{x}^{(0)}}\mathbf{n}_{\partial\Omega_{ij}}\left(\mathbf{x}^{(0)}_m\right)(\pm\varepsilon) + o\left(\pm\varepsilon\right).$$

The ignorance of the ε^2 and high order terms, the deformation of the above equation and left multiplication of $\mathbf{n}_{\partial\Omega_{ij}}$ gives

$$\mathbf{n}^T_{\partial\Omega_{ij}}(\mathbf{x}^{(0)}_{m-\varepsilon})\bullet\left[\mathbf{x}^{(i)}_{m-\varepsilon} - \mathbf{x}^{(0)}_{m-\varepsilon}\right] = \mathbf{n}^T_{\partial\Omega_{ij}}(\mathbf{x}^{(0)}_m)\bullet\left[\mathbf{x}^{(i)}_{m-} - \mathbf{x}^{(0)}_m\right] - \varepsilon G^{(i)}_{\partial\Omega_{ij}}(\mathbf{x}_m, t_m, \mathbf{p}_i, \boldsymbol{\lambda}),$$
$$\mathbf{n}^T_{\partial\Omega_{ij}}(\mathbf{x}^{(0)}_{m+\varepsilon})\bullet\left[\mathbf{x}^{(j)}_{m+\varepsilon} - \mathbf{x}^{(0)}_{m+\varepsilon}\right] = \mathbf{n}^T_{\partial\Omega_{ij}}(\mathbf{x}^{(0)}_m)\bullet\left[\mathbf{x}^{(j)}_{m+} - \mathbf{x}^{(0)}_m\right] + \varepsilon G^{(j)}_{\partial\Omega_{ij}}(\mathbf{x}_m, t_m, \mathbf{p}_j, \boldsymbol{\lambda}).$$

Because of $\mathbf{x}^{(\alpha)}_{m\pm} = \mathbf{x}^{(0)}_m = \mathbf{x}_m$, the foregoing equation becomes

$$\mathbf{n}^T_{\partial\Omega_{ij}}(\mathbf{x}^{(0)}_{m-\varepsilon})\bullet\left[\mathbf{x}^{(0)}_{m-\varepsilon} - \mathbf{x}^{(i)}_{m-\varepsilon}\right] = \varepsilon G^{(i)}_{\partial\Omega_{ij}}(\mathbf{x}_m, t_m, \mathbf{p}_i, \boldsymbol{\lambda}),$$
$$\mathbf{n}^T_{\partial\Omega_{ij}}(\mathbf{x}^{(0)}_{m+\varepsilon})\bullet\left[\mathbf{x}^{(j)}_{m+\varepsilon} - \mathbf{x}^{(0)}_{m+\varepsilon}\right] = \varepsilon G^{(j)}_{\partial\Omega_{ij}}(\mathbf{x}_m, t_m, \mathbf{p}_j, \boldsymbol{\lambda}).$$

With Eq.(10.13), the foregoing equation gives Eq.(10.12). Using Eq.(10.12), equation (10.13) can be obtained. This theorem is proved. ■

The above theorem is identical to the one in Luo (2005a, 2006b) owing to the zero-order contact between the flow and boundary. With the higher order singularity of a flow to the boundary, the $(2k_i : m_j)$-passable flow ($k_i, m_j \in \mathbb{N}$) is described as follows.

DEFINITION 10.4. For a discontinuous dynamical system in Eq.(10.1), $\mathbf{x}(t_m) \equiv \mathbf{x}_m \in \partial\Omega_{ij}$ at time t_m. Suppose $\mathbf{x}^{(i)}(t_{m-}) = \mathbf{x}_m = \mathbf{x}^{(j)}(t_{m+})$. For an arbitrarily small $\varepsilon > 0$, there are two time intervals $[t_{m-\varepsilon}, t_m)$ and $(t_m, t_{m+\varepsilon}]$. The flow $\mathbf{x}^{(i)}(t)$ is

$C^r_{[t_{m-\varepsilon},t_m)}$-continuous ($r \geq 2k_i+1$) for time t and $\| d^{r+1}\mathbf{x}^{(i)}/dt^{r+1} \| < \infty$. The flow $\mathbf{x}^{(j)}(t)$ is $C^r_{(t_m,t_{m+\varepsilon}]}$-continuous ($r \geq m_j+1$) for time t and $\| d^{r+1}\mathbf{x}^{(j)}/dt^{r+1} \| < \infty$. The flow $\mathbf{x}^{(i)}(t)$ of the $(2k_i)^{th}$-order and $\mathbf{x}^{(j)}(t)$ of the $(m_j)^{th}$-order to the boundary $\partial\Omega_{ij}$ is $(2k_i : m_j)$-*semi-passable* from domain Ω_i to Ω_j (expressed by $\overrightarrow{\partial\Omega_{ij}}$) if

$$\left.\begin{array}{l} G^{(s,i)}_{\partial\Omega_{ij}}(\mathbf{x}_m, t_{m-}, \mathbf{p}_i, \boldsymbol{\lambda}) = 0 \ \text{ for } s = 0,1,\cdots,2k_i-1 \\ G^{(2k_i,i)}_{\partial\Omega_{ij}}(\mathbf{x}_m, t_{m-}, \mathbf{p}_i, \boldsymbol{\lambda}) \neq 0, \end{array}\right\} \tag{10.14}$$

$$\left.\begin{array}{l} G^{(s,j)}_{\partial\Omega_{ij}}(\mathbf{x}_m, t_{m+}, \mathbf{p}_j, \boldsymbol{\lambda}) = 0 \ \text{ for } s = 0,1,\cdots,m_j-1 \\ G^{(m_j,j)}_{\partial\Omega_{ij}}(\mathbf{x}_m, t_{m+}, \mathbf{p}_j, \boldsymbol{\lambda}) \neq 0, \end{array}\right\} \tag{10.15}$$

$$\left.\begin{array}{ll} \text{either} & \left.\begin{array}{l} \mathbf{n}^T_{\partial\Omega_{ij}}(\mathbf{x}^{(0)}_{m-\varepsilon})\bullet\left[\mathbf{x}^{(0)}_{m-\varepsilon} - \mathbf{x}^{(i)}_{m-\varepsilon}\right] > 0 \\ \mathbf{n}^T_{\partial\Omega_{ij}}(\mathbf{x}^{(0)}_{m+\varepsilon})\bullet\left[\mathbf{x}^{(j)}_{m+\varepsilon} - \mathbf{x}^{(0)}_{m+\varepsilon}\right] > 0 \end{array}\right\} \text{for } \mathbf{n}_{\partial\Omega_{ij}} \to \Omega_j \\ \text{or} & \left.\begin{array}{l} \mathbf{n}^T_{\partial\Omega_{ij}}(\mathbf{x}^{(0)}_{m-\varepsilon})\bullet\left[\mathbf{x}^{(0)}_{m-\varepsilon} - \mathbf{x}^{(i)}_{m-\varepsilon}\right] < 0 \\ \mathbf{n}^T_{\partial\Omega_{ij}}(\mathbf{x}^{(0)}_{m+\varepsilon})\bullet\left[\mathbf{x}^{(j)}_{m+\varepsilon} - \mathbf{x}^{(0)}_{m+\varepsilon}\right] < 0 \end{array}\right\} \text{for } \mathbf{n}_{\partial\Omega_{ij}} \to \Omega_i. \end{array}\right\} \tag{10.16}$$

If $m_j = 2k_j$, the $(2k_i : 2k_j)$-passable flow can be shown in Fig.10.2. However, for $m_j = 2k_j - 1$, the $(2k_i : 2k_j - 1)$-passable flow from domain Ω_i to Ω_j is sketched in Fig.10.3(a). The tangential flow of the $(2k_j-1)^{th}$-order exists in domain Ω_j. The dotted curves represent the tangential curves to the boundary for time $t \in [t_{m-\varepsilon}, t_m)$. The starting point of the flow is $(t_{m-\varepsilon}, \mathbf{x}^{(i)}_{m-\varepsilon})$ in domain Ω_i. If the flow arrives to the point $(t_m, \mathbf{x}_m)$ of the boundary $\partial\Omega_{ij}$, the flow will follow the tangential flow in domain Ω_j. The $(2k_j : 2k_i - 1)$-passable flow from domain Ω_j to Ω_i is presented in Fig.10.3(b) with the same behavior as in Fig.10.3(a). So, a new passable flow is formed as the post-transversal, tangential flow in Luo (2005b). From the definition of the $(2k_i : m_j)$-passable flow, the corresponding necessary and sufficient conditions can be given through the following theorem.

THEOREM 10.2. *For a discontinuous dynamical system in Eq.*(10.1), $\mathbf{x}(t_m) \equiv \mathbf{x}_m \in \partial\Omega_{ij}$ *at time* t_m. *Suppose* $\mathbf{x}^{(i)}(t_{m-}) = \mathbf{x}_m = \mathbf{x}^{(j)}(t_{m+})$. *For an arbitrarily small* $\varepsilon > 0$, *there are two time intervals* $[t_{m-\varepsilon}, t_m)$ *and* $(t_m, t_{m+\varepsilon}]$. *The flow* $\mathbf{x}^{(i)}(t)$ *is*

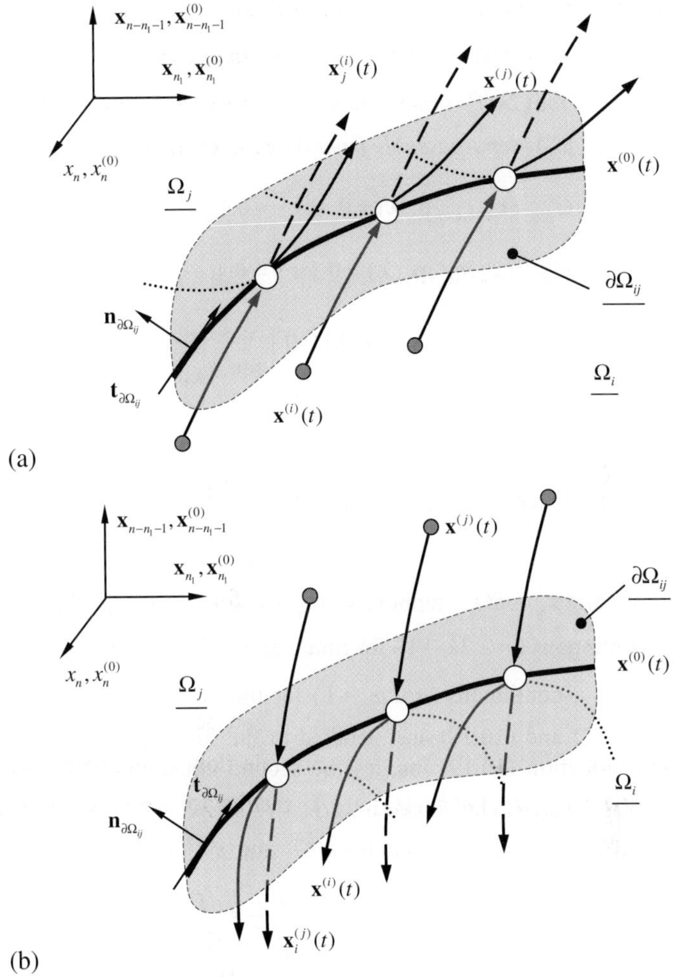

Figure 10.3. Passable flows: (a) from Ω_i to Ω_j with $(2k_i : 2k_j - 1)$-order and (b) from Ω_j to Ω_i with $(2k_j : 2k_j - 1)$-order. $\mathbf{x}^{(i)}(t)$ and $\mathbf{x}^{(j)}(t)$ represent the *real* flows in domains Ω_i and Ω_j, respectively, which are depicted by the thin solid curves. $\mathbf{x}^{(j)}_i(t)$ and $\mathbf{x}^{(i)}_j(t)$ give the *imaginary* flows in domains Ω_i and Ω_j, respectively controlled by the vector fields in Ω_j and Ω_i, which are depicted by the dashed curves. The flow on the boundary is described by $\mathbf{x}^{(0)}(t)$. The normal and tangential vectors $\mathbf{n}_{\partial\Omega_{ij}}$ and $\mathbf{t}_{\partial\Omega_{ij}}$ on the boundary are depicted. The dotted curves represent the curves before time t_{m+}. The hollow circles are the switching points and the shaded circles are the starting points.

$C^r_{[t_{m-\varepsilon},t_m)}$-continuous ($r \ge 2k_i+1$) for time t and $\| d^{r+1}\mathbf{x}^{(i)}/dt^{r+1} \| < \infty$. The flow $\mathbf{x}^{(j)}(t)$ is $C^r_{(t_m,t_{m+\varepsilon}]}$-continuous ($r \ge m_j+1$) for time t and $\| d^{r+1}\mathbf{x}^{(j)}/dt^{r+1} \| < \infty$. The flow $\mathbf{x}^{(i)}(t)$ of the $(2k_i)^{th}$-order and $\mathbf{x}^{(j)}(t)$ of the $(m_j)^{th}$-order to the boundary $\partial\Omega_{ij}$ is $(2k_i : m_j)$-semi-passable from domain Ω_i to Ω_j iff

$$G^{(s,i)}_{\partial\Omega_{ij}}(\mathbf{x}_m, t_{m-}, \mathbf{p}_i, \boldsymbol{\lambda}) = 0 \ \text{ for } s = 0,1,\cdots,2k_i-1; \tag{10.17}$$

$$G^{(s,j)}_{\partial\Omega_{ij}}(\mathbf{x}_m, t_{m+}, \mathbf{p}_j, \boldsymbol{\lambda}) = 0 \ \text{ for } s = 0,1,\cdots,m_j-1; \tag{10.18}$$

$$\left.\begin{array}{ll}\text{either} & \left.\begin{array}{l} G^{(2k_i,i)}_{\partial\Omega_{ij}}(\mathbf{x}_m, t_{m-}, \mathbf{p}_i, \boldsymbol{\lambda}) > 0 \\ G^{(m_j,j)}_{\partial\Omega_{ij}}(\mathbf{x}_m, t_{m+}, \mathbf{p}_j, \boldsymbol{\lambda}) > 0 \end{array}\right\} \text{for } \mathbf{n}_{\partial\Omega_{ij}} \to \Omega_j \\ \text{or} & \left.\begin{array}{l} G^{(2k_i,i)}_{\partial\Omega_{ij}}(\mathbf{x}_m, t_{m-}, \mathbf{p}_i, \boldsymbol{\lambda}) < 0 \\ G^{(m_j,j)}_{\partial\Omega_{ij}}(\mathbf{x}_m, t_{m+}, \mathbf{p}_j, \boldsymbol{\lambda}) < 0 \end{array}\right\} \text{for } \mathbf{n}_{\partial\Omega_{ij}} \to \Omega_i. \end{array}\right\} \tag{10.19}$$

PROOF. For a point $\mathbf{x}_m \in \partial\Omega_{ij}$, suppose $\mathbf{x}^{(i)}(t_{m-}) = \mathbf{x}_m = \mathbf{x}^{(j)}(t_{m+})$. The flow $\mathbf{x}^{(i)}(t)$ is $C^r_{[t_{m-\varepsilon},t_m)}$-continuous ($r \ge 2k_i+1$) for time t and $\| d^{r+1}\mathbf{x}^{(i)}/dt^{r+1} \| < \infty$. The flow $\mathbf{x}^{(j)}(t)$ is $C^r_{(t_m,t_{m+\varepsilon}]}$-continuous ($r \ge m_j+1$) for time t, and $\| d^{r+1}\mathbf{x}^{(j)}/dt^{r+1} \| < \infty$. Equations (10.17) and (10.18) are identical to the first equations of Eq.(10.14) and (10.15). Equation (10.19) implies the second equations of Eq.(10.14) and (10.15). For $a \in [t_{m-\varepsilon}, t_{m-})$ or $b \in (t_{m+}, t_{m+\varepsilon}]$, the Taylor series of $\mathbf{x}^{(i)}(t_{m-\varepsilon})$ and $\mathbf{x}^{(j)}(t_{m+\varepsilon})$ at $\mathbf{x}^{(i)}(a)$ and $\mathbf{x}^{(j)}(b)$ up to the ε^{2k_i+1} and ε^{m_j+1}-terms give

$$\begin{aligned}\mathbf{x}^{(i)}_{m-\varepsilon} &\equiv \mathbf{x}^{(i)}(t_{m-} - \varepsilon) = \mathbf{x}^{(i)}(a) + \sum_{s=1}^{2k_i} \tfrac{d^s}{dt^s}\mathbf{x}^{(i)}(a)(t_{m-} - \varepsilon - a)^s \\ &\quad + \tfrac{d^{2k_i+1}}{dt^{2k_i+1}}\mathbf{x}^{(i)}(a)(t_{m-} - \varepsilon - a)^{2k_i+1} + o((t_{m-} - \varepsilon - a)^{2k_i+1}),\end{aligned}$$

$$\begin{aligned}\mathbf{x}^{(j)}_{m+\varepsilon} &\equiv \mathbf{x}^{(j)}(t_{m+} + \varepsilon) = \mathbf{x}^{(j)}(b) + \sum_{s=1}^{m_j} \tfrac{d^s}{dt^s}\mathbf{x}^{(j)}(b)(t_{m+} + \varepsilon - b)^s \\ &\quad + \tfrac{d^{m_j+1}}{dt^{m_j+1}}\mathbf{x}^{(j)}(b)(t_{m+} + \varepsilon - b)^{m_j+1} + o((t_{m+} + \varepsilon - b)^{m_j+1}),\end{aligned}$$

As $a \to t_{m-}$ and $b \to t_{m+}$, taking the limit of the foregoing equations leads to

$$\begin{aligned}\mathbf{x}^{(i)}_{m-\varepsilon} &\equiv \mathbf{x}^{(i)}(t_{m-} - \varepsilon) = \mathbf{x}^{(i)}(t_{m-}) + \sum_{s=1}^{2k_i} \tfrac{d^s}{dt^s}\mathbf{x}^{(i)}(t_{m-})(-\varepsilon)^s \\ &\quad + \tfrac{d^{2k_i+1}}{dt^{2k_i+1}}\mathbf{x}^{(i)}(t_{m-})(-\varepsilon)^{2k_i+1} + o((-\varepsilon)^{2k_i+1}),\end{aligned}$$

$$\mathbf{x}_{m+\varepsilon}^{(j)} \equiv \mathbf{x}^{(j)}\left(t_{m+}+\varepsilon\right)=\mathbf{x}^{(j)}(t_{m+})+\sum_{s=1}^{m_j}\tfrac{d^s}{dt^s}\mathbf{x}^{(j)}(t_+)\varepsilon^s$$
$$+\tfrac{d^{m_j+1}}{dt^{m_j+1}}\mathbf{x}^{(j)}(t_{m+})\varepsilon^{m_j+1}+o(\varepsilon^{m_j+1}).$$

In a similar fashion, we have

$$\mathbf{x}_{m-\varepsilon}^{(0)} \equiv \mathbf{x}^{(0)}\left(t_m-\varepsilon\right)=\mathbf{x}^{(0)}(t_m)+\sum_{s=1}^{2k_i}\tfrac{d^s}{dt^s}\mathbf{x}^{(0)}(t_m)\left(-\varepsilon\right)^s$$
$$+\tfrac{d^{2k_i+1}}{dt^{2k_i+1}}\mathbf{x}^{(0)}(t_m)\left(-\varepsilon\right)^{2k_i+1}+o((-\varepsilon)^{2k_i+1}),$$

$$\mathbf{x}_{m+\varepsilon}^{(0)} \equiv \mathbf{x}^{(0)}\left(t_{m+}+\varepsilon\right)=\mathbf{x}^{(0)}(t_m)+\sum_{s=1}^{m_j}\tfrac{d^s}{dt^s}\mathbf{x}^{(0)}(t_m)\varepsilon^s$$
$$+\tfrac{d^{m_j+1}}{dt^{m_j+1}}\mathbf{x}^{(0)}(t_m)\varepsilon^{m_j+1}+o(\varepsilon^{m_j+1});$$

$$\mathbf{n}_{\partial\Omega_{ij}}(\mathbf{x}_{m-\varepsilon}^{(0)})=\mathbf{n}_{\partial\Omega_{ij}}(\mathbf{x}_m^{(0)})+\sum_{s=1}^{2k_i}D_{\mathbf{x}_m^{(0)}}^s\mathbf{n}_{\partial\Omega_{ij}}(\mathbf{x}_m^{(0)})\left(-\varepsilon\right)^s$$
$$+D_{\mathbf{x}_m^{(0)}}^{2k_i+1}\mathbf{n}_{\partial\Omega_{ij}}(\mathbf{x}_m^{(0)})\left(-\varepsilon\right)^{2k_i+1}+o((-\varepsilon)^{2k_i+1}),$$

$$\mathbf{n}_{\partial\Omega_{ij}}(\mathbf{x}_{m+\varepsilon}^{(0)})=\mathbf{n}_{\partial\Omega_{ij}}(\mathbf{x}_m^{(0)})+\sum_{s=1}^{m_j}D_{\mathbf{x}_m^{(0)}}^s\mathbf{n}_{\partial\Omega_{ij}}(\mathbf{x}_m^{(0)})\varepsilon^s$$
$$+D_{\mathbf{x}_m^{(0)}}^{m_j+1}\mathbf{n}_{\partial\Omega_{ij}}(\mathbf{x}_m^{(0)})\varepsilon^{m_j+1}+o(\varepsilon^{m_j+1}).$$

The ignorance of the ε^{2k_i+2} and ε^{m_j+2}-terms and high order terms, the deformation of the above equation and left multiplication of $\mathbf{n}_{\partial\Omega_{ij}}$ gives

$$\mathbf{n}_{\partial\Omega_{ij}}^T(\mathbf{x}_{m-\varepsilon}^{(0)})\bullet\left[\mathbf{x}_{m-\varepsilon}^{(i)}-\mathbf{x}_{m-\varepsilon}^{(0)}\right]=\mathbf{n}_{\partial\Omega_{ij}}^T(\mathbf{x}_m^{(0)})\bullet\left[\mathbf{x}_{m-}^{(i)}-\mathbf{x}_m^{(0)}\right]$$
$$+\sum_{s=0}^{2k_i}\left(-\varepsilon\right)^s G_{\partial\Omega_{ij}}^{(s-1,i)}(\mathbf{x}_m,t_{m-},\mathbf{p}_i,\boldsymbol{\lambda})$$
$$+\left(-\varepsilon\right)^{2k_i+1}G_{\partial\Omega_{ij}}^{(2k_i,i)}(\mathbf{x}_m,t_{m-},\mathbf{p}_i,\boldsymbol{\lambda}),$$

$$\mathbf{n}_{\partial\Omega_{ij}}^T(\mathbf{x}_{m+\varepsilon}^{(0)})\bullet\left[\mathbf{x}_{m+\varepsilon}^{(j)}-\mathbf{x}_{m+\varepsilon}^{(0)}\right]=\mathbf{n}_{\partial\Omega_{ij}}^T(\mathbf{x}_m^{(0)})\bullet\left[\mathbf{x}_{m+}^{(j)}-\mathbf{x}_m^{(0)}\right]$$
$$+\sum_{s=1}^{m_j}\varepsilon^s G_{\partial\Omega_{ij}}^{(s-1,j)}(\mathbf{x}_m,t_{m+},\mathbf{p}_j,\boldsymbol{\lambda})$$
$$+\varepsilon^{m_j+1}G_{\partial\Omega_{ij}}^{(m_j,j)}(\mathbf{x}_m,t_{m+},\mathbf{p}_j,\boldsymbol{\lambda}).$$

Because of $\mathbf{x}_{m\pm}^{(\alpha)}=\mathbf{x}_m^{(0)}=\mathbf{x}_m$, with Eqs.(10.15) and (10.16), we have

$$\mathbf{n}_{\partial\Omega_{ij}}^T(\mathbf{x}_{m-\varepsilon}^{(0)})\bullet\left[\mathbf{x}_{m-\varepsilon}^{(0)}-\mathbf{x}_{m-\varepsilon}^{(i)}\right]=(-1)^{2k_i+2}\varepsilon^{k_i}G_{\partial\Omega_{ij}}^{(2k_i,i)}(\mathbf{x}_m,t_{m-},\mathbf{p}_i,\boldsymbol{\lambda}),$$
$$\mathbf{n}_{\partial\Omega_{ij}}^T(\mathbf{x}_{m+\varepsilon}^{(0)})\bullet\left[\mathbf{x}_{m+\varepsilon}^{(j)}-\mathbf{x}_{m+\varepsilon}^{(0)}\right]=\varepsilon^{m_j+1}G_{\partial\Omega_{ij}}^{(m_j,j)}(\mathbf{x}_m,t_{m+},\mathbf{p}_j,\boldsymbol{\lambda}).$$

With Eq.(10.19), the foregoing equation gives Eq.(10.16). But, using Eq.(10.16), the foregoing equation gives Eq.(10.19). The proof is completed. ■

10.3. Non-passable flows

In this section, the non-passable flows will be discussed. The detailed discussion can be found in Luo (2005a, 2006b). Several non-passable flows is sketched in Fig.10.4 for helping one understand the concept of non-passable flows. The full non-passable flow of the first kind (sink flows) and the full non-passable flow of the second kind (source flows) are sketched in Fig.10.4(a) and (b). The sink flow is defined as follows.

DEFINITION 10.5. For a discontinuous dynamical system in Eq.(10.1), $\mathbf{x}(t_m) \equiv \mathbf{x}_m \in \partial\Omega_{ij}$ at time t_m. Suppose $\mathbf{x}^{(i)}(t_{m-}) = \mathbf{x}_m = \mathbf{x}^{(j)}(t_{m-})$. For an arbitrarily small $\varepsilon > 0$, there is a time interval $[t_{m-\varepsilon}, t_m)$. The flow $\mathbf{x}^{(i)}(t)$ and $\mathbf{x}^{(j)}(t)$ to the boundary $\partial\Omega_{ij}$ is *non-passable* with the first kind (or called the sink flow) if

$$\left.\begin{array}{ll}\text{either} & \left.\begin{array}{l}\mathbf{n}^T_{\partial\Omega_{ij}}(\mathbf{x}^{(0)}_{m-\varepsilon})\cdot\left[\mathbf{x}^{(0)}_{m-\varepsilon} - \mathbf{x}^{(i)}_{m-\varepsilon}\right] > 0\\ \mathbf{n}^T_{\partial\Omega_{ij}}(\mathbf{x}^{(0)}_{m-\varepsilon})\cdot\left[\mathbf{x}^{(0)}_{m-\varepsilon} - \mathbf{x}^{(j)}_{m-\varepsilon}\right] < 0\end{array}\right\}\text{for } \mathbf{n}_{\partial\Omega_{ij}} \to \Omega_j \\ \text{or} & \left.\begin{array}{l}\mathbf{n}^T_{\partial\Omega_{ij}}(\mathbf{x}^{(0)}_{m-\varepsilon})\cdot\left[\mathbf{x}^{(0)}_{m-\varepsilon} - \mathbf{x}^{(i)}_{m-\varepsilon}\right] < 0\\ \mathbf{n}^T_{\partial\Omega_{ij}}(\mathbf{x}^{(0)}_{m-\varepsilon})\cdot\left[\mathbf{x}^{(0)}_{m-\varepsilon} - \mathbf{x}^{(j)}_{m-\varepsilon}\right] > 0\end{array}\right\}\text{for } \mathbf{n}_{\partial\Omega_{ij}} \to \Omega_i.\end{array}\right\} \tag{10.20}$$

THEOREM 10.3. *For a discontinuous dynamical system in Eq.*(10.1), $\mathbf{x}(t_m) \equiv \mathbf{x}_m \in \partial\Omega_{ij}$ *at time* t_m. *Suppose* $\mathbf{x}^{(i)}(t_{m-}) = \mathbf{x}_m = \mathbf{x}^{(j)}(t_{m-})$. *For an arbitrarily small* $\varepsilon > 0$, *there is a time interval* $[t_{m-\varepsilon}, t_m)$. *The flow* $\mathbf{x}^{(i)}(t)$ *and* $\mathbf{x}^{(j)}(t)$ *are* $C^r_{[t_{m-\varepsilon}, t_m)}$-*continuous* ($r \geq 1$) *for time* t *and* $\| d^{r+1}\mathbf{x}^{(\alpha)}/dt^{r+1} \| < \infty$ ($\alpha \in \{i, j\}$). *The flow* $\mathbf{x}^{(i)}(t)$ *and* $\mathbf{x}^{(j)}(t)$ *to the boundary* $\partial\Omega_{ij}$ *is non-passable with the first kind iff*

$$\left.\begin{array}{ll}\text{either} & \left.\begin{array}{l}G^{(i)}_{\partial\Omega_{ij}}(\mathbf{x}_m, t_{m-}, \mathbf{p}_i, \boldsymbol{\lambda}) > 0\\ G^{(j)}_{\partial\Omega_{ij}}(\mathbf{x}_m, t_{m-}, \mathbf{p}_j, \boldsymbol{\lambda}) < 0\end{array}\right\}\text{for } \mathbf{n}_{\partial\Omega_{ij}} \to \Omega_j, \\ \text{or} & \left.\begin{array}{l}G^{(i)}_{\partial\Omega_{ij}}(\mathbf{x}_m, t_{m-}, \mathbf{p}_i, \boldsymbol{\lambda}) < 0\\ G^{(j)}_{\partial\Omega_{ij}}(\mathbf{x}_m, t_{m-}, \mathbf{p}_j, \boldsymbol{\lambda}) > 0\end{array}\right\}\text{for } \mathbf{n}_{\partial\Omega_{ij}} \to \Omega_i.\end{array}\right\} \tag{10.21}$$

PROOF. The proof is the same as in Theorem 10.1. ■

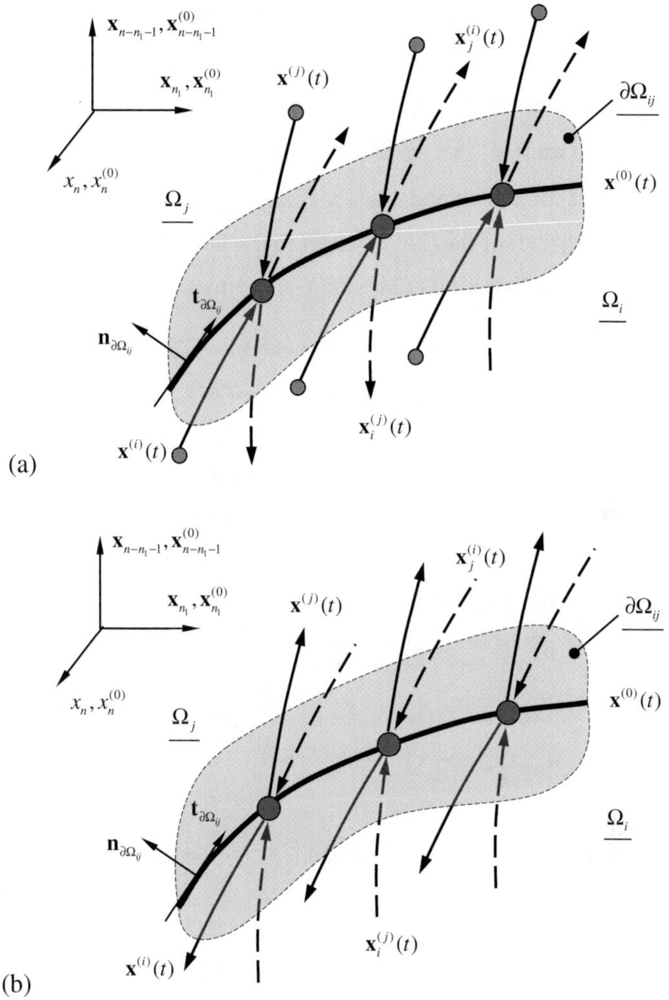

Figure 10.4. The $(2k_i : 2k_j)$ -non-passable flows of: (a) the first kind (sink flows) and (b) the second kind (source flows). $\mathbf{x}^{(i)}(t)$ and $\mathbf{x}^{(j)}(t)$ represent the *real* flows in domains Ω_i and Ω_j, respectively, which are depicted by the thin solid curves. $\mathbf{x}_i^{(j)}(t)$ and $\mathbf{x}_j^{(i)}(t)$ represent the *imaginary* flows in domains Ω_i and Ω_j, respectively controlled by the vector fields in Ω_j and Ω_i, which are depicted by the dashed curves. The flow on the boundary is described by $\mathbf{x}^{(0)}(t)$. The normal and tangential vectors $\mathbf{n}_{\partial\Omega_{ij}}$ and $\mathbf{t}_{\partial\Omega_{ij}}$ on the boundary are depicted. The hollow circles are the switching points and the shaded circles are the starting points.

DEFINITION 10.6. For a discontinuous dynamical system in Eq.(10.1), $\mathbf{x}(t_m) \equiv \mathbf{x}_m \in \partial\Omega_{ij}$ at time t_m. Suppose $\mathbf{x}^{(i)}(t_{m-}) = \mathbf{x}_m = \mathbf{x}^{(j)}(t_{m-})$. For an arbitrarily small $\varepsilon > 0$, there is a time interval $[t_{m-\varepsilon}, t_m)$. The flow $\mathbf{x}^{(i)}(t)$ is $C^r_{[t_{m-\varepsilon}, t_m)}$-continuous ($r \geq 2k_i + 1$) for time t and $\| d^{r+1}\mathbf{x}^{(i)} / dt^{r+1} \| < \infty$. The flow $\mathbf{x}^{(j)}(t)$ is $C^r_{[t_{m-\varepsilon}, t_m)}$-continuous ($r \geq 2k_j + 1$) for time t and $\| d^{r+1}\mathbf{x}^{(j)} / dt^{r+1} \| < \infty$. The flow $\mathbf{x}^{(i)}(t)$ of the $(2k_i)^{th}$-order and $\mathbf{x}^{(j)}(t)$ of the $(2k_j)^{th}$-order to the boundary $\partial\Omega_{ij}$ is $(2k_i : 2k_j)$-*non-passable* with the first kind ($(2k_i : 2k_j)$-sink flow) if

$$\left.\begin{aligned} &G^{(s,i)}_{\partial\Omega_{ij}}(\mathbf{x}_m, t_{m-}, \mathbf{p}_i, \boldsymbol{\lambda}) = 0 \quad \text{for } s = 0, 1, \cdots, 2k_i - 1 \\ &G^{(2k_i,i)}_{\partial\Omega_{ij}}(\mathbf{x}_m, t_{m-}, \mathbf{p}_i, \boldsymbol{\lambda}) \neq 0, \end{aligned}\right\} \tag{10.22}$$

$$\left.\begin{aligned} &G^{(s,j)}_{\partial\Omega_{ij}}(\mathbf{x}_m, t_{m-}, \mathbf{p}_j, \boldsymbol{\lambda}) = 0 \quad \text{for } s = 0, 1, \cdots, 2k_j - 1 \\ &G^{(2k_j,j)}_{\partial\Omega_{ij}}(\mathbf{x}_m, t_{m-}, \mathbf{p}_j, \boldsymbol{\lambda}) \neq 0, \end{aligned}\right\} \tag{10.23}$$

$$\left.\begin{aligned} &\text{either} \left.\begin{aligned} &\mathbf{n}^T_{\partial\Omega_{ij}}(\mathbf{x}^{(0)}_{m-\varepsilon}) \bullet \left[\mathbf{x}^{(0)}_{m-\varepsilon} - \mathbf{x}^{(i)}_{m-\varepsilon}\right] > 0 \\ &\mathbf{n}^T_{\partial\Omega_{ij}}(\mathbf{x}^{(0)}_{m-\varepsilon}) \bullet \left[\mathbf{x}^{(0)}_{m-\varepsilon} - \mathbf{x}^{(j)}_{m-\varepsilon}\right] < 0 \end{aligned}\right\} \text{for } \mathbf{n}_{\partial\Omega_{ij}} \to \Omega_j \\ &\text{or} \left.\begin{aligned} &\mathbf{n}^T_{\partial\Omega_{ij}}(\mathbf{x}^{(0)}_{m-\varepsilon}) \bullet \left[\mathbf{x}^{(0)}_{m-\varepsilon} - \mathbf{x}^{(i)}_{m-\varepsilon}\right] < 0 \\ &\mathbf{n}^T_{\partial\Omega_{ij}}(\mathbf{x}^{(0)}_{m-\varepsilon}) \bullet \left[\mathbf{x}^{(0)}_{m-\varepsilon} - \mathbf{x}^{(j)}_{m-\varepsilon}\right] > 0 \end{aligned}\right\} \text{for } \mathbf{n}_{\partial\Omega_{ij}} \to \Omega_i. \end{aligned}\right\} \tag{10.24}$$

THEOREM 10.4. *For a discontinuous dynamical system in Eq.*(10.1), $\mathbf{x}(t_m) \equiv \mathbf{x}_m \in \partial\Omega_{ij}$ *at time* t_m. *Suppose* $\mathbf{x}^{(i)}(t_{m-}) = \mathbf{x}_m = \mathbf{x}^{(j)}(t_{m-})$. *For an arbitrarily small* $\varepsilon > 0$, *there is a time interval* $[t_{m-\varepsilon}, t_m)$. *The flow* $\mathbf{x}^{(i)}(t)$ *is* $C^r_{[t_{m-\varepsilon}, t_m)}$-*continuous* ($r \geq 2k_i + 1$) *for time t and* $\| d^{r+1}\mathbf{x}^{(i)} / dt^{r+1} \| < \infty$. *The flow* $\mathbf{x}^{(j)}(t)$ *is* $C^r_{[t_{m-\varepsilon}, t_m)}$-*continuous* ($r \geq 2k_j + 1$) *for time t and* $\| d^{r+1}\mathbf{x}^{(j)} / dt^{r+1} \| < \infty$. *The flow* $\mathbf{x}^{(i)}(t)$ *of the* $(2k_i)^{th}$-*order and* $\mathbf{x}^{(j)}(t)$ *of the* $(2k_j)^{th}$-*order to the boundary* $\partial\Omega_{ij}$ *is* $(2k_i : 2k_j)$-*non-passable with the first kind* ($(2k_i : 2k_j)$-sink flow) *iff*

$$G^{(s,i)}_{\partial\Omega_{ij}}(\mathbf{x}_m, t_{m-}, \mathbf{p}_i, \boldsymbol{\lambda}) = 0 \quad \text{for } s = 0, 1, \cdots, 2k_i - 1; \tag{10.25}$$

$$G^{(s,j)}_{\partial\Omega_{ij}}(\mathbf{x}_m, t_{m-}, \mathbf{p}_j, \boldsymbol{\lambda}) = 0 \quad \text{for } s = 0, 1, \cdots, 2k_j - 1; \tag{10.26}$$

$$\left.\begin{array}{ll}\text{either} & \left.\begin{array}{l}G_{\partial\Omega_{ij}}^{(2k_i,i)}(\mathbf{x}_m,t_{m-},\mathbf{p}_i,\boldsymbol{\lambda})>0\\ G_{\partial\Omega_{ij}}^{(2k_j,j)}(\mathbf{x}_m,t_{m-},\mathbf{p}_i,\boldsymbol{\lambda})<0\end{array}\right\}\text{for } \mathbf{n}_{\partial\Omega_{ij}}\to\Omega_j\\ \text{or} & \left.\begin{array}{l}G_{\partial\Omega_{ij}}^{(2k_i,i)}(\mathbf{x}_m,t_{m-},\mathbf{p}_i,\boldsymbol{\lambda})<0\\ G_{\partial\Omega_{ij}}^{(2k_j,j)}(\mathbf{x}_m,t_{m-},\mathbf{p}_i,\boldsymbol{\lambda})>0\end{array}\right\}\text{for } \mathbf{n}_{\partial\Omega_{ij}}\to\Omega_i.\end{array}\right\} \tag{10.27}$$

PROOF. The proof is similar to the proof of Theorem 10.2. ■

DEFINITION 10.7. For a discontinuous dynamical system in Eq.(10.1), $\mathbf{x}(t_m)\equiv\mathbf{x}_m \in\partial\Omega_{ij}$ at time t_m. Suppose $\mathbf{x}^{(i)}(t_{m+})=\mathbf{x}_m=\mathbf{x}^{(j)}(t_{m+})$. For an arbitrarily small $\varepsilon>0$, there is a time interval $(t_m,t_{m+\varepsilon}]$. The flow $\mathbf{x}^{(i)}(t)$ and $\mathbf{x}^{(j)}(t)$ to the boundary $\partial\Omega_{ij}$ is *non-passable* with the second kind (or called the source flow) if

$$\left.\begin{array}{ll}\text{either} & \left.\begin{array}{l}\mathbf{n}_{\partial\Omega_{ij}}^T(\mathbf{x}_{m+\varepsilon}^{(0)})\bullet\left[\mathbf{x}_{m+\varepsilon}^{(i)}-\mathbf{x}_{m+\varepsilon}^{(0)}\right]<0\\ \mathbf{n}_{\partial\Omega_{ij}}^T(\mathbf{x}_{m+\varepsilon}^{(0)})\bullet\left[\mathbf{x}_{m+\varepsilon}^{(j)}-\mathbf{x}_{m+\varepsilon}^{(0)}\right]>0\end{array}\right\}\text{for } \mathbf{n}_{\partial\Omega_{ij}}\to\Omega_j\\ \text{or} & \left.\begin{array}{l}\mathbf{n}_{\partial\Omega_{ij}}^T(\mathbf{x}_{m+\varepsilon}^{(0)})\bullet\left[\mathbf{x}_{m+\varepsilon}^{(i)}-\mathbf{x}_{m+\varepsilon}^{(0)}\right]>0\\ \mathbf{n}_{\partial\Omega_{ij}}^T(\mathbf{x}_{m+\varepsilon}^{(0)})\bullet\left[\mathbf{x}_{m+\varepsilon}^{(j)}-\mathbf{x}_{m+\varepsilon}^{(0)}\right]<0\end{array}\right\}\text{for } \mathbf{n}_{\partial\Omega_{ij}}\to\Omega_i.\end{array}\right\} \tag{10.28}$$

THEOREM 10.5. *For a discontinuous dynamical system in Eq.*(10.1), $\mathbf{x}(t_m)\equiv\mathbf{x}_m \in\partial\Omega_{ij}$ *at time* t_m. *Suppose* $\mathbf{x}^{(i)}(t_{m+})=\mathbf{x}_m=\mathbf{x}^{(j)}(t_{m+})$. *For an arbitrarily small* $\varepsilon>0$, *there is a time interval* $(t_m,t_{m+\varepsilon}]$. *The flow* $\mathbf{x}^{(i)}(t)$ *is* $C^r_{(t_m,t_{m+\varepsilon}]}$*-continuous* $(r\geq 1)$ *for time t and* $\| d^{r+1}\mathbf{x}^{(i)}/dt^{r+1}\|<\infty$. *The flow* $\mathbf{x}^{(j)}(t)$ *is* $C^r_{(t_m,t_{m+\varepsilon}]}$*-continuous* $(r\geq 1)$ *for time t and* $\| d^{r+1}\mathbf{x}^{(j)}/dt^{r+1}\|<\infty$. *The flow* $\mathbf{x}^{(i)}(t)$ *and* $\mathbf{x}^{(j)}(t)$ *to the boundary* $\partial\Omega_{ij}$ *is non-passable with the second kind* (*or source flow*) *iff*

$$\left.\begin{array}{ll}\text{either} & \left.\begin{array}{l}G_{\partial\Omega_{ij}}^{(i)}(\mathbf{x}_m,t_{m+},\mathbf{p}_i,\boldsymbol{\lambda})<0\\ G_{\partial\Omega_{ij}}^{(j)}(\mathbf{x}_m,t_{m+},\mathbf{p}_j,\boldsymbol{\lambda})>0\end{array}\right\}\text{for } \mathbf{n}_{\partial\Omega_{ij}}\to\Omega_j,\\ \text{or} & \left.\begin{array}{l}G_{\partial\Omega_{ij}}^{(i)}(\mathbf{x}_m,t_{m+},\mathbf{p}_i,\boldsymbol{\lambda})>0\\ G_{\partial\Omega_{ij}}^{(j)}(\mathbf{x}_m,t_{m+},\mathbf{p}_j,\boldsymbol{\lambda})<0\end{array}\right\}\text{for } \mathbf{n}_{\partial\Omega_{ij}}\to\Omega_i.\end{array}\right\} \tag{10.29}$$

PROOF. The proof is the same as in Theorem 10.1. ■

DEFINITION 10.8. For a discontinuous dynamical system in Eq.(10.1), $\mathbf{x}(t_m) \equiv \mathbf{x}_m \in \partial\Omega_{ij}$ at time t_m. Suppose $\mathbf{x}^{(i)}(t_{m+}) = \mathbf{x}_m = \mathbf{x}^{(j)}(t_{m+})$. For an arbitrarily small $\varepsilon > 0$, there is a time interval $(t_m, t_{m+\varepsilon}]$. The flow $\mathbf{x}^{(i)}(t)$ is $C^r_{(t_m, t_{m+\varepsilon}]}$-continuous ($r \geq 2k_i + 1$) for time t and $\| d^{r+1}\mathbf{x}^{(i)}/dt^{r+1} \| < \infty$. The flow $\mathbf{x}^{(j)}(t)$ is $C^r_{(t_m, t_{m+\varepsilon}]}$-continuous ($r \geq 2k_j + 1$) for time t and $\| d^{r+1}\mathbf{x}^{(j)}/dt^{r+1} \| < \infty$. The flow $\mathbf{x}^{(i)}(t)$ of the $(2k_i)^{th}$-order and $\mathbf{x}^{(j)}(t)$ of the $(2k_j)^{th}$-order to the boundary $\partial\Omega_{ij}$ is $(2k_i : 2k_j)$-*non-passable* with the second kind (or $(2k_i : 2k_j)$-source flow) if

$$\left.\begin{aligned} &G^{(s,i)}_{\partial\Omega_{ij}}(\mathbf{x}_m, t_{m+}, \mathbf{p}_i, \boldsymbol{\lambda}) = 0 \ \text{ for } s = 0, 1, \cdots, 2k_i - 1 \\ &G^{(2k_i,i)}_{\partial\Omega_{ij}}(\mathbf{x}_m, t_{m+}, \mathbf{p}_i, \boldsymbol{\lambda}) \neq 0, \end{aligned}\right\} \tag{10.30}$$

$$\left.\begin{aligned} &G^{(s,j)}_{\partial\Omega_{ij}}(\mathbf{x}_m, t_{m+}, \mathbf{p}_j, \boldsymbol{\lambda}) = 0 \ \text{ for } s = 0, 1, \cdots, 2k_j - 1 \\ &G^{(2k_j,j)}_{\partial\Omega_{ij}}(\mathbf{x}_m, t_{m+}, \mathbf{p}_j, \boldsymbol{\lambda}) \neq 0, \end{aligned}\right\} \tag{10.31}$$

$$\left.\begin{aligned} &\text{either} \left.\begin{aligned} &\mathbf{n}^T_{\partial\Omega_{ij}}(\mathbf{x}^{(0)}_{m+\varepsilon}) \cdot \left[\mathbf{x}^{(i)}_{m+\varepsilon} - \mathbf{x}^{(0)}_{m+\varepsilon}\right] < 0 \\ &\mathbf{n}^T_{\partial\Omega_{ij}}(\mathbf{x}^{(0)}_{m+\varepsilon}) \cdot \left[\mathbf{x}^{(j)}_{m+\varepsilon} - \mathbf{x}^{(0)}_{m+\varepsilon}\right] > 0 \end{aligned}\right\} \text{for } \mathbf{n}_{\partial\Omega_{ij}} \to \Omega_j \\ &\text{or} \left.\begin{aligned} &\mathbf{n}^T_{\partial\Omega_{ij}}(\mathbf{x}^{(0)}_{m+\varepsilon}) \cdot \left[\mathbf{x}^{(i)}_{m+\varepsilon} - \mathbf{x}^{(0)}_{m+\varepsilon}\right] > 0 \\ &\mathbf{n}^T_{\partial\Omega_{ij}}(\mathbf{x}^{(0)}_{m+\varepsilon}) \cdot \left[\mathbf{x}^{(j)}_{m+\varepsilon} - \mathbf{x}^{(0)}_{m+\varepsilon}\right] < 0 \end{aligned}\right\} \text{for } \mathbf{n}_{\partial\Omega_{ij}} \to \Omega_i. \end{aligned}\right\} \tag{10.32}$$

THEOREM 10.6. *For a discontinuous dynamical system in Eq.*(10.1), $\mathbf{x}(t_m) \equiv \mathbf{x}_m \in \partial\Omega_{ij}$ *at time* t_m. *Suppose* $\mathbf{x}^{(i)}(t_{m+}) = \mathbf{x}_m = \mathbf{x}^{(j)}(t_{m+})$. *For an arbitrarily small* $\varepsilon > 0$, *there is a time interval* $(t_m, t_{m+\varepsilon}]$. *The flow* $\mathbf{x}^{(i)}(t)$ *is* $C^r_{(t_m, t_{m+\varepsilon}]}$-*continuous* ($r \geq 2k_i + 1$) *for time t and* $\| d^{r+1}\mathbf{x}^{(i)}/dt^{r+1} \| < \infty$. *The flow* $\mathbf{x}^{(j)}(t)$ *is* $C^r_{(t_m, t_{m+\varepsilon}]}$-*continuous* ($r \geq 2k_j + 1$) *for time t and* $\| d^{r+1}\mathbf{x}^{(j)}/dt^{r+1} \| < \infty$. *The flow* $\mathbf{x}^{(i)}(t)$ *of the* $(2k_i)^{th}$-*order and* $\mathbf{x}^{(j)}(t)$ *of the* $(2k_j)^{th}$-*order to the boundary* $\partial\Omega_{ij}$ *is* $(2k_i : 2k_j)$-*non-passable with the second kind (or* $(2k_i : 2k_j)$-*source flow) iff*

$$G^{(s,i)}_{\partial\Omega_{ij}}(\mathbf{x}_m, t_{m+}, \mathbf{p}_i, \boldsymbol{\lambda}) = 0 \ \text{ for } s = 0, 1, \cdots, 2k_i - 1; \tag{10.33}$$

$$G^{(s,j)}_{\partial\Omega_{ij}}(\mathbf{x}_m, t_{m+}, \mathbf{p}_j, \boldsymbol{\lambda}) = 0 \ \text{ for } s = 0, 1, \cdots, 2k_j - 1; \tag{10.34}$$

$$\left.\begin{array}{ll}\text{either} & \left.\begin{array}{l}G_{\partial\Omega_{ij}}^{(2k_i,i)}(\mathbf{x}_m,t_{m+},\mathbf{p}_i,\boldsymbol{\lambda})<0\\ G_{\partial\Omega_{ij}}^{(2k_j,j)}(\mathbf{x}_m,t_{m+},\mathbf{p}_j,\boldsymbol{\lambda})>0\end{array}\right\}\text{for }\mathbf{n}_{\partial\Omega_{ij}}\to\Omega_j\\ \text{or} & \left.\begin{array}{l}G_{\partial\Omega_{ij}}^{(2k_i,i)}(\mathbf{x}_m,t_{m+},\mathbf{p}_i,\boldsymbol{\lambda})>0\\ G_{\partial\Omega_{ij}}^{(2k_j,j)}(\mathbf{x}_m,t_{m+},\mathbf{p}_j,\boldsymbol{\lambda})<0\end{array}\right\}\text{for }\mathbf{n}_{\partial\Omega_{ij}}\to\Omega_i.\end{array}\right\} \tag{10.35}$$

PROOF. The proof is similar to the proof of Theorem 10.2. ■

The full sink and source flows to the boundary were discussed. Next, the half -non-passable flows to the boundary will be discussed. The half-sink flow to the separation boundary is sketched in Fig.10.5. The half-non-passable flow of the first kind is termed *the half-sink flow*. Such a half-sink flow in Ω_i is shown in Fig.10.5(a). Only $\mathbf{x}^{(i)}(t)$ for time $t\in[t_{m-\varepsilon},t_m)$ is a real flow, and the other imaginary flows $\mathbf{x}_j^{(i)}(t)$ for time $t\in[t_{m-\varepsilon},t_{m+\varepsilon}]$ and $\mathbf{x}_i^{(j)}(t)$ for $t\in(t_m,t_{m+\varepsilon}]$ are represented by dashed curves. To the same boundary $\partial\Omega_{ij}$, a half sink flow in Ω_j is sketched in Fig.10.5(b). The input flow $\mathbf{x}^{(j)}(t)$ for time $t\in[t_{m-\varepsilon},\varepsilon)$ is only a real flow. The strict mathematic description is given as follows.

DEFINITION 10.9. For a discontinuous dynamical system in Eq.(10.1), $\mathbf{x}(t_m)\equiv\mathbf{x}_m\in\partial\Omega_{ij}$ at time t_m. Suppose $\mathbf{x}^{(i)}(t_{m-})=\mathbf{x}_m=\mathbf{x}_i^{(j)}(t_{m\pm})$. For an arbitrarily small $\varepsilon>0$, there are two time intervals $[t_{m-\varepsilon},t_m)$ and $[t_{m-\varepsilon},t_{m+\varepsilon}]$. The flow $\mathbf{x}^{(i)}(t)$ is $C^r_{[t_{m-\varepsilon},t_m)}$-continuous ($r\geq 2k_i+1$) for t and $\|d^{r+1}\mathbf{x}^{(i)}/dt^{r+1}\|<\infty$. The imaginary flow $\mathbf{x}_i^{(j)}(t)$ is $C^r_{[t_{m-\varepsilon},t_{m+\varepsilon}]}$-continuous ($r\geq 2k_j$) and $\|d^{r+1}\mathbf{x}_i^{(j)}/dt^{r+1}\|<\infty$ for time t. The flow $\mathbf{x}^{(i)}(t)$ of the $(2k_i)^{th}$-order and $\mathbf{x}_i^{(j)}(t)$ of the $(2k_j-1)^{th}$-order to the boundary $\partial\Omega_{ij}$ is $(2k_i:2k_j-1)$-half-*non-passable* with the first kind in domain Ω_i (or $(2k_i:2k_j-1)$-half sink flow) if

$$\left.\begin{array}{l}G_{\partial\Omega_{ij}}^{(s,i)}(\mathbf{x}_m,t_{m-},\mathbf{p}_i,\boldsymbol{\lambda})=0\ \text{ for } s=0,1,\cdots,2k_i-1\\ G_{\partial\Omega_{ij}}^{(2k_i,i)}(\mathbf{x}_m,t_{m-},\mathbf{p}_i,\boldsymbol{\lambda})\neq 0,\end{array}\right\} \tag{10.36}$$

$$\left.\begin{array}{l}G_{\partial\Omega_{ij}}^{(s,j)}(\mathbf{x}_m,t_{m\pm},\mathbf{p}_j,\boldsymbol{\lambda})=0\ \text{ for } s=0,1,\cdots,2k_j-2\\ G_{\partial\Omega_{ij}}^{(2k_j-1,j)}(\mathbf{x}_m,t_{m\pm},\mathbf{p}_j,\boldsymbol{\lambda})\neq 0,\end{array}\right\} \tag{10.37}$$

$$\left.\begin{array}{ll}\text{either} & \mathbf{n}^T_{\partial\Omega_{ij}}(\mathbf{x}^{(0)}_{m-\varepsilon})\bullet\left[\mathbf{x}^{(0)}_{m-\varepsilon}-\mathbf{x}^{(i)}_{m-\varepsilon}\right]>0 \ \text{ for } \mathbf{n}_{\partial\Omega_{ij}}\rightarrow\Omega_j\\ \text{or} & \mathbf{n}^T_{\partial\Omega_{ij}}(\mathbf{x}^{(0)}_{m-\varepsilon})\bullet\left[\mathbf{x}^{(0)}_{m-\varepsilon}-\mathbf{x}^{(i)}_{m-\varepsilon}\right]<0 \ \text{ for } \mathbf{n}_{\partial\Omega_{ij}}\rightarrow\Omega_i,\end{array}\right\} \tag{10.38}$$

$$\left.\begin{array}{ll}\text{either} & \left.\begin{array}{l}\mathbf{n}^T_{\partial\Omega_{ij}}(\mathbf{x}^{(0)}_{m-\varepsilon})\bullet\left[\mathbf{x}^{(0)}_{m-\varepsilon}-\mathbf{x}^{(j)}_{i(m-\varepsilon)}\right]>0\\ \mathbf{n}^T_{\partial\Omega_{ij}}(\mathbf{x}^{(0)}_{m+\varepsilon})\bullet\left[\mathbf{x}^{(j)}_{i(m+\varepsilon)}-\mathbf{x}^{(0)}_{m+\varepsilon}\right]<0\end{array}\right\}\text{for } \mathbf{n}_{\partial\Omega_{ij}}\rightarrow\Omega_j\\ \text{or} & \left.\begin{array}{l}\mathbf{n}^T_{\partial\Omega_{ij}}(\mathbf{x}^{(0)}_{m-\varepsilon})\bullet\left[\mathbf{x}^{(0)}_{m-\varepsilon}-\mathbf{x}^{(j)}_{i(m-\varepsilon)}\right]<0\\ \mathbf{n}^T_{\partial\Omega_{ij}}(\mathbf{x}^{(0)}_{m+\varepsilon})\bullet\left[\mathbf{x}^{(j)}_{i(m+\varepsilon)}-\bar{\mathbf{x}}^{(0)}_{m+\varepsilon}\right]>0\end{array}\right\}\text{for } \mathbf{n}_{\partial\Omega_{ij}}\rightarrow\Omega_i.\end{array}\right\} \tag{10.39}$$

From the above definition, the corresponding theorem gives the necessary and sufficient conditions for such a half-nonpassable flow of the first kind (or half-sink flow).

THEOREM 10.7. *For a discontinuous dynamical system in Eq.*(10.1), $\mathbf{x}(t_m)\equiv\mathbf{x}_m \in\partial\Omega_{ij}$ *at time* t_m. *Suppose* $\mathbf{x}^{(i)}(t_{m-})=\mathbf{x}_m=\mathbf{x}^{(j)}_i(t_{m\pm})$. *For an arbitrarily small* $\varepsilon>0$, *there are two time intervals* $[t_{m-\varepsilon},t_m)$ *and* $[t_{m-\varepsilon},t_{m+\varepsilon}]$. *The flow* $\mathbf{x}^{(i)}(t)$ *is* $C^r_{[t_{m-\varepsilon},t_m)}$*-continuous* ($r\geq 2k_i+1$) *and* $\| d^{r+1}\mathbf{x}^{(i)}/dt^{r+1}\|<\infty$ *for time t. The imaginary flow* $\mathbf{x}^{(j)}_i(t)$ *is* $C^r_{[t_{m-\varepsilon},t_{m+\varepsilon}]}$*-continuous* ($r\geq 2k_j$) *and* $\| d^{r+1}\mathbf{x}^{(j)}_i/dt^{r+1}\|<\infty$ *for time t. The flow* $\mathbf{x}^{(i)}(t)$ *of the* $(2k_i)^{th}$*-order and* $\mathbf{x}^{(j)}_i(t)$ *of the* $(2k_j-1)^{th}$*-order to the boundary* $\partial\Omega_{ij}$ *is* $(2k_i:2k_j-1)$*-half-non-passable with the first kind in domain* Ω_i *(or* $(2k_i:2k_j-1)$*-half sink flow) iff*

$$G^{(s,i)}_{\partial\Omega_{ij}}(\mathbf{x}_m,t_{m-},\mathbf{p}_i,\boldsymbol{\lambda})=0 \ \text{ for } s=0,1,\cdots,2k_i-1; \tag{10.40}$$

$$G^{(s,j)}_{\partial\Omega_{ij}}(\mathbf{x}_m,t_{m\pm},\mathbf{p}_j,\boldsymbol{\lambda})=0 \ \text{ for } s=0,1,\cdots,2k_j-2; \tag{10.41}$$

$$\left.\begin{array}{ll}\text{either} & \left.\begin{array}{l}G^{(2k_i,i)}_{\partial\Omega_{ij}}(\mathbf{x}_m,t_{m-},\mathbf{p}_i,\boldsymbol{\lambda})>0\\ G^{(2k_j-1,j)}_{\partial\Omega_{ij}}(\mathbf{x}_m,t_{m\pm},\mathbf{p}_j,\boldsymbol{\lambda})<0\end{array}\right\}\text{for } \mathbf{n}_{\partial\Omega_{ij}}\rightarrow\Omega_j\\ \text{or} & \left.\begin{array}{l}G^{(2k_i,i)}_{\partial\Omega_{ij}}(\mathbf{x}_m,t_{m-},\mathbf{p}_i,\boldsymbol{\lambda})<0\\ G^{(2k_j-1,j)}_{\partial\Omega_{ij}}(\mathbf{x}_m,t_{m\pm},\mathbf{p}_j,\boldsymbol{\lambda})>0\end{array}\right\}\text{for } \mathbf{n}_{\partial\Omega_{ij}}\rightarrow\Omega_i.\end{array}\right\} \tag{10.42}$$

PROOF. The proof is similar to the proof of Theorem 10.2. ■

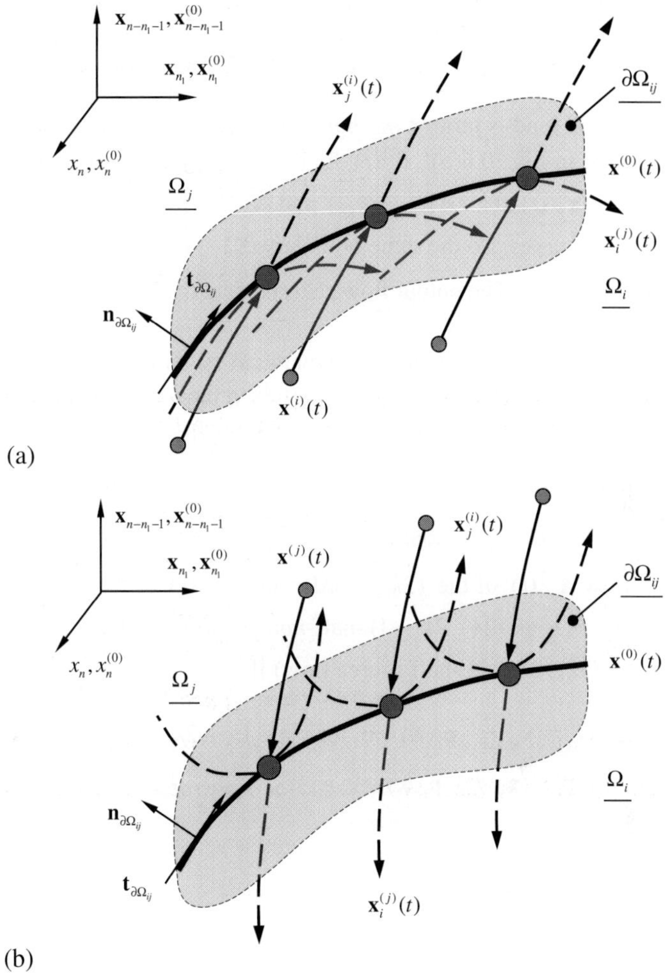

Figure 10.5. The half sink flows in: (a) in Ω_i with $(2k_i : 2k_j - 1)$ -order and (b) in Ω_j with $(2k_j : 2k_i - 1)$ -order. $\mathbf{x}^{(i)}(t)$ and $\mathbf{x}^{(j)}(t)$ represent the *real* flows in domains Ω_i and Ω_j, respectively, which are depicted by the thin solid curves. $\mathbf{x}^{(j)}_i(t)$ and $\mathbf{x}^{(i)}_j(t)$ represent the *imaginary* flows in domains Ω_i and Ω_j, respectively controlled by the vector fields in Ω_j and Ω_i, which are depicted by the dashed curves. The flow on the boundary is described by $\mathbf{x}^{(0)}(t)$. The normal and tangential vectors $\mathbf{n}_{\partial\Omega_{ij}}$ and $\mathbf{t}_{\partial\Omega_{ij}}$ on the boundary are depicted. The hollow circles are the switching points and the shaded circles are the starting points.

After the description of a non-passable flow of the first kind, the half-non-passable flow of the second kind will be discussed. Before the mathematical description is given, the intuitive illustration of the half-non-passable flow is sketched in Fig.10.6 to help one understand this new concept. The half-non-passable flow of the second kind is termed *the half-source flow*. The half-source flows in Ω_i are presented in Fig.10.6(a). $\mathbf{x}^{(i)}(t)$ for $t \in (t_m, t_{m+\varepsilon}]$ is only the real flow. The imaginary flows $\mathbf{x}_j^{(i)}(t)$ for $t \in [t_{m-\varepsilon}, t_{m+\varepsilon}]$ and $\mathbf{x}_i^{(j)}(t)$ for $t \in [t_{m-\varepsilon}, t_m)$ are represented by dashed curves. To the same boundary $\partial\Omega_{ij}$, a half-source flow in Ω_j is sketched in Fig.10.6(b). The output flow $\mathbf{x}^{(j)}(t)$ for $t \in (t_m, t_{m+\varepsilon}]$ is a real flow.

DEFINITION 10.10. For a discontinuous dynamical system in Eq.(10.1), $\mathbf{x}(t_m) \equiv \mathbf{x}_m \in \partial\Omega_{ij}$ at time t_m. For an arbitrarily small $\varepsilon > 0$, there are two time intervals $[t_{m-\varepsilon}, t_m)$ and $[t_{m-\varepsilon}, t_{m+\varepsilon}]$. Suppose $\mathbf{x}^{(i)}(t_{m+}) = \mathbf{x}_m = \mathbf{x}_i^{(j)}(t_{m\pm})$. The flow $\mathbf{x}^{(i)}(t)$ is $C^r_{[t_{m-\varepsilon}, t_m)}$-continuous ($r \geq 2k_i + 1$) and $\| d^{r+1}\mathbf{x}^{(i)}/dt^{r+1} \| < \infty$ for time *t*. The imaginary flow $\mathbf{x}_i^{(j)}(t)$ is $C^r_{[t_{m-\varepsilon}, t_{m+\varepsilon}]}$-continuous ($r \geq 2k_j$) and $\| d^{r+1}\mathbf{x}_i^{(j)}/dt^{r+1} \| < \infty$ for time *t*. The flow $\mathbf{x}^{(i)}(t)$ of the $(2k_i)^{th}$-order and $\mathbf{x}_i^{(j)}(t)$ of the $(2k_j - 1)^{th}$-order to the boundary $\partial\Omega_{ij}$ is $(2k_i : 2k_j - 1)$-half-*non-passable* with the second kind in domain Ω_i (or $(2k_i : 2k_j - 1)$-half source flow) if

$$\left.\begin{aligned} &G^{(s,i)}_{\partial\Omega_{ij}}(\mathbf{x}_m, t_{m+}, \mathbf{p}_i, \boldsymbol{\lambda}) = 0 \ \text{ for } s = 0, 1, \cdots, 2k_i - 1 \\ &G^{(2k_i,i)}_{\partial\Omega_{ij}}(\mathbf{x}_m, t_{m+}, \mathbf{p}_i, \boldsymbol{\lambda}) \neq 0, \end{aligned}\right\} \tag{10.43}$$

$$\left.\begin{aligned} &G^{(s,j)}_{\partial\Omega_{ij}}(\mathbf{x}_m, t_{m\pm}, \mathbf{p}_j, \boldsymbol{\lambda}) = 0 \ \text{ for } s = 0, 1, \cdots, 2k_j - 2 \\ &G^{(2k_j-1,j)}_{\partial\Omega_{ij}}(\mathbf{x}_m, t_{m\pm}, \mathbf{p}_j, \boldsymbol{\lambda}) \neq 0, \end{aligned}\right\} \tag{10.44}$$

$$\left.\begin{aligned} &\text{either } \mathbf{n}^T_{\partial\Omega_{ij}}(\mathbf{x}^{(0)}_{m+\varepsilon})\cdot\left[\mathbf{x}^{(i)}_{m+\varepsilon} - \mathbf{x}^{(0)}_{m+\varepsilon}\right] < 0 \ \text{ for } \mathbf{n}_{\partial\Omega_{ij}} \to \Omega_j \\ &\text{or } \quad \mathbf{n}^T_{\partial\Omega_{ij}}(\mathbf{x}^{(0)}_{m+\varepsilon})\cdot\left[\mathbf{x}^{(i)}_{m+\varepsilon} - \mathbf{x}^{(0)}_{m+\varepsilon}\right] > 0 \ \text{ for } \mathbf{n}_{\partial\Omega_{ij}} \to \Omega_i, \end{aligned}\right\} \tag{10.45}$$

$$\left.\begin{aligned} &\text{either } \left.\begin{aligned} &\mathbf{n}^T_{\partial\Omega_{ij}}(\mathbf{x}^{(0)}_{m-\varepsilon})\cdot\left[\mathbf{x}^{(0)}_{m-\varepsilon} - \mathbf{x}^{(j)}_{i(m-\varepsilon)}\right] > 0 \\ &\mathbf{n}^T_{\partial\Omega_{ij}}(\mathbf{x}^{(0)}_{m+\varepsilon})\cdot\left[\mathbf{x}^{(j)}_{i(m+\varepsilon)} - \mathbf{x}^{(0)}_{m+\varepsilon}\right] < 0 \end{aligned}\right\} \text{for } \mathbf{n}_{\partial\Omega_{ij}} \to \Omega_j \\ &\text{or } \left.\begin{aligned} &\mathbf{n}^T_{\partial\Omega_{ij}}(\mathbf{x}^{(0)}_{m-\varepsilon})\cdot\left[\mathbf{x}^{(0)}_{m-\varepsilon} - \mathbf{x}^{(j)}_{i(m-\varepsilon)}\right] < 0 \\ &\mathbf{n}^T_{\partial\Omega_{ij}}(\mathbf{x}^{(0)}_{m+\varepsilon})\cdot\left[\mathbf{x}^{(j)}_{i(m+\varepsilon)} - \overline{\mathbf{x}}^{(0)}_{m+\varepsilon}\right] > 0 \end{aligned}\right\} \text{for } \mathbf{n}_{\partial\Omega_{ij}} \to \Omega_i. \end{aligned}\right\} \tag{10.46}$$

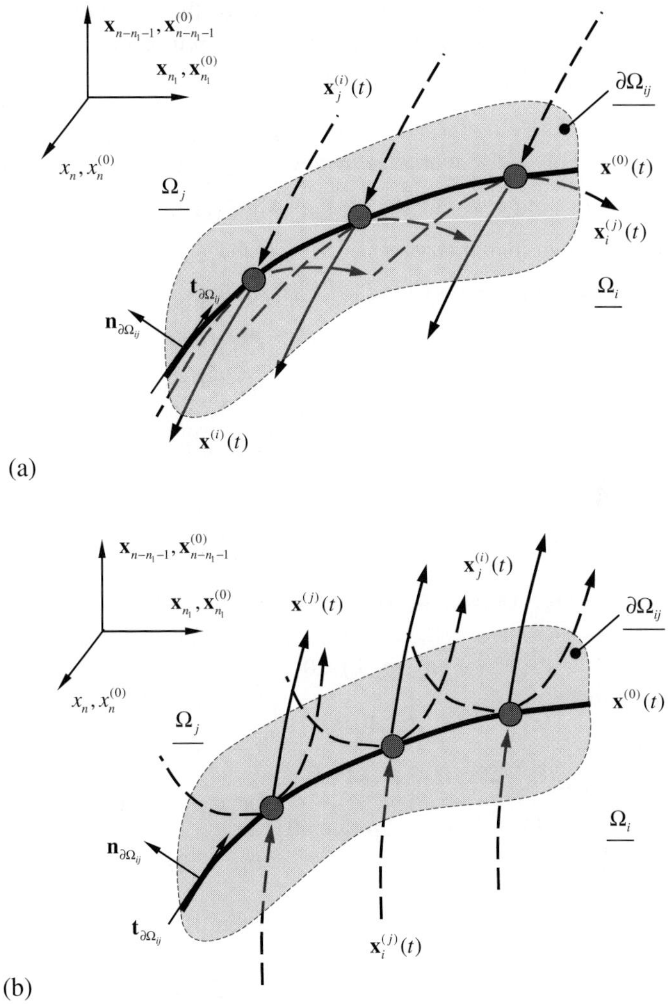

Figure 10.6. Half source flows: (a) in Ω_i with $(2k_i : 2k_j - 1)$-order and (b) in Ω_j with $(2k_i : 2k_j - 1)$-order. $\mathbf{x}^{(i)}(t)$ and $\mathbf{x}^{(j)}(t)$ represent the *real* flows in domains Ω_i and Ω_j, respectively, which are depicted by the thin solid curves. $\mathbf{x}^{(j)}_i(t)$ and $\mathbf{x}^{(i)}_j(t)$ represent the *imaginary* flows in domains Ω_i and Ω_j ,respectively controlled by the vector fields in Ω_j and Ω_i, which are depicted by the dashed curves. The boundary is described by $\mathbf{x}^{(0)}(t)$. The normal and tangential vectors $\mathbf{n}_{\partial\Omega_{ij}}$ and $\mathbf{t}_{\partial\Omega_{ij}}$ on the boundary are depicted. The hollow circles are the switching points and the shaded circles are the starting points.

From the above definition, the necessary and sufficient conditions for such a $(2k_i : 2k_j - 1)$ half-nonpassable flow of the second kind (or $(2k_i : 2k_j - 1)$ half-source flow) are stated in the following theorem.

THEOREM 10.8. *For a discontinuous dynamical system in Eq.*(10.1), $\mathbf{x}(t_m) \equiv \mathbf{x}_m \in \partial\Omega_{ij}$ *at time* t_m. *Suppose* $\mathbf{x}^{(i)}(t_{m-}) = \mathbf{x}_m = \mathbf{x}_i^{(j)}(t_{m\pm})$. *For an arbitrarily small* $\varepsilon > 0$, *there are two time intervals* $[t_{m-\varepsilon}, t_{m+\varepsilon}]$ *and* $(t_m, t_{m+\varepsilon}]$. *The flow* $\mathbf{x}^{(i)}(t)$ *is* $C^r_{[t_{m-\varepsilon}, t_m)}$*-continuous* ($r \geq 2k_i + 1$)*and* $\| d^{r+1}\mathbf{x}^{(i)}/dt^{r+1} \| < \infty$ *for time* t. *The imaginary flow* $\mathbf{x}_i^{(j)}(t)$ *is* $C^r_{[t_{m-\varepsilon}, t_{m+\varepsilon}]}$*-continuous* ($r \geq 2k_j$) *and* $\| d^{r+1}\mathbf{x}_i^{(j)}/dt^{r+1} \| < \infty$ *for time* t. *The flow* $\mathbf{x}^{(i)}(t)$ *of the* $(2k_i)^{th}$*-order and* $\mathbf{x}_i^{(j)}(t)$ *of the* $(2k_j - 1)^{th}$*-order to the boundary* $\partial\Omega_{ij}$ *is* $(2k_i : 2k_j - 1)$*-half-non-passable with the second kind in domain* Ω_i *(or* $(2k_i : 2k_j - 1)$*-source flow) iff*

$$G^{(s,i)}_{\partial\Omega_{ij}}(\mathbf{x}_m, t_{m+}, \mathbf{p}_i, \boldsymbol{\lambda}) = 0 \text{ for } s = 0, 1, \cdots, 2k_i - 1; \tag{10.47}$$

$$G^{(s,j)}_{\partial\Omega_{ij}}(\mathbf{x}_m, t_{m\pm}, \mathbf{p}_j, \boldsymbol{\lambda}) = 0 \text{ for } s = 0, 1, \cdots, 2k_j - 2; \tag{10.48}$$

$$\left.\begin{array}{ll} \text{either} & \left.\begin{array}{l} G^{(2k_i,i)}_{\partial\Omega_{ij}}(\mathbf{x}_m, t_{m+}, \mathbf{p}_i, \boldsymbol{\lambda}) < 0 \\ G^{(2k_j-1,j)}_{\partial\Omega_{ij}}(\mathbf{x}_m, t_{m\pm}, \mathbf{p}_i, \boldsymbol{\lambda}) < 0 \end{array}\right\} \text{for } \mathbf{n}_{\partial\Omega_{ij}} \to \Omega_j \\ \text{or} & \left.\begin{array}{l} G^{(2k_i,i)}_{\partial\Omega_{ij}}(\mathbf{x}_m, t_{m+}, \mathbf{p}_i, \boldsymbol{\lambda}) > 0 \\ G^{(2k_j-1,j)}_{\partial\Omega_{ij}}(\mathbf{x}_m, t_{m\pm}, \mathbf{p}_i, \boldsymbol{\lambda}) > 0 \end{array}\right\} \text{for } \mathbf{n}_{\partial\Omega_{ij}} \to \Omega_i. \end{array}\right\} \tag{10.49}$$

PROOF. The proof is similar to the proof of Theorem 10.2. ■

10.4. Tangential flows

The tangency of a flow to the boundary in discontinuous dynamical systems is the same as the global tangency of a flow to the separatrix in Chapter 3.

DEFINITION 10.11. For a discontinuous dynamical system in Eq.(10.1), $\mathbf{x}(t_m) \equiv \mathbf{x}_m \in \partial\Omega_{ij}$ at time t_m. Suppose $\mathbf{x}^{(\alpha)}(t_{m\pm}) = \mathbf{x}_m$ ($\alpha \in \{i, j\}$). For an arbitrarily small $\varepsilon > 0$, there is a time interval $[t_{m-\varepsilon}, t_{m+\varepsilon}]$. The flow $\mathbf{x}^{(\alpha)}(t)$ is $C^r_{[t_{m-\varepsilon}, t_{m+\varepsilon}]}$-continuous ($r \geq 2$) for t. A flow $\mathbf{x}^{(\alpha)}(t)$ in Ω_α is *tangential* to the boundary $\partial\Omega_{ij}$ if

$$G_{\partial\Omega_{ij}}^{(\alpha)}(\mathbf{x}_m,t_m,\mathbf{p}_\alpha,\boldsymbol{\lambda})=0 \text{ and } G_{\partial\Omega_{ij}}^{(1,\alpha)}(\mathbf{x}_m,t_m,\mathbf{p}_\alpha,\boldsymbol{\lambda})\neq 0; \tag{10.50}$$

$$\left.\begin{array}{ll}\text{either} & \left.\begin{array}{l}\mathbf{n}_{\partial\Omega_{ij}}^T(\mathbf{x}_{m-\varepsilon}^{(0)})\bullet\left[\mathbf{x}_{m-\varepsilon}^{(0)}-\mathbf{x}_{m-\varepsilon}^{(\alpha)}\right]>0\\ \mathbf{n}_{\partial\Omega_{ij}}^T(\mathbf{x}_{m+\varepsilon}^{(0)})\bullet\left[\mathbf{x}_{m+\varepsilon}^{(\alpha)}-\mathbf{x}_{m+\varepsilon}^{(0)}\right]<0\end{array}\right\}\text{for } \mathbf{n}_{\partial\Omega_{ij}}\to\Omega_\beta\\ \text{or} & \left.\begin{array}{l}\mathbf{n}_{\partial\Omega_{ij}}^T(\mathbf{x}_{m-\varepsilon}^{(0)})\bullet\left[\mathbf{x}_{m-\varepsilon}^{(0)}-\mathbf{x}_{m-\varepsilon}^{(\alpha)}\right]<0\\ \mathbf{n}_{\partial\Omega_{ij}}^T(\mathbf{x}_{m+\varepsilon}^{(0)})\bullet\left[\mathbf{x}_{m+\varepsilon}^{(\alpha)}-\overline{\mathbf{x}}_{m+\varepsilon}^{(0)}\right]>0\end{array}\right\}\text{for } \mathbf{n}_{\partial\Omega_{ij}}\to\Omega_\alpha.\end{array}\right\} \tag{10.51}$$

THEOREM 10.9. *For a discontinuous dynamical system in Eq.*(10.1), $\mathbf{x}(t_m)\equiv\mathbf{x}_m\in\partial\Omega_{ij}$ *at time* t_m. *Suppose* $\mathbf{x}^{(\alpha)}(t_{m\pm})=\mathbf{x}_m$ ($\alpha\in\{i,j\}$). *For an arbitrarily small* $\varepsilon>0$, *there is a time interval* $[t_{m-\varepsilon},t_{m+\varepsilon}]$. *The flow* $\mathbf{x}^{(\alpha)}(t)$ *is* $C^r_{[t_{m-\varepsilon},t_{m+\varepsilon}]}$*-continuous* ($r\geq 2$) *for time t, and* $\|d^{r+1}\mathbf{x}^{(\alpha)}/dt^{r+1}\|<\infty$. *A flow* $\mathbf{x}^{(\alpha)}(t)$ *in* Ω_α *is tangential to the boundary* $\partial\Omega_{ij}$ *iff*

$$G_{\partial\Omega_{ij}}^{(0,\alpha)}(\mathbf{x}_m,t_m,\mathbf{p}_\alpha,\boldsymbol{\lambda})=0; \tag{10.52}$$

$$\left.\begin{array}{ll}\text{either} & G_{\partial\Omega_{ij}}^{(1,\alpha)}(\mathbf{x}_m,t_m,\mathbf{p}_\alpha,\boldsymbol{\lambda})<0 \text{ for } \mathbf{n}_{\partial\Omega_{ij}}\to\Omega_\beta\\ \text{or} & G_{\partial\Omega_{ij}}^{(1,\alpha)}(\mathbf{x}_m,t_m,\mathbf{p}_\alpha,\boldsymbol{\lambda})>0 \text{ for } \mathbf{n}_{\partial\Omega_{ij}}\to\Omega_\alpha.\end{array}\right\} \tag{10.53}$$

PROOF. The proof is the same as in Theorem 3.5. ■

DEFINITION10.12. For a discontinuous dynamical system in Eq.(10.1), $\mathbf{x}(t_m)\equiv\mathbf{x}_m\in\partial\Omega_{ij}$ at time t_m. Suppose $\mathbf{x}^{(\alpha)}(t_{m\pm})=\mathbf{x}_m$ ($\alpha\in\{i,j\}$). For an arbitrarily small $\varepsilon>0$, there is a time interval $[t_{m-\varepsilon},t_{m+\varepsilon}]$. A flow $\mathbf{x}^{(\alpha)}(t)$ is $C^r_{[t_{m-\varepsilon},t_{m+\varepsilon}]}$-continuous ($r\geq k_\alpha+1$) and $\|d^{r+1}\mathbf{x}^{(\alpha)}/dt^{r+1}\|<\infty$ for time *t*. A flow $\mathbf{x}^{(\alpha)}(t)$ in Ω_α is *tangential* to the boundary $\partial\Omega_{ij}$ with the $(2k_\alpha-1)^{th}$-order if

$$\left.\begin{array}{l}G_{\partial\Omega_{ij}}^{(s,\alpha)}(\mathbf{x}_m,t_m,\mathbf{p}_\alpha,\boldsymbol{\lambda})=0 \text{ for } s=0,1,\cdots,2k_\alpha-2\\ G_{\partial\Omega_{ij}}^{(2k_\alpha-1,\alpha)}(\mathbf{x}_m,t_m,\mathbf{p}_\alpha,\boldsymbol{\lambda})\neq 0,\end{array}\right\} \tag{10.54}$$

$$\text{either}\quad \left.\begin{array}{l}\mathbf{n}_{\partial\Omega_{ij}}^T(\mathbf{x}_{m-\varepsilon}^{(0)})\bullet\left[\mathbf{x}_{m-\varepsilon}^{(0)}-\mathbf{x}_{m-\varepsilon}^{(\alpha)}\right]>0\\ \mathbf{n}_{\partial\Omega_{ij}}^T(\mathbf{x}_{m+\varepsilon}^{(0)})\bullet\left[\mathbf{x}_{m+\varepsilon}^{(\alpha)}-\mathbf{x}_{m+\varepsilon}^{(0)}\right]<0\end{array}\right\}\text{for } \mathbf{n}_{\partial\Omega_{ij}}\to\Omega_\beta \tag{10.55a}$$

$$\text{or} \quad \left.\begin{array}{l} \mathbf{n}^T_{\partial\Omega_{ij}}(\mathbf{x}^{(0)}_{m-\varepsilon})\bullet\left[\mathbf{x}^{(0)}_{m-\varepsilon}-\mathbf{x}^{(\alpha)}_{m-\varepsilon}\right]<0 \\ \mathbf{n}^T_{\partial\Omega_{ij}}(\mathbf{x}^{(0)}_{m+\varepsilon})\bullet\left[\mathbf{x}^{(\alpha)}_{m+\varepsilon}-\bar{\mathbf{x}}^{(0)}_{m+\varepsilon}\right]>0 \end{array}\right\} \text{for } \mathbf{n}_{\partial\Omega_{ij}} \to \Omega_\alpha. \tag{10.55b}$$

THEOREM 10.10. *For a discontinuous dynamical system in Eq.*(10.1), $\mathbf{x}(t_m) \equiv \mathbf{x}_m \in \partial\Omega_{ij}$ *at time* t_m. *Suppose* $\mathbf{x}^{(\alpha)}(t_{m\pm}) = \mathbf{x}_m$ ($\alpha \in \{i, j\}$). *For an arbitrarily small* $\varepsilon > 0$, *there is a time interval* $[t_{m-\varepsilon}, t_{m+\varepsilon}]$. *The flow* $\mathbf{x}^{(\alpha)}(t)$ *is* $C^r_{[t_{m-\varepsilon}, t_{m+\varepsilon}]}$*-continuous* ($r \ge k_\alpha + 1$) *for time t and* $\| d^{r+1}\mathbf{x}^{(\alpha)}/dt^{r+1} \| < \infty$. *A flow* $\mathbf{x}^{(\alpha)}(t)$ *in* Ω_α *is tangential to the boundary* $\partial\Omega_{ij}$ *with the* $(2k_\alpha - 1)^{th}$*-order iff*

$$G^{(s,\alpha)}_{\partial\Omega_{ij}}(\mathbf{x}_m, t_m, \mathbf{p}_\alpha, \boldsymbol{\lambda}) = 0 \text{ for } s = 0, 1, \cdots, 2k_\alpha - 2, \tag{10.56}$$

$$\left.\begin{array}{ll} \text{either} & G^{(2k_\alpha-1,\alpha)}_{\partial\Omega_{ij}}(\mathbf{x}_m, t_m, \mathbf{p}_\alpha, \boldsymbol{\lambda}) < 0 \text{ for } \mathbf{n}_{\partial\Omega_{ij}} \to \Omega_\beta \\ \text{or} & G^{(2k_\alpha-1,\alpha)}_{\partial\Omega_{ij}}(\mathbf{x}_m, t_m, \mathbf{p}_\alpha, \boldsymbol{\lambda}) > 0 \text{ for } \mathbf{n}_{\partial\Omega_{ij}} \to \Omega_\alpha. \end{array}\right\} \tag{10.57}$$

PROOF. The proof is the same as in Theorem 3.6. ■

The conditions for grazing bifurcation of a flow to the boundary in domain Ω_i can be given as in Eq.(3.35). The flow grazing bifurcation the boundary can be determined by the G-function $G^{(2k_\alpha-1,\alpha)}_{\partial\Omega_{ij}}(\mathbf{x}_m, t_m, \mathbf{p}_\alpha, \boldsymbol{\lambda})$. The conditions for a flow tangential to the boundary are $G^{(r,\alpha)}_{\partial\Omega_{ij}}(\mathbf{x}_m, t_m, \mathbf{p}_\alpha, \boldsymbol{\lambda}) = 0$ ($r = 0, 1, \cdots, 2k_\alpha - 2$) and $G^{(2k_i-1,i)}_{\partial\Omega_{ij}}(\mathbf{x}_m, t_m, \mathbf{p}_i, \boldsymbol{\lambda}) < 0$ (or $G^{(2k_i-1,i)}_{\partial\Omega_{ij}}(\mathbf{x}_m, t_m, \mathbf{p}_i, \boldsymbol{\lambda}) > 0$) for the boundary $\partial\Omega_{ij}$ with $\mathbf{n}_{\partial\Omega_{ij}} \to \Omega_j$ (or with $\mathbf{n}_{\partial\Omega_{ij}} \to \Omega_i$). To develop a uniform theory of the tangential flow with the passable and non-passable flow, the imaginary flow tangency will be introduced. In order to distinguish a real tangential flow from an imaginary tangential flow, the tangency of a real flow to the boundary can be re-stated as follows.

DEFINITION 10.13. For a discontinuous dynamical system in Eq.(10.1), $\mathbf{x}(t_m) \equiv \mathbf{x}_m \in \partial\Omega_{ij}$ at time t_m. Suppose $\mathbf{x}^{(i)}(t_{m+}) = \mathbf{x}_m = \mathbf{x}^{(j)}_i(t_{m\pm})$. For an arbitrarily small $\varepsilon > 0$, there are two time intervals $[t_{m-\varepsilon}, t_m)$ and $[t_{m-\varepsilon}, t_{m+\varepsilon}]$. The flow $\mathbf{x}^{(i)}(t)$ is $C^r_{[t_{m-\varepsilon}, t_{m+\varepsilon}]}$-continuous ($r \ge 2k_i + 1$) for time t and $\| d^{r+1}\mathbf{x}^{(i)}/dt^{r+1} \| < \infty$. The sink or source flow $\mathbf{x}^{(j)}(t)$ is $C^r_{[t_{m-\varepsilon}, t_m)}$ or $C^r_{(t_m, t_{m+\varepsilon}]}$-continuous ($r \ge 2k_j$) for time t and

$\| d^{r+1}\mathbf{x}^{(j)}/dt^{r+1} \| < \infty$. The flow $\mathbf{x}^{(i)}(t)$ of the $(2k_i-1)^{th}$-order with $\mathbf{x}^{(j)}(t)$ of the $(2k_j)^{th}$-order to the boundary $\partial\Omega_{ij}$ is a $(2k_i-1:2k_j)$-tangential flow in domain Ω_i if

$$\left.\begin{array}{l} G^{(s,i)}_{\partial\Omega_{ij}}(\mathbf{x}_m, t_{m\pm}, \mathbf{p}_i, \boldsymbol{\lambda}) = 0 \text{ for } s = 0,1,\cdots,2k_i-2 \\ G^{(2k_i,i)}_{\partial\Omega_{ij}}(\mathbf{x}_m, t_{m\pm}, \mathbf{p}_i, \boldsymbol{\lambda}) \neq 0, \end{array}\right\} \tag{10.58}$$

$$\left.\begin{array}{l} G^{(s,j)}_{\partial\Omega_{ij}}(\mathbf{x}_m, t_{m\pm}, \mathbf{p}_j, \boldsymbol{\lambda}) = 0 \text{ for } s = 0,1,\cdots,2k_j-1 \\ G^{(2k_j-1,j)}_{\partial\Omega_{ij}}(\mathbf{x}_m, t_{m\pm}, \mathbf{p}_j, \boldsymbol{\lambda}) \neq 0, \end{array}\right\} \tag{10.59}$$

$$\left.\begin{array}{ll} \text{either} & \left.\begin{array}{l} \mathbf{n}^T_{\partial\Omega_{ij}}(\mathbf{x}^{(0)}_{m-\varepsilon})\bullet\left[\mathbf{x}^{(0)}_{m-\varepsilon} - \mathbf{x}^{(i)}_{m-\varepsilon}\right] > 0 \\ \mathbf{n}^T_{\partial\Omega_{ij}}(\mathbf{x}^{(0)}_{m+\varepsilon})\bullet\left[\mathbf{x}^{(i)}_{m+\varepsilon} - \mathbf{x}^{(0)}_{m+\varepsilon}\right] < 0 \end{array}\right\} \text{for } \mathbf{n}_{\partial\Omega_{ij}} \to \Omega_j \\ \text{or} & \left.\begin{array}{l} \mathbf{n}^T_{\partial\Omega_{ij}}(\mathbf{x}^{(0)}_{m-\varepsilon})\bullet\left[\mathbf{x}^{(0)}_{m-\varepsilon} - \mathbf{x}^{(i)}_{m-\varepsilon}\right] < 0 \\ \mathbf{n}^T_{\partial\Omega_{ij}}(\mathbf{x}^{(0)}_{m+\varepsilon})\bullet\left[\mathbf{x}^{(i)}_{m+\varepsilon} - \overline{\mathbf{x}}^{(0)}_{m+\varepsilon}\right] > 0 \end{array}\right\} \text{for } \mathbf{n}_{\partial\Omega_{ij}} \to \Omega_i, \end{array}\right\} \tag{10.60}$$

$$\left.\begin{array}{l} \left.\begin{array}{ll} \text{either} & \mathbf{n}^T_{\partial\Omega_{ij}}(\mathbf{x}^{(0)}_{m+\varepsilon})\bullet\left[\mathbf{x}^{(j)}_{m+\varepsilon} - \mathbf{x}^{(0)}_{m+\varepsilon}\right] > 0 \text{ for } \mathbf{n}_{\partial\Omega_{ij}} \to \Omega_j \\ \text{or} & \mathbf{n}^T_{\partial\Omega_{ij}}(\mathbf{x}^{(0)}_{m+\varepsilon})\bullet\left[\mathbf{x}^{(j)}_{m+\varepsilon} - \overline{\mathbf{x}}^{(0)}_{m+\varepsilon}\right] < 0 \text{ for } \mathbf{n}_{\partial\Omega_{ij}} \to \Omega_i, \end{array}\right\} \\ \text{or} \\ \left.\begin{array}{ll} \text{either} & \mathbf{n}^T_{\partial\Omega_{ij}}(\mathbf{x}^{(0)}_{m-\varepsilon})\bullet\left[\mathbf{x}^{(0)}_{m-\varepsilon} - \mathbf{x}^{(j)}_{m-\varepsilon}\right] < 0 \text{ for } \mathbf{n}_{\partial\Omega_{ij}} \to \Omega_j \\ \text{or} & \mathbf{n}^T_{\partial\Omega_{ij}}(\mathbf{x}^{(0)}_{m-\varepsilon})\bullet\left[\mathbf{x}^{(0)}_{m-\varepsilon} - \mathbf{x}^{(j)}_{m-\varepsilon}\right] > 0 \text{ for } \mathbf{n}_{\partial\Omega_{ij}} \to \Omega_i. \end{array}\right\} \end{array}\right\} \tag{10.61}$$

Such a $(2k_i-1:2k_j)$-tangential flow in domain Ω_i are sketched in Fig.10.7(a) with source in domain Ω_j and (b) with sink in domain Ω_j. The $(2k_j-1:2k_i)$-tangential flow in domain Ω_j are sketched in Fig.10.8(a) with source in domain Ω_i and (b) with sink in domain Ω_i. The sink and source flows are represented by the dotted curves. The tangential flows are presented by solid curves. The dashed curves denote the imaginary flows. If the starting point is on the flow $\mathbf{x}^{(j)}(t)$ (or $\mathbf{x}^{(j)}(t)$) in Fig.10.7(b) (or Fig.10.8(a)), the passable flow from domain Ω_j to Ω_i (or Ω_j to Ω_i) is formed. From the above definition, the necessary and sufficient conditions for the tangential flow are given in the following theorem.

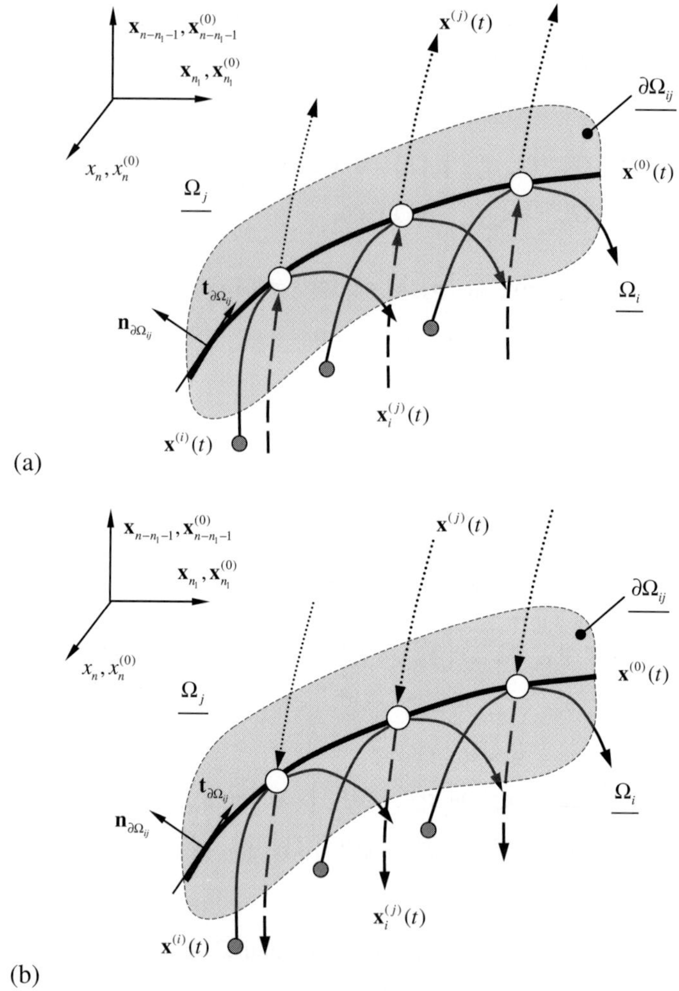

Figure 10.7. The $(2k_i - 1 : 2k_j)$ -tangential flows in Ω_i : (a) with source in Ω_j and (b) with sink in Ω_j . $\mathbf{x}^{(i)}(t)$ and $\mathbf{x}^{(j)}(t)$ represent the *real* flows in domains Ω_i and Ω_j , respectively, which are depicted by the thin solid curves. $\mathbf{x}_i^{(j)}(t)$ and $\mathbf{x}_j^{(i)}(t)$ represent the *imaginary* flows in domains Ω_i and Ω_j ,respectively controlled by the vector fields in Ω_j and Ω_i , which are depicted by the dashed curves. The flow on the boundary is described by $\mathbf{x}^{(0)}(t)$. The normal and tangential vectors $\mathbf{n}_{\partial\Omega_{ij}}$ and $\mathbf{t}_{\partial\Omega_{ij}}$ on the boundary are depicted. The dotted curves represent the real flow without the starting point . The hollow circles are the switching points and the shaded circles are the starting points.

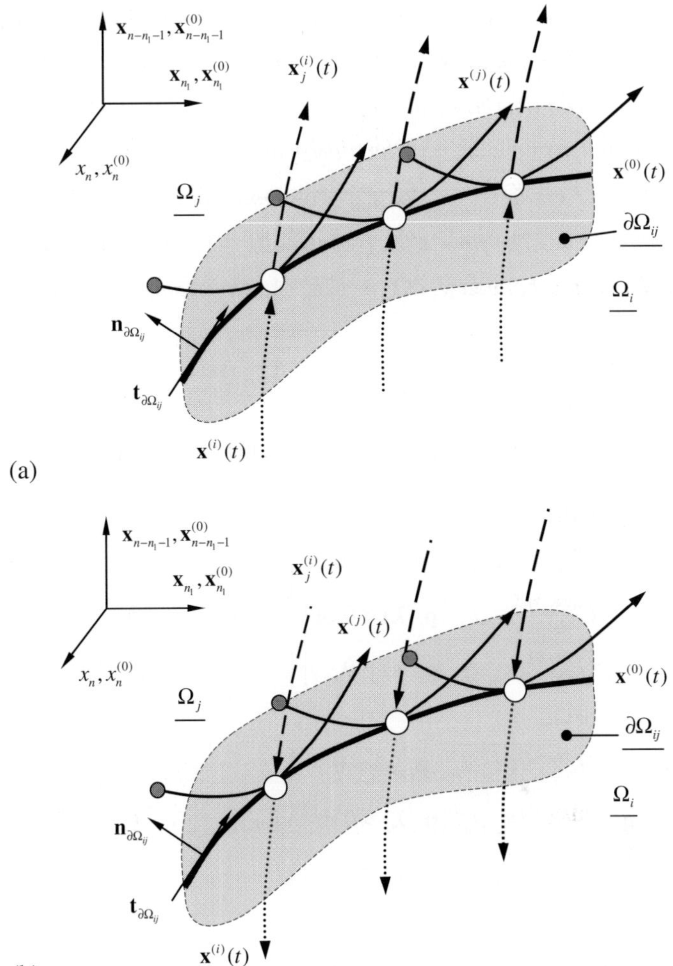

Figure 10.8. The $(2k_j - 1 : 2k_i)$ -tangential flows in Ω_j : (a) with sink in Ω_i and (b) with source in Ω_i. $\mathbf{x}^{(i)}(t)$ and $\mathbf{x}^{(j)}(t)$ represent the *real* flows in domains Ω_i and Ω_j , respectively, which are depicted by the thin solid curves. $\mathbf{x}_i^{(j)}(t)$ and $\mathbf{x}_j^{(i)}(t)$ represent two *imaginary* flows in domains Ω_i and Ω_j , respectively controlled by the vector fields in Ω_j and Ω_i , which are depicted by the dashed curves. The flow on the boundary is described by $\mathbf{x}^{(0)}(t)$. The normal and tangential vectors $\mathbf{n}_{\partial\Omega_{ij}}$ and $\mathbf{t}_{\partial\Omega_{ij}}$ on the boundary are depicted. The dotted curves represent the real flow without the starting point. The hollow circles are the switching points and the shaded circles are the starting points.

THEOREM 10.11. *For a discontinuous dynamical system in Eq.(10.1),* $\mathbf{x}(t_m) \equiv \mathbf{x}_m \in \partial\Omega_{ij}$ *at time* t_m. *Suppose* $\mathbf{x}^{(i)}(t_{m+}) = \mathbf{x}_m = \mathbf{x}_i^{(j)}(t_{m\pm})$. *For an arbitrarily small* $\varepsilon > 0$, *there are two time intervals* $[t_{m-\varepsilon}, t_m)$ *and* $[t_{m-\varepsilon}, t_{m+\varepsilon}]$. *The flow* $\mathbf{x}^{(i)}(t)$ *is* $C^r_{[t_{m-\varepsilon}, t_{m+\varepsilon}]}$*-continuous* ($r \geq 2k_i + 1$) *for time t and* $\| d^{r+1}\mathbf{x}^{(i)}/dt^{r+1} \| < \infty$. *The sink or source flow* $\mathbf{x}^{(j)}(t)$ *is* $C^r_{[t_{m-\varepsilon}, t_m)}$ *or* $C^r_{(t_m, t_{m+\varepsilon}]}$*-continuous* ($r \geq 2k_j$) *for time t and* $\| d^{r+1}\mathbf{x}^{(j)}/dt^{r+1} \| < \infty$. *The flow* $\mathbf{x}^{(i)}(t)$ *of the* $(2k_i - 1)^{th}$*-order and* $\mathbf{x}^{(j)}(t)$ *of the* $(2k_j)^{th}$*-order to the boundary* $\partial\Omega_{ij}$ *is* $(2k_i - 1 : 2k_j)$*-tangential flow in domain* Ω_i *iff*

$$G^{(s,i)}_{\partial\Omega_{ij}}(\mathbf{x}_m, t_{m\pm}, \mathbf{p}_i, \boldsymbol{\lambda}) = 0 \text{ for } s = 0, 1, \cdots, 2k_i - 2; \tag{10.62}$$

$$G^{(s,j)}_{\partial\Omega_{ij}}(\mathbf{x}_m, t_{m\pm}, \mathbf{p}_j, \boldsymbol{\lambda}) = 0 \text{ for } s = 0, 1, \cdots, 2k_j - 1; \tag{10.63}$$

$$\left.\begin{array}{l} \text{either} \\ \left.\begin{array}{l} G^{(2k_i-1,i)}_{\partial\Omega_{ij}}(\mathbf{x}_m, t_{m\pm}, \mathbf{p}_i, \boldsymbol{\lambda}) < 0 \\ \left.\begin{array}{l} G^{(2k_j,j)}_{\partial\Omega_{ij}}(\mathbf{x}_m, t_{m-}, \mathbf{p}_i, \boldsymbol{\lambda}) < 0 \text{ or} \\ G^{(2k_j,j)}_{\partial\Omega_{ij}}(\mathbf{x}_m, t_{m+}, \mathbf{p}_i, \boldsymbol{\lambda}) > 0 \end{array}\right] \end{array}\right\} \text{for } \mathbf{n}_{\partial\Omega_{ij}} \to \Omega_j \\ \text{or} \\ \left.\begin{array}{l} G^{(2k_i-1,i)}_{\partial\Omega_{ij}}(\mathbf{x}_m, t_{m\pm}, \mathbf{p}_i, \boldsymbol{\lambda}) > 0 \\ \left.\begin{array}{l} G^{(2k_j,j)}_{\partial\Omega_{ij}}(\mathbf{x}_m, t_{m-}, \mathbf{p}_i, \boldsymbol{\lambda}) > 0 \text{ or} \\ G^{(2k_j,j)}_{\partial\Omega_{ij}}(\mathbf{x}_m, t_{m+}, \mathbf{p}_i, \boldsymbol{\lambda}) < 0 \end{array}\right] \end{array}\right\} \text{for } \mathbf{n}_{\partial\Omega_{ij}} \to \Omega_i. \end{array}\right\} \tag{10.64}$$

PROOF. The proof is similar to the proof of Theorem 10.2. ■

DEFINITION 10.14. For a discontinuous dynamical system in Eq.(10.1), $\mathbf{x}(t_m) \equiv \mathbf{x}_m \in \partial\Omega_{ij}$ at time t_m. Suppose $\mathbf{x}^{(\alpha)}(t_{m\pm}) = \mathbf{x}_m = \mathbf{x}_\alpha^{(\beta)}(t_{m\pm})$ ($\alpha, \beta \in \{i, j\}$ and $\alpha \neq \beta$). For an arbitrarily small $\varepsilon > 0$, there is a time interval $[t_{m-\varepsilon}, t_{m+\varepsilon}]$. The flow $\mathbf{x}^{(\alpha)}(t)$ is $C^r_{[t_{m-\varepsilon}, t_{m+\varepsilon}]}$-continuous ($r \geq 2k_\alpha$) and $\| d^{r+1}\mathbf{x}^{(\alpha)}/dt^{r+1} \| < \infty$ for time *t*. The imaginary flow $\mathbf{x}_\alpha^{(\beta)}(t)$ is $C^r_{[t_{m-\varepsilon}, t_{m+\varepsilon}]}$-continuous ($r \geq 2k_\beta$) for time t and $\| d^{r+1}\mathbf{x}_\alpha^{(\beta)}/dt^{r+1} \| < \infty$. The flow $\mathbf{x}^{(\alpha)}(t)$ of the $(2k_\alpha - 1)^{th}$-order and $\mathbf{x}_\alpha^{(\beta)}(t)$ of the $(2k_\beta - 1)^{th}$-order to the boundary $\partial\Omega_{ij}$ is a $(2k_\alpha - 1 : 2k_\beta - 1)$-tangential flow in domain Ω_α if

$$\left.\begin{array}{l} G^{(s,\alpha)}_{\partial\Omega_{\alpha\beta}}(\mathbf{x}_m,t_{m\pm},\mathbf{p}_\alpha,\boldsymbol{\lambda})=0 \ \text{ for } s=0,1,\cdots,2k_\alpha-2 \\ G^{(2k_\alpha-1,\alpha)}_{\partial\Omega_{\alpha\beta}}(\mathbf{x}_m,t_{m\pm},\mathbf{p}_\alpha,\boldsymbol{\lambda})\neq 0, \end{array}\right\} \tag{10.65}$$

$$\left.\begin{array}{l} G^{(s,\beta)}_{\partial\Omega_{\alpha\beta}}(\mathbf{x}_m,t_{m\pm},\mathbf{p}_\beta,\boldsymbol{\lambda})=0 \ \text{ for } s=0,1,\cdots,2k_\beta-2 \\ G^{(2k_\beta-1,\beta)}_{\partial\Omega_{\alpha\beta}}(\mathbf{x}_m,t_{m-},\mathbf{p}_\beta,\boldsymbol{\lambda})\neq 0, \end{array}\right\} \tag{10.66}$$

$$\left.\begin{array}{ll} \text{either} & \left.\begin{array}{l} \mathbf{n}^T_{\partial\Omega_{ij}}(\mathbf{x}^{(0)}_{m-\varepsilon})\bullet\left[\mathbf{x}^{(0)}_{m-\varepsilon}-\mathbf{x}^{(\alpha)}_{m-\varepsilon}\right]>0 \\ \mathbf{n}^T_{\partial\Omega_{ij}}(\mathbf{x}^{(0)}_{m+\varepsilon})\bullet\left[\mathbf{x}^{(\alpha)}_{m+\varepsilon}-\mathbf{x}^{(0)}_{m+\varepsilon}\right]<0 \end{array}\right\} \text{for } \mathbf{n}_{\partial\Omega_{\alpha\beta}}\to\Omega_\beta \\ \text{or} & \left.\begin{array}{l} \mathbf{n}^T_{\partial\Omega_{ij}}(\mathbf{x}^{(0)}_{m-\varepsilon})\bullet\left[\mathbf{x}^{(0)}_{m-\varepsilon}-\mathbf{x}^{(\alpha)}_{m-\varepsilon}\right]<0 \\ \mathbf{n}^T_{\partial\Omega_{ij}}(\mathbf{x}^{(0)}_{m+\varepsilon})\bullet\left[\mathbf{x}^{(\alpha)}_{m+\varepsilon}-\overline{\mathbf{x}}^{(0)}_{m+\varepsilon}\right]>0 \end{array}\right\} \text{for } \mathbf{n}_{\partial\Omega_{\alpha\beta}}\to\Omega_\alpha, \end{array}\right\} \tag{10.67}$$

$$\left.\begin{array}{ll} \text{either} & \left.\begin{array}{l} \mathbf{n}^T_{\partial\Omega_{ij}}(\mathbf{x}^{(0)}_{m-\varepsilon})\bullet\left[\mathbf{x}^{(0)}_{m-\varepsilon}-\mathbf{x}^{(\beta)}_{\alpha(m-\varepsilon)}\right]>0 \\ \mathbf{n}^T_{\partial\Omega_{ij}}(\mathbf{x}^{(0)}_{m+\varepsilon})\bullet\left[\mathbf{x}^{(\beta)}_{\alpha(m+\varepsilon)}-\mathbf{x}^{(0)}_{m+\varepsilon}\right]<0 \end{array}\right\} \text{for } \mathbf{n}_{\partial\Omega_{\alpha\beta}}\to\Omega_\beta \\ \text{or} & \left.\begin{array}{l} \mathbf{n}^T_{\partial\Omega_{ij}}(\mathbf{x}^{(0)}_{m-\varepsilon})\bullet\left[\mathbf{x}^{(0)}_{m-\varepsilon}-\mathbf{x}^{(\beta)}_{\alpha(m-\varepsilon)}\right]<0 \\ \mathbf{n}^T_{\partial\Omega_{ij}}(\mathbf{x}^{(0)}_{m+\varepsilon})\bullet\left[\mathbf{x}^{(\beta)}_{\alpha(m+\varepsilon)}-\overline{\mathbf{x}}^{(0)}_{m+\varepsilon}\right]>0 \end{array}\right\} \text{for } \mathbf{n}_{\partial\Omega_{\alpha\beta}}\to\Omega_\alpha. \end{array}\right\} \tag{10.68}$$

The $(2k_\alpha-1:2k_\beta-1)$-tangential flows in domain Ω_α and Ω_β ($\alpha,\beta\in\{i,j\}$ and $\alpha\neq\beta$)) are sketched in Fig.10.9 with the corresponding imaginary tangential flows. The real tangential flows are presented by the solid curves. The dashed curves denote the imaginary tangential flows. The corresponding necessary and sufficient conditions for the tangential flow are given by the following theorem.

THEOREM 10.12. *For a discontinuous dynamical system in Eq.*(10.1), $\mathbf{x}(t_m)\equiv\mathbf{x}_m\in\partial\Omega_{ij}$ *at time* t_m. *Suppose* $\mathbf{x}^{(\alpha)}(t_{m\pm})=\mathbf{x}_m=\mathbf{x}^{(\beta)}_\alpha(t_{m\pm})$ ($\alpha,\beta\in\{i,j\}$ and $\alpha\neq\beta$). *For an arbitrarily small* $\varepsilon>0$, *there is a time interval* $[t_{m-\varepsilon},t_{m+\varepsilon}]$. *The flow* $\mathbf{x}^{(\alpha)}(t)$ *is* $C^r_{[t_{m-\varepsilon},t_{m+\varepsilon}]}$*-continuous* ($r\geq 2k_\alpha$) *and* $\| d^{r+1}\mathbf{x}^{(\alpha)}/dt^{r+1}\|<\infty$ *for time t. The imaginary flow* $\mathbf{x}^{(\beta)}_\alpha(t)$ *is* $C^r_{[t_{m-\varepsilon},t_{m+\varepsilon}]}$*-continuous* ($r\geq 2k_\beta$) *for time t and* $\| d^{r+1}\mathbf{x}^{(\beta)}_\alpha/dt^{r+1}\|<\infty$. *The flow* $\mathbf{x}^{(\alpha)}(t)$ *of the* $(2k_\alpha-1)^{th}$*-order and* $\mathbf{x}^{(\beta)}_\alpha(t)$ *of the* $(2k_\beta-1)^{th}$*-order to the boundary* $\partial\Omega_{ij}$ *is a* $(2k_\alpha-1:2k_\beta-1)$*-tangential flow in* *domain* Ω_α *iff*

$$G^{(s,\alpha)}_{\partial\Omega_{\alpha\beta}}(\mathbf{x}_m, t_{m\pm}, \mathbf{p}_\alpha, \boldsymbol{\lambda}) = 0 \text{ for } s = 0,1,\cdots,2k_\alpha - 2; \tag{10.69}$$

$$G^{(s,\beta)}_{\partial\Omega_{\alpha\beta}}(\mathbf{x}_m, t_{m\pm}, \mathbf{p}_\beta, \boldsymbol{\lambda}) = 0 \text{ for } s = 0,1,\cdots,2k_\beta - 2; \tag{10.70}$$

$$\left.\begin{array}{l}\text{either}\\ \left.\begin{array}{l} G^{(2k_\alpha-1,\alpha)}_{\partial\Omega_{\alpha\beta}}(\mathbf{x}_m, t_{m\pm}, \mathbf{p}_\alpha, \boldsymbol{\lambda}) < 0 \\ G^{(2k_\beta-1,\beta)}_{\partial\Omega_{\alpha\beta}}(\mathbf{x}_m, t_{m\pm}, \mathbf{p}_\beta, \boldsymbol{\lambda}) < 0 \end{array}\right\} \text{for } \mathbf{n}_{\partial\Omega_{\alpha\beta}} \to \Omega_\beta \\ \text{or} \\ \left.\begin{array}{l} G^{(2k_\alpha-1,\alpha)}_{\partial\Omega_{\alpha\beta}}(\mathbf{x}_m, t_{m\pm}, \mathbf{p}_\alpha, \boldsymbol{\lambda}) > 0 \\ G^{(2k_\beta-1,\beta)}_{\partial\Omega_{\alpha\beta}}(\mathbf{x}_m, t_{m\pm}, \mathbf{p}_\beta, \boldsymbol{\lambda}) > 0 \end{array}\right\} \text{for } \mathbf{n}_{\partial\Omega_{\alpha\beta}} \to \Omega_\alpha. \end{array}\right\} \tag{10.71}$$

PROOF. The proof is similar to the proof of Theorem 10.2. ■

DEFINITION 10.15. For a discontinuous dynamical system in Eq.(10.1), $\mathbf{x}(t_m) \equiv \mathbf{x}_m \in \partial\Omega_{ij}$ at time t_m. Suppose $\mathbf{x}^{(\alpha)}(t_{m\pm}) = \mathbf{x}_m = \mathbf{x}^{(\beta)}(t_{m\pm})$ ($\alpha, \beta \in \{i, j\}$ and $\alpha \neq \beta$). For an arbitrarily small $\varepsilon > 0$, there is a time interval $[t_{m-\varepsilon}, t_{m+\varepsilon}]$. The flow $\mathbf{x}^{(\alpha)}(t)$ is $C^r_{[t_{m-\varepsilon}, t_{m+\varepsilon}]}$-continuous ($r \geq 2k_\alpha$) for time t and $\| d^{r+1}\mathbf{x}^{(\alpha)}/dt^{r+1} \| < \infty$. The flow $\mathbf{x}^{(\beta)}(t)$ is $C^r_{[t_{m-\varepsilon}, t_{m+\varepsilon}]}$-continuous ($r \geq 2k_\beta$) and $\| d^{r+1}\mathbf{x}^{(\beta)}/dt^{r+1} \| < \infty$ for time *t*. The flow $\mathbf{x}^{(\alpha)}(t)$ of the $(2k_\alpha - 1)^{th}$-order and $\mathbf{x}^{(\beta)}(t)$ of the $(2k_\beta - 1)^{th}$-order to the boundary $\partial\Omega_{\alpha\beta}$ is a $(2k_\alpha - 1 : 2k_\beta - 1)$-double tangential flow if

$$\left.\begin{array}{l} G^{(s,\alpha)}_{\partial\Omega_{\alpha\beta}}(\mathbf{x}_m, t_{m\pm}, \mathbf{p}_\alpha, \boldsymbol{\lambda}) = 0 \text{ for } s = 0,1,\cdots,2k_\alpha - 2 \\ G^{(2k_\alpha-1,\alpha)}_{\partial\Omega_{\alpha\beta}}(\mathbf{x}_m, t_{m\pm}, \mathbf{p}_\alpha, \boldsymbol{\lambda}) \neq 0, \end{array}\right\} \tag{10.72}$$

$$\left.\begin{array}{l} G^{(s,\beta)}_{\partial\Omega_{\alpha\beta}}(\mathbf{x}_m, t_{m\pm}, \mathbf{p}_\beta, \boldsymbol{\lambda}) = 0 \text{ for } s = 0,1,\cdots,2k_\beta - 2 \\ G^{(2k_\beta-1,\beta)}_{\partial\Omega_{\alpha\beta}}(\mathbf{x}_m, t_{m-}, \mathbf{p}_\beta, \boldsymbol{\lambda}) \neq 0, \end{array}\right\} \tag{10.73}$$

$$\text{either} \quad \left.\begin{array}{l} \mathbf{n}^T_{\partial\Omega_{ij}}(\mathbf{x}^{(0)}_{m-\varepsilon}) \bullet \left[\mathbf{x}^{(0)}_{m-\varepsilon} - \mathbf{x}^{(\alpha)}_{m-\varepsilon}\right] > 0 \\ \mathbf{n}^T_{\partial\Omega_{ij}}(\mathbf{x}^{(0)}_{m+\varepsilon}) \bullet \left[\mathbf{x}^{(\alpha)}_{m+\varepsilon} - \mathbf{x}^{(0)}_{m+\varepsilon}\right] < 0 \end{array}\right\} \text{for } \mathbf{n}_{\partial\Omega_{\alpha\beta}} \to \Omega_\beta \tag{10.74a}$$

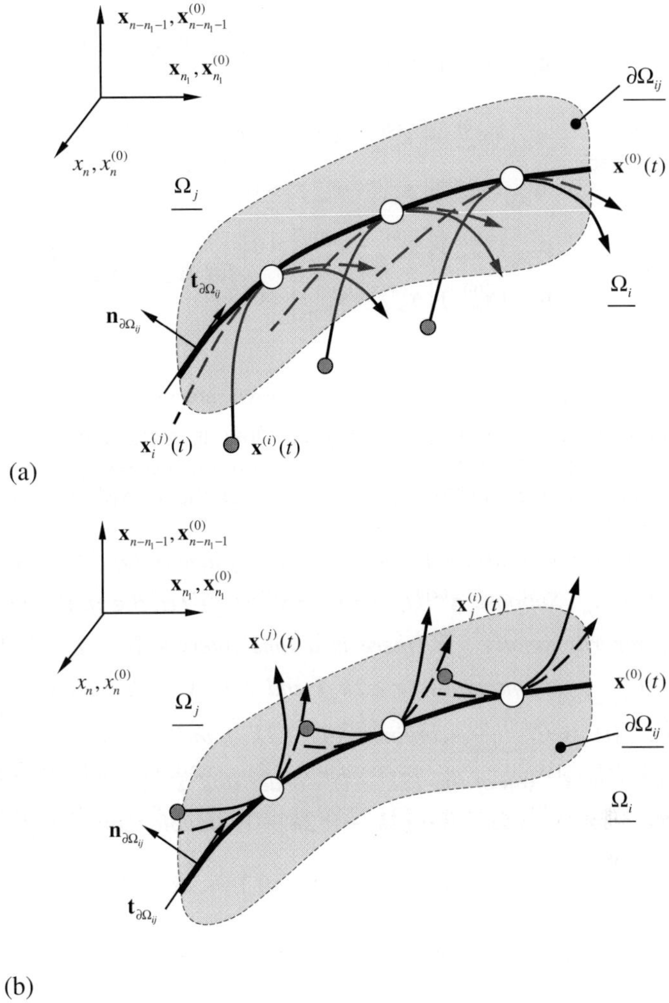

Figure 10.9. Tangential flows with imaginary flows in: (a) Ω_i with $(2k_i-1:2k_j-1)$ -order and (b) Ω_j with $(2k_j-1:2k_i-1)$ -order. $\mathbf{x}^{(i)}(t)$ and $\mathbf{x}^{(j)}(t)$ represent the *real* flows in domains Ω_i and Ω_j, respectively, which are depicted by the thin solid curves. $\mathbf{x}_i^{(j)}(t)$ and $\mathbf{x}_j^{(i)}(t)$ represent the *imaginary* flows in domains Ω_i and Ω_j ,respectively controlled by the vector fields in Ω_j and Ω_i, which are depicted by the dashed curves. The flow on the boundary is described by $\mathbf{x}^{(0)}(t)$. The normal and tangential vectors $\mathbf{n}_{\partial\Omega_{ij}}$ and $\mathbf{t}_{\partial\Omega_{ij}}$ on the boundary are depicted. The hollow circles are the switching points and the shaded circles are the starting points.

$$\text{or}\quad \left.\begin{array}{l}\mathbf{n}^T_{\partial\Omega_{ij}}(\mathbf{x}^{(0)}_{m-\varepsilon})\bullet\left[\mathbf{x}^{(0)}_{m-\varepsilon}-\mathbf{x}^{(\alpha)}_{m-\varepsilon}\right]<0\\ \mathbf{n}^T_{\partial\Omega_{ij}}(\mathbf{x}^{(0)}_{m+\varepsilon})\bullet\left[\mathbf{x}^{(\alpha)}_{m+\varepsilon}-\bar{\mathbf{x}}^{(0)}_{m+\varepsilon}\right]>0\end{array}\right\}\text{for } \mathbf{n}_{\partial\Omega_{\alpha\beta}}\to\Omega_\alpha, \tag{10.74b}$$

$$\left.\begin{array}{ll}\text{either} & \left.\begin{array}{l}\mathbf{n}^T_{\partial\Omega_{ij}}(\mathbf{x}^{(0)}_{m-\varepsilon})\bullet\left[\mathbf{x}^{(0)}_{m-\varepsilon}-\mathbf{x}^{(\beta)}_{m-\varepsilon}\right]<0\\ \mathbf{n}^T_{\partial\Omega_{ij}}(\mathbf{x}^{(0)}_{m+\varepsilon})\bullet\left[\mathbf{x}^{(\beta)}_{m+\varepsilon}-\mathbf{x}^{(0)}_{m+\varepsilon}\right]>0\end{array}\right\}\text{for } \mathbf{n}_{\partial\Omega_{\alpha\beta}}\to\Omega_\beta\\ \text{or} & \left.\begin{array}{l}\mathbf{n}^T_{\partial\Omega_{ij}}(\mathbf{x}^{(0)}_{m-\varepsilon})\bullet\left[\mathbf{x}^{(0)}_{m-\varepsilon}-\mathbf{x}^{(\beta)}_{m-\varepsilon}\right]>0\\ \mathbf{n}^T_{\partial\Omega_{ij}}(\mathbf{x}^{(0)}_{m+\varepsilon})\bullet\left[\mathbf{x}^{(\beta)}_{m+\varepsilon}-\bar{\mathbf{x}}^{(0)}_{m+\varepsilon}\right]<0\end{array}\right\}\text{for } \mathbf{n}_{\partial\Omega_{\alpha\beta}}\to\Omega_\alpha,\end{array}\right\} \tag{10.75}$$

The $(2k_\alpha-1:2k_\beta-1)$-double tangential flows are sketched in Fig.10.10(a) by the solid curves. The double tangential flow is formed by the two real tangential flows in both domains. The corresponding necessary and sufficient conditions for the tangential flows are given through the following theorem.

THEOREM 10.13. *For a discontinuous dynamical system in Eq.(10.1),* $\mathbf{x}(t_m)\equiv\mathbf{x}_m\in\partial\Omega_{ij}$ *at time* t_m. *Suppose* $\mathbf{x}^{(\alpha)}(t_{m\pm})=\mathbf{x}_m=\mathbf{x}^{(\beta)}(t_{m\pm})$ $(\alpha,\beta\in\{i,j\}$ *and* $\alpha\neq\beta)$. *For an arbitrarily small* $\varepsilon>0$, *there is a time interval* $[t_{m-\varepsilon},t_{m+\varepsilon}]$. *The flow* $\mathbf{x}^{(\alpha)}(t)$ *is* $C^r_{[t_{m-\varepsilon},t_{m+\varepsilon}]}$*-continuous* $(r\geq 2k_\alpha)$ *and* $\|d^{r+1}\mathbf{x}^{(\alpha)}/dt^{r+1}\|<\infty$ *for time t. The flow* $\mathbf{x}^{(\beta)}(t)$ *is* $C^r_{[t_{m-\varepsilon},t_{m+\varepsilon}]}$*-continuous* $(r\geq 2k_\beta)$ *and* $\|d^{r+1}\mathbf{x}^{(\beta)}/dt^{r+1}\|<\infty$ *for time* t. *The flow* $\mathbf{x}^{(\alpha)}(t)$ *of the* $(2k_\alpha-1)^{th}$*-order and* $\mathbf{x}^{(\beta)}(t)$ *of the* $(2k_\alpha-1)^{th}$*-order to the boundary* $\partial\Omega_{\alpha\beta}$ *is a* $(2k_\alpha-1:2k_\beta-1)$*-double tangential flow iff*

$$G^{(s,\alpha)}_{\partial\Omega_{\alpha\beta}}(\mathbf{x}_m,t_{m\pm},\mathbf{p}_\alpha,\boldsymbol{\lambda})=0\quad\text{for } s=0,1,\cdots,2k_\alpha-2; \tag{10.76}$$

$$G^{(s,\beta)}_{\partial\Omega_{\alpha\beta}}(\mathbf{x}_m,t_{m\pm},\mathbf{p}_\beta,\boldsymbol{\lambda})=0\quad\text{for } s=0,1,\cdots,2k_\beta-2; \tag{10.77}$$

$$\left.\begin{array}{ll}\text{either} & \left.\begin{array}{l}G^{(2k_\alpha-1,\alpha)}_{\partial\Omega_{\alpha\beta}}(\mathbf{x}_m,t_{m\pm},\mathbf{p}_\alpha,\boldsymbol{\lambda})<0\\ G^{(2k_\beta-1,\beta)}_{\partial\Omega_{\alpha\beta}}(\mathbf{x}_m,t_{m\pm},\mathbf{p}_\beta,\boldsymbol{\lambda})>0\end{array}\right\}\text{for } \mathbf{n}_{\partial\Omega_{\alpha\beta}}\to\Omega_\beta\\ \text{or} & \left.\begin{array}{l}G^{(2k_\alpha-1,\alpha)}_{\partial\Omega_{\alpha\beta}}(\mathbf{x}_m,t_{m\pm},\mathbf{p}_\alpha,\boldsymbol{\lambda})>0\\ G^{(2k_\beta-1,\beta)}_{\partial\Omega_{\alpha\beta}}(\mathbf{x}_m,t_{m\pm},\mathbf{p}_\beta,\boldsymbol{\lambda})<0\end{array}\right\}\text{for } \mathbf{n}_{\partial\Omega_{\alpha\beta}}\to\Omega_\alpha.\end{array}\right\} \tag{10.78}$$

PROOF. The proof is similar to the proof of Theorem 10.2. ■

DEFINITION 10.16. For a discontinuous dynamical system in Eq.(10.1), $\mathbf{x}(t_m) \equiv \mathbf{x}_m \in \partial\Omega_{ij}$ at time t_m. Suppose $\mathbf{x}_\beta^{(\alpha)}(t_{m\pm}) = \mathbf{x}_m = \mathbf{x}_\alpha^{(\beta)}(t_{m\pm})$ ($\alpha, \beta \in \{i, j\}$ and $\alpha \neq \beta$). For an arbitrarily small $\varepsilon > 0$, there is a time interval $[t_{m-\varepsilon}, t_{m+\varepsilon}]$. The imaginary flow $\mathbf{x}_\beta^{(\alpha)}(t)$ is $C^r_{[t_{m-\varepsilon}, t_{m+\varepsilon}]}$-continuous ($r \geq 2k_\alpha$) and $\| d^{r+1}\mathbf{x}_\beta^{(\alpha)} / dt^{r+1} \| < \infty$ for time t. The imaginary flow $\mathbf{x}_\alpha^{(\beta)}(t)$ is $C^r_{[t_{m-\varepsilon}, t_{m+\varepsilon}]}$-continuous ($r \geq 2k_\beta$) for time t and $\| d^{r+1}\mathbf{x}_\alpha^{(\beta)} / dt^{r+1} \| < \infty$. The imaginary flow $\mathbf{x}_\beta^{(\alpha)}(t)$ of the $(2k_\alpha - 1)^{th}$-order and the imaginary $\mathbf{x}_\alpha^{(\beta)}(t)$ of the $(2k_\beta - 1)^{th}$-order to the boundary $\partial\Omega_{ij}$ is a $(2k_\alpha - 1 : 2k_\beta - 1)$-double inaccessible tangential flow if

$$\left.\begin{array}{l} G_{\partial\Omega_{\alpha\beta}}^{(s,\alpha)}(\mathbf{x}_m, t_{m\pm}, \mathbf{p}_\alpha, \boldsymbol{\lambda}) = 0 \ \text{ for } s = 0, 1, \cdots, 2k_\alpha - 2 \\ G_{\partial\Omega_{\alpha\beta}}^{(2k_\alpha - 1,\alpha)}(\mathbf{x}_m, t_{m\pm}, \mathbf{p}_\alpha, \boldsymbol{\lambda}) \neq 0, \end{array}\right\} \tag{10.79}$$

$$\left.\begin{array}{l} G_{\partial\Omega_{\alpha\beta}}^{(s,\beta)}(\mathbf{x}_m, t_{m\pm}, \mathbf{p}_\beta, \boldsymbol{\lambda}) = 0 \ \text{ for } s = 0, 1, \cdots, 2k_\beta - 2 \\ G_{\partial\Omega_{\alpha\beta}}^{(2k_\beta - 1,\beta)}(\mathbf{x}_m, t_{m-}, \mathbf{p}_\beta, \boldsymbol{\lambda}) \neq 0, \end{array}\right\} \tag{10.80}$$

$$\left.\begin{array}{ll} \text{either} & \left.\begin{array}{l} \mathbf{n}_{\partial\Omega_{ij}}^T(\mathbf{x}_{m-\varepsilon}^{(0)}) \cdot \left[\mathbf{x}_{m-\varepsilon}^{(0)} - \mathbf{x}_{\beta(m-\varepsilon)}^{(\alpha)}\right] < 0 \\ \mathbf{n}_{\partial\Omega_{ij}}^T(\mathbf{x}_{m+\varepsilon}^{(0)}) \cdot \left[\mathbf{x}_{\beta(m+\varepsilon)}^{(\alpha)} - \mathbf{x}_{m+\varepsilon}^{(0)}\right] > 0 \end{array}\right\} \text{for } \mathbf{n}_{\partial\Omega_{\alpha\beta}} \to \Omega_\beta \\ \text{or} & \left.\begin{array}{l} \mathbf{n}_{\partial\Omega_{ij}}^T(\mathbf{x}_{m-\varepsilon}^{(0)}) \cdot \left[\mathbf{x}_{m-\varepsilon}^{(0)} - \mathbf{x}_{\beta(m-\varepsilon)}^{(\alpha)}\right] > 0 \\ \mathbf{n}_{\partial\Omega_{ij}}^T(\mathbf{x}_{m+\varepsilon}^{(0)}) \cdot \left[\mathbf{x}_{\beta(m+\varepsilon)}^{(\alpha)} - \bar{\mathbf{x}}_{m+\varepsilon}^{(0)}\right] < 0 \end{array}\right\} \text{for } \mathbf{n}_{\partial\Omega_{\alpha\beta}} \to \Omega_\alpha, \end{array}\right\} \tag{10.81}$$

$$\left.\begin{array}{ll} \text{either} & \left.\begin{array}{l} \mathbf{n}_{\partial\Omega_{ij}}^T(\mathbf{x}_{m-\varepsilon}^{(0)}) \cdot \left[\mathbf{x}_{m-\varepsilon}^{(0)} - \mathbf{x}_{\alpha(m-\varepsilon)}^{(\beta)}\right] > 0 \\ \mathbf{n}_{\partial\Omega_{ij}}^T(\mathbf{x}_{m+\varepsilon}^{(0)}) \cdot \left[\mathbf{x}_{\alpha(m+\varepsilon)}^{(\beta)} - \mathbf{x}_{m+\varepsilon}^{(0)}\right] < 0 \end{array}\right\} \text{for } \mathbf{n}_{\partial\Omega_{\alpha\beta}} \to \Omega_\beta \\ \text{or} & \left.\begin{array}{l} \mathbf{n}_{\partial\Omega_{ij}}^T(\mathbf{x}_{m-\varepsilon}^{(0)}) \cdot \left[\mathbf{x}_{m-\varepsilon}^{(0)} - \mathbf{x}_{\alpha(m-\varepsilon)}^{(\beta)}\right] < 0 \\ \mathbf{n}_{\partial\Omega_{ij}}^T(\mathbf{x}_{m+\varepsilon}^{(0)}) \cdot \left[\mathbf{x}_{\alpha(m+\varepsilon)}^{(\beta)} - \bar{\mathbf{x}}_{m+\varepsilon}^{(0)}\right] > 0 \end{array}\right\} \text{for } \mathbf{n}_{\partial\Omega_{\alpha\beta}} \to \Omega_\alpha, \end{array}\right\} \tag{10.82}$$

The $(2k_\alpha - 1 : 2k_\beta - 1)$-double inaccessible tangential flows are sketched in Fig.10.10(b) by the dashed curves. Such a double inaccessible flow is formed by two imaginary tangential flows to the boundary. No any flows in the two doamins can access the boundary. The corresponding necessary and sufficient conditions for the tangential flows are given as follows.

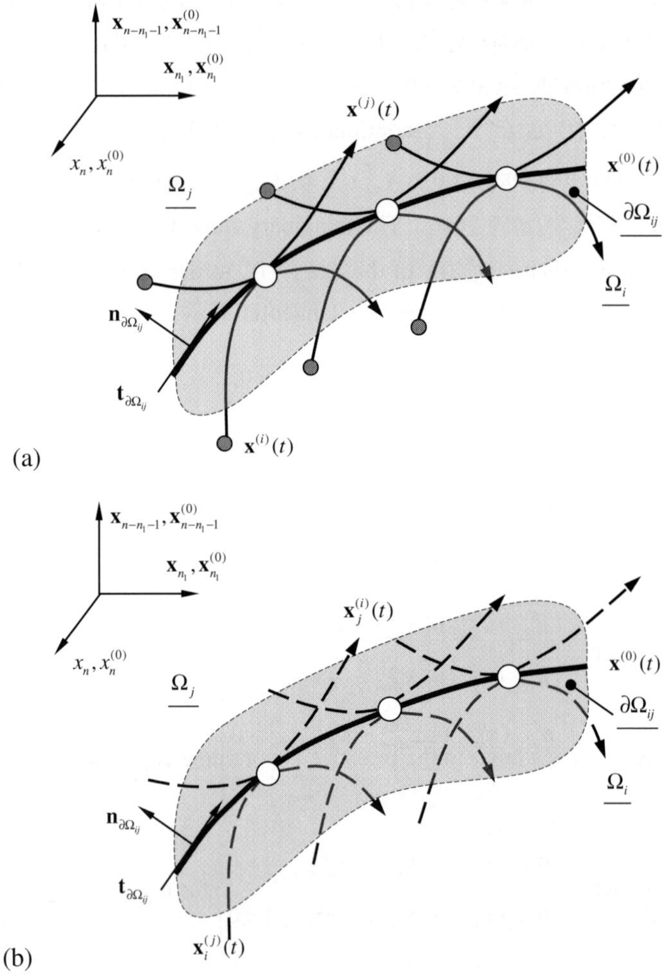

Figure 10.10. (a) Tangential flow in both Ω_i and Ω_j and (b) imaginary tangential flow in both Ω_i and Ω_j. $\mathbf{x}^{(i)}(t)$ and $\mathbf{x}^{(j)}(t)$ represent the *real* flows in domains Ω_i and Ω_j, respectively, which are depicted by the thin solid curves. $\mathbf{x}^{(j)}_i(t)$ and $\mathbf{x}^{(i)}_j(t)$ represent the *imaginary* flows in domains Ω_i and Ω_j ,respectively controlled by the vector fields in Ω_j and Ω_i, which are depicted by the dashed curves. The flow on the boundary is described by $\mathbf{x}^{(0)}(t)$. The normal and tangential vectors $\mathbf{n}_{\partial\Omega_{ij}}$ and $\mathbf{t}_{\partial\Omega_{ij}}$ on the boundary are depicted. The hollow circles are the switching points and the shaded circles are the starting points.

THEOREM 10.14. *For a discontinuous dynamical system in Eq.*(10.1), $\mathbf{x}(t_m) \equiv \mathbf{x}_m \in \partial\Omega_{ij}$ *at time* t_m. *Suppose* $\mathbf{x}_\beta^{(\alpha)}(t_{m\pm}) = \mathbf{x}_m = \mathbf{x}_\alpha^{(\beta)}(t_{m\pm})$ ($\alpha, \beta \in \{i, j\}$ *and* $\alpha \neq \beta$). *For an arbitrarily small* $\varepsilon > 0$, *there is a time interva* $[t_{m-\varepsilon}, t_{m+\varepsilon}]$. *The imaginary flow* $\mathbf{x}_\beta^{(\alpha)}(t)$ *is* $C_{[t_{m-\varepsilon}, t_{m+\varepsilon}]}^r$*-continuous* ($r \geq 2k_\alpha$) *and* $\| d^{r+1}\mathbf{x}_\beta^{(\alpha)} / dt^{r+1} \| < \infty$ *for time t. The imaginary flow* $\mathbf{x}_\alpha^{(\beta)}(t)$ *is* $C_{[t_{m-\varepsilon}, t_{m+\varepsilon}]}^r$*-continuous* ($r \geq 2k_\beta$) *for time t and* $\| d^{r+1}\mathbf{x}_\alpha^{(\beta)} / dt^{r+1} \| < \infty$. *The imaginary flow* $\mathbf{x}_\beta^{(\alpha)}(t)$ *of the* $(2k_\alpha - 1)^{th}$*-order and the imaginary* $\mathbf{x}_\alpha^{(\beta)}(t)$ *of the* $(2k_\beta - 1)^{th}$*-order to the boundary* $\partial\Omega_{ij}$ *is a* $(2k_\alpha - 1 : 2k_\beta - 1)$*-double inaccessible tangential flow iff*

$$G_{\partial\Omega_{\alpha\beta}}^{(s,\alpha)}(\mathbf{x}_m, t_{m\pm}, \mathbf{p}_\alpha, \boldsymbol{\lambda}) = 0 \text{ for } s = 0, 1, \cdots, 2k_\alpha - 2; \tag{10.83}$$

$$G_{\partial\Omega_{\alpha\beta}}^{(s,\beta)}(\mathbf{x}_m, t_{m\pm}, \mathbf{p}_\beta, \boldsymbol{\lambda}) = 0 \text{ for } s = 0, 1, \cdots, 2k_\beta - 2; \tag{10.84}$$

$$\left.\begin{array}{ll} \text{either} & \left.\begin{array}{l} G_{\partial\Omega_{\alpha\beta}}^{(2k_\alpha-1,\alpha)}(\mathbf{x}_m, t_{m\pm}, \mathbf{p}_\alpha, \boldsymbol{\lambda}) > 0 \\ G_{\partial\Omega_{\alpha\beta}}^{(2k_\beta-1,\beta)}(\mathbf{x}_m, t_{m\pm}, \mathbf{p}_\beta, \boldsymbol{\lambda}) < 0 \end{array}\right\} \text{for } \mathbf{n}_{\partial\Omega_{\alpha\beta}} \to \Omega_\beta \\ \text{or} & \left.\begin{array}{l} G_{\partial\Omega_{\alpha\beta}}^{(2k_\alpha-1,\alpha)}(\mathbf{x}_m, t_{m\pm}, \mathbf{p}_\alpha, \boldsymbol{\lambda}) < 0 \\ G_{\partial\Omega_{\alpha\beta}}^{(2k_\beta-1,\beta)}(\mathbf{x}_m, t_{m\pm}, \mathbf{p}_\beta, \boldsymbol{\lambda}) > 0 \end{array}\right\} \text{for } \mathbf{n}_{\partial\Omega_{\alpha\beta}} \to \Omega_\alpha. \end{array}\right\} \tag{10.85}$$

PROOF. The proof is similar to the proof of Theorem 10.2. ■

10.5. Flow switching bifurcations

In this section, the flow switching bifurcations from the passable to non-passable flow and the sliding fragmentation bifurcation from the non-passable to passable flow will be discussed. This section will extend the idea in Luo (2006b, 2007d). Before discussion of switching bifurcations, the product of the normal vector fields of the $(m_i : m_j)$-order on the boundary $\partial\Omega_{ij}$ is introduced as follows.

DEFINITION 10.17. For a discontinuous dynamical system in Eq.(10.1), $\mathbf{x}(t_m) = \mathbf{x}_m \in [\mathbf{x}_{m_1}, \mathbf{x}_{m_2}] \subset \partial\Omega_{ij}$ for time t_m and $\mathbf{x}^{(\alpha)}(t_{m\pm}) = \mathbf{x}_m$, $\alpha \in \{i, j\}$. For an arbitrarily small $\varepsilon > 0$, there is a time interval $[t_{m-\varepsilon}, t_{m+\varepsilon}]$. The flow $\mathbf{x}^{(i)}(t)$ is $C_{[t_{m-\varepsilon}, t_{m+\varepsilon}]}^r$-continuous ($r \geq m_i + 1$) for time t and $\| d^{r+1}\mathbf{x}^{(i)} / dt^{r+1} \| < \infty$. The flow $\mathbf{x}^{(j)}(t)$ is $C_{[t_{m-\varepsilon}, t_{m+\varepsilon}]}^r$-continuous ($r \geq m_j + 1$) for time t and $\| d^{r+1}\mathbf{x}^{(j)} / dt^{r+1} \| < \infty$. *The*

$(m_i : m_j)$-*product of G-functions* on the boundary $\partial\Omega_{ij}$ is defined as

$$\begin{aligned} L_{ij}^{(m_i:m_j)}(t_m) &\equiv L_{ij}^{(m_i:m_j)}(\mathbf{x}_m, t_m, \mathbf{p}_i, \mathbf{p}_j, \boldsymbol{\lambda}) \\ &= G_{\partial\Omega_{ij}}^{(m_i,i)}(\mathbf{x}_m, t_{m-}, \mathbf{p}_i, \boldsymbol{\lambda}) \times G_{\partial\Omega_{ij}}^{(m_j,j)}(\mathbf{x}_m, t_{m+}, \mathbf{p}_j, \boldsymbol{\lambda}) \end{aligned} \tag{10.86}$$

and for $m_i = m_j = 0$, we have $L_{ij}^{(0:0)} = L_{ij}$

$$\begin{aligned} L_{ij}(t_m) &\equiv L_{ij}(\mathbf{x}_m, t_m, \mathbf{p}_i, \mathbf{p}_j, \boldsymbol{\lambda}) \\ &= G_{\partial\Omega_{ij}}^{(i)}(\mathbf{x}_m, t_{m-}, \mathbf{p}_i, \boldsymbol{\lambda}) \times G_{\partial\Omega_{ij}}^{(j)}(\mathbf{x}_m, t_{m+}, \mathbf{p}_j, \boldsymbol{\lambda}). \end{aligned} \tag{10.87}$$

From the foregoing definition, the normal vector field products for the full passable, sink and source on the boundary $\partial\Omega_{\alpha\beta}$ are

$$\left.\begin{aligned} &L_{\alpha\beta}^{(2k_\alpha:2k_\beta)}(t_m) > 0 \text{ on } \overrightarrow{\partial\Omega}_{\alpha\beta}; \\ &L_{\alpha\beta}^{(2k_\alpha:2k_\beta)}(t_m) < 0 \text{ on } \overline{\partial\Omega}_{\alpha\beta} = \widetilde{\partial\Omega}_{\alpha\beta} \cup \widehat{\partial\Omega}_{\alpha\beta}. \end{aligned}\right\} \tag{10.88}$$

The switching bifurcation of a flow at $(t_m, \mathbf{x}_m)$ on the boundary $\partial\Omega_{\alpha\beta}$ requires

$$L_{\alpha\beta}^{(2k_\alpha:2k_\beta)}(t_m) = 0. \tag{10.89}$$

For a passable flow at $\mathbf{x}(t_m) \equiv \mathbf{x}_m \in [\mathbf{x}_{m_1}, \mathbf{x}_{m_2}] \subset \overrightarrow{\partial\Omega}_{ij}$, consider a time interval $[t_{m_1}, t_{m_2}]$ for $[\mathbf{x}_{m_1}, \mathbf{x}_{m_2}]$ on the boundary and the normal vector field product for $t_m \in [t_{m_1}, t_{m_2}]$ and $\mathbf{x}_m \in [\mathbf{x}_{m_1}, \mathbf{x}_{m_2}]$ is positive, i.e., $L_{ij}^{(2k_i:2k_j)}(t_m) > 0$. To determine the switching bifurcation, the global minimum of such a normal vector field product should be determined. Because $\mathbf{x}_m$ is a function of t_m, the two total derivatives of $L_{ij}^{(2k_i:2k_j)}(t_m)$ are introduced by

$$DL_{ij}^{(2k_i:2k_j)} = \nabla L_{ij}^{(2k_i:2k_j)}(\mathbf{x}_m, t_m, \mathbf{p}_i, \mathbf{p}_j, \boldsymbol{\lambda}) \cdot \mathbf{F}_{ij}^{(0)}(\mathbf{x}_m, t_m) + \frac{\partial L_{ij}^{(2k_i:2k_j)}(\mathbf{x}_m, t_m, \mathbf{p}_i, \mathbf{p}_j, \boldsymbol{\lambda})}{\partial t_m}, \tag{10.90}$$

$$D^r L_{ij}^{(2k_i:2k_j)} = D^{r-1}\left\{ DL_{ij}^{(2k_i:2k_j)}(\mathbf{x}_m, t_m, \mathbf{p}_i, \mathbf{p}_j, \boldsymbol{\lambda}) \right\} \tag{10.91}$$

for $r = 1, 2 \cdots$. Thus, the local minimum of $L_{ij}^{(2k_i:2k_j)}(t_m)$ is determined by

$$D^r L_{ij}^{(2k_i:2k_j)}(t_m) = 0, \quad (r = 1, 2, \cdots, 2l-1) \tag{10.92}$$

$$D^{2l} L_{ij}^{(2k_i:2k_j)}(t_m) > 0. \tag{10.93}$$

DEFINITION 10.18. For a discontinuous dynamical system in Eq.(10.1), $\mathbf{x}(t_m) = \mathbf{x}_m \in [\mathbf{x}_{m_1}, \mathbf{x}_{m_2}] \subset \overrightarrow{\partial\Omega}_{ij}$ for time t_m. Suppose $\mathbf{x}^{(i)}(t_{m\pm}) = \mathbf{x}_m = \mathbf{x}^{(j)}(t_{m\mp})$. For an arbitrarily small $\varepsilon > 0$, there are two time intervals (i.e., $[t_{m-\varepsilon}, t_m)$, $(t_m, t_{m+\varepsilon}]$). The flow $\mathbf{x}^{(i)}(t)$ is $C^r_{[t_{m-\varepsilon}, t_m)}$-continuous ($r \geq 2k_i + 1$)and $\| d^{r+1}\mathbf{x}^{(i)}/dt^{r+1} \| < \infty$ for time t. The flow $\mathbf{x}^{(j)}(t)$ is $C^r_{(t_m, t_{m+\varepsilon}]}$-continuous ($r \geq 2k_j + 1$) and $\| d^{r+1}\mathbf{x}^{(j)}/dt^{r+1} \| < \infty$ for time t. The local minimum value set of the $(2k_i : 2k_j)$-product of G-functions (i.e., $L_{ij}^{(2k_i:2k_j)}(t_m)$) is defined by

$$
{}_{\min}L_{ij}^{(2k_i:2k_j)}(t_m) = \left\{ L_{ij}^{(2k_i:2k_j)}(t_m) \left| \begin{array}{l} \text{for } t_m \in \left[t_{m_1}, t_{m_2}\right] \text{and } \mathbf{x}_m \in \left[\mathbf{x}_{m_1}, \mathbf{x}_{m_2}\right], \\ D^r L_{ij}^{(2k_i:2k_j)} = 0 \text{for } r = \{1, 2, \cdots 2l-1\}, \\ \text{and } D^{2l} L_{ij}^{(2k_i:2k_j)} > 0. \end{array} \right. \right\} \tag{10.94}
$$

From the local minimum set of $L_{ij}^{(2k_i:2k_j)}(t_m)$, the global minimum values of $L_{ij}^{(2k_i:2k_j)}(t_m)$ is defined as follows.

DEFINITION 10.19. For a discontinuous dynamical system in Eq.(10.1), $\mathbf{x}(t_m) = \mathbf{x}_m \in [\mathbf{x}_{m_1}, \mathbf{x}_{m_2}] \subset \overrightarrow{\partial\Omega}_{ij}$ for time t_m. Suppose $\mathbf{x}^{(i)}(t_{m\pm}) = \mathbf{x}_m = \mathbf{x}^{(j)}(t_{m\mp})$. For an arbitrarily small $\varepsilon > 0$, there are two time intervals (i.e., $[t_{m-\varepsilon}, t_m)$ and $(t_m, t_{m+\varepsilon}]$). The flow $\mathbf{x}^{(i)}(t)$ is $C^r_{[t_{m-\varepsilon}, t_m)}$ -continuous ($r \geq 2k_i + 1$) and $\| d^{r+1}\mathbf{x}^{(i)}/dt^{r+1} \| < \infty$ for time t. The flow $\mathbf{x}^{(j)}(t)$ is $C^r_{(t_m, t_{m+\varepsilon}]}$-continuous ($r \geq 2k_j + 1$) for time t and $\| d^{r+1}\mathbf{x}^{(j)}/dt^{r+1} \| < \infty$. The global minimum value of *the* $(2k_i : 2k_j)$*-product of* G-functions (i.e., $L_{ij}^{(2k_i:2k_j)}(\mathbf{x}_m, t_m, \mathbf{p}_i, \mathbf{p}_j, \boldsymbol{\lambda})$) is defined by

$$
{}_{G\min}L_{ij}^{(2k_i:2k_j)}(t_m) = \min_{t_m \in \left[t_{m_1}, t_{m_2}\right]} \left\{ {}_{\min}L_{ij}^{(2k_i:2k_j)}(t_m), L_{ij}^{(2k_i:2k_j)}(t_{m_1}), L_{ij}^{(2k_i:2k_j)}(t_{m_2}) \right\} \tag{10.95}
$$

To look into the switching bifurcation to the system parameter $\mathbf{q} \in \{\mathbf{p}_i, \mathbf{p}_j, \boldsymbol{\lambda}\}$, $D^r L_{ij}^{(2k_i:2k_j)}$ in Eq.(10.91) is replaced by $\frac{d^r}{d\mathbf{q}^r} L_{ij}^{(2k_i:2k_j)}$. Similarly, the maximum set of the $(2k_i : 2k_j)$-product of G-functions (i.e., $L_{ij}^{(2k_i:2k_j)}(t_m)$) can be developed as follows.

DEFINITION 10.20. For a discontinuous dynamical system in Eq.(10.1), $\mathbf{x}(t_m)=\mathbf{x}_m \in [\mathbf{x}_{m_1}, \mathbf{x}_{m_2}] \subset \overline{\partial\Omega}_{ij}$ for time t_m. Suppose $\mathbf{x}^{(\alpha)}(t_{m\pm})=\mathbf{x}_m$ ($\alpha \in \{i,j\}$). For an arbitrarily small $\varepsilon>0$, there are two time intervals ($[t_{m-\varepsilon}, t_m)$ or $(t_m, t_{m+\varepsilon}]$). The flow $\mathbf{x}^{(i)}(t)$ is $C^r_{[t_{m-\varepsilon},t_m)}$ or $C^r_{[t_{m-\varepsilon},t_{m+\varepsilon}]}$-continuous ($r \ge 2k_i+1$) and for time t $\| d^{r+1}\mathbf{x}^{(i)}/dt^{r+1} \| < \infty$. The flow $\mathbf{x}^{(j)}(t)$ is $C^r_{[t_{m-\varepsilon},t_{m+\varepsilon}]}$ or $C^r_{(t_m,t_{m+\varepsilon}]}$-continuous ($r \ge 2k_j+1$) for time t and $\| d^{r+1}\mathbf{x}^{(j)}/dt^{r+1} \| < \infty$. *The local maximum set* of the $(2k_i:2k_j)$ product of G-functions (i.e., $L_{ij}^{(2k_i:2k_j)}(\mathbf{x}_m, t_m, \mathbf{p}_i, \mathbf{p}_j, \boldsymbol{\lambda})$) is defined by

$$
{}_{\max}L_{ij}^{(2k_i:2k_j)}(t_m)=\left\{ L_{ij}^{(2k_i:2k_j)}(t_m) \left| \begin{array}{l} \text{for } t_m \in [t_{m_1}, t_{m_2}] \text{ and } \mathbf{x}_m \in [\mathbf{x}_{m_1}, \mathbf{x}_{m_2}], \\ D^r L_{ij}^{(2k_i:2k_j)} = 0 \text{ for } r=\{1,2,\cdots 2l\}, \\ \text{and } D^{2l+1} L_{ij}^{(2k_i:2k_j)} < 0. \end{array} \right. \right\} \tag{10.96}
$$

From the local maximum set of $L_{ij}^{(2k_i:2k_j)}(t_m)$, the global maximum value of $L_{ij}^{(2k_i:2k_j)}(t_m)$ is defined as follows.

DEFINITION 10.21: For a discontinuous dynamical system in Eq.(10.1), $\mathbf{x}(t_m)=\mathbf{x}_m \in [\mathbf{x}_{m_1}, \mathbf{x}_{m_2}] \subset \overline{\partial\Omega}_{ij}$ for time t_m. $\mathbf{x}^{(\alpha)}(t_{m\pm})=\mathbf{x}_m$ ($\alpha \in \{i,j\}$). For a small $\varepsilon>0$, there is a time interval ($[t_{m-\varepsilon}, t_m)$ or $(t_m, t_{m+\varepsilon}]$). The flow $\mathbf{x}^{(i)}(t)$ is $C^r_{[t_{m-\varepsilon},t_m)}$ or $C^r_{(t_m,t_{m+\varepsilon}]}$-continuous ($r \ge 2k_i+1$) for time t and $\| d^{r+1}\mathbf{x}^{(i)}/dt^{r+1} \| < \infty$. The flow $\mathbf{x}^{(j)}(t)$ is $C^r_{[t_{m-\varepsilon},t_m)}$ or $C^r_{(t_m,t_{m+\varepsilon}]}$-continuous ($r \ge 2k_j+1$) and $\| d^{r+1}\mathbf{x}^{(j)}/dt^{r+1} \| < \infty$ for time t. The global maximum of *the* $(2k_i:2k_j)$ *product of G-functions* (i.e., $L_{ij}^{(2k_i:2k_j)}(\mathbf{x}_m, t_m, \mathbf{p}_i, \mathbf{p}_j, \boldsymbol{\lambda})$) is defined by

$$
{}_{G\max}L_{ij}^{(2k_i:2k_j)}(t_m)=\max_{t_m \in [t_{m_1}, t_{m_2}]} \left\{ {}_{\max}L_{ij}^{(2k_i:2k_j)}(t_m), L_{ij}^{(2k_i:2k_j)}(t_{m_1}), L_{ij}^{(2k_i:2k_j)}(t_{m_2}) \right\}. \tag{10.97}
$$

DEFINITION 10.22. For a discontinuous dynamical system in Eq.(10.1), $\mathbf{x}(t_m)=\mathbf{x}_m \in [\mathbf{x}_{m_1}, \mathbf{x}_{m_2}] \subset \overrightarrow{\partial\Omega}_{ij}$ for time t_m. Suppose $\mathbf{x}^{(i)}(t_{m-})=\mathbf{x}_m=\mathbf{x}^{(j)}(t_{m\pm})$. For an arbitrarily small $\varepsilon>0$, there are two time intervals $[t_{m-\varepsilon}, t_m)$ and $(t_m, t_{m+\varepsilon}]$. Both flows $\mathbf{x}^{(i)}(t)$ and $\mathbf{x}^{(j)}(t)$ are $C^r_{[t_{m-\varepsilon},t_m)}$ and $C^r_{[t_{m-\varepsilon},t_{m+\varepsilon}]}$-continuous ($r \ge 2$) for time t, respectively, and $\| d^{r+1}\mathbf{x}^{(\alpha)}/dt^{r+1} \| < \infty$ ($\alpha \in \{i,j\}$). The tangential bifurcation of the flow $\mathbf{x}^{(j)}(t)$ at $\mathbf{x}_m$ on the boundary $\overrightarrow{\partial\Omega_{ij}}$ is termed *the switching bifurcation of the first kind of the non-passable flow* (or called *the sliding bifurcation*) if

$$\begin{aligned} &G^{(j)}_{\partial\Omega_{ij}}(\mathbf{x}_m, t_{m\pm}, \mathbf{p}_j, \boldsymbol{\lambda}) = 0 \text{ and } G^{(i)}_{\partial\Omega_{ij}}(\mathbf{x}_m, t_{m-}, \mathbf{p}_i, \boldsymbol{\lambda}) \neq 0 \\ &G^{(1,j)}_{\partial\Omega_{ij}}(\mathbf{x}_m, t_{m\pm}, \mathbf{p}_j, \boldsymbol{\lambda}) \neq 0 \end{aligned} \tag{10.98}$$

$$\left.\begin{aligned} &\left.\begin{aligned} &\mathbf{n}^T_{\partial\Omega_{ij}}(\mathbf{x}^{(0)}_{m-\varepsilon})\bullet\left[\mathbf{x}^{(0)}_{m-\varepsilon} - \mathbf{x}^{(j)}_{m-\varepsilon}\right] < 0,\ \mathbf{n}^T_{\partial\Omega_{ij}}(\mathbf{x}^{(0)}_{m+\varepsilon})\bullet\left[\mathbf{x}^{(j)}_{m+\varepsilon} - \mathbf{x}^{(0)}_{m+\varepsilon}\right] > 0 \\ &\mathbf{n}^T_{\partial\Omega_{ij}}(\mathbf{x}^{(0)}_{m-\varepsilon})\bullet\left[\mathbf{x}^{(0)}_{m-\varepsilon} - \mathbf{x}^{(i)}_{m-\varepsilon}\right] > 0 \text{ for } \mathbf{n}_{\partial\Omega_{ij}} \to \Omega_j, \end{aligned}\right\} \\ &\text{or} \\ &\left.\begin{aligned} &\mathbf{n}^T_{\partial\Omega_{ij}}(\mathbf{x}^{(0)}_{m-\varepsilon})\bullet\left[\mathbf{x}^{(0)}_{m-\varepsilon} - \mathbf{x}^{(j)}_{m-\varepsilon}\right] > 0,\ \mathbf{n}^T_{\partial\Omega_{ij}}(\mathbf{x}^{(0)}_{m+\varepsilon})\bullet\left[\mathbf{x}^{(j)}_{m+\varepsilon} - \mathbf{x}^{(0)}_{m+\varepsilon}\right] < 0 \\ &\mathbf{n}^T_{\partial\Omega_{ij}}(\mathbf{x}^{(0)}_{m-\varepsilon})\bullet\left[\mathbf{x}^{(0)}_{m-\varepsilon} - \mathbf{x}^{(i)}_{m-\varepsilon}\right] < 0 \text{ for } \mathbf{n}_{\partial\Omega_{ij}} \to \Omega_i. \end{aligned}\right\} \end{aligned}\right\} \tag{10.99}$$

THEOREM 10.15. *For a discontinuous dynamical system in Eq.*(10.1), $\mathbf{x}(t_m) = \mathbf{x}_m \in [\mathbf{x}_{m_1}, \mathbf{x}_{m_2}] \subset \overrightarrow{\partial\Omega_{ij}}$ *for time* t_m. *Suppose* $\mathbf{x}^{(i)}(t_{m-}) = \mathbf{x}_m = \mathbf{x}^{(j)}(t_{m\pm})$. *For an arbitrarily small* $\varepsilon > 0$, *there are two time intervals* $[t_{m-\varepsilon}, t_m)$ *and* $(t_m, t_{m+\varepsilon}]$. *The flows* $\mathbf{x}^{(i)}(t)$ *and* $\mathbf{x}^{(j)}(t)$ *are* $C^r_{[t_{m-\varepsilon}, t_m)}$ *and* $C^r_{[t_{m-\varepsilon}, t_{m+\varepsilon}]}$*-continuous* ($r \geq 3$) *for time t and* $\| d^{r+1}\mathbf{x}^{(\alpha)}/dt^{r+1} \| < \infty$ ($\alpha \in \{i, j\}$). *The bifurcation of the passable flow of* $\mathbf{x}^{(i)}(t)$ *and* $\mathbf{x}^{(j)}(t)$ *at* $\mathbf{x}_m$ *switching to the non-passable flow of the first kind on the boundary* $\overrightarrow{\partial\Omega_{ij}}$ *occurs iff*

$$\left.\begin{aligned} &G^{(j)}_{\partial\Omega_{ij}}(\mathbf{x}_m, t_{m\pm}, \mathbf{p}_j, \boldsymbol{\lambda}) = 0, \\ &G^{(i)}_{\partial\Omega_{ij}}(\mathbf{x}_m, t_{m-}, \mathbf{p}_i, \boldsymbol{\lambda}) > 0 \text{ for } \mathbf{n}_{\partial\Omega_{ij}} \to \Omega_j \\ &G^{(i)}_{\partial\Omega_{ij}}(\mathbf{x}_m, t_{m-}, \mathbf{p}_i, \boldsymbol{\lambda}) < 0 \text{ for } \mathbf{n}_{\partial\Omega_{ij}} \to \Omega_i, \end{aligned}\right\} \tag{10.100a}$$

$$\text{or } L_{ij}(\mathbf{x}_{m_2}, t_{m_2}, \mathbf{p}_i, \mathbf{p}_j, \boldsymbol{\lambda}) = 0 \text{ and } G^{(i)}_{\partial\Omega_{ij}}(\mathbf{x}_m, t_{m-}, \mathbf{p}_i, \boldsymbol{\lambda}) \neq 0, \tag{10.100b}$$

$$\text{or } {}_{G\min}L_{ij}(t_m) = 0 \text{ and } G^{(i)}_{\partial\Omega_{ij}}(\mathbf{x}_m, t_{m-}, \mathbf{p}_i, \boldsymbol{\lambda}) \neq 0; \tag{10.100c}$$

$$\left.\begin{aligned} &G^{(1,j)}_{\partial\Omega_{ij}}(\mathbf{x}_m, t_{m\pm}, \mathbf{p}_j, \boldsymbol{\lambda}) > 0 \text{ for } \mathbf{n}_{\partial\Omega_{ij}} \to \Omega_j \\ &G^{(1,j)}_{\partial\Omega_{ij}}(\mathbf{x}_m, t_{m\pm}, \mathbf{p}_j, \boldsymbol{\lambda}) < 0 \text{ for } \mathbf{n}_{\partial\Omega_{ij}} \to \Omega_i. \end{aligned}\right\} \tag{10.101}$$

PROOF. The proof is the same as in the proof of Theorem 10.1 and Theorem

10.2. This theorem can be proved. ∎

DEFINITION 10.23. For a discontinuous dynamical system in Eq.(10.1), $\mathbf{x}(t_m)=\mathbf{x}_m\in[\mathbf{x}_{m_1},\mathbf{x}_{m_2}]\subset\overrightarrow{\partial\Omega_{ij}}$ for time t_m. Suppose $\mathbf{x}^{(i)}(t_{m-})=\mathbf{x}_m=\mathbf{x}^{(j)}(t_{m\pm})$. For an arbitrarily small $\varepsilon>0$, there are two time intervals (i.e., $[t_{m-\varepsilon},t_m)$ and $(t_m,t_{m+\varepsilon}]$). The flow $\mathbf{x}^{(i)}(t)$ is $C^r_{[t_{m-\varepsilon},t_m)}$-continuous ($r\geq 2k_i+1$) and $\|d^{r+1}\mathbf{x}^{(i)}/dt^{r+1}\|<\infty$ for time t. The flow $\mathbf{x}^{(j)}(t)$ is $C^r_{[t_{m-\varepsilon},t_{m+\varepsilon}]}$-continuous ($r\geq 2k_j+1$) for time t and $\|d^{r+1}\mathbf{x}^{(j)}/dt^{r+1}\|<\infty$. The bifurcation of the $(2k_i:2k_j)$-passable flow of $\mathbf{x}^{(i)}(t)$ and $\mathbf{x}^{(j)}(t)$ at $\mathbf{x}_m$ on the boundary $\overrightarrow{\partial\Omega_{ij}}$ is termed *the switching bifurcation of the first kind of the* $(2k_i:2k_j)$*-non-passable flow* (or called *the* $(2k_i:2k_j)$*-sliding bifurcation*) if

$$\left.\begin{aligned}&G^{(s,j)}_{\partial\Omega_{ij}}(\mathbf{x}_m,t_{m\pm},\mathbf{p}_j,\boldsymbol{\lambda})=0 \text{ for } s=0,1,\cdots,2k_j\\&G^{(s,i)}_{\partial\Omega_{ij}}(\mathbf{x}_m,t_{m-},\mathbf{p}_i,\boldsymbol{\lambda})=0 \text{ for } s=0,1,\cdots,2k_i-1\\&G^{(2k_i,i)}_{\partial\Omega_{ij}}(\mathbf{x}_m,t_{m-},\mathbf{p}_i,\boldsymbol{\lambda})\neq 0 \text{ and } G^{(2k_j+1,j)}_{\partial\Omega_{ij}}(\mathbf{x}_m,t_{m\pm},\mathbf{p}_j,\boldsymbol{\lambda})\neq 0\end{aligned}\right\}\tag{10.102}$$

$$\left.\begin{aligned}&\left.\begin{aligned}&\mathbf{n}^T_{\partial\Omega_{ij}}(\mathbf{x}^{(0)}_{m-\varepsilon})\bullet\left[\mathbf{x}^{(0)}_{m-\varepsilon}-\mathbf{x}^{(j)}_{m-\varepsilon}\right]<0,\ \mathbf{n}^T_{\partial\Omega_{ij}}(\mathbf{x}^{(0)}_{m+\varepsilon})\bullet\left[\mathbf{x}^{(j)}_{m+\varepsilon}-\mathbf{x}^{(0)}_{m+\varepsilon}\right]>0\\&\mathbf{n}^T_{\partial\Omega_{ij}}(\mathbf{x}^{(0)}_{m-\varepsilon})\bullet\left[\mathbf{x}^{(0)}_{m-\varepsilon}-\mathbf{x}^{(i)}_{m-\varepsilon}\right]>0 \text{ for } \mathbf{n}_{\partial\Omega_{ij}}\to\Omega_j,\end{aligned}\right\}\\&\text{or}\\&\left.\begin{aligned}&\mathbf{n}^T_{\partial\Omega_{ij}}(\mathbf{x}^{(0)}_{m-\varepsilon})\bullet\left[\mathbf{x}^{(0)}_{m-\varepsilon}-\mathbf{x}^{(j)}_{m-\varepsilon}\right]>0,\ \mathbf{n}^T_{\partial\Omega_{ij}}(\mathbf{x}^{(0)}_{m+\varepsilon})\bullet\left[\mathbf{x}^{(j)}_{m+\varepsilon}-\mathbf{x}^{(0)}_{m+\varepsilon}\right]<0\\&\mathbf{n}^T_{\partial\Omega_{ij}}(\mathbf{x}^{(0)}_{m-\varepsilon})\bullet\left[\mathbf{x}^{(0)}_{m-\varepsilon}-\mathbf{x}^{(i)}_{m-\varepsilon}\right]<0 \text{ for } \mathbf{n}_{\partial\Omega_{ij}}\to\Omega_i.\end{aligned}\right\}\end{aligned}\right\}\tag{10.103}$$

THEOREM 10.16. *For a discontinuous dynamical system in Eq.*(10.1), $\mathbf{x}(t_m)=\mathbf{x}_m\in[\mathbf{x}_{m_1},\mathbf{x}_{m_2}]\subset\overrightarrow{\partial\Omega_{ij}}$ *for time* t_m. *Suppose* $\mathbf{x}^{(i)}(t_{m-})=\mathbf{x}_m=\mathbf{x}^{(j)}(t_{m\pm})$. *For an arbitrarily small* $\varepsilon>0$, *there are two time intervals* $[t_{m-\varepsilon},t_m)$ *and* $(t_m,t_{m+\varepsilon}]$. *The flow* $\mathbf{x}^{(i)}(t)$ *is* $C^r_{[t_{m-\varepsilon},t_m)}$*-continuous* ($r\geq 2k_i+1$) *and* $\|d^{r+1}\mathbf{x}^{(i)}/dt^{r+1}\|<\infty$ *for time t. The flow* $\mathbf{x}^{(j)}(t)$ *is* $C^r_{[t_{m-\varepsilon},t_{m+\varepsilon}]}$*-continuous* ($r\geq 2k_j+1$) *and* $\|d^{r+1}\mathbf{x}^{(j)}/dt^{r+1}\|<\infty$ *for time t. The bifurcation of the* $(2k_i:2k_j)$*-passable flow of* $\mathbf{x}^{(i)}(t)$ *and* $\mathbf{x}^{(j)}(t)$ *at* $\mathbf{x}_m$ *switching to the* $(2k_i:2k_j)$*-non-passable flow of the first kind on the boundary* $\overrightarrow{\partial\Omega_{ij}}$ iff

$$\left.\begin{aligned} &G_{\partial\Omega_{ij}}^{(s,j)}(\mathbf{x}_m,t_{m\pm},\mathbf{p}_j,\boldsymbol{\lambda})=0 \text{ for } s=0,1,\cdots,2k_j-1;\\ &G_{\partial\Omega_{ij}}^{(s,i)}(\mathbf{x}_m,t_{m-},\mathbf{p}_i,\boldsymbol{\lambda})=0 \text{ for } s=0,1,\cdots,2k_i-1; \end{aligned}\right\} \tag{10.104}$$

$$\left.\begin{aligned} &G_{\partial\Omega_{ij}}^{(2k_i,i)}(\mathbf{x}_m,t_{m-},\mathbf{p}_i,\boldsymbol{\lambda})>0 \text{ for } \mathbf{n}_{\partial\Omega_{ij}}\to\Omega_j\\ &G_{\partial\Omega_{ij}}^{(2k_i,i)}(\mathbf{x}_m,t_{m-},\mathbf{p}_i,\boldsymbol{\lambda})<0 \text{ for } \mathbf{n}_{\partial\Omega_{ij}}\to\Omega_i \end{aligned}\right\} \tag{10.105a}$$

$$G_{\partial\Omega_{ij}}^{(2k_j,j)}(\mathbf{x}_m,t_{m\pm},\mathbf{p}_j,\boldsymbol{\lambda})=0,$$

$$\text{or } L_{ij}^{(2k_i:2k_j)}\left(\mathbf{x}_m,t_m,\mathbf{p}_i,\mathbf{p}_j,\boldsymbol{\lambda}\right)=0 \text{ and } G_{\partial\Omega_{ij}}^{(2k_i,i)}(\mathbf{x}_m,t_{m-},\mathbf{p}_i,\boldsymbol{\lambda})\neq 0, \tag{10.105b}$$

$$\text{or } {}_{G\min}L_{ij}^{(2k_i:2k_j)}\left(t_m\right)=0 \text{ and } G_{\partial\Omega_{ij}}^{(2k_i,i)}(\mathbf{x}_m,t_{m-},\mathbf{p}_i,\boldsymbol{\lambda})\neq 0; \tag{10.105c}$$

$$\left.\begin{aligned} &G_{\partial\Omega_{ij}}^{(2k_j+1,j)}(\mathbf{x}_m,t_{m\pm},\mathbf{p}_j,\boldsymbol{\lambda})>0 \text{ for } \mathbf{n}_{\partial\Omega_{ij}}\to\Omega_j\\ &G_{\partial\Omega_{ij}}^{(2k_j+1,j)}(\mathbf{x}_m,t_{m\pm},\mathbf{p}_j,\boldsymbol{\lambda})<0 \text{ for } \mathbf{n}_{\partial\Omega_{ij}}\to\Omega_i. \end{aligned}\right\} \tag{10.106}$$

PROOF. The proof is the same as in the proof of Theorem 10.1 and Theorem 10.2. This theorem can be proved. ■

DEFINITION 10.24. For a discontinuous dynamical system in Eq.(10.1), $\mathbf{x}(t_m)=\mathbf{x}_m\in[\mathbf{x}_{m_1},\mathbf{x}_{m_2}]\subset\overrightarrow{\partial\Omega}_{ij}$ for time t_m. Suppose $\mathbf{x}^{(i)}(t_{m\pm})=\mathbf{x}_m=\mathbf{x}^{(j)}(t_{m+})$. For an arbitrarily small $\varepsilon>0$, there are two time intervals $[t_{m-\varepsilon},t_m)$ and $(t_m,t_{m+\varepsilon}]$. Both flows $\mathbf{x}^{(i)}(t)$ and $\mathbf{x}^{(j)}(t)$ are $C^r_{[t_{m-\varepsilon},t_{m+\varepsilon}]}$ and $C^r_{[t_{m-\varepsilon},t_m)}$-continuous ($r\geq 2$) for time t, respectively, and $\| d^{r+1}\mathbf{x}^{(\alpha)}/dt^{r+1}\|<\infty$ ($\alpha\in\{i,j\}$). The tangential bifurcation of the flow $\mathbf{x}^{(i)}(t)$ at $\mathbf{x}_m$ on the boundary $\overrightarrow{\partial\Omega}_{ij}$ is termed *the switching bifurcation of the non-passable flow of the second kind* (or called *the source flow bifurcation*) if

$$\left.\begin{aligned} &G_{\partial\Omega_{ij}}^{(j)}(\mathbf{x}_m,t_{m+},\mathbf{p}_j,\boldsymbol{\lambda})\neq 0,\\ &G_{\partial\Omega_{ij}}^{(i)}(\mathbf{x}_m,t_{m\pm},\mathbf{p}_i,\boldsymbol{\lambda})=0 \text{ and } G_{\partial\Omega_{ij}}^{(1,i)}(\mathbf{x}_m,t_{m\pm},\mathbf{p}_i,\boldsymbol{\lambda})\neq 0 \end{aligned}\right\} \tag{10.107}$$

$$\left.\begin{aligned} &\mathbf{n}^T_{\partial\Omega_{ij}}(\mathbf{x}^{(0)}_{m-\varepsilon})\cdot\left[\mathbf{x}^{(0)}_{m-\varepsilon}-\mathbf{x}^{(i)}_{m-\varepsilon}\right]>0,\ \mathbf{n}^T_{\partial\Omega_{ij}}(\mathbf{x}^{(0)}_{m+\varepsilon})\cdot\left[\mathbf{x}^{(i)}_{m+\varepsilon}-\mathbf{x}^{(0)}_{m+\varepsilon}\right]<0\\ &\mathbf{n}^T_{\partial\Omega_{ij}}(\mathbf{x}^{(0)}_{m+\varepsilon})\cdot\left[\mathbf{x}^{(j)}_{m+\varepsilon}-\mathbf{x}^{(0)}_{m+\varepsilon}\right]>0 \text{ for } \mathbf{n}_{\partial\Omega_{ij}}\to\Omega_j \end{aligned}\right\} \tag{10.108a}$$

or

$$\left.\begin{array}{l}\mathbf{n}^T_{\partial\Omega_{ij}}(\mathbf{x}^{(0)}_{m-\varepsilon})\bullet\left[\mathbf{x}^{(0)}_{m-\varepsilon}-\mathbf{x}^{(i)}_{m-\varepsilon}\right]<0,\ \mathbf{n}^T_{\partial\Omega_{ij}}(\mathbf{x}^{(0)}_{m+\varepsilon})\bullet\left[\mathbf{x}^{(i)}_{m+\varepsilon}-\mathbf{x}^{(0)}_{m+\varepsilon}\right]>0\\ \mathbf{n}^T_{\partial\Omega_{ij}}(\mathbf{x}^{(0)}_{m+\varepsilon})\bullet\left[\mathbf{x}^{(j)}_{m+\varepsilon}-\mathbf{x}^{(0)}_{m+\varepsilon}\right]<0 \text{ for } \mathbf{n}_{\partial\Omega_{ij}}\rightarrow\Omega_i.\end{array}\right\} \quad (10.108b)$$

THEOREM 10.17. *For a discontinuous dynamical system in Eq.*(10.1), $\mathbf{x}(t_m)=\mathbf{x}_m\in[\mathbf{x}_{m_1},\mathbf{x}_{m_2}]\subset\overrightarrow{\partial\Omega}_{ij}$ *for time* t_m. *Suppose* $\mathbf{x}^{(i)}(t_{m-})=\mathbf{x}_m=\mathbf{x}^{(j)}(t_{m+})$. *For an arbitrarily small* $\varepsilon>0$, *there are two time intervals* $[t_{m-\varepsilon},t_m)$ *and* $(t_m,t_{m+\varepsilon}]$. *Both flows* $\mathbf{x}^{(i)}(t)$ *and* $\mathbf{x}^{(j)}(t)$ *are* $C^r_{[t_{m-\varepsilon},t_{m+\varepsilon}]}$ *and* $C^r_{[t_{m-\varepsilon},t_m)}$*-continuous* ($r\geq 2$) *for time* t *and* $\| d^{r+1}\mathbf{x}^{(\alpha)}/dt^{r+1}\|<\infty$ ($\alpha\in\{i,j\}$)*.The bifurcation of the passable flow of* $\mathbf{x}^{(i)}(t)$ *and* $\mathbf{x}^{(j)}(t)$ *at* $\mathbf{x}_m$ *switching to the non-passable flow of the second kind on the boundary* $\overrightarrow{\partial\Omega}_{ij}$ *iff*

$$\left.\begin{array}{l}G^{(i)}_{\partial\Omega_{ij}}(\mathbf{x}_m,t_{m\pm},\mathbf{p}_i,\boldsymbol{\lambda})=0,\\ \left.\begin{array}{l}G^{(j)}_{\partial\Omega_{ij}}(\mathbf{x}_m,t_{m+},\mathbf{p}_j,\boldsymbol{\lambda})>0 \text{ for } \mathbf{n}_{\partial\Omega_{ij}}\rightarrow\Omega_j\\ G^{(j)}_{\partial\Omega_{ij}}(\mathbf{x}_m,t_{m+},\mathbf{p}_j,\boldsymbol{\lambda})<0 \text{ for } \mathbf{n}_{\partial\Omega_{ij}}\rightarrow\Omega_i,\end{array}\right\}\end{array}\right\} \quad (10.109a)$$

$$\text{or } L_{ij}\left(\mathbf{x}_{m_2},t_{m_2},\mathbf{p}_i,\mathbf{p}_j,\boldsymbol{\lambda}\right)=0 \text{ and } G^{(j)}_{\partial\Omega_{ij}}(\mathbf{x}_m,t_{m+},\mathbf{p}_i,\boldsymbol{\lambda})\neq 0, \quad (10.109b)$$

$$\text{or } {}_{G\min}L_{ij}\left(t_m\right)=0 \text{ and } G^{(j)}_{\partial\Omega_{ij}}(\mathbf{x}_m,t_{m+},\mathbf{p}_i,\boldsymbol{\lambda})\neq 0; \quad (10.109c)$$

$$\left.\begin{array}{l}G^{(1,i)}_{\partial\Omega_{ij}}(\mathbf{x}_m,t_{m\pm},\mathbf{p}_i,\boldsymbol{\lambda})<0 \text{ for } \mathbf{n}_{\partial\Omega_{ij}}\rightarrow\Omega_j\\ G^{(1,i)}_{\partial\Omega_{ij}}(\mathbf{x}_m,t_{m\pm},\mathbf{p}_i,\boldsymbol{\lambda})>0 \text{ for } \mathbf{n}_{\partial\Omega_{ij}}\rightarrow\Omega_i.\end{array}\right\} \quad (10.110)$$

PROOF. The proof is the same as in the proof of Theorem 10.1 and Theorem 10.2. This theorem can be proved. ■

DEFINITION 10.25. For a discontinuous dynamical system in Eq.(10.1), $\mathbf{x}(t_m)=\mathbf{x}_m\in[\mathbf{x}_{m_1},\mathbf{x}_{m_2}]\subset\overrightarrow{\partial\Omega}_{ij}$ for time t_m. For an arbitrarily small $\varepsilon>0$, there are two time intervals $[t_{m-\varepsilon},t_m)$ and $(t_m,t_{m+\varepsilon}]$. The flow $\mathbf{x}^{(i)}(t)$ is $C^r_{[t_{m-\varepsilon},t_{m+\varepsilon}]}$-continuous ($r\geq 2k_i+2$) for time t and $\| d^{r+1}\mathbf{x}^{(i)}/dt^{r+1}\|<\infty$. The flow $\mathbf{x}^{(j)}(t)$ is $C^r_{(t_m,t_{m+\varepsilon}]}$-continuous ($r\geq 2k_j+1$) for time t and $\| d^{r+1}\mathbf{x}^{(j)}/dt^{r+1}\|<\infty$. The tangential bifurcation of the $(2k_i:2k_j)$-passable flow of $\mathbf{x}^{(i)}(t)$ and $\mathbf{x}^{(j)}(t)$ at $\mathbf{x}_m$ on the boundary $\overrightarrow{\partial\Omega_{ij}}$ is termed *the switching bifurcation of the* $(2k_i:2k_j)$ *, non-passable flow of the second kind* (or called *the* $(2k_i:2k_j)$ *-source flow bifurcation*) if

$$\left.\begin{array}{l} G_{\partial\Omega_{ij}}^{(r,i)}(\mathbf{x}_m,t_{m\pm},\mathbf{p}_i,\boldsymbol{\lambda})=0 \text{ for } r=0,1,\cdots,2k_i \\ G_{\partial\Omega_{ij}}^{(r,j)}(\mathbf{x}_m,t_{m+},\mathbf{p}_j,\boldsymbol{\lambda})=0 \text{ for } r=0,1,\cdots,2k_j-1 \\ G_{\partial\Omega_{ij}}^{(2k_j,j)}(\mathbf{x}_m,t_{m+},\mathbf{p}_j,\boldsymbol{\lambda})\neq 0 \text{ and } G_{\partial\Omega_{ij}}^{(2k_i+1,i)}(\mathbf{x}_m,t_{m\pm},\mathbf{p}_i,\boldsymbol{\lambda})\neq 0 \end{array}\right\} \tag{10.111}$$

$$\left.\begin{array}{l} \left.\begin{array}{l} \mathbf{n}_{\partial\Omega_{ij}}^T(\mathbf{x}_{m-\varepsilon}^{(0)})\bullet\left[\mathbf{x}_{m-\varepsilon}^{(0)}-\mathbf{x}_{m-\varepsilon}^{(i)}\right]>0, \mathbf{n}_{\partial\Omega_{ij}}^T(\mathbf{x}_{m+\varepsilon}^{(0)})\bullet\left[\mathbf{x}_{m+\varepsilon}^{(i)}-\mathbf{x}_{m+\varepsilon}^{(0)}\right]<0 \\ \mathbf{n}_{\partial\Omega_{ij}}^T(\mathbf{x}_{m+\varepsilon}^{(0)})\bullet\left[\mathbf{x}_{m+\varepsilon}^{(j)}-\mathbf{x}_{m+\varepsilon}^{(0)}\right]>0 \text{ for } \mathbf{n}_{\partial\Omega_{ij}}\to\Omega_j, \end{array}\right\} \\ \text{or} \\ \left.\begin{array}{l} \mathbf{n}_{\partial\Omega_{ij}}^T(\mathbf{x}_{m-\varepsilon}^{(0)})\bullet\left[\mathbf{x}_{m-\varepsilon}^{(0)}-\mathbf{x}_{m-\varepsilon}^{(i)}\right]<0, \mathbf{n}_{\partial\Omega_{ij}}^T(\mathbf{x}_{m+\varepsilon}^{(0)})\bullet\left[\mathbf{x}_{m+\varepsilon}^{(i)}-\mathbf{x}_{m+\varepsilon}^{(0)}\right]>0 \\ \mathbf{n}_{\partial\Omega_{ij}}^T(\mathbf{x}_{m+\varepsilon}^{(0)})\bullet\left[\mathbf{x}_{m+\varepsilon}^{(j)}-\mathbf{x}_{m+\varepsilon}^{(0)}\right]<0 \text{ for } \mathbf{n}_{\partial\Omega_{ij}}\to\Omega_i. \end{array}\right\} \end{array}\right\} \tag{10.112}$$

THEOREM 10.18. *For a discontinuous dynamical system in Eq.(10.1),* $\mathbf{x}(t_m)=\mathbf{x}_m \in[\mathbf{x}_{m_1},\mathbf{x}_{m_2}]\subset\overrightarrow{\partial\Omega_{ij}}$ *for time* t_m*. Suppose* $\mathbf{x}^{(i)}(t_{m-})=\mathbf{x}_m=\mathbf{x}^{(j)}(t_{m+})$*. For an arbitrarily small* $\varepsilon>0$*, there are two time intervals* $[t_{m-\varepsilon},t_m)$ *and* $(t_m,t_{m+\varepsilon}]$*. The flow* $\mathbf{x}^{(i)}(t)$ *is* $C^r_{[t_{m-\varepsilon},t_{m+\varepsilon}]}$*-continuous* $(r\geq 2k_i+2)$ *for time t and* $\| d^{r+1}\mathbf{x}^{(i)}/dt^{r+1}\|<\infty$*. The flow* $\mathbf{x}^{(j)}(t)$ *is* $C^r_{(t_m,t_{m+\varepsilon}]}$*-continuous* $(r\geq 2k_j+1)$ *and* $\| d^{r+1}\mathbf{x}^{(j)}/dt^{r+1}\|<\infty$ *for time t. The bifurcation of the* $(2k_i:2k_j)$ *-passable flow of* $\mathbf{x}^{(i)}(t)$ *and* $\mathbf{x}^{(j)}(t)$ *at* $\mathbf{x}_m$ *switching to the* $(2k_i:2k_j)$ *- non-passable flow of the second kind on the boundary* $\overrightarrow{\partial\Omega_{ij}}$ *occurs iff*

$$\left.\begin{array}{l} G_{\partial\Omega_{ij}}^{(r,j)}(\mathbf{x}_m,t_{m+},\mathbf{p}_j,\boldsymbol{\lambda})=0 \text{ for } r=0,1,\cdots,2k_j-1; \\ G_{\partial\Omega_{ij}}^{(r,i)}(\mathbf{x}_m,t_{m\pm},\mathbf{p}_i,\boldsymbol{\lambda})=0 \text{ for } r=0,1,\cdots,2k_i-1; \end{array}\right\} \tag{10.113}$$

$$\begin{array}{l} \left.\begin{array}{l} G_{\partial\Omega_{ij}}^{(2k_j,j)}(\mathbf{x}_m,t_{m+},\mathbf{p}_j,\boldsymbol{\lambda})>0 \text{ for } \mathbf{n}_{\Omega_{ij}}\to\Omega_j \\ G_{\partial\Omega_{ij}}^{(2k_j,j)}(\mathbf{x}_m,t_{m+},\mathbf{p}_j,\boldsymbol{\lambda})<0 \text{ for } \mathbf{n}_{\Omega_{ij}}\to\Omega_i, \end{array}\right\} \\ G_{\partial\Omega_{ij}}^{(2k_i,i)}(\mathbf{x}_m,t_{m\pm},\mathbf{p}_i,\boldsymbol{\lambda})=0; \end{array} \tag{10.114a}$$

$$\text{or } L_{ij}^{(2k_i:2k_j)}\left(\mathbf{x}_m,t_m,\mathbf{p}_i,\mathbf{p}_j,\boldsymbol{\lambda}\right)=0 \text{ and } G_{\partial\Omega_{ij}}^{(2k_j,j)}(\mathbf{x}_m,t_{m+},\mathbf{p}_j,\boldsymbol{\lambda})\neq 0, \tag{10.114b}$$

$$\text{or} \quad {}_{G\min}L_{ij}^{(2k_i:2k_j)}(t_m) = 0 \text{ and } G_{\partial\Omega_{ij}}^{(2k_j,j)}(\mathbf{x}_m, t_{m+}, \mathbf{p}_j, \boldsymbol{\lambda}) \neq 0; \tag{10.114c}$$

$$\left.\begin{array}{l} G_{\partial\Omega_{ij}}^{(2k_i+1,i)}(\mathbf{x}_m, t_{m\pm}, \mathbf{p}_i, \boldsymbol{\lambda}) < 0 \text{ for } \mathbf{n}_{\Omega_{ij}} \to \Omega_j \\ G_{\partial\Omega_{ij}}^{(2k_i+1,i)}(\mathbf{x}_m, t_{m\pm}, \mathbf{p}_i, \boldsymbol{\lambda}) > 0 \text{ for } \mathbf{n}_{\Omega_{ij}} \to \Omega_i. \end{array}\right\} \tag{10.115}$$

PROOF. The proof is the same as in the proof of Theorem 10.1 and Theorem 10.2. This theorem can be proved. ■

DEFINITION 10.26. For a discontinuous dynamical system in Eq.(10.1), $\mathbf{x}(t_m) = \mathbf{x}_m \in [\mathbf{x}_{m_1}, \mathbf{x}_{m_2}] \subset \overrightarrow{\partial\Omega}_{ij}$ for time t_m. Suppose $\mathbf{x}^{(i)}(t_{m-}) = \mathbf{x}_m = \mathbf{x}^{(j)}(t_{m+})$. For an arbitrarily small $\varepsilon > 0$, there are two time intervals $[t_{m-\varepsilon}, t_m)$ and $(t_m, t_{m+\varepsilon}]$. Both flows $\mathbf{x}^{(i)}(t)$ and $\mathbf{x}^{(j)}(t)$ are $C^r_{[t_{m-\varepsilon}, t_m)}$ and $C^r_{[t_{m-\varepsilon}, t_{m+\varepsilon}]}$-continuous ($r \geq 2$) for time t, respectively, and $\| d^{r+1}\mathbf{x}^{(\alpha)}/dt^{r+1} \| < \infty$ ($\alpha \in \{i, j\}$). The tangential bifurcation of the flow $\mathbf{x}^{(i)}(t)$ and $\mathbf{x}^{(j)}(t)$ at $\mathbf{x}_m$ on the boundary $\overrightarrow{\partial\Omega}_{ij}$ is termed *the switching bifurcation of the flow from* $\overrightarrow{\partial\Omega}_{ij}$ to $\overleftarrow{\partial\Omega}_{ij}$ if

$$\left.\begin{array}{l} G_{\partial\Omega_{ij}}^{(i)}(\mathbf{x}_m, t_{m-}, \mathbf{p}_i, \boldsymbol{\lambda}) = 0 \text{ and } G_{\partial\Omega_{ij}}^{(j)}(\mathbf{x}_m, t_{m+}, \mathbf{p}_j, \boldsymbol{\lambda}) = 0, \\ G_{\partial\Omega_{ij}}^{(1,i)}(\mathbf{x}_m, t_{m-}, \mathbf{p}_i, \boldsymbol{\lambda}) \neq 0 \text{ and } G_{\partial\Omega_{ij}}^{(1,j)}(\mathbf{x}_m, t_{m+}, \mathbf{p}_j, \boldsymbol{\lambda}) \neq 0; \end{array}\right\} \tag{10.116}$$

$$\left.\begin{array}{l} \left.\begin{array}{l} \mathbf{n}^T_{\partial\Omega_{ij}}(\mathbf{x}^{(0)}_{m-\varepsilon})\bullet\left[\mathbf{x}^{(0)}_{m-\varepsilon} - \mathbf{x}^{(i)}_{m-\varepsilon}\right] > 0, \ \mathbf{n}^T_{\partial\Omega_{ij}}(\mathbf{x}^{(0)}_{m+\varepsilon})\bullet\left[\mathbf{x}^{(i)}_{m+\varepsilon} - \mathbf{x}^{(0)}_{m+\varepsilon}\right] < 0 \\ \mathbf{n}^T_{\partial\Omega_{ij}}(\mathbf{x}^{(0)}_{m-\varepsilon})\bullet\left[\mathbf{x}^{(0)}_{m-\varepsilon} - \mathbf{x}^{(j)}_{m-\varepsilon}\right] < 0, \ \mathbf{n}^T_{\partial\Omega_{ij}}(\mathbf{x}^{(0)}_{m+\varepsilon})\bullet\left[\mathbf{x}^{(j)}_{m+\varepsilon} - \mathbf{x}^{(0)}_{m+\varepsilon}\right] > 0 \end{array}\right\} \\ \text{for } \mathbf{n}_{\partial\Omega_{ij}} \to \Omega_j \\ \text{or} \\ \left.\begin{array}{l} \mathbf{n}^T_{\partial\Omega_{ij}}(\mathbf{x}^{(0)}_{m-\varepsilon})\bullet\left[\mathbf{x}^{(0)}_{m-\varepsilon} - \mathbf{x}^{(i)}_{m-\varepsilon}\right] < 0, \ \mathbf{n}^T_{\partial\Omega_{ij}}(\mathbf{x}^{(0)}_{m+\varepsilon})\bullet\left[\mathbf{x}^{(i)}_{m+\varepsilon} - \mathbf{x}^{(0)}_{m+\varepsilon}\right] > 0 \\ \mathbf{n}^T_{\partial\Omega_{ij}}(\mathbf{x}^{(0)}_{m-\varepsilon})\bullet\left[\mathbf{x}^{(0)}_{m-\varepsilon} - \mathbf{x}^{(j)}_{m-\varepsilon}\right] > 0, \ \mathbf{n}^T_{\partial\Omega_{ij}}(\mathbf{x}^{(0)}_{m+\varepsilon})\bullet\left[\mathbf{x}^{(j)}_{m+\varepsilon} - \mathbf{x}^{(0)}_{m+\varepsilon}\right] < 0 \end{array}\right\} \\ \text{for } \mathbf{n}_{\partial\Omega_{ij}} \to \Omega_i. \end{array}\right\} \tag{10.117}$$

THEOREM 10.19. *For a discontinuous dynamical system in Eq.(10.1),* $\mathbf{x}(t_m) = \mathbf{x}_m \in [\mathbf{x}_{m_1}, \mathbf{x}_{m_2}] \subset \overrightarrow{\partial\Omega}_{ij}$ *for time* t_m. *Suppose* $\mathbf{x}^{(i)}(t_{m\pm}) = \mathbf{x}_m = \mathbf{x}^{(j)}(t_{m\pm})$. *For an arbitrarily small* $\varepsilon > 0$, *there is a time interval* $[t_{m-\varepsilon}, t_{m+\varepsilon}]$. *Both flows* $\mathbf{x}^{(i)}(t)$ *and*

$\mathbf{x}^{(j)}(t)$ *are* $C^r_{[t_{m-\varepsilon},t_{m+\varepsilon}]}$*-continuous* ($r \geq 3$) *for time* t *and* $\| d^{r+1}\mathbf{x}^{(\alpha)}/dt^{r+1} \| < \infty$ ($\alpha \in \{i,j\}$). *The switching bifurcation of the flow from* $\overrightarrow{\partial\Omega}_{ij}$ *to* $\overleftarrow{\partial\Omega}_{ij}$ *occurs iff*

$$G^{(i)}_{\partial\Omega_{ij}}(\mathbf{x}_m, t_{m-}, \mathbf{p}_j, \boldsymbol{\lambda}) = 0 \text{ and } G^{(j)}_{\partial\Omega_{ij}}(\mathbf{x}_m, t_{m+}, \mathbf{p}_i, \boldsymbol{\lambda}) = 0, \tag{10.118a}$$

$$\text{or } L_{ij}(\mathbf{x}_{m_2}, t_{m_2}, \mathbf{p}_i, \mathbf{p}_j, \boldsymbol{\lambda}) = 0, \tag{10.118b}$$

$$\text{or } {}_{G\min}L_{ij}(t_m) = 0; \tag{10.118c}$$

$$\left.\begin{aligned} &G^{(1,i)}_{\partial\Omega_{ij}}(\mathbf{x}_m, t_{m-}, \mathbf{p}_i, \boldsymbol{\lambda}) < 0, G^{(1,j)}_{\partial\Omega_{ij}}(\mathbf{x}_m, t_m, \mathbf{p}_j, \boldsymbol{\lambda}) > 0 \text{ for } \mathbf{n}_{\Omega_{ij}} \to \Omega_j \\ &G^{(1,i)}_{\partial\Omega_{ij}}(\mathbf{x}_m, t_{m-}, \mathbf{p}_i, \boldsymbol{\lambda}) > 0, G^{(1,j)}_{\partial\Omega_{ij}}(\mathbf{x}_m, t_{m-}, \mathbf{p}_j, \boldsymbol{\lambda}) < 0 \text{ for } \mathbf{n}_{\Omega_{ij}} \to \Omega_i. \end{aligned}\right\} \tag{10.119}$$

PROOF. The proof is the same as in the proof of Theorem 10.1 and Theorem 10.2. This theorem can be proved. ■

DEFINITION 10.27. For a discontinuous dynamical system in Eq.(10.1), $\mathbf{x}(t_m) = \mathbf{x}_m \in [\mathbf{x}_{m_1}, \mathbf{x}_{m_2}] \subset \overrightarrow{\partial\Omega}_{ij}$ for time t_m. *Suppose* $\mathbf{x}^{(i)}(t_{m\pm}) = \mathbf{x}_m = \mathbf{x}^{(j)}(t_{m\pm})$. For an arbitrarily small $\varepsilon > 0$, there is a time interval $[t_{m-\varepsilon}, t_{m+\varepsilon}]$. The flow $\mathbf{x}^{(i)}(t)$ is $C^r_{[t_{m-\varepsilon},t_{m+\varepsilon}]}$-continuous ($r \geq 2k_i + 1$) for time t and $\| d^{r+1}\mathbf{x}^{(i)}/dt^{r+1} \| < \infty$. The flow $\mathbf{x}^{(j)}(t)$ is $C^r_{[t_{m-\varepsilon},t_{m+\varepsilon}]}$-continuous ($r \geq 2k_j + 1$) for time t and $\| d^{r+1}\mathbf{x}^{(j)}/dt^{r+1} \| < \infty$. The tangential bifurcation of the $(2k_i : 2k_j)$-passable flow of $\mathbf{x}^{(i)}(t)$ and $\mathbf{x}^{(j)}(t)$ at $\mathbf{x}_m$ on the boundary $\overrightarrow{\partial\Omega}_{ij}$ is termed *the switching bifurcation of the* $(2k_j : 2k_i)$-*passable flow from* $\overrightarrow{\partial\Omega}_{ij}$ to $\overleftarrow{\partial\Omega}_{ij}$ if

$$\left.\begin{aligned} &G^{(s,i)}_{\partial\Omega_{ij}}(\mathbf{x}_m, t_{m-}, \mathbf{p}_i, \boldsymbol{\lambda}) = 0 \text{ for } s = 0,1,\cdots,2k_i \\ &G^{(s,j)}_{\partial\Omega_{ij}}(\mathbf{x}_m, t_{m+}, \mathbf{p}_j, \boldsymbol{\lambda}) = 0 \text{ for } s = 0,1,\cdots,2k_j \\ &G^{(2k_i+1,i)}_{\partial\Omega_{ij}}(\mathbf{x}_m, t_{m-}, \mathbf{p}_i, \boldsymbol{\lambda}) \neq 0 \text{ and } G^{(2k_j+1,j)}_{\partial\Omega_{ij}}(\mathbf{x}_m, t_{m+}, \mathbf{p}_j, \boldsymbol{\lambda}) \neq 0 \end{aligned}\right\} \tag{10.120}$$

$$\left.\begin{aligned} &\mathbf{n}^T_{\partial\Omega_{ij}}(\mathbf{x}^{(0)}_{m-\varepsilon})\cdot\left[\mathbf{x}^{(0)}_{m-\varepsilon} - \mathbf{x}^{(i)}_{m-\varepsilon}\right] > 0,\ \mathbf{n}^T_{\partial\Omega_{ij}}(\mathbf{x}^{(0)}_{m+\varepsilon})\cdot\left[\mathbf{x}^{(i)}_{m+\varepsilon} - \mathbf{x}^{(0)}_{m+\varepsilon}\right] < 0 \\ &\mathbf{n}^T_{\partial\Omega_{ij}}(\mathbf{x}^{(0)}_{m-\varepsilon})\cdot\left[\mathbf{x}^{(0)}_{m-\varepsilon} - \mathbf{x}^{(j)}_{m-\varepsilon}\right] > 0,\ \mathbf{n}^T_{\partial\Omega_{ij}}(\mathbf{x}^{(0)}_{m+\varepsilon})\cdot\left[\mathbf{x}^{(j)}_{m+\varepsilon} - \mathbf{x}^{(0)}_{m+\varepsilon}\right] > 0 \end{aligned}\right\} \tag{10.121a}$$

for $\mathbf{n}_{\partial\Omega_{ij}} \to \Omega_j$ or

$$\left.\begin{array}{l}\mathbf{n}^T_{\partial\Omega_{ij}}(\mathbf{x}^{(0)}_{m-\varepsilon})\bullet\left[\mathbf{x}^{(0)}_{m-\varepsilon}-\mathbf{x}^{(i)}_{m-\varepsilon}\right]<0,\ \mathbf{n}^T_{\partial\Omega_{ij}}(\mathbf{x}^{(0)}_{m+\varepsilon})\bullet\left[\mathbf{x}^{(i)}_{m+\varepsilon}-\mathbf{x}^{(0)}_{m+\varepsilon}\right]>0\\ \mathbf{n}^T_{\partial\Omega_{ij}}(\mathbf{x}^{(0)}_{m-\varepsilon})\bullet\left[\mathbf{x}^{(0)}_{m-\varepsilon}-\mathbf{x}^{(j)}_{m-\varepsilon}\right]<0,\ \mathbf{n}^T_{\partial\Omega_{ij}}(\mathbf{x}^{(0)}_{m+\varepsilon})\bullet\left[\mathbf{x}^{(j)}_{m+\varepsilon}-\mathbf{x}^{(0)}_{m+\varepsilon}\right]<0\\ \text{for } \mathbf{n}_{\partial\Omega_{ij}}\to\Omega_i.\end{array}\right\}\quad(10.121\text{b})$$

THEOREM 10.20. *For a discontinuous dynamical system in Eq.(10.1),* $\mathbf{x}(t_m)=\mathbf{x}_m \in[\mathbf{x}_{m_1},\mathbf{x}_{m_2}]\subset\overrightarrow{\partial\Omega}_{ij}$ *for* t_m. *Suppose* $\mathbf{x}^{(i)}(t_{m-})=\mathbf{x}_m=\mathbf{x}^{(j)}(t_{m+})$. *For an arbitrarily small* $\varepsilon>0$, *there are two time intervals* $[t_{m-\varepsilon},t_m)$ *and* $(t_m,t_{m+\varepsilon}]$. *The flow* $\mathbf{x}^{(i)}(t)$ *is* $C^r_{[t_{m-\varepsilon},t_m)}$*-continuous* ($r\geq 2k_i+1$) *for time t and* $\|d^{r+1}\mathbf{x}^{(i)}/dt^{r+1}\|<\infty$. *The flow* $\mathbf{x}^{(j)}(t)$ *is* $C^r_{(t_m,t_{m+\varepsilon}]}$*-continuous* ($r\geq 2k_j+2$) *for time t and* $\|d^{r+1}\mathbf{x}^{(j)}/dt^{r+1}\|<\infty$. *The bifurcation of the* $(2k_i:2k_j)$*-passable flow of* $\mathbf{x}^{(i)}(t)$ *and* $\mathbf{x}^{(j)}(t)$ *at* $\mathbf{x}_m$ *switching to the* $(2k_i:2k_j)$*-non-passable flow of the second kind on the boundary* $\overrightarrow{\partial\Omega}_{ij}$ *occurs iff*

$$\left.\begin{array}{l}G^{(s,j)}_{\partial\Omega_{ij}}(\mathbf{x}_m,t_{m+},\mathbf{p}_j,\boldsymbol{\lambda})=0\text{ for } s=0,1,\cdots,2k_j-1;\\ G^{(s,i)}_{\partial\Omega_{ij}}(\mathbf{x}_m,t_{m-},\mathbf{p}_i,\boldsymbol{\lambda})=0\text{ for } s=0,1,\cdots,2k_i-1;\end{array}\right\}\quad(10.122)$$

$$G^{(2k_i,i)}_{\partial\Omega_{ij}}(\mathbf{x}_m,t_{m\pm},\mathbf{p}_i,\boldsymbol{\lambda})=0\text{ and } G^{(2k_j,j)}_{\partial\Omega_{ij}}(\mathbf{x}_m,t_{m\pm},\mathbf{p}_j,\boldsymbol{\lambda})=0,\quad(10.123\text{a})$$

$$\text{or}\quad L^{(2k_i:2k_j)}_{ij}\left(\mathbf{x}_m,t_m,\mathbf{p}_i,\mathbf{p}_j,\boldsymbol{\lambda}\right)=0,\quad(10.123\text{b})$$

$$\text{or}\quad {}_{G\min}L^{(2k_i:2k_j)}_{ij}\left(t_m\right)=0;\quad(10.123\text{c})$$

$$\left.\begin{array}{l}\left.\begin{array}{l}G^{(2k_i+1,i)}_{\partial\Omega_{ij}}(\mathbf{x}_m,t_{m\pm},\mathbf{p}_i,\boldsymbol{\lambda})<0\\ G^{(2k_j+1,j)}_{\partial\Omega_{ij}}(\mathbf{x}_m,t_{m\pm},\mathbf{p}_j,\boldsymbol{\lambda})>0\end{array}\right\}\text{for } \mathbf{n}_{\partial\Omega_{ij}}\to\Omega_j\\ \left.\begin{array}{l}G^{(2k_i+1,i)}_{\partial\Omega_{ij}}(\mathbf{x}_m,t_{m\mp},\mathbf{p}_i,\boldsymbol{\lambda})>0\\ G^{(2k_j+1,j)}_{\partial\Omega_{ij}}(\mathbf{x}_m,t_{m\mp},\mathbf{p}_j,\boldsymbol{\lambda})<0\end{array}\right\}\text{for } \mathbf{n}_{\partial\Omega_{ij}}\to\Omega_i.\end{array}\right\}\quad(10.124)$$

PROOF. The proof is the same as in the proof of Theorem 10.1 and Theorem 10.2. This theorem can be proved. ■

Following the definitions in DEFINITIONS 10.20-10.27, the sliding and source fragmentation bifurcations can be similarly defined.

DEFINITION 10.28. For a discontinuous dynamical system in Eq.(10.1), $\mathbf{x}(t_m) = \mathbf{x}_m \in [\mathbf{x}_{m_1}, \mathbf{x}_{m_2}] \subset \widetilde{\partial\Omega}_{ij}$ for time t_m.

(i) The tangential bifurcation of the flow $\mathbf{x}^{(j)}(t)$ at $\mathbf{x}_m$ on the boundary $\widetilde{\partial\Omega}_{ij}$ is termed *the fragmentation bifurcation of the non-passable flow of the first kind* (or called *the sliding fragmentation bifurcation*) if Eqs.(10.98) and (10.99) hold.

(ii) The tangential bifurcation of the flow $\mathbf{x}^{(i)}(t)$ with the $(2k_i)^{th}$-order and $\mathbf{x}^{(j)}(t)$ with the $(2k_j)^{th}$-order at $\mathbf{x}_m$ on the boundary $\widetilde{\partial\Omega}_{ij}$ is termed *the fragmentation bifurcation of the* $(2k_i:2k_j)$ *-non-passable flow of the first kind* (or called *the* $(2k_i:2k_j)$ *-sliding fragmentation bifurcation*) if Eqs.(10.102) and (10.103) hold.

The necessary and sufficient conditions for the sliding fragmentation bifurcation of the non-passable flow of the first kind are given by Eqs.(10.100) and (10.101) with ${}_{G\max}L_{ij}(t_m)$ replacing ${}_{G\min}L_{ij}(t_m)$. Similarly, the necessary and sufficient conditions for the sliding fragmentation bifurcation of the $(2k_i:2k_j)$-non-passable flow of the first kind are presented by Eqs.(10.104)-(10.106) with ${}_{G\max}L_{ij}^{(2k_i:2k_j)}(t_m)$ replacing ${}_{G\min}L_{ij}^{(2k_i:2k_j)}(t_m)$.

DEFINITION 10.28. For a discontinuous dynamical system in Eq.(10.1), $\mathbf{x}(t_m) = \mathbf{x}_m \in [\mathbf{x}_{m_1}, \mathbf{x}_{m_2}] \subset \widehat{\partial\Omega}_{ij}$ for time t_m.

(i) The tangential bifurcation of the flow $\mathbf{x}^{(j)}(t)$ at $\mathbf{x}_m$ on the boundary $\widehat{\partial\Omega}_{ij}$ is termed *the fragmentation bifurcation of the non-passable flow of the second kind* (or called *the source fragmentation bifurcation*) if Eqs.(10.107) and (10.108) hold.

(ii) The tangential bifurcation of the flow $\mathbf{x}^{(i)}(t)$ with the $(2k_i)^{th}$-order and $\mathbf{x}^{(j)}(t)$ with the $(2k_j)^{th}$-order at $\mathbf{x}_m$ on the boundary $\widehat{\partial\Omega}_{ij}$ is termed *the fragmentation bifurcation of the* $(2k_i:2k_j)$ *-non-passable flow of the second kind* (or called *the* $(2k_i:2k_j)$ *-source fragmentation bifurcation*) if Eqs.(10.111) and (10.112) hold.

The necessary and sufficient conditions for the source fragmentation bifurcation of the non-passable flow of the second kind are given by Eqs.(10.109) and (10.110) with ${}_{G\max}L_{ij}(t_m)$ replacing ${}_{G\min}L_{ij}(t_m)$. Similarly, the necessary and sufficient conditions for the sliding fragmentation bifurcation of the $(2k_i:2k_j)$-non-passable flow of the second kind are presented by Eqs.(10.113)-(10.115) with ${}_{G\max}L_{ij}^{(2k_i:2k_j)}(t_m)$ replacing ${}_{G\min}L_{ij}^{(2k_i:2k_j)}(t_m)$.

DEFINITION 10.30. For a discontinuous dynamical system in Eq.(10.1), $\mathbf{x}(t_m) = \mathbf{x}_m \in [\mathbf{x}_{m_1}, \mathbf{x}_{m_2}] \subset \widetilde{\partial\Omega}_{ij}$ (or $\widehat{\partial\Omega}_{ij}$) for time t_m.

(i) The tangential bifurcation of the flow $\mathbf{x}^{(i)}(t)$ and $\mathbf{x}^{(j)}(t)$ at $\mathbf{x}_m$ on the boundary $\widetilde{\partial\Omega}_{ij}$ (or $\widehat{\partial\Omega}_{ij}$) is termed *the switching bifurcation of the non-passable flow from* $\widetilde{\partial\Omega}_{ij}$ to $\widehat{\partial\Omega}_{ij}$ (or *from* $\widehat{\partial\Omega}_{ij}$ to $\widetilde{\partial\Omega}_{ij}$) if Eqs.(10.116) and (10.117) hold.

(ii) The tangential bifurcation of the flow $\mathbf{x}^{(i)}(t)$ with the $(2k_i)^{th}$-order and $\mathbf{x}^{(j)}(t)$ with the $(2k_j)^{th}$-order at $\mathbf{x}_m$ on the boundary $\widetilde{\partial\Omega}_{ij}$ (or $\widehat{\partial\Omega}_{ij}$) is termed *the switching bifurcation of the* $(2k_i:2k_j)$*-non-passable flow from* $\widetilde{\partial\Omega}_{ij}$ to $\widehat{\partial\Omega}_{ij}$ (or from $\widehat{\partial\Omega}_{ij}$ to $\widetilde{\partial\Omega}_{ij}$) if Eqs.(10.120) and (10.121) hold.

The necessary and sufficient conditions for the switching bifurcation of a non -passable flow *from* $\widetilde{\partial\Omega}_{ij}$ to $\widehat{\partial\Omega}_{ij}$ (or *from* $\widehat{\partial\Omega}_{ij}$ to $\widetilde{\partial\Omega}_{ij}$) are from Eqs.(10.118) and (10.119) with ${}_{G\max}L_{ij}(t_m)$ replacing ${}_{G\min}L_{ij}(t_m)$, However, the conditions for the switching bifurcation of the $(2k_i:2k_j)$ -non-passable flow *from* $\widetilde{\partial\Omega}_{ij}$ to $\widehat{\partial\Omega}_{ij}$ (or *from* $\widehat{\partial\Omega}_{ij}$ to $\widetilde{\partial\Omega}_{ij}$) the second kind are presented by Eqs.(10.122)-(10.124) with ${}_{G\max}L_{ij}^{(2k_i:2k_j)}(t_m)$ replacing ${}_{G\min}L_{ij}^{(2k_i:2k_j)}(t_m)$.The above conditions for the switching bifurcations of the $(2k_\alpha:2k_\beta)$-flows are summarized in Table 10.1. The following notations should be used for simplicity.

$$L_{\alpha\beta}^{(m_\alpha:m_\beta)} \equiv L_{\alpha\beta}^{(m_\alpha:m_\beta)}(\mathbf{x}_m, t_m, \mathbf{p}_\alpha, \mathbf{p}_\beta, \boldsymbol{\lambda}) \tag{10.125}$$

$$\begin{aligned} G_{\pm}^{(m_\alpha,\alpha)} &\equiv G_{\partial\Omega_{\alpha\beta}}^{(m_\alpha,\alpha)}(\mathbf{x}_m, t_{m\pm}, \mathbf{p}_\alpha, \boldsymbol{\lambda}) \\ G_{-}^{(m_\alpha,\alpha)} &\equiv G_{\partial\Omega_{\alpha\beta}}^{(m_\alpha,\alpha)}(\mathbf{x}_m, t_{m-}, \mathbf{p}_\alpha, \boldsymbol{\lambda}) \\ G_{+}^{(m_\alpha,\alpha)} &\equiv G_{\partial\Omega_{\alpha\beta}}^{(m_\alpha,\alpha)}(\mathbf{x}_m, t_{m+}, \mathbf{p}_\alpha, \boldsymbol{\lambda}) \end{aligned} \tag{10.126}$$

Because the concept of the imaginary flow is introduced, the switching bifurcations of the $(2k_\alpha:2k_\beta-1)$, $(2k_\alpha-1:2k_\beta)$ and $(2k_\alpha-1:2k_\beta)$-flows can follow the discussion on the switching bifurcation of the $(2k_\alpha:2k_\beta)$-flows. The corres-

Table 10.1 Sufficient and necessary conditions for $(2k_\alpha : 2k_\beta)$-switching bifurcations

$(2k_\alpha : 2k_\beta)$ passable flows $G_-^{(2k_\alpha)} > 0, G_+^{(2k_\beta)} > 0$ $L_{\alpha\beta}^{(2k_\alpha : 2k_\beta)} > 0$	$\overset{G_+^{(2k_\beta)}=0,\ \text{or}\ L_{\alpha\beta}^{(2k_\alpha:2k_\beta)}=0}{\longrightarrow}$ / $\underset{G_-^{(2k_\beta)}=0,\ \text{or}\ L_{\alpha\beta}^{(2k_\alpha:2k_\beta)}=0}{\longleftarrow}$ $\left.\begin{array}{l} G_\pm^{(2k_\beta+1)} > 0 \text{ for } \mathbf{n}_{\partial\Omega_{\alpha\beta}} \to \Omega_\beta; \\ G_\pm^{(2k_\beta+1)} < 0 \text{ for } \mathbf{n}_{\partial\Omega_{\alpha\beta}} \to \Omega_\alpha. \end{array}\right\}$	$(2k_\alpha : 2k_\beta)$ full sink flows $G_-^{(2k_\alpha)} > 0,\ G_-^{(2k_\beta)} < 0$ $L_{\alpha\beta}^{(2k_\alpha : 2k_\beta)} < 0$
$(2k_\alpha : 2k_\beta)$ passable flows $G_-^{(2k_\alpha)} > 0, G_+^{(2k_\beta)} > 0$ $L_{\alpha\beta}^{(2k_\alpha : 2k_\beta)} > 0$	$\overset{G_-^{(2k_\alpha)}=0,\ \text{or}\ L_{\alpha\beta}^{(2k_\alpha:2k_\beta)}=0}{\longrightarrow}$ / $\underset{G_+^{(2k_\alpha)}=0,\ \text{or}\ L_{\alpha\beta}^{(2k_\alpha:2k_\beta)}=0}{\longleftarrow}$ $\left.\begin{array}{l} G_\pm^{(2k_\alpha+1)} < 0 \text{ for } \mathbf{n}_{\partial\Omega_{\alpha\beta}} \to \Omega_\beta; \\ G_\pm^{(2k_\alpha+1)} > 0 \text{ for } \mathbf{n}_{\partial\Omega_{\alpha\beta}} \to \Omega_\alpha \end{array}\right\}$	$(2k_\alpha : 2k_\beta)$ full source flows $G_+^{(2k_\alpha)} > 0,\ G_+^{(2k_\beta)} < 0$ $L_{\alpha\beta}^{(2k_\alpha : 2k_\beta)} < 0$
$(2k_\alpha : 2k_\beta)$ passable flows $G_-^{(2k_\alpha)} > 0, G_+^{(2k_\beta)} > 0$ $L_{\alpha\beta}^{(2k_\alpha : 2k_\beta)} > 0$	$\overset{G_-^{(2k_\alpha)}=0, G_+^{(2k_\beta)}=0;\ \text{or}\ L_{\alpha\beta}^{(2k_\alpha:2k_\beta)}=0,}{\longrightarrow}$ / $\underset{G_+^{(2k_\alpha)}=0, G_-^{(2k_\beta)}=0;\ \text{or}\ L_{\alpha\beta}^{(2k_\alpha:2k_\beta)}=0}{\longleftarrow}$ $\left.\begin{array}{l} G_\pm^{(2k_\alpha+1)} < 0, G_\pm^{(2k_\beta+1)} > 0 \text{ for } \mathbf{n}_{\partial\Omega_{\alpha\beta}} \to \Omega_\beta; \\ G_\pm^{(2k_\alpha+1)} > 0, G_\pm^{(2k_\beta+1)} < 0 \text{ for } \mathbf{n}_{\partial\Omega_{\alpha\beta}} \to \Omega_\alpha \end{array}\right\}$	$(2k_\alpha : 2k_\beta)$ passable flows $G_+^{(2k_\alpha)} > 0, G_-^{(2k_\beta)} > 0$ $L_{\alpha\beta}^{(2k_\alpha : 2k_\beta)} > 0$
$(2k_\alpha : 2k_\beta)$ full sink flows $G_-^{(2k_\alpha)} > 0,\ G_-^{(2k_\beta)} < 0$ $L_{\alpha\beta}^{(2k_\alpha : 2k_\beta)} < 0$	$\overset{G_-^{(2k_\alpha)}=0, G_-^{(2k_\beta)}=0;\ \text{or}\ L_{\alpha\beta}^{(2k_\alpha:2k_\beta)}=0,}{\longrightarrow}$ / $\underset{G_+^{(2k_\alpha)}=0, G_+^{(2k_\beta)}=0;\ \text{or}\ L_{\alpha\beta}^{(2k_\alpha:2k_\beta)}=0}{\longleftarrow}$ $\left.\begin{array}{l} G_\pm^{(2k_\alpha+1)} < 0, G_\pm^{(2k_\beta+1)} > 0 \text{ for } \mathbf{n}_{\partial\Omega_{\alpha\beta}} \to \Omega_\beta; \\ G_\pm^{(2k_\alpha+1)} > 0, G_\pm^{(2k_\beta+1)} < 0 \text{ for } \mathbf{n}_{\partial\Omega_{\alpha\beta}} \to \Omega_\alpha \end{array}\right\}$	$(2k_\alpha : 2k_\beta)$ full source flows $G_+^{(2k_\alpha)} > 0,\ G_+^{(2k_\beta)} < 0$ $L_{\alpha\beta}^{(2k_\alpha : 2k_\beta)} < 0$

Table 10.2 Sufficient and necessary conditions for $(2k_\alpha : 2k_\beta - 1)$-switching bifurcations

$(2k_\alpha : 2k_\beta - 1)$ passable flows $G_-^{(2k_\alpha)} > 0, G_+^{(2k_\beta - 1)} > 0$ $L_{\alpha\beta}^{(2k_\alpha : 2k_\beta - 1)} > 0$	$\underset{G_\pm^{(2k_\beta-1)}=0,\ G_\pm^{(2k_\beta)}>0}{\overset{G_+^{(2k_\beta-1)}=0, G_+^{(2k_\beta)}<0}{\rightleftarrows}}$ $L_{\alpha\beta}^{(2k_\alpha : 2k_\beta - 1)} = 0$	$(2k_\alpha : 2k_\beta - 1)$ half sink flows $G_-^{(2k_\alpha)} > 0,\ G_\pm^{(2k_\beta - 1)} < 0$ $L_{\alpha\beta}^{(2k_\alpha : 2k_\beta - 1)} < 0$
$(2k_\alpha : 2k_\beta - 1)$ passable flows $G_-^{(2k_\alpha)} > 0, G_+^{(2k_\beta - 1)} > 0$ $L_{\alpha\beta}^{(2k_\alpha : 2k_\beta - 1)} > 0$	$\underset{G_+^{(2k_\beta)}=0, G_\pm^{(2k_\beta-1)}=0, G_\pm^{(2k_\beta)}>0}{\overset{G_-^{(2k_\alpha)}=0,\ G_+^{(2k_\beta-1)}=0, G_+^{(2k_\beta)}<0}{\rightleftarrows}}$ $L_{\alpha\beta}^{(2k_\alpha : 2k_\beta - 1)} = 0, G_\pm^{(2k_\alpha + 1)} < 0$	$(2k_\alpha : 2k_\beta - 1)$ half source flows $G_+^{(2k_\alpha)} < 0,\ G_\pm^{(2k_\beta - 1)} < 0$ $L_{\alpha\beta}^{(2k_\alpha : 2k_\beta - 1)} < 0$
$(2k_\alpha : 2k_\beta - 1)$ passable flows $G_-^{(2k_\alpha)} > 0, G_+^{(2k_\beta - 1)} > 0$ $L_{\alpha\beta}^{(2k_\alpha : 2k_\beta - 1)} > 0$	$\underset{G_+^{(2k_\alpha)}=0}{\overset{G_-^{(2k_\alpha)}=0}{\rightleftarrows}}$ $L_{\alpha\beta}^{(2k_\alpha : 2k_\beta - 1)} = 0, G_\pm^{(2k_\alpha + 1)} < 0$	$(2k_\alpha : 2k_\beta - 1)$ tangential flows $G_+^{(2k_\alpha)} < 0, G_\pm^{(2k_\beta - 1)} > 0$ $L_{\alpha\beta}^{(2k_\alpha : 2k_\beta - 1)} > 0$
$(2k_\alpha : 2k_\beta - 1)$ half sink flows $G_-^{(2k_\alpha)} > 0,\ G_\pm^{(2k_\beta - 1)} < 0$ $L_{\alpha\beta}^{(2k_\alpha : 2k_\beta - 1)} < 0$	$\underset{G_+^{(2k_\alpha)}=0}{\overset{G_-^{(2k_\alpha)}=0}{\rightleftarrows}}$ $L_{\alpha\beta}^{(2k_\alpha : 2k_\beta)} = 0, G_\pm^{(2k_\alpha + 1)} < 0$	$(2k_\alpha : 2k_\beta - 1)$ half source flows $G_+^{(2k_\alpha)} < 0,\ G_\pm^{(2k_\beta - 1)} < 0$ $L_{\alpha\beta}^{(2k_\alpha : 2k_\beta - 1)} < 0$

Table 10.3 Sufficient and necessary conditions for $(2k_\alpha : 2k_\beta - 1)$-switching bifurcations

$(2k_\alpha : 2k_\beta - 1)$ passable flows	$\underset{G_\pm^{(2k_\beta-1)}=0,\, G_\pm^{(2k_\beta)}<0}{\overset{G_+^{(2k_\beta-1)}=0,\, G_+^{(2k_\beta)}>0}{\rightleftarrows}}$	$(2k_\alpha : 2k_\beta - 1)$ half sink flows
$G_-^{(2k_\alpha)} < 0,\, G_+^{(2k_\beta-1)} < 0$ $L_{\alpha\beta}^{(2k_\alpha:2k_\beta-1)} > 0$	$L_{\alpha\beta}^{(2k_\alpha:2k_\beta-1)} = 0$	$G_-^{(2k_\alpha)} < 0,\, G_\pm^{(2k_\beta-1)} > 0$ $L_{\alpha\beta}^{(2k_\alpha:2k_\beta-1)} < 0$
$(2k_\alpha : 2k_\beta - 1)$ passable flows	$\underset{G_+^{(2k_\alpha)}=0,\, G_\pm^{(2k_\beta-1)}=0,\, G_\pm^{(2k_\beta)}<0}{\overset{G_-^{(2k_\alpha)}=0,\, G_+^{(2k_\beta-1)}=0,\, G_+^{(2k_\beta)}>0}{\rightleftarrows}}$	$(2k_\alpha : 2k_\beta - 1)$ half source flows
$G_-^{(2k_\alpha)} < 0,\, G_+^{(2k_\beta-1)} < 0$ $L_{\alpha\beta}^{(2k_\alpha:2k_\beta-1)} > 0$	$L_{\alpha\beta}^{(2k_\alpha:2k_\beta-1)} = 0,\, G_\pm^{(2k_\alpha+1)} > 0$	$G_+^{(2k_\alpha)} > 0,\, G_\pm^{(2k_\beta-1)} > 0$ $L_{\alpha\beta}^{(2k_\alpha:2k_\beta-1)} > 0$
$(2k_\alpha : 2k_\beta - 1)$ passable flows	$\underset{G_+^{(2k_\alpha)}=0,}{\overset{G_-^{(2k_\alpha)}=0,}{\rightleftarrows}}$	$(2k_\alpha : 2k_\beta - 1)$ tangential flows
$G_-^{(2k_\alpha)} < 0,\, G_+^{(2k_\beta-1)} < 0$ $L_{\alpha\beta}^{(2k_\alpha:2k_\beta-1)} > 0$	$L_{\alpha\beta}^{(2k_\alpha:2k_\beta-1)} = 0,\, G_\pm^{(2k_\alpha+1)} > 0$	$G_+^{(2k_\alpha)} > 0,\, G_\pm^{(2k_\beta-1)} < 0$ $L_{\alpha\beta}^{(2k_\alpha:2k_\beta-1)} < 0$
$(2k_\alpha : 2k_\beta - 1)$ half sink flows	$\underset{G_+^{(2k_\alpha)}=0,\, G_+^{(2k_\beta)}=0,}{\overset{G_-^{(2k_\alpha)}=0,\, G_-^{(2k_\beta)}=0}{\rightleftarrows}}$	$(2k_\alpha : 2k_\beta - 1)$ half source flows
$G_-^{(2k_\alpha)} < 0,\, G_\pm^{(2k_\beta-1)} > 0$ $L_{\alpha\beta}^{(2k_\alpha:2k_\beta-1)} < 0$	$L_{\alpha\beta}^{(2k_\alpha:2k_\beta)} = 0,\, G_\pm^{(2k_\alpha+1)} > 0$	$G_+^{(2k_\alpha)} > 0,\, G_\pm^{(2k_\beta-1)} > 0$ $L_{\alpha\beta}^{(2k_\alpha:2k_\beta-1)} > 0$

Table 10.4 Sufficient and necessary conditions for $(2k_\alpha - 1 : 2k_\beta)$-switching bifurcations

$(2k_\alpha - 1 : 2k_\beta)$ tangential flows $G_{\pm}^{(2k_\alpha-1)} < 0,\ G_{+}^{(2k_\beta)} > 0$ $L_{\alpha\beta}^{(2k_\alpha-1:2k_\beta)} < 0$	$\xrightleftharpoons[G_{\pm}^{(2k_\beta-1)}=0,G_{-}^{(2k_\beta+1)}>0]{G_{+}^{(2k_\beta)}=0,G_{+}^{(2k_\beta+1)}<0}$ $L_{\alpha\beta}^{(2k_\alpha-1:2k_\beta)} = 0$	$(2k_\alpha - 1 : 2k_\beta)$ tangential flows $G_{\pm}^{(2k_\alpha-1)} < 0,\ G_{-}^{(2k_\beta)} < 0$ $L_{\alpha\beta}^{(2k_\alpha-1:2k_\beta)} > 0$
$(2k_\alpha - 1 : 2k_\beta)$ tangential flows $G_{\pm}^{(2k_\alpha-1)} < 0,\ G_{+}^{(2k_\beta)} > 0$ $L_{\alpha\beta}^{(2k_\alpha-1:2k_\beta)} < 0$	$\xrightleftharpoons[G_{\pm}^{(2k_\alpha-1)}=0,G_{-}^{(2k_\beta)}=0;G_{-}^{(2k_\alpha)}<0,G_{\pm}^{(2k_\beta+1)}>0]{G_{\pm}^{(2k_\alpha-1)}=0,\ G_{+}^{(2k_\beta)}=0;G_{+}^{(2k_\alpha)}>0,G_{\pm}^{(2k_\beta+1)}<0}$ $L_{\alpha\beta}^{(2k_\alpha-1:2k_\beta)} = 0,$	$(2k_\alpha - 1 : 2k_\beta)$ half sink flows $G_{\pm}^{(2k_\alpha-1)} > 0,\ G_{-}^{(2k_\beta)} < 0$ $L_{\alpha\beta}^{(2k_\alpha-1:2k_\beta)} < 0$
$(2k_\alpha - 1 : 2k_\beta)$ tangential flows $G_{\pm}^{(2k_\alpha-1)} < 0,\ G_{+}^{(2k_\beta)} > 0$ $L_{\alpha\beta}^{(2k_\alpha-1:2k_\beta)} < 0$	$\xrightleftharpoons[G_{\pm}^{(2k_\alpha-1)}=0,G_{-}^{(2k_\alpha)}<0]{G_{\pm}^{(2k_\alpha-1)}=0,\ G_{+}^{(2k_\alpha)}>0}$ $L_{\alpha\beta}^{(2k_\alpha-1:2k_\beta)} = 0$	$(2k_\alpha - 1 : 2k_\beta)$ half source flows $G_{\pm}^{(2k_\alpha-1)} > 0,\ G_{+}^{(2k_\beta)} > 0$ $L_{\alpha\beta}^{(2k_\alpha-1:2k_\beta)} > 0$
$(2k_\alpha - 1 : 2k_\beta)$ half sink flows $G_{\pm}^{(2k_\alpha-1)} > 0,\ G_{-}^{(2k_\beta)} < 0$ $L_{\alpha\beta}^{(2k_\alpha-1:2k_\beta)} < 0$	$\xrightleftharpoons[G_{+}^{(2k_\beta)}=0;G_{\pm}^{(2k_\beta+1)}<0]{G_{-}^{(2k_\beta)}=0;\ G_{\pm}^{(2k_\beta+1)}>0}$ $L_{\alpha\beta}^{(2k_\alpha:2k_\beta)} = 0,$	$(2k_\alpha - 1 : 2k_\beta)$ half source flows $G_{\pm}^{(2k_\alpha-1)} > 0,\ G_{+}^{(2k_\beta)} > 0$ $L_{\alpha\beta}^{(2k_\alpha-1:2k_\beta)} > 0$

Table 10.5 Sufficient and necessary conditions for $(2k_\alpha - 1 : 2k_\beta)$-switching bifurcations

$(2k_\alpha - 1 : 2k_\beta)$ tangential flows $G_{\pm}^{(2k_\alpha - 1)} > 0$, $G_{-}^{(2k_\beta)} > 0$ $L_{\alpha\beta}^{(2k_\alpha - 1:2k_\beta)} > 0$	$\underset{G_{\pm}^{(2k_\beta)}=0;G_{-}^{(2k_\beta+1)}>0}{\overset{G_{+}^{(2k_\beta)}=0;G_{\pm}^{(2k_\beta+1)}<0}{\rightleftharpoons}}$ $L_{\alpha\beta}^{(2k_\alpha - 1:2k_\beta)} = 0$	$(2k_\alpha - 1 : 2k_\beta)$ tangential flows $G_{\pm}^{(2k_\alpha - 1)} > 0$, $G_{+}^{(2k_\beta)} < 0$ $L_{\alpha\beta}^{(2k_\alpha - 1:2k_\beta)} < 0$
$(2k_\alpha - 1 : 2k_\beta)$ tangential flows $G_{\pm}^{(2k_\alpha - 1)} > 0$, $G_{+}^{(2k_\beta)} > 0$ $L_{\alpha\beta}^{(2k_\alpha - 1:2k_\beta)} > 0$	$\underset{G_{\pm}^{(2k_\alpha-1)}=0,G_{-}^{(2k_\beta)}=0;G_{-}^{(2k_\alpha)}>0,G_{\pm}^{(2k_\beta+1)}>0}{\overset{G_{\pm}^{(2k_\alpha-1)}=0,\ G_{+}^{(2k_\beta)}=0;G_{+}^{(2k_\alpha)}<0,G_{\pm}^{(2k_\beta+1)}<0}{\rightleftharpoons}}$ $L_{\alpha\beta}^{(2k_\alpha:2k_\beta - 1)} = 0$	$(2k_\alpha - 1 : 2k_\beta)$ half source flows $G_{\pm}^{(2k_\alpha - 1)} < 0$, $G_{+}^{(2k_\beta)} < 0$ $L_{\alpha\beta}^{(2k_\alpha - 1:2k_\beta)} > 0$
$(2k_\alpha - 1 : 2k_\beta)$ tangential flows $G_{\pm}^{(2k_\alpha - 1)} > 0$, $G_{-}^{(2k_\beta)} > 0$ $L_{\alpha\beta}^{(2k_\alpha - 1:2k_\beta)} > 0$	$\underset{G_{\pm}^{(2k_\alpha-1)}=0,G_{-}^{(2k_\alpha)}>0}{\overset{G_{\pm}^{(2k_\alpha-1)}=0,\ G_{+}^{(2k_\alpha)}<0}{\rightleftharpoons}}$ $L_{\alpha\beta}^{(2k_\alpha:2k_\beta - 1)} = 0$	$(2k_\alpha - 1 : 2k_\beta)$ half sink flows $G_{\pm}^{(2k_\alpha - 1)} < 0$, $G_{-}^{(2k_\beta)} > 0$ $L_{\alpha\beta}^{(2k_\alpha - 1:2k_\beta)} < 0$
$(2k_\alpha - 1 : 2k_\beta)$ half source flows $G_{\pm}^{(2k_\alpha - 1)} < 0$, $G_{+}^{(2k_\beta)} < 0$ $L_{\alpha\beta}^{(2k_\alpha - 1:2k_\beta)} > 0$	$\underset{G_{+}^{(2k_\beta)}=0;G_{\pm}^{(2k_\beta+1)}<0}{\overset{G_{-}^{(2k_\beta)}=0;G_{\pm}^{(2k_\beta+1)}>0}{\rightleftharpoons}}$ $L_{\alpha\beta}^{(2k_\alpha:2k_\beta)} = 0,$	$(2k_\alpha - 1 : 2k_\beta)$ half sink flows $G_{\pm}^{(2k_\alpha - 1)} < 0$, $G_{-}^{(2k_\beta)} > 0$ $L_{\alpha\beta}^{(2k_\alpha - 1:2k_\beta)} < 0$

Table 10.6 Sufficient and necessary conditions for $(2k_\alpha - 1 : 2k_\beta)$-switching bifurcations

$(2k_i-1:2k_j-1)$ double tangential flows $G_\pm^{(2k_i-1)}<0\,,\,G_\pm^{(2k_j-1)}>0$ $L_{ij}^{(2k_i-1:2k_j-1)}<0$	$\overset{G_\pm^{(2k_i-1)}=0,G_\pm^{(2k_i)}>0;G_\pm^{(2k_j-1)}=0,G_\pm^{(2k_j)}<0}{\underset{G_\pm^{(2k_i-1)}=0,G_\pm^{(2k_i)}<0;G_\pm^{(2k_j-1)}=0,G_\pm^{(2k_j)}>0}{\rightleftarrows}}$ $L_{ij}^{(2k_i-1:2k_i-1)}=0$	$(2k_i-1:2k_j-1)$ double inaccessible flows $G_\pm^{(2k_i-1)}>0,\ G_\pm^{(2k_j-1)}<0$ $L_{ij}^{(2k_i-1:2k_j-1)}<0$
$(2k_i-1:2k_j-1)$ double tangential flows $G_\pm^{(2k_i-1)}<0\,,\,G_\pm^{(2k_j-1)}>0$ $L_{ij}^{(2k_i-1:2k_j-1)}<0$	$\overset{G_\pm^{(2k_j-1)}=0,G_\pm^{(2k_j)}<0}{\underset{G_\pm^{(2k_j-1)}=0,G_\pm^{(2k_j)}>0}{\rightleftarrows}}$ $L_{ij}^{(2k_i-1:2k_i-1)}=0$	$(2k_i-1:2k_j-1)$ single tangential flows in Ω_i $G_\pm^{(2k_i-1)}<0,\ G_\pm^{(2k_j-1)}<0$ $L_{ij}^{(2k_i-1:2k_j-1)}>0$
$(2k_i-1:2k_j-1)$ double tangential flows $G_\pm^{(2k_i-1)}<0\,,\,G_\pm^{(2k_j-1)}>0$ $L_{ij}^{(2k_i-1:2k_j-1)}<0$	$\overset{G_\pm^{(2k_i-1)}=0,G_\pm^{(2k_i)}>0}{\underset{G_\pm^{(2k_i-1)}=0,G_\pm^{(2k_i)}<0}{\rightleftarrows}}$ $L_{ij}^{(2k_i-1:2k_i-1)}=0$	$(2k_i-1:2k_j-1)$ single tangential flows in Ω_j $G_\pm^{(2k_i-1)}>0,\ G_\pm^{(2k_j-1)}>0$ $L_{ij}^{(2k_i-1:2k_j-1)}>0$
$(2k_i-1:2k_j-1)$ single tangential flows in Ω_i $G_\pm^{(2k_i-1)}<0,\ G_\pm^{(2k_j-1)}<0$ $L_{ij}^{(2k_i-1:2k_j-1)}>0$	$\overset{G_\pm^{(2k_i-1)}=0,G_\pm^{(2k_i)}>0;G_\pm^{(2k_j-1)}=0,G_\pm^{(2k_j)}<0}{\underset{G_\pm^{(2k_i-1)}=0,G_\pm^{(2k_i)}<0;G_\pm^{(2k_j-1)}=0,G_\pm^{(2k_j)}>0}{\rightleftarrows}}$ $L_{ij}^{(2k_i-1:2k_i-1)}=0$	$(2k_i-1:2k_j-1)$ single tangential flows in Ω_j $G_\pm^{(2k_i-1)}>0,\ G_\pm^{(2k_j-1)}>0$ $L_{ij}^{(2k_i-1:2k_j-1)}>0$

ponding necessary and sufficient conditions for such flows can be obtained. For simplicity, the conditions for the $(2k_\alpha : 2k_\beta - 1)$, $(2k_\alpha - 1 : 2k_\beta)$ and $(2k_\alpha - 1 : 2k_\beta)$ - flows are summarized in Tables 10.2-10.6, respectively. The switching bifurcations between a passable flow and half-non-passable flow, and between a passable flow and single tangential flow are presented. In addition, the switching bifurcations between a half-non-passable flow to a single tangential flow, and between a double tangential flow and a double inaccessible flow are given. All the conditions can help us understand the complexity in discontinuous dynamical systems. Compared with continuous dynamic systems, the non-passable flows and imaginary flows do not exist. If discontinuous dynamical systems possess flow barriers on the separation boundary, the corresponding passability, non-passability and tangency to the separation boundary can be discussed similarly. Further, the switching bifurcation can be further presented. Except the G-functions are different in two discontinuous dynamical systems, the corresponding conditions for the passability, non-passability and tangency to the separation boundary are similar to discontinuous dynamical system without flow barrier. Therefore, the new G-function is introduced for discontinuous dynamical systems with flow barriers.

From input and output flow barriers on the boundary $\partial\Omega_{\alpha\beta}$ in Luo (2006b), the input and output vector fields of the domain Ω_α on the boundary $\partial\Omega_{\alpha\beta}$ are defined as $\mathbf{F}_{in}^{(\alpha)}(\mathbf{x},t,\mathbf{p}_\alpha)$ and $\mathbf{F}_b^{(\alpha)}(\mathbf{x},t,\mathbf{p}_\alpha)$. The G-function for the flow barriers should be defined

$$\begin{aligned}
G_{\partial\Omega_{ij}}^{(\alpha)}(\mathbf{x}_m,t_{m-},\mathbf{p}_\alpha,\boldsymbol{\lambda}) &\equiv \mathbf{n}_{\partial\Omega_{ij}}^T(\mathbf{x}^{(0)},t,\boldsymbol{\lambda})\bullet[\mathbf{F}_{in}^{(\alpha)}(\mathbf{x}^{(\alpha)},t,\mathbf{p}_\alpha)-\mathbf{F}^{(0)}(\mathbf{x}^{(0)},t,\boldsymbol{\lambda})]\Big|_{(\mathbf{x}_m^{(0)},\mathbf{x}_m^{(\alpha)},t_{m-})}\\
&= \mathbf{n}_{\partial\Omega_{ij}}^T(\mathbf{x}^{(0)},t,\boldsymbol{\lambda})\bullet\mathbf{F}_{in}^{(\alpha)}(\mathbf{x}^{(\alpha)},t,\mathbf{p}_\alpha)+\frac{\partial\varphi_{ij}(\mathbf{x}^{(0)},t,\boldsymbol{\lambda})}{\partial t}\Big|_{(\mathbf{x}_m^{(0)},\mathbf{x}_m^{(\alpha)},t_{m-})}\\
&= \nabla\varphi_{ij}(\mathbf{x}^{(0)},t,\boldsymbol{\lambda})\bullet\mathbf{F}_{in}^{(\alpha)}(\mathbf{x}^{(\alpha)},t,\mathbf{p}_\alpha)+\frac{\partial\varphi_{ij}(\mathbf{x}^{(0)},t,\boldsymbol{\lambda})}{\partial t}\Big|_{(\mathbf{x}_m^{(0)},\mathbf{x}_m^{(\alpha)},t_{m-})}.
\end{aligned}\tag{10.127}$$

$$\begin{aligned}
G_{\partial\Omega_{ij}}^{(\alpha)}(\mathbf{x}_m,t_{m+},\mathbf{p}_\alpha,\boldsymbol{\lambda}) &\equiv \mathbf{n}_{\partial\Omega_{ij}}^T(\mathbf{x}^{(0)},t,\boldsymbol{\lambda})\bullet[\mathbf{F}_{b}^{(\alpha)}(\mathbf{x}^{(\alpha)},t,\mathbf{p}_\alpha)-\mathbf{F}^{(0)}(\mathbf{x}^{(0)},t,\boldsymbol{\lambda})]\Big|_{(\mathbf{x}_m^{(0)},\mathbf{x}_m^{(\alpha)},t_{m+})}\\
&= \mathbf{n}_{\partial\Omega_{ij}}^T(\mathbf{x}^{(0)},t,\boldsymbol{\lambda})\bullet\mathbf{F}_{b}^{(\alpha)}(\mathbf{x}^{(\alpha)},t,\mathbf{p}_\alpha)+\frac{\partial\varphi_{ij}(\mathbf{x}^{(0)},t,\boldsymbol{\lambda})}{\partial t}\Big|_{(\mathbf{x}_m^{(0)},\mathbf{x}_m^{(\alpha)},t_{m+})}\\
&= \nabla\varphi_{ij}(\mathbf{x}^{(0)},t,\boldsymbol{\lambda})\bullet\mathbf{F}_{b}^{(\alpha)}(\mathbf{x}^{(\alpha)},t,\mathbf{p}_\alpha)+\frac{\partial\varphi_{ij}(\mathbf{x}^{(0)},t,\boldsymbol{\lambda})}{\partial t}\Big|_{(\mathbf{x}_m^{(0)},\mathbf{x}_m^{(\alpha)},t_{m+})}.
\end{aligned}\tag{10.128}$$

The higher-order derivatives of the G-function for discontinuous dynamical systems with the input and output flow barriers are defined as

$$G_{\partial\Omega_{ij}}^{(k,\alpha)}(\mathbf{x}_m,t_{m-},\mathbf{p}_\alpha,\boldsymbol{\lambda})=\sum_{r=1}^{k+1}C_{k+1}^{r}D_{\bar{\mathbf{x}}}^{k+1-r}\mathbf{n}_{\partial\Omega_{ij}}^{T}(\mathbf{x}^{(0)},t,\boldsymbol{\lambda})\cdot\Big[D_{\mathbf{x}}^{r-1}\mathbf{F}_{in}^{(\alpha)}\big(\mathbf{x}^{(\alpha)},t,\mathbf{p}_\alpha\big)$$
$$-D_{\mathbf{x}^{(0)}}^{r-1}\mathbf{F}^{(0)}(\mathbf{x}^{(0)},t,\boldsymbol{\lambda})\Big]\Big|_{(\mathbf{x}_m^{(0)},\mathbf{x}_m^{(\alpha)},t_{m-})}. \tag{10.129}$$

$$G_{\partial\Omega_{ij}}^{(k,\alpha)}(\mathbf{x}_m,t_{m+},\mathbf{p}_\alpha,\boldsymbol{\lambda})=\sum_{r=1}^{k+1}C_{k+1}^{r}D_{\bar{\mathbf{x}}}^{k+1-r}\mathbf{n}_{\partial\Omega_{ij}}^{T}(\mathbf{x}^{(0)},t,\boldsymbol{\lambda})\cdot\Big[D_{\mathbf{x}}^{r-1}\mathbf{F}_{b}^{(\alpha)}\big(\mathbf{x}^{(\alpha)},t,\mathbf{p}_\alpha\big)$$
$$-D_{\mathbf{x}^{(0)}}^{r-1}\mathbf{F}^{(0)}(\mathbf{x}^{(0)},t,\boldsymbol{\lambda})\Big]\Big|_{(\mathbf{x}_m^{(0)},\mathbf{x}_m^{(\alpha)},t_{m+})}. \tag{10.130}$$

Further the dynamics of discontinuous dynamical systems with flow barriers can be investigated. Once the afore-mentioned flow barriers exist, the conditions for the non-passable flows should be measured by the new G-function.

10.6. First integral quantity increment

Consider the vector field in Eq.(10.1) for domain Ω_i as

$$\dot{\mathbf{x}}^{(i)}\equiv\mathbf{F}^{(i)}\big(\mathbf{x}^{(i)},t,\mathbf{p}_i\big)=\mathbf{f}^{(i)}\big(\mathbf{x}^{(i)},\boldsymbol{\mu}_i\big)+\mathbf{g}(\mathbf{x}^{(i)},t,\boldsymbol{\pi}_i),\ \ \mathbf{x}^{(i)}\in\Omega_i. \tag{10.131}$$

As in Eq.(2.2), the time-independent and integrable dynamical system in domain Ω_i is given by

$$\dot{\bar{\mathbf{x}}}^{(i)}=\mathbf{f}^{(i)}\big(\bar{\mathbf{x}}^{(i)},\boldsymbol{\mu}_i\big),\ \ \bar{\mathbf{x}}^{(i)}\in\Omega_i. \tag{10.132}$$

The first integral quantity is given by

$$F^{(i)}\big(\bar{\mathbf{x}}^{(i)},\boldsymbol{\mu}_i\big)=E^{(i)}. \tag{10.133}$$

Once the first integrable quantity is determined for sub-domains, the flow in the corresponding sub-domains can be measured. The intuitive illustration is given in Fig.10.11. $\mathbf{x}^{(i)}(t)$ and $\mathbf{x}^{(j)}(t)$, as depicted by the dark solid curves, represent flows in domains Ω_i and Ω_j, respectively. $\bar{\mathbf{x}}^{(i)}(t)$ and $\bar{\mathbf{x}}^{(j)}(t)$, as depicted by the light solid curves, represent integrable flows in domains Ω_i and Ω_j. The hollow circles are switching points and the shaded circles are starting and ending points. Because the dynamical system in each domain is continuous, the increment of the first integral quantity in each domain can be computed as in Chapter 4.

DEFINITION 10.31. For $t\in[t_k,t_{k+1}]$, the increment of the first integral quantity in domain Ω_α is defined as

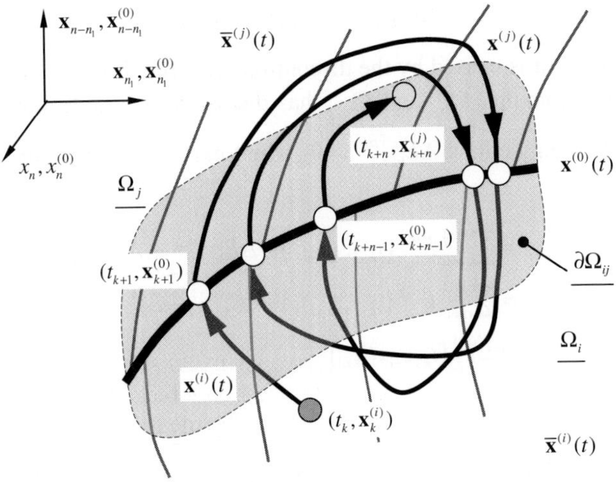

Figure 10.11. The flow measured by different integral flows in the different domains. $\mathbf{x}^{(i)}(t)$ and $\mathbf{x}^{(j)}(t)$, as depicted by the dark solid curves, represent the flows in domains Ω_i and Ω_j, respectively. $\overline{\mathbf{x}}^{(i)}(t)$ and $\overline{\mathbf{x}}^{(j)}(t)$ as depicted by the light solid curves, represent the integrable flows in domains Ω_i and Ω_j. The hollow circles are the switching points and the shaded circles are the starting and ending points.

$$L^{(\alpha)}(t_k, t_{k+1}) \equiv \int_{t_k}^{t_{k+1}} \nabla F(\mathbf{x}^{(\alpha)}, \boldsymbol{\mu}_\alpha) \cdot \mathbf{g}(\mathbf{x}^{(\alpha)}, t, \boldsymbol{\pi}_\alpha)\, dt. \tag{10.134}$$

If the investigated flow $\mathbf{x}^{(\alpha)}(t)$ for $t \in [t_k, \infty)$ is in domain Ω_α, the first integral quantity is computed for time interval $[t_k, t]$ by

$$L^{(\alpha)}(t_k, t) \equiv \int_{t_k}^{t} \nabla F(\mathbf{x}^{(\alpha)}, \boldsymbol{\mu}_\alpha) \cdot \mathbf{g}(\mathbf{x}^{(\alpha)}, t, \boldsymbol{\pi}_\alpha)\, dt. \tag{10.135}$$

However, for a discontinuous system, the flow will exist in finite domains Ω_α ($\alpha \in \{\alpha_1, \alpha_2, \cdots, \alpha_n\}$). The flow starts from domain Ω_{α_1} at time t_k and ending at time t_{k+n}. For time t_{k+i}, there are two points $(t_{k+i}, \mathbf{x}_{k+i}^{(\alpha_i)})$ and $(t_{k+i}, \mathbf{x}_{k+i+1}^{(\alpha_{i+1})})$. At the two points, we have

$$F^{(\alpha_i)}(\mathbf{x}_{k+i}^{(\alpha_i)}, \boldsymbol{\mu}_{\alpha_i}) = E_{k+i}^{(\alpha_i)} \text{ and } F^{(\alpha_{i+1})}(\mathbf{x}_{k+i}^{(\alpha_{i+1})}, \boldsymbol{\mu}_{\alpha_{i+1}}) = E_{k+i}^{(\alpha_{i+1})}. \tag{10.136}$$

The switching increment of the first integral quantity is determined at time t_{k+i} by

$$\Delta E_{k+i}^{\alpha_i \uparrow \alpha_{i+1}} = E_{k+i}^{(\alpha_{i+1})} - E_{k+i}^{(\alpha_i)}. \tag{10.137}$$

Such an increment is caused by the discontinuity. If the discontinuous dynamical system in two domains Ω_{α_i} and $\Omega_{\alpha_{i+1}}$ has the same integrable vector field with $\mathbf{x}_{k+i}^{(\alpha_i)} = \mathbf{x}_{k+i+1}^{(\alpha_{i+1})}$, then we have $\Delta E_{k+i}^{\alpha_i \uparrow \alpha_{i+1}} = 0$. The total integral quantity increment is computed by

$$\begin{aligned} L^{(\alpha_1 \cdots \alpha_n)}(t_k, t_{k+n}) &= \sum_{i=0}^{n-1} L^{(\alpha_i)}\left(t_{k+i}, t_{k+i+1}\right) + \Delta E_{k+i}^{\alpha_i \uparrow \alpha_{i+1}} + L^{(\alpha_n)}\left(t_{k+n-1}, t_{k+n}\right) \\ &= \sum_{i=0}^{n-1} L^{(\alpha_i)}\left(t_{k+i}, t_{k+i+1}\right) + (E_{k+i}^{(\alpha_{i+1})} - E_{k+i}^{(\alpha_i)}) \\ &\quad + L^{(\alpha_n)}\left(t_{k+n-1}, t_{k+n}\right) \end{aligned} \tag{10.138}$$

where for $i = 1, 2, \cdots, n$

$$L^{(\alpha_i)}\left(t_{k+i-1}, t_{k+i}\right) \equiv \int_{t_{k+i-1}}^{t_{k+i}} \nabla F\left(\mathbf{x}^{(\alpha_i)}, \boldsymbol{\mu}_{\alpha_i}\right) \bullet \mathbf{g}\left(\mathbf{x}^{(\alpha_i)}, t, \boldsymbol{\pi}_{\alpha_i}\right) dt. \tag{10.139}$$

If a flow for $t \in [t_k, t_{k+n}]$ is periodic, then the total first integral quantity increment should be zero, i.e.,

$$L^{(\alpha_1 \cdots \alpha_n)}(t_k, t_{k+n}) = 0. \tag{10.140}$$

For chaotic motion, $L^{(\alpha_1 \cdots \alpha_n)}(t_k, t_{k+n}) \neq 0$ can be used to develop an iterative relations. Further, the chaotic motion can be investigated.

For a 2-D perturbed Hamiltonian system, equation (10.139) becomes

$$\begin{aligned} L^{(\alpha_i)}\left(t_{k+i}, t_{k+i+1}\right) &\equiv \int_{t_{k+i}}^{t_{k+i+1}} \nabla F\left(\mathbf{x}^{(\alpha_i)}, \boldsymbol{\mu}_{\alpha_i}\right) \bullet \mathbf{g}\left(\mathbf{x}^{(\alpha_i)}, t, \boldsymbol{\pi}_{\alpha_i}\right) dt \\ &= \int_{t_{k+i}}^{t_{k+i+1}} \Big[f_2\left(\mathbf{x}^{(\alpha_i)}, \boldsymbol{\mu}_{\alpha_i}\right) \bullet g_1\left(\mathbf{x}^{(\alpha_i)}, t, \boldsymbol{\pi}_{\alpha_i}\right) \\ &\quad - f_1\left(\mathbf{x}^{(\alpha_i)}, \boldsymbol{\mu}_{\alpha_i}\right) \bullet g_2\left(\mathbf{x}^{(\alpha_i)}, t, \boldsymbol{\pi}_{\alpha_i}\right) \Big] dt. \end{aligned} \tag{10.141}$$

Once Eq.(10.141) is substituted into Eq.(10.138), the $L^{(\alpha_1 \cdots \alpha_n)}(t_k, t_{k+n})$-function can be obtained. In addition, for a special Poincare surface, such an L-function can be developed rather than the Melnikov functions in discontinuous dynamical systems.

References

1. Abdullaev, S.S. and Zaslavsky, G.M., 1995, Self-similarity of stochastic magnetic field lines near the X-point, *Physics of Plasmas*, **2**, 4533-4541.
2. Abdullaev, S.S. and Zaslavsky, G.M., 1996, Application of the separatrix map to study perturbed magnetic field lines near the separatrix, *Physics of Plasmas*, **3**, 516-528.
3. Ahn, T., Kim, G. and Kim, S., 1996, Analysis of the separatrix map in Hamiltonian systems, *Physica D*, **89**, 315-328.
4. Anosov, D.A., 1962, Structural stability of geodesic flows on compact Riemannian manifolds of negative curvature. *Doklady Akademic Nauk SSSR*, **145**, 707-709.
5. Ao, P., 2004, Potential in stochastic differential equations: Novel construction, *Journal of Physics A: Mathematical and General*, **37**, L25-L30.
6. Ao, P., 2005, Laws in Darwinian Evolutionary Theory, *Physics of Life Reviews,* **2**(2), 117-156.
7. Arnold, V.I., 1963, Small denominators and the problem of stability of motion in classical and celestial mechanics, Russian Mathematical Surveys, **18**, 85-192.
8. Arnold, V.I., 1964, Instability of dynamical systems with several degrees of freedom, *Soviet Mathematics Doklody*, **5**, 581-585.
9. Arnold, V.I., 1989, *Mathematical Methods of Classical Mechanics*, Springer: New York.
10. Bartler, T., 1999, *Lyapunov Exponents and Chaos Investigation*, Ph.D. thesis, Department of Aerospace Engineering, the University of Cincinnati, Cincinnati, Ohio, USA.
11. Benettin, G., Galgani, L. and Strelcyn, J.M., 1976, Kolmogorov entropy and numerical experiments, *Physical Review*, **A14**, 2338-2345.
12. Benettin, G., Galgani, L., Giorgilli, A. and Strelcyn, J.M., 1980a, Lyapunov characteristic exponents for smooth dynamical systems and for Hamiltonian systems: a method for computing all of them-1.Theory, *Meccanica*, **15**, 9-21
13. Benettin, G., Galgani, L., Giorgilli, A. and Strelcyn, J.M., 1980b, Lyapunov characteristic exponents for smooth dynamical systems and for Hamiltonian systems: a method for computing all of them-2. Numerical Applications, *Meccanica*, **15**, 9-21

14. Birkhoff, G.D., 1913, Proof of Poincare's geometric theorem, *Transactions on American Mathematical Society*, **14**, 14-22.

15. Birkhoff, G.D., 1927, *Dynamical Systems*, American Mathematical Society, New York,.

16. Chirikov, B.V., 1960, Resonance processes in magnetic traps, *Journal of Nuclear Energy, Part C Plasma*, **1**, 253-260.

17. Chirikov, B.V., 1979, A universal instability of many-dimensional oscillator systems, *Physics Reports*, **52**, 263-379.

18. Dieci, L., Russell, R. and Velk, E.V., 1994, Unitary integrators and application to continuous orthogonalization techniques, *SIAM Journal of Numerical Analysis*, **31**, 261-281.

19. Dieci, L.and Velk, E.V., 1995, Computation of a few Lyapunov exponents for continuous and discrete dynamical systems, *Applied Numerical Mathematics*, **34**, 403-423.

20. Dieci, L., Russell, R. and Velk, E.V., 1997, On the computation of Lyapunov exponents for continuous dynamical systems, *SIAM Journal of Numerical Analysis*, **34**, 403-423.

21. Escande, D.F. and Doveil, F., 1981, Renormalization method for the onset of stochasticity in a Hamiltonian system. *Physical Letter*, **83A**, 307-310.

22. Escande, D.F., 1985, Stochasticity in classic Hamiltonian systems: universal aspects, *Physics Reports*, **121**, 165-261.

23. Feng, K. and Qin, M.Z., 1991, Hamiltonian algorithms for Hamiltonian systems and a comparative numerical study, *Computer Physics Communications*, **65**, 173-187.

24. Filonenko, N.N., Sagdeev, R.Z. and Zaslavsky, G.M., 1967, Destruction of magnetic surfaces by magnetic field irregularities: Part II, *Nuclear Fusion*, **7**, 253-266.

25. Ford, J., Stoddard, D. and Turner, J.S.,1973, On the integrability of the Toda lattice, *Progress of Theoretical Physics*, **50**, 1547-1560.

26. Gelfreich,V.G., Lazutkin,V.F. and Tabanov,M.B.,1991, Exponentially small splitting in Hamiltonian systems, *Chaos,* **1**, 137-142.

27. Gelfreich, V.G., Lazutkin,V.F. and Svanidze, N.V., 1994, A refined formula for the separatrix splitting for the standard map, *Physica D*, **71**, 82-101.

28. Glimm, J., 1963, Formal stability of Hamiltonian systems. *Communications on Pure and Applied Mathematics*, **16**, 509-526.

29. Goldhirsch, I., Sulem, P. and Orszag, S., 1987, Stability and Lyapunov stability of dynamical system: a differential approach and a numerical method, *Physica D*, **27**, 311-337.

30. Goldstein, H., Poole Jr. C. and Safko, J.L., 2002, *Classic Mechanics*, 3rd edition, Addison-Wesley: San Francisco.

31. Greene, J.M., 1968, Two-dimensional measure-preserving mappings, *Journal of Mathematical Physics*, **9**, 760-768.

32. Greene, J.M., 1979, A method for computing the stochastic transition, *Journal of Mathematical Physics*, **20**, 1183-1201.

33. Greenspan, B.D.,1981, Bifurcations in periodically forced oscillations: subharmonics and homoclinic orbits, *Ph.D. thesis*, Center for Applied Mathematics, Cornell University, USA.

34. Greenspan, B.D. and Holmes, P.J., 1983, Homoclinic orbits, subharmonics and global bifurcations in forced oscillations, In *Nonlinear Dynamics and Turbulence*, G.Barenblatt, G. Iooss, and D.D. Joseph (eds.) London, Pitman.

35. Guckenheimer, J. and Holmes, P., 1983, *Nonlinear oscillations, Dynamical systems, and Bifurcations of Vector Fields*, Springer-Verlag: New York.

36. Han, R.P.S. and Luo, A.C.J., 1998, Resonant layers in nonlinear dynamics, *ASME Journal of Applied Mechanics*, **65**, 727-736.

37. Hayashi, C., 1964, *Nonlinear Oscillations in Physical Systems*, McGraw-Hill: New York.

38. Henon, M. and Heiles, C., 1964, The applicability of the third integral motion: some numerical experiments, *Astronomical Journal,* **69**, 73-79.

39. Hopf, E., 1937, *Ergodentheorie*, Ergebnisse der Mathematics und ihrer Grenzgebiete, Springer: Berlin.

40. Holmes, P.J., Marsden, J.E. and Scheurle, J.,1988, Exponentially small splittings of separatrices with applications to KAM theory and degenerate bifurcations, *Contemporary Mathematics*, **81**, 213-244.

41. Iomin, A. and Fishman, S., 1996, Semiclassical quantization of a separatrix map, *Physical Review* E, **54**, R1-R5.

42. Izrailev, F.M. and Chirikov, B.V., 1966, Statistical properties of a nonlinear string, *Doklady Akademic Nauk SSSR,* **166**, 57-59.

43. Kolmogorov, A.N., 1954, The conservation of conditionally periodic motions under a small perturbations of the Hamiltonian, *Doklady Akademy Nauk SSSR*, **98**, 527-530.

44. Kreyszig, E., 1959, *Differential Geometry*, University of Toronto Press: Toronto.

45. Langrage, J.L.,1788, *Mecanique Analytique*, (edition Albert Blanchard, Paris, 1965).

46. Lazutkin, V.F., Schachmannski, I.G. and Tabanov, M.B., 1989, Splitting of separatrices for standard and semistandard mappings, *Physica D*, **40**, 235-

248.

47. Lichtenberg,/A.J. and Lieberman, M.A., 1992, *Regular and Chaotic Dynamics*, 2nd Edition, Springer-Verlag: New York.
48. Lin, W.A. and Reichl, L.E., 1986, External field induced chaos in an infinite square well potential, *Physica D*, **19**, 145-152.
49. Lu, C., 2007, Chaos of a parametrically excited undamped pendulum, *Communications in Nonlinear Science and Numerical Simulation*, **12**, 45-57.
50. Lu, J., Yang, G., Oh, H. and Luo, A.C.J., 2005, Computing Lyapunov exponents of continuous dynamical systems:Method of Lyapunov vectors, *Chaos, Solitons and Fractals,* **23**, 1879-1892.
51. Luo, A.C.J., 1995, Analytical modeling of bifurcations, chaos and multifractals in nonlinear dynamics. *Ph.D. Dissertation*, University of Manitoba, Winnipeg, Manitoba, Canada.
52. Luo, A.C.J., 2001a, Resonant-overlap phenomena in stochastic layers of nonlinear Hamiltonian systems with periodical excitations, *Journal of Sound and Vibration*, **240**(5), 821-836.
53. Luo, A.C.J., 2001b, Resonance and stochastic layer in a parametrically excited pendulum, *Nonlinear Dynamics*, **26**, 355-367.
54. Luo, A.C.J., 2001c, Chaotic motions in the generic separatrix band of the Mathieu-Duffing oscillator, *Journal of Sound and Vibration*, **248**(3), 521-532.
55. Luo, A.C.J., 2002, Resonant layers in a parametrically excited pendulum. *International Journal of Bifurcation and Chaos*, **12**(2), 409-419.
56. Luo, A.C.J., 2004a, Nonlinear Dynamics theory of stochastic layers in nonlinear Hamiltonian, *Applied Mechanics Review,* **57**, 161-172.
57. Luo, A.C.J., 2004b, Chaotic motion in the resonant separatrix bands of a Mathieu-Duffing oscillator with a Twin-well Potential, *Journal of Sound and Vibrations*, **273**, 653-666.
58. Luo, A.C.J., 2005a, A theory for nonsmooth dynamic systems on the connectable domains, *Communications in Nonlinear Science and Numerical Simulation*, **10**, 1-55.
59. Luo, A.C.J., 2005b, Chaotic motions in traveling plates, IMeCHe, Part K, *Journal of Multi-body Dynamics*, **219**, 237-247.
60. Luo, A.C.J., 2005c, Imaginary, sink and source flows in the vicinity of separatrix of non-smooth dynamical systems, *Journal of Sound and Vibration* **285**, 443-456.
61. Luo, A.C.J., 2006a, Predictions of quasi-periodic and chaotic motions in nonlinear Hamiltonian systems, *Chaos, Solitons and Fractals* **28**, 627-649.

62. Luo, A.C.J., 2006b, *Singularity and Dynamics on Discontinuous Vector Fields*, Elsevier: Amsterdam.
63. Luo, A.C.J., 2006c, Chaos and quasi-periodic motion on the separatrix surface in nonlinear Hamiltonian system, *ASME Journal of Computational and Nonlinear Dynamics*, **1**, 135-142.
64. Luo, A.C.J., 2006d, On grazing and strange attractor fragmentation in non-smooth dynamical systems, *Communications in Nonlinear Science and Numerical Simulation* **11**, 922-933.
65. Luo, A.C.J., 2007a, A theory for *n*-dimensional, nonlinear dynamics on continuous vector fields, *Communications in Nonlinear Science and Numerical Simulation* **12**, 117-194.
66. Luo, A.C.J. 2007b, On global transversality and chaos in two-dimensional dynamical systems, *ASME Journal of Computational and Nonlinear Dynamics*, **2**, 242-248.
67. Luo, A.C.J., 2007c, Global tangency and transversality of periodic flows and chaos in a periodically forced, damped Duffing oscillator, *International Journal of Bifurcations and Chaos*, in press.
68. Luo, A.C.J., 2007d, On flow switching bifurcations in discontinuous dynamical systems, *Communications in Nonlinear Science and Numerical Simulation*, **12**, 100-116.
69. Luo, A.C.J. and Gu, K., 2003, On Resonant separatrix bands of a Duffing oscillator with twin-well potentials, *Chaos, Solitons and Fractals,* **15**, 771-782.
70. Luo, A.C.J., Gu, K. and Han, R.P.S., 1999, Resonant-separatrix webs in stochastic layers of the twin-well Duffing oscillator, *Nonlinear Dynamics*, **19**, 37-48.
71. Luo, A.C.J. and Han, R.P.S., 1999, Analytical predictions of chaos in a nonlinear rod, *Journal of Sound and Vibration*, **227 (3)**, 523-544.
72. Luo, A.C.J. and Han, R.P.S., 2000a, Investigations of stochastic layers in nonlinear dynamics, *Journal of Vibration and Acoustics*, **122**, 36-41.
73. Luo, A.C.J. and Han, R.P.S., 2000b, The dynamics of resonant and stochastic layers in a periodically-driven pendulum, *Chaos, Solitons and Fractals*, **11**, 2349-2359.
74. Luo, A.C.J. and Han,R.P.S.,2001, The resonance theory for stochastic layers in nonlinear dynamical systems, *Chaos, Solitons and Fractals*, **12**, 2493-2508.
75. Luo, A.C.J., Han, R.P.S. and Xiang, Y.M., 1995, Chaotic analysis of subharmonic resonant waves in undamped and damped strings, *Journal of*

Hydrodynamics, **7B**, 92-104.

76. Luo, A.C.J. and Wang, F.Y., 2002, Chaotic motion in a micro-electro-mechanical system with non-linearity from capacitors, *Communications in Nonlinear Science and Numerical Simulation*, 7, 31-49.
77. Luo, A.C.J. and Wang, F.Y., 2004, Nonlinear dynamics of a micro-electro-mechanical system with time-varying capacitors, *ASME Journal of Vibration and Acoustics,* **126**, 77-83.
78. Lyapunov,A.M., 1907, The general problem of the stability of motion. *Communications in Society of Mathematics Kharkov*. Translation from Russian to French in 'Probleme general de la stabilite du mouvement', Annual Fac. Sci. Toulouse (2), **9**, 203-475; reprinted in Annals Mathematical Studies **17**, Princeton University Press, 1947.
79. McLachlan, R. and Atela, P., 1992, The accuracy of symplectic integrators, *Nonlinearity*, **5**, 541-562.
80. Melnikov, V.K., 1962, On the behavior of trajectories of system near to autonomous Hamiltonian systems, *Soviet Mathematics-Doklady,* **3**, 109-112.
81. Melnikov, V.K., 1963, On the stability of the center for time periodic perturbations, *Transaction Moscow Mathematical Society*, **12**, 1-57.
82. Minorsky, N., 1962, *Nonlinear Oscillations*, Van Nostrand: New York.
83. Moser, J., 1962, On the invariant curves of area-preserving mapping on an annulus, Nachr, Akad. Wiss. Goettingen Math. Phys. *Kl.*, **1**,1-20.
84. Neishtadt, A.I., 1984, The separation of motion in systems with rapidly rotating phase,(PMM) *Journal of Applied Mathematics and Mechanics*, **48**, 133-139.
85. Nekhoroshev, N.N., 1977, An exponential estimate of the time of stability of nearly-integrable Hamiltonian systems, *Russian Mathematics Surveys*,**32(6)**, 1-65.
86. Poincare, H., 1890, Sur La Probleme des Trois Corps et les Equations de la Dynamique, *Acta Mathematica*, 13,1-271.
87. Pioncare, H.,1892, *Les Methods Nouvelles de la Mechanique Celeste*, Gauthier-Villars, Paris.
88. Reichl, L.E., 1992, *The Transition to Chaos in Conservative Classic System: Quantum Manifestations*, Springer-Verlag, New York.
89. Reichl, L.E. and Zheng, W.M., 1984a, Field-Induced barrier penetration in the quadratic potential, *Physical Review*, **29A**, 2186-2193.
90. Reichl, L.E. and Zheng, W.M., 1984b, Perturbed double-well system: The pendulum approximation and low-frequency effect, *Physical Review*, **30A**, 1068-1077.

91. Rom-Kedar, V., 1990, Transport rates of a class of two-dimensional maps and flow, *Physica D*, **43**, 229-268.

92. Rom-Kedar, V., 1994, Homoclinic tangles-classification and applications, *Nonlinearity*, 7, 441-473.

93. Rom-Kedar, V., 1995, Secondary homoclinic bifurcation theorems, *Chaos*, **5**, 385-401.

94. Rosenbluth, M.N., Sagdeev, R.Z. , Taylor, J.B. and Zalavski, G.M., 1966, Destruction of magnetic surfaces by magnetic field irregularities, *Nuclear Fusion*, **6**, 297-300.

95. Sinai, Y.G., 1976, *Introduction to Ergodic Theory*, Princeton University Press: Princeton.

96. Slutskin, A. A., 1964, Motion of a one-dimensional nonlinear oscillator under adiabatic conditions, *Soviet Physics JETP*, **18**, 676-682.

97. Smale, S., 1967, Differentiable Dynamical Systems. Bull. American Mathematical Society, **73**, 747-817.

98. Stoker, J.J., 1950, *Nonlinear Vibrations*, Wiley, New York.

99. Tabor, M., 1989, *Chaos and Integrability in Nonlinear Dynamics: An Introduction*, John Wiley and Sons, New York.

100. Treschev, D.V., 1995, An averaging method for Hamiltonian systems, exponentially close to integrable ones, *Chaos*, **6**, 6-14.

101. Treschev, D.V., 1998, Width of stochastic layers in near-integrable two-dimensional symplectic maps, *Physica D,* **116**, 21-43.

102. Vecheslavov, V.V., 1996, Motion in the vicinity of the separatrix of a nonlinear resonance in the presence of high-frequency excitations, *Soviet Physics JEPT*, **82**, 1190-1195.

103. Udwadia, F. and von Bremen, H., 2001, An efficient and stable approach for computation of Lyapunov characteristic exponents of continuous dynamical systems, *Applied Mathematical Computation*, **121**, 219-58.

104. Ueda, Y., 1980, Steady motions exhibited by Duffing's equation: A picture book of regular and chaotic motion, In *New Approaches to Nonlinear Problems in Dynamics*, P.J. Holmes (ed.) SIAM, Philadelphia, PA. 311-322.

105. Walker, G.H. and Ford, J.,1969, Amplitude instability and ergodic behavior for conservative nonlinear oscillator systems, *Physical Review*, **188**, 416-432.

106. Wolf, A., Swift, J., Swinney, S. and Vastano, J., 1985, Determining Lyapunov exponents from a time series, *Physica D*, **16**, 285-317.

107. Yang, G., Lu, J. and Luo, A.C.J., 2005, On the computation of Lyapunov exponents for forced vibration of a Lennard-Jones Oscillator, *Chaos,*

Solitons and Fractals, **23**, 833-841.

108. Yin, L. and Ao, P., 2006, Existence and construction of dynamical potential in nonequilibrium process without detailed balance, *Journal of Physics A: Mathematical and General*, **39**, 8593-8601.

109. Zaslvasky, G.M., 1998, *Physics of Chaos in Hamiltonian Systems*, Imperial Colleage Press.

110. Zaslavsky,G.M. and Chirikov,B.V., 1964, On the mechanism of one-dimensional Fermi acceleration, Doklady of Akademy Nauk **159**, 306-309 (translation: Soviet Physics-Doklady 1965, **9**, 989-992)

111. Zaslavsky, G.M. and Filonenko, N.N., 1968, Stochastic instability of trapped particles and conditions of application of the quasi-linear approximation, *Soviet Physics JETP*, **27**, 851-857.

112. Zaslavsky,G.M. and Chirikov,B.V., 1972, Stochastic Instability of nonlinear oscillations, *Soviet Physics USP*, **52**, 263-272.

113. Zaslavsky, G.M. and Abdullaev, S.S., 1995, Scaling properties and anomalous transport of particles inside the stochastic layer, *Physical Review E*, **51**, 3901-3910.

Subject Index